HANDBOOK OF QUANTITATIVE SUPPLY CHAIN ANALYSIS:
Modeling in the E-Business Era

Recent titles in the
INTERNATIONAL SERIES IN
OPERATIONS RESEARCH & MANAGEMENT SCIENCE
Frederick S. Hillier, Series Editor, *Stanford University*

Ramík, J. & Vlach, M. / *GENERALIZED CONCAVITY IN FUZZY OPTIMIZATION AND DECISION ANALYSIS*
Song, J. & Yao, D. / *SUPPLY CHAIN STRUCTURES:* Coordination, Information and Optimization
Kozan, E. & Ohuchi, A. / *OPERATIONS RESEARCH/ MANAGEMENT SCIENCE AT WORK*
Bouyssou et al. / *AIDING DECISIONS WITH MULTIPLE CRITERIA:* Essays in Honor of Bernard Roy
Cox, Louis Anthony, Jr. / *RISK ANALYSIS:* Foundations, Models and Methods
Dror, M., L'Ecuyer, P. & Szidarovszky, F. / *MODELING UNCERTAINTY:* An Examination of Stochastic Theory, Methods, and Applications
Dokuchaev, N. / *DYNAMIC PORTFOLIO STRATEGIES:* Quantitative Methods and Empirical Rules for Incomplete Information
Sarker, R., Mohammadian, M. & Yao, X. / *EVOLUTIONARY OPTIMIZATION*
Demeulemeester, R. & Herroelen, W. / *PROJECT SCHEDULING: A Research Handbook*
Gazis, D.C. / *TRAFFIC THEORY*
Zhu, J. / *QUANTITATIVE MODELS FOR PERFORMANCE EVALUATION AND BENCHMARKING*
Ehrgott, M. & Gandibleux, X. /*MULTIPLE CRITERIA OPTIMIZATION:* State of the Art Annotated Bibliographical Surveys
Bienstock, D. / *Potential Function Methods for Approx. Solving Linear Programming Problems*
Matsatsinis, N.F. & Siskos, Y. / *INTELLIGENT SUPPORT SYSTEMS FOR MARKETING DECISIONS*
Alpern, S. & Gal, S. / *THE THEORY OF SEARCH GAMES AND RENDEZVOUS*
Hall, R.W./*HANDBOOK OF TRANSPORTATION SCIENCE - 2nd Ed.*
Glover, F. & Kochenberger, G.A. / *HANDBOOK OF METAHEURISTICS*
Graves, S.B. & Ringuest, J.L. / *MODELS AND METHODS FOR PROJECT SELECTION:* Concepts from Management Science, Finance and Information Technology
Hassin, R. & Haviv, M./ *TO QUEUE OR NOT TO QUEUE:* Equilibrium Behavior in Queueing Systems
Gershwin, S.B. et al/ *ANALYSIS & MODELING OF MANUFACTURING SYSTEMS*
Maros, I./ *COMPUTATIONAL TECHNIQUES OF THE SIMPLEX METHOD*
Harrison, T., Lee, H. & Neale, J./ *THE PRACTICE OF SUPPLY CHAIN MANAGEMENT:* Where Theory And Application Converge
Shanthikumar, J.G., Yao, D. & Zijm, W.H./ *STOCHASTIC MODELING AND OPTIMIZATION OF MANUFACTURING SYSTEMS AND SUPPLY CHAINS*
Nabrzyski, J., Schopf, J.M., Węglarz, J./ *GRID RESOURCE MANAGEMENT:* State of the Art and Future Trends
Thissen, W.A.H. & Herder, P.M./ *CRITICAL INFRASTRUCTURES:* State of the Art in Research and Application
Carlsson, C., Fedrizzi, M., & Fullér, R./ *FUZZY LOGIC IN MANAGEMENT*
Soyer, R., Mazzuchi, T.A., & Singpurwalla, N.D./ *MATHEMATICAL RELIABILITY:* An Expository Perspective
Talluri, K. & van Ryzin, G./ *THE THEORY AND PRACTICE OF REVENUE MANAGEMENT*
Kavadias, S. & Loch, C.H./*PROJECT SELECTION UNDER UNCERTAINTY:* Dynamically Allocating Resources to Maximize Value
Sainfort, F., Brandeau, M.L., Pierskalla, W.P./ *HANDBOOK OF OPERATIONS RESEARCH AND HEALTH CARE: Methods and Applications*
Cooper, W.W., Seiford, L.M., Zhu, J./ *HANDBOOK OF DATA ENVELOPMENT ANALYSIS: Models and Methods*
Sherbrooke, C.C./ *OPTIMAL INVENTORY MODELING OF SYSTEMS:* Multi-Echelon Techniques, Second Edition
Chu, S.-C., Leung, L.C., Hui, Y. V., Cheung, W./ *4th PARTY CYBER LOGISTICS FOR AIR CARGO*

* *A list of the early publications in the series is at the end of the book* *

HANDBOOK OF QUANTITATIVE SUPPLY CHAIN ANALYSIS:
Modeling in the E-Business Era

Edited by

DAVID SIMCHI-LEVI
Massachusetts Institute of Technology

S. DAVID WU
Lehigh University

ZUO-JUN (MAX) SHEN
University of Florida

Kluwer Academic Publishers
Boston/Dordrecht/London

Distributors for North, Central and South America:
Kluwer Academic Publishers
101 Philip Drive
Assinippi Park
Norwell, Massachusetts 02061 USA
Telephone (781) 871-6600
Fax (781) 871-6528
E-Mail <kluwer@wkap.com>

Distributors for all other countries:
Kluwer Academic Publishers Group
Post Office Box 322
3300 AH Dordrecht, THE NETHERLANDS
Telephone 31 78 6576 000
Fax 31 78 6576 474
E-Mail <orderdept@wkap.nl>

 Electronic Services <http://www.wkap.nl>

Library of Congress Cataloging-in-Publication

Simchi-Levi, David/ Wu, S. David/ Shen, Zuo-Jun
Handbook of Supply Chain Analysis: Modeling in an E-Business Era
ISBN HB   1-4020-7952-4
ISBN E-book   1-4020-7953-2

Copyright © 2004 by Kluwer Academic Publishers

All rights reserved. No part of this publication may be reproduced, stored in a retrieval system or transmitted in any form or by any means, electronic, mechanical, photo-copying, microfilming, recording, or otherwise, without the prior written permission of the publisher, with the exception of any material supplied specifically for the purpose of being entered and executed on a computer system, for exclusive use by the purchaser of the work.
Permissions for books published in the USA: permissions@wkap.com
Permissions for books published in Europe: permissions@wkap.nl
Printed on acid-free paper.

Printed in the United States of America

Contents

Chapter 1
Supply Chain Analysis and E-Business: An Overview — 1
David Simchi-Levi, S. David Wu and Z. Max Shen
 1. Introduction — 1
 2. Main Components of the Handbook — 3
  2.1 Emerging Paradigms for Supply Chain Analysis — 3
  2.2 Auctions and Bidding — 4
  2.3 Supply Chain Coordinations in E-Business — 5
  2.4 Multi-Channel Coordination — 7
  2.5 Network Design, IT, and Financial Services. — 8
 3. Conclusions — 9
Acknowledgments — 9

Part I Emerging Paradigms for Supply Chain Analysis

Chapter 2
Game Theory in Supply Chain Analysis — 13
Gérard P. Cachon and Serguei Netessine
 1. Introduction — 13
  1.1 Scope and relation to the literature — 14
 2. Non-cooperative static games — 14
  2.1 Game setup — 15
  2.2 Best response functions and the equilibrium of the game — 17
  2.3 Existence of equilibrium — 21
  2.4 Uniqueness of equilibrium — 29
  2.5 Multiple equilibria — 35
  2.6 Comparative statics in games — 36
 3. Dynamic games — 40
  3.1 Sequential moves: Stackelberg equilibrium concept — 40
  3.2 Simultaneous moves: repeated and stochastic games — 41
  3.3 Differential games — 45
 4. Cooperative games — 48
  4.1 Games in characteristic form and the core of the game — 49
  4.2 Shapley value — 50
  4.3 Biform games — 51
 5. Signaling, Screening and Bayesian Games — 52
  5.1 Signaling Game — 53
  5.2 Screening — 56
  5.3 Bayesian games — 57

6.	Summary and Opportunities		58
References			59

Chapter 3
Supply Chain Intermediation: A Bargaining Theoretic Framework — 67
S. David Wu

1.	Introduction		67
2.	Supply Chain Intermediation		70
3.	Supply Chain Intermediary Theory		73
	3.1	The Basic Settings	73
	3.2	Mechanism Design and the Revelation Principle	74
	3.3	Models of Supply Chain Intermediation	76
4.	Bilateral Bargaining with Complete Information		77
	4.1	Bilateral Bargaining to Divide the System Surplus	79
	4.2	A Bilateral Supply-Chain Bargaining Model	81
	4.3	The Subgame Perfect Equilibrium	82
	4.4	Analysis of the Bargaining Game	85
	4.5	Intermediary's Role in Price Setting, Searching, and Matching	87
5.	Bilateral Bargaining with Incomplete Information		89
	5.1	The Basic Setting	90
	5.2	The Direct Revelation Mechanism	92
6.	Multilateral Trade with Incomplete Information		97
	6.1	Multilateral Trade with Vertical Integration	97
	6.2	Multilateral Trade with Markets	101
7.	Related Work in the Supply Chain Literature and Research Opportunities		105
References			112

Chapter 4
Decentralized Decision Making in Dynamic Technological Systems: The Principal-Agent Paradigm — 117
Stefanos Zenios

1.	Background and Motivation		117
2.	The Static Principal-Agent Model		119
3.	The Dynamic Principal-Agent Model		121
	3.1	The Principal's Problem	123
	3.2	Step 1: The Static Problem.	125
	3.3	Step 2: The Dynamic Problem.	126
4.	A Model with Multiple Agents		127
5.	A Production-Inventory System		129
6.	A Service Delivery System		133
7.	The Past and the Future		137
8.	Concluding Remarks		138
Acknowledgments			138
References			139

Part II Auctions and Bidding

Chapter 5
Auctions, Bidding and Exchange Design — 143

Jayant Kalagnanam and David C. Parkes

1.	Introduction		143
	1.1	A framework for auctions	144
	1.2	Outline	146
2.	Economic Considerations		146
	2.1	Preliminaries	147
	2.2	Mechanism Design	148
	2.3	Competitive Equilibrium	157
	2.4	Indirect Revelation Mechanisms	159
3.	Implementation Considerations		164
	3.1	Bidding Languages	164
	3.2	Winner-Determination Complexity	166
4.	Interactions between Computation and Incentives		179
	4.1	Strategic Complexity	180
	4.2	Communication Complexity	181
	4.3	Valuation Complexity	182
	4.4	Implementation Complexity	183
5.	Specific Market Mechanisms		186
	5.1	Combinatorial Auctions	186
	5.2	Multi-unit Auctions	189
	5.3	Multiattribute Auctions	192
	5.4	Procurement Reverse Auctions	194
	5.5	Capacity constrained allocation mechanisms	196
	5.6	Double Auctions and Exchanges	197
6.	Discussion		200
	References		204

Chapter 6
Auctions and Pricing in E-Marketplaces 213
Wedad Elmaghraby

1.	B2B E-Marketplaces	213
2.	Current State of Pricing in B2B Marketplaces	215
3.	Customizing Auctions - A Case Study of FreeMarkets	219
	3.1 Auction Formats	221
	3.2 Combating Collusion	224
4.	Bidder Support in Auctions - Manugistics' NetWORKS Target Pricing^{TM}	227
5.	Future Directions for Research in Auction Theory	230
6.	Precision Pricing - Manugistics' Networks Precision Pricing	232
7.	Future Directions for Research in Price Discrimination	238
8.	Conclusion	240
	References	243

Chapter 7
Design of Combinatorial Auctions 247
Sven de Vries and Rakesh V. Vohra

1.	Introduction	247
2.	Mechanism Design Perspective	249
3.	Optimal Auctions	250
4.	Efficient Auctions	254
	4.1 The VCG Auction	255

5.	Implementing an Efficient Auction		257
	5.1 Winner Determination		257
	5.2 Supposed Problems with VCG		261
	5.3 Ascending Implementations of VCG		264
	5.4 Threshold and Collusion Problems		276
	5.5 Other Ascending Auctions		277
	5.6 Complexity of Communication		278
6.	Interdependent Values		278
7.	Two Examples		281
	7.1 The German UMTS-Auction		281
	7.2 Logistics Auctions		284
References			287

Part III Supply Chain Coordinations in E-Business

Chapter 8

The Marketing-Operations Interface — 295

Sergio Chayet, Wallace J. Hopp and Xiaowei Xu

1. Product Development		298
1.1	Conjoint Analysis For Concept Development	299
1.2	System Level and Detailed Design	301
1.3	Prototyping and Testing	304
1.4	Macro-Level Research	305
2. Sales		306
2.1	Pricing	307
2.2	Lead Time Quoting	309
2.3	Quality Management	311
2.4	Product Variety	312
3. Production/Delivery		313
4. Service		316
4.1	Previous Research	316
4.2	Research Opportunities	317
5. Conclusions		319
References		322

Chapter 9

Coordination of Pricing and Inventory Decisions: A Survey and Classification — 335

L. M. A. Chan, Z. J. Max Shen, David Simchi-Levi and Julie L. Swann

1. Introduction		335
1.1	Motivation	335
1.2	Scope	336
1.3	Classification and Outline	338
2. Single Period Models		340
3. Multiple Period Models		343
3.1	Models to Explain Price Realizations	343
3.2	General Pricing and Production Models	344
3.3	Retail, Clearance, and Promotion	351
3.4	Fixed Pricing	358
4. Extension Areas		359
4.1	Multiple Products, Classes, and Service levels	359
4.2	Capacity as a Decision	366

	4.3	Supply Chain Coordination	367
	4.4	Competition	369
	4.5	Demand Learning and Information	373
5.	Industry		376
	5.1	Dynamic Pricing Practice	376
	5.2	Related Research	376
	5.3	Price Discrimination in Practice	377
	5.4	Potential problems with Dynamic Pricing	378
6.	Conclusions and Future Research		379
References			382

Chapter 10
Collaborative Forecasting and its Impact on Supply Chain Performance 393
Yossi Aviv

1.	Notation and Preliminaries		402
2.	Common Approaches for Modeling Demand Uncertainty and Forecast Evolution in the Inventory Management Literature		404
	2.1	Demand models with unknown parameters	404
	2.2	A Markov-modulated demand process	406
	2.3	A linear state-space model	407
3.	Common Types of Single-Location Inventory Control Policies		410
	3.1	Dynamic models for inventory management	410
	3.2	Heuristic policies	416
	3.3	An adaptive replenishment policy for the linear state-space model	419
4.	Models for Decentralized Forecasting Processes		421
	4.1	The orders generated by the retailer	422
	4.2	Enriched information structures	425
	4.3	Assessment of the benefits of information sharing	428
5.	An inventory model of collaborative forecasting		429
	5.1	Installation-based inventory systems	430
	5.2	Echelon-based inventory systems	432
6.	Cost analysis		435
	6.1	Cost assessment	435
	6.2	Policy coordination in the supply chain	437
	6.3	Decoupled two-level inventory systems	439
	6.4	Results from the study of CF	440
7.	Summary		442
References			443

Chapter 11
Available to Promise 447
Michael O. Ball, Chien-Yu Chen and Zhen-Ying Zhao

1.	Introduction		447
	1.1	Push-Pull Framework	448
	1.2	Available to Promise (ATP)	449
2.	Business Examples		449
	2.1	Overview of Conventional ATP	449
	2.2	Toshiba Electronic Product ATP System	451
	2.3	Dell Two-stage Order Promising Practice	452
	2.4	Maxtor ATP Execution for Hard Disk Drive	453
	2.5	ATP Functionality in Commercial Software	454
3.	ATP Modelling Issues		455

		3.1	ATP Implementation Dimensions	455
		3.2	Factors Affecting ATP Implementations	456
		3.3	Push vs. Pull ATP Models	459
	4.	Push-Based ATP Models		461
		4.1	Push ATP Rules and Policy Analysis	462
		4.2	Deterministic Optimization-Based Push ATP Models	463
		4.3	Stochastic Push ATP Models	467
	5.	Pull-Based ATP Models		469
		5.1	Pull-Based ATP Models for an MTS Production Environment	470
		5.2	Real Time Order Promising and Scheduling	471
		5.3	Optimization-Based Batch ATP Models	473
		5.4	Experimental Implementation	477
	6.	Conclusions		480
	Acknowledgments			480
	References			481

Chapter 12

Due Date Management Policies 485

Pınar Keskinocak and Sridhar Tayur

	1.	Characteristics of a Due Date Management Problem		487
		1.1	Due Date Management Decisions	487
		1.2	Dimensions of a Due Date Management Problem	489
		1.3	Objectives of Due Date Management	490
		1.4	Solution Approaches for Due Date Management Problems	494
	2.	Scheduling Policies in Due Date Management		494
	3.	Offline Models for Due Date Management		499
		3.1	Equal Order Arrival Times	500
		3.2	Distinct Order Arrival Times	501
	4.	Online Models for Due Date Management		503
		4.1	Due-Date Setting Rules	506
		4.2	Choosing the Parameters of Due Date Rules	520
		4.3	Mathematical Models for Setting Due Dates	521
	5.	Due Date Management with Service Constraints		523
	6.	Due Date Management with Price and Order Selection Decisions		530
		6.1	Due Date Management with Order Selection Decisions (DDM-OS)	531
		6.2	Due Date Management with Price and Order Selection Decisions (DDM-P)	536
	7.	Conclusions and Future Research Directions		542
	References			547

Part IV Multi-Channel Coordination

Chapter 13

Modeling Conflict and Coordination in Multi-channel Distribution Systems: A Review 557

Andy A. Tsay and Narendra Agrawal

| | 1. | Introduction | | 557 |
| | | 1.1 | Business Setting | 557 |

		1.2	Scope of Discussion	559
	2.		Related Literature	561
		1.3	Contribution	562
		2.1	Descriptive research	562
		2.2	Analytical research	564
	3.		Analytical Research on Conflict and Coordination in Multi-Channel Systems With Both Manufacturer-Owned And Intermediated Channels	571
		3.1	Manufacturer-owned channel is direct sales	574
		3.2	Manufacturer-owned channel contains physical stores	582
		3.3	Discussion	585
	4.		Research Opportunities	586
		4.1	Representing channel characteristics	586
		4.2	Evaluating distribution strategies	594
		4.3	Concluding remarks	595
	Acknowledgments			595
	References			597

Chapter 14
Supply chain structures on the Internet and the role of marketing-operations interaction 607
Serguei Netessine and Nils Rudi

1.	Introduction	607
2.	Literature survey	611
3.	Notation and modeling assumptions	614
4.	Supply chain models without coordination	616
	4.1 Model I - vertically integrated supply chain	616
	4.2 Model T - traditional supply chain	618
	4.3 Model D – drop-shipping	619
	4.4 Comparative analysis of the stationary policies	625
5.	Supply chain coordination	629
6.	Numerical experiments	632
7.	Conclusions and discussion	636
References		639

Chapter 15
Coordinating Traditional and Internet Supply Chains 643
Kyle D. Cattani, Wendell G. Gilland and Jayashankar M. Swaminathan

1.	Introduction	643
	1.1 Overview of Research	644
	1.2 Procurement	644
	1.3 Pricing	646
	1.4 Distribution / Fulfillment	647
2.	Procurement	648
	2.1 Coordinating Traditional and Internet Procurement	648
	2.2 Formation of Consortia	655
3.	Pricing	656
	3.1 Independent Competition	656
	3.2 Bricks and Clicks	660
	3.3 Forward Integration	665
	3.4 Full Integration	668
4.	Distribution / Fulfilment	668

		4.1	Direct channel as outlet	669
		4.2	Direct channel as Service buffer	671
	5.	Conclusion		674
	References			675

Part V Network Design, IT, and Financial Services

Chapter 16

Using a Structural Equations Modeling Approach to Design and Monitor Strategic International Facility Networks — 681
Panos Kouvelis and Charles L. Munson

	1.	Introduction		681
		1.1	A conceptual framework to Classify Global Network Structures	681
		1.2	Literature Review	682
		1.3	Research Contributions	685
	2.	Structural Equations Model		686
		2.1	Problem Statement	686
		2.2	Motivation for the Structural Equations Modeling Approach	686
		2.3	Framework Measures (Independent Variables)	687
		2.4	Proxies for the Global Network Structure Dimensions (Dependent Variables)	689
		2.5	Model Development	689
		2.6	Model Validation	695
		2.7	Other Products	696
	3.	Government Incentives: Tax Holidays and Subsidized Loans		696
	4.	Conclusion		701
	Acknowledgments			702
	Appendix: MIP Model for Facility Location			702
	References			708

Chapter 17

Integrated Production and Distribution Operations: Taxonomy, Models, and Review — 711
Zhi-Long Chen

	1.	Introduction	711
	2.	Model Classification	713
	3.	Production - Transportation Problems	715
	4.	Joint Lot Sizing and Finished Product Delivery Problems	717
	5.	Joint Raw Material Delivery and Lot Sizing Problems	720
	6.	General Tactical	721
	7.	Joint Job Processing and Finished Job Delivery Problems	729
	8.	Directions for Future Research	732
	Acknowledgments		734
	References		735

Chapter 18

Next Generation ERP Systems: Scalable and Decentralized Paradigms — 747
Paul M. Griffin and Christina R. Scherrer

	1.	Introduction	747

2.	A Brief History		749
3.	Current ERP Functionality		751
4.	Implementation Issues		755
5.	Scalability and ERP		759
	5.1	New Developments in Enterprise Scalability	759
	5.2	Standards in ERP	763
6.	Decentralized ERP		766
	6.1	e-Market Intermediaries	767
	6.2	Outsourced ERP Systems	770
7.	Current Enterprise Issues: ERPII and ECM		772
8.	Conclusions		775
References			778

Chapter 19
Delivering e-banking services: An emerging internet business model and a case study   783

Andreas C. Soteriou and Stavros A. Zenios

1.	Introduction		783
2.	The changing landscape of demand for financial services		785
3.	The changing landscape of supply of financial services		787
4.	An Emerging e-Banking Model		790
5.	The design of a web-based personal asset allocation system		794
	5.1	The integrated decision support system	796
	5.2	Business plan for the deployment of the system	797
	5.3	Bringing it altogether: Lessons from the implementation of the system	800
6.	Concluding Remarks		802
References			803

Index   805

David Simchi-Levi, S. David Wu, and Z. Max Shen (Eds.)
Handbook of Quantitative Supply Chain Analysis:
Modeling in the E-Business Era
©2004 Kluwer Academic Publishers

Chapter 1

SUPPLY CHAIN ANALYSIS AND E-BUSINESS
An Overview

David Simchi-Levi
Department of Civil and Environmental Engineering
Massachusetts Institute of Technology
dslevi@mit.edu

S. David Wu[*]
Department of Industrial and Systems Engineering
Lehigh University
david.wu@lehigh.edu

Z. Max Shen
Department of Industrial and Systems Engineering
University of Florida
shen@ise.ufl.edu

1. Introduction

Supply chain analysis is the study of quantitative models that characterize various economic tradeoffs in the supply chain. The field has made significant strides in both theoretical and practical fronts. On the theoretical front, supply chain analysis inspires new research ventures that blend operations research, game theory, and microeconomics. These ventures result in an unprecedented amalgamation of prescriptive, descriptive, and predictive models characteristic of each subfield. On the practical front, supply chain analysis offers solid foundations for strategic positioning, policy setting, and decision making. Over the past two decades, not only has supply chain analysis become a strategic fo-

---

[*] S. David Wu is supported by NSF Grants DMI-0075391 and DMI-0121395.

cus of leading firms, it has also spawned an impressive array of research that brings together diverse research communities. Adding to this diversity and intellectual energy is the emergence of E-Business. E-Business creates new competitive dimensions that are fast-paced, ever-changing, and risk-prone, dimensions where innovation, speed, and technological savvy often define success. Most importantly, E-Business challenges the premises and expands the scope of supply chain analysis.

The research community has responded to the E-Business challenge. Despite the infamous dot-com bust in the early 2000's, scores of research initiatives, workshops, technical papers, and special journal issues have been devoted to the subject. E-Business remains a critical subject not only in the research community, but also in corporate boardrooms. Instead of the revolution that would replace every facet of business, the rise of E-Business might be viewed as the emergence of new economic intermediaries that offer opportunities for innovation. These new intermediaries offer different means to respond to market demands (e.g., Internet vs. traditional channels), to facilitate sourcing, procurement, and price discovery (e.g., electronic auctions), and to develop new mechanisms for coordination and execution (e.g., dynamic pricing, revenue management, and collaborative forecasting).

The area intersecting supply chain analysis and E-Business is in its infancy; it is still taking shape and emerging. Indeed, there are still debates and contentions as to whether E-Business offers any fundamentally new research dimensions. We thought that this might be the right moment to put together a book that takes a close look at what has been done in the field of supply chain analysis that may be relevant to the emerging environment of E-Business. We set out to edit a research handbook that pays as much attention to looking back as to looking forward. The handbook is intended as reference material for researchers, graduate students, and practitioners alike who are interested in pursuing research and development in this area, and who need both a comprehensive view of the existing literature, and a multitude of ideas for future directions. The handbook should serve quite nicely as supplementary material for advanced graduate level courses in operations research, industrial engineering, operations management, supply chain management, or applied economics.

The handbook contains 18 chapters organized in five main parts as follows:

1. Emerging Paradigms for Supply Chain Analysis,

2. Auctions and Bidding,

3. Supply Chain Coordinations in E-Business,

4. Multi-Channel Coordination, and

5. Network Design, IT, and Financial Services.

Each chapter was written by one or more leading researchers in the area. These authors were invited on the basis of their scholarly expertise and unique insights in each particular subarea. An outline of the chapter was first submitted to us; we tried to coordinate the contents and foci of the chapters for better overall coverage. Since this is a handbook rather than a collection of research papers, we encouraged the authors to position their chapters broadly so as to provide a comprehensive overview or survey of an area of interest. On the other hand, we also encouraged the authors to focus on quantitative models and analytical insights provided by these models. Individual authors ultimately determined the emphasis and perspectives of each chapter. All chapters were reviewed by us and at least one outside referee.

We acknowledge that supply chain management and E-Business are represented by a wide variety of disciplines and research communities. The categorization and coverage of this handbook is neither complete nor exhaustive. However, we do believe that the 18 chapters represent an excellent sample of research topics on the intersection of quantitative supply chain analysis and E-Business. Through these topics, we are able to touch and connect a key body of literature that forms the basic ingredients and theoretical underpinnings for future developments; we hope this provides a foundation that helps to perpetuate next generation research in quantitative supply chain analysis and E-Business.

2. Main Components of the Handbook

In the following sections, we provide an overview for each of the five parts of the handbook.

2.1 Emerging Paradigms for Supply Chain Analysis

The chapters in this part explore three emerging paradigms that are of increasing importance in quantitative supply chain analysis: game theory, bargaining theory, and agency theory. In **Chapter 2**, Cachon and Netessine provide a comprehensive survey of game theory in supply chain analysis. Game theory has become an essential tool in the analysis of supply chains, which often involves multiple agents with conflicting objectives. This chapter surveys the applications of game theory to supply chain analysis and outlines game-theoretic concepts that have potential for future application. They discuss both non-cooperative and cooperative game theory in static and dynamic settings. Careful attention is given to techniques for demonstrating the existence and uniqueness of equilibrium in non-cooperative games. A newsvendor game is employed throughout to demonstrate the application of various tools. In **Chapter 3**, Wu provides an overview of supply chain intermediation using the framework of bargaining theory. Supply chain intermediaries are those economic agents who coordinate and arbitrate transactions between suppliers and customers. For any set of agents in the supply chain who desire to form supplier-buyer relationships, they may choose to do so through direct bargaining or through some form of intermediation. Thus, the merit of an intermediary must be justified by the outcome of its competing bargaining game. In this context, he examines bilateral and multilateral bargaining games under complete and incomplete information. In **Chapter 4**, Zenios introduces the principal-agent (or agency) paradigm for supply chain analysis. A principal has a primary stake in the performance of a system but delegates operational control to one or more agents. The agency theory has been adopted in various fields such as investment management, personnel economics, and managerial accounting. He first introduces the classical static principal-agent model, and he then introduces the dynamic principal-agent model and presents a solution methodology based on dynamic programming. Next, he presents an extension of the basic dynamic model in a setting where the decision processes are controlled by multiple agents with potentially conflicting objectives. Following this, he presents applications in a decentralized inventory system and a service delivery system.

2.2 Auctions and Bidding

The game-theoretic and bargaining-theoretic discussions extend naturally to auctions theory, the subject of the second part of the handbook. Auction is among the oldest and most widely used market mechanisms. There is an extensive literature on auctions theory that is both theoretically rich and practically significant. It is not our intent to survey the massive literature in auctions theory; the handbook focuses on the emergence of auctions as an intermediary in E-Business transactions, serving functions such as price discovering, revenue management, resource allocation, multilateral negotiations, and sourcing/procurement. We start out with **Chapter 5**, where Kalagnanam and Parkes provide a comprehensive survey of auctions from the perspective of decentralized resource allocation. This perspective is important because it brings in the dimensions of optimization and computing that have been previously overlooked in the auctions literature but that have critical importance for E-Business applications, such as procurement auctions. Procurement auctions motivate the study of theoretical and computational roadblocks involved in multi-unit, multi-attribute, sequential, and combinatorial auctions. Kalagnanam and Parkes provide details for some of the most interesting designs from the literature, establish state-of-the-art results, and identify emerging research directions. In **Chapter 6**, Elmaghraby focuses on the roles of auctions and precision pricing in B2B e-marketplaces. Using in-depth case studies of FreeMarkets' customized auction services and Manugistics' Network Target

Overview

Pricing (NTP) system, she explores the design dimensions and implementation challenges associated with the use of auctions in B2B markets. She discusses the decisions that FreeMarkets faces when designing a procurement auction, and then discusses the design and use of NTP, a bidding support software developed by Manugistics to aid suppliers in submitting bids. She then turns her attention to an alternative pricing mechanism in B2B e-marketplaces, precision pricing. She presents a detailed case study of Manugistics' Precision Pricing (P2) system, discussing the challenges in properly implementing precision pricing in B2B contexts. In **Chapter 7**, de Vries and Vohra survey the state of knowledge on the design of combinatorial auctions. Combinatorial auctions involve the sale of assets that demonstrate complementarities, e.g., radio spectrums, where the bidders are allowed to bid on combinations or bundles of different assets that reflect their preferences. The authors focus on the revenue maximization and economic efficiency of such auctions. They present various integer programming formulations of the winner determination problem, and give an in-depth discussion of efficient mechanisms such as the VCG auction and its variations.

2.3 Supply Chain Coordinations in E-Business

In the new environment of E-Business, the mechanisms by which firms coordinate different stages in the supply chain are undergoing profound changes. New technologies have emerged to acquire instantaneous feedback from customers and markets, to enable information-sharing with suppliers, and to collaborate decision-making throughout the supply chain. This enhanced level of coordination significantly increases the dependencies among retailers, suppliers, manufacturers, and intermediaries, posing significant challenges in strategic positioning, planning, and execution. This part of the handbook is dedicated to different aspects of supply chain coordinations that attempt to address the emerging needs of E-Business. In **Chapter 8**, Chayet, Hopp, and Xu focus on the marketing-operations interface. In specific, they address the changing customer contact mechanisms and information flows due to the Internet. They examine the stages of the "customer contact chain" to identify areas in which the Internet presents new management issues. By reviewing the literatures traditionally related to Marketing and Operations Management, they suggest new research avenues for addressing the marketing-operations challenges posed by E-Business. In **Chapter 9**, Chan, Shen, Simchi-Levi, and Swann address the coordination of pricing and inventory decisions in the supply chain. Pricing is among the most significant issues in E-Business, especially when it is integrated with production/inventory decisions. The authors provide an in-depth survey of the literature where price is a decision variable, where customer demand depends on the price chosen, and where production/inventory decisions can be coordinated with pricing. They consider a variety of pricing strategies, including dynamic pricing, simultaneous pricing of multiple products, and pricing across customer segments; they also consider different forms of price coordination, including price-coordination across multiple channels and price-coordination with production lead time decisions and capacity investment decisions. They devise a taxonomy which classifies the literature in terms of nine key elements. Collaborative forecasting is another significant issue in supply chain coordination. In **Chapter 10**, Aviv provides an in-depth examination of collaborative forecasting models and their impact on supply chain performance. He describes models using the *linear state-space framework* that integrates the demand process with forecasting, inventory control, and collaboration processes. He surveys demand models that can be used in an integrative collaborative forecasting model. He then discusses the dynamic inventory control problem for each of the demand models. An integrative model in a two-stage supply chain is presented in detail, and the author describes the evolution of information and replenishment policies along with the value of information sharing in the context of this model. The author further examines the benefits of collaborative forecasting processes in a certain auto-regressive demand environment and their impact to the inventory cost performance. In **Chapter 11**, Ball, Chen, and Zhao give an overview on the Available to Promise (ATP) business process. ATP is the set of capabilities that supports the response to customer order requests. Traditionally ATP refers to a simple database lookup into the Master Production Schedule. As the variety and complexity of product offerings increase, ATP has become a key E-Business function that requires sophisticated modelling and IT support. The authors provide an overview of ATP-related research as well as of business practices. They classify ATP research into two main categories: push-based models that allocate resources and prepare responses based on forecasted demand and pull-based models that generate responses based on actual customer orders. Directly related to ATP is the issue of lead-time quotation, or more generally, due date management. In **Chapter 12**, Keskinocak and Tayur provide an extensive survey of the due date management (DDM) literature. Due date management decisions typically include order acceptance (demand management), due date quotation, and sequencing and scheduling. The authors survey a wide variety of DDM models, including off-line models that assume the demand is known *a priori*, on-line models that consider dynamic arrival of orders, models that consider service level constraints, and models that consider order acceptance and pricing decisions. They point to several important research directions directly relevant to the E-Business environment.

2.4 Multi-Channel Coordination

Integrating traditional and Internet channels is among the most significant E-Business developments that have a direct impact on supply chain management. For instance, an emerging retail structure known as "bricks-and-clicks" allows retailers such as Best Buy and Circuit City to maintain market presence through the Internet and the physical stores. However, the retailers must face significant operational challenges in effectively coordinating the dual channels. After all, the Internet stores require no physical inventory, and it is common for the wholesalers to stock and own the inventory and ship directly to the customers at the retailers' request. In **Chapter 13**, Tsay and Agrawal provide a review of quantitative models for multi-channel coordination. They focus on the issue of channel conflict where one channel may object to the actions taken by another (e.g., pricing, tax, shipping charges). They note that the fear of alienating incumbent channel intermediaries is among the most cited reasons why many manufacturers choose to avoid direct (Internet or otherwise) sales. The authors emphasize the implications of Internet sales for supply chain distribution strategies and suggest potential research opportunities. In **Chapter 14**, Netessine and Rudi present an in-depth analysis of drop-shipping Internet channel, in which the retailer handles customer acquisition, but the wholesaler takes inventory risk and performs fulfillment. They conduct a game-theoretic analysis aiming to compare three distinct supply chain settings: (1) a vertically integrated supply chain, (2) a traditionally structured supply chain where the retailer assumes the inventory risk, and (3) a drop-shipping supply chain. They consider three different power structures for the drop-shipping supply chain: a powerful wholesaler, a powerful retailer, and an equally powerful wholesaler and retailer. They demonstrate how decision power affects the decision variables and profits, and they show that both the traditional and drop-shipping supply chains are system sub-optimal. They further show that inefficiencies arising in the drop-shipping channel are different from these in the traditional channel and propose a new mechanism that coordinates the former. In **Chapter 15**, Cattani, Gilland, and Swaminathan provide a literature survey on multi-channel coordination. They construct the survey based on three channel coordination opportunities along the supply chain: procurement, pricing, and distribution. In this order, they consider research in e-procurement and electronic marketplaces, pricing coordination between traditional and Internet channels, and distribution/fullfillment strategies. The survey focuses on quantitative models based on four basic settings: independent competition, bricks-and-clicks, forward integration, and full integration.

2.5 Network Design, IT, and Financial Services.

Fundamental to the operations and execution in a supply chain are the issues of network design and information technology. E-Business potentially increases the complexity of and changes the design criteria for these systems. In this part of the handbook, we devote three chapters to the state-of-the-art concerning supply chain network design, production/distribution integration, and ERP systems. While significant supply chain infrastructures have been put into place in the retail and manufacturing sectors, the use of supply chain and E-Business concepts is only beginning to penetrate service sectors such as health care and financial services. We will devote the final chapter to the impact of supply chain and E-Business on financial institutions. In **Chapter 16**, Kouvelis and Munson describe a conceptual framework for analyzing strategic international facility network structures. They suggest ways to operationalize the conceptual framework as a managerial tool for the design and monitoring of a facility network. Three main dimensions are being considered: market focus, plant focus, and network dispersion. The authors use global sensitivity analysis to develop a structural equations model based on a mixed integer program that captures essential design tradeoffs of global networks and explicitly incorporates government subsidies, trade tariffs, and taxation issues. The resulting structural equations model classifies a firm's network structure according to the authors' conceptual framework via the calculation of a few key independent variables. The chapter exemplifies current research efforts in building comprehensive models that capture the complex facets of global supply networks, taking into consideration factors such as tariffs, transportation costs, and subsidized financing opportunities. Focusing on a similar modeling paradigm, Chen provides in **Chapter 17** a comprehensive survey of integrated production and distribution models in the supply chain literature; his survey focuses on mathematical programming models. He suggests a taxonomy that classifies existing models into five classes based on the level of decisions, the structure of production-distribution integration, and problem parameters such as infinite/finite horizon, single/multiple period, and constant/dynamic demands. He reviews each class of problems and suggests important directions for future research. In **Chapter 18**, Griffin and Scherrer provide a survey of next generation Enterprise Resource Planning (ERP) systems. They suggest that current ERP systems have not lived up to their initial promise, particularly in the E-Business environments. They explore some of the reasons for this unsatisfactory performance and discuss new development trends in next generation ERP systems. The authors focus their discussion on the issues of scalability and decentralization and discuss research needs in these areas. In **Chapter 19**, Soteriou and Zenios offer their unique perspectives on the impact of supply chain and E-Business concepts to financial services. Specifically, they build on previous literature in this area by discussing e-banking services; they present a new business model that focuses on the market potentials of e-banking and explore factors for a successful e-banking strategy. The authors review some major changes in the world of financial services in the E-Business era, and examine both supply and demand forces that shape Internet-based financial services. They present a case study base on a personal financial planning system deployed by several Italian banks, and analyze some of its components in light of their business model.

3. Conclusions

The area of quantitative supply chain analysis is undergoing profound changes due to the fast emergence of E-Business practices in the industry. This handbook is an attempt to gauge the broader impact of E-Business on supply chain research. We hope that this is not the end but the beginning of an era where new ideas and innovation opportunities brought by E-Business provide the catalyst for next generation development of supply chain strategies. We believe that the body of fundamental research summarized in this handbook will play a pivotal role in shaping the future of this development.

Acknowledgments

We are grateful to Lian Qi and Jia Ren who completed the typesetting of this handbook. We acknowledge the supports from Kluwer Academic Publishers throughout the preparation of the manuscript.

Part I

EMERGING PARADIGMS FOR SUPPLY CHAIN ANALYSIS

David Simchi-Levi, S. David Wu, and Z. Max Shen (Eds.)
Handbook of Quantitative Supply Chain Analysis:
Modeling in the E-Business Era
©2004 Kluwer Academic Publishers

Chapter 2

GAME THEORY IN SUPPLY CHAIN ANALYSIS

Gérard P. Cachon
The Wharton School
University of Pennsylvania
Philadelphia, PA 19104
cachon@wharton.upenn.edu

Serguei Netessine
The Wharton School
University of Pennsylvania
Philadelphia, PA 19104
netessine@wharton.upenn.edu

Keywords: Game theory, non-cooperative, cooperative, equilibrium concepts

1. Introduction

Game theory (hereafter GT) is a powerful tool for analyzing situations in which the decisions of multiple agents affect each agent's payoff. As such, GT deals with interactive optimization problems. While many economists in the past few centuries have worked on what can be considered game-theoretic models, John von Neumann and Oskar Morgenstern are formally credited as the fathers of modern game theory. Their classic book "Theory of Games and Economic Behavior", von Neumann and Morgenstern (1944), summarizes the basic concepts existing at that time. GT has since enjoyed an explosion of developments, including the concept of equilibrium by Nash (1950), games with imperfect information by Kuhn (1953), cooperative games by Aumann (1959) and Shubik (1962) and auctions by Vickrey (1961), to name just a few. Citing Shubik (2002), "In the 50s ... game theory was looked upon as a curiosum not to be taken seriously by any behavioral scientist. By the late 1980s, game theory in the new industrial organization has taken over ... game theory has proved its success in many disciplines."

This chapter has two goals. In our experience with GT problems we have found that many of the useful theoretical tools are spread over dozens of papers and books, buried among other tools that are not as useful in supply chain management (hereafter SCM). Hence, our first goal is to construct a brief tutorial through which SCM researchers can quickly locate GT tools and apply GT concepts. Due to the need for short explanations, we omit all proofs, choosing to focus only on the intuition behind the results we discuss. Our second goal is to provide ample but by no means exhaustive references on the specific applications of various GT techniques. These references offer an in-depth understanding of an application where necessary. Finally, we intentionally do not explore the implications of GT analysis on supply chain management, but rather, we emphasize the means of conducting the analysis to keep the exposition short.

1.1 Scope and relation to the literature

There are many GT concepts, but this chapter focuses on concepts that are particularly relevant to SCM and, perhaps, already found their applications in the literature. We dedicate a considerable amount of space to the discussion of static non-cooperative, non-zero sum games, the type of game which has received the most attention in the recent SCM literature. We also discuss cooperative games, dynamic/differential games and games with asymmetric/incomplete information. We omit discussion of important GT concepts covered in other chapters in this book: auctions in Chapters 5 and 7; and principal-agent models in Chapter 4.

The material in this chapter was collected predominantly from Friedman (1986), Fudenberg and Tirole (1991), Moulin (1986), Myerson (1997), Topkis (1998) and Vives (1999). Some previous surveys of GT models in management science include Lucas's (1971) survey of mathematical theory of games, Feichtinger and Jorgensen's (1983) survey of differential games and Wang and Parlar's (1989) survey of static models. A recent survey by Li and Whang (2001) focuses on application of GT tools in five specific OR/MS models.

2. Non-cooperative static games

In non-cooperative static games the players choose strategies simultaneously and are thereafter committed to their chosen strategies, i.e., these are simultaneous move, one-shot games. Non-cooperative GT seeks a rational prediction of how the game will be played in practice.[1] The solution concept for these games was formally introduced by John Nash (1950) although some instances of using similar concepts date back a couple of centuries.

2.1 Game setup

To break the ground for the section, we introduce basic GT notation. A warning to the reader: to achieve brevity, we intentionally sacrifice some precision in our presentation. See texts like Friedman (1986) and Fudenberg and Tirole (1991) if more precision is required.

Throughout this chapter we represent games in the normal form. A game in the normal form consists of (1) *players* indexed by $i = 1, ..., n$, (2) *strategies* or more generally a set of strategies denoted by x_i, $i = 1, ..., n$ available to each player and (3) *payoffs* $\pi_i(x_1, x_2, ..., x_n)$, $i = 1, ..., n$ received by each player. Each strategy is defined on a set X_i, $x_i \in X_i$, so we call the Cartesian product $X_1 \times X_2 \times ... \times X_n$ the *strategy space*. Each player may have a unidimensional strategy or a multi-dimensional strategy. In most SCM applications players have unidimensional strategies, so we shall either explicitly or implicitly assume unidimensional strategies throughout this chapter. Furthermore, with the exception of one example, we will work with continuous strategies, so the strategy space is R^n.

A player's strategy can be thought of as the complete instruction for which actions to take in a game. For example, a player can give his or her strategy to a person that has absolutely no knowledge of the player's payoff or preferences and that person should be able to use the instructions contained in the strategy to choose the actions the player desires. As a result, each player's set of feasible strategies must be independent of the strategies chosen by the other players, i.e., the strategy choice by one player is not allowed to limit the feasible strategies of another player. (Otherwise the game is ill defined and any analytical results obtained from the game are questionable.)

In the normal form players choose strategies simultaneously. Actions are adopted after strategies are chosen and those actions correspond to the chosen strategies.

As an alternative to the one-shot selection of strategies in the normal form, a game can also be designed in the extensive form. With the extensive form actions are chosen only as needed, so sequential choices are possible. As a result, players may learn information between the selection of actions, in particular, a player may learn which actions were previously chosen or the outcome of a random event. Figure 2.1 provides an example of a simple extensive form

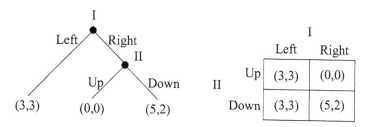

Figure 2.1. Extensive vs normal form game representation

game and its equivalent normal form representation: there are two players, player I chooses from {Left,Right} and player II chooses from {Up, Down}. In the extensive form player I chooses first, then player II chooses after learning player I's choice. In the normal form they choose simultaneously. The key distinction between normal and extensive form games is that in the normal form a player is able to commit to all future decisions. We later show that this additional commitment power may influence the set of plausible equilibria.

A player can choose a particular strategy or a player can choose to randomly select from among a set of strategies. In the former case the player is said to choose a *pure strategy* whereas in the latter case the player chooses a *mixed strategy*. There are situations in economics and marketing that have used mixed strategies: see, e.g., Varian (1980) for search models and Lal (1990) for promotion models. However, mixed strategies have not been applied in SCM, in part because it is not clear how a manager would actually implement a mixed strategy. For example, it seems unreasonable to suggest that a manager should "flip a coin" among various capacity levels. Fortunately, mixed strategy equilibria do not exist in games with a unique pure strategy equilibrium. Hence, in those games attention can be restricted to pure strategies without loss of generality. Therefore, in the remainder of this chapter we consider only pure strategies.

In a *non-cooperative* game the players are unable to make binding commitments before choosing their strategies. In a *cooperative* game players are able to make binding commitments. Hence, in a cooperative game players can make side-payments and form coalitions. We begin our analysis with non-cooperative static games.

In all sections, except the last one, we work with the games of *complete information*, i.e., the players' strategies and payoffs are common knowledge to all players.

As a practical example throughout this chapter, we utilize the classic newsvendor problem transformed into a game. In the absence of competition each newsvendor buys Q units of a single product at the beginning of a single selling season. Demand during the season is a random variable D with distribution function F_D and density function f_D. Each unit is purchased for c and sold on the market for $r > c$. The newsvendor solves the following optimization problem

$$\max_Q \pi = \max_Q E_D\left[r \min(D, Q) - cQ\right],$$

with the unique solution

$$Q^* = F_D^{-1}\left(\frac{r-c}{r}\right).$$

Goodwill penalty costs and salvage revenues can easily be incorporated into the analysis, but for our needs we normalized them out.

Now consider the GT version of the newsvendor problem with two retailers competing on product availability. Parlar (1988) was the first to analyze this problem, which is also one of the first articles modeling inventory management in a GT framework. It is useful to consider only the two-player version of this game because then graphical analysis and interpretations are feasible. Denote the two players by subscripts i and j, their strategies (in this case stocking quantities) by Q_i, Q_j and their payoffs by π_i, π_j.

We introduce interdependence of the players' payoffs by assuming the two newsvendors sell the same product. As a result, if retailer i is out of stock, all unsatisfied customers try to buy the product at retailer j instead. Hence, retailer i's total demand is $D_i + (D_j - Q_j)^+$: the sum of his own demand and the demand from customers not satisfied by retailer j. Payoffs to the two players are then

$$\pi_i(Q_i, Q_j) = E_D\left[r_i \min\left(D_i + (D_j - Q_j)^+, Q_i\right) - c_i Q_i\right],\ i,j = 1,2.$$

2.2 Best response functions and the equilibrium of the game

We are ready for the first important GT concept: *best response functions*.

DEFINITION 1. *Given an n-player game, player i's best response (function) to the strategies x_{-i} of the other players is the strategy x_i^* that maximizes* player i's payoff $\pi_i(x_i, x_{-i})$:

$$x_i^*(x_{-i}) = \arg\max_{x_i} \pi_i(x_i, x_{-i}).$$

($x_i^*(x_{-i})$ is probably better described as a correspondence rather than a function, but we shall nevertheless call it a function with an understanding that we are interpreting the term "function" liberally.) If π_i is quasi-concave in x_i the best response is uniquely defined by the first-order conditions of the payoff functions. In the context of our competing newsvendors example, the best response functions can be found by optimizing each player's payoff functions w.r.t. the player's own decision variable Q_i while taking the competitor's strategy Q_j as given. The resulting best response functions are

$$Q_i^*(Q_j) = F^{-1}_{D_i + (D_j - Q_j)^+}\left(\frac{r_i - c_i}{r_i}\right), \ i, j = 1, 2.$$

Taken together, the two best response functions form a *best response mapping* $R^2 \to R^2$ or in the more general case $R^n \to R^n$. Clearly, the best response is the best player i can hope for given the decisions of other players. Naturally, an outcome in which all players choose their best responses is a candidate for the non-cooperative solution. Such an outcome is called a Nash Equilibrium (hereafter NE) of the game.

DEFINITION 2. *An outcome* $(x_1^*, x_2^*, ..., x_n^*)$ *is a Nash equilibrium of the game if* x_i^* *is a best response to* x_{-i}^* *for all* $i = 1, 2, ..., n$.

Going back to competing newsvendors, NE is characterized by solving *a system* of best responses that translates into the system of first-order conditions:

$$Q_1^*(Q_2^*) = F^{-1}_{D_1 + (D_2 - Q_2^*)^+}\left(\frac{r_1 - c_1}{r_1}\right),$$
$$Q_2^*(Q_1^*) = F^{-1}_{D_2 + (D_1 - Q_1^*)^+}\left(\frac{r_2 - c_2}{r_2}\right).$$

When analyzing games with two players it is often helpful to graph the best response functions to gain intuition. Best responses are typically defined implicitly through the first-order conditions, which makes analysis difficult. Nevertheless, we can gain intuition by finding out how each player reacts to an increase in the stocking quantity by the other player (i.e., $\partial Q_i^*(Q_j)/\partial Q_j$) through employing implicit differentiation as follows:

$$\frac{\partial Q_i^*(Q_j)}{\partial Q_j} = -\frac{\frac{\partial^2 \pi_i}{\partial Q_i \partial Q_j}}{\frac{\partial^2 \pi_i}{\partial Q_i^2}} = -\frac{r_i f_{D_i + (D_j - Q_j)^+ | D_j > Q_j}(Q_i) \Pr(D_j > Q_j)}{r_i f_{D_i + (D_j - Q_j)^+}(Q_i)} < 0.$$

(2.1)

The expression says that the *slopes* of the best response functions are negative, which implies an intuitive result that each player's best response is monotonically decreasing in the other player's strategy. Figure 2.2 presents this result for the symmetric newsvendor game. The equilibrium is located on the intersection of the best responses and we also see that the best responses are, indeed, decreasing.

One way to think about a NE is as a *fixed point* of the best response mapping $R^n \to R^n$. Indeed, according to the definition, NE must satisfy the system of equations $\partial \pi_i / \partial x_i = 0$, all i. Recall that a fixed point x of mapping $f(x)$, $R^n \to R^n$ is any x such that $f(x) = x$. Define $f_i(x_1, ..., x_n) = \partial \pi_i / \partial x_i + x_i$. By the definition of a fixed point,

$$f_i(x_1^*, ..., x_n^*) = x_i^* = \partial \pi_i(x_1^*, ..., x_n^*)/\partial x_i + x_i^* \to \partial \pi_i(x_1^*, ..., x_n^*)/\partial x_i = 0, \text{ all } i.$$

Hence, x^* solves the first-order conditions if and only if it is a fixed point of mapping $f(x)$ defined above.

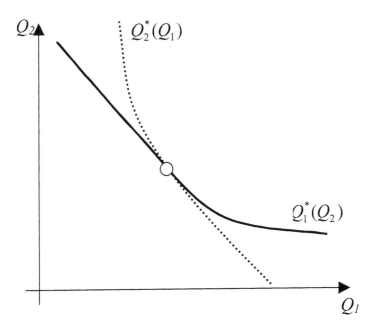

Figure 2.2. Best responses in the newsvendor game

The concept of NE is intuitively appealing. Indeed, it is a self-fulfilling prophecy. To explain, suppose a player were to guess the strategies of the other players. A guess would be consistent with payoff maximization and therefore would be reasonable only if it presumes that strategies are chosen to maximize every player's payoff given the chosen strategies. In other words, with any set of strategies that is not a NE there exists at least one player that is choosing a non payoff maximizing strategy. Moreover, the NE has a self-enforcing property: no player wants to unilaterally deviate from it since such behavior would lead to lower payoffs. Hence NE seems to be the necessary condition for the prediction of any rational behavior by players[2].

While attractive, numerous criticisms of the NE concept exist. Two particularly vexing problems are the non-existence of equilibrium and the multiplicity of equilibria. Without the existence of an equilibrium, little can be said regarding the likely outcome of the game. If there are multiple equilibria, then it is not clear which one will be the outcome. Indeed, it is possible the outcome is not even an equilibrium because the players may choose strategies from different equilibria. For example, consider the normal form game in Figure 2.1. There are two Nash equilibria in that game {Left,Up} and {Right,Down}: each is a best response to the other player's strategy. However, because the players choose their strategies simultaneously it is possible that player I chooses Right (the 2nd equilibrium) while player II choose Up (the 1st equilibrium), which results in {Right,Up}, the worst outcome for both players.

In some situations it is possible to rationalize away some equilibria via a refinement of the NE concept: e.g., trembling hand perfect equilibrium by Selten (1975), sequential equilibrium by Kreps and Wilson (1982) and proper equilibria by Myerson (1997). These refinements eliminate equilibria that are based on non-credible threats, i.e., threats of future actions that would not actually be adopted if the sequence of events in the game led to a point in the game in which those actions could be taken. The extensive form game in Figure 2.1 illustrates this point. {Left, Up} is a Nash equilibrium (just as it is in the comparable normal form game) because each player is choosing a best response to the other player's strategy: Left is optimal for player I given player II plans to play Up and player II is indifferent between Up or Down given player I chooses Left. But if player I were to choose Right, then it is unreasonable to assume player II would actually follow through with UP: UP yields a payoff of 0 while Down yields a payoff of 2. Hence, the {Left, Up} equilibrium is supported by a non-credible threat by player II to play Up. Although these refinements are viewed as extremely important in economics (Selten was awarded the Nobel prize for his work), the need for these refinements has not yet materialized in

Game Theory in Supply Chain Analysis 21

the SCM literature. But that may change as more work is done on sequential/dynamic games.

An interesting feature of the NE concept is that the system optimal solution (i.e., a solution that maximizes the sum of players' payoffs) need not be a NE. Hence, decentralized decision making generally introduces inefficiency in the supply chain. There are, however, some exceptions: see Mahajan and van Ryzin (1999b) and Netessine and Zhang (2003) for situations in which competition may result in the system-optimal performance. In fact, a NE may not even be on the *Pareto frontier*: the set of strategies such that each player can be made better off only if some other player is made worse off. A set of strategies is *Pareto optimal* if they are on the Pareto frontier; otherwise a set of strategies is *Pareto inferior*. Hence, a NE can be Pareto inferior. The Prisoner's Dilemma game (see Fudenberg and Tirole 1991) is the classic example of this: only one pair of strategies when both players "cooperate" is Pareto optimal, and the unique Nash equilibrium is when both players "defect" happens to be Pareto inferior. A large body of the SCM literature deals with ways to align the incentives of competitors to achieve optimality. See Cachon (2002) for a comprehensive survey and taxonomy. See Cachon (2003) for a supply chain analysis that makes extensive use of the Pareto optimal concept.

2.3 Existence of equilibrium

A NE is a solution to a system of n first-order conditions, so an equilibrium may not exist. Non-existence of an equilibrium is potentially a conceptual problem since in this case it is not clear what the outcome of the game will be. However, in many games a NE does exist and there are some reasonably simple ways to show that at least one NE exists. As already mentioned, a NE is a fixed point of the best response mapping. Hence fixed point theorems can be used to establish the existence of an equilibrium. There are three key fixed point theorems, named after their creators: Brouwer, Kakutani and Tarski, see Border (1999) for details and references. However, direct application of fixed point theorems is somewhat inconvenient and hence generally not done. For exceptions see Lederer and Li (1997) and Majumder and Groenevelt (2001a) for existence proofs that are based on Brouwer's fixed point theorem. Alternative methods, derived from these fixed point theorems, have been developed. The simplest and the most widely used technique for demonstrating the existence of NE is through verifying concavity of the players' payoffs.

THEOREM 1 (Debreu 1952). *Suppose that for each player the strategy space is compact[3] and convex and the payoff function is continuous and quasi-concave with respect to each player's own strategy. Then there exists at least one pure strategy NE in the game.*

If the game is symmetric in a sense that the players' strategies and payoffs are identical, one would imagine that a symmetric solution should exist. This is indeed the case, as the next Theorem ascertains.

THEOREM 2. *Suppose that a game is symmetric and for each player the strategy space is compact and convex and the payoff function is continuous and quasi-concave with respect to each player's own strategy. Then there exists at least one symmetric pure strategy NE in the game.*

To gain some intuition about why non-quasi-concave payoffs may lead to non-existence of NE, suppose that in a two-player game, player 2 has a bi-modal objective function with two local maxima. Furthermore, suppose that a small change in the strategy of player 1 leads to a shift of the global maximum for player 2 from one local maximum to another. To be more specific, let us say that at x_1' the global maximum $x_2^*(x_1')$ is on the left (Figure 2.3 left) and at x_1'' the global maximum $x_2^*(x_2'')$ is on the right (Figure 2.3 right). Hence, a small change in x_1 from x_1' to x_1'' induces a jump in the best response of player 2, x_2^*. The resulting best response mapping is presented in Figure 2.4 and there is no NE in pure strategies in this game. In other words, best response functions do not intersect anywhere. As a more specific example, see Netessine and Shumsky (2001) for an extension of the newsvendor game to the situation in which product inventory is sold at two different prices; such a game may not have a NE since both players' objectives may be bimodal. Furthermore, Cachon and Harker (2002) demonstrate that pure strategy NE may not exist in two other important settings: two retailers competing with cost functions described by the Economic Order Quantity (EOQ) or two service providers competing with service times described by the $M/M/1$ queuing model.

The assumption of a compact strategy space may seem restrictive. For example, in the newsvendor game the strategy space R_+^2 is not bounded from above. However, we could easily bound it with some large enough finite number to represent the upper bound on the demand distribution. That bound would not impact any of the choices, and therefore the transformed game behaves just as the original game with an unbounded strategy space. (However, that bound cannot depend on any player's strategy choice.)

To continue with the newsvendor game analysis, it is easy to verify that the newsvendor's objective function is concave and hence quasi-concave w.r.t. the stocking quantity by taking the second derivative. Hence the conditions of Theorem 1 are satisfied and a NE exists. There are virtually dozens of papers employing Theorem 1. See, for example, Lippman and McCardle (1997)

Game Theory in Supply Chain Analysis

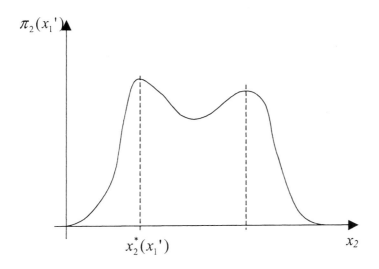

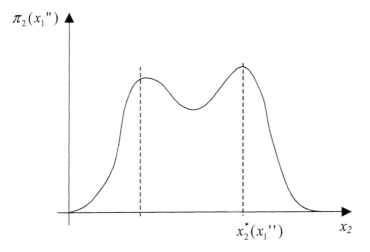

Figure 2.3. Example with a bi-modal objective function

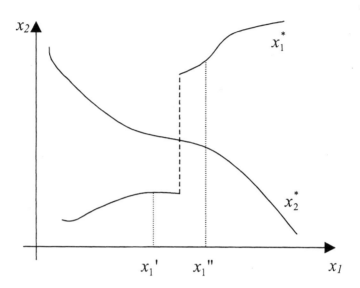

Figure 2.4. Non-existence of NE

for the proof involving quasi-concavity, Mahajan and van Ryzin (1999a) and Netessine et al. (2002) for the proofs involving concavity. Clearly, quasi-concavity of each player's objective function only implies uniqueness of the best response but does not imply a unique NE. One can easily envision a situation where unique best response functions cross more than once so that there are multiple equilibria (see Figure 2.5).

If quasi-concavity of the players' payoffs cannot be verified, there is an alternative existence proof that relies on Tarski's (1955) fixed point theorem and involves the notion of supermodular games. The theory of supermodular games is a relatively recent development introduced and advanced by Topkis (1998).

DEFINITION 3. *A twice continuously differentiable payoff function $\pi_i(x_1, ..., x_n)$ is supermodular (submodular) iff $\partial^2 \pi_i / \partial x_i \partial x_j \geq 0$ (≤ 0) for all x and all $j \neq i$. The game is called supermodular if the players' payoffs are supermodular.*

Supermodularity essentially means complementarity between any two strategies and is not linked directly to either convexity, concavity or even continuity. (This is a significant advantage when forced to work with discrete strategies, e.g., Cachon 2001.) However, similar to concavity/convexity, supermodular-

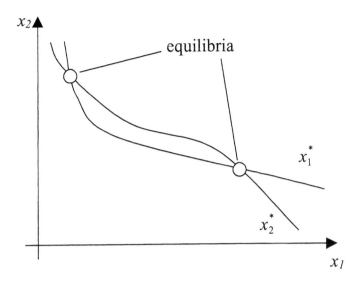

Figure 2.5. Non-uniqueness of the equilibrium

ity/submodularity is preserved under maximization, limits and addition and hence under expectation/integration signs, an important feature in stochastic SCM models. While in most situations the positive sign of the second derivative can be used to verify supermodularity (using Definition 3), sometimes it is necessary to utilize supermodularity-preserving transformations to show that payoffs are supermodular. Topkis (1998) provides a variety of ways to verify that the function is supermodular and some of his results are used in Cachon (2001), Cachon and Lariviere (1999), Netessine and Shumsky (2001) and Netessine and Rudi (2003). The following theorem follows directly from Tarski's fixed point result and provides another tool to show existence of NE in non-cooperative games:

THEOREM 3. *In a supermodular game there exists at least one NE.*

Coming back to the competitive newsvendors example, recall that the second-order cross-partial derivative was found to be

$$\frac{\partial^2 \pi_i}{\partial Q_i \partial Q_j} = -r_i f_{D_i+(D_j-Q_j)^+|D_j>Q_j}(Q_i)\Pr(D_j > Q_j) < 0,$$

so that the newsvendor game is submodular and hence existence of equilibrium cannot be assured. However, a standard trick is to re-define the ordering of the players' strategies. Let $y = -Q_j$ so that

$$\frac{\partial^2 \pi_i}{\partial Q_i \partial y} = -r_i f_{D_i + (D_j + y)^+ | D_j > Q_j}(Q_i) \Pr(D_j > -y) > 0,$$

and the game becomes supermodular in (x_i, y) so existence of NE is assured. Notice that we do not change either payoffs or the structure of the game, we only alter the ordering of one player's strategy space. Obviously, this trick only works in two-player games, see also Lippman and McCardle (1997) for the analysis of the more general version of the newsvendor game using a similar transformation. Hence, we can state that in general NE exists in games with decreasing best responses (submodular games) with two players. This argument can be generalized slightly in two ways that we mention briefly, see Vives (1999) for details. One way is to consider an n-player game where best responses are functions of aggregate actions of all other players, that is, $x_i^* = x_i^* \left(\sum_{j \neq i} x_j \right)$. If best responses in such a game are decreasing, then NE exists. Another generalization is to consider the same game with $x_i^* = x_i^* \left(\sum_{j \neq i} x_j \right)$ but require symmetry. In such a game, existence can be shown even with non-monotone best responses provided that there are only jumps up but on intervals between jumps best responses can be increasing or decreasing.

We now step back to discuss the intuition behind the supermodularity results. Roughly speaking, Tarski's fixed point theorem only requires best response mappings to be non-decreasing for the existence of equilibrium and does not require quasi-concavity of the players' payoffs and allows for jumps in best responses. While it may be hard to believe that non-decreasing best responses is the only requirement for the existence of a NE, consider once again the simplest form of a single-dimensional equilibrium as a solution to the fixed point mapping $x = f(x)$ on the compact set. It is easy to verify after a few attempts that if $f(x)$ is non-decreasing but possibly with jumps up then it is not possible to derive a situation without an equilibrium. However, when $f(x)$ jumps down, non-existence is possible (see Figure 2.6).

Hence, increasing best response functions is the only major requirement for an equilibrium to exist; players' objectives do not have to be quasi-concave or even continuous. However, to describe an existence theorem with non-continuous payoffs requires the introduction of terms and definitions from lattice theory. As a result, we restricted ourselves to the assumption of continuous payoff functions, and in particular, to twice-differentiable payoff functions.

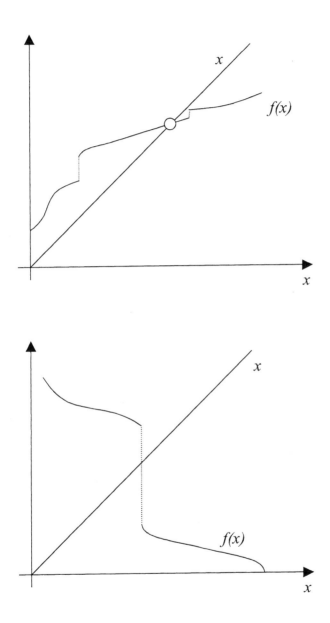

Figure 2.6. Increasing (left) and decreasing (right) mappings

Although it is now clear why increasing best responses ensure existence of an equilibrium, it is not immediately obvious why Definition 3 provides a sufficient condition, given that it only concerns the sign of the second-order cross-partial derivative. To see this connection, consider separately the continuous and the dis-continuous parts of the best response $x_i^*(x_j)$. When the best response is continuous, we can apply the Implicit Function Theorem to find its slope as follows

$$\frac{\partial x_i^*}{\partial x_j} = -\frac{\frac{\partial^2 \pi_i}{\partial x_i \partial x_j}}{\frac{\partial^2 \pi_i}{\partial x_i^2}}.$$

Clearly, if x_i^* is the best response, it must be the case that $\partial^2 \pi_i / \partial x_i^2 < 0$ or else it would not be the best response. Hence, for the slope to be positive it is sufficient to have $\partial^2 \pi_i / \partial x_i \partial x_j > 0$ which is what Definition 3 provides. This reasoning does not, however, work at discontinuities in best responses since the Implicit Function Theorem cannot be applied. To show that only jumps up are possible if $\partial^2 \pi_i / \partial x_i \partial x_j > 0$ holds, consider a situation in which there is a jump down in the best response. As one can recall, jumps in best responses happen when the objective function is bi-modal (or more generally multi-modal). For example, consider a specific point $x_j^\#$ and let $x_i^1\left(x_j^\#\right) < x_i^2\left(x_j^\#\right)$ be two distinct points at which first-order conditions hold (i.e., the objective function π_i is bi-modal). Further, suppose $\pi_i\left(x_i^1\left(x_j^\#\right), x_j^\#\right) < \pi_i\left(x_i^2\left(x_j^\#\right), x_j^\#\right)$ but $\pi_i\left(x_i^1\left(x_j^\# + \varepsilon\right), x_j^\# + \varepsilon\right) > \pi_i\left(x_i^2\left(x_j^\# + \varepsilon\right), x_j^\# + \varepsilon\right)$. That is, initially $x_i^2\left(x_j^\#\right)$ is a global maximum but as we increase $x_j^\#$ infinitesimally, there is a *jump down* and a smaller $x_i^1\left(x_j^\# + \varepsilon\right)$ becomes the global maximum. For this to be the case, it must be that

$$\frac{\partial \pi_i\left(x_i^1\left(x_j^\#\right), x_j^\#\right)}{\partial x_j} > \frac{\partial \pi_i\left(x_i^2\left(x_j^\#\right), x_j^\#\right)}{\partial x_j},$$

or, in words, the objective function rises faster at $\left(x_i^1\left(x_j^\#\right), x_j^\#\right)$ than at $\left(x_i^2\left(x_j^\#\right), x_j^\#\right)$. This, however, can only happen if $\partial^2 \pi_i / \partial x_i \partial x_j < 0$ at least somewhere on the interval $\left[x_i^1\left(x_j^\#\right), x_i^2\left(x_j^\#\right)\right]$ which is a contradiction. Hence, if $\partial^2 \pi_i / \partial x_i \partial x_j > 0$ holds then only jumps up in the best response are possible.

2.4 Uniqueness of equilibrium

From the perspective of generating qualitative insights, it is quite useful to have a game with a unique NE. If there is only one equilibrium, then one can characterize equilibrium actions without much ambiguity. Unfortunately, demonstrating uniqueness is generally much harder than demonstrating existence of equilibrium. This section provides several methods for proving uniqueness. No single method dominates; all may have to be tried to find the one that works. Furthermore, one should be careful to recognize that these methods assume existence, i.e., existence of NE must be shown separately. Finally, it is worth pointing out that uniqueness results are only available for games with continuous best response functions and hence there are no general methods to prove uniqueness of NE in supermodular games.

2.4.1 Method 1. Algebraic argument.

In some rather fortunate situations one can ascertain that the solution is unique by simply looking at the optimality conditions. For example, in a two-player game the optimality condition of one of the players may have a unique closed-form solution that does not depend on the other player's strategy and, given the solution for one player, the optimality condition for the second player can be solved uniquely. See Hall and Porteus (2000) and Netessine and Rudi (2001) for examples. In other cases one can assure uniqueness by analyzing geometrical properties of the best response functions and arguing that they intersect only once. Of course, this is only feasible in two-player games. See Parlar (1988) for a proof of uniqueness in the two-player newsvendor game and Majumder and Groenevelt (2001b) for a supply chain game with competition in reverse logistics. However, in most situations these geometrical properties are also implied by the more formal arguments stated below. Finally, it may be possible to use a contradiction argument: assume that there is more than one equilibrium and prove that such an assumption leads to a contradiction, as in Lederer and Li (1997).

2.4.2 Method 2. Contraction mapping argument.

Although the most restrictive among all methods, the contraction mapping argument is the most widely known and is the most frequently used in the literature because it is the easiest to verify. The argument is based on showing that the best response mapping is a contraction, which then implies the mapping has a unique fixed point. To illustrate the concept of a contraction mapping, suppose we would like to find a solution to the following fixed point equation:

$$x = f(x), x \in R^1.$$

To do so, a sequence of values is generated by an iterative algorithm, $\{x^{(1)}, x^{(2)}, x^{(3)}, ...\}$ where $x^{(1)}$ is arbitrarily picked and $x^{(t)} = f\left(x^{(t-1)}\right)$.

The hope is that this sequence converges to a unique fixed point. It does so if, roughly speaking, each step in the sequence moves closer to the fixed point. One could verify that if $|f'(x)| < 1$ in some vicinity of x^* then such an iterative algorithm converges to a unique $x^* = f(x^*)$. Otherwise, the algorithm diverges. Graphically, the equilibrium point is located on the intersection of two functions: x and $f(x)$. The iterative algorithm is presented in Figure 2.7.

The iterative scheme in Figure 2.7 left is a contraction mapping: it approaches the equilibrium after every iteration.

DEFINITION 4. *Mapping $f(x)$, $R^n \to R^n$ is a contraction iff* $\|f(x_1) - f(x_2)\| \le \alpha \|x_1 - x_2\|, \forall x_1, x_2, \alpha < 1$.

In words, the application of a contraction mapping to any two points strictly reduces (i.e., $\alpha = 1$ does not work) the distance between these points. The norm in the definition can be any norm, i.e., the mapping can be a contraction in one norm and not a contraction in another norm.

THEOREM 4. *If the best response mapping is a contraction on the entire strategy space, there is a unique NE in the game.*

One can think of a contraction mapping in terms of iterative play: player 1 selects some strategy, then player 2 selects a strategy based on the decision by player 1, etc. If the best response mapping is a contraction, the NE obtained as a result of such iterative play is *stable* but the opposite is not necessarily true, i.e., no matter where the game starts, the final outcome is the same. See also Moulin (1986) for an extensive treatment of stable equilibria.

A major restriction in Theorem 4 is that the contraction mapping condition must be satisfied everywhere. This assumption is quite restrictive because the best response mapping may be a contraction locally, say in some not necessarily small ε-neighborhood of the equilibrium but not outside of it. Hence, if iterative play starts in this ε-neighborhood, then it converges to the equilibrium, but starting outside that neighborhood may not lead to the equilibrium (even if the equilibrium is unique). Even though one may wish to argue that it is reasonable for the players to start iterative play some place close to the equilibrium, formalization of such an argument is rather difficult. Hence, we must impose the condition that the entire strategy space be considered. See Stidham (1992) for an interesting discussion of stability issues in a queuing system.

While Theorem 4 is a starting point towards a method for demonstrating uniqueness, it does not actually explain how to validate that a best reply mapping is a contraction. Suppose we have a game with n players each endowed

Game Theory in Supply Chain Analysis

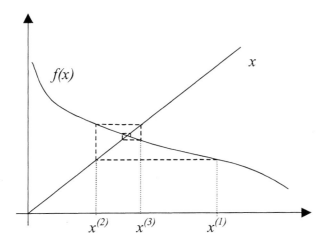

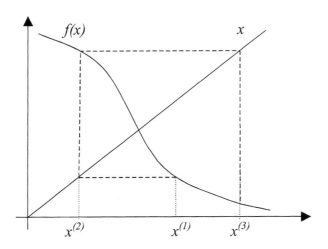

Figure 2.7. Converging (left) and diverging (right) iterations

with the strategy x_i and we have obtained the best response functions for all players, $x_i = f_i(x_{-i})$. We can then define the following matrix of derivatives of the best response functions:

$$A = \begin{vmatrix} 0 & \frac{\partial f_1}{\partial x_2} & \cdots & \frac{\partial f_1}{\partial x_n} \\ \frac{\partial f_2}{\partial x_1} & 0 & \cdots & \frac{\partial f_2}{\partial x_2} \\ \cdots & \cdots & \cdots & \cdots \\ \frac{\partial f_n}{\partial x_1} & \frac{\partial f_n}{\partial x_2} & \cdots & 0 \end{vmatrix}.$$

Further, denote by $\rho(A)$ the spectral radius of matrix A and recall that the spectral radius of a matrix is equal to the largest absolute eigenvalue $\rho(A) = \{\max |\lambda| : Ax = \lambda x, x \neq 0\}$, see Horn and Johnson (1996).

THEOREM 5. *The mapping $f(x)$, $R^n \to R^n$ is a contraction if and only if $\rho(A) < 1$ everywhere.*

Theorem 5 is simply an extension of the iterative convergence argument we used above into multiple dimensions, and the spectral radius rule is an extension of the requirement $|f'(x)| < 1$. Still, Theorem 5 is not as useful as we would like it to be: calculating eigenvalues of a matrix is not trivial. Instead, it is often helpful to use the fact that the largest eigenvalue and hence the spectral radius is bounded above by any of the matrix norms, see Horn and Johnson (1996). So instead of working with the spectral radius itself, it is sufficient to show $\|A\| < 1$ for any one matrix norm. The most convenient matrix norms are the maximum column-sum and the maximum row-sum norms (see Horn and Johnson 1996 for other matrix norms). To use either of these norms to verify the contraction mapping, it is sufficient to verify that no column sum or no row sum of matrix A exceeds one,

$$\sum_{i=1}^{n} \left|\frac{\partial f_k}{\partial x_i}\right| < 1, \text{ or } \sum_{i=1}^{n} \left|\frac{\partial f_i}{\partial x_k}\right| < 1, \forall k.$$

Netessine and Rudi (2003) used the contraction mapping argument in this most general form in the multiple-player variant of the newsvendor game described above.

A challenge associated with the contraction mapping argument is finding best response functions because in most SC models best responses cannot be found explicitly. Fortunately, Theorem 5 only requires the derivatives of the best response functions, which can be done using the Implicit Function Theorem (from now on, IFT, see Bertsekas 1999). Using the IFT, Theorem 5 can

be re-stated as
$$\sum_{i=1, i\neq k}^{n} \left|\frac{\partial^2 \pi_k}{\partial x_k \partial x_i}\right| < \left|\frac{\partial^2 \pi_k}{\partial x_k^2}\right|, \forall k. \tag{2.2}$$

This condition is also known as "diagonal dominance" because the diagonal of the matrix of second derivatives, also called the Hessian, dominates the off-diagonal entries:

$$H = \begin{vmatrix} \frac{\partial^2 \pi_1}{\partial x_1^2} & \frac{\partial^2 \pi_1}{\partial x_1 \partial x_2} & \cdots & \frac{\partial^2 \pi_1}{\partial x_1 \partial x_n} \\ \frac{\partial^2 \pi_2}{\partial x_2 \partial x_1} & \frac{\partial^2 \pi_2}{\partial x_2^2} & \cdots & \frac{\partial^2 \pi_1}{\partial x_2 \partial x_n} \\ \cdots & \cdots & \cdots & \cdots \\ \frac{\partial^2 \pi_n}{\partial x_n \partial x_1} & \frac{\partial^2 \pi_n}{\partial x_n \partial x_2} & \cdots & \frac{\partial^2 \pi_n}{\partial x_n^2} \end{vmatrix}. \tag{2.3}$$

Contraction mapping conditions in the diagonal dominance form have been used extensively by Bernstein and Federgruen (2000, 2001a, 2001c, 2003). As has been noted by Bernstein and Federgruen (2002), many standard economic demand models satisfy this condition.

In games with only two players the condition in Theorem 5 simplifies to

$$\left|\frac{\partial f_1}{\partial x_2}\right| < 1 \text{ and } \left|\frac{\partial f_2}{\partial x_1}\right| < 1, \tag{2.4}$$

i.e., the slopes of the best response functions are less than one. This condition is especially intuitive if we use the graphical illustration (Figure 2.2). Given that the slope of each best response function is less than one everywhere, if they cross at one point then they cannot cross at an additional point. A contraction mapping argument in this form was used by van Mieghem (1999) and by Rudi et al. (2001).

Returning to the newsvendor game example, we have found that the slopes of the best response functions are

$$\left|\frac{\partial Q_i^*(Q_j)}{\partial Q_j}\right| = \left|\frac{f_{D_i + (D_j - Q_j)^+ | D_j > Q_j}(Q_i) \Pr(D_j > Q_j)}{f_{D_i + (D_j - Q_j)^+}(Q_i)}\right| < 1.$$

Hence, the best response mapping in the newsvendor game is a contraction and the game has a unique and stable NE.

2.4.3 Method 3. Univalent mapping argument.
Another method for demonstrating uniqueness of equilibrium is based on verifying that the best response mapping is one-to-one: that is, if $f(x)$ is a $R^n \to R^n$ mapping, then $y = f(x)$ implies that for all $x' \neq x, y \neq f(x')$. Clearly, if

the best response mapping is one-to-one then there can be at most one fixed point of such mapping. To make an analogy, recall that, if the equilibrium is interior[4], the NE is a solution to the system of the first-order conditions: $\partial \pi_i / \partial x_i = 0$, $\forall i$, which defines the best response mapping. If this mapping is single-dimensional $R^1 \to R^1$ then it is quite clear that the condition sufficient for the mapping to be one-to-one is quasi-concavity of π_i. Similarly, for the $R^n \to R^n$ mapping to be one-to-one we require quasi-concavity of the mapping which translates into quasi-definiteness of the Hessian:

THEOREM 6. *Suppose the strategy space of the game is convex and all equilibria are interior. Then if the determinant $|H|$ is negative quasi-definite (i.e., if the matrix $H + H^T$ is negative definite) on the players' strategy set, there is a unique NE.*

Proof of this result can be found in Gale and Nikaido (1965) and some further developments that deal with boundary equilibria are found in Rosen (1965). Notice that the univalent mapping argument is somewhat weaker than the contraction mapping argument. Indeed, the re-statement (2.2) of the contraction mapping theorem directly implies univalence since the dominant diagonal assures us that H is negative definite. Hence, it is negative quasi-definite. It immediately follows that the newsvendor game satisfies the univalence theorem. However, if some other matrix norm is used, the relationship between the two theorems is not that specific. In the case of just two players the univalence theorem can be written as, according to Moulin (1986),

$$\left| \frac{\partial^2 \pi_2}{\partial x_2 \partial x_1} + \frac{\partial^2 \pi_1}{\partial x_1 \partial x_2} \right| \leq 2\sqrt{\left| \frac{\partial^2 \pi_1}{\partial x_1^2} \cdot \frac{\partial^2 \pi_2}{\partial x_2^2} \right|}, \forall x_1, x_2.$$

2.4.4 Method 4. Index theory approach..
This method is based on the Poincare-Hopf index theorem found in differential topology, see, e.g., Gillemin and Pollack (1974). Similarly to the univalence mapping approach, it requires a certain sign from the Hessian, but this requirement need hold only at the equilibrium point.

THEOREM 7. *Suppose the strategy space of the game is convex and all payoff functions are quasi-concave. Then if $(-1)^n |H|$ is positive whenever $\partial \pi_i / \partial x_i = 0$, all i, there is a unique NE.*

Observe that the condition $(-1)^n |H|$ is trivially satisfied if $|H|$ is negative definite which is implied by the condition (2.2) of contraction mapping, i.e., this method is also somewhat weaker than the contraction mapping argument. Moreover, the index theory condition need only hold at the equilibrium. This makes it the most general, but also the hardest to apply. To gain some intuition about why the index theory method works, consider the two-player game. The condition of Theorem 7 simplifies to

$$\begin{vmatrix} \frac{\partial^2 \pi_1}{\partial x_1^2} & \frac{\partial^2 \pi_1}{\partial x_1 \partial x_2} \\ \frac{\partial^2 \pi_2}{\partial x_2 \partial x_1} & \frac{\partial^2 \pi_2}{\partial x_2^2} \end{vmatrix} > 0 \;\forall x_1, x_2 : \frac{\partial \pi_1}{\partial x_1} = 0, \frac{\partial \pi_2}{\partial x_2} = 0,$$

which can be interpreted as meaning the multiplication of the slopes of best response functions should not exceed one at the equilibrium:

$$\frac{\partial f_1}{\partial x_2} \frac{\partial f_2}{\partial x_1} < 1 \; at \; x_1^*, x_2^*. \tag{2.5}$$

As with the contraction mapping approach, with two players the Theorem becomes easy to visualize. Suppose we have found best response functions $x_1^* = f_1(x_2)$ and $x_2^* = f_2(x_1)$ as in Figure 2.2. Find an inverse function $x_2 = f_1^{-1}(x_1)$ and construct an auxiliary function $g(x_1) = f_1^{-1}(x_1) - f_2(x_1)$ that measures the distance between two best responses. It remains to show that $g(x_1)$ crosses zero only once since this would directly imply a single crossing point of $f_1(x_1)$ and $f_2(x_2)$. Suppose we could show that every time $g(x_1)$ crosses zero, it does so *from below*. If that is the case, we are assured there is only a single crossing: it is impossible for a continuous function to cross zero more than once from below because it would also have to cross zero from above somewhere. It can be shown that the function $g(x_1)$ crosses zero only from below if the slope of $g(x_1)$ *at the crossing point* is positive as follows

$$\frac{\partial g(x_1)}{\partial x_1} = \frac{\partial f_1^{-1}(x_1)}{\partial x_1} - \frac{\partial f_2(x_1)}{\partial x_1} = \frac{1}{\frac{\partial f_2(x_2)}{\partial x_2}} - \frac{\partial f_2(x_1)}{\partial x_1} > 0,$$

which holds if (2.5) holds. Hence, in a two-player game condition (2.5) is sufficient for the uniqueness of the NE. Note that condition (2.5) trivially holds in the newsvendor game since each slope is less than one and hence the multiplication of slopes is less than one as well *everywhere*. Index theory has been used by Netessine and Rudi (2001b) to show uniqueness of the NE in a retailer-wholesaler game when both parties stock inventory and sell directly to consumers and by Cachon and Kok (2002) and Cachon and Zipkin (1999).

2.5 Multiple equilibria

Many games are just not blessed with a unique equilibrium. The next best situation is to have a few equilibria. The worst situation is either to have an infinite number of equilibria or no equilibrium at all. The obvious problem with multiple equilibria is that the players may not know which equilibrium will prevail. Hence, it is entirely possible that a non-equilibrium outcome results because one player plays one equilibrium strategy while a second player chooses a strategy associated with another equilibrium. However, if a game is repeated, then it is possible that the players eventually find themselves in one particular equilibrium. Furthermore, that equilibrium may not be the most desirable one.

If one does not want to acknowledge the possibility of multiple outcomes due to multiple equilibria, one could argue that one equilibrium is more reasonable than the others. For example, there may exist only one symmetric equilibrium and one may be willing to argue that a symmetric equilibrium is more focal than an asymmetric equilibrium. (See Mahajan and van Ryzin 1999a for an example). In addition, it is generally not too difficult to demonstrate the uniqueness of a symmetric equilibrium. If the players have unidimensional strategies, then the system of n first-order conditions reduces to a single equation and one need only show that there is a unique solution to that equation to prove the symmetric equilibrium is unique. If the players have m-dimensional strategies, $m > 1$, then finding a symmetric equilibrium reduces to determining whether a system of m equations has a unique solution (easier than the original system, but still challenging).

An alternative method to rule out some equilibria is to focus only on the Pareto optimal equilibrium, of which there may be only one. For example, in supermodular games the equilibria are Pareto rankable under an additional condition that each players' objective function is increasing in other players' strategies, i.e., there is a most preferred equilibrium by every player and a least preferred equilibrium by every player. (See Wang and Gerchak 2002 for an example). However, experimental evidence exists that suggests players do not necessarily gravitate to the Pareto optimal equilibrium as is demonstrated by Cachon and Camerer, 1996. Hence, caution is warranted with this argument.

2.6 Comparative statics in games

In GT models, just as in the non-competitive SCM models, many of the managerial insights and results are obtained through comparative statics such as monotonicity of the optimal decisions w.r.t. some parameter of the game.

2.6.1 The Implicit Functions Theorem approach.
This approach works for both GT and single decision-maker applications as will become evident from the statement of the next theorem.

Game Theory in Supply Chain Analysis

THEOREM 9. *Consider the system of equations*

$$\frac{\partial \pi_i(x_1, ..., x_n, a)}{\partial x_i} = 0, \ i = 1, ..., n,$$

defining $x_1^*, ..., x_n^*$ *as implicit functions of parameter* a. *If all derivatives are continuous functions and the Hessian (2.3) evaluated at* $x_1^*, ..., x_n^*$ *is non-zero, then the function* $x^*(a)$, $R^1 \to R^n$ *is continuous on a ball around* x^* *and its derivatives are found as follows:*

$$\begin{vmatrix} \frac{\partial x_1^*}{\partial a} \\ \frac{\partial x_2^*}{\partial a} \\ ... \\ \frac{\partial x_n^*}{\partial a} \end{vmatrix} = - \begin{vmatrix} \frac{\partial^2 \pi_1}{\partial x_1^2} & \frac{\partial^2 \pi_1}{\partial x_1 \partial x_2} & ... & \frac{\partial^2 \pi_1}{\partial x_1 \partial x_n} \\ \frac{\partial^2 \pi_2}{\partial x_2 \partial x_1} & \frac{\partial^2 \pi_2}{\partial x_2^2} & ... & \frac{\partial^2 \pi_1}{\partial x_2 \partial x_n} \\ ... & ... & ... & ... \\ \frac{\partial^2 \pi_n}{\partial x_n \partial x_1} & \frac{\partial^2 \pi_n}{\partial x_n \partial x_2} & ... & \frac{\partial^2 \pi_n}{\partial x_n^2} \end{vmatrix}^{-1} \begin{vmatrix} \frac{\partial \pi_1}{\partial x_1 \partial a} \\ \frac{\partial \pi_1}{\partial x_2 \partial a} \\ ... \\ \frac{\partial \pi_1}{\partial x_n \partial a} \end{vmatrix}. \quad (2.6)$$

Since the IFT is covered in detail in many non-linear programming books and its application to the GT problems is essentially the same, we do not delve further into this matter. In many practical problems, if $|H| \neq 0$ then it is instrumental to multiply both sides of the expression (2.6) by H^{-1}. That is justified because the Hessian is assumed to have a non-zero determinant to avoid the cumbersome task of inverting the matrix. The resulting expression is a system of n linear equations which have a closed form solution. See Netessine and Rudi (2001b) for such an application of the IFT in a two-player game and Bernstein and Federgruen (2000) in n-player games.

The solution to (2.6) in the case of two players is

$$\frac{\partial x_1^*}{\partial a} = -\frac{\frac{\partial^2 \pi_1}{\partial x_1 \partial a}\frac{\partial^2 \pi_2}{\partial x_2^2} - \frac{\partial^2 \pi_1}{\partial x_1 \partial x_2}\frac{\partial^2 \pi_2}{\partial x_2 \partial a}}{|H|}, \quad (2.7)$$

$$\frac{\partial x_2^*}{\partial a} = -\frac{\frac{\partial^2 \pi_1}{\partial x_1^2}\frac{\partial^2 \pi_2}{\partial x_2 \partial a} - \frac{\partial^2 \pi_1}{\partial x_1 \partial a}\frac{\partial^2 \pi_2}{\partial x_2 \partial x_1}}{|H|}. \quad (2.8)$$

Using our newsvendor game as an example, suppose we would like to analyze sensitivity of the equilibrium solution to changes in r_1 so let $a = r_1$. Notice that $\partial \pi_2 / \partial Q_2 \partial r_1 = 0$ and also that the determinant of the Hessian is positive. Both expressions in the numerator of (2.7) are positive as well so that $\partial Q_1^* / \partial r_1 > 0$. Further, the numerator of (2.8) is negative so that $\partial Q_2^* / \partial r_1 < 0$. Both results are intuitive.

Solving a system of n equations analytically is generally cumbersome and one may have to use Kramer's rule or analyze an inverse of H instead, see Bernstein and Federgruen (2000) for an example. The only way to avoid this complication is to employ supermodular games as described below. However, the IFT method has an advantage that is not enjoyed by supermodular games: it can handle constraints of any form. That is, any constraint on the players' strategy spaces of the form $g_i(x_i) \leq 0$ or $g_i(x_i) = 0$ can be added to the objective function by forming a Lagrangian:

$$L_i(x_1, ..., x_n, \lambda_i) = \pi_i(x_1, ..., x_n) - \lambda_i g_i(x_i).$$

All analysis can then be carried through the same way as before with the only addition being that the Lagrange multiplier λ_i becomes a decision variable. For example, let's assume in the newsvendor game that the two competing firms stock inventory at a warehouse. Further, the amount of space available to each company is a function of the total warehouse capacity C, e.g., $g_i(Q_i) \leq C$. We can construct a new game where each retailer solves the following problem:

$$\max_{Q_i \in \{g_i(Q_i) \leq C\}} E_D\left[r_i \min\left(D_i + (D_j - Q_j)^+, Q_i\right) - c_i Q_i\right], i = 1, 2.$$

Introduce two Lagrange multipliers, λ_i, $i = 1, 2$ and re-write the objective functions as

$$\max_{Q_i, \lambda_i} L(Q_i, \lambda_i, Q_j)$$
$$= E_D\left[r_i \min\left(D_i + (D_j - Q_j)^+, Q_i\right) - c_i Q_i - \lambda_i \left(g_i(Q_i) - C\right)\right].$$

The resulting four optimality conditions can be analyzed using the IFT the same way as has been demonstrated previously.

2.6.2 Supermodular games approach.
In some situations, supermodular games provide a more convenient tool for comparative statics.

THEOREM 11. *Consider a collection of supermodular games on R^n parameterized by a parameter a. Further, suppose $\partial^2 \pi_i / \partial x_i \partial a \geq 0$ for all i. Then the largest and the smallest equilibria are increasing in a.*

Roughly speaking, a sufficient condition for monotone comparative statics is supermodularity of players' payoffs in strategies *and* a parameter. Note that, if there are multiple equilibria, we cannot claim that every equilibrium is monotone in a; rather, a set of all equilibria is monotone in the sense of Theorem 10. A convenient way to think about the last Theorem is through the augmented Hessian:

$$\begin{vmatrix} \frac{\partial^2 \pi_1}{\partial x_1^2} & \frac{\partial^2 \pi_1}{\partial x_1 \partial x_2} & \cdots & \frac{\partial^2 \pi_1}{\partial x_1 \partial x_n} & \frac{\partial^2 \pi_1}{\partial x_1 \partial a} \\ \frac{\partial^2 \pi_2}{\partial x_2 \partial x_1} & \frac{\partial^2 \pi_2}{\partial x_2^2} & \cdots & \frac{\partial^2 \pi_1}{\partial x_2 \partial x_n} & \frac{\partial^2 \pi_1}{\partial x_2 \partial a} \\ \cdots & \cdots & \cdots & \cdots & \cdots \\ \frac{\partial^2 \pi_n}{\partial x_n \partial x_1} & \frac{\partial^2 \pi_n}{\partial x_n \partial x_2} & \cdots & \frac{\partial^2 \pi_n}{\partial x_n^2} & \frac{\partial^2 \pi_n}{\partial x_n \partial a} \\ \frac{\partial^2 \pi_1}{\partial x_1 \partial a} & \frac{\partial^2 \pi_1}{\partial x_2 \partial a} & \cdots & \frac{\partial^2 \pi_n}{\partial x_n \partial a} & \frac{\partial^2 \pi_n}{\partial a^2} \end{vmatrix}.$$

Roughly, if all *off-diagonal* elements of this matrix are positive, then the monotonicity result holds (signs of diagonal elements do not matter and hence concavity is not required). To apply this result to competing newsvendors we will analyze sensitivity of equilibrium inventories $\left(Q_i^*, Q_j^*\right)$ to r_i. First, transform the game to strategies (Q_i, y) so that the game is supermodular and find cross-partial derivatives

$$\begin{aligned} \frac{\partial^2 \pi_i}{\partial Q_i \partial r_i} &= \Pr\left(D_i + (D_j - Q_j)^+ > Q_i\right) \geq 0, \\ \frac{\partial \pi_j}{\partial y \partial r_i} &= 0 \geq 0, \end{aligned}$$

so that (Q_i^*, y^*) are both increasing in r_i or Q_i^* is increasing and Q_j^* is decreasing in r_i just as we have already established using the IFT.

The simplicity of the argument (once supermodular games are defined) as compared to the machinery required to derive the same result using the IFT is striking. Such simplicity has attracted much attention in SCM and has resulted in extensive applications of supermodular games. Examples include Cachon (2001), Corbett and DeCroix (2001), Netessine and Shumsky (2001) and Netessine and Rudi (2001b), to name just a few. There is, however, an important limitation to the use of Theorem 10: it cannot handle many constraints as IFT can. Namely, the decision space must be a lattice to apply supermodularity i.e., it must include its coordinate-wise maximum and minimum. Hence, a constraint of the form $x_i \leq b$ can be handled but a constraint $x_i + x_j \leq b$ cannot since points $(x_i, x_j) = (b, 0)$ and $(x_i, x_j) = (0, b)$ are within the constraint but the coordinate-wise maximum of these two points (b, b) is not. Notice that to avoid dealing with this issue in detail we stated in the theorems that the strategy space should all be R^n. Since in many SCM applications there are constraints on the players' strategies, supermodularity must be applied with care.

3. Dynamic games

While many SCM models are static, including all newsvendor-based models, a significant portion of the SCM literature is devoted to dynamic models in which decisions are made over time. In most cases the solution concept for these games is similar to the backwards induction used when solving dynamic programming problems. There are, however, important differences as will be clear from the discussion of repeated games. As with dynamic programming problems, we continue to focus on the games of complete information, i.e., at each move in the game all players know the full history of play.

3.1 Sequential moves: Stackelberg equilibrium concept

The simplest possible dynamic game was introduced by Stackelberg (1934). In a Stackelberg duopoly model, player 1, the Stackelberg leader, chooses a strategy first and then player 2, the Stackelberg follower, observes this decision and makes his own strategy choice. Since in many SCM models the upstream firm, e.g., the wholesaler, possesses certain power over the typically smaller downstream firm, e.g., the retailer, the Stackelberg equilibrium concept has found many applications in SCM literature. We do not address the issues of who should be the leader and who should be the follower.

To find an equilibrium of a Stackelberg game which often called the Stackelberg equilibrium we need to solve a dynamic multi-period problem via backwards induction. We will focus on a two-period problem for analytical convenience. First, find the solution $x_2^*(x_1)$ for the second player as a response to any decision made by the first player:

$$x_2^*(x_1) : \frac{\partial \pi_2(x_2, x_1)}{\partial x_2} = 0.$$

Next, find the solution for the first player anticipating the response by the second player:

$$\frac{d\pi_1(x_1, x_2^*(x_1))}{dx_1} = \frac{\partial \pi_1(x_1, x_2^*)}{\partial x_1} + \frac{\partial \pi_1(x_1, x_2)}{\partial x_2} \frac{\partial x_2^*}{\partial x_1} = 0.$$

Intuitively, the first player chooses the best possible point on the second player's best response function. Clearly, the first player can choose a NE, so the leader is always at least as well off as he would be in NE. Hence, if a player were allowed to choose between making moves simultaneously or being a leader in a game with complete information he would always prefer to be the leader. However, if new information is revealed after the leader makes a play, then it is not always advantageous to be the leader.

Whether the follower is better off in the Stackelberg or simultaneous move game depends on the specific problem setting. See Netessine and Rudi (2001a) for examples of both situations and comparative analysis of Stackelberg vs NE; see also Wang and Gerchak (2002) for a comparison between the leader vs follower roles in a decentralized assembly model. For example, consider the newsvendor game with sequential moves. The best response function for the second player remains the same as in the simultaneous move game:

$$Q_2^*(Q_1) = F_{D_2+(D_1-Q_1)^+}^{-1}\left(\frac{r_2 - c_2}{r_2}\right).$$

For the leader the optimality condition is

$$\frac{d\pi_1(Q_1, Q_2^*(Q_1))}{dQ_1} = r_1 \Pr\left(D_1 + (D_2 - Q_2)^+ > Q_1\right) - c_1$$

$$-r_1 \Pr\left(D_1 + (D_2 - Q_2)^+ < Q_1, D_2 > Q_2\right)\frac{\partial Q_2^*}{\partial Q_1}$$

$$= 0,$$

where $\partial Q_2^*/\partial Q_1$ is the slope of the best response function found in (2.1). Existence of a Stackelberg equilibrium is easy to demonstrate given the continuous payoff functions. However, uniqueness may be considerably harder to demonstrate. A sufficient condition is quasi-concavity of the leader's profit function, $\pi_1(x_1, x_2^*(x_1))$. In the newsvendor game example, this implies the necessity of finding derivatives of the density function of the demand distribution as is typical for many problems involving uncertainty. In stochastic models this is feasible with certain restrictions on the demand distribution. See Lariviere and Porteus (2001) for an example with a supplier that establishes the wholesale price and a newsvendor that then chooses an order quantity and Cachon (2003) for the reverse scenario in which a retailer sets the wholesale price and buys from a newsvendor supplier. See Netessine and Rudi (2001a) for a Stackelberg game with a wholesaler choosing a stocking quantity and the retailer deciding on promotional effort. One can further extend the Stackelberg equilibrium concept into multiple periods, see Erhun et al. (2000) and Anand et al. (2002) for examples.

3.2 Simultaneous moves: repeated and stochastic games

A different type of dynamic game arises when both players take actions in multiple periods. Since inventory models used in SCM literature often involve inventory replenishment decisions that are made over and over again, multi-period games should be a logical extension of these inventory models. Two major types of multiple-period games exist: without and with time dependence.

In the multi-period game without time dependence the exact same game is played over and over again hence the term *repeated* games. The strategy for each player is now a sequence of actions taken in all periods. Consider one repeated game version of the competing newsvendor game in which the newsvendor chooses a stocking quantity at the start of each period, demand is realized and then leftover inventory is salvaged. In this case, there are no links between successive periods other than the players' memory about actions taken in all the previous periods. Although repeated games have been extensively analyzed in economics literature, it is awkward in a SCM setting to assume that nothing links successive games; typically in SCM there is some transfer of inventory and/or backorders between periods. As a result, repeated games thus far have not found many applications in the SCM literature. Exceptions are Debo (1999), Taylor and Plambeck (2003) and Ren et al. (2003) in which reputational effects are explored as means of supply chain coordination in place of the formal contracts.

A fascinating feature of repeated games is that the set of equilibria is much larger than the set of equilibria in a static game and may include equilibria that are not possible in the static game. At first, one may assume that the equilibrium of the repeated game would be to play the same static NE strategy in each period. This is, indeed, an equilibrium but only one of many. Since in repeated games the players are able to condition their behavior on the observed actions in the previous periods, they may employ so-called *trigger strategies*: the player will choose one strategy until the opponent changes his play at which point the first player will change the strategy. This *threat* of reverting to a different strategy may even induce players to achieve the best possible outcome, i.e., the centralized solution, which is called an *implicit collusion*. Many such threats are, however, non-credible in the sense that once a part of the game has been played, such a strategy is not an equilibrium anymore for the remainder of the game, as is the case in our example in Figure 2.1. To separate out credible threats from non-credible, Selten (1965) introduced the notion of a *subgame-perfect equilibrium*. See Hall and Porteus (2000) and van Mieghem and Dada (1999) for solutions involving subgame-perfect equilibria in dynamic games.

Subgame-perfect equilibria reduce the equilibrium set somewhat. However, infinitely-repeated games are still particularly troublesome in terms of multiplicity of equilibria. The famous Folk theorem[5] proves that any convex combination of the feasible payoffs is attainable in the infinitely repeated game as an equilibrium, implying that "virtually anything" is an equilibrium outcome [6]. See Debo (1999) for the analysis of a repeated game between the wholesaler setting the wholesale price and the newsvendor setting the stocking quantity.

In time-dependent multi-period games players' payoffs in each period depend on the actions in the previous as well as current periods. Typically the payoff structure does not change from period to period (so called stationary payoffs). Clearly, such setup closely resembles multi-period inventory models in which time periods are connected through the transfer of inventories and backlogs. Due to this similarity, time-dependent games have found applications in SCM literature. We will only discuss one type of time-dependent multi-period games, *stochastic games* or *Markov games*, due to their wide applicability in SCM. See also Majumder and Groenevelt (2001b) for the analysis of deterministic time-dependent multi-period games in reverse logistics supply chains. Stochastic games were developed by Shapley (1953a) and later by Sobel (1971), Kirman and Sobel (1974) and Heyman and Sobel (1984). The theory of stochastic games is also extensively covered in Filar and Vrieze (1996).

The setup of the stochastic game is essentially a combination of a static game and a Markov Decisions Process: in addition to the set of players with strategies which is now a vector of strategies, one for each period, and payoffs, we have a set of states and a transition mechanism $p(s'|s,x)$, probability that we transition from state s to state s' given action x. Transition probabilities are typically defined through random demand occurring in each period. The difficulties inherent in considering non-stationary inventory models are passed over to the game-theoretic extensions of these models, so a standard simplifying assumption is that demands are independent and identical across periods. When only a single decision-maker is involved, such an assumption leads to a unique stationary solution (e.g., stationary inventory policy of some form: order-up-to, S-s, etc.). In a GT setting, however, things get more complicated; just as in the repeated games described above, non-stationary equilibria, e.g., trigger strategies, are possible. A standard approach is to consider just one class of equilibria – e.g., stationary – since non-stationary policies are hard to implement in practice and they are not always intuitively appealing. Hence, with the assumption that the policy is stationary the stochastic game reduces to an equivalent static game and equilibrium is found as a sequence of NE in an appropriately modified single-period game. Another approach is to focus on "Markov" or "state-space" strategies in which the past influences the future through the state variables but not through the history of the play. A related equilibrium concept is that of *Markov Perfect Equilibrium* (MPE) which is simply a profile of Markov strategies that yields a Nash equilibrium in every subgame. The concept of MPE is discussed in Chapter 9 of Fudenberg and Tirole (1991). See also Tayur and Yang (2002) for the application of this concept.

To illustrate, consider an infinite-horizon variant of the newsvendor game with lost sales in each period and inventory carry-over to the subsequent period, see Netessine et al. 2002 for complete analysis. The solution to this problem in a non-competitive setting is an order-up-to policy. In addition to unit-revenue r and unit-cost c we introduce inventory holding cost h incurred by a unit carried over to the next period and a discount factor β. Also denote by x_i^t the inventory position at the beginning of the period and by y_i^t the order-up-to quantity. Then the infinite-horizon profit of each player is

$$\pi_i\left(x^1\right) = E\sum_{t=1}^{\infty}\beta_i^{t-1}\left[r_i\min\left(y_i^t, D_i^t + (D_j^t - y_j^t)^+\right)\right.$$
$$\left. -h_i\left(y_i^t - D_i^t - (D_j^t - y_j^t)^+\right)^+ - c_iQ_i^t\right],$$

with the inventory transition equation

$$x_i^{t+1} = \left(y_i^t - D_i^t - (D_j^t - y_j^t)^+\right)^+.$$

Using the standard manipulations from Heyman and Sobel (1984), this objective function can be converted to

$$\pi_i\left(x^1\right) = c_ix_i^1 + \sum_{t=1}^{\infty}\beta_i^{t-1}G_i^t\left(y_i^t\right), \ i = 1, 2,$$

where $G_i^t\left(y_i^t\right)$ is a single-period objective function

$$G_i^t(y_i^t) = E\left[(r_i - c_i)\left(D_i^t + (D_j^t - y_j^t)^+\right)\right.$$
$$- (r_i - c_i)\left(D_i^t + (D_j^t - y_j^t)^+ - y_i^t\right)^+$$
$$\left. - (h_i + c_i(1 - \beta_i))\left(y_i^t - D_i^t - (D_j^t - y_j^t)^+\right)^+\right],$$

$i = 1, 2, \ t = 1, 2, ...$

Assuming demand is stationary and independently distributed across periods $D_i = D_i^t$ we further obtain that $G_i^t(y_i^t) = G_i(y_i^t)$ since the single-period game is the same in each period. By restricting consideration to the stationary inventory policy $y_i = y_i^t$, $t = 1, 2, ...$, we can find the solution to the multi-period game as a sequence of the solutions to a single-period game $G_i(y_i)$ which is

$$y_i^* = F_{D_i + (D_j - y_j^*)^+}^{-1}\left(\frac{r_i - c_i}{r_i + h_i - c_i\beta_i}\right), \ i = 1, 2.$$

With the assumption that the equilibrium is stationary, one could argue that stochastic games are no different from static games; except for a small change in the right-hand side reflecting inventory carry-over and holding costs, the solution is essentially the same. However, more elaborate models capture some effects that are not present in static games but can be envisioned in stochastic games. For example, if we were to introduce backlogging in the above model, a couple of interesting situations would arise: a customer may backlog the product with either the first or with the second competitor he visits if both are out of stock. These options introduce the behavior that is observed in practice but cannot be modeled within the static game (see Netessine at al. 2002 for detailed analysis) since firms' inventory decisions affect their demand in the future. Among other applications of stochastic games are papers by Cachon and Zipkin (1999) analyzing a two-echelon game with the wholesaler and the retailer making stocking decisions, Bernstein and Federgruen (2002) analyzing price and service competition, Netessine and Rudi (2001a) analyzing the game with the retailer exerting sales effort and the wholesaler stocking the inventory and van Mieghem and Dada (1999) studying a two-period game with capacity choice in the first period and production decision under the capacity constraint in the second period.

3.3 Differential games

So far we have described dynamic games in discrete time, i.e., games involving a sequence of decisions that are separated in time. *Differential games* provide a natural extension for decisions that have to be made continuously. Since many SC models rely on continuous-time processes, it is natural to assume that differential games should find a variety of applications in SCM literature. However, most SCM models include stochasticity in one form or another. At the same time, due to the mathematical difficulties inherent in differential games, we are only aware of deterministic differential GT models in SCM. Although theory for stochastic differential games does exist, applications are quite limited, see Basar and Olsder (1995). Marketing and economics have been far more successful in applying differential games since deterministic models are standard in these areas. Hence, we will only briefly outline some new concepts necessary to understand the theory of differential games.

The following is a simple example of a differential game taken from Kamien and Schwartz (2000). Suppose two players indexed by $i = 1, 2$ are engaged into production and sales of the same product. Firms choose production levels $u_i(t)$ at any moment of time and incur total cost $C_i(u_i) = cu_i + u_i^2/2$. The price in the market is determined as per Cournot competition. Typically, this would mean that $p(t) = a - u_1(t) - u_2(t)$. However, the twist in this problem is that if the production level is changed, price adjustments are not instantaneous. Namely, there is a parameter s, referred to as the speed of price adjustment, so that the price is adjusted according to the following differential equation:

$$p'(t) = s\left[a - u_1(t) - u_2(t) - p(t)\right], \ p(0) = p_0.$$

Finally, each firm maximizes

$$\pi_i = \int_0^\infty e^{-rt}\left(p(t)u_i(t) - C_i(u_i(t))\right) dt, \ i = 1, 2.$$

The standard tools needed to analyze differential games are the calculus of variations or optimal control theory, see Kamien and Schwartz (2000). In a standard optimal control problem a single decision-maker sets the control variable that affects the state of the system. In contrast, in differential games several players select control variables that may affect a common state variable and/or payoffs of all players. Hence, differential games can be looked at as a natural extension of the optimal control theory. In this section we will consider two distinct types of player strategies: *open-loop* and *closed-loop* which is also sometimes called *feedback*. In the open-loop strategy the players select their decisions or control variables once at the beginning of the game and do not change them so that the control variables are only functions of time and do not depend on the other players' strategies. Open-loop strategies are simpler in that they can be found through the straightforward application of optimal control which makes them quite popular. Unfortunately, an open-loop strategy may not be subgame-perfect. On the contrary, in a closed-loop strategy the player bases his strategy on current time and the states of both players' systems. Hence, feedback strategies are subgame-perfect: if the game is stopped at any time, for the remainder of the game the same feedback strategy will be optimal, which is consistent with the solution to the dynamic programming problems that we employed in the stochastic games section. The concept of a feedback strategy is more satisfying, but is also more difficult to analyze. In general, optimal open-loop and feedback strategies differ, but they may coincide in some games.

Since it is hard to apply differential game theory in stochastic problems, we cannot utilize the competitive newsvendor problem to illustrate the analysis. Moreover, the analysis of even the most trivial differential game is somewhat involved mathematically so we will limit our survey to stating and contrasting optimality conditions in the cases of open-loop and closed-loop NE. Stackelberg equilibrium models do exist in differential games as well but are rarer, see Basar and Olsder (1995). Due to mathematical complexity, games with more than two players are hardly ever analyzed. In a differential game with two players, each player is endowed with a control $u_i(t)$ that the player uses to maximize the objective function π_i

$$\max_{u_i(t)} \pi_i(u_i, u_j) = \max_{u_i(t)} \int_0^T f_i\left(t, x_i(t), x_j(t), u_i(t), u_j(t)\right) dt,$$

where $x_i(t)$ is a state variable describing the state of the system. The state of the system evolves according to the differential equation

$$x_i'(t) = g_i\left(t, x_i(t), x_j(t), u_i(t), u_j(t)\right),$$

which is the analog of the inventory transition equation in the multi-period newsvendor problem. Finally, there are initial conditions $x_i(0) = x_{i0}$.

The open-loop strategy implies that each players' control is only a function of time, $u_i = u_i(t)$. A feedback strategy implies that each players' control is also a function of state variables, $u_i = u_i(t, x_i(t), x_j(t))$. As in the static games, NE is obtained as a fixed point of the best response mapping by simultaneously solving a system of first-order optimality conditions for the players. Recall that to find the optimal control we first need to form a Hamiltonian. If we were to solve two individual non-competitive optimization problems, the Hamiltonians would be $H_i = f_i + \lambda_i g_i$, $i = 1, 2$, where $\lambda_i(t)$ is an adjoint multiplier. However, with two players we also have to account for the state variable of the opponent so that the Hamiltonian becomes

$$H_i = f_i + \lambda_i^1 g_i + \lambda_i^2 g_j, \; i, j = 1, 2.$$

To obtain the necessary conditions for the open-loop NE we simply use the standard necessary conditions for any optimal control problem:

$$\frac{\partial H_1}{\partial u_1} = 0, \frac{\partial H_2}{\partial u_2} = 0, \tag{2.9}$$

$$\frac{\partial \lambda_1^1}{\partial t} = -\frac{\partial H_1}{\partial x_1}, \frac{\partial \lambda_1^2}{\partial t} = -\frac{\partial H_1}{\partial x_2}, \tag{2.10}$$

$$\frac{\partial \lambda_2^1}{\partial t} = -\frac{\partial H_2}{\partial x_2}, \frac{\partial \lambda_2^2}{\partial t} = -\frac{\partial H_2}{\partial x_1}. \tag{2.11}$$

For the feedback equilibrium the Hamiltonian is the same as for the open-loop strategy. However, the necessary conditions are somewhat different:

$$\frac{\partial H_1}{\partial u_1} = 0, \frac{\partial H_2}{\partial u_2} = 0, \tag{2.12}$$

$$\frac{\partial \lambda_1^1}{\partial t} = -\frac{\partial H_1}{\partial x_1} - \frac{\partial H_1}{\partial u_2}\frac{\partial u_2^*}{\partial x_1}, \frac{\partial \lambda_1^2}{\partial t} = -\frac{\partial H_1}{\partial x_2} - \frac{\partial H_1}{\partial u_2}\frac{\partial u_2^*}{\partial x_2}, \tag{2.13}$$

$$\frac{\partial \lambda_2^1}{\partial t} = -\frac{\partial H_2}{\partial x_2} - \frac{\partial H_2}{\partial u_1}\frac{\partial u_1^*}{\partial x_2}, \frac{\partial \lambda_2^2}{\partial t} = -\frac{\partial H_2}{\partial x_1} - \frac{\partial H_2}{\partial u_1}\frac{\partial u_1^*}{\partial x_1}. \tag{2.14}$$

Notice that the difference is captured by an extra term on the right when we compare (2.10) and (2.13) or (2.11) and (2.14). The difference is due to the fact that the optimal control of each player under the feedback strategy depends on $x_i(t)$, $i = 1, 2$. Hence, when differentiating the Hamiltonian to obtain equations (2.13) and (2.14) we have to account for such dependence (note also that two terms disappear when we use (2.12) to simplify).

As we mentioned earlier, there are numerous applications of differential games in economics and marketing, especially in the area of dynamic pricing, see, e.g., Eliashberg and Jeuland (1986). Eliashberg and Steinberg (1987), Desai (1992) and Desai (1996) use the open-loop Stackelberg equilibrium concept in a marketing-production game with the manufacturer and the distributor. Gaimon (1989) uses both open and closed-loop NE concepts in a game with two competing firms choosing prices and production capacity when the new technology reduces firms' costs. Mukhopadhyay and Kouvelis (1997) consider a duopoly with firms competing on prices and quality of design and derive open and closed-loop NE.

4. Cooperative games

The subject of cooperative games first appeared in the seminal work of von Neumann and Morgenstern (1944). However, for a long time cooperative game theory did not enjoy as much attention in the economics literature as non-cooperative GT. Papers employing cooperative GT to study SCM had been scarce, but are becoming more popular. This trend is probably due to the prevalence of bargaining and negotiations in SC relationships.

Cooperative GT involves a major shift in paradigms as compared to non-cooperative GT: the former focuses on the outcome of the game in terms of the value created through cooperation of a subset of players but does not specify the actions that each player will take, while the latter is more concerned with the specific actions of the players. Hence, cooperative GT allows us to model outcomes of complex business processes that otherwise might be too difficult to describe, e.g., negotiations, and answers more general questions, e.g., how well is the firm positioned against competition (see Brandenburger and Stuart 1996). However, there are also limitations to cooperative GT, as we will later discuss.

In what follows, we will cover transferable utility cooperative games (players can share utility via side payments) and two solution concepts: the core of the game and the Shapley value, and also biform games that have found several applications in SCM. Not covered are alternative concepts of value, e.g., nucleous and the σ-value, and games with non-transferable utility that have not yet found application in SCM. Material in this section is based mainly on Stuart (2001) and Moulin (1995). Perhaps the first paper employing cooperative games in SCM is Wang and Parlar (1994) who analyze the newsvendor game with three players, first in a non-cooperative setting and then under cooperation with and without Transferable Utility.

4.1 Games in characteristic form and the core of the game

Recall that the non-cooperative game consists of a set of players with their strategies and payoff functions. In contrast, the cooperative game, which is also called the game in characteristic form, consists of the set of players N with subsets or *coalitions* $S \subseteq N$ and a *characteristic function* $v(S)$ that specifies a (maximum) value (which we assume is a real number) created by any subset of players in N, i.e., the total pie that members of a coalition can create and divide. The specific actions that players have to take to create this value are not specified: the characteristic function only defines the total value that can be created by utilizing all players' resources. Hence, players are free to form any coalitions that are beneficial to them and no player is endowed with power of any sort. Furthermore, the value a coalition creates is independent of the coalitions and actions taken by the non-coalition members. This decoupling of payoffs is natural in political settings (e.g., the majority gets to choose the legislation), but it far more problematic in competitive markets. For example, in the context of cooperative game theory, the value HP and Compaq can generate by merging is independent of the actions taken by Dell, Gateway, IBM, Ingram Micron, etc.[7]

A frequently used solution concept in cooperative GT is the *core of the game*:

DEFINITION 5: *The utility vector $\pi_1, ..., \pi_N$ is in the core of the cooperative game if $\forall S \subset N$, $\sum_{i \in S} \pi_i \geq v(S)$ and $\sum_{i \in N} \pi_i \geq v(N)$*

A utility vector is in the core if the total utility of every possible coalition is at least as large as the coalition's value, i.e., there does not exist a coalition of players that could make all of its members at least as well off and one member strictly better off.

As is true for NE, the core of the game may not exist, i.e., it may be empty, and the core is often not unique. Existence of the core is an important issue because with an empty core it is difficult to predict what coalitions would form and what value each player would receive. If the core exists, then the core typically specifies a range of utilities that a player can appropriate, i.e., competition alone does not fully determine the players' payoffs. What utility each player will actually receives is undetermined: it may depend on the details of the residual bargaining process, a source of criticism of the core concept. (Biform games, described below, provide one possible resolution of this indeterminacy.)

In terms of specific applications to the SCM, Hartman et al. (2000) considered the newsvendor centralization game, i.e., a game in which multiple retailers decide to centralize their inventory and split profits resulting from the benefits of risk pooling. Hartman at al. (2000) further show that this game has a non-empty core under certain restrictions on the demand distribution. Muller et al. (2002) relax these restrictions and show that the core is always non-empty. Further, Muller et al. (2002) give a condition under which the core is a singleton.

4.2 Shapley value

The concept of the core, though intuitively appealing, also possesses some unsatisfying properties. As we mentioned, the core might be empty or indeterministic[8]. As it is desirable to have a unique NE in non-cooperative games, it is desirable to have a solution concept for cooperative games that results in a unique outcome. Shapley (1953b) offered an axiomatic approach to a solution concept that is based on three axioms. First, the value of a player should not change due to permutations of players, i.e., only the role of the player matters and not names or indices assigned to players. Second, if a player's added value to the coalition is zero then this player should not get any profit from the coalition, or, in other words, only players generating added value should share the benefits. (A player's added value is the difference between the coalitions value with that player and without that player.) Those axioms are intuitive, but the third is far less so. The third axiom requires additivity of payoffs: if v_1 and v_2 are characteristic functions in any two games, and if q_1 and q_2 are a player's Shapely value in these two games, then the player's Shapely value in the composite game, $v_1 + v_2$, must be $q_1 + q_2$. This is not intuitive because it is not clear by what is meant by a composite game. Nevertheless, Shapley (1953b) demonstrates that there is a unique value for each player, called the Shapley value, that satisfies all three axioms.

THEOREM 11. *The Shapley value, π_i, for player i in an N-person non-cooperative game with transferable utility is* :

$$\pi_i = \sum_{S \subseteq N \setminus i} \frac{|S|!(|N|-|S|-1)!}{|N|!} \left(v\left(S \cup \{i\} \right) - v(S) \right).$$

The Shapley value assigns to each player his marginal contribution $(v(S \cup \{i\} - v(S))$ when S is a random coalition of agents preceding i and the ordering is drawn at random. To explain further, (see Myerson 1997), suppose players are picked randomly to enter into a coalition. There are $|N|!$ different orderings for all players, and for any set S that does not contain player i there are $|S|!(|N| - |S| - 1)!$ ways to order players so that all of the players in S are picked ahead of player i. If the orderings are equally likely, there is a probability of $|S|!(|N| - |S| - 1)!/|N|!$ that when player i is picked he will find S players in the coalition already. The marginal contribution of adding player i to coalition S is $(v(S \cup \{i\}) - v(S))$. Hence, the Shapley value is nothing more than a marginal expected contribution of adding player i to the coalition.

Because the Shapley value is unique, it has found numerous applications in economics and political sciences. So far, however, SCM applications are scarce: except for discussion in Granot and Sosic (2001) we are not aware of any other papers employing the concept of the Shapley value. Although uniqueness of the Shapely value is a convenient feature, caution should surely be taken with Shapley value: the Shapley value need not be in the core, hence, although the Shapely is appealing from the perspective of fairness, it may not be a reasonable prediction of the outcome of a game (i.e., because it is not in the core, there exists some subset of players that can deviate and improve their lots).

4.3   Biform games

From the SCM point of view, cooperative games are somewhat unsatisfactory in that they do not explicitly describe the equilibrium actions taken by the players that is often the key in SC models. *Biform games*, developed by Brandenburger and Stuart (2003), compensate to some extent for this shortcoming.

A biform game can be thought of as a non-cooperative game with cooperative games as outcomes and those cooperative games lead to specific payoffs. Similar to the non-cooperative game, the biform game has a set of players N, a set of strategies for each player and also a cost function associated with each strategy (cost function is optional – we include it since most SCM applications of biform games involve cost functions). The game begins by players making choices from among their strategies and incurring costs. After that, a cooperative game occurs in which the characteristic value function depends on the chosen actions. Hopefully the core of each possible cooperative game is non-empty, but it is also unlikely to be unique. As a result, there is not specific outcome of the cooperative sub-game, i.e., it is not immediately clear what value each player can expect. The proposed solution is that each player is assigned a confidence index, $\alpha_i \in [0, 1]$, and the α_is are common knowledge.

Each player then expects to earn in each possible cooperative game a weighted average of the minimum and maximum values in the core, with α_i being the weight. For example, if $\alpha_i = 0$, then the player earns the minimum value in the core and if $\alpha_i = 1$ then the player earns the maximum value in the core. Once a specific value is assigned to each player for each cooperative sub-game, the first stage non-cooperative game can be analyzed just like any other non-cooperative game.

Biform games have been successfully adopted in several SCM papers. Anupindi et al. (2001) consider a game where multiple retailers stock at their own locations as well as at several centralized warehouses. In the first (non-cooperative) stage retailers make stocking decisions. In the second (cooperative) stage retailers observe demand and decide how much inventory to transship among locations to better match supply and demand and how to appropriate the resulting additional profits. Anupindi et al. (2001) conjecture that a characteristic form of this game has an empty core. However, the biform game has a non-empty core and they find the allocation of rents based on dual prices that is in the core. Moreover, they find an allocation mechanism that is in the core and that allows them to achieve coordination, i.e., the first-best solution. Granot and Sosic (2001) analyze a similar problem but allow retailers to hold back the residual inventory. In their model there are actually three stages: inventory procurement, decision about how much inventory to share with others and finally the transshipment stage. Plambeck and Taylor (2001a, 2001b) analyze two similar games between two firms that have an option of pooling their capacity and investments to maximize the total value. In the first stage, firms choose investment into effort that affects the market size. In the second stage, firms bargain over the division of the market and profits.

5. Signaling, Screening and Bayesian Games

So far we have considered only games in which the players are on "equal footing" with respect to information, i.e., each player knows every other player's expected payoff with certainty for any set of chosen actions. However, such ubiquitous knowledge is rarely present in supply chains. One firm may have a better forecast of demand than another firm, or a firm may possess superior information regarding its own costs and operating procedures. Furthermore, a firm may know that another firm may have better information, and therefore choose actions that acknowledge this information shortcoming. Fortunately, game theory provides tools to study these rich issues, but, unfortunately, they do add another layer of analytical complexity. This section briefly describes three types of games in which the information structure has a strategic role: signaling games, screening games and Bayesian games. Detailed methods for the analysis of these games are not provided. Instead, a general description is provided along with specific references to supply chain management papers that study these games.

5.1 Signaling Game

In its simplest form, a signaling game has two players, one of which has better information than the other and it is the player with the better information that makes the first move. For example, Cachon and Lariviere (2001) consider a model with one supplier and one manufacturer. The supplier must build capacity for a key component to the manufacturer's product, but the manufacturer has a better demand forecast than the supplier. In an ideal world the manufacturer would truthfully share her demand forecast with the supplier so that the supplier could build the appropriate amount of capacity. However, the manufacturer always benefits from a larger installed capacity in case demand turns out to be high but it is the supplier that bears the cost of that capacity. Hence, the manufacturer has an incentive to inflate her forecast to the supplier. The manufacturer's hope is that the supplier actually believes the rosy forecast and builds additional capacity. Unfortunately, the supplier is aware of this incentive to distort the forecast, and therefore should view the manufacturer's forecast with skepticism. The key issue is whether there is something the manufacturer should do to make her forecast convincing, i.e., credible.

While the reader should refer to Cachon and Lariviere (2001) for the details of the game, some definitions and concepts are needed to continue this discussion. The manufacturer's private information, or type, is her demand forecast. There is a set of possible types that the manufacturer could be and this set is known to the supplier, i.e., the supplier is aware of the possible forecasts, but is not aware of the manufacturer's actual forecast. Furthermore, at the start of the game the supplier and the manufacturer know the probability distribution over the set of types. We refer to this probability distribution as the supplier's belief regarding the types. The manufacturer chooses her action first which in this case is a contract offer and a forecast, the supplier updates his belief regarding the manufacturer's type given the observed action, and then the supplier chooses his action which in this case is the amount of capacity to build. If the supplier's belief regarding the manufacturer's type is resolved to a single type after observing the manufacturer's action (i.e., the supplier assigns a 100% probability that the manufacturer is that type and a zero probability that the manufacturer is any other type) then the manufacturer has signaled a type to the supplier. The trick for the supplier is to ensure that the manufacturer has signaled her actual type.

While we are mainly interested in the set of contracts that credibly signal the manufacturer's type, it is worth beginning with the possibility that the manufacturer does not signal her type. In other words, the manufacturer chooses an action such that the action does not provide the supplier with additional information regarding the manufacturer's type. That outcome is called a *pooling equilibrium*, because the different manufacturer types behave in the same way, i.e., the different types are pooled into the same set of actions. As a result, Bayes' rule does not allow the supplier to refine his beliefs regarding the manufacturer's type.

A pooling equilibrium is not desirable from the perspective of supply chain efficiency because the manufacturer's type is not communicated to the supplier. Hence, the supplier does not choose the correct capacity given the manufacturer's actual demand forecast. However, this does not mean that both firms are disappointed with a pooling equilibrium. If the manufacturer's demand forecast is worse than average, then that manufacturer is quite happy with the pooling equilibrium because the supplier is likely to build more capacity than he would if he learned the manufacturer's true type. It is the manufacturer with a higher than average demand forecast that is disappointed with the pooling equilibrium because then the supplier is likely to underinvest in capacity.

A pooling equilibrium is often supported by the belief that every type will play the pooling equilibrium and any deviation from that play would only be done by a manufacturer with a low demand forecast. This belief can prevent the high demand manufacturer from deviating from the pooling equilibrium: a manufacturer with a high demand forecast would rather be treated as an average demand manufacturer (the pooling equilibrium) than a low demand manufacturer (if deviating from the pooling equilibrium). Hence, a pooling equilibrium can indeed be a NE in the sense that no player has a unilateral incentive to deviate given the strategies and beliefs chosen by the other players.

While a pooling equilibrium can meet the criteria of a NE, it nevertheless may not be satisfying. In particular, why should the supplier believe that the manufacturer is a low type if the manufacturer deviates from the pooling equilibrium? Suppose the supplier were to believe a deviating manufacturer has a high demand forecast. If a high type manufacturer is better off deviating but a low type manufacturer is not better off, then only the high type manufacturer would choose such a deviation. The key part in this condition is that the low type is not better off deviating. In that case it is not reasonable for the supplier to believe the deviating manufacturer could only be a high type, so the supplier should adjust his belief. Furthermore, the high demand manufacturer should then deviate from the pooling equilibrium, i.e., this reasoning, which is called the intuitive criterion, breaks the pooling equilibrium, see Kreps (1990).

The contrast to a pooling equilibrium is a *separating* equilibrium which is also called a *signaling* equilibrium. With a separating equilibrium the different manufacturer types choose different actions, so the supplier is able to perfectly refine his belief regarding the manufacturer's type given the observed action. The key condition for a separating equilibrium is that only one manufacturer type is willing to choose the action designated for that type. If there is a continuum of manufacturer types, then it is quite challenging to obtain a separating equilibrium: it is difficult to separate two manufacturers that have nearly identical types. However, separating equilibria are more likely to exist if there is a finite number of discrete types.

There are two main issues with respect to separating equilibria: what actions lead to separating equilibrium and does the manufacturer incur a cost to signal, i.e., is the manufacturer's expected profit in the separating equilibrium lower than what it would be if the manufacturer's type were known to the supplier with certainty? In fact, these two issues are related: an ideal action for a high demand manufacturer is one that costlessly signals her high demand forecast. If a costless signal does not exist, then the goal is to seek the lowest cost signal.

Cachon and Lariviere (2001) demonstrate that whether a costless signal exists depends upon what commitments the manufacturer can impose on the supplier. For example, suppose the manufacturer dictates to the supplier a particular capacity level in the manufacturer's contract offer. Furthermore, suppose the supplier accepts that contract and by accepting the contract the supplier has essentially no choice but to build that level of capacity since the penalty for noncompliance is too severe. They refer to this regime as forced compliance. In that case there exist many costless signals for the manufacturer. However, if the manufacturer's contract is not iron-clad, so the supplier could potentially deviate, which is referred to as voluntary compliance, then the manufacturer's signaling task becomes more complex.

One solution for a high demand manufacturer is to give a sufficiently large lump sum payment to the supplier: the high demand manufacturer's profit is higher than the low demand manufacturer's profit, so only a high demand manufacturer could offer that sum. This has been referred to as signaling by "burning money": only a firm with a lot of money can afford to burn that much money.

While burning money can work, it is not a smart signal: burning one unit of income hurts the high demand manufacturer as much as it hurts the low demand manufacturer. The signal works only because the high demand manufacturer has more units to burn. A better signal is a contract offer that is costless to a high demand manufacturer but expensive to a low demand manufacturer. A good example of such a signal is a minimum commitment. A minimum commitment is costly only if realized demand is lower than the commitment, because then the manufacturer is forced to purchase more units than desired. That cost is less likely for a high demand manufacturer, so in expectation a minimum commitment is costlier for a low demand manufacturer. Interestingly, Cachon and Lariviere (2001) show that a manufacturer would never offer a minimum commitment with perfect information, i.e., these contracts may be used in practice solely for the purpose of signaling information.

5.2 Screening

In a screening game the player that lacks information is the first to move. For example, in the screening game version of the supplier-manufacturer game described by Cachon and Lariviere (2001) the supplier makes the contract offer. In fact, the supplier offers a menu of contracts with the intention of getting the manufacturer to reveal her type via the contract selected in the menu. In the economics literature this is also referred to as *mechanism design*, because the supplier is in charge of designing a mechanism to learn the manufacturer's information. See Porteus and Whang (1999) for a screening game that closely resembles this one.

The space of potential contract menus is quite large, so large that it is not immediately obvious how to begin to find the supplier's optimal menu. For example, how many contracts should be offered and what form should they take? Furthermore, for any given menu the supplier needs to infer for each manufacturer type which contract the type will choose. Fortunately, the *revelation principle* (Kreps 1990) provides some guidance.

The revelation principle begins with the presumption that a set of optimal mechanisms exists. Associated with each of these mechanisms is a NE that specifies which contract each manufacturer type chooses and the supplier's action given the chosen contract. With some of these equilibria it is possible that some manufacturer type chooses a contract that is not designated for that type. For example, the supplier intends the low demand manufacturer to choose one of the menu options, but instead the high demand manufacturer chooses that option. Even though this does not seem desirable, it is possible that this mechanism is still optimal in the sense that the supplier can do no better on average. The supplier ultimately cares only about expected profit, not the means by which that profit is achieved. Nevertheless, the revelation principle states that an optimal mechanism that involves deception (the wrong manufacturer chooses a contract) can be replaced by a mechanism that does not involve deception, i.e., there exists an equivalent mechanism that is truth-telling. Hence, in the hunt for an optimal mechanism it is sufficient to consider the set of revealing mechanisms: the menu of contracts is constructed such that each option is designated for a type and that type chooses that option.

Even though an optimal mechanism may exist for the supplier, this does not mean the supplier earns as much profit as he would if he knew the manufacturer's type. The gap between what a manufacturer earns with the menu of contracts and what the same manufacturer would earn if the supplier knew her type is called an information rent. A feature of these mechanisms is that separation of the manufacturer types goes hand in hand with a positive information rent, i.e., a manufacturer's private information allows the manufacturer to keep some rent that the manufacturer would not be able to keep if the supplier knew her type. Hence, even though there may be no cost to information revelation with a signaling game, the same is not true with a screening game.

There have been a number of applications of the revelation principle in the supply chain literature: e.g., Chen (2001) studies auction design in the context of supplier procurement contracts; Corbett (2001) studies inventory contract design; Baiman et al. (2003) study procurement of quality in a supply chain.

5.3 Bayesian games

With a signaling game or a screening game actions occur sequentially so information can be revealed through the observation of actions. There also exist games with private information that do not involve signaling or screening. Consider the capacity allocation game studied by Cachon and Lariviere (1999). A single supplier has a finite amount of capacity. There are multiple retailers, and each knows his own demand but not the demand of the other retailers. The supplier announces an allocation rule, the retailers submit their orders and then the supplier produces and allocates units. If the retailers' total order is less than capacity, then each retailer receives his entire order. If the retailers' total order exceeds capacity, the supplier's allocation rule is implemented to allocate the capacity. The issue is the extent to which the supplier's allocation rule influences the supplier's profit, the retailer's profit and the supply chain's profit.

In this setting the firms that have the private information (the retailers) choose their actions simultaneously. Therefore, there is no information exchange among the firms. Even the supplier's capacity is fixed before the game starts, so the supplier is unable to use any information learned from the retailers' orders to choose a capacity. However, it is possible that correlation exists in the retailers' demand information, i.e., if a retailer observes his demand type to be high, then he might assess the other retailers' demand type to be high as well (if there is a positive correlation). Roughly speaking, in a Bayesian game each player uses Bayes' rule to update his belief regarding the types of the other players. An equilibrium is then a set of strategies for each type that is optimal given the updated beliefs with that type and the actions of all other types. See Fudenberg and Tirole (1991) for more information on Bayesian games.

6. Summary and Opportunities

As has been noted in other reviews, Operations Management has been slow to adopt GT. But because SCM is an ideal candidate for GT applications, we have recently witnessed an explosion of GT papers in SCM. As our survey indicates, most of these papers utilize only a few GT concepts, in particular the concepts related to non-cooperative static games. Some attention has been given to stochastic games but several other important areas need additional work: cooperative, repeated, differential, signaling, screening and Bayesian games.

The relative lack of GT applications in SCM can be partially attributed to the absence of GT courses from the curriculum of most doctoral programs in operations research/management. One of our hopes with this survey is to spur some interest in GT tools by demonstrating that they are intuitive and easy to apply for a person with traditional operations research training.

With the invention of the Internet, certain GT tools have received significant attention: web auctions gave a boost to auction theory, and numerous web sites offer an opportunity to haggle, thus making bargaining theory fashionable. In addition, the advent of relatively cheap information technology has reduced transaction costs and enabled a level of disintermediation that could not be achieved before. Hence, it can only become more important to understand the interactions among independent agents within and across firms. While the application of game theory to supply chain management is still in its infancy, much more progress will soon come.

Notes

1. Some may argue that GT should be a tool for choosing how a manager should play a game, which may involve playing against rational or semi-rational players. In some sense there is no conflict between these descriptive and normative roles for GT, but this philosophical issue surely requires more in-depth treatment than can be afforded here.
2. However, an argument can also be made that to predict rational behavior by players it is sufficient that players not choose dominated strategies, where a dominated strategy is one that yields a lower payoff than some other strategy (or convex combination of other strategies) for all possible strategy choices by the other players.
3. Strategy space is compact if it is closed and bounded.
4. Interior equilibrium is the one in which first-order conditions hold for each player. The alternative is boundary equilibrium in which at least one of the players select the strategy on the boundary of his strategy space.
5. The name is due to the fact that its source is unknown and dates back to 1960; Friedman (1986) was one of the first to treat Folk Theorem in detail.
6. A condition needed to insure attainability of an equilibrium solution is that the discount factor is large enough. The discount factor also affects effectiveness of trigger and many other strategies.
7. One interpretation of the value function is that it is the minimum value a coalition can guarantee for itself assuming the other players take actions that are most damaging to the coalition. But that can be criticized as overly conservative.
8. Another potential problem is that the core might be very large. However, as Brandenburger and Stuart (2003) point out, this may happen for a good reason: to interprete such situations, one can think of competition as not having much force in the game, hence the division of value will largely depend on the intangibles involved.

References

Anand, K., R. Anupindi and Y. Bassok. 2002. Strategic inventories in procurement contracts. Working Paper, University of Pennsylvania.

Anupindi, R., Y. Bassok and E. Zemel. 2001. A general framework for the study of decentralized distribution systems. *Manufacturing & Service Operations Management*, Vol.3, 349-368.

Aumann, R. J. 1959. Acceptable Points in General Cooperative N-Person Games, pp. 287-324 in *"Contributions to the Theory of Games"*, Volume IV, A. W. Tucker and R. D. Luce, editors. Princeton University Press.

Baiman, S., S. Netessine and H. Kunreuther. 2003. Procurement in supply chains when the end-product exhibits the weakest link property. Working Paper, University of Pennsylvania.

Basar, T. and G.J. Olsder. 1995. *Dynamic noncooperative game theory*. SIAM, Philadelphia.

Bernstein, F. and A. Federgruen. 2000. Comparative statics, strategic complements and substitute in oligopolies. Forthcoming in *Journal of Mathematical Economics*.

Bernstein, F. and A. Federgruen. 2001a. Decentralized supply chains with competing retailers under Demand Uncertainty. Forthcoming in *Management Science*.

Bernstein, F. and A. Federgruen. 2003. Pricing and Replenishment Strategies in a Distribution System with Competing Retailers. *Operations Research*, Vol.51, No.3, 409-426.

Bernstein, F. and A. Federgruen. 2001c. A General Equilibrium Model for Decentralized Supply Chains with Price- and Service- Competition. Working Paper, Duke University.

Bernstein, F. and A. Federgruen. 2002. Dynamic inventory and pricing models for competing retailers. Forthcoming in *Naval Research Logistics*.

Bertsekas, D. P. 1999. *Nonlinear Programming*. Athena Scientific.

Border, K.C. 1999. *Fixed point theorems with applications to economics and game theory*. Cambridge University Press.

Brandenburger, A. and H.W. Stuart, Jr. 1996. Value-based business strategy. *Journal of Economics and Management Strategy*, Vol.5, No.1, 5-24.

Brandenburger, A. and H.W. Stuart, Jr. 2003. Biform games. Working Paper, Columbia University.

Cachon, G. 2001. Stock wars: inventory competition in a two-echelon supply chain. *Operations Research*, Vol.49, 658-674.

Cachon, G. 2002. Supply chain coordination with contracts. Forthcoming in "*Handbooks in Operations Research and Management Science: Supply Chain management*", S. Graves and T. de Kok, editors.

Cachon, G. 2003. The allocation of inventory risk in a supply chain: push, pull and advanced purchase discount contracts. University of Pennsylvania working paper. Conditionally accepted, *Management Science*.

Cachon, G. and C. Camerer. 1996. Loss avoidance and forward induction in coordination games. *Quarterly Journal of Economics*. Vol.112, 165-194.

Cachon, G. and P. T. Harker. 2002. Competition and outsourcing with scale economies. *Management Science*, Vol.48, 1314-1333.

Cachon, G. and M. Lariviere. 1999. Capacity choice and allocation: strategic behavior and supply chain performance. *Management Science*. Vol.45, 1091-1108.

Cachon, G. and M. Lariviere. 2001. Contracting to assure supply: how to share demand forecasts in a supply chain. *Management Science*. Vol.47, 629-646.

Cachon, G.P. and P.H. Zipkin. 1999. Competitive and cooperative inventory policies in a two-stage supply chain. *Management Science*, Vol.45, 936-953.

Cachon, G. and G. Kok. 2002. Heuristic equilibrium in the newsvendor model with clearance pricing. Working Paper, University of Pennsylvania.

Chen, F. 2001. Auctioning supply contracts. Working paper, Columbia University.

Ren, J., Cohen, M., T. Ho., and C. Terwiesch. 2003. Sharing forecast information in a long-term supply chain relationship. Working Paper, University of Pennsylvania.

Corbett, C. 2001. Stochastic inventory systems in a supply chain with asymmetric information: cycle stocks, safety stocks, and consignment stock. *Operations Research*. Vol.49, 487-500.

Corbett C. J. and G. A. DeCroix. 2001. Shared-Savings Contracts for Indirect Materials in Supply Chains: Channel Profits and Environmental Impacts. *Management Science*, Vol.47, 881-893.

Debo, L. 1999. Repeatedly selling to an impatient newsvendor when demand fluctuates: a supergame framework for co-operation in a supply chain. Working Paper, Carnegie Mellon University.

Debreu, D. 1952. A social equilibrium existence theorem. *Proceedings of the National Academy of Sciences*, Vol.38, 886-893.

Desai, V.S. 1992. Marketing-production decisions under independent and integrated channel structures. *Annals of Operations Research*, Vol.34, 275-306.

Desai, V.S. 1996. Interactions between members of a marketing-production channel under seasonal demand. *European Journal of Operational Research*, Vol.90, 115-141.

Eliashberg, J. and A.P. Jeuland. 1986. The impact of competitive entry in a developing market upon dynamic pricing strategies. *Marketing Science*, Vol.5, 20-36.

Eliashberg, J. and R. Steinberg. 1987. Marketing-production decisions in an industrial channel of distribution. *Management Science*, Vol.33, 981-1000.

Erhun F., P. Keskinocak and S. Tayur. 2000. Analysis of capacity reservation and spot purchase under horizontal competition. Working Paper, Georgia Institute of Technology.

Feichtinger, G. and S. Jorgensen. 1983. Differential game models in management science. *European Journal of Operational Research*, Vol.14, 137-155.

Filar, J. and K. Vrieze. 1996. *Competitive Markov decision processes*. Springer-Verlag.

Friedman, J.W. 1986. *Game theory with applications to economics*. Oxford University Press.

Fudenberg, D. and J. Tirole. 1991. *Game theory*. MIT Press.

Gaimon, C. 1989. Dynamic game results of the acquisition of new technology. *Operations Research*, Vol.3, 410-425.

Gale, D. and H. Nikaido. 1965. The Jacobian matrix and global univalence of mappings. *Mathematische Annalen*, Vol.159, 81-93.

Gillemin V. and A. Pollak. 1974. *Differential Topology*. Prentice Hall, NJ.

Granot, D. and G. Sosic. 2001. A three-stage model for a decentralized distribution system of retailers. Forthcoming, *Operations Research*.

Hall, J. and E. Porteus. 2000. Customer service competition in capacitated systems. *Manufacturing & Service Operations Management*, Vol.2, 144-165.

Hartman, B. C., M. Dror and M. Shaked. 2000. Cores of inventory centralization games. *Games and Economic Behavior*, Vol.31, 26-49.

Heyman, D. P. and M. J. Sobel. 1984. *Stochastic models in Operations Research*, Vol.II: Stochastic Optimization. McGraw-Hill.

Horn, R.A. and C.R. Johnson. 1996. *Matrix analysis*. Cambridge University Press.

Kamien, M.I. and N.L. Schwartz. 2000. *Dynamic optimization: the calculus of variations and optimal control in economics and management*. North-Holland.

Kirman, A.P. and M.J. Sobel. 1974. Dynamic oligopoly with inventories. *Econometrica*, Vol.42, 279-287.

Kreps, D. M. 1990. *A Course in Microeconomic Theory*. Princeton University Press.

Kreps, D. and R. Wilson. 1982. Sequential equilibria. *Econometrica*, Vol.50, 863-894.

Kuhn, H. W. 1953. Extensive Games and the Problem of Information. pp. 193-216 in *"Contributions to the Theory of Games"*, Volume II, H. W. Kuhn and A. W. Tucker, editors. Princeton University Press.

Lal, R. 1990. Price promotions: limiting competitive encroachment. *Marketing Science*, Vol.9, 247-262.

Lariviere, M.A. and E.L. Porteus. 2001. Selling to the newsvendor: an analysis of price-only contracts. *Manufacturing & Service Operations Management*, Vol.3, 293-305.

Lederer, P. and L. Li. 1997. Pricing, production, scheduling, and delivery-time competition. *Operations Research*, Vol.45, 407-420.

Li, L. and S. Whang. 2001. Game theory models in operations management and information systems. In *"Game theory and business applications"*, K. Chatterjee and W.F. Samuelson, editors. Kluwer Academic Publishers.

Lippman, S.A. and K.F. McCardle. 1997. The competitive newsboy. *Operations Research*, Vol.45, 54-65.

Lucas, W.F. 1971. An overview of the mathematical theory of games. *Management Science*, Vol.18, 3-19.

Mahajan, S and G. van Ryzin. 1999a. Inventory competition under dynamic consumer choice. *Operations Research*, Vol.49, 646-657.

Mahajan, S and G. van Ryzin. 1999b. Supply chain coordination under horizontal competition. Working Paper, Columbia University.

Majumder, P. and H. Groenevelt. 2001a. Competition in remanufacturing. *Production and Operations Management*, Vol.10, 125-141.

Majumder, P. and H. Groenevelt. 2001b. Procurement competition in remanufacturing. Working Paper, Duke University.

Moulin, H. 1986. *Game theory for the social sciences*. New York University Press.

Moulin, H. 1995. *Cooperative microeconomics: a game-theoretic introduction*. Princeton University Press.

Muller, A., M. Scarsini and M. Shaked. 2002. The newsvendor game has a nonempty core. *Games and Economic Behavior*, Vol.38, 118-126.
Mukhopadhyay, S.K. and P. Kouvelis. 1997. A differential game theoretic model for duopolistic competition on design quality. *Operations Research*, Vol.45, 886-893.
Myerson, R.B. 1997. *Game theory*. Harvard University Press.
Nash, J. F. 1950. Equilibrium Points in N-Person Games. *Proceedings of the National Academy of Sciences of the United States of America*, Vol.36, 48-49.
Netessine, S. and N. Rudi. 2001a. Supply chain structures on the Internet and the role of marketing-operations interaction. Forthcoming in *"Supply chain analysis in e-business era"*, D. Simchi-Levi, S.D. Wu and M. Shen, Edts., Kluwer.
Netessine, S. and N. Rudi. 2001b. Supply Chain choice on the Internet. Working Paper, University of Pennsylvania. Available at http://www.netessine.com.
Netessine, S. and N. Rudi. 2003. Centralized and competitive inventory models with demand substitution. *Operations Research*, Vol.53, 329-335.
Netessine, S. and R. Shumsky. 2001. Revenue management games: horizontal and vertical competition. Working Paper, University of Pennsylvania. Available at http://www.netessine.com.
Netessine, S., N. Rudi and Y. Wang. 2002. Dynamic inventory competition and customer retention. Working Paper, University of Pennsylvania, available at http://www.netessine.com.
Netessine, S. and F. Zhang. 2003. The impact of supply-side externalities among downstream firms on supply chain efficiency. Working Paper, University of Pennsylvania, available at http://www.netessine.com.
Parlar, M. 1988. Game theoretic analysis of the substitutable product inventory problem with random demands. *Naval Research Logistics*, Vol.35, 397-409.
Plambeck, E. and T. Taylor. 2001a. Sell the plant? The impact of contract manufacturing on innovation, capacity and profitability. Working Paper, Stanford University.
Plambeck, E. and T. Taylor. 2001b. Renegotiation of supply contracts. Working Paper, Stanford University.
Porteus, E. and S. Whang. 1999. Supply chain contracting: non-recurring engineering charge, minimum order quantity, and boilerplate contracts. Working paper, Stanford University.
Rosen, J.B. 1965. Existence and uniqueness of equilibrium points for concave N-person games. *Econometrica*, Vol.33, 520-533.
Rudi, N., S. Kapur and D. Pyke. 2001. A two-location inventory model with transshipment and local decision making. *Management Science*, Vol.47, 1668-1680.

Selten, R. 1965. Spieltheoretische behaundlung eine oligopolmodells mit nachfragetragheit. *Zeitschrift fur die gesamte staatswissenschaft.* Vol.12, 301-324.

Selten, R. 1975. Reexamination of the perfectness concept for equilibrium points in extensive games. *International Journal of Game Theory*, Vol.4, 25-55.

Shapley, L. 1953a. Stochastic games. *Proceedings of the National Academy of Sciences*, Vol.39, 1095-1100.

Shapley, L. 1953b. A value for $n-$person game. pp. 307-317 in "*Contributions to the Theory of Games*", Volume II, H. W. Kuhn and A. W. Tucker, editors. Princeton University Press.

Shubik, M. 1962. Incentives, decentralized control, the assignment of joint costs and internal pricing. *Management Science,* Vol.8, 325-343.

Shubik, M. 2002. Game theory and operations research: some musings 50 years later. *Operations Research*, Vol.50, 192-196.

Sobel, M.J. 1971. Noncooperative stochastic games. *Annals of Mathematical Statistics*, Vol.42, 1930-1935.

Stackelberg, H. von. 1934. *Markform and Gleichgewicht*. Vienna: Julius Springer.

Stidham, S. 1992. Pricing and capacity decisions for a service facility: stability and multiple local optima. *Management Science.* Vol.38, 1121-1139.

Stuart, H. W., Jr. 2001. Cooperative games and business strategy. In "*Game theory and business applications*", K. Chatterjee and W.F. Samuelson, editors. Kluwer Academic Publishers.

Tarski, A. 1955. A lattice-theoretical fixpoint theorem and its applications. *Pacific Journal of Mathematics*, Vol.5, 285-308.

Taylor, T. A. and E. L. Plambeck. 2003. Supply chain relationships and contracts: the impact of repeated interaction on capacity investment and procurement. Working paper, Columbia University.

Tayur, S. and W. Yang. 2002. Equilibrium analysis of a natural gas supply chain. Working Paper, Carnegie Mellon University.

Topkis, D. M. 1998. *Supermodularity and complementarity*. Princeton University Press.

van Mieghem, J. 1999. Coordinating investment, production and subcontracting. *Management Science*, Vol.45, 954-971.

van Mieghem, J. and M. Dada. 1999. Price versus production postponement: capacity and competition. *Management Science*, Vol.45, 1631-1649.

Varian, H. 1980. A model of sales. *American Economic Review*, Vol.70, 651-659.

Vickrey W. 1961. Counterspeculation, auctions, and competitive sealed tenders. *Journal of Finance*, Vol.16, 8-37.

Vives, X. 1999. *Oligopoly pricing: old ideas and new tools*. MIT Press.

von Neumann, J. and O. Morgenstern. 1944. *Theory of games and economic behavior*. Princeton University Press.

Wang, Y. and Y. Gerchak. 2003. Capacity games in assembly systems with uncertain demand. *Manufacturing & Service Operations Management,* Vol.5, No.3, 252-267.

Wang, Q. and M. Parlar. 1989. Static game theory models and their applications in management science. *European Journal of Operational Research*, Vol.42, 1-21.

Wang, Q. and M. Parlar. 1994. A three-person game theory model arising in stochastic inventory control theory. *European Journal of Operational Research*, Vol.76, 83-97.

David Simchi-Levi, S. David Wu, and Z. Max Shen (Eds.)
Handbook of Quantitative Supply Chain Analysis:
Modeling in the E-Business Era
©2004 Kluwer Academic Publishers

Chapter 3

SUPPLY CHAIN INTERMEDIATION:
A Bargaining Theoretic Framework

S. David Wu*
Department of Industrial and Systems Engineering
Lehigh University
david.wu@lehigh.edu

Keywords: Bargaining Theory, Game Theory, Auctions, Supply Chain Intermediary, Supply Chain Coordination

1. Introduction

This chapter explores the theory of supply chain intermediation. Using a bargaining theoretic framework, we set out to examine why intermediaries exist, different forms they operate, and the way they influence supply chain efficiency. The notion of intermediary has its root in the economics literature, referring to those economic agents who coordinate and arbitrate transactions in between a group of suppliers and customers. Distinctions are often drawn between a "market maker" and a "broker" intermediary Resnick et al., 1998. The former buys, sells, and holds inventory (e.g., retailers, wholesales), while the latter provides services without owning the goods being transacted (e.g., insurance agents, financial brokage). Sarkar et al. (1995) offer a list of various intermediation services. They distinguish the services that benefit the customers (e.g. assistance in search and evaluation, needs assessment and product matching, risk reduction, and product distribution/delivery) and those that benefit the suppliers (e.g. creating and disseminating product information). Taking a step further, Spulber (1996) views intermediary as the fundamental building

---

*S. David Wu is supported by NSF Grants DMI-0075391 and DMI-0121395

block of economic activities. He proposes the *intermediation theory of the firm* which suggests that the very existence of firms is due to the needs for intermediated exchange between a group of suppliers and customers. A firm is created when "the gains from intermediated exchange exceed the gains from direct exchange (between the supplier and the customer)." He also suggests that "with intermediated exchange, firms select prices, clear markets, allocate resources, and coordinate transactions." By this definition, firms *are* intermediaries which establish and operate markets.

Much of the earlier debate regarding the social/economic impact of Internet surrounds the possible "disintermediation" of traditional entities (c.f., Wigand and Benjamin 1996) and the formation of new intermediaries Kalakota and Whinston, 1997; Bollier, 1996. Disintermediation occur when an intermediary is removed from a transaction. The term was first used with regard to the financial services industry in the late 1960's to describe the trend for small investors to invest directly in financial instruments such as money market funds rather than through the traditional intermediary, a bank savings account Gellman, 1996. Popular discussions suggest that efficiencies in B2B e-commerce are obtained by disintermediation: that is, by cutting out "middlemen" and supplanting presumably costly intermediaries with direct transactions between the suppliers and buyers Hoffman, 1995; Imparato and Harari, 1995; Schiller and Zellner, 1994. On the other side of the debate, Fox (1999), Lu (1997a; 1997b), Crowston (1996), and Sarkar et al. (1995) show that intermediaries are still essential in electronic commerce, and argue that only the form of intermediation changes (reintermediation). Bailey (1998) suggests that both intermediation and disintermediation hypotheses are correct under different circumstances. He considers three basic transaction structures: *disintermediated* (direct exchange), *market* (where each intermediary carries all products from all suppliers, and the consumer only needs to visit one intermediary for these products), *hierarchy* (where each supplier chooses exactly one intermediary as in a distribution channel, and the consumer must choose among all intermediaries for different products). He shows that the preferred market structure to minimize transaction costs dependents on the number of suppliers. If the number is very small, a disintermediated market is preferred. As the number of suppliers increases, the market is preferred. After a point when the suppliers become numerous, the hierarchy is preferred.

The economics literature in market intermediation, agency theory, and bargaining theory offer rich and solid foundations for the study of intermediaries and their role in the supply chain. Spulber (1999) proposed the intermediary theory as a means to understanding market microstructure. The theory offers powerful explanation for why intermediaries exist, their advantage over direct exchange, and their roles in price setting, transaction costs, and the nature of competition. He suggests that markets reach equilibrium through strategic pricing and contracting by intermediaries. Intermediaries serve the critical functions of reducing transaction costs, pooling and diversifying risk, lowering costs of matching and searching, and alleviating adverse selection. Financial market literature also offers significant insights in the role of intermediation and market design. Campbell, et al. (1999) provides a comprehensive survey on the econometrics of financial markets. O'Hara (1995), and Frankel et al. (1996) offer significant insights of the theory of financial market microstructures. Harker and Zenios (2000) investigates main performance drivers in financial institutions and the roles of intermediations in that context.

Bargaining theory provides a powerful tool for the analysis of intermediaries. As stated above, the intermediary must offer intermediated trade that is no worse than the outcome expected from direct negotiation. Bargaining theory helps to characterize expected outcome from direct negotiation in various situations. In the seminal work of Nash (1950), he defines the bargaining problem as "two individuals who have the opportunity to collaborate for mutual benefits in more than one way. (p. 155)." There have been two main streams of research on bargaining theory: 1) axiomatic (cooperative game) models, and 2) strategic (non-cooperative game) models. Nash (1950 and 1953) lays the framework for the axiomatic Nash Bargaining Solution where he first defines the basic axioms that any bargaining solution should "naturally" satisfy, he then shows that the solution of the so called Nash product uniquely satisfies the stated axioms. Kalai and Smorodinsky (1975) replace a controversial axiom from the original Nash proposal and revise the unique solution. Binmore (1987) summarizes the efforts over the years that either relaxes or adds to the Nash axioms and gives further analysis of the Nash's bargaining model. An important characteristic of the axiomatic approach is that it leaves out the actual process of negotiations while focusing on the expected outcome based on pre-specified solution properties. In this chapter, we will focus on non-corporative models of bargaining. Ståhl (1972) is among the first who investigates a non-cooperative, sequential bargaining process by explicitly modelling bargaining as a sequence of offers and counter offers. Using the notion of sequential bargaining, Rubinstein (1982) lays out the framework for non-cooperative bargaining models. He proposes an alternating-offer bargaining procedure where the agents take turns in making offers and counter offers to one another until an agreement is reached. The agents face time-discounted gain (a "shrinking pie") which provide them the incentive to compromise. An intuitive comparison between the axiomatic and strategic bargaining theory can be found in Sutton, 1986.

A majority of the earlier bargaining literature focuses on bilateral bargaining with complete information. There is a significant and growing literature on sequential bargaining with incomplete information (c.f., Roth (1985), Wilson (1987)). In this setting, the players involve in the bargaining situation has only incomplete information about the opponent's valuation. Rubinstein (1985a,b) proposes an alternating-offer model with incomplete information where player-one's valuation is known but player-two's cost takes one of two values, with a certain probability. He develops the concept of sequential equilibrium and shows that many sequential equilibria may exist, unless additional assumptions are made about the player's beliefs. Myerson and Satterthwaite (1983) propose a mechanism design framework for bilateral bargaining where incomplete information is represented in the form of a distribution function with known supports. The mechanism design framework is more general than that of non-cooperative bargaining theory, and it provides a means to analyzing situations in multilateral settings. The latter has important implications in the context of supply chain intermediation, which we will also explore in this chapter.

The rest of the chapter is organized as follows: in Section 2, we define the scope and set up the context for the theory of supply chain intermediation. In Section 3, we outline a modelling framework starting from bilateral bargaining with complete information, to bilateral bargaining with incomplete information, then multilateral bargaining with incomplete information. In Sections 4 to 6 we discuss each of these models in some detail. In Section 7 we conclude the chapter by pointing to related work in the supply chain literature and outlining future research opportunities.

2. Supply Chain Intermediation

Many situations may arise in the supply chain where a group of suppliers and buyers find beneficial to seek the service of a third party agent as an *intermediary*. We may consider intermediaries in two broad categories: *transactional intermediaries* who improve the efficiency of a certain supply chain transactions (e.g., the wholesaler who facilitates the transactions between a group of manufacturers and retailers), and *informational intermediaries* who alleviate inefficiencies due to information asymmetry (e.g., an arbitrator, an auditor, an insurance agency). In either case, the intermediary must devise proper mechanisms (e.g., a long-term contract, a partnership agreement, auctions, etc.) to facilitate her operation. *Supply chain intermediation* refers to the coordination and arbitration functions provided by the intermediary. In the following, we summarize supply chain intermediation by the above categorization.

Transactional Intermediary. Consider supply chain transactions from the customers, retailers, wholesaler/distributor, manufacturer, to the raw material suppliers. Each supply chain player can be viewed as a intermediary between her upstream suppliers and downstream customers. Over the long run, a supply chain player is only engaged when she creates value from such intermediation, she would be disengaged (disintermediated) otherwise. For instance, in a three-tier supply chain with retailers, wholesalers, and manufacturers, the wholesaler serves as an intermediary between the retailers and the manufacturers. Operationally, the wholesaler may create value by holding inventory for the manufacturers such that just-in-time delivery could be made to the retailers. Over time, the wholesaler may help reducing the manufacturer's risk by aggregating demands from multiple retailers, or reducing the retailer's shortage risk by offering alternative products from multiple manufacturers. Over the long-run, a certain manufacturers, wholesalers, and retailers may form strategic alliance to further improve efficiency by streamlining their transactions electronically, by joint forecasting and inventory planning, etc. While providing the service as an intermediary, the wholesaler incurs intermediation costs (i.e., overhead plus her own profit) for the manufacturers and retailers. As market condition changes, the intermediation costs may not be justified by the reduction in transaction costs when comparing to direct exchange, or an alternative form of intermediation. In this case, disintermediation and/or reintermediation will eventually occur, i.e., a retailer may choose a new intermediary, say, a buy-side procurement auction for some of her products, while ordering directly from the manufacturer for other products. In general, a transactional intermediary may serve the following functions:

- reducing uncertainty by setting and stabilizing prices,
- reducing the costs associated with searching and matching,
- providing immediacy by holding inventory or reserving capacity, and
- aggregating supply or demand to achieve economy of scale.

Informational Intermediary. While at the transactional level a supply chain may operate with a high level of transparency, at the tactical and strategic level it typically operates under incomplete or asymmetric information. Financial incentives represented by the buyer's willingness-to-pay level and the supplier's opportunity cost tend to be private information subject to distortion. The buyer and supplier may both have outside options that influence their bargaining positions, therefore their valuations. This information asymmetry could significantly complicate the supplier-buyer interaction, leading to inefficiency known as adverse selection (i.e., players making misinformed decisions due to information distortion). This creates the needs for a third-party trust agent (an informational intermediary) who either acts as a broker between the trading parties, or as an arbitrator who regulates the trade in some way. In either case, the intermediary may devise mechanisms that elicit private information from the players, thereby improving trade efficiency. Similar to a transactional intermediary, an informational intermediary incurs her own costs and must create (net) value in order to justify her existence. In general, we may characterize informational intermediation as follows:

- avoiding adverse selection by administrating coordination mechanisms,
- creating a trusted institution thereby reducing the needs for direct negotiation, thus the transaction overhead, and
- synthesizing dispersed information to reduce information asymmetry.

If one is curious about the utilities of supply chain intermediary theory, it may be helpful to consider the perspective of a supply chain "integrator." A supply chain integrator represents the leader of a vertically integrated supply chain, or a certain collective effort in the supply chain to improve overall efficiency. To the supply chain integrator, the transactional and information intermediaries are strategic instruments who can be used to improve a certain aspect of supply chain efficiency. For instance, the integrator may want to instigate different classes of service in the supply chain, where a buyer may set up "preferred" status for a certain subset of suppliers. A preferred supplier is given a guaranteed sourcing percentage (of a product) in exchange for better quality and favorite pricing. However, neither the supplier nor the buyer is willing to share information openly. Thus, the buyer may have no way to verify if the quality and pricing offered by a particular supplier is truly favorable (relative to other buyers), and the supplier may have no way to verify the sourcing split the buyer actually uses (across all suppliers). In this case, the integrator may create an informational intermediary to facilitate the preferred supplier program. The intermediary is to make sure that the buyer correctly ranks the suppliers based on her established criteria, and the preferred supplier program satisfies basic requirements of an efficient mechanism.

As another example, suppose the leader of a vertically-integrated supply chain is to explore new strategies to integrate her Internet and traditional retail channels. The Internet channel operates most efficiently using *drop shipping*, where the wholesaler stocks and owns the inventory and ships products directly to the customers at the retailers' request (see Chapter 14). On the other hand, retailers in the traditional distribution channels must stock and own their inventory for shelf display. To successfully integrate the two channels, it may be necessary to replace the existing wholesaler with a new intermediary (reintermeidation), who implements mechanisms that reconcile the conflicting goals and different operational requirements of the two channels. In this context, the new intermediary plays a critical role, addressing issues ranging from demand management and inventory ownership, to stocking decision rights.

More generally, the supply chain integrator may consider strategically placing intermediaries in the supply chain to improve efficiency. To be economically viable, an intermediary must create intermediated trades that are more profitable than (1) direct exchange between the suppliers and buyers, and (2) other competing forms of intermediary. The intermediary creates value by improving transaction efficiency and/or reducing the effects of information asymmetry, while creating a system surplus that benefit all players involved. The value-creation is accomplished by overcoming obstacles that hamper profitable trades and by preventing inefficient trades from taking place. In the following section, we establish the basic framework for supply chain intermediary theory, focusing on the roles of the intermediary in dividing system surplus and regulating trades.

3. Supply Chain Intermediary Theory

3.1 The Basic Settings

To establish a framework for supply chain intermediary analysis, we focus on the economic incentives of three types of players: suppliers, buyers, and intermediaries. All players are self interested, profit seeking, and risk neutral. In the simplest form, each supplier has an opportunity cost s, each buyer has a willingness to pay level v that could be public or private information depending on the model assumptions. The intermediary offers an asked price w to the supplier and a bid price p to the buyer while creating a non-negative bid-ask spread $(p - w)$ to support her operation. The intermediary has the authority to determine whether a particular trade is to take place using control β. Adopting some mechanism $\Gamma(\beta, p, w)$, the intermediary optimizes her own profit.

The setting above describes the key elements we use to define supply chain intermediation. To further characterize supply chain intermediation in different settings and scopes, we consider models distinguished by two main factors: information symmetry (complete vs. incomplete information), and cardinality of interaction (bilateral vs. multilateral). Under the multilateral setting, we further distinguish vertically integrated channels and matching markets. This characterization suggests the following simple taxonomy that we will use to structure the remainder of the chapter.

Supply Chain Intermediation Models:

1. Complete Information
    - Bilateral Bargaining (Section 4)

2. Incomplete Information
    - Bilateral Bargaining (Section 5)

- Multilateral Trade (Section 6)
  - Vertical Integration (Section 6.1)
  - Markets (Section 6.2)

As hinted above, all supply chain intermediation models discussed in this chapter assume a profit-maximizing intermediary. The intermediary, either serving as a mediator or an arbitrator, always has an explicit interest in profit. This assumption provides a simple and unifying view between supply chain coordination and intermediation. Simply stated, the intermediary must ensure that sufficient system surplus is generated from the intermediated trade such that (1) the players are no worse off participating in the trade compared to their other options, (2) in the case of incomplete information, the player has the incentive to reveal her true valuations, and the trade is ex post efficient, and (3) the player receives non-negative profit. When any of the above conditions are not satisfied, the intermediary has the option of calling off the trade. In a more generalized case, the intermediary may choose to subsidize a short-term trade (violating condition (3)) for long-term profit, but we do not consider this extension. Thus, after providing necessary funds in support of the trade, the intermediary keeps the remaining system surplus. The profitability of the intermediary symbolizes the strength of intermediated trade, while the opposite signals the eventual fate of disintermediation. This draws contrast to the existing supply chain coordination literature, where the system surplus is divided among the players depending on the coordination mechanism (e.g., the specific form of a contract), the result typically favors the leader of the channel who has the first-move advantage.

We will introduce an analytical model for each of cases listed above. The models help to characterize the role of intermediation, to determine when should they exist, and to understand how could they extract profit while sustaining the trade efficiency. For the incomplete information cases we need to make use of the *mechanism design* framework and the *revelation principle*, which we will briefly summarize in the following section. In Section 3.3 we will summarize the settings of the four supply chain intermediation models.

3.2 Mechanism Design and the Revelation Principle

To carry out transactions at a lower cost, the intermediary must design an efficient mechanism that offers the service. We now introduce a mechanism design framework to characterize the main components of supply chain intermediation. Consider the base model of *bilateral* bargaining under *incomplete information*. There is a significant literature on strategic sequential bargaining models with incomplete information. Roth (1985) and Wilson (1987) provide excellent surveys of this literature. A subset of the literature is concerned about the design of mechanisms that carry out the bargaining process. This mechanism design literature contributes two important concepts that are fundamental to bargaining analysis with incomplete information. First, the *revelation principle* Myerson, 1979 states that regardless of the actual mechanism constructed by the intermediary, given the Bayesian-Nash equilibrium outcome of the mechanism we can construct an equivalent direct mechanism where the buyer and the supplier reveal their respective valuation to the intermediary, and the intermediary determines if the trade is to take place. This allows the study of a large class of bargaining games without the need to specify each of the games in detail. Second, *ex post efficiency* requires that when all the information is revealed, the players' payoffs resulting from the bargaining process are Pareto efficient. It can be shown that if there exists a bargaining mechanism where the corresponding bargaining game has a Bayesian-Nash Equilibrium that generates an ex post efficient outcome, then the bargaining mechanism can be ex post efficient. A mechanism is *incentive compatible* if it is the best strategy for the players to reveal their true valuations. It is *individually rational* if the players are no worse-off participating in the game than not participating. In summary, when putting into a mechanism design framework, it is sufficient for the supply chain intermediary to consider a direct revelation mechanism that is *incentive compatible, individually rational*, and *ex post efficient*. In other words, regardless of the actual mechanism being constructed, it is sufficient to consider a direct mechanism as follows:

Step 1. To the intermediary, the buyer reveals her valuation v. The supplier reveals her valuation s.

Step 2. Based on the players' reports, the intermediary specifies a mechanism $\Gamma(\beta, p, w)$ as follows:

    a. The intermediary determines $\beta(s, v)$ which specifies if the current trade is to take place:
$$\beta(s,v) = 1, \quad \text{if a certain creiteria are satisfied}$$
$$= 0, \quad \text{Otherwise.}$$

    b. If the trade is to take place ($\beta(s, v) = 1$), the intermediary collects asked price p from the buyer and pay the bid price w to the supplier. The intermediary determines the bid-ask spread (p, w) to maximize a certain well-fare function, subject to *incentive compatibility*, and *individually rationality* constraints.

    c. If the trade is not to take place ($\beta(s, v) = 0$), the players take their outside options.

In general, the intermediary ensures that mechanism $\Gamma(\beta, p, w)$ is ex post efficient while balancing the budget (otherwise, the intermediary is to call off the trade). The above procedure offers a general mechanism design framework for supply chain intermediation. In the following section, we use this framework to consider a few different settings.

3.3 Models of Supply Chain Intermediation

We summarize four basic models for supply chain intermediation according to the taxonomy established earlier. The simplest model is the complete information case based on bilateral bargaining with complete information. This is followed by three incomplete information cases. Figure 3.1 illustrates the schematics for the four different models of supply chain intermediation.

1. In **bilateral bargaining with complete information** (Figure 3.1-(a)), the supplier's opportunity cost is s and the buyer's willingness to pay is v. The intermediary determines if the trade is to take place (β) based on the cost information. If so, she collects asked price p from the buyer and pay bid price w to the supplier. The intermediary determines β, p, w to maximize a certain function subject to individual rationality (see Section 4).

2. In **bilateral bargaining with incomplete information** (Figure 3.1-(b)), the supplier and the buyer hold private information $s \in [s_1, s_2], v \in [v_1, v_2]$ as defined above with ex ante distributions $F(s)$ and $G(v)$, respectively. The intermediary decides if the trade is to take place ($\beta(s, v)$). If so, she collects $p(s, v)$ from the buyer and pay $w(s, v)$ to the supplier. The intermediary determines β, p, w to maximize a certain function subject to incentive compatibility, individual rationality, and ex post efficiency (see Section 5).

3. In **multilateral trade with vertical integration** (Figure 3.1-(c)) there is one supplier and m buyers (bidders). Each bidder i holds private information about her valuation $v_i \in [a_i, b_i]$ which is known to the others in the form of distribution function $G_i : [a_i, b_i] \to [0, 1]$. Each bidder i reports her valuation v_i to the intermediary, the supplier reports her opportunity cost s. Given s and the vector of the reported valuations $v = (v_1, ..., v_m)$, the intermediary determines the probability $\beta_i(s, v)$ that bidder i will get the object (i.e., determines which bidder gets the object), collect the amount $p_i(s, v)$ from each bidder i, and pay $w(s, v)$ to the supplier. The intermediary determines (β, p, w) to maximize a certain function subject to incentive compatibility and individual rationality (see Section 6.1).

Supply Chain Intermediation

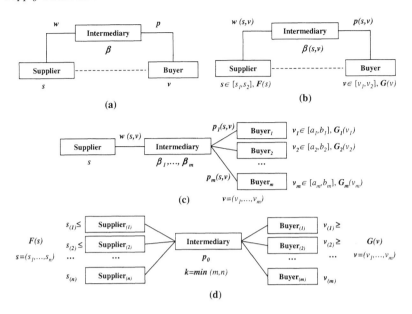

Figure 3.1. Models of Supply Chain Intermediation

4. In **multilateral trade with markets** (Figure 3.1-(d)) there are m buyers and n suppliers, each buyer i has a private valuation v_i for a single unit of good, and each supplier j has a privately known cost s_j for the good she sells. It is the common believe of all traders that each buyer's value is distributed according to $G(v)$, and each supplier's cost is distributed according to $F(s)$. The buyers and the suppliers report their valuations to the intermediary, who finds the efficient trade quantity $k \leq \min(m, n)$ and determines the market clearing price p_0 subject to budget balanceness (see Section 6.2).

4. Bilateral Bargaining with Complete Information

We now present a simple supply chain intermediary model based on the setting of bilateral bargaining with complete information. There is one supplier who is to provide a certain product to a buyer. The supplier produces (or acquires) the product at a unit cost of s. The buyer is willing to pay v for each unit of the good. The supplier and the buyer may choose to trade directly, in which case a transaction cost T incurs. Suppose there are positive net gains π^d

from the trade, then:
$$\pi^d = v - s - T > 0. \tag{3.1}$$

Suppose that an intermediary can purchase the goods from the supplier at unit price w and sell it to the buyer at unit price p, while incurring a transaction cost of K. The intermediary posts the bid-asked prices based on a certain criteria. Based on the posted price, the supplier and buyer may choose to trade directly, or through the intermediary. Clearly the intermediated trade will occur if and only if it offers a lower transaction cost, i.e., $K \le T$. For a typical trade, the sequence of events is as follows:

Step 1. The intermediary makes a binding offer of an asked price p and a bid price w.

Step 2. After observing p and w, the buyer and the supplier decide whether to trade directly with one another, or to accept the intermediary's offer.

Step 3. If the supplier and buyer are to transact via the intermediary, trade takes place at p and w with a transaction cost of K. If they trade directly, they must bargain over the allocation of the gain π^d, while incurring a transaction cost T.

The above posted price model can be described in the form of a direct mechanism as specified in Section 3.2, where the intermediary determines if the trade is to take place based on the following criteria:

$$\begin{aligned}\beta(s,v) &= 1, \quad \text{if } v \ge p \text{ and } w \ge s \\ &= 0, \quad \text{Otherwise.}\end{aligned}$$

This simple model captures the basic decisions faced by the supplier and the buyer: to use direct, or intermediated trade based on the unit price. In the former case, the supplier and the buyer must split the net gain through bilateral bargaining. Suppose the bargaining results in a split $\alpha \in [0,1]$, such that the buyer receives $\alpha \cdot \pi^d$ and the supplier receives $(1-\alpha) \cdot \pi^d$. In order for the intermediary to attract the supplier and the buyer to the intermediated trade, she must offer an asked price p and a bid price w such that

$$v - p = \alpha \cdot \pi^d \tag{3.2}$$
$$w - s = (1 - \alpha) \cdot \pi^d \tag{3.3}$$

The intermediary sets the bid-ask spread $(p - w)$, which is equal to the transaction cost T according to (3.2) and (3.3):

$$(p - w) = (v - \alpha \cdot \pi^d) - ((1 - \alpha) \cdot \pi^d - s) = v - s - \pi^d = T \tag{3.4}$$

The intermediary's profit is generated from the bid-ask spread after taking out the transaction cost, i.e., $(p - w - K) = T - K$. Thus, the intermediary can only extract profit if she could offer a more efficient transaction with $K < T$.

In the following section, we present a supply chain bargaining model which further characterizes how the system surplus (the gain of trade π^d) is divided, and how the players' bargaining power influence the surplus division.

4.1 Bilateral Bargaining to Divide the System Surplus

One important function for the intermediary is to provide a shortcut to the otherwise lengthy, and possibly costly negotiations between the supplier and the buyer, while at the same time achieving the expected benefit brought by direct bargaining. Bilateral bargaining provides the basis for an intermediary to design an efficient trade and to determine a bid-ask spread that is sufficiently attractive from the players' perspectives.

Economists (c.f., Rubinstein and Wolinsky, 1985, 1990) use models of bargaining and searching to present markets as decentralized mechanisms with pairwise interactions of buyers and suppliers. In this context, intermediaries could either increase the likelihood of matching, or improve the terms of trade relative to direct exchange. Rubinstein (1982) lays out an alternating offer bargaining procedure where the agents face time-discounted gain, and in each iteration, an agent must decide to either (1) accept the opponent's offer (in which case the bargaining ends), or (2) propose a counter offer. Binmore and Herrero (1988) propose a third option where an agent may decide to leave the current negotiation and opt for her "outside options" (e.g., previously quoted deals). Ponsati and Sakovics (1998) also consider outside options as part of the Rubinstein model. Muthoo (1995) considers outside options in the form of a search in a bargaining search game. An important aspect of the extended bargaining model is to allow the possibility for the negotiation to breakdown. Binmore et. al (1986) study a version of the alternating offer model with breakdown probability. In this model, there is no time pressure (time-discounted gain), but there is a probability that a rejected offer is the last offer made in the game, meaning that the negotiation breaks down.

The supply chain literature takes a different perspective on supplier-buyer interaction. The most well known model of pairwise supplier-buyer interaction is in supply chain contracting. The scope of the contract is typically limited to the two agents involved in the negotiation at a particular point in time with the assumption that they have agreed to coordinate via some form of contract. Cachon (2002) describes the typical sequence of events as follows: "the supplier offers the retailer a contract; the retailer accepts or rejects the contract; assuming the retailer accepts the contract, the retailer submits an order quantity, q, to the supplier; the supplier produces and delivers to the retailer before the selling season; season demand occurs; and finally transfer payments are made between the firms based upon the agreed contract. If the retailer rejects the contract, the game ends and each firm earns a default payoff." A typical goal for supply chain contracting is to design "channel coordinated" contracts (i.e., contracts where the players' Nash equilibrium coincides with the supply chain optimum), while at the same time satisfies *individual rationality* and *incentive compatibility* constraints. So long as that is the case, the agents are thought to be justified to accept the contract terms. The channel surplus created by the coordination contract is split arbitrarily, typically in favor of the "leader" who initiates the contract design.

The above approach encounters two basic problems when considered in the broader context of supply chain coordination: (1) there is no guarantee that either agent involved in the current negotiation should necessarily accept the "channel coordinated" contract when other outside options are easily accessible, and (2) rather than settling for a predetermined split of the channel surplus, both players may desire to negotiate for a (hopefully) larger share of the surplus. Outside options play a role here, shaping the agent's perception of her bargaining power. Ertogral and Wu (2001) show that the dynamics of supplier-buyer contract negotiation would change fundamentally if the agents were to enter a repeated, alternating-offer bargaining game on the contract surplus, and the equilibrium condition for the bargaining game may not coincide with contract stipulation. The bargaining model offers an alternative view of supply chain interaction as follows: first, contract negotiation is generalized to a *bilateral bargaining* over the expected channel surplus; second, instead of assuming the contract terms would be accepted in one offer, an alternating-offer bargaining process takes place before a final agreement is reached; third, the players' corresponding bargaining power, not the pre-determined contract stipulation, determines the ultimate split of the channel surplus. As we will argue throughout this chapter, the viewpoints offered by supply chain intermediary theory and bargaining theory broaden the scope for supply chain coordination and allow for additional versatility in modelling.

In the following section, we model the pair-wise supply chain interaction as a bilateral bargaining game with complete information. We will summarize main results derived in Ertogral and Wu (2001), and introduce a bargaining theoretic perspective which help to analyze the tradeoff between direct and intermediated exchanges.

4.2 A Bilateral Supply-Chain Bargaining Model

Consider a bargaining situation between a pair of suppliers and buyers who set out to negotiate the terms associated with a certain system surplus, say $\pi = \pi^d$. The supplier and the buyer are to make several offers and counter offers before settling on a final agreement. Before entering negotiation, the supplier and buyer each have recallable outside options W_s and W_b, respectively. We limit the definition of outside options to tangibles known at the point of negotiation, e.g., negotiations a player previously carried out with other agents in the market, which she could fall back on. Intangibles such as an anticipated future deal is not considered an outside option. We assume that the total maximum surplus generated from the current trade is greater than or equal to the sum of the outside options. This is reasonable since otherwise at least one of the players will receive a deal worse than her outside option, and would have no incentive to participate in the first place. We further assume that when an agent is indifferent between accepting the current offer or waiting for future offers, she will choose to accept the current offer. The sequence of events in the bargaining game is as follows:

1. With equal probability, either the supplier or the buyer makes an offer that yields a certain split of the system surplus π

2. The other agent either

   - accepts the offer (the negotiation ends).
   - rejects the offer and waits for the next round offer.

3. With a certain probability, $(1 - \psi)$, the negotiation breaks down and the agents take their outside options, W_s and W_b.

4. If the negotiation continues, the game restarts from step 1.

The above bargaining game is similar to Rubinstein's alternating-offer bargaining model with three additional elements: (1) both players are equally likely to make the next offer, (2) the negotiation breaks down with a certain probability, and (3) each player has an outside option. The first treatment allows us to view each iteration of the bargaining processes independently regardless of who makes the previous offer. The breakdown probability characterizes the stability of the bargaining situations, which could be influenced by either player's anticipation of a more attractive future deal, non-perfectly rational players, and other intangibles that can not be measured by monetary gains (e.g., trust and goodwill, or there lack of). The breakdown probability is defined exogenously here but could perceivably be modelled endogenously with some added complexity. The outside option is important as a player's bargaining power is a combination of her ability to influence the breakdown probability and her outside options. The player with higher valuation on her outside option is more likely to receive a larger share of the surplus.

4.3 The Subgame Perfect Equilibrium

The bargaining game outlined above iterates until one of the agents accepts the offer (Step 2), or when the negotiation breaks down (Step 3). The subgame perfect equilibrium (SPE) strategies are the ones that constitute the Nash equilibrium in every iteration of the game (the subgame). In a perfect equilibrium, an agent would accept a proposal if it offered at least as much as what she expected to gain in the future, given the strategy set of the other agent. In this bargaining game, each subgame starts with the same structure: either it is initiated by the supplier or the buyer. Thus, the perfect equilibrium strategies of the agents are symmetrical in each subgame. We will analyze the game in a time line of offers to find the subgame perfect equilibrium, similar to the approach taken in Shaked and Sutton (1984) and Sutton (1986). We introduce the following additional notations:

$M_b(M_s)$: The maximum share the buyer (the supplier) could receive in a subgame perfect equilibrium for any subgame initiated with the buyer's (the supplier's) offer.

$m_b(m_s)$: The minimum share the buyer (the supplier) could receive in a subgame perfect equilibrium for any subgame initiated with the buyer's (the supplier's) offer.

ψ : The probability that negotiations will continue to the next round.

The subgame equilibrium analysis proceeds as follows: we first assume that in subgame perfect equilibrium there is an infinite number of solutions leading to gains ranging from m_b to M_b for the buyer, and m_s to M_s for the supplier. We then show that the player's share under each of the four extreme cases m_b, M_b, m_s and M_s can be derived from an event-tree structure shown in Figure 3.2. Given the derived shares, we can then determine if there is a unique SPE solution for the players where $m_b = M_b$ and $m_s = M_s$.

We now derive the best-case scenario for the buyer where she initiates the subgame and receives the maximum possible share M_b in SPE. This best-case scenario is illustrated in Figure 3.1. The root node represents that the buyer makes the initial offer, with probability $(1 - \psi)$ the bargaining breaks down. With probability ψ the bargaining continues to the next round, where the buyer and the supplier have equal probability ($\psi/2$) to make the next offer. If the buyer makes the next offer, the subtree repeats same structure. If the supplier makes the next offer, again, with probability $(1 - \psi)$ the bargaining breaks down. With probability ψ the bargaining continues to the next round, where

Supply Chain Intermediation

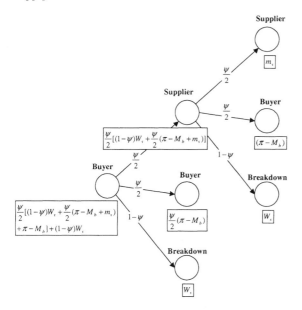

Figure 3.2. A tree defining the largest share the buyer could obtain in a subgame perfect equilibrium

the buyer and the supplier have equal probability ($\psi/2$) to make the next offer. The subtree from this point on repeats the same structure. For convenience, the nodal label in Figure 3.2 represents the share the *supplier* would receive in perfect equilibrium. Thus, the buyer's maximum gain would be labelled in terms of the supplier's share $\pi - M_b$, while the supplier's minimum gain is labelled m_s.

To derive the SPE condition we evaluate the event tree backward from the leaf nodes. Since the event tree in the figure represents the case where the buyer gets the largest possible perfect equilibrium share, when the supplier makes the offer, she settles for the minimum perfect equilibrium share m_s. When the buyer makes the offer, she receives the maximum gain possible and leaves $\pi - M_b$ to the supplier. In case the bargaining breaks down, the supplier receives her outside option W_s. The offers at the next tier follows the same logic. When the buyer makes the offer, she leaves $\pi - M_b$ to supplier as before. If the supplier makes the offer, she settles for the least amount she expects to gain in the future, which is equal to

$$(1 - \psi)W_s + \frac{\psi}{2}(\pi - M_b + m_s) \tag{3.5}$$

Going back one offer to the root node, we can see that the supplier would expect to gain, in the perfect equilibrium, a minimum share of

$$\frac{\psi}{2}\left[(1-\psi)W_s + \frac{\psi}{2}(\pi - M_b + m_s) + \pi - M_b\right] + (1-\psi)W_s \quad (3.6)$$

Therefore, the maximum share the buyer could gain in a SPE is as follows:

$$M_b = \pi - \left[\frac{\psi}{2}\left[(1-\psi)W_s + \frac{\psi}{2}(\pi - M_b + m_s) + \pi - M_b\right] + (1-\psi)W_s\right] \quad (3.7)$$

With slight modification, we can also find the minimum SPE share that the buyer would receive in a subgame starting with the buyer's offer. In specific, we only need to replace M_b with m_b, and m_s with M_s in equation (3.7). Thus, the minimum share the buyer would receive in the subgame is as follows:

$$m_b = \pi - \left[\frac{\psi}{2}\left[(1-\psi)W_s + \frac{\psi}{2}(\pi - m_b + M_s) + \pi - m_b\right] + (1-\psi)W_s\right] \quad (3.8)$$

Since the roles of the supplier and buyer are completely symmetrical in the game, we can write the expressions for M_s and m_s by simply changing the indices, and we end up with four linear equations with four unknowns. The following proposition summarizes the subgame perfect equilibrium description.

PROPOSITION 1 *The following system of equations defines the subgame perfect equilibrium for the bilateral bargaining between the supplier and the buyer.*

$$M_b = \pi - \left[\frac{\psi}{2}\left[(1-\psi)W_s + \frac{\psi}{2}(\pi - M_b + m_s) + \pi - M_b\right] + (1-\psi)W_s\right]$$

$$m_b = \pi - \left[\frac{\psi}{2}\left[(1-\psi)W_s + \frac{\psi}{2}(\pi - m_b + M_s) + \pi - m_b\right] + (1-\psi)W_s\right]$$

$$M_s = \pi - \left[\frac{\psi}{2}\left[(1-\psi)W_b + \frac{\psi}{2}(\pi - M_s + m_b) + \pi - M_s\right] + (1-\psi)W_b\right]$$

$$m_s = \pi - \left[\frac{\psi}{2}\left[(1-\psi)W_b + \frac{\psi}{2}(\pi - m_s + M_b) + \pi - m_s\right] + (1-\psi)W_b\right]$$

We may solve the linear equations in Proposition 1 and find the exact expressions for M_b, m_b, M_s and m_s. We can now specify the subgame perfect equilibrium strategies of the players as follows.

PROPOSITION 2 *The unique subgame perfect equilibrium strategy for the players is as follows: if the buyer (supplier) initiates the offer, she should ask* for X_b (X_s) share of the system surplus π, where

$$X_b = (\pi - W_s) - \frac{\psi^2}{2(2-\psi)}(\pi - W_b - W_s) \qquad (3.9)$$

$$X_s = (\pi - W_b) - \frac{\psi^2}{2(2-\psi)}(\pi - W_b - W_s) \qquad (3.10)$$

Proof: By solving the system of equations given in Proposition 1, we may conclude that $M_b = m_b = X_b$ and $M_s = m_s = X_s$. ◊

Since the maximum and the minimum SPE shares are equal for a given player, the SPE strategy is unique. We may interpret from expressions (3.9) and (3.10) that when a buyer (supplier) makes an offer, she ask for the difference between the system surplus π and the supplier's (buyer's) outside option W_s (W_b) minus a "risk premium" equals to $-\frac{\psi^2}{2(2-\psi)}(\pi - W_b - W_s)$. Note that the risk premium is a fraction of the "mutual gain" $(\pi - W_b - W_s)$, that the players could share if they reach an agreement. If they do not reach an agreement and continue with the bargaining, there is a risk that the process will break down and they receive only their respective outside options (thus the *mutual gain* would be lost). Proposition 2 says that in equilibrium the initiating party would offer a fraction of the *mutual gain* to the opponent that is sufficient to neutralize the opponent's desire to continue with the bargaining process.

4.4 Analysis of the Bargaining Game

Under complete information, we may conclude from Proposition 2 that the bargaining process will end in one iteration when either the supplier or the buyer initiates the negotiation with the SPE offer, and the opponent would accept the offer. This is true since the SPE offer makes the opponent indifferent between accepting the current offer and waiting for future offers. One important issue remains is whether there exists a first-mover advantage in the game. We attend to this issue in the following proposition.

PROPOSITION 3 *The first-mover advantage exists in the alternating offer bargaining game. The advantage diminishes as the probability of breakdown decreases, and goes to zero if the probability of breakdown is zero.*

Proof: If we take the difference between the SPE shares of the two players we get the following:

$$X_b - (\pi - X_s) = X_s - (\pi - X_b) = \frac{(\pi - W_b - W_s)(2 - \psi^2 - \psi)}{2 - \psi} \qquad (3.11)$$

Since $2 \geq \psi^2 + \psi$ and $(\pi - W_b - W_s) \geq 0$, the above expression always yields a value greater than or equal to zero. It is zero when $\psi = 1$, or equivalently when the breakdown probability $(1 - \psi)$ is zero. In the following, we show that the players' *individual rationally* conditions are satisfied under SPE. ⋄

PROPOSITION 4 *The player who initiates, and the player who accepts the SPE offer both gain no less than their respective outside options.*

Proof: For the player who initiates the offer, the difference between her SPE share and her outside option is as follows:

$$X_b - W_b = X_s - W_s = \frac{1}{2}\frac{(\pi - W_b - W_s)(4 - \psi^2 - 2\psi)}{2 - \psi} \quad (3.12)$$

The expression above is always positive. Hence the player who initiates the SPE offer will gain no less than her outside option. For the player who accepts the SPE offer, the difference between her SPE share and her outside option is as follows:

$$(\pi - X_s) - W_b = (\pi - X_b) - W_s = \frac{1}{2}\frac{(\pi - W_b - W_s)\psi^2}{2 - \psi} \quad (3.13)$$

This expression is always positive as well. Therefore, in SPE both players gain no less than their outside options, regardless of who initiate the offer. ⋄

In the following, we further specify the relationship between the breakdown probability $(1 - \psi)$ and the SPE share of the initiating player.

PROPOSITION 5 *The SPE share of the initiating player is linearly increasing (for $\psi > 0$) in her outside option, and linearly decreasing in her opponent's outside option.*

Proof: Taking the first and second derivatives of the buyer's SPE offer with respect to the outside options W_b and W_s, we see that

$$\frac{\partial X_b}{\partial W_b} = \frac{\psi^2}{2(2 - \psi)} \geq 0,$$

$$\frac{\partial^2 X_b}{\partial W_b^2} = 0,$$

$$\frac{\partial X_b}{\partial W_s} = \frac{-(4 - \psi^2 - 2\psi)}{2(2 - \psi)} < 0,$$

$$\frac{\partial^2 X_b}{\partial W_s^2} = 0.$$

It should be clear that the case when the supplier initiates the SPE offer would lead to similar results. ⋄

Another interesting aspect of the bargaining game is that, the offering party obtains the maximum share when the breakdown probability approaches 1 as described in the following proposition.

Supply Chain Intermediation

PROPOSITION 6 *The SPE share of the initiating player is maximized when the breakdown probability approaches 1 ($(1-\psi) \to 1$), where the share equals to the system surplus π less the opponent's outside option.*

Proof: Suppose the buyer is the offering player; taking the first derivative of the buyer's SPE offer with respect to ψ gives:

$$\frac{\partial X_b}{\partial \psi} = \frac{1}{2} \frac{\psi[(\psi - 4)(\pi - W_b - W_s)]}{(2 - \psi)^2} \qquad (3.14)$$

Since $\frac{\partial X_b}{\partial \psi} \leq 0$ for $0 \leq \psi \leq 1$, and we have assumed that $\pi - W_b - W_s \geq 0$, we may conclude that X_b is maximized at $\psi = 0$.

It should be clear that the case when the supplier initiates the SPE offer would lead to similar results. ⋄

Proposition 6 is intuitive in that if both players know that the negotiation is likely to breakdown ($(1 - \psi) \to 1$), the player initiating the bargaining would know that her opponent (in anticipation of the breakdown) is willing to accept an offer equivalent to the outside option. Thus, there is no reason for the offering player to offer more than the opponent's outside option.

4.5 Intermediary's Role in Price Setting, Searching, and Matching

Given the bilateral supply chain bargaining model and the supply chain intermediary theory described above, we will now examine the role of supply chain intermediary in setting prices, and in matching suppliers with buyers. Using the categorization in Section 2, these are transactional intermediation aiming to "reducing uncertainty by setting and stabilizing prices." and "reducing the costs associated with searching and matching."

We first establish the pricing criteria for a supply chain intermediary. Recall that a supply chain intermediary is economically viable if she can carry out transactions at a lower cost than (1) direct exchange between the suppliers and buyers, and (2) other competing intermediaries. If we use the result of bilateral bargaining to represent the expected gain the supplier and the buyer would expected from direct exchange, we may establish the role of any supply chain intermediary between the buyer and the supplier as follows.

THEOREM 1 *A supply chain intermediary is viable if she can operate with a transaction cost no more than*

$$(v - s) - W_b - W_s - \frac{\psi^2}{(2 - \psi)}(\pi - W_b - W_s) \qquad (3.15)$$

Proof: From Proposition 2 we know that the supplier and the buyer would expect from direct bargaining a payoff no less than $(\pi - X_b)$ and $\pi - X_s$, respectively. To attract the supplier and the buyer from direct bargaining, a supply chain intermediary must offer an asked price $\acute{p}$ and a bid price $\acute{w}$ that satisfy the following conditions:

$$v - \acute{p} \geq \pi - X_s \quad (3.16)$$
$$\acute{w} - s \geq \pi - X_b$$

Thus, we have

$$\begin{aligned}
\acute{p} &\leq v - \pi + X_s \\
&= v - \pi + (\pi - W_b) - \frac{\psi^2}{2(2-\psi)}(\pi - W_b - W_s) \\
&= v - W_b - \frac{\psi^2}{2(2-\psi)}(\pi - W_b - W_s)
\end{aligned}$$

$$\begin{aligned}
\acute{w} &\geq \pi - X_b + s \\
&= \pi - [(\pi - W_s) - \frac{\psi^2}{2(2-\psi)}(\pi - W_b - W_s)] + s \\
&= W_s + s + \frac{\psi^2}{2(2-\psi)}(\pi - W_b - W_s)
\end{aligned}$$

Therefore, the intermediary's bid-ask spread has an upper bound as follows:

$$\begin{aligned}
\acute{p} - \acute{w} &\leq [v - W_b - \frac{\psi^2}{2(2-\psi)}(\pi - W_b - W_s)] \\
&\quad - [W_s + s + \frac{\psi^2}{2(2-\psi)}(\pi - W_b - W_s)] \\
&= (v - s) - W_b - W_s - \frac{\psi^2}{(2-\psi)}(\pi - W_b - W_s)
\end{aligned}$$

Moreover, under Bertrand price competition, the market price will equal to marginal cost and the intermediary will earn zero profit. Thus, she must offer a bid-ask spread no more than

$$Min\{K, (v-s) - W_b - W_s - \frac{\psi^2}{(2-\psi)}(\pi - W_b - W_s)\}$$

In order to stay viable (non-negative profit), the intermediary must be able to operate with a transaction cost K no more than the upper bound of the bid-ask spread (the second term). ◇

From the above theorem, and more specifically from (3.15), note that the intermediary needs to be concerned about the supplier and the buyer's bargaining power. As the players' *outside options* increase, it will become increasing difficult for the intermediary to stay viable, and disintermediation will eventually occur. Moreover, the breakdown probability $(1 - \psi)$ plays a role. In general, the *higher* the breakdown probability the *easier* it is for the intermediary to stay viable. Specifically, when the breakdown probability is zero ($\psi = 1$), the intermediary must offer a transaction cost no more than $(v - s - \pi)$. When the breakdown probability is 1 ($\psi = 0$), she may offer a transaction cost up to $(v - s - W_b - W_s)$.

The above analysis provides the following insights concerning supply chain intermediation:

> When both the supplier and the buyer are in weak bargaining positions (limited outside options), or when direct trade is expected to be volatile (as characterized by the breakdown probability), intermediated trade will be desirable. Conversely, when either the supplier or the buyer is in a strong bargaining position, or when direct trade is expected to be stable, disintermediation is likely to occur.

Note that the above insights are derived entirely from marginal cost analysis under complete information, and no consideration are given concerning information asymmetry. This is the subject of discussion in the remainder of the chapter.

5. Bilateral Bargaining with Incomplete Information

The supply chain intermediary theory takes the viewpoint that supplier-buyer interaction could be either direct or intermediated. If the interaction is direct, it can be modelled explicitly as a bargaining process. If it is intermediated, the intermediary must convince the players that they are not worse off than they would be with direct bargaining. Thus, the bargaining theoretic analysis does not suggest that every supplier-buyer negotiation is actually taking place as a bilateral bargaining game. Rather, the bargaining-theoretic outcomes provide the rationale for the third-party intermediary to perform her intermediation functions. Specifically, the intermediary is to carry out the expected bargaining outcome via an efficient mechanism, while eliminating the needs for bilateral bargaining to actually take place. In Sections 3 and 4 we combine the theoretic foundation established by Spulber (1999) and Rubinstein (1982) to define a posted-price model of the supply chain intermediary theory. In this model, the intermediary posts the bid-ask spread, and the trade takes place if and only if the buyer and the supplier agree to the ask and bid prices, respectively. To establish the bid-ask spread, the intermediary must offer prices such that the players are no worse off than bargaining directly with one another. The Rubinstein (1982) model allows us to consider a richer set of bargaining parameters such as bargaining power, breakdown probability, etc. This analysis is based entirely on marginal costs, which is sufficient if we assume the players and the intermediary have complete information. We demonstrate that the bargaining power of supply chain participants determines the nature of their interactions. We show that players' relative bargaining power could be used to characterize when an intermediated trade is viable and when disintermediation is likely.

In this section, we consider the case when players are subject to asymmetric information, i.e., each player may hold private information on her valuation of the object, her outside options, or her quality level/expectation, which directly influence the bargaining process. Specifically, we are interested in the case where the supplier holds private information on her opportunity cost s, and the buyer holds private information on her willingness to pay level v. Acting on this information, the supply chain intermediary establishes intermediated trade via a mechanism. We introduce the analytic framework established by Myerson (1982), and Myerson and Satterthwaite (1983) that lays out the foundation for intermediated trades under incomplete information. We then introduce potential research topics using this perspective.

There is a significant and growing literature on bargaining with incomplete information. For the alternating-offer bargaining game described above, the offer and counter offers not only express a player's willingness to settle on the deal, they also serve as signals by which the players communicate their private information. Such signals may not be truthful as both parties may have incentive to distort the signal if doing so could increase their gains. Earlier literature in this area uses the notion of a sequential equilibrium due to Kreps and Wilson (1982) by reducing the bargaining situation to Harsanyi's (1967) game with imperfect information. To further refine the notion of sequential equilibrium in bargaining, Rubinstein (1985a,b)Rubinstein, 1985a; Rubinstein, 1985b introduces the alternating-offer model with incomplete information. He shows that many sequential equilibria may exist, and he defines unique equilibrium outcomes by adding conjectures on the way players rationalize their opponents' bargaining power.

5.1 The Basic Setting

We consider a one-buyer, one-supplier basic model where the players could either trade through an intermediary, or via a direct matching market (their outside option). The players hold private information on their costs, established based on their respective outside options. The *supplier* holds private information on her opportunity cost $\tilde{s}$, which takes values on the interval $[s_1, s_2]$ with a prior probability density function $f(s)$, and cumulative distribution function $F(s)$. Similarly, the *buyer* holds private information on her willingness to pay level $\tilde{v}$, taking on the interval $[v_1, v_2]$, with cumulative distribution G, and density g. Each player knows her own valuation at the time of trade, but considers the other's valuation a random variable, distributed as above.

In competition with the players' outside options, the intermediary must offer an intermediated trade that is "more attractive" to the buyer and the supplier. The intermediary is subject to the same information asymmetry in the market as the market participants, however, the intermediary has two main advantages: (1) she has access to aggregate information gained by dealing with multiple buyers and suppliers over time, and (2) she has the freedom to design an intermediated trading mechanism that taxes, subsidizes, or calls off individual transactions. The latter is important due to the *impossibility theory* by Vickrey (1961), which states that it is impossible to design a mechanism that satisfies incentive compatibility, budget balanceness, and ex post Pareto efficiency at the same time. Since the intermediary does not need to balance the budget in every single transaction as is required in a direct matching market, Myerson and Satterthwaite (1983) show that it is possible to design an incentive compatible mechanism that is ex post efficient. For instance, a profit maximizing intermediary could tax the market by setting a bid-ask spread, rejecting an unprofitable trade, or subsidizing the trade while achieving budget balance (and profit) over the long run.

By the *revelation principle* (Section 3.2), it is sufficient to consider an incentive compatible direct mechanism. In other words, regardless of the mechanism constructed by the intermediary, given the equilibrium of the mechanism, we can construct an equivalent incentive compatible direct mechanism, where the buyer and the supplier report their respective valuations to the intermediary, and the intermediary determines if the trade is to take place. If so, she determines the buyer's payment and the suppliers' revenue. Otherwise, the players take their outside options in a direct matching market. Let $\Gamma(\beta, p, w)$ represents the direct revelation mechanism, where $\beta(s, v)$ is the probability that the trade will take place, $p(s, v)$ is the expected payment to be made by the buyer to the intermediary (the asked price), and $w(s, v)$ is the expected payment from the intermediary to the supplier (the bid price), where s and v are the valuations given by the supplier and buyer, respectively. As mentioned above, the intermediary is aware of the buyer and the supplier's outside options as random variables characterized by distributions G and F, respectively. Based on this information the intermediary establishes the buyer's virtual willingness to pay $\Psi_b(v)$ as follows:

$$\Psi_b(v) = v - \frac{1 - G(v)}{g(v)} \tag{3.17}$$

Similarly, the intermediary establishes the supplier's virtual opportunity cost $\Psi_s(s)$ as follows:

$$\Psi_s(s) = s + \frac{F(s)}{f(s)} \tag{3.18}$$

Given a direct mechanism $\Gamma(\beta, p, w)$, we define the following quantities. The expected payment from the buyer to the intermediary (given that her willingness-to-pay level $\tilde{v} = v$) is as follows:

$$\tilde{p}(v) = \int_{s_1}^{s_2} p(\tau_s, v) f(\tau_s) d\tau_s \qquad (3.19)$$

The probability for the buyer to complete the trade given that her opportunity cost $\tilde{v} = v$ is

$$\tilde{\beta}_b(v) = \int_{s_1}^{s_2} \beta(\tau_s, v) f(\tau_s) d\tau_s \qquad (3.20)$$

Similarly, the supplier's expected payment from the intermediary given that her opportunity cost $\tilde{s} = s$ is as follows:

$$\tilde{w}(s) = \int_{v_1}^{v_2} w(s, \tau_b) g(\tau_b) d\tau_b \qquad (3.21)$$

The probability for the supplier to complete the trade given that her opportunity cost $\tilde{s} = s$ is:

$$\tilde{\beta}_s(s) = \int_{v_1}^{v_2} \beta(s, \tau_b) g(\tau_b) d\tau_b \qquad (3.22)$$

Thus, from (3.19) to (3.22), the buyer's and the supplier's expected gain from the intermediated trade can be defined as follows:

$$\pi_b(v) = v\tilde{\beta}_b(v) - \tilde{p}(v) \qquad (3.23)$$
$$\pi_s(s) = \tilde{w}(s) - s\tilde{\beta}_s(s) \qquad (3.24)$$

The direct mechanism Γ is said to be *incentive compatible* if reporting the truthful valuation is the preferred strategy for the players:

$$\pi_b(v) \geq v\tilde{\beta}_b(\acute{v}) - \tilde{p}(\acute{v}), \ \forall v, \acute{v} \in [v_1, v_2] \qquad (3.25)$$
$$\pi_s(s) \geq \tilde{w}(\acute{s}) - s\tilde{\beta}_s(\acute{s}), \ \forall s, \acute{s} \in [s_1, s_2] \qquad (3.26)$$

In other words, there is no incentive for the players to report $\acute{v}$ and $\acute{s}$ when their true valuations are v and s, respectively. The mechanism is said to be *individually rational* if it offers each player an expected gain that is non-zero.

$$\pi_b(v) \geq 0, \ \forall v, \acute{v} \in [v_1, v_2] \qquad (3.27)$$
$$\pi_s(s) \geq 0, \ \forall s, \acute{s} \in [s_1, s_2] \qquad (3.28)$$

5.2 The Direct Revelation Mechanism

In the general framework of supply chain intermediation (Section 3.2), an intermediary can be characterized in a mechanism design framework using the *revelation principle*, requiring the specification of a direct revelation mecha-

Supply Chain Intermediation

nism that is *individually rational, incentive compatible,* and *ex post efficient.* We know that any Bayesian-Nash equilibrium of any trading game with intermediary can be simulated by an equivalent incentive compatible direct mechanism. Following the supply chain intermediary framework we may specify intermediation under bilateral bargaining with incomplete information as follows:

Step 1. To the intermediary, the buyer reveals her valuation, her outside options, and her quality requirements, characterized by v. The supplier reveals her valuation, her outside options, and her quality type, characterized by s.

Step 2. The intermediary is subject to the same information asymmetry as the players. Based on the players' reports, and the probability distributions G and F characterizing the asymmetric information, the intermediary constructs a virtual willingness to pay $\psi_b(v)$ for the buyer a virtual opportunity cost $\psi_s(s)$ for the supplier, in reference to the trade at hand.

Step 3. The intermediary specifies a mechanism $\Gamma(\beta, p, w)$ as follows:

1. The intermediary determines $\beta(s, v)$ which specifies if the current trade is to take place:

$$\beta(s,v) = 1, \quad \text{if } \Theta(\Psi_b(v), \Psi_s(s)) \text{ is satisfied}$$
$$= 0, \quad \text{Otherwise.}$$

where $\Theta(x, y)$ specifies the relationship between x and y.

2. If the trade is to take place ($\beta(s, v) = 1$), the intermediary specifies a bid-ask spread (p, w) to maximize her own expected profit

$$\pi_I(s,v) = \int_{v_1}^{v_2} \int_{s_1}^{s_2} (p(\tau_s, \tau_b) - w(\tau_s, \tau_b)) f(\tau_s) g(\tau_b) d\tau_s d\tau_b \tag{3.29}$$

subject to *incentive compatibility,* and *individually rationality* constraints.

3. If the trade is not to take place ($\beta(s, v) = 0$), the players take their outside options.

4. The intermediary must ensure that mechanism $\Gamma(\beta, p, w)$ is ex post efficient.

We further illustrate the construct of this framework in the remainder of this section. The expected gain of trade from the buyer's (supplier's) perspectives is π_b (π_s) as defined in (3.23) ((3.24)). Thus, the total expected gains from trade for the buyer and the supplier are as follows:

$$\int_{v_1}^{v_2} \pi_b(\tau_b)g(\tau_b)d\tau_b + \int_{s_1}^{s_2} \pi_s(\tau_s)f(\tau_s)d\tau_s \qquad (3.30)$$

By definition, the total expected gain from trade is as follows:

$$\pi_T = \int_{v_1}^{v_2}\int_{s_1}^{s_2} (\tau_b - \tau_s)\beta(\tau_s, \tau_b)f(\tau_s)g(\tau_b)d\tau_s d\tau_b \qquad (3.31)$$

Since the expected gains for the buyer and supplier must equal to the expected gains from trade minus the expected net profit to the intermediary, we have the following relationship.

$$\pi_T - \pi_I = \int_{v_1}^{v_2} \pi_b(\tau_b)g(\tau_b)d\tau_b + \int_{s_1}^{s_2} \pi_s(\tau_s)f(\tau_s)d\tau_s \qquad (3.32)$$

Furthermore, Myerson and Satterthwaite (1983) presents the following important theorem:

THEOREM 2 *For any incentive-compatible mechanism with an intermediary, $\tilde{\beta}_s(s)$ is nonincreasing, $\tilde{\beta}_b(v)$ is nondecreasing, and*

$$\pi_I + \pi_b(v_1) + \pi_s(s_2) = \pi_I + \min_{v \in [v_1,v_2]} \pi_b(v) + \min_{s \in [s_1,s_2]} \pi_s(s)$$
$$= \int_{v_1}^{v_2}\int_{s_1}^{s_2} (\Psi_b(\tau_b) - \Psi_s(\tau_s))\beta(\tau_s, \tau_b))f(\tau_s)g(\tau_b)d\tau_s d\tau_b$$

To streamline the discussion we will only outline the main component of the proof as follows. First of all, by incentive compatibility (3.25) and (3.26), it is straightforward to show that $\tilde{\beta}_s(s)$ is nonincreasing, $\tilde{\beta}_b(v)$ is nondecreasing. Furthermore, from relationship (3.32), we have

$$\int_{v_1}^{v_2}\int_{s_1}^{s_2} (\tau_b - \tau_s)\beta(\tau_s, \tau_b)f(\tau_s)g(\tau_b)d\tau_s d\tau_b$$
$$= \pi_I + \int_{v_1}^{v_2} \pi_b(\tau_b)g(\tau_b)d\tau_b + \int_{s_1}^{s_2} \pi_s(\tau_s)f(\tau_s)d\tau_s$$
$$= \pi_I + \pi_b(v_1) + \int_{v_1}^{v_2}\int_{v_1}^{\tau_2} \tilde{\beta}_b(\tau_b)d\tau_b f(\tau_2)d\tau_2$$
$$+ \pi_s(s_2) + \int_{s_1}^{s_2}\int_{\tau_1}^{s_2} \tilde{\beta}_s(\tau_s)d\tau_s f(\tau_1)d\tau_1$$
$$= \pi_I + \pi_b(v_1) + \pi_s(s_2) + \int_{s_1}^{s_2} F(\tau_s)\tilde{\beta}_s(\tau_s)d\tau_s + \int_{v_1}^{v_2} G(\tau_b)\tilde{\beta}_b(\tau_b)d\tau_b$$
$$= \pi_I + \pi_b(v_1) + \pi_s(s_2)$$
$$+ \int_{v_1}^{v_2}\int_{s_1}^{s_2} (F(\tau_s)g(\tau_b) + (1 - G(\tau_b)f(\tau_s))\beta(\tau_s, \tau_b)d\tau_s d\tau_b \qquad (3.33)$$

Thus, we have the following relationship:

$$\pi_I + \pi_b(v_1) + \pi_s(s_2)$$
$$= \int_{v_1}^{v_2} \int_{s_1}^{s_2} (\tau_b - \tau_s)\beta(\tau_s, \tau_b) f(\tau_s) g(\tau_b) d\tau_s d\tau_b$$
$$- \int_{v_1}^{v_2} \int_{s_1}^{s_2} (F(\tau_s)g(\tau_b) + (1 - G(\tau_b))f(\tau_s))\beta(\tau_s, \tau_b) d\tau_s d\tau_b$$

Rewriting the above relationship using the definition of $\Psi_b(.)$ and $\Psi_s(.)$ in (3.17) and (3.18) gives us the equation stated in the theorem.

A mechanism is ex post efficient iff the buyer gets the object whenever her valuation v is higher than the supplier's cost s, otherwise the trade is not taking place. Using Theorem 2, Myerson and Satterthwaite (1983) shows that it is possible to construct an ex post efficient mechanism so long as the trade is subsidized by an intermediary as needed. Specifically,

$$\pi_I + \pi_b(v_1) + \pi_s(s_2) = -\int_{v_1}^{s_2} (1 - G(\tau))F(\tau) d\tau \quad (3.34)$$

Thus, $\int_{v_1}^{s_2}(1 - G(\tau))F(\tau)d\tau$ is the minimum subsidy required from the intermediary. However, a profit-minded intermediary may want to optimize her profit over a longer time horizon, or design a trading mechanism that would maximize her profit in each individual trade. The former requires enhanced knowledge of the market which presents an interesting research topic to be discussed further. The latter could be done by a mechanism which only allow *profitable* (while individually rational) trades to take place. We describe the construct of such a mechanism in the following.

With Theorem 2, we may rewrite the intermediary's profit function as follows:

$$\pi_I = \int_{v_1}^{v_2} \int_{s_1}^{s_2} (\Psi_b(\tau_b) - \Psi_s(\tau_s))\beta(\tau_s, \tau_b)) f(\tau_s) g(\tau_b) d\tau_s d\tau_b - \pi_b(v_1) - \pi_s(s_2)$$
(3.35)

Based on the supply chain intermediation framework outlined above, the intermediary devises a direct mechanism $\Gamma(\beta, p, w)$ to maximize her profit (3.35), subject to incentive compatibility and individual rationality. First, the intermediary must determine $\beta(s, v)$ which specifies whether the trade is to take place given the reported s, v and her knowledge of their virtual opportunity costs and virtual willingness to pay. Given the simple form of the profit function (3.35) it is straightforward to find a profit maximizing β subject to individual rationality as follows:

$$\begin{aligned}\beta(s,v) &= 1, \quad \text{if } \Psi_b(v) \geq \Psi_s(s) \\ &= 0, \quad \text{Otherwise.}\end{aligned} \quad (3.36)$$

and $\pi_b(v_1) = \pi_s(s_2) = 0$. In other words, we simply define the condition $\Theta(\Psi_b(v), \Psi_s(s)) \equiv \Psi_b(v) \geq \Psi_s(s)$. Moreover, while the intermediary satisfies the individual rationality constraint she offers no additional surplus to the players. Under this mechanism the intermediary's profit is determined by the difference between the buyer's virtual willingness to pay and the supplier's virtual opportunity cost. If the trade is to take place (i.e., $\Psi_b(v) \geq \Psi_s(s)$), the intermediary must specify an asked price $p(s, v)$ and a bid price $w(s, v)$ that satisfy the incentive compatibility constraints (3.25) and (3.26). From Theorem 2, it can be shown that if $\Psi_b(.)$ and $\Psi_s(.)$ are monotone functions, one possible solution is to set $p(s, v) = v_1$ and $w(s, v) = s_2$. In other words, to attract the players from their outside options to the intermediated trade, the intermediary asks the lowest willingness to pay level the buyer could have quoted, while paying the supplier the highest possible opportunity cost. Of course, in order for the trade to occur in the first place, it must be the case that $\Psi_b(v_1) \geq \Psi_s(s_2)$. We now state the following theorem (Myerson and Satterthwaite (1983)).

THEOREM 3 *Suppose $\Psi_b(.)$ and $\Psi_s(.)$ are monotone increasing functions in $[v_1, v_2]$ and $[s_1, s_2]$, respectively. Then among all individually rational mechanisms, the intermediary's expected profit is maximized by a mechanism in which the trade takes place iff $\Psi_b(\tilde{v}) \geq \Psi_s(\tilde{s})$.*

In essence, to maximize her own profit the intermediary must restrict the trade to "profitable" situations, as indicated by the difference between the buyer's virtual willingness to pay and the supplier's virtual opportunity cost. For instance, suppose $\tilde{s}$ and $\tilde{v}$ are both uniformly distributed on the unit interval. Then, based on the above mechanism the trade takes place if and only if $\Psi_b(\tilde{v}) = 2\tilde{v} - 1 \geq 2\tilde{s} = \Psi_s(\tilde{s})$. Or equivalently, $\tilde{v} - \tilde{s} \geq \frac{1}{2}$, i.e., the trade takes place iff the buyer's valuation exceeds the supplier's valuation by $\frac{1}{2}$.

The insights provided by the above analysis is important in that it illustrates another important role played by the supply chain intermediary. In theory, the intermediary must subsidize trade as needed, but it is possible for the intermediary to regulate trades based on the players' virtual valuations such that only trades expected to be profitable are actually taking place. Note that under the criteria specified in (3.36) the intermediary's profit (based on (3.35)) is always non-negative. In other words, one way for the intermediary to manage is by regulating when the trade is to take place, while never sponsoring any trade that she is expected to subsidize. However, this results in a fairly conservative policy for intermediated trade, e.g., in the above example, the buyer's valuation must exceed the supplier's by $\frac{1}{2}$. An important extension for this line of research is to model the situation where the intermediary subsidizes a certain unprofitable trades with the goal of maximizing gains over a longer horizon.

6. Multilateral Trade with Incomplete Information

Our analysis has so far focused on the role of intermediary in bilateral bargaining situations. We now turn our attention to multilateral trades. In the supply chain, multilateral trade could occur in at least two different settings. First, in a **vertically integrated** setting, a pre-established supply chain structure dictates the set of suppliers a wholesaler deals with, or the set of retailers a supplier sells to. What is left to be determined is the particular term of trade (e.g., price, quality, delivery date). At any one time, a supplier may face a particular set of buyers. In this setting, the intermediary creates value by devising efficient mechanisms that help the supplier to elicit willingness to pay information from the buyers and to identify the most lucrative trade. Similarly, in a buyer-centric environment, the intermediary may devise mechanisms that help the buyer to elicit cost information from a preestablished set of suppliers, identifying the most desirable supplier for the trade. In Section 6.1, we characterize multilateral trade with vertical integration using the basic framework of Myerson (1981) and Bulow and Roberts (1989). We will show that from the perspective of supply chain intermediation, this multilateral trading environment is directly linked to the bilateral bargaining framework established before.

The second setting is in a **matching markets**. In this setting, there is no preestablished supply structure, the buyers and suppliers come to a central exchange (e.g., an eCommerce site, a procurement auction) and the intermediary functions as a coordinator of the exchange. A matching market emerges since it may be costly for the buyers and the suppliers to seek out each other directly. However, there are costs involved in setting up a central place (e.g., infrastructure costs) and there are variable costs associate with the transactions (e.g., communication of price, quality, and product specifications). The intermediary creates value by continuously shaping the portfolio of suppliers (customers) that best match the needs (market potential) of the customers (suppliers). The intermediary makes a profit by creating a nonzero bid-ask spread which clears the market. In Section 6.2, we characterize the role of an intermediary in multilateral trade with markets using the framework by McAfee (1992). Similarly, from the viewpoint of supply chain intermediation, the model associated to matching markets is directly linked to the bilateral bargaining model.

6.1 Multilateral Trade with Vertical Integration

In this section, we characterize the role of intermediary in multilateral trades where one supplier faces multiple buyers or one buyer faces multiple suppliers. Without the lost of generality, we consider a supplier facing m buyers so that $M = (1, ..., m)$. The trade under consideration consists of one particular bundle of goods and services that can be considered a single object. The intermediary faces a mechanism design problem with the goal of eliciting buyers' willingness to pay for the object. The supplier's opportunity cost s is common knowledge. Each buyer i holds private information about her valuation $v_i \in [a_i, b_i]$ that is known to the others in the form of distribution function $G_i(v_i)$ and density $g_i(v_i)$. Each buyer i holds a vector of value estimates $v_{-i} = (v_1, ... v_{i-1}, v_{i+1}, ..., v_m)$ for other buyers. All players are influenced by other buyers' valuations, which result in a quasi-linear valuation $u_i(v)$ for each buyer i. This problem has been examined extensively in the context of optimal auctions (c.f., Myerson 1981; Maskin and Riley 1984; Milgrom and Weber 1982). Riley and Samuleson (1981) shows that for a broad family of auction rules, expected seller (supplier) revenue is maximized if the seller announces a certain reserve price (the minimum bid she would accept). They show that this reserve price is independent of the number of buyers and it is strictly greater than the supplier's opportunity cost s. Myerson (1981) proposes the optimal auction design problem: the supplier chooses, among all possible mechanisms, one that would maximize her expected net revenue. This perspective is useful in that it helps us to define the role of the intermediary in the one-supplier, multiple-buyer setting. As before, it is sufficient to consider an incentive compatible direct mechanism that will carry out the trade. To the intermediary, the supplier reports her opportunity cost s, and each buyer i reports her valuation v_i. Thus, the intermediary holds a vector of value estimates $v = (v_1, ..., v_m)$. The intermediary establishes the buyer's virtual willingness to pay $\Psi_i(v_i)$ as follows:

$$\Psi_i(v_i) = v_i - \frac{1 - G_i(v_i)}{g_i(v_i)} \tag{3.37}$$

Since the supplier's opportunity cost s is known, the intermediary establishes the supplier's opportunity cost $\Psi_s(s) = s$. However, the intermediary may announce a reserve price $u_s(v)$ based on her knowledge of the vector v and that $u_s(v) \geq \Psi_s(s) = s$. It may be convenient to think that the intermediary determines a reserve price for the object at $u_s(v)$, and she submits a bid of $u_s(v)$ such that if none of the bids received from the buyers are above $u_s(v)$, the intermediary keeps the object (the trade fails to take place).

Let $\Gamma(\beta, p, w)$ represents the direct revelation mechanism, where $\beta_i(v)$ is the probability that buyer i will get the object, $p_i(v)$ is the expected payment from buyer i to the intermediary, and $w(v)$ is the expected payment from the intermediary to the supplier. Given mechanism $\Gamma(\beta, p, w)$, knowledge of her own valuation v_i, and other bidders $j \neq i$ valuations in terms of $g_j(v_j)$ buyer i's expected gain from the trade is as follows:

$$\pi_i(\beta, p, v_i) = \int_{T_{-i}} (u_i(v)\beta_i(v) - p_i(v))g_{-i}(\tau_{-i})d\tau_{-i} \tag{3.38}$$

where $d\tau_{-i} = d\tau_1, ..., d\tau_{i-1}, d\tau_{i+1}, ..., d\tau_m$ and $T_{-i} = [a_1, b_1] \times ... [a_{i-1}, b_{i-1}] \times [a_{i+1}, b_{i+1}] \times ... [a_m, b_m]$. Similarly, the intermediary's expected gain for the trade as follows:

$$\pi_I(\beta, p, w) = \int_T [u_s(\tau)(1 - \sum_{j=1}^m \beta_j(\tau)) + \sum_{j=1}^m (p_j(\tau))] g(\tau) d\tau \quad (3.39)$$

where $d\tau = d\tau_1, ..., d\tau_m$ and $T = [a_1, b_1] \times ... \times [a_m, b_m]$. The intermediary is to maximize (3.39) subject to the following constraints:
individual rationality:

$$\pi_i(\beta, p, v_i) \geq 0 \quad \forall i \in M \quad (3.40)$$

incentive compatibility:

$$\pi_i(\beta, p, v_i) \geq \pi_i(\beta, p, \acute{v}_i) \quad \forall v_i, \acute{v}_i \in [a_i, b_i] \quad (3.41)$$

(where $\acute{v}_i$ is the valuation reported by buyer i) and the probability conditions:

$$\beta_i(v) \geq 0 \text{ and } \sum_{j=1}^m \beta_i(v) \leq 1, \quad \forall i \in M \quad \forall v \in T \quad (3.42)$$

Using similar techniques as described in the bilateral bargaining analysis (Section 5), we may rewrite the intermediary's maximization function similar to that of (3.35) as follows:

$$\pi_I(\beta, p) = \int_T (\sum_{i \in M} (\Psi_i(v_i) - \Psi_s) \beta_i(\tau)) g(\tau) d\tau - \sum_{i \in M} \pi_i(\beta, p, v_i) \quad (3.43)$$

Similar to the bilateral case, the intermediary could satisfy the individual rationality constraint by offering no additional surplus to the players, i.e.,

$$\sum_{i \in M} \pi_i(\beta, p, v_i) = 0, \psi_s = w = s. \quad (3.44)$$

Thus, the intermediary's profit is determined by the difference between the buyers' virtual willingness to pay and the supplier's opportunity cost. If the trade is to take place, there must be at least one buyer i such that $\Psi_i(v_i) \geq \Psi_s = s$). The intermediary could ensure that this is the case by stating a reserve price $u_s(v) \geq s$. It can be shown that if $\Psi_i(v_i)$ is a monotone strictly increasing function of v_i, for every $i \in M$, one possible solution to the above auction maximization problem is as follows:

$$\begin{aligned} \beta_i(v) &= 1, \text{ if } \Psi_i(v_i) = \max_{j \in M} \Psi_j(v_j) \geq u_s(v) \\ &= 0, \text{ Otherwise.} \end{aligned} \quad (3.45)$$

In other words, the intermediary offers the object to the buyer with the highest virtual valuation $\Psi_i(v_i)$ so long as it is above the reserve price. While the above model was developed in the context of auction optimization, it provides a general framework of analysis for multilateral trade with vertical integration. Similar to the bilateral bargaining case, the intermediary plays the important role of regulating the trade such that it is profitable. Moreover, the intermediary may screen out buyers by setting the reserve price, i.e., buyers whose willingness to pay $v_i < u_s(v)$ has no incentive to participate in the trade. Since the intermediary only need to pay the supplier her opportunity cost s, she could generate profit from the difference $(u_s(v) - s)$, where $u_s(v)$ represents her knowledge of the market.

An important insight from the above analysis is that the intermediary matches the supplier with the buyer with the highest virtual willingness to pay, which may not be the buyer with the highest willingness to pay. This is because Myserson's model assumes asymmetric buyers (i.e., v_i's are draw from independent, but not necessarily identical distributions). Bulow and Roberts (1989) offers an insightful interpretation of *virtual willingness to pay* as follows.

Define the X axis as the probability that the buyer's value exceeds a certain value, $1 - G_i(v_i) = q$, and the Y axis as value v. For each buyer i, graph the inverse of her cumulative distribution function G_i (where $G_i(a_i) = 0, G_i(b_i) = 1$) (see Figure 3.3). This represents the buyer's demand curve. The buyer's revenue is qv_i, where $v_i = G_i^{-1}(1 - q)$. From the demand curve for each buyer, we may compute the buyer's marginal revenue as follows, i.e.,

$$\begin{aligned}
\frac{dq \cdot v_i}{dq} &= \frac{dq G_i^{-1}(1-q)}{dq} \\
&= G_i^{-1}(1-q) + \frac{dG_i^{-1}(1-q)}{dq} \\
&= G_i^{-1}(1-q) - \frac{q}{g_i(G_i^{-1}(1-q))} \\
&= v_i - \frac{1 - G_i(v_i)}{g_i(v_i)}
\end{aligned}$$

Clearly, the buyer's marginal revenue is identical to her virtual valuation Ψ_i (3.37). In setting up the reserve price, the intermediary may be thought of as a buyer with a value and marginal revenue of zero. Thus, the intermediary offers the object to the buyer with the highest marginal revenue so long as it is positive (above her own).

Using the above interpretation, Bulow and Roberts (1989) show that Myerson's optimal auction problem can be described as the third-degree monopoly price discrimination problem where instead of m independent bidders, there are m independent markets. The monopolist allocates the object(s) to the

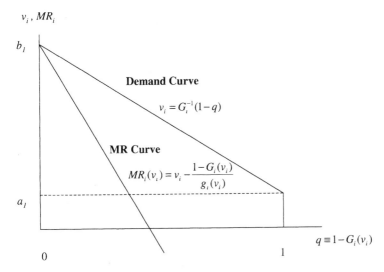

Figure 3.3. Interpretation of the Bidder i's Virtual Willingness to Pay (Bulow and Roberts (1989)

buyer(s) with the highest marginal revenue. The only crucial assumption being made here is that marginal revenue is downward sloping in quantity within each market. Moreover, in the context of bilateral bargaining with incomplete information (5), Bulow and Roberts (1989) offer a similar interpretation for the buyer's *virtual willingness to pay* Ψ_b (3.17) and the supplier's *virtual opportunity cost* Ψ_s (3.18), showing that Ψ_b and Ψ_s are equivalent to the buyer's *marginal revenue* and the supplier's *marginal cost*, respectively. Thus, the intermediary's function is to make sure the trade only takes place when the (buyer's) *marginal revenue* is greater than the (supplier's) *marginal cost*.

From the above discussion, it is interesting to note that the theoretical underpinning of multilateral trade is directly linked to the bilateral bargaining situation, and both can be interpreted in the context of pricing problem in microeconomics theory.

6.2 Multilateral Trade with Markets

In this section, we further extend the insights derived from bilateral bargaining in a multilateral setting. We introduce a multilateral trading model that captures the basic essence of intermediation in an exchange setting. There are m buyers and n suppliers, each buyer i has a private valuation v_i for a single unit of good, and each supplier j has a privately known cost s_j for the good she sells. It is the common believe of all traders that each buyer's value is distributed according to $G(v)$, and each supplier's cost is distributed according to $F(s)$. A well known form of trading in this environment is a *call market*, which describe the basic operation of the New York Stock Exchange (NYSE). In a call market, the intermediary collects bids from the buyers and offers (asks) from the suppliers, constructs supply and demand curves, determines a market-clearing price, and executes the trade. Several mechanisms have being proposed to model the call market. Wilson (1985) initiated the study of double auction as a means to model multilateral trading with incomplete information. Extending the results from Myerson (1981), Myerson and Satterthwaite (1983), and Gresik and Stterthwaite (1989) he shows that a sealed-tender double action is incentive efficient if the number of traders are sufficiently large. In such a double auction, all trades are made at a single market clearing price. Rustichini et al. (1994) models the call market as a *k-double auction*, where the buyers' bids and the suppliers' offers are aggregated to form (discrete) supply and demand curves. The crossing of their graphs determines an interval $[a, b]$ from which a market clearing price p_0 is defined as $p_0 = (1-k)a + kb$. The choice of $k \in [0, 1]$ defines a specific mechanism. Trades occur among buyers who bid at least p_0 and sellers who offer no more than p_0. Hagerty and Rogerson (1985) discusses a fixed-price mechanism where trades occur among buyer and sellers who indicate their willingness to trade at a fixed-price p_0^*, with traders on the long side of the market randomly given the right to trade.

McAfee (1992) proposes a double auction model that explicitly considers the role of an intermediary who intervenes in the trade and keeps track of supply and demand at asked and bid prices. Like the market specialist in NYSE, the intermediary makes a profit by regulating the trade using a certain mechanism. In the following, we describe this double auction as a direct revelation mechanism.

Step 1. The buyers report their willingness-to-pay to the intermediary, who ranks them as $v_{(1)} \geq v_{(2)} \geq ... \geq v_{(m)}$). Similarly, the suppliers report their opportunity costs and are ranked as $s_{(1)} \leq s_{(2)} ... \leq s_{(n)}$. Where index (i) represents the ith highest valuation buyer or the ith lowest cost supplier. Further, we define $v_{(m+1)} = sup\{v : G(v) = 0\}$ (the lowest possible value) and $s_{(n+1)} = inf\{s : F(s) = 1\}$ (the highest possible cost).

Step 2. The intermediary finds the efficient *trading quantity* $k \leq \min m, n$ satisfying $v_k \geq s_k$ and $v_k + 1 < s_k + 1$.

Step 3. The intermediary determines the *market clearing price* p_0 as follows:

$$p_0 = \frac{1}{2}(v_{(k+1)} + s_{(k+1)}) \tag{3.46}$$

Step 4. To execute the market with budget balanceness, if $p_0 \in [s_{(k)}, v_{(k)}]$, all buyers and suppliers (1) through (k) trade at the market clearing price $p = w = p_0$. The intermediary makes zero profit $(p - w = 0)$. Otherwise, the $k - 1$ highest value buyers and trade with the $k - 1$ lowest cost suppliers, where buyers pay $p = v_{(k)}$ and suppliers receive $w = s_{(k)}$, and the intermediary keeps the total bid-ask spread $(k - 1)(p - w)$.

The final step is key for the intermediary to maintain budget balanceness while making a profit (sometimes) by: (1) charging the buyers a higher price then the sellers receive, i.e., buying at the asked price $w = s_{(k)}$ and selling at the bid price $p = v_{(k)}$, thus creating a profit $(k-1)(p-w) = (k-1)(v_{(k)} - s_{(k)})$, and (2) preventing the least profitable trade (the trade between the lowest value buyer and highest cost seller) from taking place. McAfee (1992) shows that for the direct mechanism described above, it is a dominant strategy for the traders to truthful reporting their valuations. This is important as it eliminates the needs to consider strategic behavior of the traders, which is the a complication found in the double auction analysis.

The simple model above describes the main role of the intermediary in multilateral trading. In addition to coordinating the multilateral exchange, the intermediary designs (bidding) mechanisms for customers and suppliers that reveals their willingness to pay levels and opportunity costs. The intermediary sets bid and asked prices to maximize profit and balances the purchases and the sales. As mentioned above, the intermediary sets up a central place for the suppliers and buyers to trade, while continuously selecting the portfolio of suppliers (buyers) that best match the needs (market potential) of the buyers (suppliers). This selection process occurs naturally in the above mechanism as only (up to) k most efficient trades take place between the most compatible pairs of suppliers and buyers. All buyers with a below threshold willingness-to-pay level and all suppliers with an above threshold opportunity cost will be excluded from the exchange. The above selection function can be further illustrated by the oral double auction described in McAfee (1992).

Milgrom and Weber (1982) describes a variant of the English Auction where the price is posted electronically. All bidders are active at price zero. The price is raised continuously, and a bidder who wishes to remain active at the current price must depress a button. When she releases the button, she is dropped out of the auction. No bidders who has dropped out can become active again. After any bidder withdraws, all remaining bidders know the price at which she drops out. When there is only one bidder left in the room, the auction ends. McAfee (1992) proposes an oral double auction work in a similar fashion, but with multiple buyers and sellers. In the following, we use the oral double auction model to characterize the basic functions of exchange coordination carried out by a market intermediary (see Figure 3.4).

Step 0. Buyers and suppliers enter a central trading space operated by an intermediary. The intermediary keeps track of the state of the system $(m(t), n(t), p(t), w(t))$ in continuous time t; where $m(t)$ and $n(t)$ are the number of active buyers and suppliers, $p(t)$ and $w(t)$ are the bid and asked prices, respectively. At $t = 0$, $m(0) = m$, $n(0) = n$, and the bid and asked prices are set at the most favorable levels, i.e.,

$$\begin{align} p(0) &= inf\{v : G(v) > 0\} \\ w(0) &= sup\{s : F(s) < 1\}. \end{align} \quad (3.47)$$

The buyers and suppliers may choose to leave the trading space anytime during the process, thus become inactive. An inactive trader can not be active again during the trading process.

Step 1. At any time t during the trading process, the intermediary updates the bid and asked prices based on the number of active buyer and suppliers. Specifically, she raises the bid price at a unit rate if there are more buyers than suppliers; she reduces the asked price at a unit rate if there are more suppliers than buyers, i.e.,

$$\begin{align} p'(t) &= 1, \quad \text{if } m(t) \geq n(t) \\ &= 0, \quad \text{if } m(t) < n(t) \\ w'(t) &= 1, \quad \text{if } n(t) \geq m(t) \\ &= 0, \quad \text{if } n(t) < m(t) \end{align} \quad (3.48)$$

Step 2. The trading process completes at the first time T when the number of buyers equals the number of suppliers, and the bid price is no less than the asked price, i.e.,

$$\begin{align} m(T) &= n(T) \\ p(T) &\geq w(T) \end{align}$$

Step 3. The trades take place. The intermediary collects the bid price $p(T)$ from the buyers and pay the asked price $w(T)$ to the suppliers. The intermediary keeps the difference $m(T)(p(T) - w(T))$.

The decision for the buyers and the suppliers are quite simple. They only need to decide whether to stay active in the trade or not. It is a dominant strategy for a buyer with willingness-to-pay of v to remain active in the intermediated trade as long as $v > p(t)$, and for a supplier with opportunity cost s to stay active as long as $s < w(t)$. The above procedure is a stylized implementation of the direct mechanism described earlier. Suppose there are more suppliers than buyers initially entering the intermediated trade ($m \leq n$). The

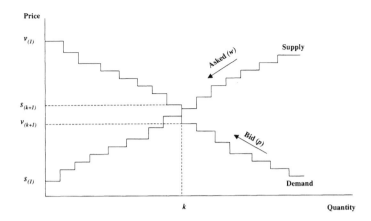

Figure 3.4. Oral Double Auction Proposed by McAfee (1992)

asked price decreases until $n - m$ (highest cost) suppliers drop out. The trading process now continues with the equal number of buyer and suppliers, but $p(t) < w(t)$. If a supplier leaves, asked prices freeze and bid prices rise until a buyer leave. Similarly, if a buyer leaves, bid prices freeze and asked prices decline until a supplier leaves. In this way, the gap between ask and bid prices decrease until $p(t) \geq w(t)$.

The general settings of the double auction assume that any supplier can be matched with any buyer. In the context of industrial procurement, Kalagnanam et al. (2000) propose a double auction mechanism allowing additional "assignment constraints" that specify which bids can be matched with which asks. It is possible that a buyer's demand can be met by the a subset of suppliers, and vice versa. They propose efficient network flow algorithms to find the market clearing allocation which satisfies the assignment constraints. They further examine the case where demand is indivisible. In this case, finding the market clearing allocation requires the solution of a combinatorial optimization problem. For an overview of various aspects of combinatorial auction such as this, please refer to Chapter 7.

7. Related Work in the Supply Chain Literature and Research Opportunities

In this chapter, we explore a modelling paradigm for supply chain coordination using the notion of supply chain intermediary. Our goal is to complement the view offered by the supply chain contracting literature that uses long-term coordination contracts as a primary mechanism to achieve channel efficiency. We propose to look at the problem of supply chain coordination somewhat differently: for any subset of players in the supply chain who desire to establish supplier-buyer relationships, they may choose to do so directly or through some form of intermediation. A transactional intermediary improves operational efficiencies by serving as an intermediate supply chain player or as a third party service provider. An informational intermediary alleviates the effects of information asymmetry by serving as a broker, an arbitrator, or a mediator. An intermediary utilizes a variety of mechanisms such as bilateral contracts, alliances, coalitions, and auctions to facilitate her coordination/arbitration functions. To streamline the thinking, we consider the intermediary a profit-seeking entity. This provides an unambiguous way to divide the system surplus generated from coordination, i.e., after providing the players proper incentives to participate and to truthfully reveal their private information, the intermediary keeps the remainder of the system surplus (if non-negative). Thus, the intermediary's profit is the net surplus from trade minus the transaction costs she has incurred. One may consider an intermediary's "profits" as the performance measure of an intermediated trade. When an intermediated trade is unprofitable, one may infer that either an alternative form of intermediation is needed, or disintermediation may be unavoidable. Since we allow the intermediary to call off the trade, trades where the surplus is insufficient to cover the incentive costs and/or the transaction cost may not take place at all.

We establish the analysis of supply chain intermediation using a bargaining-theoretic framework. We introduce four basic settings: bilateral bargaining with complete information, bilateral bargaining with incomplete information, multilateral trade with vertically integration, and multilateral trade with markets. Each setting captures a different aspect of transactional/informational intermediation. As we have demonstrated in the chapter, the direct-mechanism framework introduced in the context of bilateral bargaining with incomplete information forms the basis for multilateral analysis. This theoretical connection establishes a convenient analytical framework for the study of supply chain intermediaries, thus the study of supply chain coordination.

A steam of research has appeared in the supply chain literature which also considers the use of bargaining theoretic models to expand the view of negotiation and coordination in the supply chain. Chod and Rudy (2003) consider a situation where two firms unilaterally decide on the investment levels on resources (capacity or inventory) based on imperfect market forecasts. As new information becomes available, the firms update their forecasts and they have the option to trade excess resources. Chod and Rudy (2003) formulate this problem as a non-cooperative stochastic investment game in which the payoffs depend on an embedded Nash bargaining game. Deng and Yano (2002) consider the situation where a buyer first orders from the supplier (at the contracted price) before market demand is observed, then places additional orders (at the spot price) after the demand is observed. They consider the setting of the spot price the result of a bargaining process. They show that the players' bargaining power regarding the spot prices depends upon the outcomes of demand. Taylor and Plambeck (2003) argue that writing binding contracts that specify the price (and possibly capacity) prior to the supplier's capacity investment may be difficult or impossible. Instead of formal, court-enforceable contracts, they study informal agreements that are sustained by repeated interaction. These relational contracts are analyzed in as a repeated bargaining game where the buyer's demand and the supplier's capacity are private information. Kohli and Park (1989) analyze quantity discounts as a cooperative bilateral bargaining problem. Instead of the typical setting where the seller dictates the quantity discount scheme, the order quantity and its corresponding price (discount) are determined through seller-buyer negotiation. They show that joint efficiency can be achieved through the bargaining process, and they study the effect of bargaining power on the bargaining outcome. Nagarajan and Bassok (2002) consider a cooperative, multilateral bargaining game in a supply chain, where n suppliers are selling complementary components to a buyer (an assembler). They propose a three-stage game: the suppliers form coalitions, the coalitions compete for a position in the bargaining sequence, then the coalitions negotiate with the assembler on the wholesale price and the supply quantity. They show that each player's payoff is a function of the player's negotiation power, the negotiation sequence, and the coalitional structure. Plambeck and Taylor (2001) consider the situation where two independent, price-setting OEM's are investing in demand-stimulating innovations. The OEMs may outsource their productions to an independent contract manufacturer (CM), and the negotiation of the outsourcing contract is modelled as a bargaining game. They show that the bargaining outcome induces the CM to invest in the system-optimal capacity level and to allocate capacity optimally among the OEMs. In a subsequent paper, Plambeck and Taylor (2002) consider the situation where two OEMs sign quantity flexible (QF) contracts with the CM *before* they invest in the innovation. In case the CM has excess capacity after the demands are observed, the three parties (two OEMs and the CM) bargain over the allocation of this capacity. They shown that this "renegotiation" could significantly increases system profit if it is anticipated in the supply contract. Van Mieghem (1999) considers two asymmetric firms, a subcontractor and a manufacturer, who invest non-cooperatively in capacity under demand uncertainty. After demand is realized, the manufacturer may purchase some of the excess capacity from the subcontractor. He models the problem as a multivariate, multidimensional, competitive newsvendor problem. He argues that ex ante contracts may be too expensive or impossible to enforce, while the supplier's investments (in quality, IT infrastructure, and technology innovation) may be non-contractible. Thus, he analyzes *incomplete contracts* Hart and Moore, 1988 where the players leave some contract parameters unspecified ex ante while agreeing to negotiate ex post. Modelled as a bilateral bargaining game, the incomplete contract allows the consideration of the player's bargaining power. Anupindi et al. (2001) consider a decentralized distribution system with n independent retailers. Facing stochastic demands, the retailers may hold stocks locally and/or at centralized locations. Each retailer chooses her own inventory level based on a certain stocking policy. After demand is realized, the retailers bargain cooperatively over the transshipment of excess inventories to meet additional demands. They propose a cooperative inventory allocation and transshipment mechanism that induces the retailers to choose the Nash equilibrium inventory levels that maximize the system profit.

As discussed throughout the chapter, the viewpoints offered by the supply chain intermediary theory and the bargaining theory could potentially broaden the scope for *supply chain coordination*. In the following, we outline a few research opportunities offered by the proposed paradigm.

Supply Chain Coordination and the Division of Surplus . Most supply chain contracts split the channel surplus arbitrarily, which invariably favor the channel-leader who designs the contract. Bargaining theory offers a generalized view of the negotiation process, taking into account the influence of bargaining power on the division of systems surplus. Both cooperative (c.f., Nagarajan and Bassok 2002) and non-cooperative (c.f., Ertogral and Wu 2001) bargaining games could be used to model the negotiation involved in splitting the channel surplus. In any case, a supply chain intermediary may devise a mechanism (e.g., post a bid and an asked prices) and eliminates the need for bilateral bargaining (contract negotiation) to actually taking place. The bargaining theory and the intermediary theory suggest that the mechanism offered by the intermediary reflects supply chain players' bargaining power. Suppose a supply chain has a retailer significantly more powerful than the manufacturers. One would expect that the power structure is reflected in the bid-ask spread set by the wholesales and distributors. In other words, a supply chain player's bargaining power ultimately determines the offers she receives from the intermediary, thus the share of the system surplus she receives. A variety of supply contracts in the literature (e.g., buy-back, QF, profit-sharing) could be examined based on the players' bargaining power, and new insights could be gained on the actual division of system surplus. While it is quite straightforward to incorporate outside options in the complete information setting (as shown in Section 4), it is significantly more challenging when the players' outside options (therefore bargaining power) are defined endogenously as private information. Moreover, it might be interesting to model the signaling game during the negotiation process where the players choose to reveal a certain portion of their outside options at a certain point in time during the negotiation. Computing Baysian-Nash equilibrium under these conditions are known to be very challenging.

Strategic Supply Chain Design . As suggested earlier in the chapter, we may take the viewpoint of a supply chain integrator who uses intermediaries as a strategic tool to improve overall supply chain design. Suppose a high-tech OEM has a traditional, third party distribution channel for her electronic products. The OEM is interested in developing an Internet channel incorporating the drop-shipping model, i.e., the manufacturer fills customer orders directly without going through intermediaries such as the retailers and distributors. In order to integrate the traditional and the Internet channels, the OEM may have to address issues such as the following: How to configure transactional intermediaries who could support the growth of both traditional and Internet channels? How to configure informational intermediaries who could overcome information asymmetry (e.g., demand, pricing) across and within the channels? How to reintermidate exiting intermediaries in the traditional channel for the integrated channel? What incentives are there to facilitate the transition from traditional to integrated channels? The above example represents opportunities to study strategic supply chain design focusing on placing new intermediaries, disengaging existing intermediaries, evaluating the interdependencies of intemediation/disintermeidation, and assessing the impact of different intermediation strategies to overall supply chain efficiency.

Intermediary to foster Information Sharing . The need for information intermediation arises when supply chain players recognize their limited abilities to share information. For instance, supply chain partnership agreements are typically built on the basis that the buyer desires favorite pricing and responsive supply, while the supplier wants improved demand visibility and stability. However, as discussed earlier in the chapter, the players may not want to share private information such as pricing, quality level, and sourcing strategies, thus a third party auditor is typically called in to monitor the partnership. In this context, the auditor is an informational intermediary who alleviates the inefficiency due to asymmetric information. In fact, the success of any supply chain partnership hinges on the intermediation mechanism used to handle information sharing. The design of such informational intermediation presents significant research opportunities. Suppose the supply chain partners are to develop a joint demand management process, but there is information asymmetry since only the buyer can observe the demand. The demand sharing process may be described as a simple signaling game as follows:

(1) Nature draws demand d_t for period t from a distribution D, (2) the buyer observes d_t and then choose a message $\hat{d}_t$ for the supplier, (3) The supplier observes $\hat{d}_t$ (but not d_t) and use it to determine the production level k_t, (4) The buyer and the supplier each receives a payoff as a function of $d_t, \hat{d}_t$, and k_t. The game repeats for each period t. This signaling game is a dynamic game with incomplete information, and it is known that the perfect Bayesian equilibrium outcome leads to inefficiency. Similar to the bilateral bargaining game with incomplete information, it is possible to develop a direct mechanism such that it is incentive compatible and individually rational for the players to align their demand signals thus avoiding adverse selection. Designing an informational intermediary in this context involves the selection of a payoff function, a demand signaling scheme, and a fundamental understanding of the players' behaviors under perfect Bayesian equilibrium.

Emerging Supply Chain Microstructure It is beneficial to broaden the view of multilateral trade from the discussion in Section 6 to consider the case where the intermediary competes with a direct matching market (c.f., Rubinstein and Wolinsky, 1987; Gehrig 1993), i.e., markets where the supplier and buyer meet directly without any form of intermediation. Thus, a buyer must choose from (1) entering the matching market, in which case she searches and meets the supplier directly and bargain over the price and (2) transacting through an intermediary who offers a bid-ask spread by aggregating price/capacity information from many buyers and suppliers. In the latter case, the buyer may get to choose from competing intermediaries who offer alternative price/performance trade-offs. When the buyer bargains directly with a supplier in the matching market, the bargaining may breakdown since the supplier's capacity may not be sufficient to accommodate the buyer's demands, or the buyer's demands may not be sufficient to justify the supplier's investment on capacity. When the buyer enters an intermediated trade, the intermediary may be able to increase the probability of a successful trade since she possesses aggregated information concerning the buyers and suppliers in the market, and she expects to transact with many buyers and suppliers over time. In general, the buyer chooses to transact with an intermediary when the expected cost of searching and bargaining in the matching market exceeds the intermediary's offer. Given the choice, the buyer transacts with the intermediary who offers the most attractive price/performacne trade-off. The general setting described here allows us to study the microstructure in emerging supply chain environments. For instance, contract manufacturers in the high-tech industry such as Solectron, Flextronics, and Celestica maybe considered emerging intermediaries between the the brand-carrying OEM such as HP, Dell, and Nokia and the upstream supply chain capacity. By consolidating demands from different OEMs, and by investing and developing highly flexible processes, these contract manufacturers are able to realize a much higher utilization on their equipment and, therefore, reduce the unit costs. On the other hand, by consolidating the component procurement for different customers, contract manufacturers are able to enjoy economies of scale from upstream suppliers. Thus, the contract manufactures are intermediaries who could offer the OEMs greater variety of products at a significantly lower cost, while offering the component suppliers greater stability in their demands. However, the contract manufacturers must compete with one another on different price/performance tradeoffs. The key for a contract manufacturer's competitiveness is in her ability to sustain highly flexible processes in a dynamically changing environment of technological innovations. As a result, the price/performance profile of a contract manufacturer changes over time as the technological life-cycle marches over time, leading to a challenging and dynamic research problem.

References

Anupindi, R., Bassok, Y., and Zemel, E. (2001). A general framework for the study of decentralized distribution systems. *Manufacturing and Service Operations Management*, 3:349–368.

Bailey, J. P. (1998). *Intermediation and electronic markets: aggregation and pricing in Internet commerce.* PhD thesis, Massachusetts Institute of Technology, Cambridge, MA.

Binmore, K., Rubinstein, A., and Wolinsky, A. (1986). The Nash bargaining solution in economic modelling. *Rand J. Econom.*, 17(2):176–188.

Binmore, K. G. (1987). Nash bargaining theory I, II, III. In Binmore, K. G. and Dasgupta, P., editors, *The economics of bargaining*, pages 27–46, 61–76, 239–256. Blackwell, Oxford, UK.

Binmore, K. G. and Herrero, M. J. (1988). Matching and bargaining in dynamic markets. *Rev. Econom. Stud.*, 55(1):17–31.

Bollier, D. (1996). In *The future of electronic commerce: a report of the fourth annual Aspen Institute roundtable on information technology*, Washington, DC. The Aspen Institute.

Bulow, J. and Roberts, J. (1989). The simple economics of optimal auctions. *Journal of Political Economy*, 97(5):1060–1090.

Cachon, G. P. (2002). Supply chain coordination with contracts. Technical report.

Campbell, J., Lo, A. W., and McKinlay, A. C. (1999). *The econometrics of fiancial markets*. Princeton University Press, Princeton, NJ.

Chod, J. and Rudi, N. (2003). Strategic investments, trading and pricing under forecast updating. Technical report, University of Rochester, Rochester, NY.

Crowston, K. (1996). Market-enabling internet agents. International Conference on Information Systems.

Deng, S. and Yano, A. C. (2002). On the role of a second purchase opportunity in a two-echelon supply chain. Technical report, University of California, Berkeley, CA.

Ertogral, K. and Wu, S. D. (2001). A bargaining game for supply chain contracting. Technical report, Lehigh Univeristy, Department of Industrial and Systems Engineering, Bethlehem, PA.

Fox, B. (1999). Does the world still need retailers? RTMagazine.

Frankel, J., Galli, G., and Giovannini, A. The microstructure of foreign exchange markets.

Gehrig, T. (1993). Intermediation in search markets. *Journal of Economics and Management Strategy*, 2:97–120.

Gellman, R. (1996). Disintermediaiton and the internet. *Government Information Quarterly*, 13(1):1–8.

Gresik, T. and Satterthwaite, M. A. (1989). The rate at which a simple market becomes efficient as the number of traders increases: An asymptotic result for optimal trading mechanisms. *Journal of Economic Theory*, 48:304–332.

Hagerty, K. and Rogerson, W. (1985). Robust trading mechanisms. *Journal of Economic Theory*, 42:94–107.

Harker, P. and Zenios, S., editors (2000). *Performance of Financial Institutions; Efficiency, Innovation, Regulations*. Cambridge University Press, UK.

Harsanyi, J. C. (1967). Games with incomplete information played by "Bayesian" players. I. The basic model. *Management Sci.*, 14:159–182.

Hart, O. and Moore, J. (1988). Incomplete contracts and renegotiation. *Econometrica*, 56:755–785.

Hoffman, T. (1995). No more middlemen. Computerworld.

Imparato, N. and Harari, O. (1995). Squeezing the middleman. The Futurist.

Kalagnanam, J. R., Davenport, A. J., and Lee, H. S. (2000). Computational aspects of clearing continuous call double auctions with assignment constraints and indivisible demand. Technical Report IBM Research Report RC21660(97613), IBM T. J. Watson Research Center, Yorktown Heights, NY.

Kalai, E. and Smorodinsky, M. (1975). Other solutions to Nash's bargaining problem. *Econometrica*, 43:513–518.

Kalakota, R. and Whinston, A. B., editors (1997). *Readings in electronic commerce*. Addison-Wesley, Reading, MA.

Kohli, R. and Park, H. (1989). A cooperative game theory model of quantity discounts. *Management Science*, 35(6):693–707.

Kreps, D. M. and Wilson, R. (1982). Sequential equilibria. *Econometrica*, 50(4):863–894.

Lu, J. (1997a). Lemons in cyberspace: A call for middlemen. Paper read at Global Networking 97.

Lu, J. (1997b). Middlemen in cyberspace. Paper read at Global Networking 97.

Maskin, E. S. and Riley, J. G. (1984). Optimal auction with risk averse buyers. *Econometrica*, 52:1473–1518.

McAfee, R. P. (1992). A dominant strategy double auction. *J. Econom. Theory*, 56(2):434–450.

Milgrom, P. R. and Weber, R. J. (1982). A theory of auction and competitive bidding. *Econometrica*, 50(5):1089–1122.

Muthoo, A. (1995). On the strategic role of outside options in bilateral bargaining. *Oper. Res.*, 43(2):292–297.

Myerson, R. B. (1979). Incentive compatibility and the bargaining problem. *Econometrica*, 47(1):61–73.

Myerson, R. B. (1981). Optimal auction design. *Mathematics of Operations Ressearch*, 6(1):58–73.

Myerson, R. B. (1982). Optimal coordination mechanisms in generalized principal-agent problems. *J. Math. Econom.*, 10(1):67–81.

Myerson, R. B. and Satterthwaite, M. A. (1983). Efficient mechanisms for bilateral trading. *J. Econom. Theory*, 29(2):265–281.

Nagarajan, M. and Bassok, Y. (2002). A bargaining framework in supply chains-the assembly problem. Technical report, Marshall School of Business, University of Southern California, Los Angeles, CA.

Nash, Jr., J. F. (1950). The bargaining problem. *Econometrica*, 18:155–162.

Nash, Jr., J. F. (1953). Two-person cooperative games. *Econometrica*, 21:128–140.

O'Hara, M. (1995). *Market microstructure theory*. Blackwell, Cambridge, MA.

Plambeck, E. L. and Taylor, T. A. (2001). Sell the plant? the impact of contract manufacturing on innovation, capacity and profitability. Technical report, Graduate School of Business, Stanford University, Palo Alto, CA.

Plambeck, E. L. and Taylor, T. A. (2002). Renegotiation of supply contracts. Technical report, Graduate School of Business, Stanford University, Palo Alto, CA.

Ponsatí, C. and Sákovics, J. (1998). Rubinstein bargaining with two-sided outside options. *Econom. Theory*, 11(3):667–672.

Resnick, P., Zechhauser, R., and Avery, C. (1998). Role for electronic brokers. Technical report, Havard Kennedy Schol of Goverment, Cambridge, MA.

Riley, J. G. and Samuelson, W. F. (1981). Optimal auctions. *American Economic Review*, 71:381–392.

Roth, A. E. (1985). A note on risk aversion in a perfect equilibrium model of bargaining. *Econometrica*, 53(1):207–211.

Rubinstein, A. (1982). Perfect equilibrium in a bargaining model. *Econometrica*, 50(1):97–109.

Rubinstein, A. (1985a). A bargaining model with incomplete information about time preferences. *Econometrica*, 53(5):1151–1172.

Rubinstein, A. (1985b). Choice of conjectures in a bargaining game with incomplete information. In Roth, A. E., editor, *Game-theoretic models of bargaining*, pages 99–114. Cambridge University Press, Cambridge.

Rubinstein, A. and Wolinsky, A. (1985). Equilibrium in a market with sequential bargaining. *Econometrica*, 53(5):1133–1150.

Rubinstein, A. and Wolinsky, A. (1987). Middlemen. *Quaterly Journal of Economics*, 102:63–78.

Rubinstein, A. and Wolinsky, A. (1990). Decentralized trading, strategic behavior and teh walrasian outcome. *Review of Economic Studies*, 57:63–78.

Rustichini, A., Satterthwaite, M. A., and Williams, S. R. (1994). Convergence to efficiency in a simple market with incomplete information. *Econometrica*, 62(5):1041–1063.

Sarker, M., Butler, B., and Steinfeld, C. (1995). Intermediaries and cybermediairies: a continuing role for mediating players in the electronic marketplace. *Journal of Computer-Mediated Communication*, 1(3).

Schiller, Z. and Zellner, W. (1994). Making the middleman an endangered species. Business Week.

Shaked, A. and Sutton, J. (1984). Involuntary unemployment as a perfect equilibrium in a bargaining model. *Econometrica*, 52(6):1351–1364.

Spulber, D. F. (1996). Market microstructure and intermediation. *Journal of Economic Perspectives*, 10(3):135–152.

Spulber, D. F. (1999). *Market microstructure*. Cambridge University Press, Cambridge, U.K.

Stå hl, I. (1972). *Bargaining theory*. Stockholm Research Institute, Stockholm School of Economics, Stockholm.

Sutton, J. (1986). Noncooperative bargaining theory: an introduction. *Rev. Econom. Stud.*, 53(5):709–724.

Taylor, T. A. and Plambeck, E. L. (2003). Supply chain relationships and contracts: The impact of repeated interaction on capacity investment and procurement. Technical report, Graduate School of Business, Stanford University, Palo Alto, CA.

Van Mieghem, J. (1999). Coordinating investment, production and subcontracting. *Management Science*, 45(7):954–971.

Vickrey, W. (1961). Counterspeculation, auctions, and competitive sealed tenders. *Journal of Finance*, 16:8–37.

Wigand, R. T. and Benjamin, R. I. (1996). Electronic commerce: effects on electronic markets. *Journal of Computer-Mediated Communication*, 1(3).

Wilson, R. B. (1987). On equilibria of bid-ask markets. In *Arrow and the ascent of modern economic theory*, pages 375–414. New York Univ. Press, New York.

David Simchi-Levi, S. David Wu, and Z. Max Shen (Eds.)
Handbook of Quantitative Supply Chain Analysis:
Modeling in the E-Business Era
©2004 Kluwer Academic Publishers

Chapter 4

DECENTRALIZED DECISION MAKING IN DYNAMIC TECHNOLOGICAL SYSTEMS

The Principal-Agent Paradigm

Stefanos Zenios
Graduate School of Business
Stanford University, Stanford, CA 94305

Keywords: Principal-Agent Models; Decentralized Control; Dynamic Programming; Queueing Systems.

1. Background and Motivation

Operations Management (OM) is concerned with the study of technological systems that produce goods or services through the repetitive execution of standardized tasks. Although many of these systems are decentralized with an owner (or principal) delegating operational control to one or more noncooperative decision makers (or agents), traditional OM models suppress that feature for analytical tractability. Yet, the behavior of the noncooperative agents is also important and should be captured in OM models. Numerous recent research efforts have attempted to do that, but to maintain tractability they suppress most of the system complexity which is the landmark of OM models; see Anupindi, Bassok, and Zemel (1999), Cachon and Lariviere (1997), Lippman and Mc-Cardle (1997), Van Mieghem (1998). In addition, the few notable exceptions that capture both the processing dynamics and the agents' behavior utilize sophisticated models whose tractability relies on specialized assumptions; see Bassok and Anupindi (1999), George (1999), Hall and Porteus (1999), Lee and Whang (1996).

Decentralized systems are plagued by the uncoordinated actions of the multiple noncooperative agents and, thus, can be inefficient. However, the agents' actions can be coordinated either by adopting specific incentive contracts that motivate them to behave in a way that is desirable from the system's perspective, or else by promoting competition between them. In either case, to effectively manage such systems, one needs to determine mechanisms, in the form of contracts, cost-sharing rules, etc., to coordinate the actions of the noncooperative agents and to enhance overall system performance. These coordination mechanisms are commonly referred to as *incentive efficient* because they maximize the system's performance subject to the *incentive compatibility* constraint that each agent acts according to his own self-interest (see Milgrom and Roberts, 1992).

The principal-agent (or agency) paradigm, in which a principal (to be called "her") has a primary stake in the performance of a system but delegates operational control to one or more agents (to be called "him"), can provide the departing point for the analysis of incentive efficient mechanisms. The agency paradigm has found many applications in numerous fields of managerial relevance, such as investment management (Admati and Pfeiffer, 1997), personnel economics (Lazear, 1995), managerial accounting (Lambert, 1987), and taxation and oil drilling (Wolfson, 1985). However, agency models are underrepresented in the OM literature because the classical principal-agent models do not capture the rich dynamics that appear even in stylized OM models. The paper by Porteus and Whang (1991) is one of the first to adopt agency models in manufacturing operations, and more recently Chen (1998) uses an agency model to study salesforce incentives, and Cohen (1999) utilizes agency models to study Vendor Managed Inventory. Van Ackere (1993) provides a general survey of the principal-agent paradigm and describes several promising applications in operations management and management sciences. Gilbert and Whang (1998) utilize this paradigm to examine the effect of incentives on resource pooling and demonstrate that such effects favor non-consolidating queues.

In this chapter, we present a parsimonious modeling framework for the analysis of dynamic and stochastic decentralized systems with delegated control. The framework's foundational block is provided by the *dynamic principal-agent model* described in Plambeck and Zenios (2000). In this model, a single agent controls a Markov Decision Process and has access to perfect capital markets. The principal wishes to identify an incentive efficient compensation scheme to reward the agent for his unobservable controls. We will start with a description of the classical static principal-agent model (section 2), and then we will review the dynamic principal-agent model and present the solution methodology that is based on the recursive logic of dynamic programming (section 3). We will then present an extension of the basic dynamic model in a setting where the transitions of the Markov Decision Processes are controlled by multiple agents with potentially conflicting objectives (section 4). Following that, we will present two applications of the basic model: one in a decentralized production inventory system (section 5) and another one in the context of a service delivery system (6). We will then present a more general outline of potential applications of the principal-agent paradigm in Operations Management highlighting the opportunities that are generated by internet-based communications (see section 7) and we conclude with general remarks (section 6).

2. The Static Principal-Agent Model

In order to better appreciate and comprehend the dynamic principal-agent model, it is worthwhile to start with a review of the classical static model. The model assumes that the principal hires an agent for an one-time project and the project's profits are partially affected by the agent's actions. The agent's actions are assumed to be unobservable by the principal and thus, the principal devises a performance-based compensation scheme that links the agent's remuneration to observed performance outcomes such as profits. The analysis focuses on the derivation of an incentive efficient compensation schemes that will entice the agent to adopt actions that are desirable from the principal's point of view; see Grossman and Hart (1983).

To make this more concrete, let us consider a simple example. Let $a \in \Re$ denote the agent's action and let the principal's revenue be $\pi = a + e$ where e is $N(0, \sigma^2)$. The agent's cost for his action is $\frac{1}{2}ca^2$. The principal will pay the agent a compensation $s(\pi)$ that is based on the observable revenue π, and for simplicity we will consider linear compensation systems

$$s(\pi) = s_0 + s_1 \pi. \qquad (4.1)$$

The agent will take an action that will maximize his expected utility which is given by

$$E[s(\pi)] - \frac{1}{2}\beta Var[s(\pi)] - \frac{1}{2}ca^2; \qquad (4.2)$$

the coefficient β captures the agent's risk-attitude. The principal wishes to determine the linear payment system that will maximize her expected profit

$$E[\pi - s(\pi)] \qquad (4.3)$$

subject to two constraints: The incentive compatibility constraint which requires that the agent will take the action that maximizes his own expected utility, and the individual rationality constraint (also called the participation constraint) that the agent's expected utility under the optimal action is non-negative.

In order to solve this problem, we first consider the agent's response to the generic compensation system $s(\pi) = s_0 + s_1 \pi$. His expected utility under action a is

$$s_0 + s_1 a - \frac{1}{2}\beta s_1^2 \sigma^2 - \frac{1}{2} c a^2. \tag{4.4}$$

Thus, his optimal action is

$$a^* = \frac{s_1}{c}, \tag{4.5}$$

and his optimal expected utility is

$$s_0 + \frac{s_1^2}{2c} - \frac{1}{2}\beta s_1^2 \sigma^2. \tag{4.6}$$

Consequently, the principal's expected profit under the agent's optimal action is

$$\frac{s_1}{c} - \frac{s_1^2}{c} - s_0. \tag{4.7}$$

Now, since the principal can anticipate the agent's optimal response to any linear compensation system, her problem can be formulated as follows: Determine the payment components s_0 and s_1 to maximize

$$\frac{s_1}{c} - \frac{s_1^2}{c} - s_0 \tag{4.8}$$

subject to the so-called participation constraint that the agent's expected utility is non-negative

$$s_0 + \frac{s_1^2}{2c} - \frac{1}{2}\beta s_1^2 \sigma^2 \geq 0 \tag{4.9}$$

The reader will observe that in this formulation, the incentive compatibility constraint has been incorporated into the objective function (4.8), and the only constraint left is the participation constraint.

The analysis of the optimization problem (4.8)-(4.9) is straightforward. First, it can be easily shown that the participation constraint is binding. Hence,

$$s_0 = \frac{1}{2}\beta s_1^2 \sigma^2 - \frac{s_1^2}{2c}. \tag{4.10}$$

Substitution of (4.10) into (4.8) implies that the principal's problem is to determine the coefficient s_1 that maximizes

$$\frac{s_1}{c} - \frac{s_1^2}{2c} - \frac{1}{2}\beta s_1^2 \sigma^2. \tag{4.11}$$

Thus, the optimal choice for s_1 is

$$s_1^* = \frac{1}{1 + \beta \sigma^2 c}, \tag{4.12}$$

and s_0^* can be derived from (4.10).

The expression for s_1^* highlights the main economic tensions captured by the principal-agent model. In order to motivate the agent to work hard, the principal promises him a share of the revenue. The larger the share, the stronger the incentives. However, the agent finds large revenue shares unappealing because of his aversion to risk. The expression states that the principal must balance her desire to provide strong incentives for action with the agent's aversion to the risk generated by such incentives. The share of revenue paid to the agent decreases as the agent becomes more risk-averse, or the "noise" in the performance measure (revenue) increases.

The preceding analysis also suggests a backwards solution methodology. First, the principal considers the agent's response to a generic compensation system. Second, given this response, she optimizes over the choices of compensation systems. This methodology will be revisited in our description of the dynamic principal-agent model below.

Before we conclude this section, it is also worthwhile to introduce and summarize the terminology we will be using in the remainder of this chapter. A compensation system is said to induce or implement an action if that action represents the agent's optimal choice. Such action is called incentive compatible. A pair that consists of a compensation system and an action is called a contract. Such a contract is feasible if the prescribed action is incentive compatible and if the compensation system satisfies the participation constraint. The optimal contract is the pair (not necessarily unique) of a compensation system and incentive compatible action that maximizes the principal's expected payoff and satisfies the participation constraint. The action prescribed in the optimal contract is called "second-best" as opposed to the so-called "first-best" action that is optimal when the agent's action is observable by the principal.

3. The Dynamic Principal-Agent Model

The repeated counterpart of the static model in which the agency relationship extends over multiple periods and the profit outcomes in each period are represented as independent replications of the same process is also well studied. Although the dynamics of the repeated model are still simplistic for our purposes, its analysis reveals an important difference between static and dynamic agency models: Memory plays a critical role in the dynamic model, and history-dependent compensation schemes are, in general, more efficient than memoryless schemes (Rogerson, 1985); that is, these contracts generate a higher expected payoff for the principal than their memoryless counterparts. Furthermore, the role of memory depends on whether the agent consumes all payments he receives in each period in that period, or whether he can spread his consumption over time. When the agent has access to banking and he can smooth his consumption over time, memoryless schemes can be as effective as history-dependent schemes (Fudenberg, Holmstrom, and Milgrom, 1990).

In Plambeck and Zenios (2000), a dynamic principal-agent model which extends the repeated principal-agent paradigm to dynamic systems with stochastic dependency between periods is developed. The main contribution of that paper is summarized in the following statement: *Consider a centralized stochastic dynamic system that can be analyzed using dynamic programming. Then there exist well-defined conditions under which the decentralized counterpart of this system can also be analyzed using dynamic programming techniques.* This result establishes an analogy between centralized and decentralized systems and can provide the departing point for the analysis of incentive problems in dynamic systems.

The basic model assumes that the relationship between the principal and a single agent extends over multiple periods. To capture the process dynamics, the model assumes that the agent controls a Markov Decision Process. And to eliminate the complications that arise from history-dependent policies, it assumes that the agent has access to frictionless capital markets and he can smooth his consumption over time; in a frictionless capital market there are no transaction costs and the interest rate on negative bank balances is the same as the interest rate on positive balances. Overall, the model consists of three major components: a physical component for process dynamics, an economic component for the preferences of the two parties, and an information component.

The physical model assumes that there are T epochs, or periods, indexed by $t = 1, ..., T$. Let X_{t-1} denote the state of the Markov chain at the beginning of period t and let X_0 denote the initial state; the state set is assumed to be finite. The states of the Markov chain are observable both by the principal and the agent. The state history of the Markov chain until the beginning of period t is denoted by $X^{t-1} = (X_0, X_1, ..., X_{t-1})$. Within each period, the following sequence of events takes place: First, the agent selects an action $a_t(X^{t-1})$ from the set of all permissible actions in state X_{t-1}, $A(X_{t-1})$. As a result of his actions, the agent incurs a cost $g_{X_{t-1}}(a_t(X^{t-1}))$. Next, a transition to a new state X_t is observed. These transitions occur according to the stationary transition matrix $P_{X_{t-1}, X_t}(a_t(X^{t-1})) = P\left(X_t | X_{t-1}, a_t(X^{t-1})\right)$. Then, the principal, who cannot observe the agent's actions, receives a state-dependent reward $\pi(X_t)$ and makes a transfer payment to the agent according to the history-dependent compensation scheme $s_t(X^t)$. This terminates the sequence of events for period t. In the last period T, the same sequence of events occurs and then the relationship terminates.

The economic model assumes that the agent solves a consumption-effort problem in which he chooses not only an action strategy $a = \left(a_t(X^{t-1})\right)_{t=1,...,T}$ but also a consumption stream $c = (c_t(X^{t-1}))_{t=1,...,T}$ to maximize his expected utility. His utility is assumed additively separable and exponential:

$$U(a,c) = -\sum_{t=1}^{T} \alpha^{t-1} \exp\left[-r\left(c_t(X^{t-1})\right)\right]$$
$$-\frac{\alpha^T}{(1-\alpha)} \exp\left[-r(w + (1-\alpha)W_T)\right]; \qquad (4.13)$$

where r is the coefficient of risk-aversion, w is a retirement wage received for every period after T, and W_T is the agent's terminal wealth. The agent's wealth in period t is denoted W_{t-1}, it is assumed to be unobservable by the principal but observable by the agent, and it evolves according to the cash flow equation:

$$W_t = \frac{1}{\alpha}\left[W_{t-1} - c_t(X^{t-1}) - g_{X_{t-1}}(a_t(X^{t-1}))\right] + s_t(X^t); \qquad (4.14)$$

where $\alpha^{-1} - 1$ is the interest rate ($0 < \alpha < 1$). The consumption-effort problem was first articulated in Fudenberg, Holmstrom and Milgrom (1990). Smith (1998) and Plambeck and Zenios (2000) provide additional arguments supporting it. For the principal, the economic model assumes that she is risk-neutral and evaluates her reward and compensation scheme with their net present value,

$$\sum_{t=1}^{T} \alpha^t \left[\pi(X_t) - s_t(X^t)\right] + \alpha^{T+1}\Pi(X_T); \qquad (4.15)$$

where $\Pi(X_T)$ is the principal's terminal reward. The assumption that the agent is risk-averse is common in agency models and captures the agent's aversion to stochastic payment streams. When $r = 0$, the agent is risk-neutral.

Lastly, the information model assumes that both parties have complete knowledge of each other's utility functions and cost structures. Furthermore, they both know how the agent's actions affect the transition probabilities. Figure 4.1 summarizes the key features of the model.

3.1 The Principal's Problem

The next step is to formulate the problem that will be analyzed in this setting. The formulation imitates the one for the static agency model and is as follows: Determine an incentive efficient compensation scheme $s = (s_t(X^t))_{t=1,\ldots,T}$, and instructions for the agent in the form of an action strategy $a = (a_t(X^{t-1}))_{t=1,\ldots,T}$ and consumption strategy $c = (c_t(X^{t-1}))_{t=1,\ldots,T}$ to maximize

$$E_{a,c}\left[\sum_{t=1}^{T} \alpha^t \left[\pi(X_t) - s_t(X^t)\right] + \alpha^{T+1}\Pi(X_T) \,|\, X_0\right] \qquad (4.16)$$

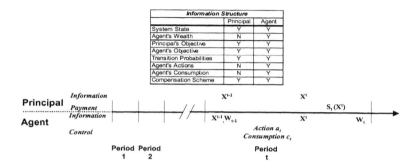

Figure 4.1. Sequence of events and information structure for the dynamic principal-agent model. (reproduced by permission, M&SOM journal)

subject to the *incentive compatibility constraints*

$$E_{a,c}\left[U(a,c)\,|\,X_0,W_0\right] \geq E_{\tilde{a},\tilde{c}}\left[U(\tilde{a},\tilde{c})\,|\,X_0,W_0\right]; \qquad (4.17)$$

for every admissible $\tilde{a} = (\tilde{a}_t(X^t))_{t=1,\ldots,T}$ and $\tilde{c} = (\tilde{c}_t(X^t))_{t=1,\ldots,T}$

and the *participation constraints*

$$E_{a,c}\left[U(a,c)\,|\,X_0,W_0\right] \geq -\frac{1}{1-\alpha}\exp\left[-r\left(w + (1-\alpha)W_0\right)\right]. \qquad (4.18)$$

The right-hand side of (4.18) gives the agent's *reservation utility* in a competitive market offering him a fixed wage w per period. The incentive compatibility constraint states that the agent will voluntarily adopt the instructions proposed by the principal for they represent his best response to the offered compensation scheme. The participation constraint states that the market for agents is competitive and thus the compensation scheme should guarantee that the agent receives the reservation utility that prevails in the market.

The analysis of problem (4.16)-(4.18) determines an optimal *long-term contract* which consists of an optimal compensation scheme $s^* = (s_t^*(X^t))_{t=1,\ldots,T}$, and the optimal action and consumption strategies $a^* = (a_t^*(X^t))_{t=1,\ldots,T}$ and $c^* = (c_t^*(X^t))_{t=1,\ldots,T}$ adopted by the agent in response to this compensation scheme. The action strategy a^* and consumption strategy c^* can be viewed as instructions for the agent: they cannot be enforced, but the agent will voluntarily adopt them if they represent his optimal response to the compensation scheme s^*.

The optimal long-term contract can be derived using a two-step dynamic programming procedure. The first step analyzes the static analogue of the dynamic principal-agent model for each possible initial state j and action $a \in$

$A(j)$, and obtains compensation rates $s^*_{jk}(a) : k = 1, ..., N$ which reward the agent based on observed single-period transitions and implement action a at a minimum cost to the principal. The expected single-period compensation cost is denoted $z^*_j(a)$. The second step solves a dynamic programming recursion to derive the action strategy in the long-term contract, and constructs the long-term compensation scheme from the single-period compensation rates.

3.2 Step 1: The Static Problem.

The static problem considers the single-period analogue of the multiperiod problem in which the agent's utility is equal to his single-period reservation utility. This problem can then be used to construct the solution to the multi-period problem as it will be explained below. The problem is formulated as follows: For each state $j = 1, ..., N$ and action $a \in A(j)$ determine a set of compensation rates $s^*_{jk}(a) : k = 1, ..., N$ that solve the following convex optimization problem:

$$\min \sum_{k=1}^{N} P_{jk}(a) s_{jk}(a) \tag{4.19}$$

$$\sum_{k=1}^{N} P_{jk}(a) \left(-\exp\left[-r(1-\alpha)\left(s_{jk}(a) - \alpha^{-1} g_j(a)\right)\right]\right) \tag{4.20}$$

$$= -\exp\left[-r\alpha^{-1}(1-\alpha)w\right]$$

$$\sum_{k=1}^{N} P_{jk}(a') \left(-\exp\left[-r(1-\alpha)\left(s_{jk}(a) - \alpha^{-1} g_j(a')\right)\right]\right) \tag{4.21}$$

$$\leq -\exp\left[-r\alpha^{-1}(1-\alpha)w\right] \qquad \text{for all } a' \in A(j);$$

the compensation rates $s^*_{jk}(a)$ represent a single-period payment to the agent when a transition to state k is observed. The solution to this optimization problem, which will be referred to as the Single Period Optimization Problem and will be denoted by $SP(j, a)$, gives the compensation scheme that implements action $a \in A(j)$ at a minimum cost to the principal when $T = 1$ and the initial state is j. Constraint (4.20) is the participation constraint, which states that if the agent takes action a, then his expected single-period utility will be equal to his reservation utility; this constraint is binding because of the exponential utility and banking assumptions. Constraint (4.21) is the incentive compatibility constraint which states that the agent's expected single-period utility under each alternative action $a' \in A(j)$ is lower than or equal to his reservation utility, which is attained under action a; it is also assumed that if the agent's utility is maximized by another action in addition to the desired action a, the agent will cooperate with the principal and choose the desired action a. Thus, action a represent's the agent's best single-period response to the compensation rates $s_{jk}(a)$. The optimal value of this optimization problem is denoted by $z_j^*(a)$, and the set of compensation rates that attain the optimum is denoted $s_{jk}^*(a) : k = 1, ..., N$. Although the optimum may not always be attainable, the following proposition characterizes conditions under which it is (all the proofs for the results in this section are available in Plambeck and Zenios, 2000):

PROPOSITION 1 *If the Optimization Problem $SP(j,a)$ has a non-empty feasible set, then there exists a single-period compensation scheme that implements action $a \in A(j)$ in the single-period problem at a minimum cost to the principal. This compensation scheme solves the optimization problem $SP(j,a)$ and makes a transfer payment $s_{jk}^*(a)$ to the agent, whenever a transition to state k is observed. The expected single-period cost of compensation is $z_j^*(a)$.*

At this point, it is natural to define the set of implementable policies $A^*(j)$ as the set of all policies $a \in A(j)$ such that the single-period optimization problem $SP(j,a)$ has a feasible solution. The solutions to the single-period problem (4.19)-(4.21) can be used to construct compensation schemes that implement a given action strategy at a minimum cost to the principal. This is explained in the following theorem:

THEOREM 4 *The principal can implement an admissible action strategy $a = \left(a_t(X^{t-1})\right)_{t=1,...,T}$ if and only if $a_t(X^{t-1}) \in A^*(X_{t-1})$ almost surely. If the latter condition holds, then the following compensation scheme implements this action strategy at a minimum cost to the principal:*

$$s_t(X^t) = s_{X_{t-1},X_t}^*(a_t(X^t)). \tag{4.22}$$

The expected present value in period $t = 1$ of all the payments made to the agent under this compensation scheme is

$$E_a\left[\sum_{t=1}^T \alpha^t z_{X_{t-1}}^*(a_t(X^{t-1}))|X^0\right]. \tag{4.23}$$

3.3 Step 2: The Dynamic Problem.

The next step is to determine an implementable action strategy that maximizes the principal's expected profit among all feasible implementable strategies. This involves substituting expression (4.23) into the principal's objective function (4.16) and solving for the optimal strategy a. Simple algebra shows that this is equivalent to the following problem: Determine the action strategy

a that maximizes

$$E_a \left[\sum_{t=1}^{T} \alpha^t \left[\pi(X_t) - z^*_{X_{t-1}}(a_t(X^{t-1})) \right] + \alpha^{T+1} \Pi(X_T) | X_0 \right]. \quad (4.24)$$

From this it follows that the second-best action strategy can be derived using a dynamic programming recursion and it is history-independent:

THEOREM 5 *(a) There exists an optimal long-term contract in which the action strategy a^* is memoryless. This strategy is given by the actions that maximize the right-hand side of the dynamic programming recursion:*

$$\begin{aligned} V^*_{T+1}(j) &= \Pi_{T+1}(j) \\ V^*_t(j) &= \alpha \max_{a \in A^*(j)} \left\{ -z^*_j(a) + \sum_{k=1}^{N} P_{jk}(a) \left(\pi(k) + V^*_{t+1}(k) \right) \right\} \end{aligned} \quad (4.25)$$

The compensation scheme that implements this action strategy is derived from the single-period optimal compensation schemes $s^*_{jk}(a)$ as follows:

$$s^*_t(X^t) = s_{X_{t-1}, X_t}(a^*_t(X_{t-1})). \quad (4.26)$$

(b) This optimal contract is not only optimal in the initial state and period, but it is also optimal in every subsequent period and state.

The proof is long and rich in technical and economic subtleties but the main idea is simple and intuitive: Because the agent's utility is additively separable and exponential and because the agent has access to frictionless capital markets, it follows that the principal can redistribute her payments to the agent across periods so that the agent's value function is equal to the reservation utility $-\frac{1}{1-\alpha} \exp\left[-r\left(w + (1-\alpha)W_t\right)\right]$ in each period t and state X_{t-1}. Therefore, the multiperiod incentive compatibility constraint (4.17) decomposes into single period constraints that generate the static problem, and Theorem 1 follows from simple algebra.

4. A Model with Multiple Agents

While the dynamic principal-agent model assumes a single agent, most systems of interest involve multiple noncooperative agents who control different aspects of the technology. A typical example is the decentralized multiechelon supply chain of Lee and Whang (1996) in which the inventory replenishment decisions at each stage of the supply chain are delegated to a different noncooperative agent. A natural question that arises in these systems is *what mechanisms will induce the noncooperative agents to adopt actions that will maximize overall system efficiency?*

In order to provide an answer to this question, we consider a *generalized dynamic principal-agent* model with two agents. The two agents will be indexed $i = 1, 2$, $A^i(j)$ will denote agent i's action set, a^i will denote agent i's actions, and $U^i(a^i, c^i)$ agent i's utility. The cost of effort for agent i in state j is $g^i_j(a)$. The multiagent model will differ from its single-agent counterpart in three important respects. First, a long-term contract will now consist of a compensation scheme, $s^i(X^t)$, and instructions for consumption and action strategies c^i and a^i, for each agent i. Second, the participation constraint will now require that the consumption and action strategies suggested in the contract guarantee that each agent receives his reservation utility. And lastly, the incentive compatibility constraint will now be formulated in terms of the equilibrium strategies of the two agents. That is, the incentive compatibility constraint will require that the consumption and action strategies specified in the contract are a subgame perfect Nash equilibrium, in which each agent will choose action and consumption strategies that represent his best response to the other agent's strategies in each period and state; a subgame perfect Nash equilibrium is the dynamic analogue of the well-known Nash equilibrium (see Gibbons, 1992).

Although one would anticipate that the incentive compatibility constraint will create difficulties, the proof of Theorem 1 suggests the following plan of action: First, it can be established that, in equilibrium, the principal can redistribute the payments in each sample path so that each agent's equilibrium expected future utility in each state will be equal to his reservation utility; this will generalize the analogous result for the single-agent problem. Second, it can be shown that the multiperiod incentive compatibility constraint decomposes into single-period constraints, and thus the solution to the multiperiod problem can be extracted from the solution to its single-period counterpart. The single-period problem can then be formulated as follows: For each state $j = 1, ..., N$ and action-pair $(a^1, a^2) \in A^1(j) \times A^2(j)$ determine a set of compensation rates $s^1_{jk}(a^1, a^2) : k = 1, ..., N$ and $s^2_{jk}(a^1, a^2) : k = 1, ..., N$ that solve the following mathematical program with equilibrium constraints:

$$\min \sum_{k=1}^{N} P_{jk}(a^1, a^2)[s^1_{jk}(a^1, a^2) + s^2_{jk}(a^1, a^2)] \quad (4.27)$$

$$\sum_{k=1}^{N} P_{jk}(a^1, a^2) \left(-\exp\left[-r(1-\alpha) \left(s^i_{jk}(a) - \alpha^{-1} g^i_j(a) \right) \right] \right)$$
$$= -\exp\left[-r\alpha^{-1}(1-\alpha)w \right] \text{ for } i = 1, 2 \quad (4.28)$$

$$\sum_{k=1}^{N} P_{jk}(a', a^2) \left(-\exp\left[-r(1-\alpha) \left(s^1_{jk}(a^1, a^2) - \alpha^{-1} g^1_j(a') \right) \right] \right)$$
$$\leq -\exp\left[-r\alpha^{-1}(1-\alpha)w \right]; \quad a' \in A^1(j); \quad (4.29)$$

$$\sum_{k=1}^{N} P_{jk}(a^1, a') \left(-\exp\left[-r(1-\alpha) \left(s^2_{jk}(a^1, a^2) - \alpha^{-1} g^2_j(a') \right) \right] \right)$$
$$\leq -\exp\left[-r\alpha^{-1}(1-\alpha)w \right]; \quad a' \in A^2(j). \quad (4.30)$$

The incentive compatibility constraints (4.29)-(4.30) state that the pair (a^1, a^2) is a Nash equilibrium for the single-period game in which each agent receives a terminal payoff equal to his reservation utility at the end of the period.

The analysis of the remainder of the problem follows predictable lines and the second-best action strategy, which now represents an equilibrium profile, can be obtained using dynamic programming as before. If we let $z^*_j(a^1, a^2)$ denote the optimal objective value for the optimization problem (4.27)-(4.30), then the second-best equilibrium action strategy can now be obtained by solving the following dynamic programming recursion:

$$V^*_{T+1}(j) = \Pi_{T+1}(j)$$
$$V^*_t(j) = \alpha \max_{(a^1, a^2) \in A^1(j) \times A^2(j)} \left\{ -z^*_j(a^1, a^2) + \sum_{k=1}^{N} P_{jk}(a^1, a^2) \left(\pi(k) + V^*_{t+1}(k) \right) \right\} \quad (4.31)$$

The compensation scheme that implements this action strategy is derived from the single-period optimal compensation schemes $s^1_{jk}(a^1, a^2)$, $s^2_{jk}(a^1, a^2)$ as follows:

$$s^1_t(X^t) = s^1_{X_{t-1}, X_t}(a^1_t(X_{t-1}), a^2_t(X_{t-1})), \quad (4.32)$$
$$s^2_t(X^t) = s^2_{X_{t-1}, X_t}(a^1_t(X_{t-1}), a^2_t(X_{t-1})). \quad (4.33)$$

5. A Production-Inventory System

Consider a production-inventory system where production is delegated from a principal to an agent. The time required to produce one unit of the product (from an unlimited supply of raw materials) is exponentially distributed, with the rate dynamically controlled by the agent. The agent incurs a cost of production that is an increasing and convex function of the production rate. Implicitly, the agent turns away alternative business (beginning with the least lucrative) and/or has an increasing marginal cost of effort as he ramps up the rate of production for the principal. Finished goods are immediately stored in a finished goods inventory used to serve customer demand. Unfilled demand is backordered. The principal incurs linear inventory costs per unit of time for backordering and for holding finished goods, and her objective is to minimize the expected total discounted inventory cost plus the total discounted cost of any compensation for the agent. Furthermore, she cannot monitor the agent's choice of production rate, but does observe the level of finished goods inventory. By observing the timing of the agent's output (jumps in the inventory level) the principal can draw inference about his choice of production rate, and by making payments to the agent contingent on the inventory process, the principal can create incentives for the agent to control the production rate in the manner she desires. In particular, the principal would like for the agent to be responsive, increasing his production rate as the inventory level falls. This will increase the agent's long run average production cost, but reduce the principal's own inventory holding and backorder costs.

This system is described in more detail in Plambeck and Zenios (2002). Its dynamics can be captured by the dynamic principal-agent model and hence in the remainder we will outline the solution to the principal's problem by utilizing the two-step solution methodology developed in section 3. It will be assumed that the preferences of the two parties are reflected by the utilities presented in section 3 and, thus, only the dynamics of the system and its cost and reward structure require some specification.

The dynamics of this system are as follows: Finished goods are produced at a single production server, and stored in inventory to meet demand. The agent controls the instantaneous production rate at time t, $\mu(t)$, and the resulting number of units of finished good produced up to time t is given by the nonhomogeneous Poisson process $N\left(\int_0^t \mu(s)ds\right)$, where $N(\cdot)$ is a standard Poisson process. The cumulative demand up until time t is an independent Poisson process $D(t)$, with rate λ. To describe the system dynamics, let $X(t)$ denote the number of units (possibly negative) in inventory at time t. Then, the inventory level obeys the following dynamics:

$$X(t) = X(0) + N\left(\int_0^t \mu(s)ds\right) - D(t) \quad \text{for } 0 \leq t \leq T; \qquad (4.34)$$

where the production rate $\mu(t)$ must satisfy:

$$0 \leq \mu(t) \leq \overline{\mu} \tag{4.35}$$

$$\mu(t) \text{ is non-anticipating with respect to } X(t). \tag{4.36}$$

Therefore, the inventory process $X(\cdot)$ is a birth-death Markov chain. In the remainder of the paper, the term 'job completion' will be used to refer to the production of a unit of finished good inventory. These model dynamics are based on George and Harrison (1999).

The cost and reward structure for each party are as follows: The principal is assumed to incur an inventory cost based on the function

$$f(x) \equiv h\,[x]^+ + b\,[x]^-;$$

where h represents the unit holding cost per unit time and b represents the unit backorder cost per unit time. She also incurs the cost of any payments to the agent. These payments consist of discrete transfer payments $S(t)$ to be made if a job is completed at time t (strictly speaking the payment is made at time $t+$), and a continuous payment *rate* $s(t)$ at time t. The terms of compensation at time t are allowed to be history-dependent and they must be non-anticipating with respect to the inventory process X; for future reference, let X^t denote all the information about the inventory process observed up to time t. The reader will note that the compensation scheme does not include payments for jumps in the demand process $D(t)$. Such payments are unnecessary because they increase the agent's risk without providing any incentives for productive output; see Holmstrom (1978).

The agent receives the compensation transferred by the principal and incurs a variable production cost at a rate $g(\mu)$ whenever the production rate is μ. The function $g(\mu)$ is assumed to be increasing and convex with $g(0) = 0$. This convexity may reflect, for example, an increasing marginal cost of labor associated with overtime or that as the agent shifts production capacity away from his alternative customers, he withdraws capacity from the least lucrative customers first.

The main result identifies conditions under which a given production policy can be induced, and shows that the cheapest payment scheme to induce it consists of a differential piece rate and an inventory penalty. This is summarized in the following Theorem (proven in Plambeck and Zenios, 2002):

THEOREM 6 *Consider an admissible production policy* $\mu = (\mu(t))_{0 \leq t \leq T}$ *that specifies the desired production rate over time. The principal can induce this policy if and only if the following condition is true almost surely for each* $t \in [0, T]$:

$$g'(\mu(t)) < \frac{1}{\alpha r}. \tag{4.37}$$

If this condition holds, then the payment scheme inducing policy μ at minimum expected cost to the principal consists of a payment rate

$$s(t) = w + g(\mu(t)) - \mu(t)g'(\mu(t)) \tag{4.38}$$

at time t, and a discrete payment of

$$S(t) = \begin{cases} -(\alpha r)^{-1} \ln\left[1 - \alpha r g'(\mu(t))\right] & \text{if } \mu(t) > 0 \\ 0 & \text{if } \mu(t) = 0 \end{cases} \tag{4.39}$$

made at time t+, if a job is completed at time t.

Condition (4.37) implies that not every production rate remains feasible when production control is delegated from a principal to the agent. Moreover, the maximum inducible production rate, given by

$$\sup\left\{\mu \in [0, \bar{\mu}] : g'(\mu) < \frac{1}{\alpha r}\right\}, \tag{4.40}$$

is non-increasing in the agent's coefficient of risk-aversion and discount rate. That is, the range of inducible production rates shrinks as the agent becomes increasingly risk averse and impatient.

Theorem 6 provides the stepping stone leading to the second step in the analysis. Specifically, the theorem implies that there exists an "induced cost function"

$$z(\mu, r) \equiv \begin{cases} w + g(\mu) - \mu g'(\mu) - \frac{\mu}{\alpha r} \ln\left[1 - \alpha r g'(\mu(t))\right] & \text{if } g'(\mu) < \frac{1}{\alpha r} \\ +\infty & \text{otherwise} \end{cases} \tag{4.41}$$

which gives the principal's instantaneous expected cost for inducing μ using the theorem's payment mechanism. Therefore, the dynamic principal-agent problem can be reformulated as an equivalent single-party dynamic optimization problem: The principal determines a second best production policy $\mu(t)$, which will be induced using the payment scheme from Theorem 6, and which minimizes her expected total cost,

$$E\left[\int_0^T e^{-\alpha t}\left[z(\mu(t), r) + h[X(t)]^+ + b[X(t)]^-\right] dt + e^{-\alpha T}\Pi(X(T)) \mid X(0)\right]. \tag{4.42}$$

subject to the inventory dynamics

$$X(t) = X(0) + N\left(\int_0^t \mu(s)ds\right) - D(t) \quad \text{for } 0 \le t \le T. \tag{4.43}$$

This dynamic optimization problem can be solved using a standard dynamic programming recursion, and for brevity we will only consider the infinite horizon case $T \to \infty$. Let $V(x, r)$ denote the principal's infinite horizon optimal cost function when the inventory level is x and the agent's coefficient of risk aversion is r. Then, $V(x,r)$ satisfies the dynamic programming recursion

$$V(x,r) = (\alpha + \overline{\mu} + \lambda)^{-1} \min_{0 \le \mu \le \overline{\mu}} \begin{array}{l} \{f(x) + z(\mu,r) + \lambda V(x-1,r) \\ + \mu V(x+1,r) + (\overline{\mu} - \mu) V(x,r) \}. \end{array}$$
(4.44)

Moreover, if we let $\mu^*(x,r)$ denote the production rate that maximizes the right hand side of (4.44), then the second best production policy $\mu^*(t)$ solving the dynamic optimization problem (4.44) is given by:

$$\mu^*(t) = \mu^*(X(t), r). \tag{4.45}$$

Then, the optimal payment scheme is characterized as the one that induces the second best production policy $\mu^*(t)$ according to the differential piece rate and inventory penalty of Theorem 6. If the production cost function $g(\mu)$ is strictly convex, then μ^* is the *unique* optimal production plan for the agent. If $g(\mu)$ is linear, then the second best policy μ^* is an optimal response for the agent, but not the unique optimal response; following standard practice in the economics literature, we assume that the agent will cooperate with the principal by selecting the second best policy. Structural properties of the second best production policy and optimal compensation scheme are discussed in Plambeck and Zenios (2002).

6. A Service Delivery System

Consider now a service delivery system with delegated control. In service operations one is typically interested in customer retention. Therefore, an appropriate model should focus on the impact of service efforts on customer retention and thus, it should capture the customer dynamics. Kelly (1979) argues that the infinite server queue provides a natural building block for the analysis of customer dynamic systems. This motivates the analysis of the following infinite server principal-agent queueing system: Customers arrive exogenously at a rate λ and join an infinite server queue. While they are in the queue, they generate income for the principal at a constant rate. Depending on the endogenously determined service rate (which literally represents customer service), customers may experience either a minor service failure which can be rectified with additional cost, or a major catastrophic failure which results in customer loss. The service effort is determined by a single agent that incurs a cost-of-effort. As before, the objective is to design an incentive efficient contract in which the principal motivates the agent to provide an adequate service level to her customers.

We consider a simple model in which decisions are made at discrete times and the planning horizon is infinite. The model is described in more detail in Fuloria and Zenios (2001). The preferences for the two parties and information structure is described in section 3. Here we present the model dynamics and cost structure. Periods, indexed by $t = 1, 2, \ldots$, represent the minimum length of time between decisions and/or payments. The state of the system at the beginning of period t, denoted by Q_{t-1}, represents the number of customers requiring service in that period.

The events that affect the state of the system in each period are as follows: First, the agent chooses a one-dimensional service level decision $a_t \in \Omega$, where Ω is assumed to be the closed interval $[\underline{a}, \bar{a}]$. The agent incurs a "cost of service" $g(a_t)Q_{t-1}$, where the cost function $g(\cdot)$ is strictly increasing and convex. Service level affects stochastically the number of customers who will experience a minor service failure, denoted H_t, as well as the number of major service failures in the same period, denoted D_t. The two random variables H_t and D_t are observed next. The probability distribution for H_t is assumed to be Binomial with number of trials Q_{t-1} and probability of minor service failure for each trial given by a proportional hazards function $p_H(a_t) = \bar{p}_H \exp(-\theta_H a_t)$, where $\theta_H > 0$; $p_H(a)$ denotes the per-period minor-failure hazard for a customer receiving service a. Similarly, the probability distribution for D_t is also Binomial with Q_{t-1} trials and probability of major failure for each trial given by $p_D(a_t) = \bar{p}_D \exp(-\theta_D a_t)$, where $\theta_D > 0$. The variables H_t and D_t are assumed to be statistically independent. Minor service failures are associated with a reputation cost V_H incurred by the principal. In addition, each customer generates a reward κ per period for the principal. In each period, the principal makes a transfer payment s_t to the agent which is restricted to be linear according to the formula,

$$s_t = s^0 + s^F Q_{t-1} + s^D D_t + s^H H_t; \tag{4.46}$$

the payment coefficients s^0, s^F, s^D and s^H will be specified during contract negotiation. The final event within period t is the arrival of new customers. These occur according to a Poisson process and the total number of new arrivals, denoted by A_t, has a Poisson distribution with rate λ. The state of the system is then updated according to

$$Q_t = Q_{t-1} - D_t + A_t. \tag{4.47}$$

The first step in the analysis is to obtain the payment system that induces an action strategy for a minimum expected cost to the agent. Because of the linear payment structure and exponential utility it can be shown that the payment system induces a stationary action strategy in which the quality of service provided by the agent is state-independent. To induce a state-dependent strategy, the payment system should utilize state-dependent coefficients.

The following proposition (proven in Fuloria and Zenios, 2001) characterizes the optimal payment system.

PROPOSITION 2 (a) *Let* $s^0(a), s^F(a), s^H(a)$ *and* $s^D(a)$ *denote the coefficients of the payment system that induces service level a at a minimum cost to the principal. Let* $x^H(a)$ *and* $x^D(a)$ *denote the solution to the non-linear equations,*

$$\frac{dg(a)}{da} = \frac{\alpha}{r(1-\alpha)} \left[\theta_H (1 - x^H(a)) + \theta_D (1 - x^D(a)) \right] \qquad (4.48)$$

$$\frac{\theta_H x^H(a) \left[p_H(a) x^H(a) + (1 - x^H(a)) \right]}{(1 - p_H(a))(1 - x^H(a))} =$$

$$\frac{\theta_D x^D(a) \left[p_D(a) x^D(a) + (1 - x^D(a)) \right]}{(1 - p_D(a))(1 - x^D(a))}. \qquad (4.49)$$

Then,

$$s^F(a) = \frac{g(a)}{\alpha} - \frac{1}{r(1-\alpha)} \left[\log x^H(a) + \log x^D(a) \right] \qquad (4.50)$$

$$s^H(a) = -\frac{1}{r(1-\alpha)} \log \left[\frac{\frac{1}{x^H(a)} - (1 - p_H(a))}{p_H(a)} \right] \qquad (4.51)$$

$$s^D(a) = -\frac{1}{r(1-\alpha)} \log \left[\frac{\frac{1}{x^D(a)} - (1 - p_D(a))}{p_D(a)} \right], \qquad (4.52)$$

and

$$s^0(a) = \frac{w}{\alpha}. \qquad (4.53)$$

(b) For a given service level $a \in \Omega$, *the simultaneous equations (4.48)-(4.49) have a feasible solution if and only if*

$$\frac{dg(a)}{da} \le \frac{\alpha}{r(1-\alpha)} (\theta_H + \theta_D). \qquad (4.54)$$

The reader can confirm that $s^F(a) > \alpha^{-1} g(a)$, $s^H(a) < 0$ and $s^D(a) < 0$. Therefore, the linear payment system includes a carrot and a stick: the penalties for adverse outcomes $s^H(a)$ and $s^D(a)$ are the stick, and the prospective payment $s^F(a)$, which exceeds the agent's cost of effort for action a, is the carrot. Finally, $s^0(a)$ rewards the provider for the opportunity cost associated with his participation in the contract with the purchaser.

An important implication of Proposition 2 is that not all service levels can be induced by linear payment systems: service levels for which the agent's marginal cost cannot be equated to his marginal benefit from linear payments are non-inducible. The expression in the right-hand side of (4.54) gives an (attainable) upper bound on the agent's marginal benefit from any linear payment. This bound depends on three factors: 1) the sum $\theta_H + \theta_D$, representing the strength of the association between the observed outcomes utilized in the payment system and the unobserved service-level decisions; 2) the agent's tolerance for risk; and 3) the cost of transferring wealth across periods as reflected in the discount rate α. The range of inducible actions increases as the association between outcomes and unobserved actions is strengthened, as the agent's aversion to risk decreases, and as the cost of transferring wealth across periods diminishes.

The next step in the analysis is to determine the service level a that maximizes

$$E_a[\sum_{t=1}^{\infty} \alpha^t(\kappa(Q_{t-1} - D_t) - V_H H_t - s^F(a)Q_{t-1}$$
$$- s^H(a)H_t - s^D(a)D_t)|Q_0 = n]; \quad (4.55)$$

this expression gives the principal's total payoff when action a is induced with the payment systems identified in Proposition 2. Simple algebra shows that the solution to this problem, denoted a^*, is independent of the initial state Q_0 and is obtained by maximizing the principal's payoff from a single customer,

$$\frac{\alpha\kappa(1 - p_D(a)) - \alpha V_H p_H(a) - \alpha s^F(a) - \alpha p_H(a)s^H(a) - \alpha p_D(a)s^D(a)}{1 - \alpha(1 - p_D(a))}.$$
(4.56)

To explore the effect of information asymmetry on service level, we can compare the *second-best* service intensity a^*, which is optimal when the agent's actions are not observable, to the *first-best* intensity, a^{FB}, that maximizes the principal's objective when the agent's actions are observable; the first-best service strategy delivers the same service level, the so-called first-best level, in each period and state. The first-best service level is obtained by maximizing the principal's payoff per customer when the cost of service is the actual cost $g(a)$ and not the "induced cost" $\delta\left[s^F(a) + p_H(a)s^H(a) + p_D(a)s^D(a)\right]$ used in (4.56):

$$\frac{\alpha\kappa(1 - p_D(a)) - \alpha V_H p_H(a) - g(a)}{1 - \alpha(1 - p_D(a))}. \quad (4.57)$$

Furthermore it can be shown that

$$\delta\left[s^F(a) + p_H(a)s^H(a) + p_D(a)s^D(a)\right] \geq g(a), \quad (4.58)$$

and that

$$a^* \leq a^{FB}. \quad (4.59)$$

In addition, the first-best service level coincides with the second-best one either when $r = 0$, or when $\alpha = 1$. In both cases, the tension between incentives and insurance disappears either because the agent is risk-neutral ($r = 0$), or because he can costlessly eliminate statistical fluctuations in his income through his access to banking ($\alpha = 1$).

There are numerous systems where the infinite server principal-agent model is relevant. Cachon and Harker (1999) provide two nice examples, one in PC banking and one in Dell's customer service. A third example is in health services where Health Maintenance Organizations (HMO) or government insurance agencies outsource the delivery of clinical services to specialized providers, and they wish to structure incentive contracts that will motivate the private providers to adopt cost-effective clinical decisions that maximize patient survival (Zweifel and Breyer, 1997).

7. The Past and the Future

The last ten years have witnessed a considerable growth in the number of publications that utilize techniques from economics in the study of production and operations system. A central tool in these studies is the principal-agent model which focused on problems of decentralization when there are information asymmetries: when one party is better informed than a second party. However, most of the OM problems that are of considerable interest do not fit neatly into the principal-agent models utilized in economic theory. Specifically, most theoretical economic models assume that the information asymmetries are either of hidden action (moral hazard) or of hidden system parameters (adverse selection). Further, these models focus almost exclusively in static economic environments or very simplified dynamic environments. In real-life operational systems, information asymmetry cannot be neatly categorized as in the theoretical models, and the dynamic environment is, in general, complex and multifaceted.

Further, the development of the internet as a convenient communication tool magnifies the complexity of the interactions and the nature of the information asymmetries that are possible. This clearly generates some intriguing possibilities for future research. As an example consider the emergence of business-to-business exchanges: While these exchanges are typically used for the procurement of standardized commodities, one anticipates that these exchanges will also be used for the procurement of complex products or services. When the nature of the procurement relationship is complex, incentive problems of the type studied in the principal-agent models appear. Then a fascinating research question is how to integrate the auction mechanisms the prevail in an exchange with incentive contracts that will facilitate the cost-effective procurement of complex products and services. The answer to this question will depend on the exact characteristics of the underlying product and on the dynamic environment in which production takes places. The internet facilitates the execution of complex and innovative transactions that facilitate the more efficient production of goods and services. The principal-agent model can be used to study these transactions and to understand the new complexities that emerge.

8. Concluding Remarks

This chapter presents both the classical static and a novel dynamic principal-agent model. The models provide the setting for the study of problems of delegated control in decentralized technological systems. Just like any model, these models rely on several restrictive assumptions. The most notable ones are the following: a) Both parties have complete knowledge of the underlying technology; b) There is an unambiguous performance measure that is observable and contractible; c) There is only a single principal; d) The agent's preferences assume a frictionless ideal for banking; e) System dynamics are markovian. Existing models in the literature provide some insights about the effect of these assumptions (see Salanie (1987) for the first four assumptions), but to the best of our knowledge, models that eliminate these assumptions and retain rich-enough system dynamics are lacking. The development and analysis of such models is an exciting topic for future research.

Acknowledgments

The author's research has been supported in part by the National Science Foundation grant SBER-9982446.

References

ADMATI A.R AND PFLEIDERER P. 1997. Does it All Add Up? Benchmarks and the Compensation of Active Portfolio Managers. *Journal of Business* (70) 323-350

ANUPINDI R., BASSOK Y. AND ZEMEL, E. 1999. A General Framework for the Study of Decentralized Distribution Systems. *Working Paper, Kellogg School of Business.*

BASSOK Y. AND ANUPINDI R. 1999. Analysis of Supply Contracts with Commitments and Flexibility. To appear *M&SOM.*

CACHON, G.P. AND LARIVIERE, M.A. 1997. Contracting to Assure Supply or What Did the Supplier Know and When Did He Know It? *Working Paper, The Fuqua School of Business.*

CACHON, G.P. AND HARKER, P.T. 1999. Service Competition, Outsourcing and Co-Production in a Queueing Game. *Working Paper, Fuqua School of Business.*

CHEN F. 1998. Salesforce Incentives and Inventory Management. *Working Paper, Graduate School of Business, Columbia University.*

COHEN S.L. 1999. Asymmetric Information in Vendor Managed Inventory Systems. *Report of Thesis Work in Progress, Graduate School of Business, Stanford University.*

FUDENBERG, D., HOLMSTROM, B. AND MILGROM, P. 1990, Short-Term Contracts and Long-Term Agency Relationships. *Journal of Economic Theory* 51, 1-31.

FULORIA, P.C. AND ZENIOS, S.A. 2001. Outcomes-Adjusted Reimbursement in a Health Care Delivery System. *Management Science*, 47, 735-751.

GEORGE, J. 1999. *A Queue with Controllable Service Rates: Optimization and Incentive Compatibility.* Stanford University Graduate School of Business, unpublished Ph.D. thesis.

GIBBONS, R. 1992. *Game Theory for Applied Economists.* Princeton University Press, Princeton, NJ.

GILBERT, S. M. AND Z. K. WENG. Incentive Effects Favor Nonconsolidating Queues in a Service System: The Principal-Agent Perspective. *Management Science* **44** 1662-1669.

GROSSMAN, S. J. AND O. D. HART. 1983. An Analysis of the Principal-Agent Problem. *Econometrica* 51, 7-45.

HALL J.M. AND PORTEUS, E.L. 1999. Customer Service Competition in Capacitated Systems. *Working Paper, Graduate School of Business, Stanford University.*

HOLMSTROM, B. 1982. Moral Hazards in Teams. *The Bell Journal of Economics*, 324-340.

KELLY, F.P. 1979. *Reversibility and Stochastic Networks.* John Wiley & Sons, New York, NY.

LAMBERT, R. 1987. An Analysis of the Use of Accounting and Market Measures of Performance in Executive Compensation Contracts, *Journal of Accounting Research,* Supplement.

LAZEAR, E. P. 1995. *Personnel Economics.* MIT Press, Cambridge, MA.

LEE H. AND WHANG S. 1996. Decentralized Multi-Echelon Supply Chains: Incentives and Information. *Working Paper, Graduate School of Business, Stanford University.*

LIPPMAN S.A. AND MCCARDLE, K.F. 1997. The Competitive Newsboy. *Operations Research* (45) 54-65.

MAS-COLELL, A., M. D. WHINSTON AND J. R. GREEN. 1995. *Microeconomic Theory.* Oxford University Press, New York, NY.

MILGROM, P. AND ROBERTS, J. 1992. *Economics, Organization and Management.* Prentice Hall, Englewood Cliffs, NJ.

PLAMBECK E.L. AND ZENIOS, S.A. 2000. Performance-Based Incentives in a Dynamic Principal-Agent Model. *M&SOM, 2,* 240-263.

PLAMBECK, E.L. AND ZENIOS, S.A. 2002. Incentive Efficient Control of a Make-to-Stock Production System. To appear in *Operations Research.*

PORTEUS E.L AND WHANG, S. 1991. On Manufacturing/Marketing Incentives. *Management Science* (37) 1166-1181.

ROGERSON, W.P. 1985. Repeated Moral Hazard, *Econometrica* 53:69-76.

SALANIE, B. 1997. The Economics of Contracts: A Primer. MIT Press.

SMITH, J.E. 1998. Evaluating Income Streams: A Decision Analysis Approach. *Management Science* **44** 1690-1708.

VAN ACKERE, A. 1993. The principal/agent paradigm: Its relevance to functional fields. *European Journal of Operational Research* **70** 83-103.

VAN MIEGHEM, J.A. 1998. Capacity Investment under Demand Uncertainty: Coordination and the Option Value of Subcontracting. *Submitted to Management Science.*

WOLFSON M. 1985. Incentive Problems in Oil and Gas Shelter Programs. In *Principals and Agents: The Structure of Business,* Pratt and Zeckhauser (eds.), Harvard Business School Press, Boston, MA.

ZWEIFEL P. AND BREYER, F. 1997. *Health Economics.* Oxford University Press, NY, 1997.

Part II

AUCTIONS AND BIDDING

David Simchi-Levi, S. David Wu, and Z. Max Shen (Eds.)
*Handbook of Quantitative Supply Chain Analysis:
Modeling in the E-Business Era*
©2004 Kluwer Academic Publishers

Chapter 5

AUCTIONS, BIDDING AND EXCHANGE DESIGN

Jayant Kalagnanam
*IBM Research Division
T.J.Watson Research Center,
P.O.Box 218,
Yorktown Heights, NY 1058*
jayant@us.ibm.com

David C. Parkes
*Division of Engineering and Applied Sciences,
Harvard University,
Cambridge MA 02138*
parkes@eecs.harvard.edu

1. Introduction

Auctions have found widespread use in the last few years as a technique for supporting and automating negotiations on the Internet. For example, eBay now serves as a new selling channel for individuals, and small and big enterprises. Another use for auctions is for industrial procurement . In both these settings traditional auction mechanisms such as the English , Dutch , First (or Second) price Sealed-Bid auctions are now commonplace. These auctions types are useful for settings where there is a single unit of an item being bought/sold. However, since procurement problems are business-to-business they tend to be more complex and have led to the development and application of advanced auction types that allow for negotiations over multiple units of multiple items, and the configuration of the attributes of items. At the heart of auctions is the problem of decentralized resource allocation.

A general setting for decentralized allocation is one with multiple agents with utility functions for various resources. The allocation problem for the decision maker, or *intermediary*, is to allocate these resources in an optimal way. A key difference from the classical optimization perspective is that the utility function of the agents is private information, and not explicitly known to the decision maker. In addition, standard methods in decentralized optimization fail because of the self-interest of participants. Therefore the design of decentralized allocation mechanisms must provide incentives for agents to reveal their true preferences in order to solve for the optimal allocation with respect to the true utility functions. Thus, the behavioral aspects of agents must be explicitly considered in the design. It is common in the economic mechanism design literature to assume rational, game-theoretic, agents. Another common assumption is that agents behave as myopic price-takers, that are rational in the current round of negotiation but not necessarily with respect to the final outcomes at the end of the negotiation.

In settings where the allocation problem itself is hard even if the decision maker knows the "true" utility function of each agent, the issues of incentive compatibility makes the design of an appropriate auction mechanism even more challenging.

The focus of this chapter is to provide an overview of the different auction mechanisms commonly encountered both in practice and in the literature. We will initially provide a framework for classifying auction mechanisms into different types. We will borrow a systems perspective (from the literature) to elucidate this framework.

1.1 A framework for auctions

We develop a framework for classifying auctions based on the requirements that need to be considered to set up an auction. We have identified these core components below:

Resources The first step is to identify the set of resources over which the negotiation is to be conducted. The resource could be a single item or multiple items, with a single or multiple units of each item. An additional consideration common in real settings is the type of the item, i.e. is this a standard commodity or multiattribute commodity. In the case of multiattribute items, the agents might need to specify the non-price attributes and some utility/scoring function to tradeoff across these attributes.

Market Structure An auction provides a mechanism for negotiation between buyers and sellers. In *forward auctions* a single seller is selling resources to multiple buyers. Alternately, in *reverse auctions*, a single buyer is sourcing resources from multiple suppliers, as is common in procurement. Auctions with multiple buyers and sellers are called *dou-*

ble auctions or *exchanges*, and these are commonly used for trading securities and financial instruments and increasingly within the supply chain.

Preference Structure The preference structure of agents in an auction is important and impacts some of the other factors. The preferences define an agent's utility for different outcomes. For example, when negotiating over multiple units agents might indicate a decreasing marginal utility for additional units. An agent's preference structure is important when negotiation over attributes for an item, for designing scoring rules used to signal information.

Bid Structure The structure of the bids allowed within the auction defines the flexibility with which agents can express their resource requirements. For a simple single unit, single item commodity, the bids required are simple statements of willingness to pay/accept. However, for a multi-unit identical items setting bids need to specify price and quantity. Already this introduces the possibility for allowing volume discounts, where a bid defines the price as a function of the quantity. With multiple items, bids may specify all-or-nothing bids with a price on a basket of items. In addition, agents might wish to provide several alternative bids but restrict the choice of bids.

Matching Supply to Demand A key aspect of auctions is matching supply to demand, also referred to as market clearing, or *winner determination*. The main choice here is whether to use *single-sourcing*, in which pairs of buyers and sellers are matched, or *multi-sourcing*, in which multiple suppliers can be matched with a single buyer, or vice-versa. The form of matching influences the complexity of winner determination, and problems range the entire spectrum from simple sorting problems to NP-hard optimization problems.

Information Feedback Another important aspect of an auction is whether the protocol is a *direct* mechanism or an *indirect* mechanism. In a direct mechanism, such as the first price sealed bid auction, agents submit bids without receiving feedback, such as price signals, from the auction. In an indirect mechanism, such as an ascending-price auction, agents can adjust bids in response to information feedback from the auction. Feedback about the state of the auction is usually characterized by a *price signal* and a *provisional allocation*, and provides sufficient information about the bids of other agents to enable an agent to refine its bids. In complex settings, such as multi-item auctions with bundle d bids, a direct mechanism can require an exponential number of bids to specify an agent's preference structure. In comparison, indirect mechanisms allow incremental revelation of preference information, on a "as required basis". The focus in the design of indirect mechanisms is to identify how much preference information is sufficient to achieve desired economic properties and how to implement informationally-efficient mechanisms. A related strand of research is to provide compact bidding languages for direct mechanisms.

Each of the six dimensions that we have identified provide a vector of choices that are available to set up the auction. Putting all of these together generates a matrix of auction types. The choices made for each of these dimensions will have a major impact on the complexity of the analysis required to characterize the market structure that emerges, on the complexity on agents and the intermediary to implement the mechanism, and ultimately on our ability to design mechanisms that satisfy desirable economic and computational properties.

1.2 Outline

In Section 2 we first introduce the economic literature on *mechanism design* and identify the economic properties that are desirable in the design of auction mechanisms. Continuing, in Section 3 we introduce the associated computational complexities that arise in the implementation of optimal mechanisms. Section 4 provides an extended discussion of the subtle interaction between computation and incentives. Then, in Section 5 we pick off a few specific mechanisms that are interesting both from a practical point of view and also because they illustrate some of the emerging research directions. Finally, Section 6 concludes with a brief discussion of the role of mechanism design in the design of electronic markets and considers a new computational approach, termed *experimental* mechanism design.

2. Economic Considerations

The basic economic methodology used in the design of electronic intermediaries first models the preferences, behavior, and information available to agents, and then designs a mechanism in which agent strategies result in outcomes with desirable properties. We consider two approaches to modeling agent behavior:

Game-theoretic/mechanism design The first model of agent behavior is *game-theoretic* and relates to *mechanism design theory*. In this model the equilibrium state is defined by the condition that agents play a best-response strategy to each other and cannot benefit from a unilateral deviation to an alternative strategy.

Price-taking/competitive equilibrium The second model of agent behavior is *price-taking*, or *myopic best-response*, and relates to *competitive equilibrium theory*. In this model the equilibrium state is defined by the condition that an agent plays a best-response to the *current* price and allocation in the market, without modeling either the strategies of other agents or the effect of its own actions on the future state of the market.

Mechanism design theory and game-theoretic modeling is most relevant when one or both of the following conditions hold: (a) the equilibrium solution concept makes weak game-theoretic assumptions about agent behavior, such as when a mechanism can be designed with a *dominant strategy* equilibrium, in which agents have a single strategy that is always optimal *whatever* the strategies and preferences of other agents; or (b) there are a small number of agents and it is reasonable to expect agents to be rational and well-informed about the likely preferences of other agents. On the other hand, competitive equilibrium theory and price-taking modeling is most relevant in large systems in which the effect of an agent's own strategy on the state of a market is small, or when there is considerable uncertainty about agent preferences and behaviors and no useful mechanism with a dominant strategy equilibrium.

We begin with a description of mechanism design theory and competitive equilibrium theory. Then, we outline a primal-dual approach to the design of indirect mechanisms (such as ascending price-based auctions), that unifies the mechanism design and competitive equilibrium approaches. Essentially, one can view an ascending-price auction as a method to implement a primal-dual algorithm for a resource allocation problem. Terminating with competitive equilibrium prices that also implement the outcome of a mechanism brings the price-taking behavior that is assumed in classic competitive equilibrium behavior into a game-theoretic equilibrium.

2.1 Preliminaries

Our presentation is limited to the *private value* model, in which the value to an agent for an outcome is only a function of its own private information. This is quite reasonable in the procurement of goods for direct consumption, unless there are significant opportunities for resale or unless there is significant uncertainty about the quality of goods. *Correlated* and *common* value models may be more appropriate in these settings, and the prescriptions for mechanism design can change [83].

Consider $\mathcal{I} = (1, \ldots, N)$ agents, a discrete outcome space $\mathcal{K}$, and payments $p = (p_1, \ldots, p_N) \in \mathbb{R}^N$, where p_i is the payment from agent i to the mechanism. The private information associated with agent i, which defines its value for different outcomes, is denoted with *type*, $\theta_i \in \Theta_i$. Given type θ_i, then agent i has value $v_i(k, \theta_i) \in \mathbb{R}$ for outcome $k \in \mathcal{K}$. It is useful to use

$\theta = (\theta_1, \ldots, \theta_N)$ to denote a type vector, and $\Theta = \Theta_1 \times \ldots \times \Theta_N$ for the joint type space. In simple cases in which an agent's valuation function can be represented by a single number, for example in a single-item allocation problem, it is convenient to write $v_i = \theta_i$.

We assume risk neutral agents, with *quasilinear utility functions*, $u_i(k, p, \theta_i) = v_i(k, \theta_i) - p$. This is a common assumption across the auction and mechanism design literature. Although an agent knows its own type, there is incomplete information about the types of other agents. Let $f_i(\theta_i) \in [0,1]$ denote the probability density function over the type, θ_i, of agent i, and let $F_i(\theta_i) \in [0,1]$ denote the corresponding cumulative distribution function. We assume that the types of the agents are independent random variables, and that there is *common knowledge* of these distributions, such that agent i knows $f_j(\cdot)$ for every other agent $j \neq i$, agent j knows that agent i knows, etc. We assume that the mechanism designer has the same information as the agents.

2.2 Mechanism Design

The mechanism design approach to solving distributed allocation problems with self-interested agents formulates the design problem as an *optimization* problem. Mechanism design addresses the problem of implementing solutions to distributed problems despite the fact that agents have private information about the quality of different solutions and that agents are self-interested and happy to misreport their private information if that can improve the solution in their favor. A mechanism takes information from agents and makes a decision about the outcome and payments that are implemented. It is useful to imagine the role of a mechanism designer as that of a game designer, able to determine the rules of the game but not the strategies that agents will follow.

A mechanism defines a set of feasible *strategies*, which restrict the kinds of messages that agents can send to the mechanism, and makes a commitment to use a particular *allocation rule* and a particular *payment rule* to select an outcome and determine agent payments, as a function of their strategies.[1] Game theoretic methods are used to analyze the properties of a mechanism, under the assumption that agents are rational and will follow expected-utility maximizing strategies in equilibrium.

Perhaps the most successful application of mechanism design has been to the theory of auctions. In recent years auction theory has been applied to the design of a number of real-world markets [64]. There are two natural design goals in the application of mechanism design to auctions and markets. One goal is *allocative efficiency*, in which the mechanism implements a solution that maximizes the total valuation across all agents. This is the *efficient* mechanism design problem. Another goal is *payoff maximization*, in which the mechanism implements a solution that maximizes the payoff to a particular agent. This is the *optimal* mechanism design problem. One can imagine many other variations, including settings in which the goal is to maximize the total payoff across a subset of agents, or settings in which the fairness of an allocation matters.

In particular settings, such as when there is an efficient after-market, then the optimal mechanism is also an efficient mechanism [6], but in general there exists a conflict between efficiency and optimality [65]. Competition across marketplaces can also promote goals of efficiency, with the efficient markets that maximize the total payoff surviving in the long-run [32]. Payoff maximization for a single participant is most appropriate in a setting in which there is asymmetric market power, such as in the automobile industry when market power within the supply chain is held by the big manufacturers [25, 12].

The efficient mechanism design problem has proved more tractable than the optimal mechanism design problem, with optimal payoff-maximizing mechanisms known only in quite restrictive special cases.

2.2.1 Direct Revelation Mechanisms.

The space of possible mechanisms is huge, allowing for example for multiple rounds of interaction between agents and the mechanism, and for arbitrarily complex allocation and payment rules. Given this, the problem of determining the *best* mechanism from the space of all possible mechanisms can appear impossibly difficult. The *revelation principle* [44, 47, 65] allows an important simplification. The revelation principle states that it is sufficient to restrict attention to *incentive compatible direct-revelation* mechanisms. In a direct-revelation mechanism (DRM) each agent is simultaneously asked to report its type. In an incentive-compatible (IC) mechanism each agent finds it in their own best interest to report its type truthfully. The mechanism design problem reduces to defining functions that map types to outcomes, subject to constraints that ensure that the mechanism is incentive-compatible. To understand the revelation principle, consider taking a complex mechanism, $\mathcal{M}$, and constructing a DRM, $\mathcal{M}'$, by taking reported types and simulating the equilibrium of mechanism $\mathcal{M}$. If a particular strategy, $s^*(\theta)$, is in equilibrium in $\mathcal{M}$, given types θ, then truthful reporting of types is in equilibrium in $\mathcal{M}'$ because this induces strategies $s^*(\theta)$ in the simulated mechanism.

Care should be taken in interpreting the revelation principle. First, the revelation principle does not imply that "incentive-compatibility comes for free". In fact, a central theme of mechanism design is that there is a cost to the elicitation of private information. The mechanism design literature is peppered with impossibility results that characterize sets of desiderata that are impossible to achieve simultaneously because it is necessary to incent agents to participate in a mechanism [53]. Rather, the revelation principle states that *if* a particular set of properties can be implemented in the equilibrium of some mechanism, then the properties can also be implemented in an incentive-compatible mechanism. Second, the revelation principle ignores computation and communication complexity, and should not be taken as a statement that "only direct revelation mechanisms matter in practical mechanism design". In many cases indirect mechanisms are preferable for reasons unmodeled in classic mechanism design theory, for example because they decentralize computation to participants, and can economize on preference elicitation while achieving more transparency than direct mechanisms. We return to this topic in Section 3.

The beauty of the revelation principle is that it allows theoretical impossibility and possibility results to be established in the space of direct mechanisms, and carried over to apply to *all* mechanisms. For example, an indirect mechanism can be constructed with a particular set of properties only if a direct mechanism can also be constructed with the same set of properties.

2.2.2 Efficient Mechanism Design.

In efficient mechanism design, the goal is to implement the choice, $k^* \in \mathcal{K}$, that maximizes that total value across all agents given agent types, $\theta \in \Theta$. By the revelation principle we can focus on incentive-compatible DRMs. Each agent is asked to report its type, $\hat{\theta}$, possibly untruthfully, and the mechanism chooses the outcome and the payments. The mechanism defines an allocation rule, $g : \Theta \to \mathcal{K}$, and a payment rule, $p : \Theta \to \mathbb{R}^N$. Given reported types, $\hat{\theta}$, then choice $g(\hat{\theta})$ is implemented and agent i makes payment $p_i(\hat{\theta})$.[2]

The goal of efficiency, combined with incentive-compatibility, pins down the allocation rule:

$$g_{\text{eff}}(\theta) = \arg\max_{k \in \mathcal{K}} \sum_{i \in \mathcal{I}} v_i(k, \theta_i) \quad \text{(EFF)}$$

for all $\theta \in \Theta$. The remaining mechanism design problem is to choose a payment rule that satisfies IC, along with any additional desiderata. Popular additional criteria include:

(IR) individual-rationality. An agent's expected payoff is greater than its payoff from non-participation.

(BB) budget-balance. Either *strong*, such that the total payments made by agents equal zero, or *weak*, such that the total payments made by agents are non-negative.

(revenue) maximize the total expected payments by agents.

Given payment rule, $p(\cdot)$, and allocation rule, $g(\cdot)$, let $m_i(p, \hat{\theta}_i)$, $V_i(g, \hat{\theta}_i \mid \theta_i)$, and $U_i(g, p, \hat{\theta}_i \mid \theta_i)$ denote (respectively) the expected payment, expected valuation, and expected payoff to agent i when reporting type, $\hat{\theta}_i$, assuming the

Auctions, Bidding and Exchange Design 151

other agents are truthful. It is convenient to leave the dependence of $m_i(\cdot)$ on $g(\cdot)$ and the dependence of $V_i(\cdot)$ on $p(\cdot)$ implicit.

$$m_i(p, \hat{\theta}_i) = E_{\theta_{-i}}[p_i(\hat{\theta}_i, \theta_{-i})] \qquad \text{(\textit{interim} payment)}$$
$$V_i(g, \hat{\theta}_i \mid \theta_i) = E_{\theta_{-i}}[v_i(g(\hat{\theta}_i, \theta_{-i}), \theta_i)] \qquad \text{(\textit{interim} valuation)}$$
$$U_i(g, p, \hat{\theta}_i \mid \theta_i) = V_i(g, \hat{\theta}_i \mid \theta_i) - m_i(p, \hat{\theta}_i) \qquad \text{(\textit{interim} payoff)}$$

Notation $\theta_{-i} = (\theta_1, \ldots, \theta_{i-1}, \theta_{i+1}, \ldots, \theta_N)$ denotes the type vector without agent i. The expectation is taken with respect to the joint distribution over agent types, θ_{-i}, implied by marginal probability distribution functions, $f_i(\cdot)$. Assuming IC, then $m_i(p, \theta_i), V_i(g, \theta_i \mid \theta_i)$ and $U_i(g, p, \theta_i \mid \theta_i)$ are the expected payment, valuation, and payoff to agent i in equilibrium. These are also referred to as the *interim* payments, valuations, and payoffs, because they are computed once an agent knows its own type but before it knows the types of the other agents. It is often convenient to suppress the dependence on the specific mechanism rules (g, p) and write $m_i(\theta_i), V_i(\theta_i)$ and $U_i(\theta_i)$. Finally, let $m_i(p) = E_{\theta_i}[m_i(p, \theta_i)]$ denote the expected *ex ante* payment by agent i, before its own type is known.

The efficient mechanism design problem is formulated as an optimization problem across payment rules that satisfy IC, as well as other constraints such as IR and BB. These constraints define the space of *feasible* payment rules. A *selection criteria*, $y(m_1, \ldots, m_N) \in \mathbb{R}$, defined over expected payments, can be used to choose a particular payment rule from the space of feasible rules. A typical criteria is to maximize the total expected payments, with $y(m_1, \ldots, m_N) = \sum_i m_i$. Formally, the efficient mechanism design problem [EFF] is:

$$\max_{p(\cdot)} \quad y(m_1(p), \ldots, m_N(p)) \qquad \text{[EFF]}$$
$$\text{s.t.} \quad U_i(g_{\text{eff}}, p, \theta_i \mid \theta_i) \geq U_i(g_{\text{eff}}, p, \hat{\theta}_i \mid \theta_i), \quad \forall i, \forall \theta_i \in \Theta_i \qquad \text{(IC)}$$
$$\text{additional constraints} \qquad \text{(IR),(BB),etc.}$$

where $g_{\text{eff}}(\cdot)$ is the efficient allocation rule.

The IC constraints require that when other agents truthfully report their types an agent's best response is to truthfully report its own type, for all possible types. In technical terms, this ensures that truth-revelation is a Bayesian-Nash equilibrium, and we say that the mechanism is *Bayesian-Nash incentive-compatible*. In a *Bayesian-Nash* equilibrium every agent is plays a strategy that is an expected utility maximizing response to its beliefs over the *distribution* over the strategies of other agents. An agent need not play a best-response to the *actual* strategy of another agent, given its actual type. This equilibrium is strengthened in a dominant strategy equilibrium, in which truth-revelation is the best-response for an agent whatever the strategies and preferences of other agents. A dominant strategy and IC mechanism is simply called a *strategyproof* mechanism. Formally:

$$v_i(g(\theta_i, \theta_{-i}), \theta_i) - p(\theta_i, \theta_{-i}) \geq v_i(g(\hat{\theta}_i, \theta_{-i}), \theta_i) - p(\hat{\theta}_i, \theta_{-i}), \quad \forall i, \forall \theta_i, \forall \theta_{-i} \tag{SP}$$

Strategyproofness is a useful property because agents can play their equilibrium strategy without game-theoretic modeling or counterspeculation about other agents.

Groves [48] mechanisms completely characterize the class of efficient and strategyproof mechanisms [47]. The payment rule in a Groves mechanism is defined as:

$$p_{\text{groves},i}(\hat{\theta}) = h_i(\hat{\theta}_{-i}) - \sum_{j \neq i} v_j(g_{\text{eff}}(\hat{\theta}))$$

where $h_i(\cdot) : \Theta_{-i} \to \mathbb{R}$ is an arbitrary function on the reported types of every agent except i, or simply a constant. To understand the strategyproofness of the Groves mechanisms, consider the utility of agent i, $u_i(\hat{\theta}_i)$, from reporting type $\hat{\theta}_i$, given $g_{\text{eff}}(\cdot)$ and $p_{\text{groves}}(\cdot)$, and fix the reported types, θ_{-i}, of the other agents. Then, $u_i(\hat{\theta}_i) = v_i(g_{\text{eff}}(\hat{\theta}_i, \theta_{-i}), \theta_i) - p_{\text{groves},i}(\hat{\theta}_i, \theta_{-i})$, and substituting for $p_{\text{groves}}(\cdot)$, we have $u_i(\hat{\theta}_i) = v_i(g_{\text{eff}}(\hat{\theta}_i, \theta_{-i}), \theta_i) + \sum_{j \neq i} v_j(g_{\text{eff}}(\hat{\theta}_i, \theta_{-i}), \theta_j) - h_i(\theta_{-i})$. Reporting $\hat{\theta}_i = \theta_i$ maximizes the sum of the first two terms by construction, and the final term is independent of the reported type. This holds for all θ_{-i}, and strategyproofness follows. The Groves payment rule internalizes the externality placed on the other agents in the system by the reported preferences of agent i. This aligns an agent's incentives with the system-wide goal of allocative-efficiency.

The uniqueness of Groves mechanisms provides an additional simplification to the efficient mechanism design problem when *dominant strategy* implementations are required. It is sufficient to consider the family of Groves mechanisms, and look for functions $h_i(\cdot)$ that provide Groves payments that satisfy all of the desired constraints. The Vickrey-Clarke-Groves (VCG) mechanism is an important special case, so named because it reflects the seminal ideas due to Vickrey [100] and Clarke [26]. The VCG mechanism maximizes expected revenue across all strategyproof efficient mechanisms, subject to *ex post* individual-rationality (IR) constraints. *Ex post* IR provides:

$$v_i(g(\theta_i, \theta_{-i}), \theta_i) - p_i(\theta_i, \theta_{-i}) \geq 0, \quad \forall i, \forall \theta_i, \forall \theta_{-i} \qquad (\textit{ex post IR})$$

This is an *ex post* condition, because it requires that the equilibrium payoff to an agent is always non-negative at the outcome of the mechanism, whatever the types of other agents. To keep things simple we assume that an agent has zero payoff for non-participation. The VCG mechanism defines payment:

$$p_{\text{vcg},i}(\hat{\theta}) = \sum_{j \neq i} v_j(g_{\text{eff}}(\hat{\theta}_{-i})) - \sum_{j \neq i} v_j(g_{\text{eff}}(\hat{\theta}))$$

where $g_{\text{eff}}(\hat{\theta}_{-i})$ is the efficient allocation as computed with agent i removed from the system.

It is natural to ask whether greater revenue can be achieved by relaxing strategyproofness to Bayesian-Nash IC. In fact, the VCG mechanism maximizes the expected revenue across *all* efficient and *ex post* IR mechanisms, even allowing for Bayesian-Nash implementation [57]. This equivalence result yields a further simplification to the efficient mechanism design problem, beyond that provided by the revelation principle. Whenever the additional constraints (in addition to IR and IC) are *interim* or *ex ante* in nature[3] in an efficient mechanism design problem, then it is sufficient to consider the family of Groves mechanisms in which the arbitrary $h_i(\cdot)$ functions are replaced with constants [103]. Not only is the allocation rule, $g(\cdot)$, pinned down, but so is the functional form of the payment rule, $p(\cdot)$, and the mechanism design problem reduces to optimization over a set of constants.

This analysis of the revenue properties of VCG mechanisms follows from a fundamental *payoff equivalence* result [57, 103]. The payoff equivalence result states that

$$U_i(\theta_i) = U_i(\underline{\theta}_i) + \int_C \frac{\mathrm{d}V_i(\theta_i)}{\mathrm{d}\theta_i}\bigg|_{\theta_i = \tau} \mathrm{d}\tau \qquad \text{(equiv)}$$

for all efficient mechanisms, where $\underline{\theta}_i$ is the minimal type of agent i, and C is a smooth curve from $\underline{\theta}_i$ to θ_i within Θ_i. By definition (*interim* valuation), the *interim* valuation, $V_i(\theta_i)$, in an IC mechanism depends only on the allocation rule. Therefore payoff equivalence (equiv) states that the equilibrium payoff from any two IC mechanisms with the same allocation rule, $g(\cdot)$, are equal up to an additive constant, i.e. its payoff at some particular type $\underline{\theta}_i$. A consequence of payoff equivalence is that all IC mechanisms with the same allocation rule are *revenue equivalent* up to an additive constant, which is soon pinned down by additional constraints such as IR.[4]

Finally, this characterization of the VCG mechanism provides a unified perspective on many areas of mechanism design theory, and provides a simple and direct proof of a number of impossibility results in the literature [57]. As an example, we can consider the Myerson-Satterthwaite [66] impossibility result, which demonstrates a conflict between efficiency and budget-balance in a simple two-sided market. There is one seller and one buyer, a single item to trade, and agent preferences such that both no-trade and trade can be efficient *ex ante*.

The Myerson-Satterthwaite result states that there does not exist an efficient, weak budget-balanced, and IR mechanism in this setting and any efficient exchange with voluntary participation must run at a budget deficit. Recalling that the VCG mechanism maximizes expected payments from agents across all efficient and IR mechanisms, there is a simple constructive method to prove this negative result. One simply shows that the VCG mechanism in this setting runs at a deficit.

2.2.3 Optimal Mechanism Design.

In optimal mechanism design the goal is to maximize the expected payoff of *one* particular agent (typically the seller). Recall that the primary goal in efficient mechanism design is to maximize the total payoff across *all* agents. The agent receiving this special consideration in the context of optimal auction design is often the *seller*, although this need not be the role of the agent. We find it convenient to refer to this agent as the seller in our discussion, and indicate this special agent with index 0. In optimal mechanism design the goals of the designer are aligned with the seller, and it is supposed that we have complete information about the seller's type. The mechanism design problem is formulated over the remaining agents, to maximize the expected payoff of the seller subject to IR constraints.

Myerson [65] first introduced the problem of optimal mechanism design, in the context of an auction for a single item with a seller that seeks to maximize her expected revenue. We will provide a general formulation of the optimal mechanism design problem, to parallel the formulation of the efficient mechanism design problem. However, *analytic* solutions to the optimal mechanism design problem are known only for special cases.

In this section we allow randomized allocation and payment rules. The allocation rule, $g : \Theta \to \Delta(\mathcal{K})$, defines a probability distribution over choices given reported types, and the payment rule, $p : \Theta \to \mathbb{R}^N$, defines expected payments. The ability to include non-determinism in the allocation rule allows the mechanism to break ties at random, amongst other things. Let $V_0(g, p)$ denote the expected *ex ante* valuation of the seller for the outcome, in equilibrium given the payment and allocation rules and beliefs about agent types.

By the revelation principle we can restrict attention to IC DRMs, and immediately express the optimal mechanism design problem [OPT] as

$$\max_{g(\cdot), p(\cdot)} \quad V_0(g, p) + \sum_i m_i(p) \qquad \text{[OPT]}$$

$$\text{s.t.} \quad U_i(g, p, \theta_i \mid \theta_i) \geq U_i(g, p, \hat{\theta}_i \mid \theta_i), \quad \forall i, \forall \theta_i \in \Theta_i \qquad \text{(IC)}$$

additional constraints $\hspace{4cm}$ (IR),(BB),etc.

where $m_i(p)$ is the expected equilibrium payment made by agent i, $U_i(g, p, \hat{\theta}_i \mid \theta_i)$ is the expected equilibrium payoff to agent i with type θ_i for reporting type $\hat{\theta}_i$. The objective is to maximize the payoff of the seller. In comparison with the efficient mechanism design problem, we have no longer pinned down the allocation rule and the optimization is performed over the entire space of allocation and payment rules.

One approach to compute an optimal mechanism is to decompose the problem into a master problem and a subproblem. The subproblem takes a particular allocation rule, $g'(\cdot)$, and computes the optimal payment rule given $g'(\cdot)$, subject to IC constraints. The masterproblem is then to determine an allocation rule to maximize the value of the subproblem. However, as discussed by de Vries & Vohra in Chapter 7, the set of allocation rules need not be finite or countable, and this is a hard problem without additional structure. Solutions are known for special cases, including a single-item allocation problem [65], and also a simple multiattribute allocation problem [25].

As an illustration, we provide an overview of optimal mechanism design for the single-item allocation problem. Let $\pi_{g,i}(\hat{\theta}) \geq 0$ denote the probability that agent i receives the item, given reported types $\hat{\theta}$ and allocation rule $g(\cdot)$. We also write, $v_i(k_i, \theta_i) = \theta_i$, for the choice, k_i, in which agent i receives the item, and 0 otherwise, so that an agent's type corresponds to its value for the item. Let θ_0 denote the seller's value.

Call a mechanism (g, p) *feasible* if IC and *interim* IR hold. The first step in the derivation of the optimal auction reduces IC and *interim* IR to the following conditions on (g, p):

$$Q_i(g, \theta_1) \leq Q_i(g, \theta_2), \quad \forall i \in \mathcal{I}, \forall \theta_1 < \theta_2, \forall \theta_1, \theta_2 \in \Theta_i \tag{5.1}$$

$$U_i(g, p, \theta_i) = U_i(g, p, \underline{\theta}_i) + \int_{\tau = \underline{\theta}_i}^{\theta_i} Q_i(g, \tau) d\tau, \quad \forall i \in \mathcal{I}, \forall \theta_i \in \Theta_i \tag{5.2}$$

$$U_i(g, p, \underline{\theta}_i) \geq 0, \quad \forall i \in \mathcal{I} \tag{5.3}$$

where $\underline{\theta}_i$ represents the lowest possible value that i might assign to the item, and $Q_i(g, \hat{\theta}_i)$ denotes the conditional probability that i will get the item when reporting type, $\hat{\theta}_i$, given that the other agents are truthful, i.e. $Q_i(g, \hat{\theta}_i) = E_{\theta_{-i}}[\pi_{g,i}(\hat{\theta}_i, \theta_{-i})]$.

The key to this equivalence is to recognize that IC can be expressed as:

$$U_i(g, p, \theta_i \mid \theta_i) \geq U_i(g, p, \hat{\theta}_i \mid \hat{\theta}_i) + (\theta_i - \hat{\theta}_i) Q_i(g, \hat{\theta}_i), \quad \forall \hat{\theta}_i \neq \theta_i \tag{5.4}$$

in this single-item allocation problem by a simple substitution for $U_i(g, p, \hat{\theta}_i \mid \theta_i)$. Given this, condition (5.1), which states that an agent's probability of getting the item must decrease if it announces a lower type, together with (5.2) implies condition (5.4), and IR follows from (5.2) and (5.3).

Continuing, once the payoff to an agent with type $\underline{\theta}$ is pinned down, then the interim payoff (5.2) of an agent is independent of the payment rule because $Q_i(g,\tau)$ is the conditional probability that agent i receives the item given type τ and allocation rule g. This allows the optimal mechanism design problem to be formulated as an optimization over just the *allocation rule*, with the effect of computing an optimal solution to the payoff-maximizing subproblem for a given allocation rule folded into the masterproblem, and IR constraints allowing the seller's expected payoff to be expressed in terms of the expected payoffs of the other agents. Integration of Q_i between $\underline{\theta}_i$ and θ_i yields a simplified formulation:

$$\max_{g(\cdot)} E_\theta \left[\sum_{i \in \mathcal{I}} (J_i(\theta_i) - \theta_0) \pi_{g,i}(\theta) \right] \quad \text{[OPT']}$$

s.t. $Q_i(g, \theta_1) \leq Q_i(g, \theta_2), \quad \forall i \in \mathcal{I}, \forall \theta_1 < \theta_2, \forall \theta_1, \theta_2 \in \Theta_i$ (5.1)

where the value, $J_i(\theta_i)$, is the *priority level* of agent i, and computed as:

$$J_i(\theta_i) = \theta_i - \frac{1 - F_i(\theta_i)}{f_i(\theta_i)}$$

Recall that $f_i(\cdot)$ is the probability distribution over the type of agent i, and $F_i(\cdot)$ the cumulative distribution. This priority level, sometimes called the *virtual valuation*, is less than an agent's type by the expectation of the second-order statistic of the distribution over its type. Economically, one can imagine that this represents the "information rent" of an agent, the expected payoff that an agent can extract from the private information that it has about its own type.

The optimal allocation rule, $g_{\text{opt}}(\cdot)$, requires the seller to *keep the item if $\theta_0 > \max_i J_i(\hat{\theta}_i)$ and award it to the agent with the highest $J_i(\hat{\theta}_i)$ otherwise, breaking ties at random*. It is immediate that this rule maximizes the objective [OPT']. A technical condition, *regularity*, ensures that this allocation rules satisfies (5.1). Regularity requires that the priority, $J_i(\theta_i)$, is a monotone strictly increasing function of θ_i for every agent. Myerson [65] also derives a general solution for the non-regular case. The remaining problem, given g_{opt}, is to solve for the payment rule. The optimal payment rule given a particular allocation rule is computed as:

$$p_i(\theta) = \pi_{g,i}(\theta)\theta_i - \int_{\tau=\underline{\theta}_i}^{\theta_i} \pi_{g,i}(\tau, \theta_{-i}) d\tau \quad (5.5)$$

where $\pi_{g,i}(\theta)$ is the probability that i gets the item given g and types θ. Given allocation rule, g_{opt}, this simplifies to

$$p_i(\theta) = \begin{cases} \inf\{\hat{\theta}_i \mid J_i(\hat{\theta}_i) \geq \theta_0, J_i(\hat{\theta}_i) \geq J_j(\theta_j), \forall j \neq i\} & \text{, if } \pi_{g_{\text{opt}},i}(\theta) = 1 \\ 0 & \text{, otherwise.} \end{cases}$$

where θ_0 is the value of the seller for the item. In words, only the winner makes a payment, and the payment is the smallest amount the agent could have bid and still won the auction. This payment rule makes truth-revelation a Bayesian-Nash equilibrium of the auction.

The optimal auction is a Vickrey auction with a reservation price in the special case that all agents are symmetric and the $J_i(\cdot)$ functions are strictly increasing. The seller places a reservation price, $p_0 = J^{-1}(\theta_0)$, given her value, θ_0, and the item is sold to the highest bidder for the second-highest price whenever the highest bid is greater than the reservation price. The optimal auction in this symmetric special case is strategyproof. The effect of the seller's reservation price is to increase the payment made whenever the seller's price is between the second-highest and highest bid from outside bidders, at the risk of missing a trade when the highest outside bid is lower than the seller's reservation price but higher than the seller's true valuation. Notice that the optimal auction is not *ex post* efficient.

In the general case of asymmetric bidders the optimal auction may not even sell to the agent whose value for the item is the highest. In this asymmetric case the optimal auction is *not* a Vickrey auction with a reservation price. The agent with the highest priority level gets the item, and the effect of adjusting for the prior beliefs $f_i(\cdot)$ about the type of an agent is that the optimal auction discriminates against bidders that *a priori* are expected to have higher types. This can result in an agent with a higher type having a lower priority level than an agent with a lower type. One can imagine that the optimal auction price-discriminates across buyers based on beliefs about their types.

2.3 Competitive Equilibrium

Competitive equilibrium theory is built around a model of agent price-taking behavior. At its heart is nothing more than linear-programming duality theory. One formulates a primal problem to represent an efficient allocation problem, and a dual problem to represent a pricing problem. Competitive equilibrium conditions precisely characterizes complementary-slackness conditions between an allocation and a set of prices, and implies that the allocation is optimal and therefore efficient. Competitive equilibrium conditions are useful because they can be evaluated based on myopic best-response bid information from agents, and without requiring complete information about agent valuations. This is the sense in which prices can decentralize decision-making in resource allocation problems.

The modeling assumption of price-taking behavior states that agents will take prices as given and demand items that maximize payoff given their valuations and the current prices. This is commonly described as *price-taking* or *myopic best-response* behavior. In the language of mechanism design, this can be considered a form of *myopic*, or bounded, incentive-compatibility.

To illustrate competitive equilibrium (CE) prices we will impose some structure on choice set $\mathcal{K}$ and define the *combinatorial allocation problem* (CAP). Let $\mathcal{G}$ define a set of items, and $S \subseteq \mathcal{G}$ a subset, or *bundle*, of items. A choice, $k \in \mathcal{K}$ defines a feasible allocation of bundles to agents. Introduce variables, $x_i(S) \in \{0,1\}$, to indicate that agent i receives bundle S in a particular allocation. Agent i has value $v_i(S)$ for bundle S.[5] Assume for the purpose of exposition that we have knowledge of agent valuations. The CAP can be formulated as the following integer program:

$$\max_{x_i(S)} \sum_{S \subseteq \mathcal{G}} \sum_{i \in \mathcal{I}} x_i(S) v_i(S) \quad \text{[CAP]}$$

$$\text{s.t.} \quad \sum_S x_i(S) \leq 1, \quad \forall i \in \mathcal{I}$$

$$\sum_{S \ni j} \sum_{i \in \mathcal{I}} x_i(S) \leq 1, \quad \forall j \in \mathcal{G}$$

$$x_i(S) \in \{0, 1\} \quad (5.6)$$

where $S \ni j$ indicates that bundle S contains item j. Later, we find it useful to use CAP($\mathcal{I}$) to denote the CAP with all agents $\mathcal{I}$ and CAP($\mathcal{I} \setminus j$) to denote the CAP formulated without agent $j \in \mathcal{I}$.

To apply linear-programming duality theory we must relax this IP formulation, and construct an integral LP formulation. Consider [LP$_1$] in which eq. (5.6) is relaxed to $x_i(S) \geq 0$. Then, the dual is simply written as:

$$\min_{\pi_i, p(j)} \sum_i \pi_i + \sum_j p(j) \quad \text{[DLP}_1\text{]}$$

$$\text{s.t.} \quad \pi_i + \sum_{j \in S} p(j) \geq v_i(S), \quad \forall i \in \mathcal{I}, \forall S \subseteq \mathcal{G} \quad \text{(DLP}_1\text{-1)}$$

$$\pi_i, p(j) \geq 0, \quad \forall i, j$$

The dual introduces variables $p(j) \geq 0$, for items $j \in \mathcal{G}$, which we can interpret as prices on items. Given prices, $p(j)$, the optimal dual solution sets $\pi_i = \max_S \left\{ v_i(S) - \sum_{j \in S} p(j), 0 \right\}$. This is the maximal payoff to agent i given the prices. The dual problem computes prices on items to minimize the sum of the payoffs across all agents. These are precisely a set of CE prices when the primal solution is integral.

A technical condition, *gross substitutes* [55] (or simply substitutes) on agent valuations is sufficient for integrality. Unit demand preferences, in which each agent wants at most one item (but can have different values for different items), is a special case. With this substitutes condition, LP$_1$ is integral and the complementary-slackness (CS) conditions on a feasible primal, x, and feasible dual, p, solution define conditions for competitive equilibrium:

$$\pi_i > 0 \Rightarrow \sum_S x_i(S) = 1, \quad \forall i \tag{5.7}$$

$$p(j) > 0 \Rightarrow \sum_{S \ni j} \sum_i x_i(S) = 1, \quad \forall j \tag{5.8}$$

$$x_i(S) > 0 \Rightarrow \pi_i + \sum_{j \in S} p(j) = v_i(S), \quad \forall i, \forall S \tag{5.9}$$

These have a natural economic interpretation. Conditions (5.7) and (5.9) state that the allocation must maximize the payoff for every agent at the prices. Condition (5.8) states that the seller must sell every item with a positive price, and maximize the payoff to the seller at the prices. The prices are said to *support* the efficient allocation. A seller can announce an efficient allocation and CE prices, and let every agent verify that the allocation maximizes its own payoff at the prices. In practice we will need an auction to provide incentives for agents to reveal the information about their valuations, and to converge towards a set of CE prices.

The linear program formulation, LP_1, is *not* integral in general instances of CAP and the item prices will not support the efficient allocation. Instead one needs to consider prices on bundles of items. Bikhchandani & Ostroy [17] provide a hierarchy of strengthened LP formulations to capture these generalizations, in which the variables in the dual problems correspond with *non-linear* and then non-linear and *non-anonymous* prices. Non-linear prices, $p(S) \geq 0$, on bundles $S \subseteq \mathcal{G}$, allow $p(S) \neq p(S_1) + p(S_2)$ for $S = S_1 \cup S_2$ and $S_1 \cap S_2 = \emptyset$. Non-anonymous prices, $p_i(S) \geq 0$, on bundles S to agent i, allow $p_i(S) \neq p_j(S)$ for $i \neq j$. We will return to these extended formulations and enriched price space in Section ?? in reference to the design of an ascending-price combinatorial auction.

2.4 Indirect Revelation Mechanisms

In this section, we tie together the mechanism design approach and the competitive equilibrium approach. The basic idea is to construct efficient ascending-price auctions that terminate with the outcome of the VCG mechanism. With this, price-taking behavior is a game-theoretic equilibrium of the auction despite the effect that an agent's bids might have on future price dynamics. The auctions provide a dynamic method to compute a set of competitive equilibrium prices, from which allocative efficiency follows.

To understand the motivation for the design of iterative, price-based mechanisms, we need to begin to consider the computational considerations in the implementation of useful auction mechanisms. Although the revelation principle focuses attention on incentive-compatibility, it hides all implementation and computational concerns. In particular, a direct revelation mechanism (such as the VCG) requires every agent to provide complete and exact information about its valuation over all possible outcomes. This is often unreasonable, for example in the setting of a combinatorial auction to allocate the rights to operate the 700+ bus routes in London.[6] In comparison, an agent in an iterative price-based auction can reveal its preference information as necessary.

These kinds of mechanisms, in which agents are *not* required to submit (and compute) complete and exact information about their private valuations, are referred to as *indirect* mechanisms. Indirect mechanisms, such as those based on prices, also go some way to distributing the calculation of the outcome of a mechanism across agents rather than requiring the mechanism infrastructure (such as the auctioneer) to compute the winners and the payments.

Examples of indirect mechanisms include *ascending-price auctions*, in which agents submit bids in response to prices and the auctioneer maintains a provisional allocation and adjusts prices. For example, the English auction is an ascending-price auction for a single item in which the price increases until there is only one bidder left in the auction [83]. The English auction implements the outcome of the Vickrey auction (and is allocatively-efficient). However, only two agents must bid to make progress towards the outcome, and agents can follow equilibrium strategies with lower- and upper-bounds on their values [73, 28]. In comparison, the equilibrium in the Vickrey auction requires every agent to compute, and reveal, exact information about its value.

We describe a general methodology to design iterative mechanisms that leverages a fundamental connection between linear programming, competitive equilibrium, and VCG payments. The approach has been used in recent years to design and analyze efficient ascending auctions for the assignment problem [36], combinatorial auctions [79, 81], multiattribute auctions [77], and multi-unit auctions [15]. The interested reader is also referred to Bikhchandani & Ostroy [16] and Parkes [74] for an extended discussion.

In outline, the two steps in a primal-dual approach to the design of efficient ascending auctions are:

1 Assume myopic best-response strategies. Formulate a linear program (LP) for the efficient allocation problem. The LP should be integral, such that it computes feasible solutions to the allocation problem, and have appropriate economic content. This economic content requires that the dual formulation computes competitive equilibrium prices that support the efficient allocation, *and* that there is a solution to the dual problem that provides enough information to compute VCG payments.

2 Design a primal-dual algorithm that maintains a feasible primal and dual solution, and terminates with solutions that satisfy complementary-

slackness conditions and also satisfy any additional conditions necessary to compute the VCG payments. The algorithm should not assume complete access to agent valuations, but rather access to myopic best-response bids from agents.

Note carefully that termination in VCG payments is sufficient to bring myopic best-response into a game-theoretic equilibrium. As such, the first assumption is used to leverage the primal-dual design methodology, but is not limiting from an incentive perspective. Technically, MBR is an *ex post* Nash equilibrium, in which there is no better strategy for an agent whatever the preferences of other agents, so long as the other agents also follow myopic best-response [49, 81]. *Ex post* Nash is a useful solution concept because agents can play the equilibrium without any information about the types of the other agents. All that is required is that the other agents are rational and play equilibrium strategies.

In the special case of *agents-are-substitutes*, then the *minimal* CE prices support the VCG payments to each agent. The minimal CE prices are a set of prices that *minimize* the revenue to the seller (or equivalently, maximize the total payoff to the buyers) across all CE prices (and need not be unique in general). Let *coalitional value*, $w(L) \geq 0$, denote the value of the efficient allocation for agents $L \subseteq \mathcal{I}$, i.e. the solution to $CAP(\mathcal{L})$. Agents-are-substitutes places the following constraints on the coalitional values:

$$w(\mathcal{I}) - w(\mathcal{I} \setminus K) \geq \sum_{i \in K} [w(\mathcal{I}) - w(\mathcal{I} \setminus i)], \quad \forall K \subseteq \mathcal{I}$$

This is known to be the widest class of preferences for which the VCG outcome can be supported in a competitive equilibrium, even with non-linear and non-anonymous prices [17]. It holds, for example when we have *gross substitutes* preferences [7], and in the unit-demand problem. When agents-are-substitutes fails, the minimal CE prices are not unique, and moreover they do not support the VCG outcome.

However, the primal-dual methodology does *not* require that the VCG payments are supported in a single set of CE prices. Rather, it requires that the price space is rich enough to support *Universal* competitive equilibrium (UCE) prices. From these prices, we can compute the VCG payments [81]. UCE prices are defined as CE prices that are simultaneously CE prices for $CAP(\mathcal{I})$, and also for $CAP(\mathcal{I} \setminus j)$ without each agent $j \in \mathcal{I}$. UCE prices always exist, for example setting $p_i(S) = v_i(S)$ for all agents provides UCE prices. In the case of agents-are-substitutes, then minimal CE prices are UCE prices. Given UCE prices, p_{uce}, we can compute the VCG payments to each agent as:

$$p_{\text{vcg},i} = p_{\text{uce}}(S_i^*) - [\Pi(p_{\text{uce}}, \mathcal{I}) - \Pi(p_{\text{uce}}, \mathcal{I} \setminus i)] \quad (5.10)$$

where $\Pi(p, L)$ is the maximal revenue that the auctioneer can achieve, given prices p, across all feasible allocations to agents in set $L \subseteq \mathcal{I}$.

We return to UCE prices in Section 5.1, in the context of an ascending-price combinatorial auction in which agents are interested in bundles of items. The primal-dual analysis is performed with respect to the hierarchy of extended LP formulations described in Section 2.3.

2.4.1 Example: The English Auction.
To illustrate the primal-dual methodology, we derive the English auction, which is an efficient and strategyproof auction for the single-item allocation problem. Let v_i denote agent i's value for the item. The efficient allocation problem is:

$$\max_{x_i} \sum_i v_i x_i \quad [\text{IP}_{\text{single}}]$$
$$\text{s.t.} \quad \sum_i x_i \leq 1$$
$$x_i \in \{0, 1\}$$

where $x_i = 1$ if and only if agent i is allocated the item, i.e. the goal is to allocate the item to the agent with the highest value. [LP$_{\text{single}}$] is an integral linear-program formulation with suitable economic properties.

$$\max_{x_i, y} \sum_i v_i x_i \quad [\text{LP}_{\text{single}}]$$
$$\text{s.t.} \quad \sum_i x_i + y \leq 1$$
$$x_i \leq 1, \quad \forall i$$
$$x_i, y \geq 0$$

Variable, $y \geq 0$, is introduced, with $y = 1$ indicating that the seller decided to make no allocation. The dual formulation, [DLP$_{\text{single}}$], is:

$$\min_{p, \pi_i} p + \sum_i \pi_i \quad [\text{DLP}_{\text{single}}]$$
$$\text{s.t.} \quad \pi_i \geq v_i - p, \quad \forall i$$
$$p \geq 0$$
$$p, \pi_i \geq 0$$

in which dual variable, $p \geq 0$, represents the price of the item. Given a price, p, the optimal values for π_i are $\pi = \max(0, v_i - p)$, which is the maximal payoff to agent i at the price. The CS conditions are:

$$p > 0 \Rightarrow \sum_i x_i + y = 1 \quad \text{(CS-1)}$$

$$\pi_i > 0 \Rightarrow x_i = 1, \quad \forall i \quad \text{(CS-2)}$$

$$x_i > 0 \Rightarrow \pi_i = v_i - p, \quad \forall i \quad \text{(CS-3)}$$

$$y > 0 \Rightarrow p = 0 \quad \text{(CS-4)}$$

In words, if the price is positive then the item must be allocated to an agent by (CS-1) and (CS-4); the price must be less than the value of the winning agent by (CS-3) and feasibility ($\pi \geq 0$); and the price must be greater than the value of all losing agents, so that $\pi_i = 0$ for those agents (CS-2).

The English auction maintains an ask price on the item, initially equal to zero. In each round an agent can bid at the current price or leave the auction. An agent's myopic best-response (MBR) strategy is to bid while the price is less than its value. As long as two or more agents bid in a round, the ask price is increased by the minimal bid increment, ϵ. An agent is selected from the agents that bid in each round to receive the item in the provisional allocation. The bid from the agent in the provisional allocation is retained in the next round. The auction terminates as soon as only one agent submits a bid. The agent receives the item for its final bid price.

We have just described a primal-dual algorithm. The ask price defines a feasible dual solution, the provisional allocation defines a feasible primal solution. The CS conditions hold when the auction terminates, and the final allocation is an optimal primal solution and efficient. Suppose the provisional allocation assigns the item to agent $\hat{i}$. Construct a feasible primal solution with $y = 0$, $x_{\hat{i}} = 1$ and $x_i = 0$ for all $i \neq \hat{i}$. Given ask price, p_{ask}, consider a feasible dual solution with $p = p_{\text{ask}}$. This is feasible as long as $p_{\text{ask}} \geq 0$, with the optimal dual solution given this price completed with payoffs, $\pi_i = \max(0, v_i - p)$. Conditions (CS-1,CS-3) and (CS-4) are maintained in each round. Condition (CS-2) holds on termination, because $\pi_i = 0$ for all agents except the winning agent, otherwise another agent would have bid by MBR.

The English auction also terminates with a price that implements the Vickrey payment. The *minimal* CE prices, or the dual solution that maximizes the payoff to the winning agent across all solutions, sets $p = \max_{i \neq i^*} v_i$, where i^* is the agent with the highest value. This is the payment by the winner in the equilibrium of the VCG mechanism in this setting, which is the second-price sealed-bid (Vickrey) auction. The English auction terminates with an ask price, p^*, that satisfies $p^* \geq \max_{i \neq i^*} v_i$ and $p^* - \epsilon < \max_{i \neq i^*}$, and implements the Vickrey outcome as $\epsilon \to 0$. In this simple auction this is sufficient to make MBR a *dominant* strategy for an agent together with a rule that prevents jump bids.

3. Implementation Considerations

In this section, we continue to discuss some of the computational considerations that must be addressed in taking a mathematical specification of a mechanism and building a working system. There often remains a large gap between the mathematical specification and a reasonable computational implementation. We have already introduced the idea of *indirect mechanisms* that can reduce the amount of information that agents must provide about their preferences. Here, in laying out some of the additional details that must be considered in closing this gap, we consider the choice of a language to represent agent preferences (Section 3.1) and the complexity of the winner-determination problem (Section 3.2). The winner-determination complexity can also be impacted by side constraints that represent business rules (Section 3.2.2.0).

3.1 Bidding Languages

The structure of the bidding language in an auction is important because it can restrict the ability of agents to express their preferences. In addition, the expressiveness allowed also has a big impact of the the properties of the auction. This has prompted research that examines bidding languages and their expressiveness and the impact on winner determination [20, 19, 13]. In this section we will outline two aspects of bidding languages that are central to auctions: (i) the structure of bids allowed, and (ii) the rules specified by the bid that restrict the choice of bids by the seller.

The structure of bids that are allowed are closely related to the market structure. For example, in markets where multiple units are being bought or sold it becomes necessary to allows bids that express preferences over multiple units. Some common bid structures examined in the literature are:

- divisible bids with price-quantity pairs that specify per-unit prices and allow any amount less than specified quantity can be chosen.

- divisible bids with a price schedule, for example volume discounted bids

- indivisible bids with price-quantity pairs, where the price is for the total amount bid and this is to be treated as an all-or-nothing bid.

- bundled bids with price-quantity pairs, where the bid is indivisible and the price is over the entire basket of different items and is to be treated as an all-or-nothing bid.

- configurable bids for multiattribute items that allow the bidder to specify a bid function sensitive to attribute levels chosen.

With multiple items or multiattribute items the preference structure of agents can be exponentially large. For example, if there are n items and the agent has super-additive preferences then in general the agent could specify 2^n bids. Multiattribute items with n binary attributes leads to similar informational complexity. Therefore an additional consideration is to provide a compact bid representation language that allows agents to implicitly specify their bid structure.

Several researchers have proposed mechanisms for specifying bids logically. Boutilier and Hoos [20] provide a nice overview of logical bidding languages for combinatorial auctions. These bidding languages have two flavors: (i) logical combinations of goods as formulae ($\mathbb{L}_G$), and (ii) logical combinations of bundles as formulae ($\mathbb{L}_B$).

$\mathbb{L}_G$ [20, 51] languages allow bids that are logical formulae where goods (items) are taken as atomic propositions and combined using logical connectives and a price is attached to the formula expressing the amount that the bidder is willing to pay for satisfaction of this formula. $\mathbb{L}_G$ captures perfect substitutes with disjunctions in a single formula, however imperfect substitutes might require multiple formulae to capture the agent's preferences.

$\mathbb{L}_B$ [91, 67] languages use bundles with associated prices as atomic propositions and combines them using logical connectives. One bid language is **additive-or**, $\mathbb{L}_B^{OR}$, in which one or more bids can be accepted, and the total bid price is the sum of the individual bid prices. This language is compact for particular valuations (e.g. linear-additive across items), but not expressive for general valuations. The canonical language is the **exclusive-or**, $\mathbb{L}_B^{XOR}$, language in which bids state that *at most one* bid can be accepted. The total bid price is the value of the maximal bid price across the component bundles when multiple bundles are accepted. One can also consider nested languages, such as OR-of-XORs and XOR-of-ORs, and a generalization, $\mathbb{L}_B^{OR*}$, in which dummy goods are used within atomic bids to provide expressiveness with more compactness than either OR-of-XOR or XOR-of-OR. Nisan [68] provides a discussion of the relative merits of these languages. More recent work [20] introduced $\mathbb{L}_{GB}$ for generalized logical bids that allows a combination of both items and bundles as atomic propositions within a single formula. These generalized logical bids inherit the advantages of both the atomic bid and the atomic bundle approaches and allows concise specification of utility.

Similar issues of concise representation of preferences over multiattribute items/goods is explored in Bichler *et al.* [14]. A bid can specify the values that are allowed for each attribute and an associated markup price over the base levels. In addition, an atomic proposition is associated with each value for each attribute and horn clauses are used to specify configurations that are not allowed or to specify promotions associated with certain feature sets.

3.2 Winner-Determination Complexity

In the context of auctions, the problem of computing an allocation is often referred to as the *winner-determination* problem. In simple designs (such as the English, Vickrey etc.) where only a single winner is permitted in the allocation, the optimization problem can be solved in a straightforward fashion. However, in settings where the allocation rule permits multiple winners, the optimization problem that needs to be solved can become quite computationally complex depending on the market and bid structures. In this section we outline the different settings and the associated complexity of the winner determination problem.

3.2.1 Multi-Unit Auctions.

Consider an auction for multiple units of the same type of item, and in particular the *reverse* auction setting where the focus is to minimize the cost subject to bid requirements. We will consider four cases: (i) divisible bids, (ii) indivisible bids with XOR bid structures, (iii) price schedules, which can be viewed as a compact representation for a generalized XOR with indivisible bids. Suppose in all of these cases that a buyer wants to buy Q identical units of the same item.

Divisible Bids. In the simple case, each bidder submits a price-quantity pair (p_i, q_i), to indicate that it will sell up to q_i units for a unit price of p_i/q_i. The optimal allocation can simply be identified by sorting the bids in increasing order of unit price and picking the cheapest bids until the demand for Q is satisfied. In general, the last chosen bid might get a partial allocation.

Indivisible Bids. Now, suppose that bidders specify *all-or-nothing* constraints on the bids and state that the bids are indivisible. In addition, suppose that the bidders also submit multiple bids with an XOR bidding language. Let M_i denote the number of bids from supplier i, and N denote the number of suppliers. The winner determination problem can be formulated as a knapsack problem, introducing $x_{ij} \in \{0, 1\}$ to indicate that bid j from bidder i is accepted [13].

$$\min_{x_{ij}} \sum_{i=1}^{N} \sum_{j=1}^{M_i} p_{ij} x_{ij}$$

$$\text{s.t.} \quad \sum_{i=1}^{N} \sum_{j=1}^{M_i} q_{ij} x_{ij} \geq Q$$

$$\sum_{j=1}^{M_i} x_{ij} \leq 1, \quad \forall i$$

$$x_{ij} \in \{0, 1\}$$

The special case where each bidder has a single bid reduces to a knapsack problem which is NP-hard [61]. In order to write this as a knapsack problem use the transformation $y_{ij} = 1 - x_{ij}$ and rewrite the formulation as a maximization problem.

Price Schedules. If the bids incorporate price schedules (such as volume discounts) then the winner determination can be modeled as a generalization of the multiple choice knapsack problem. The key issue is whether the price schedule is nonlinear or piecewise linear. Piecewise linear approximations are commonly used to model nonlinear functions [33, 92]. Therefore, we will focus on a model with piecewise linear price schedules.

Each supplier responds with a price schedule that consists of a list of M_i price quantity pairs, $\{(p_{i1}, [\underline{q}_{i1}, \overline{q}_{i1}]), \ldots (p_{iM_i}, [\underline{q}_{iM_i}, \overline{q}_{iM_i}])\}$. Each price quantity pair $(p_{ij}, [\underline{q}_{ij}, \overline{q}_{ij}])$, specifies the per-unit price, p_{ij}, that supplier i is willing to provide for marginal items in the interval, $[\underline{q}_{ij}, \overline{q}_{ij}]$. The ranges in the volume discount must be contiguous. Let z_{ij} denote the number of units sourced above $\underline{q}_{ij}$ from supplier i, with $z_{ij} \leq \overline{q}_{ij} - \underline{q}_{ij}$. The total price for quantity $(z_{ij} + \underline{q}_{ij})$ is:

$$p(z_{ij}) = p_{ij} z_{ij} + \sum_{\hat{j}=1}^{j-1} p_{i\hat{j}} (\overline{q}_{i\hat{j}} - \underline{q}_{i\hat{j}})$$

The price schedule incorporates an infinite large number of potential indivisible bids from each of the intervals with an XOR constraint across these possible bids.

Associate a decision variable, $x_{ij} \in \{0, 1\}$, with each level j of each price schedule i which takes the value 1 if the number of units sourced to supplier i is in the interval $[\underline{q}_{ij}, \overline{q}_{ij}]$, and continuous variable z_{ij} that specifies the exact number of units sourced above $\underline{q}_{ij}$ from supplier i. Constraints ensure that

$z_{ij} > 0 \Rightarrow x_{ij} > 0$. The winner determination formulation for this problem is:

$$\min_{x_{ij}, z_{ij}} \sum_{i=1}^{N} \sum_{j=1}^{M_i} p_{ij} z_{ij} + x_{ij} C_{ij}$$

s.t. $\quad z_{ij} - (\bar{q}_{ij} - \underline{q}_{ij}) x_{ij} \leq 0, \quad \forall i, \forall j$

$$\sum_j x_{ij} \leq 1, \quad \forall i$$

$$\sum_i \sum_j (z_{ij} + x_{ij} \underline{q}_{ij}) \geq Q$$

$$x_{ij} \in \{0, 1\}, z_{ij} \geq 0$$

where the coefficient C_{ij} computes the total price for all the items purchased up to and include $\underline{q}_{ij}$:

$$C_{ij} = \sum_{\hat{j}=1}^{j-1} p_{i\hat{j}} (\bar{q}_{i\hat{j}} - \underline{q}_{i\hat{j}})$$

A special case of this formulation where each interval in the schedule is a point interval reduces to the multiple choice knapsack problem which is NP-hard [61]. Once again the we need to use a change of variables $y_{ij} = 1 - x_{ij}$ to get the canonical maximization form.

Recently, Kothari et al. [56] have proposed a fully polynomial-time approximation scheme (FPTAS) for a variation on this price-schedule problem in which the cost functions are piecewise and *marginal-decreasing* and each supplier has a *capacity constraint*. The approach is to construct a 2-approximation to a generalized knapsack problem, which can then be used to scale a dynamic-programming algorithm and compute an $(1 + \epsilon)$ approximation in worst-case time $T = O((nc)^3/\epsilon)$, for n bidders and with a maximum of c pieces in each bid.[7]

3.2.2 Multi-Item Auctions.

In this subsection we introduce multi-item auctions for multiple heterogenous items. This is the well known combinatorial auction problem, in which we allow bidders to have arbitrary valuations over bundles of items.

Following the notation in Section 2.3, let $\mathcal{G} = (1, \ldots, N)$ denote the set of items for sale. The bidders are allowed to specify bundles $S \subseteq \mathcal{G}$, with a single price on the entire bundle and submit bids for multiple bundles via an XOR bidding language. We formulate this problem by introducing a decision variable $x_i(S)$ for each bundle S offered by bidder i. Each bidder provides a bid set $B_i \subseteq 2^{\mathcal{G}}$. Let $p_i(S)$ denote the price offered by agent i for bundle S, and consider bids in an exclusive-or language.

Forward auction. For the forward auction case of a *single seller with multiple buyers*, the winner-determination problem can be written as:

$$\max_{x_i(S)} \sum_{S \in B_i} \sum_i x_i(S) p_i(S)$$

$$\text{s.t.} \sum_{S \in B_i} x_i(S) \leq 1, \quad \forall i$$

$$\sum_{S \in B_i, S \ni j} \sum_i x_i(S) \leq 1, \quad \forall j$$

$$x_i(S) \in {0, 1}, \quad \forall i, S$$

This is a set packing formulation and is NP-hard [89]. There are special cases under which the structure of this problem simplifies and allows for polynomial time solutions. Many special cases arise out of constraints that reduce the constraint matrix to be totally unimodular [35]. A common example is the case where adjacent plots of land are being sold and bidders might want multiple plots but they need to be adjacent. However, real world problems will often not satisfy the fairly severe restrictions that provide a totally unimodular constraint matrix. Moreover, if the bidding language is not expressive then this can interact with the incentive properties of an auction because a bidder is not able to express her true valuation, even if that would be her equilibrium strategy. We wait until Section 4 for an extensive discussion of the interaction between computational constraints and incentives.

Reverse auction. Combinatorial auctions are also proposed for procurement problems in markets with one buyer and multiple sellers. The reverse combinatorial auction is formulated as a set covering problem rather than a set packing problem. An interesting (and complicating) issue that arises in this setting is that there are various business rules that are used to constrain the choice of winners. These business rules appear as side constraints in the winner determination problem. The winner determination problem with no side constraints can be written as:

$$\min_{x_i(S)} \sum_{S \in B_i} \sum_i x_i(S) p_i(S) \qquad (5.11)$$

$$\sum_{S \in B_i, S \ni j} \sum_i x_i(S) \geq 1, \quad \forall j$$

$$\sum_{S \in B_i} x_i(S) \leq 1, \quad \forall i$$

$$x_i(S) \in {0, 1}, \quad \forall i, S$$

for bids in set $B_i \subseteq 2^{\mathcal{G}}$ with prices $p_i(S)$ on $S \in B_i$. This is posed as a cost minimization problem with a demand covering constraint. In this formulation the problem is procure a single unit of each good, but this can be generalized by increasing the RHS of the first set of constraints.

Business Rules as Side Constraints. In a real world setting there are several considerations beside cost minimization. These considerations often arise from business practice and/or operational considerations and are specified as a set of constraints that need to be satisfied while picking a set of winning suppliers. Recent work [33, 94, 13] in this area provides a comprehensive overview of the constraint types that are possible. We discuss some of the main constraint classes here, and provide some example MIP formulations in the context of (5.11). In general, the specific form of these side constraints depends on the market structure.

Number of Winning Suppliers An important consideration while choosing winning bids is to make sure that the entire supply is not sourced from too few suppliers, since this creates a high exposure if some of them are not able to deliver on their promise. On the other hand, having too many suppliers creates a high overhead cost in terms of managing a large number of supplier relationships. These considerations introduce constraints on the minimum, L_s, and maximum, U_s, number of winning suppliers in the solution to the winner determination problem.

$$y_i \leq \sum_{S \in B_i} x_i(S) \leq K y_i, \quad \forall i \in N$$

$$L_s \leq \sum_i y_i \leq U_s$$

Budget Limits on Trades A common constraint that is often placed is a upper limit on the total volume of the transaction with a particular supplier. These limits could either be on the total spend or on the total quantity that is sourced to a supplier. These types of constraints are largely motivated (in a procurement setting) by considerations that the dependency on any particular supplier is managed. Similarly, often constraints are placed on the minimum amount or minimum spend on any transaction, i.e. if a supplier is picked for sourcing then the transaction should be of a minimum size. Such constraints reduce the overhead of managing a large number of very small contracts.

Marketshare Constraints Another common consideration, especially in situations where the relationships are longterm, is to restrict the market share that any supplier is awarded. The motivations are similar to the previous case.

Reservation Prices A reservation price allows the buyer to place an additional constraint on the most she will pay for some items. This can arise, for example, due to a fall-back option such as an external commodity market. If the reservation prices are specified over bundles then

$$x_i(S)p_i(S) \leq r(S)$$

where $r(S)$ is the reservation price. Alternately, reservation prices can be specified for each item, with $r(S) = \sum_{j \in S} r(j)$, and $r(j)$ to define the price on item j.

Representation Constraints These specify additional requirements such as "at least one minority supplier" is included in the set of winners. A generalization is to specify the number of winners that are required from different supplier types.

The interesting aspect of these side constraints is how they impact the computational complexity of the winner determination problem. For example, introducing either of the following constraint classes will transform even a tractable problem (e.g. with a totally unimodular structure) into a hard problem:

- Budget constraints with integrality requirements for the choice of bids lead to a knapsack type constraints and lead a NP-hard problems.
- Minimum/Maximum number of winning supplier requirements introduce integral counts (for those suppliers who have winning bids versus those who do not) and lead to a set-cover type of constraint that make winner determination NP-hard.

Volume Discounts. So far we have treated the price on a bundle as an all-or-nothing bid. An alternative is to consider *supply curves*, in which the price function is explicitly specified in terms of volume discounts for each item over which a total price is constructed for a given bundle. It is useful to assume that the supply curves are additive separable; that is,

$$p_i(A^s) = p_i(a_1^s, .., a_M^s) = \sum_{j \in \mathcal{G}} p_{ij}(a_i^s)$$

where p_{ij} are individual price curves for the commodity j from supplier i, and a_i^s is the quantity for commodity j. We can also assume that each individual curve is a piece-wise linear function.

The winner determination for supply curve auctions can also be written as a set covering problem as shown Eq. (5.11) using a Dantzig-Wolfe type decomposition [39]. To use a set covering model we introduce the concept of supply patterns. A supply pattern A^s is a vector of length M specifying the amount supplied for each of the commodities $A^s = (a_1^s, a_2^s, ..., a_k^s)$. The cost of a supply pattern for a particular supplier is computed as $p_i(A^s)$. A supply pattern is feasible for a supplier if she is able to sell the given amount from each of the commodities, and meet additional side constraints. The set of feasible supply patterns for supplier i is denoted by $\mathcal{D}_i$. Note that there could be an exponential number of feasible supply patterns for each supplier. In the mathematical model we introduce a decision variable for each feasible supply pattern of each supplier: y^s is a decision variable indicating whether pattern s is selected or not, $s \in \bigcup_i \mathcal{D}_i$.

Let Q_j denote the quantity of commodity j demanded in the procurement problem. The basic constraints of this optimization problem will ensure that the demand is met and that at most one pattern is chosen for each supplier:

$$\sum_i \sum_{s \in \mathcal{D}_i} a_j^s y^s \geq Q_j, \quad \forall j$$

$$\sum_{s \in \mathcal{D}_i} y^s \leq 1, \quad \forall i$$

A lower, L_s, and upper limit, U_s, for the total number of accepted suppliers, can again be imposed by the following constraint:

$$L_s \leq \sum_i \sum_{s \in \mathcal{D}_i} y^s \leq U_s$$

On the other hand, lower and upper limits on the amount of goods supplied by any particular supplier can be encoded in the patterns. Assume that l_{ij} and u_{ij} are such limits for a particular supplier and commodity and that L_i and U_i are limits for a supplier across all commodities. Then any feasible pattern $s \in \mathcal{D}_i$ for supplier j must satisfy the following constraints:

$$l_{ij} \leq a_j^s \leq u_{ij}, \quad \forall j$$
$$L_i \leq \sum_j a_j^s \leq U_i$$

The objective function of minimizing the procurer's cost completes the mathematical model:

$$z = \min \sum_i \sum_{s \in \mathcal{D}_i} p_i(A^s) y^s$$

3.2.3 Double auctions and Exchanges.

Double auctions are settings with multiple buyers and sellers. There exist two main institutions for double auctions: (i) the continuous double auction, which clears continuously,[8] and (ii) the clearinghouse or *call* auction, which clears periodically. In this Chapter we focus on call markets. Call markets have received more attention in the literature, and are perhaps more appropriate when bids and asks are combinatorial because collecting a number of bids before clearing the market can improve the ability to construct useful matches between complex bids and asks.

The computational aspects of market clearing depends on the market structure [54]. The aspects of market structure that have an impact on winner determination are as follows:

- *Aggregation*: The role of the market-maker in disassembling and re-assembling bundles of items. Possibilities include buy-side aggregation, sell-side aggregation or both. If no aggregation is allowed then each bid can be matched to exactly one ask.

- *Divisibility*: The ability to allocate fractions of items, and the ability to satisfy a fraction of agents' bids and asks. When an agent wants its bid or nothing, then its bid is called *indivisible*.

- *Homogeneous/Heterogeneous Goods*: Homogenous goods imply that all the goods being exchanged are all exactly the same and interchangeable (e.g. an auction for a particular financial stocks). If the goods are differentiated, or heterogeneous, then any given ask can only match with a subset of the bids.

The appropriate level of aggregation depends on the *physical attributes* of the good. For example, pieces of steel can be cut but not very easily joined (buy-side aggregation). Conversely, computer memory chips can be combined but not split (sell-side aggregation). Note that aggregation *does not* imply that the exchange must take physical possession of goods, trades can still be executed directly between agents.

Similarly, goods can have multiple attributes, and must often be considered as heterogeneous goods. For example, steel coils may differ in the grade or surface quality. Very often *substituting* a higher quality item for a lower quality item is acceptable, e.g. a bid for 10 units of 1.0GHz processors can be substituted with 10 units of 1.2GHz processors with additional cost. In contrast, in some situations the heterogenous good might *complement* each other and provide greater value as a bundle rather than separately. For example an offer for all legs of an itinerary is valuable than a set of disjointed legs.

Our discussion considers the case of bids and asks for multiple units of the same item, that we term *homogeneous bids*, and the *combinatorial exchanges* in which bids and asks can bundle together multiple units of multiple different items.

Homogeneous Bids. First, we consider the case of bids and asks for multiple units of the same item, but without allowing bundle bids. Moreover, we assume that multiple bids and asks submitted by the same bidder are connected with *additive-or* logic. For the moment we also assume that bids and asks are divisible, so that a fraction of a bid can be matched with an ask (or multiple fractions with multiple asks). We provide a general formulation of the winner determination problem in this setting. The formulation captures different market structures, in terms of aggregation and differentiation.

Consider a set of bids B and a set of asks A. Each bid, $b_i \in B$ is associated with a single type of good, and provides a unit bid price, p_i, and a quantity demanded, q_i. Similarly, each ask, $a_j \in A$, is associated with a single type of good, and provides a unit ask price, p_j, and a quantity offered, q_j. Let $0 \leq x_{ij} \leq 1$ denote the fraction of the demand q_i from bid b_i allocated to ask a_j. For any given bid b_i we also specify a set of asks $A_i \subseteq A$ to which it can be feasibly matched. Similarly, for each ask a_j we specify the set of bids $B_j \subseteq B$ to which it can be feasibly matched. These assignment restrictions model the feasibility requirements imposed by the heterogeneity of goods. We do not specify the constraints in any more detail because this depends on the structure of the market. Then, the winner-determination problem, to clear the exchange to maximize surplus, is written as:

$$\max_{x_{ij}} \sum_{i \in A} \sum_{j \in B_j} (p_i - p_j) q_i x_{ij}$$

s.t.
$$\sum_{i \in B_j} q_i x_{ij} \leq q_j, \quad \forall j \in A \quad (5.12)$$

$$\sum_{j \in A_i} x_{ij} \leq 1, \quad \forall i \in B \quad (5.13)$$

$$0 \leq x_{ij} \leq 1, \quad \forall i, j \quad (5.14)$$

In the simplest case of homogeneous goods we can drop the assignment restrictions, and set $A_i = A$ and $B_j = B$. Assuming divisibility, then x_{ij} indicates the fraction of the available quantity in bid b_i allocated to ask a_j. The matching problem can be solved by sorting the bids in decreasing order of price and offers in increasing price. The crossover point, p^* is the clearing price and bids with price above p^* and asks below p^* are matched.

We can include assignment restrictions (for example to capture the case of heterogeneous goods) and still use a LP to solve the matching problem as long as bids are divisible and the logic connecting bids is additive-or. The linear program has a network structure which can be exploited to solve the problem efficiently. Any type of aggregation is allowed without impacting the computational complexity of the problem.

Auctions, Bidding and Exchange Design

However, if the bids are indivisible we just define the decision variable, $x_{ij} \in \{0,1\}$, as a binary variable that takes a value 1 if bid b_i is assigned to ask a_j and zero otherwise and replace equation (5.14) with $x_{ij} \in \{0,1\}$. Still, if we now restrict the exchange so as not to allow any aggregation then the winner-determination problem is an assignment problem which can be solved very efficiently in polynomial time [1]. Consider a bipartite graph with asks on one side (the asks are differentiated by price and seller) and the bids on the other. The constraint (5.12) can be replaced with $\sum_{i \in B_j} x_{ij} \leq 1$ But, in general, for example with aggregation on the sell side, the constraint (5.13) with integrality restricts bids to be assigned to at most one ask and the problem becomes the generalized assignment problem which is known to be NP-hard [61]. The reader is referred to Kalagnanam et al. [54] for a detailed discussion of these issues.

Combinatorial Exchanges. In a combinatorial exchange we allow bids and asks on bundles of heterogeneous items, and allow a bidder to connect multiple bids and asks with an exclusive-or bidding language. We choose to formulate the problem for agents that either act exclusively as sellers or exclusively as buyers, but this is not necessary in general. The formulation of the market clearing problem generalizes the winner-determination problem for the one-sided combinatorial auction to allow multiple buyers and sellers.

Let $\mathcal{B}$ denote the set of buyers and $\mathcal{S}$ denote the set of sellers. We allow sellers to submit asks for multiple bundles $S \subseteq \mathcal{G}$, with an ask price, $m_i(S)$, on each bundle. Similarly, we allow buyers to submit bids for multiple bundles, with a bid price, $p_i(S)$ on each bundle. Let $i \in \mathcal{B} \cup \mathcal{S}$ index both buyers and sellers, and let $B_i \subseteq 2^{\mathcal{G}}$ denote the set of bundles that receive a bid (or ask) from agent i. Finally, variable $x_i(S) = 1$ indicates the bid on bundle S from buyer i is accepted, and $y_i(S) = 1$ indicates that the ask on bundle S from seller i is accepted. Given this, we can formulate the market clearing problem as:

$$\max_{x_i(S), y_i(S)} \sum_{S \in B_i} \sum_{i \in \mathcal{B} \cup \mathcal{S}} (x_i(S)p_i(S) - y_i(S)m_i(S))$$

$$\text{s.t.} \quad \sum_{S \in B_i} x_i(S) \leq 1, \quad \forall i \in \mathcal{B}$$

$$\sum_{S \in B_i} y_i(S) \leq 1, \quad \forall i \in \mathcal{S}$$

$$\sum_{S \in B_i, S \ni j} \sum_i (y_i(S) - x_i(S)) \geq 0, \quad \forall j$$

$$x_i(S) \in {0,1}, y_i(S) \in \{0,1\}$$

Although the general problem is NP-hard, the special case in which there is no aggregation is still equivalent to the assignment problem, and solvable via an LP formulation. For each bundle from a supplier we allow exactly one match to a bundle requested by the bidder. Similarly, each bundled bid form a bidder is restricted to match exactly one bundled offer. This reduces to an assignment problem. However, since each agent can bid a power set $S \subseteq \mathcal{G}$ the assignment problem can become exponential in the number of bids.

3.2.4 Multiattribute Auctions.

Multiattribute auctions relate to items that can be differentiated on several non-price attributes such as quality, delivery date etc. In order to evaluate different offers for a item with different attribute levels we need to appeal to multiattribute utility theory to provide a tradeoff across these different attributes. One common approach assumes *preferential independence*, and supposes that an agent's valuation for a bundle of attribute levels is a linear-additive sum across the attributes. Another more general approach captures nonlinear valuations. It is also interesting to consider both *single sourcing*, in which the buyer chooses a single supplier, and *multiple sourcing*, in which there are multiple items to procure and the buyer is willing to consider a solution that aggregates across multiple suppliers.

Let $\mathcal{J}$ denote the set of attributes of an item, with K_j to denote the domain of attribute j (assumed discrete), and $K = K_1 \times \ldots \times K_m$ denote the joint domain, with $m = |\mathcal{J}|$. Consider a reverse auction setting, and write $v^B(x) \geq 0$ and $c_i(x) \geq 0$ to denote the buyer's value and the cost of seller i for attribute bundle $x \in K$. Of course, enumerating these valuations and cost functions over the cross-product of attribute levels can be costly for participants in a market.

Thus, it is useful to consider the *preferential-independence* (PI), in which an agent can state its value for different levels of a particular attribute irrespective of the levels of another attribute. In this special case, the valuation, $v^B(x) = \sum_{j \in \mathcal{J}} v_j^B(x_j)$, where $v_j^B(x_j) \geq 0$ is the buyer's valuation for level $x_j \in K_j$ of attribute j. Similarly, the cost, $c_i(x) = \sum_{j \in \mathcal{J}} c_{ij}(x_j)$, where $c_{ij}(q_j)$ is the supplier's cost for level $x_j \in K_j$ of attribute j.

We will first consider the *single-sourcing* problem, in which a single winning bid is selected to satisfy the demand. Then, we will consider the *multiple-sourcing* problem, in which there are multiple units to procure and a buyer is interested in purchasing from multiple sellers.

Single Sourcing. In a single-sourcing setting only a single winning bid is picked to satisfy the demand. Consider the case of preferential-independence, and let b_{ijk} denote the ask price from supplier i on level $k \in K_j$ of attribute j. Similarly, let v_{jk} denote the reported value of the buyer on level

k of attribute j. The winner-determination problem is

$$\max_{x_{ijk}, y_i} \sum_{i \in \mathcal{I}} \sum_{j \in \mathcal{J}} \sum_{k \in K_j} x_{ijk}(v_{jk} - b_{ijk})$$

$$\text{s.t.} \quad \sum_{k \in K_j} x_{ijk} \leq y_i, \quad \forall i, \forall j$$

$$\sum_{i \in \mathcal{I}} y_i \leq 1$$

$$x_{ijk}, y_i \in \{0, 1\}$$

where variable y_i is used to indicate that supplier i is selected in the winning allocation. A straightforward method to solve this problem computes the best attribute values for each supplier, and then chooses the best supplier.

A more interesting setting is when the bid structure is more expressive, and in addition to specifying markup prices for attribute levels as in the preferential-independence bidding language, a supplier can provide *configuration rules* to indicate which combinations of attributes are infeasible. Similarly, promotions to encourage certain attribute levels can be specified as rules. Propositional logic can be used to capture these rules and these rules can be parsed into linear inequalities and added as side constraints to the winner determination formulations.

An interesting aspect of this configurable setting is that even in the simple case of single sourcing with a budget constraint, the identification of the optimal feature set is NP-hard [13]. Consider the simplest setting where the buyer attempts to identify the best configurations from a configurable offer from a single supplier, subject to a budget-constraint, B. Identifying the best configuration can be modeled as a variation of the multiple-choice knapsack problem. Again, let $x_{jk} = 1$ indicate that level k of attribute j is selected. Let p_b denote the base price, for a base feature set, and μ_{jk} be the markup associated with choosing level k for attribute j. Assuming an separable additive utility function, then the optimal feature set can be identified as:

$$\max_{x_{jk}, p} \sum_{j \in \mathcal{J}} \sum_{k \in K_j} v_{jk} x_{jk} - p$$

$$\text{s.t.} \quad \sum_{k \in K_j} x_{jk} = 1, \quad \forall j \in \mathcal{J}$$

$$\sum_{j \in \mathcal{J}} \sum_{k \in K_j} \mu_{jk} x_{jk} + p_b \leq p$$

$$p \leq B,$$

$$x_{jk} \in \{0, 1\}, \quad \forall j, k$$

Bichler et al. [14] provide a detailed discussion of this configurable offers problem with multiple sourcing and other side constraints.

Multiple Sourcing. There are settings where it might be necessary to source to more than one supplier either because none of the suppliers are large enough to satisfy the demand or business rules may requires a minimum number of suppliers. Let Q denote the buyer demand and let q_i denote the supply of seller i. We will use the same notation as for the single sourcing case. If the bids are divisible then identifying the optimal bids is straightforward. The bids are sorted in descending-order of surplus (value - cost), and then the optimal set of bids are picked from this sorted list until $\sum_i q_i = Q$. Notice that the last bid may be chosen fractionally.

However, if the bids are indivisible then the winner determination problem reduces to a knapsack problem and becomes NP-hard. The winner determination problem can be written as follows:

$$\max_{x_{ijk}, y_i} \sum_{i \in \mathcal{I}} \sum_{j \in \mathcal{J}} \sum_{k \in K_j} x_{ijk}(v_{jk} - b_{ijk})$$

s.t.
$$\sum_{k \in K_j} x_{ijk} \leq y_i, \quad \forall i, \forall j$$

$$\sum_{i \in \mathcal{I}} q_i y_i \geq Q$$

$$x_{ijk}, y_i \in \{0, 1\}$$

In practice it might be more realistic to impose an acceptable range for the demand.

Multi-sourcing for multiattribute items can also require special consideration of *homogeneity constraints* when picking winners, such that all the winning bids must have the same value for some attribute/s. For example, if chairs are being bought from 3 different suppliers for an auditorium, then it is important that the color for all chairs be the same. Such constraints can be generalized to allow selection of winning bids such that for an attribute of interest all bids have values adjacent to each other. In order to capture such a requirement we can introduce an indicator variable z_{jk} that takes a value 1 if any bids are chosen at level k for attribute j. Let T_{jk} denote the set of bids at level k for attribute j, then we can capture this requirement as follows:

$$z_{jk} \leq \sum_{i \in T_{jk}} x_{ijk} \leq |T_{jk}| z_{jk} \quad \forall j, k$$

$$0 \leq \sum_k z_{jk} \leq 1 \quad \forall j$$

Notice that these constraints have to be applied for each attribute level. The reader is referred to Bichler and Kalagnanam [13] for more details.

4. Interactions between Computation and Incentives

Up to this point we have considered the computational complexity of exact implementations of economic mechanisms for market-based allocation problems. However, sometimes there is no reasonable implementation of the exact mechanism. In such cases, computational considerations must be introduced explicitly during the mechanism design process itself. This will be the focus of this section.

Limited computational resources, both at agents and within the mechanism infrastructure, and limited communication bandwidth, can often necessitate the introduction of explicit approximations and restrictions within mechanism and market designs, or at least careful design to provide good computational properties in addition to good economic properties. Introducing approximations, for example to the allocation rule in a mechanism, can fundamentally change the economic properties of a mechanism. For example, many approximations to the functions $g_{\text{eff}}(\cdot)$ in the Groves mechanism payment and allocation rules break strategyproofness. We focus in this section on *interactions* between computational considerations and incentive considerations in mechanism design. Just as classic mechanism design introduces IC constraints to restrict the space of feasible mechanisms, computational constraints further restrict the space of feasible mechanisms. We divide our discussion into the following areas:

strategic complexity how much computation is required by agents to compute the game-theoretic equilibrium of a mechanism?

communication complexity how much communication is required between agents and the mechanism to implement the outcome of the mechanism?

valuation complexity how much computation is required by agents to compute, or elicit, enough information about their type to be able to compute the game-theoretic equilibrium?

implementation complexity how much computation is required to compute the outcome of a mechanism from agent strategies?

In addition to identifying tractable special cases, for example for a subset of a larger type space, and developing fast algorithms, computational considerations often make it necessary to impose explicit constraints, for example to restrict the expressiveness of a bidding language or to restrict the range of outcomes considered by the mechanism.

4.1 Strategic Complexity

The *strategic complexity* is the complexity of the game-theoretic problem facing an agent. Mechanism design uses a rational model of agent behavior, in which agents compute and play equilibrium strategies given information about the mechanism and given beliefs about the preferences, rationality, and beliefs of other agents. But agents must be able to compute equilibrium strategies to play equilibrium strategies, or at the least the mechanism designer must be able to compute equilibrium strategies and provide a certificate to allow agents to verify that strategies are in equilibrium.

Although the general question of how complex it is to construct a Nash equilibrium in a game remains open [71] a number of hardness results have been established for computing equilibria with particular properties [45]. Given this, it is important to consider the strategic complexity of the particular non-cooperative game induced by a mechanism, and for the appropriate solution concept, such as Bayesian-Nash or dominant strategy. We choose to focus on strategic complexity in incentive-compatible DRMs, which are the mechanisms for which issues of strategic complexity have received most attention.

A first approach is to design mechanisms with tractable strategic problems, such as the class of *strategyproof* mechanisms in which truth-revelation is a dominant strategy equilibrium and optimal for every agent irrespective of the types and strategies of other agents. Most work in *algorithmic mechanism design* [69] focuses on this class of strategyproof mechanisms and addresses the remaining problems of communication complexity and implementation complexity.

A second approach is to perform mechanism design with respect to explicit assumptions about the computational abilities of agents, such as restricting attention to mechanisms with polynomial-time computable equilibrium. For example, Nisan & Ronen [68] introduce the concept of a *feasible best-response*, which restricts the strategies an agent in computing its best-response to a *knowledge set*, which can be a subset of the complete strategy space. Mechanism analysis is performed with respect to a *feasible-dominant* equilibrium, in which there is a dominant-strategy in the restricted strategy space defined by agent knowledge sets. In other work, combinatorial exchange mechanisms (see Section 5.6) are proposed that make small deviations away from truthfulness unbeneficial to agents [78], and the mechanism design problem has been considered with respect to an ϵ-strategyproofness [96].

It is interesting that limited computational resources can be used as a *positive* tool within mechanism design, for example designing mechanisms in which the only computable equilibria are "good" from the perspective of system-wide design goals. As an example, the problem of strategic manipulation in voting protocols is known to be NP-hard [11], and it is possible to use randomization within a mechanism to make manipulation hard without making the implementation problem for the mechanism hard [30].

4.2   Communication Complexity

The *communication complexity* of a mechanism considers the size of messages that must be sent between agents and the mechanism to implement the outcome of a mechanism. To motivate this problem, recall that mechanism design often makes an appeal to the revelation-principle and considers direct mechanisms. However, direct mechanisms require agents to report complete and exact information about their type, which is often unreasonable in problems such as combinatorial auctions. In the worst-case the VCG mechanism for a combinatorial auction requires each agent to submit 2^M numbers, given M items, to report its complete valuation function.

A first approach to address the problem of communication complexity is to implement *indirect mechanisms* (see Section 2.4), which do not require the complete revelation of an agent's type. Instead, an agent must report its strategy to the mechanism *along the equilibrium path*. As an example, whereas the VCG mechanism for a combinatorial auction requires complete revelation of an agent's valuation function, an agent must only provide *best-response* bid information in response to prices in an ascending-price combinatorial auction. Although all mechanisms have the same *worst-case* communication complexity in the combinatorial auction setting [70], indirect mechanisms reduce the communication required in many instances of the problem [74, chapter 8].

A second approach introduces *compact* representations of agent preferences via the careful design of bidding languages (see Section 3.1). Nisan [67] notes a tradeoff between the *compactness* of a language, which measures the size of messages required to state an agent's preferences, and the *simplicity* of a language, which considers the computation required to evaluate the value of any particular outcome given a message in the language. At one extreme, one could allow agents to submit *valuation programs* [67], that provide the mechanism with a method to compute an agent's value for an outcome on-the-fly, as demanded by the implementation of the mechanism. Valuation programs can be useful when the *method* used to compute an agent's valuation for different outcomes can be described more compactly than an explicit enumeration of value for all possible outcomes. However, in practice, valuation programs require considerable trust, for example that a program is faithfully executed by a mechanism and that valuable and sensitive information is not shared with an agent's competitors.

A third approach is to restrict the expressiveness of a bidding language within a mechanism to provide compactness. In restricting the expressiveness of a bidding language it is important to consider the effect on the equilibrium properties of a mechanism [87]. For example, a VCG-based mechanism in which agents are restricted to bidding on particular bundles can prevent truthful bidding and break strategyproofness. Holzman et al. [50] describe necessary and sufficient conditions on the structure of bundles in the language to maintain strategyproofness[9] and an *ex post* no-regret property that states that at termination no agent wants to provide any information about its value that was not already permitted within the language. Related work has considered mechanism design within a class of mechanisms in which severe bounds are imposed on the amount of communication permitted between agents and the mechanism [18].

4.3 Valuation Complexity

The valuation complexity of a mechanism considers the complexity of the problem facing an agent that must determine its type. There are many settings in which it is *costly* to provide complete and exact information value information, for all possible outcomes. This valuation cost can arise for *computational* reasons [90], for example in a setting in which an agent's value for a particular procurement outcome is the solution to a hard optimization problem. Consider a logistics example, in which a firm seeks to procure a number of trucks to deliver goods to its customers. The value that the trucks bring to the firm depends on the value of the optimal solution to a truck scheduling problem. This valuation cost can also arise for *informational* reasons, because an agent must elicit preference information from a user to determine the value for a particular outcome [7].

Indirect mechanisms provide one approach to address the problem of valuation complexity. Unlike an incentive-compatible DRM, in which an agent must compute and provide complete information about its preferences to the mechanism, an agent can often compute its optimal strategy in an ascending-price auction from approximate information about preferences. Indirect mechanisms allow incremental revelation of preference information through agent bids, with feedback through prices and provisional solutions to guide the valuation computation of agents [74, chapter 8]. One can imagine that prices in an ascending-price auction structure a sequence of preference-elicitation queries, such as "what is your best-response to these prices?" When myopic best-response is an equilibrium, and when agents play that equilibrium, then each response from an agent provides additional information about an agent's preferences, refining the space of preferences that are consistent with the agent's strategy.

Experimental results demonstrate the advantages of indirect over direct mechanisms for a model of the valuation problem in which an agent can refine *bounds* on its value for bundles during an auction [73, 76]. Related work presents experimental analysis to compare the preference-elicitation costs of different schemes to elicit agent preferences in indirect implementations of combinatorial auctions [29, 52]. Recent theoretical results demonstrate the benefits of indirect vs. direct auctions in the equilibrium of a single-item auction, with a simple valuation model and agents that can choose to refine their valuations during the auction [28], and derive necessary and sufficient conditions on information about agent preferences to be able to compute the VCG outcome in a combinatorial auction [75].

4.4 Implementation Complexity

The implementation complexity of a mechanism considers the complexity of computing the outcome of a mechanism from agent strategies. For example, in a DRM this is the complexity of the problem to compute the outcome from reported agent values. In an indirect mechanism this is the complexity to update the state of the mechanism in response to agent strategies, for example to update the provisional allocation and ask prices in an ascending-price auction. We choose to focus on the issues of implementation complexity in direct mechanisms, which are the mechanisms in which this has received most attention.

One approach is to characterize restrictions on the type space in which the implementation problem is tractable. For example, the winner-determination problem in the VCG mechanism for a combinatorial auction can be solved in polynomial time with particular assumptions about the structure of agent valuations [89, 35]. A number of fast algorithms have also been developed to solve the winner-determination problem in combinatorial auctions, even though the problem remains theoretically intractable [93, 42, 2]. Recent experimental work illustrates the effectiveness of embedding the *structure* of agent valuations within mixed-integer programming formulations of the winner-determination problem [19].

Sometimes it is necessary to impose explicit restrictions and approximations in order to develop a mechanism with reasonable implementation complexity [69]. This problem is interesting because introducing approximation algorithms can often change the equilibrium strategies within mechanisms. For example, the strategyproofness of the VCG mechanism relies on the optimality of the allocation rule. Recall that the utility to agent i in the Groves mechanism is:

$$u_i(\theta) = v_i(g(\hat{\theta}_i, \hat{\theta}_{-i}), \theta_i) + \sum_{j \neq i} v_j(g(\hat{\theta}_i, \hat{\theta}_{-i}), \hat{\theta}_j) - h_i(\cdot)$$

where $\hat{\theta}$ are reported types, $g(\cdot)$ is the efficient allocation rule, and $h_i(\cdot)$ is an arbitrary function of the announced types of the other agents. Truth revelation,

$\hat{\theta}_i = \theta_i$, maximizes the payoff of agent i, so that the mechanism implements $g(\theta_i, \hat{\theta}_{-i})$, and maximizes the sum of the first two terms. Now, with an approximate solution, $\hat{g}(\cdot)$, in place of $g(\cdot)$, and information about the reported types, $\hat{\theta}_{-i}$, of the other agents, the agent should announce a type, $\hat{\theta}_i$, to solve

$$\max_{\hat{\theta}_i \in \Theta_i} v_i(k, \theta_i) + \sum_{j \neq i} v_j(k, \hat{\theta}_j)$$
$$\text{s.t.} \quad x = \hat{g}(\hat{\theta}_i, \hat{\theta}_{-i})$$

The agent chooses its announced type to correct the error in the approximation algorithm, $\hat{g}(\cdot)$, and improve the choice made with respect to its *true* type and the reported types of the other agents.

It is useful to retain strategyproofness, but allow for a tractable approximation to the efficient function, $g(\cdot)$. Nisan & Ronen [68] derive necessary and sufficient conditions for VCG-based mechanisms to maintain the useful property of strategyproofness. Let $R(g, \Theta)$ denote the *range* of the allocation algorithm used within a VCG-based mechanism, i.e. $k \in R(g, \Theta) \Leftrightarrow \exists \theta \in \Theta$ s.t. $\hat{g}(\theta) = k$. A VCG mechanism is *maximal-in-range* if the algorithm, $\hat{g}(\cdot)$ satisfies:

$$\hat{g}(\theta) = \max_{k \in R(\hat{g})} \sum_i v_i(k, \theta_i), \quad \forall \theta \in \Theta$$

When this property holds, there is nothing that an agent can do to correct the approximation error, because this would require changing the range of the algorithm.

Nisan & Ronen use this characterization to demonstrate a negative result for the performance of any range-restricted variation on the VCG mechanism. One can show that any truthful and tractable VCG mechanism for the combinatorial auction must have unreasonable worst-case allocative-efficiency, by constructing a set of preferences for which the efficient allocation is outside the range of the mechanism and that all allocations inside have low values. However, this worst-case bad performance may not be very important in practice, especially in a setting in which the range is carefully selected to provide good performance in most instances that occur in practice. From a positive perspective, the sufficiency of maximal-in-range provides a powerful *constructive* method to build truthful mechanisms with tractable implementation problems: choose a range of outcomes; provide agents with a bidding language that is expressive enough to state their preferences across outcomes in the range; and implement an optimal algorithm with respect to the bidding language and the range.

A number of interesting tractable and strategyproof mechanisms have been suggested for problems in which the VCG mechanism is intractable. For example, Lehmann et al. [60] propose a truthful and feasible mechanism for a combinatorial auction problem with *single-minded* bidders, each with value for one particular bundle of items. The optimal winner-determination problem remains intractable, even in this single-minded setting. Bartal *et al.* [10] have proposed a truthful and feasible mechanism for the multi-unit combinatorial allocation problem, in which each bidder is restricted to demand a small fraction of the available units of each good.

One can also try to *distribute* the computation across the agents that participate within a mechanism. For example, consider providing agents with an opportunity to provide better solutions to the winner-determination problem [89, 22]. Recent work in theoretical computer science, in the broad area of *distributed algorithmic mechanism design* [41], considers the computational and communication complexity of distributed implementations of mechanisms. Broad research goals include developing appropriate notions of hardness and complexity classes, and designing mechanisms with good distributed computational properties *and* good incentives [40]. A key challenge when computation is distributed across participants is to make sure that it is incentive-compatible for agents to implement the algorithm truthfully [97]. This extends the consideration of truthful information revelation, present in classic mechanism design, to also require incentives for truthful information processing and computation [98].

We can also consider relaxed strategic models, in which the goal of complete incentive-compatibility is relaxed. We briefly outline a taxonomy of strategic relaxations [41], and provide some examples of their use in the literature.

almost-strategyproofness *Approximate the strategic properties of a mechanism, perhaps along with other goals.* A particular example is the concept of ϵ-strategyproofness, in which an agent can gain at most ϵ by following some non-truthful strategy whatever the strategies of other agents. ϵ-strategyproofness has been considered by Kothari *et al.* [56] for a multi-unit auction problem and Archer *et al.* [4] for a setting of *known* single-minded bidders in which the auctioneer knows the bundles demanded by agents and only the value is private to each bidder.

tolerable manipulability *The kinds of manipulations are well characterized, and have tolerable effects on overall design goals.* This concept has been considered by Archer et al. [3], in a multicast cost-sharing setting.

feasible strategyproofness *There are beneficial manipulations available to agents, but they cannot compute them because of limited computational power.* Nisan & Ronen [68] have explored this in the context of the VCG mechanism, where they consider the equilibrium behavior of computationally limited agents within the context of approximate implementations of the VCG mechanism.

5. Specific Market Mechanisms

In this section we pick a few mechanisms that are interesting, both from a practical point of view and because they illustrate some of the emerging research directions in the design of electronic auctions, markets and intermediaries. Many of the mechanisms are indirect, with agents providing progressive information about their types and information feedback from the mechanism to guide agent strategies. This observation reinforces the importance of indirect mechanisms in practice. Many of the mechanisms also implement the outcome of the VCG mechanism (or variations), which reinforces the importance of Groves mechanisms in the design of practical mechanisms.

5.1 Combinatorial Auctions

Combinatorial auctions are characterized by the ability for agents to submit bids on *bundles* of items. This can be important in settings in which items are complements, e.g. "I only want A if I also get B," because bundle bids allow agents to express explicit contingencies across items. The applications of combinatorial auctions are numerous, including procurement [32], logistics [59, 38], and in resource allocation settings [63, 84]. Let $\mathcal{G}$ denote a set of discrete items and $\mathcal{I}$ denote a set of agents. Each agent has a valuation, $v_i(S) \geq 0$, for bundles $S \subseteq \mathcal{G}$ and quasilinear utility functions. The efficient mechanism design problem has received the most attention. Indeed, no general solution is known for the *optimal* (revenue-maximizing) combinatorial auction.

The VCG mechanism provides an efficient *sealed-bid* auction, in which agents submit reported valuation functions in a single-shot auction. Given a suitably expressive bidding language, which allows agent i to describe its valuation, v_i, this is an efficient and strategyproof solution. However it is often unreasonable to expect agents to provide valuations on all possible bundles of items. The valuation problem for a single bundle can often be time-consuming, and more difficult to automate than other processes such as winner-determination and bidding.

Given these objections to one-shot combinatorial auctions there has been considerable interest in the design of *iterative* combinatorial auctions, which can reduce the valuation work required by agents because optimal strategies must only be computed along the equilibrium path of the auction.

Proposals for iterative auctions can be described along the following two directions:

bidding language Auctions such as RAD [37] and AUSM [9] allow participants to submit *additive-or* bids, while other auctions [72, 79, 49, 7] allow participants to submit *exclusive-or* bids.

information feedback Auctions such as AUSM, the proposed combinatorial design for FCC auction #31,[10] and the Chicago GSB auction [46], provide *linear-price* feedback along with the provisional allocation. Auctions such as *i*Bundle [79] and A*k*BA [105] provide *non-linear price* feedback along with the provisional allocation. Other proposals, such as AUSM and an ascending-proxy design [7] are described without explicit price feedback.

Early theoretical results exist for particular restrictions on agent valuations. For example, the DGS auction [36] solves the unit-demand (or *assignment*) problem, in which each agent wants at most one item. Another special case of a combinatorial auction for multiple identical items and decreasing marginal valuations is solved with Ausubel's auction [5] (see Section 5.2.2).

More recently, Parkes & Ungar [79] proposed an ascending auction, *i*Bundle, which is efficient for the case of *buyer-submodular* preferences. Buyer-submodular is slightly stronger than agents-are-substitutes.[11] As before, let $w(L)$ denote the coalitional value for agents $L \subseteq \mathcal{I}$. Buyer-submodular requires:

$$w(L) - w(L \setminus K) \geq \sum_{i \in K} [w(L) - w(L \setminus i)], \quad \forall K \subset L, \forall L \subseteq \mathcal{I}$$

Straightforward MBR bidding is an *ex post* Nash equilibrium in this case, with each agent choosing to bid in each round for the set of bundles that maximize its payoff given the current prices. This is proved by Ausubel & Milgrom [7] in their analysis of the closely-related ascending-proxy auction.

*i*Bundle proceeds in rounds. In each round agents can submit XOR bids on multiple bundles, at or above the current ask price. The auctioneer maintains non-linear, and perhaps non-anonymous ask prices. The auctioneer collects bids and solves a winner-determination problem, computing a provisional allocation to maximize revenue given the bids. Finally, prices are updated based on the bids from losing agents. The auction terminates when there are no losing bidders still active in the auction.

*i*Bundle Extend & Adjust (*i*BEA) [81] uses the concept of *Universal* competitive equilibrium prices (see Section 2.4) to extend *i*Bundle beyond its normal termination round and collect just enough additional information about agent preferences to compute VCG payments. Although the final payments by agents are computed as a discount from the final prices in the auction, *i*BEA is best viewed as an ascending-price auction because agents face increasing prices while actively bidding in the auction. The discount to agents is computed at the end of the auction, based on the final ask prices and following eq. (5.10).

Ausubel & Milgrom [7] have described a proxy-agent variation on *i*Bundle, in which bidders must submit preferences to proxy agents that submit ascending bids to an auction. The auction terminates with the VCG payments for the case of buyer-submodular values, and straightforward MBR is in an *ex post* Nash equilibrium for that case (just as in *i*Bundle). However, the ascending-proxy auction will in general terminate with agent payments that are greater than their payments in the VCG outcome, at least when agents choose to reveal their true values to the proxy agents. The ascending-proxy auction satisfies a *bidder-monotonicity property* that is not satisfied by the VCG. Bidder monotonicity states that revenue to the auctioneer increases with the number of agents and provides robustness against collusive bidding strategies. One potential drawback of the ascending-proxy auction, in comparison with a VCG-based mechanism such as *i*BEA, is that there are many equilibrium in ascending-proxy when buyer-submodular does not hold and agents must solve an implicit bargaining problem to implement equilibrium outcomes.

Many iterative combinatorial auctions, including the DGS auction [36] and the *i*Bundle and *i*BEA auctions can be interpreted within the primal-dual design methodology described in Section 2.4. The following three steps are important in extending the primal-dual framework described in Section 2.4 to the combinatorial allocation problem:

a) choose an extended formulation of the efficient allocation problem so that the optimal LP relaxation is integral and so that the dual problem can be interpreted as defining the space of competitive equilibrium prices

b) implement a primal-dual algorithm for the efficient allocation problem, using myopic best-response information from agents in each round to adjust prices and the provisional allocation

c) terminate with *Universal* CE prices, to enable the computation of VCG payments and bring myopic best-response into equilibrium.

Step a) can be answered by appealing to the hierarchy of formulations for CAP provided by Bikhchandani & Ostroy [17], each of which strengthens the formulation and has the effect of enriching the price space (to include both non-linear and non-anonymous prices). In fact, *i*Bundle tries to introduce non-anonymous prices only when necessary and can be thought of as implementing a primal-dual algorithm for a hybrid formulation for the CAP.

Step b) can be answered by paying careful attention to an economic interpretation of the complementary-slackness (CS) conditions. The goal is to demonstrate that the auction terminates with an allocation and prices that satisfy the CS conditions. We provide an outline of the proof of the properties of *i*Bundle, focusing on the special case (*i*Bundle(2)) that prices are still *anonymous*. See Parkes & Ungar [79] for additional details. Let $x_i(S) = 1$ denote that bundle S is allocated to agent i, and let π_i denote the *maximal payoff* to agent i at prices $p(S)$. The first important CS condition is:

$$x_i(S) > 0 \Rightarrow \pi_i + p(S) = v_i(S), \quad \forall i, S \quad \text{(CS-1a)}$$

In words, this condition states that any bundle allocated to an agent in the provisional allocation in a particular round should maximize its payoff at the prices. This is maintained (to within 2ϵ) in *i*Bundle when agents follow MBR strategies because the provisional allocation can only include bundles that receive bids from agents.

Let $y(k) = 1$ denote that partition $k \in \Gamma$ in the space of all feasible partitions is selected by the auctioneer. Partition $k \in Gamma$ divides goods into bundles, but does not specify an agent assignment for the bundles. Notation $S \in k$ indicates that bundle S is in the partition. Let Π denote the maximal possible revenue to the auctioneer in the current round across all feasible allocations at the current prices (i.e. irrespective of the bids submitted by agents). The second important CS condition is:

$$y(k) > 0 \Rightarrow \Pi - \sum_{S \in k} p(S) = 0, \quad \forall k \in \Gamma \qquad \text{(CS-2a)}$$

In words, this condition states that the allocation must maximize the auctioneer's revenue at prices $p(S)$. Recall that the provisional allocation is selected to maximize revenue *given bids*, so it is necessary to show that the provisional allocation is maximal across all possible allocations despite this restriction. An important concept, known as *safety* is used to maintain this condition will hold (within $\min\{M, N\}\epsilon$) in all rounds of the auction. Essentially, safety ensures that the potential maximal increase in revenue due to price increases across rounds will always be matched by new (or renewed) bids from agents. Safety fails when the *exclusive-or* bids from a losing bidder are disjoint, at which point non-anonymous prices are introduced for that bidder.

Step c) can be achieved by continuing to adjust prices beyond the first round in which CE prices are determined for $CAP(\mathcal{I})$. Instead, *i*BEA continues to adjust prices until the prices are also CE prices for all subproblems without each agent in the efficient allocation. Agents are kept ignorant of this second phase, and their bids serve to provide information that eventually discounts the final payments of other agents in order to implement the VCG payments.

5.2 Multi-unit Auctions

In a multi-unit auction there is a set of K indivisible and homogeneous items, and agents have valuations, $v_i(m)$, for $m \geq 0$ units of the item. This is a special-case of the combinatorial auction problem, in which the items are identical. Useful auction designs for this problem do not simply introduce an identifier for each item and use combinatorial auctions. Rather, a useful auction allows agents to submit bids and receive price feedback expressed in terms of the quantity of units of the item.

The *efficient* multi-unit auction problem has received some attention. We consider the forward auction problem, with one seller and multiple buyers, and distinguish between two simple cases: *unit-demand* valuations and *marginal-decreasing* valuations. Both cases can be solved with iterative auctions that maintain a *single* unit price. The first case is quite straightforward, but illustrative of the primal-dual auction design method. The second case requires an innovative "clinching" auction, in which price discounts are maintained for agents during the auction [5]. This latter auction also has a direct interpretation as a primal-dual implementation for an appropriate LP formulation of the allocation problem [15].

5.2.1 Unit Demand.

In the unit demand setting each buyer demands a single item of the good and there are K items for sale. Let v_i denote agent i's value for one unit. The following LP is integral, and the solution to its dual corresponds to a competitive equilibrium price. Let $x_i \geq 0$ denote the number of units assigned to agent i.

$$\max_{x_i} \sum_{i \in \mathcal{I}} x_i v_i$$

$$\text{s.t.} \quad \sum_{i \in \mathcal{I}} x_i \leq K$$

$$0 \leq x_i \leq 1, \quad \forall i \in \mathcal{I}$$

Introducing dual variables, p and π_i, to correspond with the primal constraints, the dual formulation is:

$$\min_{p, \pi_i} \sum_{i \in \mathcal{I}} \pi_i + Ky$$

$$\text{s.t.} \quad p + \pi_i \geq v_i, \quad \forall i \in \mathcal{I}$$

$$p, \pi_i \geq 0$$

Variable, p, can be interpreted as the price on a unit, and π_i as the maximal payoff to agent i, with optimal values given price p computed as $\pi_i = \max(0, v_i - p)$. CS condition, $p > 0 \Rightarrow \sum_i x_i = K$, requires that the outcome maximizes the seller's payoff at the price and implies that the price must be zero unless all items are sold. CS conditions, $\pi_i > 0 \Rightarrow x_i = 1$ and $x_i > 0 \Rightarrow p + \pi_i = v_i$ imply that agents must receive an item if and only if the item has positive payoff at the price. These are familiar conditions for competitive equilibrium.

Moreover, as in the single-item auction in Section 2.4, the VCG outcome is implemented at the minimal CE price, which is the smallest p that corresponds to an optimal dual solution. A simple ascending auction implements a primal-dual algorithm for this problem, terminating in the VCG outcome, and with MBR a dominant strategy equilibrium [15]. The auction maintains a single ask price, p_{ask}, and allows agents to bid for a single unit in each round at the ask price. While more bids are received than there are items, K agents are selected in the provisional allocation, and the ask price is increased. The auction terminates as soon as fewer than K bids are received, with items allocated to the agents still bidding and remaining items allocated to agents that were active in the previous round, breaking ties at random.

5.2.2 Marginal-Decreasing Values.

With marginal-decreasing values, each buyer has a decreasing value for each additional unit that it receives in the assignment. We describe Ausubel's [5] efficient ascending-price auction. There is no single price that supports the efficient allocation in competitive equilibrium for this marginal-decreasing case. However, the auction is able to maintain a single price along with enough additional information to compute discounts to each agent at the end of the auction, such that the final price faced by each agent is its VCG payment. This brings a simple MBR strategy into an *ex post* Nash equilibrium. The discounts are completely analogous to the discounts in *i*BEA.

Ausubel's auction maintains a price, p_{ask}, and agents submit bids for a quantity, $q_i(p)$, in each round. The auction terminates as soon as the total quantity demanded is less than or equal to K, and otherwise the price is increased. As the price is increased and demand drops agents can "clinch" units of the item. Clinching a unit in round t locks in the price that the agent must pay for that particular unit to the current ask price. An agent clinches a unit of the item in the first round in which the demand from the other agents is low enough that the agent is sure to win the item at its current bid price, assuming that the demand from other agents will only fall as the price continues to increase. In essence, the future increase in ask price before the auction terminates represents the *discount* awarded to the agent for that item.

Table 5.1. Marginal-decreasing Multi-unit Example.

agent	$v_i(1)$	$v_i(2)$	$v_i(3)$
1	123	236	329
2	75	80	83
3	125	250	299
4	85	150	157
5	45	70	75

Consider the simple example illustrated in Table 5.1, taken from Ausubel [5]. There are 5 agents and 5 units of an item, and agent i has value, $v_i(m)$, for m units. The auction proceeds until $p_{\text{ask}} = 65$, at which time the MBR bids are $q(65) = (3, 1, 2, 1, 0)$, from agents 1,...,5 respectively. Let $c_i \geq 0$

denote the number of items clinched by agent i, initially set to zero. Agent 1 clinches the first unit, at the current price, because $c_1 + \sum_{i \neq 1} q_i(p) < K$ ($0 + 1 + 2 + 1 < 5$). The auction proceeds until $p_{\text{ask}} = 75$, at which time the MBR bids are $q(75) = (3, 0, 2, 1, 0)$. Agent 1 clinches a second unit, at the current price, because $c_1 + \sum_{i \neq 1} q_i(p) < K$ ($1 + 2 + 1 < 5$). Agent 3 clinches its first unit, at the current price, because $c_3 + \sum_{i \neq 3} q_i(p) < K$ ($0 + 3 + 1 < 5$). The auction proceeds until $p_{\text{ask}} = 85$, at which time the MBR bids are $q(85) = (3, 0, 2, 0, 0)$, and agents 1 and 2 both clinch one more unit each, at the current price. Finally, agent 1 receives 3 units, for total payment $65 + 75 + 85 = 225$, its VCG payment, and agent 3 receives 2 units, for total payment $75 + 85 = 160$, its VCG payment.

5.3 Multiattribute Auctions

Multiattribute auctions [25] extend the traditional auction setting to allow negotiation over price and attributes, with the final characteristics of the item, as well as the price, determined dynamically through agents' bids. For example, in a procurement problem, a multiattribute auction can allow different suppliers to compete over both attributes values and price.

The single-item multiattribute auction, with single-sourcing, has received the most attention in the literature. In the efficient multiattribute auction problem the goal is to find the configuration of the item and seller that maximizes the total payoff across all participants. In the optimal multiattribute auction problem the goal is to find the configuration of the item, and a price, that maximizes the payoff to the buyer across all possible auction mechanisms. We retain the notation introduced in Section 3.2.4.

5.3.1 Efficient Multiattribute Auctions.

In general, there can be *no* efficient multiattribute auction that does not run at a deficit. This follows from the Myerson-Satterthwaite impossibility result (see Section 2.2) because there is private information on two-sides of the market (for the buyer and for the sellers). In particular, the VCG mechanism runs at a deficit.

Instead, it is standard to consider a one-sided refinement to the VCG mechanism, that is budget-balanced but only approximately-efficient. In the one-sided VCG mechanism the allocation is computed as in the VCG mechanism. The only difference is that VCG payments are only provided on the sell-side, while the buyer is expected to make a payment equal to the VCG payment of the winning seller. Essentially, this leaves the buyer and the winning seller engaged in a *bargaining game* for a division of the total surplus $V(\mathcal{I})$. If the buyer reports a truthful cost function then the payoff division between the buyer and the winning seller is $[w(\mathcal{I} \setminus i^*), w(\mathcal{I}) - w(\mathcal{I} \setminus i^*)]$, where $w(L)$ is the maximal surplus with sellers $L \subseteq \mathcal{I}$ and i^* denotes the winning seller. A well-informed buyer can misstate her cost function and extract all the available surplus, to provide a payoff division of $[w(\mathcal{I}), 0]$. This difference in payoff, $w(\mathcal{I}) - w(\mathcal{I} \setminus i^*)$, places a simple and useful bound on the maximal benefit available to the buyer from deviating from a truthful strategy. As the auction becomes more competitive, the marginal product of any single seller becomes negligible and this maximal benefit tends to zero.

Parkes & Kalagnanam [77] have proposed a descending-price auction to implement the outcome of the modified VCG mechanism with MBR strategies. As with combinatorial auctions, the main advantage claimed for these price-based auctions is that they allow for incremental revelation of valuation and cost information, with price feedback guiding that process. MBR is an *ex post* Nash equilibrium strategy for the sellers against a *non-adaptive* (but perhaps untruthful) buyer strategy, and the same bounds on possible gains from manipulation are available for the buyer as in the one-sided VCG.

Of particular interest is auction ADDITIVE& DISCRETE (AD), which is designed for the special-case of preferential independence. Auction AD employs a compact price space, with linear-additive prices on attribute levels and an additional non-linear price term that applies to the bids from all suppliers. Experimental results for a simulated environment demonstrate that this price space provides a significant reduction in information revelation over an auction in which the price space maintains explicit prices on *bundles* of attribute levels.

Vulkan & Jennings [101] also describe an iterative multiattribute auction design for this problem. However, their auction is not price based and the buyer– but not the seller –must reveal her complete valuation function to the auction.

5.3.2 Optimal Multiattribute Auctions.

Che [25] has proposed optimal multiattribute auctions for the special case of seller cost functions that are defined in terms of a *single* unknown parameter. Che's auctions are direct-revelation mechanisms, and he considers both first-price and second-price variations. The second-price variation is exactly the one-sided VCG mechanism, and Che is able to derive the optimal "scoring function" (or reported cost function) that the buyer should state to maximize her payoff in equilibrium. Branco [21] extends the analysis to consider the case of correlated seller cost functions. No optimal multiattribute auction is known for a more general formulation of the problem, for example for the case of preferential-independence.

Recently, Beil & Wein [12] have proposed an *iterative* payoff-maximizing auction procedure, again for a class of parameterized cost and valuation functions with known functional forms and *naive* suppliers. The buyer restarts an auction a number of times to estimate the seller costs functions deterministically, restarting the auction with a different scoring function each time. For the final round, Beil & Wein design a scoring function that maximizes the buyer payoff by essentially reporting the same score (within ϵ) for the top two suppliers. They allow the buyer scoring function to change across rounds. The main assumption made about seller behavior is that a seller will continue to follow a straightforward bidding strategy, even though the auction is restarted many times and even though the information is used to learn their cost functions and finally extract the maximal possible payoff from the winning seller. This is clearly not a realistic equilibrium proposition.

5.4 Procurement Reverse Auctions

Reverse auctions are now routinely used for enterprise procurement. Simple single item single sourcing auctions such as the first-price sealed bid and English auctions have found use in the procurement of maintenance, repair and operations services. Increasingly, more complex formats such as combinatorial and volume discount reverse auctions are being introduced in strategic sourcing decisions for material and services used in the production process of an enterprise.

Davenport *et al.* [32] have studied the use of reverse multi-unit auctions with volume discounts in a procurement setting. Winner determination formulations are provided in Section 3.2.2, where we also discuss the introduction of business rules as side constraints, which is an important consideration in practical electronic markets. A combinatorial auction is used for single units (lot) of multiple items, with all-or-nothing bids are allowed and nonlinear prices are used to feedback information. Volume discount auctions are also used, when multiple units of multiple items are being procured and for the restricted case of bids that are separable across items. Another consideration in procurement auctions is that the outcome should be such that the final prices should be profitable for both the buyer and the suppliers, i.e. a win-win outcome. The competitive equilibrium property can be used to operationalize this notion of achieving a win-win outcome [32].

The main technical challenge in analyzing the properties of such procurement reverse auctions is that the allocation problems are integer programs and a direct appeal to primal-dual algorithms cannot be made. If an appropriate extended formulation can be identified that has the integrality property then an appeal can be made once again to primal-dual algorithms to show competitive equilibrium, as in the case of forward combinatorial auctions (see Section 5.1). The reverse combinatorial auction is fundamentally different from the forward auction because: (i) set covering formulation rather than set packing, and (ii) the inclusion of business rules as side constraints that complicate the formulation. However, extended formulations can still be derived for the reverse combinatorial auction with side constraints. This extended formulation yields dual prices on bundles of items, just in the case of the forward auction.

In this section we outline an extended formulation that can be used to provide a primal-dual based iterative descending price auction for a procurement setting. Consider an extended formulation of the allocation problem corresponding to the set covering problem (5.11). The extended formulation is defined over the space of all feasible partitions of items with agent assignments, denoted Γ_{feas}. This time, we use $k \in \Gamma_{\text{feas}}$ to denote a partitioning of items into bundles *and* an assignment to agents. Let $y(k) = 1$ iff allocation k is selected, with $k \ni (S, i)$ to denote that the allocation assigns bundle S to agent i. $v_i(S)$ is the *cost* of allocating the bundle S to agent i, and variable $x_i(S)$ takes the value 1 if the bundle S is allocated to agent i. Note that the set Γ_{feas} is chosen such that only feasible partition-assignments are allowed, respecting side-constraints in addition to item feasibility constraints. For example a partition-assignment k that violates the maximum number of winning suppliers is not considered in Γ_{feas}.

The minimization formulation shown below is integral (proof not provided here). Notice that all integrality requirements on $x_i(S)$ have been relaxed.

$$\min \sum_{S \subseteq \mathcal{G}} \sum_i v_i(S) x_i(S)$$

s.t.

$$x_i(S) \geq \sum_{k \in \Gamma_{\text{feas}}, k \ni (S,i)} y(k), \quad \forall i, \forall S$$

$$\sum_{S \subseteq \mathcal{G}} x_i(S) \geq 0 \quad \forall i$$

$$\sum_{k \in \Gamma_{\text{feas}}} y(k) \geq 1$$

$$x_i(S), y(k) \geq 0$$

The first constraint ensures that for each bundle and each agent the total allocation is at least as large as the partitions chosen containing the bundle with assignments to agent i. The second constraint ensures that the total allocation to agent i is non-negative. The third constraint required that at least one partition-assignment be chosen to satisfy demand. Now we present the dual of this formulation:

$$\max \Pi$$

s.t.

$$p_i(S) + \pi_i \leq v_i(S) \quad \forall i, \forall S$$

$$\Pi \leq \sum_{(S,i) \in k} p_i(S) \quad \forall k \in \Gamma_{\text{feas}}$$

$$\Pi, \pi_i, p_i(S) \geq 0$$

The dual variable $p_i(S)$ corresponds to the first equation in primal and similarly π_i and Π correspond to the second and third equations in primal. Now if we choose values for these dual variables as follows:

$$\pi_i = \max\left\{0, \max_{S \subseteq \mathcal{G}}(v_i(S) - p_i(S))\right\}$$

Each agent i chooses a bundle that maximizes the his/her profit, and

$$\Pi = \min_{k \in \Gamma_{\text{feas}}} \sum_{(S,i) \in k} p_i(S)$$

the buyer makes allocations to minimize cost of procurement. Notice that these choices of dual correspond to the conditions of competitive equilibria.

Choosing the dual variables in this way satisfies the complementary-slackness conditions.

$$x_i(S) > 0 \Rightarrow \pi_i + p_i(S) = v_i(S) \qquad \text{(CS-1b)}$$

$$y(k) > 0 \Rightarrow \pi_S = \sum_{(S,i) \in k} p_i(S) \qquad \text{(CS-2b)}$$

Notice that the extended formulation uses a variable for each partition-assignment pair thereby introducing price discrimination across agents. Now a descending price auction with prices on each bundle can be used to reach the competitive equilibria following a primal-dual type algorithm [32]. The bundle prices in each round converge towards a competitive equilibrium when the suppliers following a straightforward bidding strategy.

5.5 Capacity constrained allocation mechanisms

An emergent research direction is the examination of mechanisms for decentralized allocation (for multi-item procurement) in the presence of capacity constraints at the suppliers. We discuss two mechanisms that have been proposed in the literature. Both are reverse auctions with a single buyer and multiple suppliers but differ in (i) bid structure that they support and (ii) the feedback that is provided. Both mechanisms assume that a partial allocation against a bid is acceptable to the bidders.

Gallien and Wein [43] propose an iterative mechanism where the suppliers bid the unit costs for each item (strategically) and their capacity constraint (truthfully). The buyer uses this information to find a cost minimizing allocation and provides private feedback regarding potential allocation to each supplier. In addition, a *bid suggestion device* is provided by the intermediary that computes the profit maximizing bid for the supplier assuming that all other bids remain the same and that the supplier is willing to share actual unit production costs with the trusted intermediary. An important rule imposed on bidding behavior is *non-reneging* on the price of *each* item to ensure efficiency of the mechanism. An assumption made about the bidding behavior of suppliers is that they are myopic best responders (MBR), i.e. they bid to optimize profits in the next round based on the information about other bids in the current round. Under these assumptions they provide convergence bounds and an ex ante bound on the procurement cost. They use numerical simulations to show that suppliers are incented to reveal true production costs (to the intermediary) under appropriate penalties for capacity overloading.

An alternate iterative approach has been proposed recently by Dawande *et al.* [34]. They use a similar setting as Gallien and Wein but relax two fundamental assumptions: (i) they do not impose a non-reneging rule on the price for each commodity, instead require all new bids from suppliers should decrease the total procurement cost by some decrement, and (ii) they provide an oracle that is able to determine (for each supplier) whether a revised bid satisfies the cost decrement without requiring the revelation of production costs or capacity constraints explicitly. They show that that for each supplier, generating a profit maximizing bid that decreases the procurement cost for the buyer by at least δ can be done in polynomial time. This implies that in designs where the bids are not common knowledge, then each supplier and the buyer can engage in an *algorithmic protocol* to identify such proposals in polynomial number of steps. In addition, they show that such a mechanism converges to an competitive equilibrium solution where all suppliers are at their profit maximizing solution given the cost and the required cost decrement δ.

5.6 Double Auctions and Exchanges

Finally, it is interesting to consider the design of double auctions and exchanges in which their are multiple buyers and multiple sellers, present simultaneously in a market. As with multiattribute auctions, the problem that immediately arises is one of the economic impossibility of efficiency with budget-balance (by Myerson-Satterthwaite), again because both the buyers and sellers have private information about their preference structure.

In this section, we discuss settings with multi-unit homogenous item and multiple buyers and sellers, and combinatorial exchanges, with multiple heterogeneous items and bids on bundles of items. Our examples are all from the *call market* paradigm in which bids and asks are collected and then cleared periodically. We ignore any temporal aspects to the problem.

5.6.1 Double Auctions.
In a *double auction* there are multiple buyers and sellers, all interested in trading multiple units of the same item. The clearing and payment problem can be analyzed as follows. Assume that bids are sorted in descending order, such that $B_1 \geq B_2 \geq \ldots \geq B_n$, while asks are sorted in ascending order, with $A_1 \leq A_2 \leq \ldots \leq A_m$. The efficient trade is to accept the first $l \geq 0$ bids and asks, where l is the maximal index for

Table 5.2. Double auction mechanisms. The *traded* column indicates the number of trades executed, where l is the efficient number of trades. The *equil* column indicates whether the mechanism implements a dominant strategy or Bayesian-Nash equilibrium (BNE).

Name	traded	p_{buy}	p_{ask}	(EFF)	(BB)	equil	(IC)
VCG	l	$\max(A_l, B_{l+1})$	$\min(A_{l+1}, B_l)$	Yes	No	dom	yes
k-DA	l	$kA_l + (1-k)B_l$	$kA_l + (1-k)B_l$	No	Yes	BNE	no
TR-DA	$l-1$	B_l	A_l	No	Yes	dom	yes
McAfee-DA	l or $l-1$	$(A_{l+1} + B_{l+1})/2$ or B_l	$(A_{l+1} + B_{l+1})/2$ or A_l	No	Yes	dom	yes

which $B_l \geq A_l$. The problem is to determine which trade is implemented, and agent payments. In the VCG mechanism for the double auction the successful buyers make payment $\max(A_l, B_{l+1})$ and the successful sellers receive payment $\min(A_{l+1}, B_l)$. In general the VCG payments are not budget-balanced, for example with $A_{l+1} < B_l$ and $B_{l+1} > A_l$ and $A_{l+1} > B_{l+1}$.

Table 5.2 provides a summary of some of the double auction mechanisms (DAs) known in the literature. In terms of high-level properties of budget-balanced DAs, the following two characteristics are mutually-exclusive:

a) clearing the double auction to maximize the surplus given the bids and asks of agents

b) achieving incentive-compatibility, such that truthful bidding is an equilibrium strategy.

Properties a) and b) together would provide for an efficient and balanced DA, and violate the Myerson-Satterthwaite result. Notice that the TR-DA [8] and McAfee-DA [62] are truthful (in a dominant strategy), but deliberately clear the exchange to implement an inefficient trade. In comparison, the k-DA auction [104, 24, 95] clears to maximize reported surplus but is not incentive-compatible. Of course, we know that the VCG mechanism is efficient but not balanced. The parameter $k \in [0, 1]$ in k-DA is chosen before the auction begins, with the clearing price faced by all agents calculated as $kA_l + (1-k)B_l$.

The McAfee DA computes price $p^* = (A_{l+1} + B_{l+1})/2$, and implements this price if $p^* \in [A_l, B_l]$ and trades l units, otherwise $l-1$ units are traded for price B_l to buyers and A_l to sellers. The TR-DA auction implements the fall-back option of McAfee's DA. An additional DA, the α-reduction DA [8], uses the TR-DA rule with probability α, and the VCG rule with probability $1 - \alpha$. Parameter α can be chosen to make the mechanism *ex ante* balanced, and the α-reduction DA retains strategyproofness by mixing two strategyproof DAs.

5.6.2 Combinatorial Exchanges.

Parkes *et al.* [78] have suggested a family of VCG-based exchanges in which the exchange is cleared to implement the trade that maximizes reported value (or surplus). Naturally, the exchange is not incentive-compatible. The pricing problem is formulated as an LP, to constructs payments that minimize the distance to VCG payments for some metric, subject to IR and BB constraints. A number of possible distance functions are proposed, which lead to simple budget-balanced payment schemes. The authors derive some theoretical properties that hold for the rules, and present experimental results.

The pricing problem is to use the available surplus, V^*, computed at value-maximizing trade λ^*, to allocate discounts to agents that have good incentive properties while ensuring (IR) and (BB). Let V^* denote the available surplus when the exchange clears, before any discounts; let $I^* \subseteq \mathcal{I}$ denote the set of agents that trade. The pricing problem is to choose discounts, $\mathbf{\Delta} = (\Delta_1, \ldots, \Delta_I)$, to minimize the distance $\mathbf{L}(\mathbf{\Delta}, \mathbf{\Delta}_{\text{vick}})$ to Vickrey discounts, for a suitable distance function $\mathbf{L}$.

$$\min_{\mathbf{\Delta}} \quad \mathbf{L}(\mathbf{\Delta}, \mathbf{\Delta}_{\text{vick}}) \qquad \text{[PP]}$$

$$\text{s.t.} \quad \sum_{i \in I^*} \Delta_i \leq V^* \qquad \text{(BB')}$$

$$\Delta_i \leq \Delta_{\text{vick},i} \quad , \forall i \in I^* \qquad \text{(VD)}$$

$$\Delta_i \geq 0 \quad , \forall i \in I^* \qquad \text{(IR')}$$

Notice that the discounts are *per-agent*, not per bid or ask, and therefore apply to a wide range of bidding languages. Each agent may submit multiple buys and sells, depending on its bids and asks and on the bids and asks of other agents. Constraints (BB') provide worst-case (or ex post) budget-balance, and can be strengthened to allow the market-maker to take a sliver of the surplus (or inject some money into the exchange). One can also substitute an expected surplus $\overline{V}^*$ for V^* and implement average-case budget-balance.

The (IR') constraints ensure that truthful bids and asks are (ex post) individual-rational for an agent, such that an agent has non-negative utility for participation whatever the bids and asks received by the exchange. Constraints (VD) ensure that no agent receives more than its Vickrey discount. The authors consider a variety of distance functions, including standard metrics such as $\mathbf{L}_2(\mathbf{\Delta}, \mathbf{\Delta}_{\text{vick}}) = \sum_i (\Delta_{\text{vick},i} - \Delta_i)^2$ and $\mathbf{L}_\infty(\mathbf{\Delta}, \mathbf{\Delta}_{\text{vick}}) = \max_i(\Delta_{\text{vick},i} - \Delta_i)$. The $\mathbf{L}_1$ metric is not interesting, providing no distributional information because any complete allocation of surplus is optimal. Each distance function leads to a simple parameterized *payment rule*. The payment rules are presented in Table 5.3.

Each payment rule is parameterized, for example the Threshold rule, $\Delta_i^*(C_t) = \max(0, \Delta_{\text{vick},i} - C_t)$, which corresponds to L_2 and L_∞ requires a "threshold parameter", C_t. The final column in Table 5.3 summarizes the method to select the optimal parameterization for each rule. For example, the optimal Thresh-

Table 5.3. Distance Functions, Payment Rules, and optimal parameter selection methods. Constraint (BB') states that $\sum_i \Delta_i^* \leq V^*$, and $|I^*|$ (used in the Equal rule) is the number of agents that participate in the trade.

Distance Function	Name	Definition	Parameter Selection
L_2, L_∞	Threshold	$\max(0, \Delta_{\text{vick},i} - C_t^*)$.	min C_t s.t. (BB')
L_{RE}	Small	$\Delta_{\text{vick},i}$, if $\Delta_{\text{vick},i} < C_s^*$ 0 otherwise	max C_s s.t. (BB')
L_{RE2}	Fractional	$\mu^* \Delta_{\text{vick},i}$	$\mu^* = V^* / \sum_i \Delta_{\text{vick},i}$
L_{WE}	Large	$\Delta_{\text{vick},i}$, if $\Delta_{\text{vick},i} > C_l^*$ 0 otherwise	min C_l s.t. (BB')
L_Π	Reverse	$\min(\Delta_{\text{vick},i}, C_r^*)$	max C_r s.t. (BB')
-	No-Discount	0	-
-	Equal	$V^* / |I^*|$	-

old parameter, C_t^*, is selected as the smallest C_t for which the solution satisfies BB'. The optimal parameter for any particular rule is typically not the optimal parameter for another rule.

Based on analytic and experimental results, a partial ordering {Large, Threshold} ≻ Fractional ≻ Reverse ≻ {Equal, Small} is derived, with respect to the allocative-efficiency of the rules. These results were first reported by Parkes et al. [78], with the experimental results computed through a simple approximation to a restricted Bayesian Nash equilibrium. The experimental results have since received additional validation by Krych's analysis [58], in which an exact Bayesian-Nash equilibrium is computed, although for a very restricted set of agent strategies. Although Large generates slightly less manipulation and higher allocative efficiency than Threshold in the experimental tests, the Threshold discounts are quite well correlated with the Vickrey discounts, which points to the face that an agent's discount in Large is very sensitive to its bid and suggests that Large is likely to be less robust than Threshold in practice.

6. Discussion

Our goal in this chapter has been to provide a cohesive overview of the different auction mechanisms that are commonly encountered in practice and in the literature. We have emphasized a theoretical approach, and we have taken a cue from traditional mechanism design (for example via an appeal to the revelation principle) in order to understand the problem of designing incentives to solve optimization problems in settings with rational agents. We have often introduced computational techniques, for example via carefully structured bidding languages and primal-dual algorithms, in order to also achieve useful computational properties.

In closing, we find it useful to take a step back and consider the role of theoretical MD in the design of auctions and exchanges for electronic markets. Certainly, mechanism design is a powerful tool which has produced some very interesting results, both positive and negative. The standard approach to mechanism design first makes assumptions about the behavior of agents, and about the information available to agents, and then formulates the design problem as an analytic optimization problem to select the optimal mechanism subject to these assumptions. Thus, MD provides the designer with a useful "model and optimize" mindset. However, mechanism design can fail for any of the following reasons:

problem difficulty The analysis problem can be too difficult to solve analytically. Open problems include the optimal (revenue-maximizing) combinatorial auction, and the most efficient combinatorial exchange amongst the class of budget-balanced exchanges.

inadequacy of direct mechanisms Direct mechanisms are not practical in many settings. Moreover, although primal-dual methods can be used to construct indirect implementations of VCG-based mechanisms, there are no general methodologies to develop indirect implementations for other classes of direct mechanisms.

ignorance of computational considerations The analytic approach ignores the strategic, valuation, communication, and implementation complexity of mechanisms. For example, perfect rationality assumptions are implicit within the inclusion of incentive-compatibility constraints in a model.

It seems possible that computational methods can be used to begin to address the first problem, that of the difficulty of the mechanism design problem.

Conitzer & Sandholm [31] propose *automated mechanism design*, in which a computational method is used to design mechanisms with respect to highly-enumerative description of the function space and agent type space. The challenge in this automated MD approach is to develop structured representations of the problem to constrain the input size to the optimization. However, automated MD cannot solve the wider issues presented by the second two problems because it is only applicable to direct revelation mechanisms and because it continues to ignore computational considerations in the formulation of the problem.

Experimental mechanism design (e.g. [82]) presents an alternative paradigm, in which computational methods are used to determine the performance of a particular mechanism design. The design problem can then be formulated as a search through a structured space, itself chosen to be expressive enough to capture a class of interesting mechanisms. Experimental MD remains within the spirit of classic mechanism design because it continues to seek to maximize performance with respect to beliefs about the way that self-interested agents will participate. However, the analysis of a mechanism is done through explicit computation instead of the imposition of incentive-compatibility constraints at design time.

The methodology of experimental MD mirrors that of experimental economics, in which experiments with human subjects in carefully controlled laboratory settings are used to test theoretical predictions and to assess the robustness of a mechanism to unmodeled behaviors [[64, chapter 1]]. There is a growing recognition of the importance of experimental methodologies within the study of human economies. Indeed, Al Roth, an economist involved in the design of real-world markets such as those used in the medical resident matching program has recently advocated an "economics as engineering" approach [88] to market design. It seems interesting to turn to computational methods to test and design mechanisms that will be used by computational trading agents.

Experimental MD offers a number of benefits over both analytic MD and automated MD. First, it extends to indirect mechanisms, because the design space is limited only to mechanisms for which equilibrium behavior can be computed through computational methods. Second, considerations of agent bounded-rationality and off-equilibrium play can also be considered explicitly within the model, again because computational methods are used to design and evaluate the mechanism. In particular, we are not limited to incentive-compatible design in the framework of experimental MD. However, experimental MD presents three main challenges. Briefly, the challenges (and some initial directions) are those of:

evaluation Take a mechanism description and compute the performance of the mechanism with respect to models of agent self-interest and rationality. Current approaches proposed for this problem include using *genetic programming* primitives to evolve agent trading strategies [82], and methods to compute a restricted Bayesian-Nash equilibrium, including *replicator dynamics* [102], *fictitious play* [107], and *best-response dynamics* [58, 85].

optimization Implement the mechanism optimizer, which searches in the space of the mechanism description language for a good mechanism. One preliminary approach proposed for this problem is to allow *genetic programming* to evolve the rules of a parameterized family of mechanisms [82, 27].

description Define a mechanism description language, to provide the interface between the mechanism optimizer and the black box mechanism evaluator. Wurman *et al.* [106] have proposed a parameterization of the auction design space, and there has been some work to develop a declarative approach to the specification of negotiation structures and auction rules [86] and to develop ontologies for automated negotiation [99].

Although each of these problems has received some attention in isolation in recent years there certainly remain significant computational difficulties if progress is to be made in this agenda of experimental mechanism design. In the meantime it seems likely that continued progress will be made in introducing computational considerations into the mechanism design program, and in the constructive application of mechanism design theory to electronic market design.

Notes

1. A mechanism must be able to make a commitment to use these rules. Without this commitment ability the equilibrium of a mechanism can quickly unravel. For example, if an auctioneer in a second-price auction cannot commit to selling the item at the second-price than the auction looks more like a first-price auction [83].

2. Later, in discussion of optimal mechanism design, we will fall back on the more general framework of randomized allocation rules and expected payments. For now we choose to stick with deterministic allocations and payments to keep the notation as simple as possible.

3. *Ex ante* and *interim* refer to timing within the mechanism. *Ex ante* constraints are defined in expectation, before agent types are known. *Interim* constraints are defined relative to the type of a particular agent, but in expectation with respect to the types of other agents.

4. As a special case, we get the celebrated *revenue-equivalence theorem* [100, 65], which states that the most popular auction formats, i.e. English, Dutch, first-price sealed-bid and second-price sealed-bid, all yield the same price on average in a single item allocation problem with symmetric agents. This is an immediate consequence because these auctions are all efficient in the simple private values model.

5. This implies a valuation $v_i(k, \theta_i) = v_i(S')$ on allocation k, where $S' = \cup_{(S,i) \in k} S$ is the union of all bundles allocated to agent i in allocation k.

6. Indeed, a *limited* combinatorial auction, in which operators can submit a restricted number of bids, has operated for the competitive tendering of London bus routes since 1997 [23].

7. Interestingly, it is also possible to carefully combine two dynamic programming tableaus in order to approximate the VCG payments for an asymptotic cost of $O(T \log n)$, instead of the typical additional worst-case cost of a factor of $O(n)$.

8. For homogeneous items, the continuous double auction maintains a queue of bids from buyers sorted in increasing order of price and a queue of offers from the sellers in decreasing order of price. Whenever the offer price is lower than the bid price the bid and ask are matched and the difference is usually kept by the market maker. This requires maintaining a sorted list of asks and bids which is of $O(N \log N)$ where N is the number of active asks/bids.

9. Truth-revelation is defined as a bid in which an agent reports value $v_i(S) = \max_{S' \subseteq S} v_i(S')$ for all bundles S permitted in the language.

10. The Federal Communications Commission maintains a URL with documents and discussion on the design of auction #31 at http://wireless.fcc.gov/auctions/31/releases.

11. A simple extension, *i*Bundle & Adjust [80, 74], brings MBR into equilibrium for the agents-are-substitutes case.

References

[1] K Ahuja, T L Magnanti, and J B Orlin. *Network flows*. Prentice-Halls, Englewood Cliffs, 1993.

[2] Arne Andersson, Mattias Tenhunen, and Fredrik Ygge. Integer programming for auctions with bids for combinations. In *Proc. 4th International Conference on Multi-Agent Systems (ICMAS-00)*, 2000.

[3] A Archer and E Tardos. Frugal path mechanisms. In *Proc. 13th Symp. on Discrete Algorithms*, pages 991–999, 2002.

[4] Aaron Archer, Christos Papadimitriou, Kunal Talwar, and Eva Tardos. An approximate truthful mechanism for combinatorial auctions with single parameter agents. In *Proc. 14th ACM Symposium on Discrete Algorithms*, 2003.

[5] Lawrence M Ausubel. An efficient ascending-bid auction for multiple objects. *American Economic Review*, 2002. Forthcoming.

[6] Lawrence M Ausubel and Peter Cramton. The optimality of being efficient. Technical report, University of Maryland, 1999.

[7] Lawrence M Ausubel and Paul R Milgrom. Ascending auctions with package bidding. *Frontiers of Theoretical Economics*, 1:1–42, 2002.

[8] Moshe Babaioff and Noam Nisan. Concurrent auctions across the supply chain. In *Proc. 3rd ACM Conference on Electronic Commerce*, pages 1–10. ACM Press, 2001.

[9] J S Banks, J O Ledyard, and D Porter. Allocating uncertain and unresponsive resources: An experimental approach. *The Rand Journal of Economics*, 20:1–25, 1989.

[10] Y Bartal, R Gonen, and N Nisan. Incentive compatible multi unit combinatorial auctions. Technical report, The Hebrew University of Jerusalem, 2003.

[11] J J Bartholdi. The computatonal difficulty of manipulating an election. *Social Choice and Welfare*, 6:227–241, 1989.

[12] Damian R Beil and Lawrence M Wein. An inverse-optimization-based auction mechanism to support a multi-attribute RFQ process. Technical report, Operations Research Center, MIT, 2001.

[13] Martin Bichler and Jayant Kalagnanam. Winner determination problems in multi-attribute auctions. Technical report, IBM Research report, 2002.

[14] Martin Bichler, Jayant Kalagnanam, and Ho Soo Lee. RECO: Representation and Evaluation of Configurable Offers. Technical Report RC 22288, Jan, IBM Research report, 2002. in Proceedings of ICS 2003.

[15] Sushil Bikhchandani, Sven de Vries, James Schummer, and Rakesh V Vohra. Linear programming and Vickrey auctions. In Brenda Dietrich and Rakesh Vohra, editors, *Mathematics of the Internet: E-Auction and Markets*, pages 75–116. IMA Volumes in Mathematics and its Applications, Springer-Verlag, 2001.

[16] Sushil Bikhchandani and Joseph M Ostroy. Ascending price Vickrey auctions. *Games and Economic Behavior*, 2002. Forthcoming.

[17] Sushil Bikhchandani and Joseph M Ostroy. The package assignment model. *Journal of Economic Theory*, 2002. Forthcoming.

[18] Liad Blumrosen and Noam Nisan. Auctions with severely bounded communication. In *Proc. 43rd Annual Symposium on Foundations of Computer Science*, 2002.

[19] Craig Boutilier. Solving concisely expressed combinatorial auction problems. In *Proc. 18th National Conference on Artificial Intelligence (AAAI-02)*, July 2002.

[20] Craig Boutilier and Holger Hoos. Bidding languages for combinatorial auctions. In *Proc. 17th International Joint Conference on Artificial Intelligence (IJCAI-01)*, 2001.

[21] F Branco. The design of multidimensional auctions. *RAND Journal of Economics*, 28:63–81, 1997.

[22] Paul J Brewer. Decentralized computation procurement and computational robustness in a smart market. *Economic Theory*, 13:41–92, 1999.

[23] Estelle Cantillon and Martin Pesendorfer. Combination bidding in multi-unit auctions. Technical report, HBS and LSE, 2003.

[24] K Chatterjee and W Samuelson. Bargaining under incomplete information. *Operations Research*, 31:835–851, 1983.

[25] Y K Che. Design competition through multidimensional auctions. *RAND Journal of Economics*, 24:668–680, 1993.

[26] E H Clarke. Multipart pricing of public goods. *Public Choice*, 11:17–33, 1971.

[27] D Cliff. Evolution of market mechanism through a continuous space of auction types. Technical Report Technical Report HPL-2001-326, Hewlett Packard Labs, 2001.

[28] Olivier Compte and Philippe Jehiel. On the virtues of the ascending price auction: New insights in the private value setting. Technical report, CERAS and UCL, 2000.

[29] Wolfram Conen and Tuomas Sandholm. Preference elicitation in combinatorial auctions. In *Proc. 3rd ACM Conf. on Electronic Commerce (EC-01)*, pages 256–259. ACM Press, New York, 2001.

[30] V Conitzer and T Sandholm. Complexity of manipulating elections with few candidates. In *Proc. 18th National Conference on Artificial Intelligence (AAAI-02)*, July 2002. .

[31] V Conitzer and T Sandholm. Complexity of mechanism design. In *Proc. 18th Conf. on Uncertainty in Artificial Intelligence (UAI'02)*, 2002.

[32] Andrew Davenport, Abigail Hohner, and Jayant Kalagnanam. Combinatorial and quantity discount procurement auctions with mutual benefits at Mars, Inc. *Interfaces*, 2002. Forthcoming.

[33] Andrew Davenport and Jayant Kalagnanam. Price negotiations for procurement of direct inputs. In Brenda Dietrich and Rakesh Vohra, editors, *Mathematics of the Internet: E-Auction and Markets*. IMA Volumes in Mathematics and its Applications, Springer-Verlag, 2001.

[34] Milind Dawande, R. Chandrasekharan, and Jayant Kalagnanam. On a question in linear programming and its application in decentralized allocation. Technical report, IBM research report, Jul 2002.

[35] Sven de Vries and Rakesh V Vohra. Combinatorial auctions: A survey. *Informs Journal on Computing*, 2002. Forthcoming.

[36] Gabrielle Demange, David Gale, and Marilda Sotomayor. Multi-item auctions. *Journal of Political Economy*, 94(4):863–872, 1986.

[37] C DeMartini, A M Kwasnica, J O Ledyard, and D Porter. A new and improved design for multi-object iterative auctions. Technical Report SSWP 1054, California Institute of Technology, 1998. Revised March 1999.

[38] Wedad Elmaghraby and Pinar Keskinocak. Combinatorial auctions in procurement. Technical report, National University of Singapore, 2002.

[39] Marta Eso, Soumyadip Ghosh, Jayant R Kalagnanam, and Laszlo Ladanyi. Bid evaluation in procurement auctions with piece-wise linear supply curves. Technical Report RC 22219, IBM TJ Watson Research Center, 2001.

[40] Joan Feigenbaum, Arvind Krishnamurthy, Rahul Sami, and Scott Shenker. Approximation and collusion in multicast cost sharing. In *Proc. of the 3rd Conference on Electronic Commerce*, pages 253–255, 2001.

[41] Joan Feigenbaum and Scott Shenker. Distributed Algorithmic Mechanism Design: Recent Results and Future Directions. In *Proceedings of the 6th International Workshop on Discrete Algorithms and Methods for Mobile Computing and Communications*, pages 1–13, 2002.

[42] Yuzo Fujishima, Kevin Leyton-Brown, and Yoav Shoham. Taming the computational complexity of combinatorial auctions: Optimal and approximate approaches. In *Proc. 16th International Joint Conference on Artificial Intelligence (IJCAI-99)*, pages 548–553, 1999.

[43] Jeremie Gallien and Lawrence Wein. Design and analysis of a smart market for industrial procurement. Technical report, Operations Research Center, MIT, 2000.

[44] Alan Gibbard. Manipulation of voting schemes: A general result. *Econometrica*, 41:587–602, 1973.

[45] I Gilboa and E Zemel. Nash and correlated equilibria: Some complexity considerations. *Games and Economic Behavior*, pages 213–221, 1989.

[46] Robert L Graves, Linus Schrage, and Jayaram Sankaran. An auction method for course registration. *Interfaces*, 23(5):81–92, 1993.

[47] Jerry Green and Jean-Jacques Laffont. Characterization of satisfactory mechanisms for the revelation of preferences for public goods. *Econometrica*, 45:427–438, 1977.

[48] Theodore Groves. Incentives in teams. *Econometrica*, 41:617–631, 1973.

[49] Faruk Gul and Ennio Stacchetti. The English auction with differentiated commodities. *Journal of Economic Theory*, pages 66–95, 2000.

[50] R Holzman, N Kfir-Dahav, D Monderer, and M Tennenholtz. Bundling equilibrium in combinatorial auctions. *Games and Economic Behavior*, 2001.

[51] Holger H Hoos and Craig Boutilier. Solving combinatorial auctions with stochastic local search. In *Proc. 17th National Conference on Artificial Intelligence (AAAI-00)*, July 2000.

[52] B Hudson and T Sandholm. Effectiveness of preference elicitation in combinatorial auctions. In *Proc. Agent-Mediated Electronic Commerce (AMEC'IV) workshop at AAMAS'02*, 2002.

[53] Matthew O. Jackson. Mechanism theory. In *The Encyclopedia of Life Support Systems*. EOLSS Publishers, 2000.

[54] Jayant R Kalagnanam, Andrew J Davenport, and H S Lee. Computational aspects of clearing continous call double auctions with assignment constraints and indivisible demand. *Electronic Commerce Journal*, 1(3):221–238, 2001.

[55] Alexander S Kelso and Vincent P Crawford. Job matching, coalition formation, and gross substitutes. *Econometrica*, 50:1483–1504, 1982.

[56] Anshul Kothari, David C. Parkes, and Subhash Suri. Approximately-strategyproof and tractable multi-unit auctions. In *Fourth ACM Conf. on Electronic Commerce (EC'03)*, pages 166–175, 2003.

[57] Vijay Krishna and Motty Perry. Efficient mechanism design. Technical report, Pennsylvania State University, 1998. Available at: http://econ.la.psu.edu/~vkrishna/vcg18.ps.

[58] David Krych. Calculation and analysis of Nash equilibria of Vickrey-based payment rules for combinatorial exchanges. Technical report, Harvard University, 2003. Undergraduate Thesis. Available from http://www.eecs.harvard.edu/econcs.

[59] John O Ledyard, Mark Olson, David Porter, Joseph A Swanson, and David P Torma. The first use of a combined value auction for transportation services. *Interfaces*, 2000. Forthcoming.

[60] Daniel Lehmann, Liadan Ita O'Callaghan, and Yoav Shoham. Truth revelation in approximately efficient combinatorial auctions. *Journal of the ACM*, 49(5):577–602, September 2002.

[61] Silvano Martello and Paulo Toth. *Knapsack Problems*. WileyIntersciences Series in Discrete Mathematics and Optimization, 1980.

[62] R Preston McAfee. A dominant strategy double auction. *J. of Economic Theory*, 56:434–450, 1992.

[63] John McMillan. Selling spectrum rights. *Journal of Economic Perspectives*, 8:145–62, 1994.

[64] Paul Milgrom. *Putting Auction Theory to Work*. MIT Press, 2002.

[65] Robert B Myerson. Optimal auction design. *Mathematics of Operation Research*, 6:58–73, 1981.

[66] Robert B Myerson and Mark A Satterthwaite. Efficient mechanisms for bilateral trading. *Journal of Economic Theory*, 28:265–281, 1983.

[67] Noam Nisan. Bidding and allocation in combinatorial auctions. In *Proc. 2nd ACM Conf. on Electronic Commerce (EC-00)*, pages 1–12, 2000.

[68] Noam Nisan and Amir Ronen. Computationally feasible VCG mechanisms. In *Proc. 2nd ACM Conf. on Electronic Commerce (EC-00)*, pages 242–252, 2000.

[69] Noam Nisan and Amir Ronen. Algorithmic mechanism design. *Games and Economic Behavior*, 35:166–196, 2001.

[70] Noam Nisan and I Segal. The communication complexity of efficient allocation problems. Technical report, Hebrew University and Stanford University, 2002.

[71] C H Papadimitriou. Algorithms, games and the Internet. In *Proc. 33rd Annual ACM Symp. on the Theory of Computing*, pages 749–753, 2001.

[72] David C Parkes. *i*Bundle: An efficient ascending price bundle auction. In *Proc. 1st ACM Conf. on Electronic Commerce (EC-99)*, pages 148–157, 1999.

[73] David C Parkes. Optimal auction design for agents with hard valuation problems. In *Proc. IJCAI-99 Workshop on Agent Mediated Electronic Commerce*, pages 206–219, July 1999. Stockholm.

[74] David C Parkes. *Iterative Combinatorial Auctions: Achieving Economic and Computational Efficiency*. PhD thesis, Department of Computer and Information Science, University of Pennsylvania, May 2001. http://www.cis.upenn.edu/~dparkes/diss.html.

[75] David C. Parkes. Price-based information certificates for minimal-revelation combinatorial auctions. In *Agent Mediated Electronic Commerce IV: Designing Mechanisms and Systems*, volume 2531 of *Lecture Notes in Artificial Intelligence*, pages 103–122. 2002.

[76] David C Parkes. Auction design with costly preference elicitation. Submitted for publication, 2003.

[77] David C Parkes and Jayant Kalagnanam. Iterative Multiattribute Vickrey Auctions. Submitted for publication, 2003.

[78] David C Parkes, Jayant R Kalagnanam, and Marta Eso. Achieving budget-balance with Vickrey-based payment schemes in exchanges.

In *Proc. 17th International Joint Conference on Artificial Intelligence (IJCAI-01)*, 2001.

[79] David C Parkes and Lyle H Ungar. Iterative combinatorial auctions: Theory and practice. In *Proc. 17th National Conference on Artificial Intelligence (AAAI-00)*, pages 74–81, July 2000.

[80] David C Parkes and Lyle H Ungar. Preventing strategic manipulation in iterative auctions: Proxy agents and price-adjustment. In *Proc. 17th National Conference on Artificial Intelligence (AAAI-00)*, pages 82–89, July 2000.

[81] David C Parkes and Lyle H Ungar. An ascending-price generalized Vickrey auction. Technical report, Harvard University, 2002.

[82] S Phelps, P McBurney, S Parsons, and E Sklar. Co-evolutionary auction mechanism design: A preliminary report. In *Proc. Agent-Mediated Electronic Commerce (AMEC'IV) workshop at AAMAS'02*, 2002.

[83] R Preston McAfee and John McMillan. Auctions and bidding. *Journal of Economic Literature*, 25:699–738, June 1987.

[84] S J Rassenti, V L Smith, and R L Bulfin. A combinatorial mechanism for airport time slot allocation. *Bell Journal of Economics*, 13:402–417, 1982.

[85] Daniel M. Reeves and Michael P. Wellman. Computing equilibrium strategies in infinite games of incomplete information. In *Proc. of Game-theoretic and Decision-theoretic Workshop at AAMAS'03*, 2003.

[86] Daniel M Reeves, Michael P Wellman, and Benjamin N Grosof. Automated negotiation from declarative descriptions. In *Proc. 5th International Conference on Autonomous Agents (AGENTS-01)*. 2001.

[87] Amir Ronen. Mechanism design with incomplete languages. In *Proc. 3rd ACM Conf. on Electronic Commerce (EC'01)*, 2001.

[88] A E Roth. The economist as engineer. *Econometrica*, 70:1341–1378, 2002.

[89] Michael H Rothkopf, Aleksandar Pekeč, and Ronald M Harstad. Computationally manageable combinatorial auctions. *Management Science*, 44(8):1131–1147, 1998.

[90] Tuomas W Sandholm. Issues in computational Vickrey auctions. *International Journal of Electronic Commerce*, 4:107–129, 2000.

[91] Tuomas W Sandholm. eMediator: A next generation electronic commerce server. *Computational Intelligence*, 18:656–676, 2002.

[92] Tuomas W Sandholm and Subhash Suri. Market clearability. In *Proc. 17th Int. Joint Conf. on Artificial Intelligence*, 2001.

[93] Tuomas W Sandholm, Subhash Suri, Andrew Gilpin, and David Levine. CABOB: A fast optimal algorithm for combinatorial auctions. In *Proc. 17th Int. Joint Conf. on Artificial Intelligence*, 2001.

[94] Tuomas W Sandholm, Subhash Suri, Andrew Gilpin, and David Levine. Winner determination in combinatorial auction generalizations. In *Proc. International Conference on Autonomous Agents, Workshop on Agent-based Approaches to B2B*, pages 35–41, 2001.

[95] Mark A Satterthwaite and Steven R Williams. Bilateral trade with the sealed bid k-double auction: Existence and efficiency. *Journal of Economic Theory*, 48:107–133, 1989.

[96] James Schummer. Almost dominant strategy implementation. Technical report, MEDS Department, Kellogg Graduate School of Management, 2002.

[97] Jeff Shneidman and David C. parkes. Rationality and self-interest in peer to peer networks. In *2nd Int. Workshop on Peer-to-Peer Systems (IPTPS'03)*, 2003.

[98] Jeffrey Shneidman and David C. Parkes. Using redundancy to improve robustness of distributed mechanism implementations. In *Fourth ACM Conf. on Electronic Commerce (EC'03)*, 2003. (Poster paper). Extended version at http://www.eecs.harvard.edu/~parkes/pubs/redundancy.pdf.

[99] Valentina Tamma, Michael Wooldridge, and Ian Dickinson. An ontology based approach to automated negotiation. In *Proc. Agent-Mediated Electronic Commerce (AMEC'IV) workshop at AAMAS'02*, 2002.

[100] William Vickrey. Counterspeculation, auctions, and competitive sealed tenders. *Journal of Finance*, 16:8–37, 1961.

[101] N Vulkan and N R Jennings. Efficient mechanisms for the supply of services in multi-agent environments. *Decision Support Systems*, 28:5–19, 2000.

[102] Michael P Wellman, Amy Greenwald, Peter Stone, and Peter R Wurman. The 2001 Trading Agent Competition. In *Proc. 14th Conf. on Innovative Applications of Art. Intell.*, pages 935–941, 2002.

[103] Steven R Williams. A characterization of efficient, Bayesian incentive-compatible mechanisms. *Economic Theory*, 14:155–180, 1999.

[104] R Wilson. Incentive efficiency of double auctions. *Econometrica*, 53:1101–1115, 1985.

[105] Peter R Wurman and Michael P Wellman. AkBA: A progressive, anonymous-price combinatorial auction. In *Second ACM Conference on Electronic Commerce*, pages 21–29, 2000.

[106] Peter R. Wurman, Michael P. Wellman, and William E. Walsh. Specifying rules for electronic auctions. *AI Magazine*, 2002.

[107] Weili Zhu and Peter R. Wurman. Structural leverage and fictitious play in sequential auctions. In *Proc. 18th National Conference on Artificial Intelligence (AAAI-02)*, pages 385–390, July 2002.

David Simchi-Levi, S. David Wu, and Z. Max Shen (Eds.)
Handbook of Quantitative Supply Chain Analysis:
Modeling in the E-Business Era
©2004 Kluwer Academic Publishers

Chapter 6

AUCTIONS AND PRICING IN E-MARKETPLACES

Wedad Elmaghraby
Georgia Institute of Technology
Atlanta, GA 30332
wedad@isye.gatech.edu

1. B2B E-Marketplaces

By offering high-speed communication and tight connectivity, advances in information technology have opened new venues for companies to create flexible supply chains. Today, many companies, from the electronics, pharmaceutical, to the automotive industry, are focusing on their core competencies and outsourcing significant portions of their business operations. As supply chains become more decentralized upstream, the pricing of intermediate goods is no longer a formality used between departments - but rather the key to a company's survival. Strangely enough, while companies have spent millions of dollars to help them reduce their operating costs, the majority continue to use crude techniques in deciding what price to charge for their products Anthes, 1999.

The increased adoption of revenue management pricing policies, along with the emergence of Internet enabled marketplaces, referred to as e-marketplaces, is one of two separate but interrelated phenomenom that is helping to change business practices. Sophisticated capacity allocation strategies, the essence of current revenue management tools, are burgeoning on the business-to-consumer (B2C) front. Designed for markets with perishable products and limited supply, revenue management tools aid companies in *dynamically* selecting the optimal mix of products to offer to customers. This is done by allowing customers to make advance bookings for different products (fare classes) and dynamically changing the availability of each product McGill and van Ryzin, 1999 Phillips, 2003. While revenue management tools focus on capacity allocation decision, they have generally ignored related pricing decision. That is, revenue management tools take the prices for different products as given,[1] rather than simultaneously determining the optimal price for each product with the capacity allocation.

E-marketplaces allow companies to go one step beyond traditional revenue management techniques, by offering a natural medium to optimally develop sophisticated pricing policies. Via e-marketplaces, suppliers (and buyers) can reach larger markets, dynamically change prices as the need and opportunity arise, and, most importantly, gain vital information about demand elasticity to incorporate into their pricing decisions.

While companies have embraced revenue management and e-marketplaces as essential to their future growth on the B2C front, they have been slow to alter their current business-to-business (B2B) practices. In particular, companies have been slow to use e-marketplaces as a transaction medium, due to the slow pace of technology diffusion and the complexity of most B2B transactions.[2] While B2C transactions can generally be characterized as involving spot transactions of a known (small) size, B2B transactions are generally governed by contracts over an extended period time with uncertain demand and entail identifying multi-attributes in addition to price, e.g., transportation, quality, ordering and replenishment procedures, etc. Therefore, except for the sale of some commodity parts and excess inventory, e-marketplaces have played a minimal role in most B2B transactions Latham, 2000. The fact that many companies do not currently use the Internet to help them with their B2B pricing decisions does not mean that there is no future potential nor enormous gains from doing so. It is anticipated that the value of products and services sold in B2B markets through electronic channels in the United States will reach $4.7T by 2005 AMR, 2001.

The purpose of this paper is to bring to the surface the current pricing practices in B2B e-marketplaces and the challenges in successfully implementing pricing policies in practice. In particular, we focus on two pricing strategies that are natural candidates for B2B e-marketplaces; *auctions* and *precision pricing* (referred to as 3rd degree price discrimination in the economics literature). Auctions, whereby prices are determined via a bidding process, have already made some headway in B2B e-marketplaces. For example, retailers in footware, home products and fashion are using GlobalNetXchange private auction exchange Rosenblum, 2002, auto manufacturers are using Covisint's auctions capabilities Prouty, 2001 and GE uses its own Global Exchange Services Barling, 2001 to help procure goods more effectively from suppliers. Precision pricing entails quoting each customer-type a different take-it-or-leave-it price based on relevant and observable characteristics. Precision pricing's presence is more sparse in B2B e-marketplaces, but is increasingly being adopted in off-line sales channels with an eye towards adoption in e-marketplaces.

By presenting case studies of companies that design and implement auctions and precision pricing tools, we hope to gain a better understanding as to the future role and contributions the ORMS community can have in developing and executing these pricing policies. We begin by providing an overview of the pricing policies used currently in e-marketplaces.

2. Current State of Pricing in B2B Marketplaces

To better understand the current pricing practices and future potential for pricing policies in e-marketplaces, it is helpful to categorize market environments according to (i) the *size* and *frequency* of the transaction, (ii) the ability for arbitrage and (iii) the information asymmetry in the marketplace. The *size* and *frequency* of the transaction determines whether or not it is cost-effective to tailor a pricing policy to that specific transaction. For large and infrequent transactions, it is typically worthwhile to take the time and money to gather information about the marketplace and design a pricing policy accordingly. On the other hand, for small and frequent transactions, it is too costly and inefficient to specifically design a pricing policy to the particular transaction. Rather, a pricing policy should dynamically response to underlying changes in the marketplace in a (semi-)automated manner. The ability for *arbitrage* and the *information asymmetry* in the marketplace critically influence the degree to which the seller can practice price discrimination, i.e., charge different prices to different customers. It is difficult for the seller to price discriminate if a product can be easily sold in a resale market (for the market segment who buys at a lower price can sell to other market segments), if customers can easily acquire information concerning the true value/cost of a product, or if the seller has little information about customer valuations. However, if different market segments can be identified across which resale is difficult, then a seller can charge different prices to each group.

During the late 1990's, a host of B2B e-marketplaces emerged in both the service and goods industries. Companies in the transportation, plastics, chemical, and paper were among the first to build and use B2B e-marketplaces; in these industries we saw the creation of Cargonow.com, The PlasticsExchange.com, ChemConnect.com and Paperloop.com.

One of the first uses for these e-marketplaces was for the sale of excess inventory (region A in figure 6.1). Sellers in these markets who found themselves with unsold units eagerly turned to e-marketplaces as a medium via which they could reach a larger and previously untapped market and derive a higher salvage value for their items. In effect, e-marketplaces served as 'garage sales' for these sellers. These garage sales were initially operated as a bulletin board, whereby sellers post their products and request price. With a bulletin board, a transaction takes place as soon as one customer who is willing to pay the seller's requested price contacts the seller. This method, while reaching a larger customer base, does not allow sellers to use competition to drive up the selling price of their products. Therefore, the second type of pricing policy used in these garage sales is an auction. The most commonly used auction format is an ascending price English auction, whereby the auction terminates when bidding ceases and the product(s) is awarded to the highest bidder(s) at the highest bid price. In an ascending price English auction, interested buyers submit increasingly higher bids. A key attribute of these garage sales (posted price or auction) is that they are managed by the seller, i.e., the e-marketplace merely serves as a host site.

Alongside the sale of excess inventory via garage sales, e-marketplaces began to build and run double auctions for commodities, similar in nature to the Chicago Mercantile Exchange (region B in figure 6.1). It is often difficult to limit arbitrage with pure commodities, therefore a pricing mechanism that yields the same price to all customers at any one point in time is most appropriate. In a double auction, buyers (sellers) submit their bid (ask) prices indicating the price they are willing to pay (receive) for a specified quantity. The market maker aggregates the supply and demand and determines a market price that clears the market and reflects the 'true value' of the product. With commodity products, there are few if any buyer-supplier specific attributes aside from price that influence the transaction decision. This is in sharp contrast to customized products, for which identifying compatible suppliers as well as supplier specific characteristics such as quality and service can play an equally if not more important role than price. Double auctions are actively managed by the e-marketplace.[3]

The third type of pricing policy now entering e-marketplaces are customized auctions (region C in figure 6.1). An estimated 80-90% of all B2B transactions are done via long-term contracts Economist, 2001. Given the vast sums of money that are contracted upon and the multi-attribute nature of many B2B transactions, it is clear that these less frequent and larger transactions cannot be handled by standard double auctions nor self-managed garage sales. Rather, they need a pricing mechanism that is rich in its dimensions so that it can be specifically tailored to the situation at hand. That is, the pricing mechanism should be more 'hands on' and allow for the proper specification of the product, be flexible enough to reflect various market participant characteristics and adapt to changing circumstances in the marketplace. Customized auctions fit the above description and are being used to help buyers establish relationships with suppliers. This vast sum of untapped money has not gone unnoticed by makers of e-marketplaces; customized procurement auctions for large and infrequent transactions require significant auction and industry expertise in their design. In response to this need, we are beginning to see companies such as

Auctions and Pricing in E-Marketplaces 217

FreeMarkets, Logistics.com[4], ChemConnect.com and CombineNet offer companies support in the design and implementation of these auctions.[5]

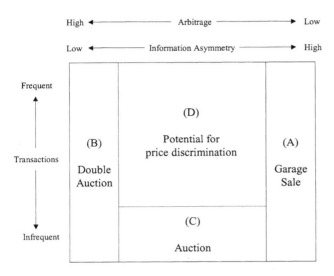

Figure 6.1. Current pricing practices in B2B e-marketplaces

Figure 6.1 illustrates the placement of each of the current pricing practices along these three market dimensions; arbitrage, information asymmetry and frequency of transactions, while figure 6.2 provides an example of the pricing policies for the transportation industry. In the transportation industry, Logistics.com aids shippers who wish to procure long-term transportation contracts with multiple carriers via combinatorial auctions.[6] With their OptiBid procurement system, Logistics.com aids shippers design and run these large (fully personalized) procurement auctions, by helping them create appropriate transportation networks, send auction information to and receive bids from the carriers, and optimize which carriers to select given the submitted bids (see region A of figure 6.2). Although most shippers and carriers enter into long-term contracts for the bulk of their business, imbalances in supply and demand often occur, giving rise to either empty carrier capacity or excess shipments that need to be transported. The National Transportation Exchange (NTE) helps to fill in these imbalances in supply and demand by offering a dynamic, automated transportation marketplace, NTE Public Exchange, where carriers can submit empty or partially filled capacity, and shippers can submit excess de-

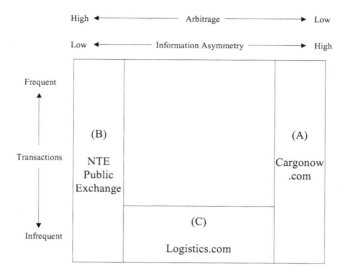

Figure 6.2. E-Marketplaces currently being used in the transportation industry

mand (region B in figure 6.2). The NTE Public Exchange is in essence a double auction over multiple products (a product being an origin-to-destination route); it matches the submitted demand and supply and determines a transaction price for each match obtained. A shipper or carrier that prefers to post its own spot requirements (and circumvent the transaction fee associated with using a public exchange such as NTE's) can use Cargonow.com's website (region A in figure 6.2).

Currently, e-marketplaces have captured the extremes of the market: There are auctions (region C) for large and infrequent transactions, double auctions (region B) for the sale of high arbitrage/ low information asymmetry (commodity) products and garage sales (region A) for the sale of high information asymmetry products such as excess inventory. What we do not observe are e-marketplaces and pricing policies for medium sized transactions where arbitrage is difficult and the seller has some information about customer valuations (region D).

In selecting a pricing policy for region D, a seller must tradeoff two main factors: (i) the *simplicity* to develop and administer the pricing policy and (ii) the *profitability* of the pricing policy. As a general rule, the more complex is a market environment, the greater the potential rewards from the use of a sophisticated pricing strategy. While double auctions for commodities and 'garage sales' are relatively simple to develop and administer, they do not take advantage of customer heterogeneity and seller information by price discrimination. One particular pricing policy that lends itself to region D market transactions is *precision pricing*. Precision pricing allows a seller to tailor his prices to different segments of his market, thereby exploiting differences in customer valuations by charging them different prices.

Where a company is able to use individualized prices and/or is able to effectively segment customers, we foresee large infrequent transactions remaining in the realm of auctions, while precision pricing will be employed for smaller more frequent transactions. Precision pricing, along with customized auctions, offer its user the promise of increased profitability. But the promise of increased profitability does not come without serious challenges. A key lesson learned from early entrants in B2B e-marketplaces is that the interconnectiveness of the Internet is not sufficient to make on-line pricing a successful venture. What is necessary is the know-how behind selecting the appropriate pricing strategy and the expertise to properly design and implement it.

To better understand the design dimensions and implementation challenges present in the use of auctions and precision pricing, we present below a few pricing 'success stories'. In section 3 we discuss the multiple decisions that FreeMarkets, a premiere customized auctions company, faces when designing a procurement auction.[7] In section 4, we discuss the design and use of a bidding support software, developed by Manugistics, to aid bidders participating in a request for quotes. In section 6, we present a description of Manugistic's precision pricing tool, the challenges in properly implementing third degree price discrimination. We conclude each topic with future directions of research for the field of ORMS.

3. Customizing Auctions - A Case Study of FreeMarkets

In the 'new' economy, customized auction market makers see themselves as transformers of supply chains. Where previously supply chains were rigid with fixed suppliers and uncompetitive procurement practices, auction market makers infuse change by breaking down barriers that obscure information, bringing new suppliers, technology and expertise to monitor the supply chain, and streamlining the procurement process, thereby allowing it to be more agile and transparent, and further rewarding efficient and competitive suppliers. One of the leading customized auction market makers in B2B e-marketplaces is FreeMarkets.

As noted by Jason Busch, Senior Manager of Strategic Marketing at FreeMarkets.[8]

"The Center for Advanced Purchasing Studies and AT Kearney did a study of all the potential ways that savings can be generated from procurement. Small gains can be achieved by running operations more efficiently. There are savings to be had in reducing inventory, cutting purchasing headcount, and ensuring all MRO buys are made on contract. But the real bang for the buck is reducing the costs of direct and indirect purchases in an organization...73% of purchasing costs are in sourcing; i.e., in what is being bought...Every $1 saved in sourcing is worth an estimated $5-$25 in increased sales."

Started in 1995, FreeMarkets was one of the first companies to offer customized auction *services* to industrial customers on a large scale. The emphasis on 'services' is clear to anyone who has designed or participated in a large industrial auction. The actual auction is the final step of the auction design process; in order for it to be a success, a substantial amount of information and work must be done beforehand in preparing both the buyers and suppliers. FreeMarkets' main focus is on helping companies through the entire sourcing process, starting from adequately specifying what product(s) the buyer wishes to purchase, identifying potential suppliers on a global scale, and designing the appropriate auction mechanism for the market transaction. Their primary product, Full Source, is a full-service product which includes FreeMarkets expertise in 8 vertical industries (Electronics, Engineering and Construction, Metals, Paper & Packaging, Plastics, Raw Materials, Maintenance Repairs Operations & Services, and Transportation), as well as an auction software which brings together supplier information and market making services in e-space. Currently, FreeMarkets has 125 customers, including Tier 1 and 2 automotive companies and oil companies such as BP Amoco.

While there are several sizeable challenges to making one of their auctions work, FreeMarkets has enjoyed considerable success in the auctions business. To their customer's procurement process they bring transparency, a list of 14,000 suppliers all over the globe, and cost competitiveness. In addition, they bring a reduction in the time it takes to procure. In a world where time is money, FreeMarkets' ability to significantly reduce procurement cycle times allows their customers to bring products to their market more quickly. For example, one of FreeMarkets' clients, Emerson, was able to reduce its procurement cycle time reduction from 12-21 weeks down to 8-12 weeks, with the use of FreeMarkets' Full-Source. The savings from these cycle time reductions can range anywhere from 5-15% based on the industry and products being auctioned.[9]

Before holding an on-line auction event, FreeMarkets must go through four critical preparation steps: (i) fully specifying all dimensions to a buyer's entire order, (ii) standardizing information across suppliers, (iii) monitoring and identifying supply market trends, and (iv) selecting an appropriate auction format. Surprisingly, many buyers will initiate a procurement process without a clear and definite idea of what it is they wish to purchase. While they may know the general characterization of the product, they often have not thought through clearly all the products' dimensions and supply specifications, such as delivery dates, order size, and required service levels. An ill-defined order can lead to misunderstanding and uncompetitive bidding. In order to make their sourcing process a success, FreeMarkets spends a considerable amount of time *detailing* the design of the product, i.e., pinpointing the exact specifications of the product the customer wishes to procure. On the opposite side of the market, the supplier base is often quite diverse, be it with respect to geographic location, technology, production capabilities and/or quality.[10] Therefore, FreeMarkets must *standardize* the product information, e.g. by defining a minimum threshold level of service and scope so as to level the playing field for the suppliers and allow the bids in the auction to be meaningful. Given the dynamic nature of the industries in which FreeMarkets operates, they carefully *monitor* and *identify* market trends so as to identify eligible suppliers to invite to submit bids and develop a better handle on their cost structures and capacity availability. For example, suppliers may change in their product/service offerings, the geographical markets in which they are operating, and their performance levels. Finally, once the product and potential suppliers have been specified, auction makers, who understand the supplier market, work with engineering experts to select the optimal auction format. With various options with respect to auction format (e.g., open or sealed-bid, qualifying round), bid format, and feedback format (discussed further below), FreeMarkets has over 30 auction formats. Their strength and success lies in their ability to properly match the market setting with an auction format.

3.1 Auction Formats

If an auction is designed "properly", it holds the promise of achieving a buyer's objectives of cutting down procurement costs and/or awarding business to "healthy" suppliers with whom she can enter into long-term strategic partnerships. Many of the companies (including United Technologies, Quaker Oats, General Electric, and Owens Corning) who have turned to FreeMarkets to help them in reducing their procurement costs have reported significant savings as a result of the on-line customized auction. In order to attain these savings, FreeMarkets has had to tailor the format of each auction to the situation at hand. There are four main auction design dimensions that must be carefully selected for the success of an auction: (i) how to *bundle* the buyer's demand and construct bidding lots, (ii) how to *sequence* the lots' auctions, (iii) what type of feedback to provide to bidders during the auction, referred to as *feedback to the marketplace* and (iv) what type of bid format, or *bid calculation form*, to use.

Given the multi-unit/product nature of most of their buyer's demand, FreeMarkets must decide which units/products to *bundle* together and have suppliers bid on as an entire package, or lot. It is often advantageous to bundle pieces of an order together either because the items experience synergies in production or because packaging allows a desirable job to be bundled with an undesirable one.[11] Due to supplier capacity constraints FreeMarkets can not typically bundle all of a buyer's order together into one package. FreeMarkets must trade off increasing the size of the bundle and allowing for more economies of scale in production with reducing the number of suppliers who have sufficient capacity to supply the entire bundle and hence can compete in the auction. Asymmetric supplier capacities further exacerbate the bundling decision. In order to determine the best number and size of lots to create, the auction maker must have a solid understanding of the bidding suppliers' production capabilities (technology), available capacity, and cost structure.

Once the components of the bundles are determined, FreeMarkets must decide on the *sequencing* of the auction. The lots on which suppliers are submitting bids typically consist of large complex orders that require the supplier to collect a substantial amount of information and resources so as to submit an appropriate bid. FreeMarkets has observed that, typically, suppliers dislike participating in more than one auction simultaneously; they prefer instead to focus their bidding on one bundle at a time, with the winner in each auction being notified at the end of the auction. As a result, FreeMarkets staggers the closing of auctions and faces the challenge of deciding in which order to sequence them. Although they do not have a general policy for ordering, FreeMarkets typically closes the largest (major) lots last.[12]

The amount of information that is observed by the auction participants can play a pivotal role in combating supplier collusion during an auction and getting reluctant suppliers to participate in the on-line auction. Therefore, the *feedback to the marketplace* is a critical auction dimension. The majority of FreeMarkets' on-line auctions are open or iterative auctions as opposed to sealed-bid auctions. The preference for open/iterative auctions stems from the belief that the psychology of an open auction induces suppliers to bid more aggressively. In special cases where FreeMarkets is procuring from a very small and collusive supplier base, e.g. the sugar market where there are only 2 main suppliers, FreeMarkets reverts to the use of a sealed-bid auction in order to reduce the information exchange and the ability of suppliers to collude and maintain high prices in the auction.

In open/iterative auctions, FreeMarkets has four feedback formats; (i) full disclosure, (ii) rank, (iii) next horse, and (iv) starting gate. Ideally, FreeMarkets would like to always provide the bidders with *full disclosure*, where bidders can see all the current bids as well as who has placed them.[13] Their preference for full disclosure is based on their experience that, in a competitive supplier market, an auction performs best when as much information as possible is made known to the bidders. Unfortunately, many supplier markets, such as metals, ocean freight, construction and air parcel, are not very competitive. In these markets, suppliers are aware of the opportunities for collusion in a full disclosure auction and the possible repercussions associated with participating in an auction.[14][15] Therefore, in an effort to combat collusion and encourage supplier participation in their auctions, FreeMarkets uses an alternative feedback format, a *rank* feedback auction. A rank auction allows all the bidders to see their and their competitors' rankings, but not the submitted bids. Rank and next horse feedback are used when the buyer is purchasing from a supplier market that is particularly disparate in cost or quality. Under *next horse feedback* format, bidders can only see the bid that is directly in front of (below) them. In addition, FreeMarkets will often usually use a *starting gate feedback* form to supplement the above three feedback forms. With starting gate feedback, bidders are disallowed from seeing anything until they place a bid. This is used to avoid suppliers from promising that they will participate in the auction, only to use the auction as a means to gain information about the marketplace without ever placing a bid.

The selection of *bid calculation* format depends on whether the suppliers are competing to sell a commodity part for which the identity of the supplier is of little importance, or if the suppliers, and their products, are heterogenous. The four bid calculation formats that are used most frequently are (i) index, (ii) descending, (iii) transformational and (iv) net present value (NPV). An *index* bid is commonly used in the sourcing of a raw material. Often raw materials are traded on a public exchange, such as NYMEX, and hence have a publicly observable market price. With an index bid, suppliers bid how much of a discount below the index they are willing to sell the product. A *descending* bid auction is the format most commonly associated with a procurement auction - suppliers must submit increasingly lower bids to win. This bid calculation format is appropriate when the suppliers' products are fairly homogeneous and hence can be compared solely on price. A *transformational* bid is a means of standardizing bids across heterogenous suppliers. For example, the suppliers may differ depending on location, technology capabilities, and/or quality. With a transformational bid, each supplier has associated with it a transformation factor for each of the criteria deemed important in the supplier selection process. These criteria and transformational factors are agreed upon by FreeMarkets and the buyer in advance of the on-line auction, and the transformational factors are either added or multiplied to the supplier's bid, so as to create comparable bids. While each supplier is made aware of the criteria that will be taken into consideration with a transformation bid, they do not know the exact individual-supplier-level factor associated with each criterion. *NPV* bid format is typically used when the buyer and supplier will enter into a longer-term contract (e.g. 3 years), and the supplier wishes/needs to change its bid over time. For example, a buyer and supplier may enter into a contract for a good where there is learning by doing in production. NPV allows the supplier to reflect any potential cost reductions into its bid by incorporating a discount rate; the suppliers' asking price is then discounted accordingly over time.

In addition to the four auction dimensions mentioned above, FreeMarkets must decide whether or not to hold a qualifying bid round, impose individual reserve prices, the size of bid decrements, and the appropriate stopping rule. In the *qualifying round*, suppliers submit bids on individual items and/or lots via a binding one-shot sealed bid auction conducted off-line. A qualifying round is typically used when there is some uncertainty in the marketplace which FreeMarkets and the buyer were unable to resolve during the preparation period. For example, a qualifying round may be used to help FreeMarkets identify (i) the degree of competition for a particular product market via the level of interest, i.e., the number and level of bids submitted; (ii) which suppliers to invite to participate in the on-line auction; (iii) which items are best grouped together into a lot; and (iii) appropriate (possibly individual-supplier) reserve prices for the on-line auction. In selecting the size of the *bid decrements*, the auction maker must trade-off the rate at which the auction progresses versus the ability of suppliers' to reflect their true costs.[16] Generally, the auction's *stopping rule* is that only a price reduction on the lowest bid will extend the auction into overtime. However, when the lowest price is not visible to all market participants, such as in the rank feedback form, FreeMarkets may relax the stopping rule to any of the first n^{th} lowest bids can extend the auction's closing time.

3.2 Combating Collusion

One of the biggest challenge that FreeMarkets faces is designing an auction in a small and potentially collusive supplier market. In these markets, on-line and semi-transparent auctions offer a great opportunity for reducing a buyer's procurement costs. But in order for the cost savings to be realized, FreeMarkets must convince all or most eligible suppliers to participate. FreeMarkets' recent experiences with designing auctions for the ocean freight and air parcel industries illustrate the many hurdles encountered in designing auctions for potentially collusive supplier markets, with the outcome not always being a success.

FreeMarkets recently conducted an auction for a large global customer who needed ocean carrier services over routes stretching all over North America, Europe, Asia and Africa. The ocean carrier(s) would transport a variety of products for the buyer as well as handle many of the services associated with international trade, such as customs clearance. The ocean freight market is dominated by a few large (global) carriers, who handle anywhere from 30-40% of the global ocean freight business, but have much higher concentrations on a few key niche routes, and many smaller local carriers. The presence of the largest three to five Tier 1 ocean carriers was deemed as necessary for the auction's success. While the buyer would need to procure the services of multiple carriers, due to the size and scope of its demand, it wished to minimize the number of carriers with whom it had to deal.

Upon initial invitation (40 carriers were invited to participate), the majority of the Tier 1 carriers declined to participate in the on-line auction. The carriers were opposed to an on-line semi-transparent auction for multiple reasons. Their main concerns were that (i) the transparency of the bids on the auction would carry over and result in negative (more competitive) effects on their other business, and (ii) an on-line auction would result in suppliers being selected on price alone, and hence result in a commoditization of their services. In the ocean freight industry, multiple criteria factor into the performance and suitability of a supplier. Carriers may differ in the speed in which they clear customs, their turn-around time in a port, and the priority they assign a buyers' containers (when capacity is tight). The Tier 1 carriers feared that competing against Tier 2 low-cost, low-service carriers would not allow them to enjoy a justifiable price mark-up for their superior services.

By carefully designing the auction format, and speaking to the carriers individually, FreeMarkets was finally able to convince all of the Tier 1 carriers to participate. Since information transparency was one of the biggest concerns of the carriers, FreeMarkets signed a non-disclosure agreement with the carriers, stating that they would not reveal who participated in the auction. For the auction, they decided to use a rank feedback format to minimize the price information that was fed back to the marketplace and a transformation bid to take into account the service dimensions to an ocean carrier. FreeMarkets spent a considerable amount of time explaining to the carriers how a transformation bid would be calculated, and stated upfront the various criteria that would be considered in the bid calculation. All carriers were informed that their past performance with regards to the number of damaged goods, number of days of container free time, and industry-adopted service levels would be used to scale their bid. In addition, before the carriers committed to participate, FreeMarkets ran a mock (non-binding) event to demonstrate how the auction would work.

Once the carriers agreed to participate, FreeMarkets held a qualifying round to determine the supplier interest and competitive landscape on all of the lanes. The sheer number of lanes being auctioned (over 1500 lanes) complicated the auction lot decisions. While FreeMarkets had spent a couple of months reviewing the buyer's aggregate ocean traffic data over the previous year to establish traffic patterns and lanes (origin-destination points) requirements, they still did not have a solid understanding of which lanes/routes were attractive to suppliers and competitive vs. non-competitive. Previously, the buyer had delegated their freight requirement decisions to regional office managers. As a result, the buyer did not have a good understanding of its own aggregate demand requirements, and possible synergies in combining lanes. In the qualifying round, all of the lanes were made available for bidding on an individual basis. Based on the observed bids, FreeMarkets developed a lotting strategy for the various lanes and established carrier-specific reserve prices to be used in the on-line auction. The rationale behind carrier-specific prices was that using the lowest submitted bid as a general reserve price for all bidders would not take into consideration non-price factors, such as quality. As a general rule, anyone who is invited to participate in the qualifying round could participate in the on-line auction. However, if a carrier submitted a bid that was an outlier, it was not invited to participate in the on-line auction.

For the on-line auction, given the vast number of lanes being auctioned, FreeMarkets held three separate auction events; one for the North America, European and Asian lanes, respectively. Each auction event lasted two days. Within each auction, the sequencing of the lots' closing times was not believed to be of critical importance. Each lot was assigned a scheduled closing time; however, bidding on the lot could go into overtime if there was any change in the rank at the first and second level. If the rank changed, then the auction was extended by 60 seconds.

The on-line auction was deemed a success by the buyer. Although the rank feedback format and transformational bids took out much of the transparency out of the auction, they were necessary for the Tier 1 carriers to agree to participate, which, in turn, was necessary for the auction to be a success. By carefully selecting the criteria used in the transformational bids, FreeMarkets was able to bring price competition into a potentially collusive supplier base, while incorporating the numerous non-price attributes that critically impact the buyer-supplier relationship.

While removing some transparency from the auction was sufficient to persuade the top carriers in ocean freight to participate, it was not sufficient for FreeMarkets' experience with the air parcel industry. The US domestic air parcel industry is dominated by two companies, FedEx and UPS. When FreeMarkets was approached by a global buyer to design an auction for their air parcel requirements, FreeMarkets knew that the key to the auction's success would be attracting FedEx and UPS both to participate. The buyer preferred to procure services from only one company, so as to facilitate the tracking of packages. FedEx and UPS did agreed to participate in a qualifying round. To further entice both companies to participate in the on-line auction, FreeMarkets stated upfront that, if FedEx or UPS chose not to participate in the on-line auction, then their bids from the qualifying round would be used as the (general) reserve price for the on-line auction. As with the ocean freight event, FreeMarkets designed a transformational bid, rank feedback auction. Despite the buyer's large presence in the US (30% of its expenditure in the US), and FreeMarkets' reserve price policy, the buyer's business represented only a small fraction of FedEx's or UPS's total market and hence FedEx and UPS refused to participate in the on-line auction. FreeMarkets' experiences with the ocean freight and air parcel industry illustrate that, while customizing an auction to the particular event can help make it a success, an auction event can easily fail due to supplier wariness of auctions and reluctance to participate.

4. Bidder Support in Auctions - Manugistics' NetWORKS Target Pricing^{TM}

Given that the desired supplier base will participate in an auction, it is imperative that the suppliers have a solid understanding of their costs, capabilities, and the competition they face in order to make an auction an ultimate success. While FreeMarkets aids buyers in designing an auction, it is also important to help suppliers with their bid preparation process. For large and infrequent purchases, it is often worthwhile for the suppliers to hire consultants to aid them in identifying their optimal bid(s). For example, in the recent FCC spectrum auctions, many of the large telecommunication companies submitting bids hired leading game theorists and economic consulting companies to help them in their bid formation process. But for small(-er) and more frequent purchases, it behooves the suppliers to have access to a (semi-automated) bidding decision support system. As we discuss below, the need for bidding decision support is very real and is the focus of Manugistics' NetWORKS Target Pricing^{TM} software.

Manugistics, a leader in supply chain management solutions software, acquired Talus, a pricing and revenue optimization company, in 2001. The reason behind this purchase was very clear: In order to better serve their customers, Manugistics needed to be able to offer them pricing services to complement the existing planning and supply chain optimization solutions. Using advanced mathematical algorithms, Manugistics now helps customers in industrial, technology, shipping and automotive industries to determine what prices to offer to which customer segments and how to maximize profitability by managing demand in response to customer and market differences and to supply constraints. Manugistics' product, NetWORKS Target Pricing^{TM} (NTP), aids suppliers in putting together a bid for an auction. NTP was developed for use in industries where the business practice was to make price offers on a bundle of products in a contractual arrangement, one customer at a time. NTP takes into consideration the specific buyer, the buyer's order, the supplier's history with the buyer, and most importantly, competitors for the buyer's business, their previous bidding behavior and winning record when calculating an appropriate bid (or range of bids) for an auction.

NTP attempts to simplify the clients bidding problem by aiding them in systematically evaluating their *win probability* given a particular bid with a particular buyer. The decision support software assumes that if a bid is submitted, the supplier has sufficient capacity to supply the buyer's entire order and that the order utilizes only a small fraction of the supplier's capacity, i.e., NTP does not incorporate any supply constraints.[17] Rather the software focuses on determining the optimal range of prices from which the supplier should submit a bid, with each price corresponding to a different win probability.

It is a complex problem for the supplier to decide what offer to make to a particular customer. While the supplier's decision problem is theoretically simple, i.e., to determine the price offer that will maximize the expected contribution per unit = Pr(winning the bid)*(unit contribution), the difficulty lies in accurately predicting the probability of winning. Except in the case of commodity products, suppliers are almost never evaluated on price alone. Rather, a list of supplier characteristics, such as size, geographic location, whether or not the buyer has done business with the supplier before, etc., are taken into account in conjunction with price to determine which supplier is best suited for a buyer. In order to make a bidder support tool operational, a supplier must be able to estimate and maintain a parameterized win probability equation that predicts win probability as a function of the characteristics of the offer itself, the customer and the market.

This process of determining the appropriate probability functions is comprised of two distinct steps. Initially, Manugistics receives from its client a list of its (the client's) customers, previous auctions in which it participated, the bids it submitted, which of its competitors were present and their bids (if available) and the auctions' outcomes. From this list, the first step is to identify the best set of predictors of win probability and their interactions. That is, Manugistics must identify attributes of the buyer and market that are important predictors of the probability of winning and for which there exists the least intra-segment heterogeneity in outcomes (both wins and losses), and the greatest inter-segment heterogeneity. These attributes can include characterization of the buyer itself, such as size, the buyer's valuation on various supplier attributes and its current supplier(s), as well as characterizations of the buyer's marketplace, such as the number of competing suppliers and their historical bidding behavior and winning record. In addition to buyer attributes, product attributes that influence the probability of winning are also identified and their effect quantified, such as the length of the contract and which package of items is being requested by the buyer.

Once the key attributes are identified, the second step is to estimate the parameters of the win probability model, using binary dependent variable regression techniques. The output of the regression model is a coefficient for each attribute, measuring the effect of that attribute on the probability of winning. Specifying the predictor attributes and estimating coefficients are undertaken at the time that the model is initialized; then periodically the steps are repeated to ensure that the model specification and parameterization remains fresh. With the repeated use of NTP, it is possible to test and refine its prediction accuracy. This is done by comparing the aggregated predictions of the model to the percentage of wins (orders actually won via an auction) actually observed. In order to make a valid comparison, the predictions are computed at the prices actually offered. By tracking differences in the predicted win probability at offered prices to the observed outcomes, the managers of the tool can detect the need to update the model coefficients.

Once NTP is installed and the prediction model is configured, a supplier can use NTP to compute the optimal bid price for a specific customer. The software has the flexibility to incorporate optional side constraints, such as constraining the price offer to be within a certain range, perhaps a range defined in terms of the price which that particular customer paid on his last contract; or constraining the win probability to be no less that some value.

NTP is currently used by a firm that supplies shipping services to businesses. The shipping firm makes tens of thousands of offers per year to thousands of different customers and prospects. There are a relatively small number of products that are offered to each customer, differing by the type of shipping (e.g. bulk versus priority documents); the destinations (e.g. domestic versus international); speed (e.g. overnight versus two-day); and ancillary services (e.g. delivery notification). The shipping firm makes over 60,000 offers per year to over 35,000 different customers and prospects. Customers differ from each other in terms of the mix of products and the volume of shipping services that they require. The also differ in terms of their own characteristics, such as the valuation that they put on reliability, their current supplier of shipping services and the competitive alternatives that are available to them. The challenge facing the shipping firm is which bid price to submit for a particular bundle of products for a particular customer.

At the time an offer is formulated, the firm has information about the customer, including the product mix and volume requested, and certain characteristics of the customer and the market. The customer characteristics include the customer's line of business, its size, and its history of buying shipping services from the firm. The market's characteristics include the firm's competitors and an estimate of their likely offers. Since the offers are private, and there is little risk of arbitrage amongst customers, the firm is essentially free to customize the price offer that it makes to each single customer, based on this information. When a salesperson makes a call on a prospect, he determines the prospect's product mix and volume requirements. These are supplied to NTP through the firm's order management system, which retrieves relevant characteristics of the prospect and the market. Based on this information, the models in NTP will return the optimal price offer to make to that customer, and ancillary information including the expected likelihood of winning the bid.

In addition to the financial benefits from offering better bids, Manugistic's customers have reported both (i) a dramatic reduction in the frequency of appeals to higher authority in the marketing organization by the sales staff and (ii) a dramatic reduction in the time it takes to quote a customer a price when using NTP. The range of prices supplied by NTP, where each price corresponding to a different win probability, provides sales agents with the flexibility and support to bid more or less aggressively for a customer's business without needing to get re-approval for price changes. Although NTP has been primarily used in sealed-bid request-for-quote settings, its input and output structure readily lend itself for use in open auction formats.

While the benefits of such a pricing tool are potentially great, there remain many challenges to its proper implementation. A critical step in determining an appropriate bid is the correct determination of the win probability function. Currently, there are large information gaps that hinder the proper specification of the probability function. Through their various implementations of NTP, Manugistics has observed time and again that companies maintain poor records of their costs and orders. In particular, poor record-keeping is prevalent when a supplier's offer loses to a competing bid. Without 'clean' data and a clear understanding of what their product is, in which product market space it lies, and who the potential competitors are, a customer-specific pricing model is severely weakened - and can potentially hurt profits.

5. Future Directions for Research in Auction Theory

Customized auctions, when designed properly, have the potential to help buyers significantly reduce their procurement costs. The growing area of research in auctions is helping us develop a better grasp on what are the critical factors when deciding on the design of the auction. There remain, however, many disconnects between the practice of designing and implementing 'real' auctions and the bulk of the auction literature.[18] These disconnects leave several interesting directions for future research: Listed below are some of the directions we believe to be the most pressing areas for research.

Product Specification A big challenge that customized auction makers face in delivering to their customer a successful sourcing experience is understanding the buyer's specifications, its effect on the potential supplier market and the optimal auction design. The bulk of the auction literature examines auctions in an environment where the product is standard and well-identified and suppliers' costs are exogenously determined. However, a buyer can influence a supplier's cost by the way that she specifies the product(s) she wishes to procure. For example, the terms of delivery, frequency of replenish, and quality of the products are all factors that can influence a supplier's cost. A serious omission from the procurement auctions literature is the important interaction between the detailing of the product, its impact on the eligible supplier base, suppliers' costs and the appropriate design of an auction.

Bundling In addition to product specification, a particularly interesting challenge posed by these B2B auctions is that they typically involve the exchange of multiple products/goods. For example, in timber, pharmaceutical, industrial parts, and electronic auctions, buyers typically wish to procure multiple units of a product, and more often than not, many different types of products via a single auction event. Unfortunately, the research in auction theory has traditionally focused on the use of single-unit auctions. One might be tempted to generalize the results of the single-object auction to the multi-object framework; however, this trust in generalizations would be misplaced. Recent work in the area of multi-object auctions has indicated that they are fundamentally different than single-object auctions and hence generalizations are invalid Ausubel and Cramton, 2002 Back and Zender, 1993 Bikhchandani, 1999 Engelbrecht-Wiggans and Kahn, 1998. In settings where suppliers experience synergies across multiple products or units of a product, the auctioneer must decide how to bundle the units together into lots. As noted earlier, the manner in which products are bundled may affect suppliers' costs as well as their ability to bid on any particular bundle. For example, by what criterion should a buyer decide which products to bundle together; competitiveness of the supplier base, cost complementarities, and/or profitability of the product? Given the multi-unit/product nature of most B2B transactions, and potential asymmetries in suppliers' ability to compete for different products, we must gain a better understanding of the impact of product bundling on supplier bidding behavior. [19]

Transformational Factors and Scoring Rules One of the biggest challenges in designing and auction is bringing into consideration the multiple factors that influence a buyer-supplier relationship.[20] Only in special cases (e.g. purchase of a commodity product) is a buyer wise to judge a supplier on the basis of price alone. Typically, many other criteria are important in the supplier selection process; for example, a supplier's quality, reliability, size, degree of vertical integration, and available capacity. Once a buyer has identified which supplier characteristics are of utmost importance to her, she must decide on how much of her preferences to reveal to suppliers and an appropriate way to judge the suppliers on the multiple attributes. Cognizant of the multi-attribute nature of suppliers and the supplier selection process, both FreeMarkets' and Manugistics' tools try to incorporate relevant supplier attributes into the pricing decision process. There exists a small but growing literature on multidimensional auctions and scoring rules, used to transform the multiple attributes to a single number or 'score' is small Athey and Levin, 1998 Bichler, 2000 Bushnell and Oren, 1994 Chao and Wilson, 2001 Che, 1993 Koppius and van Heck, 2002. Given the importance of non-price factors in the selection of suppliers, there is a great need for research that studies how the scaling and scoring of non-price attributes can affect the performance of an auction, particularly when it is difficult to observe and verify these non-price attributes.

Repeat Interaction of Suppliers and Collusion Suppliers who are part of a relatively small supplier base are bound to compete against one another over time. However, with few exceptions Elmaghraby, 2003b Katzman, 1999 Klotz and Chatterjee, 1995 Milgrom, 2000, the auction literature almost always assumes a one-shot framework, ignoring the strategic factors that may arise in a repeated interaction framework. If an auction is to be conducted in a small supplier base, then it must take into account the presence of future auctions on a supplier's bidding behavior in the current auction and the possible presence of collusion or bidding-rings. For example, if the buyer's order constitutes a significant portion of a supplier's capacity, the supplier may not be able to participate in future auctions. Furthermore, if the suppliers are part of a bidding-ring, then they may not participate under certain auction formats (or may be uncompetitively), as experienced by FreeMarkets.[21] Therefore, an important area for future research is the design and bidding behavior of suppliers who repeatedly interact over time.

Auctions in Supply Chains An additional complicating factor in B2B auctions is that buyers and sellers typically interact across multiple channels in the supply chain. For example, a supplier may sell to a distributor (the buyer in the auction) as well as compete with the distributor in selling to the final customers. To the best of our knowledge, the research in auction theory has ignored the competing interests of different echelons in a supply chain, and its effect on the optimal design and performance of an auction.

6. Precision Pricing - Manugistics' Networks Precision Pricing

While the majority of revenue management tools continue to focus on capacity allocation, research has already begun to explore the role of simultaneous price and inventory optimization (please see Elmaghraby and Keskinocak, 2003 Bitran and Caldentey, 2002 for reviews of the literature). The general approach in this area of research is to assume that *any* customer arriving a particular point in time will see the same price; the question then becomes how to set price given existing inventory conditions, future demand, and the possibility (or lack thereof) of replenishment. While valuable and necessary building blocks for increasing our understanding of pricing in the presence of inventory constraints, this line of research ignores the fact that a seller may want to price discriminate across *customer types*, in addition to across time.

Many sellers operate in an environment in which they face heterogeneous customer types, i.e., their willingness-to-pay and demand elasticity are sufficiently heterogeneous, and hence wish to develop customer-type specific prices. If the seller receives buyer inquiries on a very frequent basis, an auction is a costly and time consuming pricing mechanism. In such an environment, an appropriate pricing policy should be dynamic, not incur large costs each time a transaction takes place, and should factor in the relevant characteristics of the marketplace (such as customer types) in determining the optimal prices. One pricing policy that possesses all of these characteristics is precision pricing.

Precision pricing entails simultaneously quoting each customer-type a different price based on relevant and observable characteristics and supply conditions. In order to successfully implement precision pricing, it is imperative that (i) the seller has some information about a customer-type's willingness to pay (ii) the customer market is comprised of more than one customer type (where customers' demands differ between types), (iii) the seller has a low-cost way to identify in which group a particular customer belongs and (iv) the seller is able to prevent or limit arbitrage across customer groups.

Precision pricing is a widely practiced and accepted pricing policy on the B2C front; for example, clothing stores charge a lower price at their outlet stores (charging a different price based on sales channel) and hotels offer a discounted rate to AAA members (charging a different price based on customer membership to a group). While some companies have practiced precision pricing in their B2B transactions, they have typically done so without any decision support system. Rather, they have relied on the expertise of their marketing departments to make the 'right deal' and quote the 'right price'. Given the numerous products companies typically sell, often in the thousands, the different types of customers in the marketplace, the duration of B2B contracts and the uncertainty surrounding a buyer's actual demand, a seller's pricing problem is extremely complex. Hence, the absence of decision support systems has typically left 'money on the table'. The practice of precision pricing poses several challenges, and a considerable amount of expertise is needed both in selecting a pricing policy and in periodically (or preferably continuously) evaluating its appropriateness and modifying it accordingly. Currently, Manugistics has developed a pricing solutions software, Networks Precision Pricing (P2) to aid companies in their practice of precision pricing. P2 was developed for use in industries where the business practice was to post and maintain prices that are available to any customer that qualified (as opposed to determined on a customer by customer basis).

P2, similar to traditional revenue management tools, is designed to be used in an environment where a seller experiences a stochastic arrival of heterogeneous customer orders. Both types of tools are built on the premise that there is sufficient customer heterogeneity in the buyers' market, the ability to limit arbitrage, as well as sufficient information to accurately sort customers into different segments. The main difference between P2 and revenue management tools is that P2 *explicity* incorporates the relationship between price and demand, and hence prices, not capacity allocations, are dynamically updated over time to maximize profits. Using historical information on customers' purchasing behavior, current production commitments and expected future demand, P2 helps companies develop and maintain over time prices for *each segment* of their customers types. It does so by including an optimization program (a static linear program) that determines a schedule of spot prices (one for each customer segment) that will maximize total expected profit, given each segments' estimated sensitivity to price, the expected demand and supply over the planning horizon, as well as the current (committed) production orders and the company' costs. The spot price is valid until the program is run again and an alternative optimal price schedule is determined; this updating may occur daily or weekly. Optional side constraints may be present, such as constraining the price or the expected sales to be within specified limits, or constraining the relationship between the different segment prices for a given product. In addition to supplying a price recommendation, P2 supplies price-sensitive demand forecasts and provides a means for analysts to interact with the model parameters and price recommendation.

There are two demand-side models that must be parameterized and maintained in order to operationalize the software for a particular client. The first model is the price sensitivity model which captures demand response, by market segment, to changes in each product's price. This price sensitivity model is a composite demand model that reflects general market conditions, including the presence of competitors and their influence on demand elasticity. In order to estimate price sensitivity, Manugistics and the client must first agree upon the proper customer segmentation. This is equivalent to determining how many different price points the client can support and how the client can easily 'qualify' customers for a particular price. The customer segments should reflect classes of customers whose underlying demand elasticity is different and for which eligibility can be easily determined.[22] In a supply chain setting, the size of the customers, their place in the supply chain, their end markets are all examples of observable characteristics that may correspond to different market segments. In designing the customer segments, the ability of customers to arbitrage, i.e., resell the products to customers who face a higher price, must also be taken into account.

Once the proper customer segmentation has been decided upon, the price sensitivity of each customer segment to the various products must be calculated. This is equivalent to determining the shape of each segment's demand curve for a product. In addition to price-elasticity differing across customer segments, it can also differ across products. In order to determine the demand curve, relevant market characteristics must be identified. For example, if some of the products are "near-commodities", i.e., traded in a highly competitive market, then customers will be very price sensitive. Likewise, if the product is a non-commodity, then the seller will have some market power and customers will be less price sensitive. Once the seller has selected a segmentation for their customers and products, it is crucial to validate the segmentation with historical data. Manugistics runs statistical tests to see how consistent the client's classification of customers and products is with the data. Similar to revenue management tools, P2 does not explicitly take into account the presence of competitors, but rather includes competitors presence in their estimation of demand elasticity.

The second model which must be parameterized is the time-series forecast model. This is equivalent to determining the location (placement) of the demand curve for each customer segment for each product. The placement of the demand curve is of critical importance, since it represents the potential size of the market and whether or not supply constraints will bind. If demand from a market segment is large relative to supply, then the optimal prices are adjusted to reflect the opportunity cost associated with limited supply.

One of the main challenges in correctly using P2 and implementing 3rd degree price discrimination is sparsity of data. Given the typically large number of products sold by the client, many of which have very sparse transaction histories, it is not possible to obtain robust estimates of the sensitivity parameter independently for each product. In order to deal with this, P2 makes use of market structure characteristics of each product to pool transactions for the purposes of estimation. The estimates obtained from this group of products are refined for each individual product, based on reliability measures from the estimation.

In addition to properly characterizing the demand side of the market, Manugistics must help its clients correctly capture their supply side. This involves two components: (i) helping its clients settle on the correct definition of cost and what their *marginal* costs of production are; and (ii) identifying how much of each product the firm is able to sell in each period, i.e., its 'capable to promise' supply forecasts. Traditionally, suppliers have a loose understanding of what the marginal cost to produce an item is, due to inadequate cost tracking procedures. In addition, they often incorrectly factor in fixed (overhead) costs into their prices when submitting a bid. Manugistics aids their clients in properly identifying the costs that are variable, and hence the actual marginal cost associated with production. Capable to promise supply forecasts tell the seller how much of each product it can sell within each price time horizon. This quantity is important to specify when there are constrained, multi-use resources consumed in the production process. Currently, P2 takes the "capable to promise" quantity as input; this quantity is generally computed from a complementary manufacturing optimization software solution, where demand is modeled as fixed (no own or cross price sensitivity considerations).

P2 is generally used in an environment where there is a limited amount of supply; however the products are not necessarily perishable and the seller may have the opportunity to replenish supply in a timely fashion, i.e., supply is not fixed. To accurately capture the supply conditions, it is important for the client to select the proper planning horizon for pricing decisions. If the client is able to replenish its inventory quickly (e.g. within one week), then a short planning horizon whereby the client need only take into account demand in the very near future when selecting a current price, is most appropriate. However, if, as in many manufacturing settings, the time to replenish inventory is long, then the planning horizon must also be long, so as to correctly capture supply constraints and the relevant trade-offs of selling a product today versus keeping it to satisfy future demand.

P2 is currently used by a firm that supplies high-technology components to original equipment manufacturers (OEMs) and other manufacturers in the semi-conductor industry. P2 is used by the firm to maintain the price lists in its order management system. The firm sells its products through various channels, including inside and outside sales forces and a network of distributors. There are a very large number of products, on the order of 20,000 base products and upwards of 50,000 products considering special-purpose packaging. A large portion of the firm's business is derived from a few main (large) customers, with whom it negotiates prices in a contractual arrangement. The remainder of the firm's sales are derived from spot purchases.[23] Customers differ from each other in terms of the mix of the firm's product that they require for their own production and the lead time with which they place their orders. Demand is highly variable, with customers from different end-markets operating on different business cycles. Production is also global with complex supply and production processes, yielding a variable supply base as well.

The pricing problem for the firm is to decide what spot prices to post for each product and how the spot prices should be changed from one period to the next. At a point in time, the firm has information about previous demand expressed by all of its customers, certain customer characteristics, the current backlog on its order books, its own production plans, expectations regarding future demand and the characteristics of the markets in which its products are sold. Historically in the semi-conductor industry, customers were quoted prices according to their size (the total dollars spent in the semiconductor market) and whether the customer was a direct customer or a distributor. For these reasons, Manugistics's client choose to segment the market into four groups: small direct customer, medium direct customer, large direct customer and distributor. The decision to select only three size groups was a compromise between the desire to have as many groups as possible to allow for more precise pricing and the lack of sufficient data. Distributors were selected as a fourth and final group due to the different nature of their supply chain. The market characteristics used to forecast demand for each product included the *sector* in which each product is used (e.g. wireless telecommunication versus industrial process control) and the *degree of competition* that each product faces (e.g. commodity product versus exclusive, patent-protected technology). Furthermore, expectations regarding future demand were obtained from various third-party information sources and forecasts.

P2 operates in a batch mode, in which it is loaded with recent transactions and changes in supply conditions every week. Once the spot prices are decided, they are posted in the firm's order management system and become visible to the sales force and to its customers. The price sensitivity and forecast parameters are re-estimated each week with a rolling set of transactions; the results from these re-estimations are used to generate new forecasts and re-optimize the prices for every product. These new prices are reviewed by the firm's pricing analysts and, subject to their approval, loaded into the order management system. When a customer asks a salesperson for a price, the salesperson looks up the product in the order management system and, based on the segment to which the customer belongs, quotes the price. If accepted, an order is created and scheduled and its details are passed along to the fulfillment processes.

In order to do proper forecasting of demand, historical transactions are price-normalized to account for effects of price changes on observed demand. Then products are pooled based on the sectors into which they are sold and time series forecasts are generated for each sector. The sector forecasts are then distributed back to each product. A key issue that arises with the use of such a tool is quality assurance. It is important to know that the predictions that are made by the model are accurate, especially the predictions of customer response to price changes. This is assessed primarily by comparing the aggregated sales forecasts to actual sales. In order to make a valid comparison, the forecasts of sales at actual prices (as opposed to forecasts at optimal, but not yet implemented prices) are compared to observed sales. By tracking the forecast errors over time, the managers of the tool can detect the need to revise pooling rules or introduce persistent forecast adjustments.

The firm has obtained several benefits have been achieved by the use of this technology. Prior to employing P2, the firm suffered from problems of regional arbitrage. Each of its regional offices are able to quote customers pricing decisions. Given that several of their customers are large/global firms, the customers took advantage of the regional price differences. The firm used P2 as a catalyst towards centrally determined prices to help circumvent arbitrage and improve price quotes. The implementation is still new enough that there has been no comprehensive analysis of increased margin benefits, although preliminary investigation has set expectations around an increase of several percent. Additionally, the firm has been able to increase the frequency with which they evaluate and update spot prices from quarterly, at best, to weekly.

7. Future Directions for Research in Price Discrimination

Since 1920, when Pigou Pigou, 1920 laid forth the concept of 3rd degree price discrimination, (hitherto referred to as precision pricing) several questions concerning its use have been posed. Does 3rd degree price discrimination increase or decrease prices? Does price discrimination always make a seller off? When should we expect to see price discrimination occur? As the practice of 3rd degree price discrimination enters into various layers of a supply chain's echelons, new questions are posed. How does practicing price discrimination upstream in the supply chain differ from practicing it with only the end-customers? What effect will price discrimination have on the inventory in the supply chain? How does competition in various levels of the supply chain effect the practice of price discrimination?

While there has been a sizable amount of research done on the use of price discrimination on the B2C front, there has been relatively done on the use of price discrimination in B2B markets. One might believe that the welfare and price results for a B2C markets should carry over to B2B markets; however, the literature in this area has found this by-and-large to be not true Katz, 1987 DeGraba, 1990 Yoshida, 2000.

Precision pricing, as with auctions, is posed to take advantage of the Internet's connectivity and information gathering abilities to aid companies in the B2B transactions. As companies gain more information about their customers' behavior and demand, companies can better identify and segment their customers and price accordingly. Given the apparently different nature of B2B price discrimination for B2C, it is important that we gain a better understanding of how and when it should be implemented, its benefits and its shortcomings.

Price Discrimination in a Supply Chain In B2B environments, the customers are themselves manufacturers or retailers who potentially differ along multiple attributes. In addition to factors that influence their willingness to pay, such as the end market in which they sell and the type of product they sell, these customers may also differ in (i) their order size, (ii) the specific combination or specification of products they wish to purchase and (iii) the urgency with which they need their order, i.e., maximum acceptable leadtime. The current literature and practice of precision pricing either assumes that the seller has a fixed quantity of product to sell (takes it as input into the model, as in the case of Manugistic's P2), or can produce instantaneously at a constant variable cost Katz, 1987 DeGraba, 1990 Yoshida, 2000, an assumption seldomly satisfied in reality. A supplier's true costs are rarely comprised of only a constant variable cost. Future work on precision pricing would greatly benefit from the consideration of how pricing will affect the stream of orders a supplier receives, and in turn, how that affects the supplier's cost of meeting his demand obligations, given that a seller can (partially) observe a customer's previous purchasing history. That is, the pricing problem should be set in a multi-period horizon that incorporates possible production constraints and fixed costs. By doing so, we will be able to answer important questions such as: Should the seller develop a price schedule that will result in steady use of his capacity? Should the seller set prices so as to be able to produce his economic order quantity? How much and what type of information should the seller use about customers when selecting his prices? How far in advance should the seller accept orders (for delivery at a future date)?

In addition, further research should be done on the impact of price discrimination on the inventory in a supply chain. For example, how will downstream companies respond to different price discrimination policies? Is the seller better off implementing a static pricing policy, or should he dynamically determine the optimal prices?[24]

Price Discrimination under Competition The ability to price discriminate does not imply that a firm *should* price discriminate. Offering different prices to different customers can lead to loss of customer goodwill. One need only look at the Amazon debacle to comprehend that while tailoring prices to different market segments may theoretically boost prices, in practice it can lead to a backlash of angry customers and lost sales [25]. While this type of response to price discrimination is more likely to occur in B2C markets rather than B2B markets, it does lead to the question, 'Is price discrimination always a profitable policy for a seller?' In addition to the ill-will price discrimination can cause, it is not clear that it is always a profit boosting pricing strategy. Obviously, a monopolist who price discriminates will always increase his profits above the uniform-price profit level, for offering the same price to all customers is always an option. However, if a firm is not the only seller in the market, price discrimination may backfire and cause the firm to offer all of its customers a lower price than uniform pricing, yielding it lower profits [26]. Conversely, it may make all of the sellers better off. For example, if one were to carefully examine all of the markets in which price discrimination takes place, many of these markets could be characterized as fairly competitive, e.g. magazine and journal subscriptions, prescriptions at drugstores, restaurants, and hotels.

Therefore the impact of competition on the practice of price discrimination is an important area of future research.

Behavioral Price Discrimination One of the greatest benefits of employing precision pricing on the Internet will be the ability to track and gather information about customers' purchasing behavior. Currently, there are a number of companies on the B2C front who are taking advantage of the Internet's information possibilities (e.g. Double Click and I-Behavior). The possibility for similar information tracking is just as great on the B2B front, with even larger rewards. As demonstrated in the case of Manugistic's P2 software, companies are aware that whether or not a customer has made a previous purchase will influence their willingness to pay. The reasoning behind this is typically that there are switching costs associated with moving to a new supplier or product. Once a customer has made a purchase from a particular buyer, her demand becomes more inelastic, and the seller can exploit this to increase his prices. Therefore, the optimal practice of behavior price discrimination in B2B markets is an research area worthy of immediate attention.[27]

8. Conclusion

The purpose of this chapter is to bring to the surface the current pricing practices in B2B e-marketplaces. We discussed two pricing policies in particular; *customized auctions* is currently used in e-marketplaces, *precision pricing* is a natural direction forward for many companies exploring revenue management practices in their B2B markets. Where a seller is able to use individualized prices and/or is able to effectively segment customers, we foresee large infrequent transactions remaining in the realm of auctions, while precision pricing will be employed for smaller more frequent transactions. Precision pricing and customized auctions both offers their users the promise of increase profitability. But this promise does not come without serious challenges. To better understand the design dimensions and implementation challenges of using auctions and precision pricing, we presented a few pricing 'success stories'. The challenges faced by FreeMarkets and Manugistics are not unique to these companies, but are likely to be faced by any company who wishes to implement sophisticated pricing policies in e-marketplaces. The ability of these pricing policies to improve company profitability will rest on the models used to capture and represent the market and market participant environments. We believe that the ORMS community can have a large role in improving the design and use of these and other sophisticated pricing mechanisms in e-marketplaces.

Notes

1. These prices are generally determined by a marketing department in a separate division of the company from where the revenue management tools and decisions are made.
2. This observation is based on discussions with FreeMarkets, Manugistics, and numerous other dynamic pricing software solution providers.
3. Friedman and Rust Friedman and Rust, 1993 provide and excellent survey of the double auction literature.
4. Logistics.com was recently bought out by Manhattan Associates.
5. Another pricing mechanism appropriate in these situations is negotiation. While negotiations tend to be held between a single buyer and seller, auctions allow multiple sellers to compete for the buyer's business simultaneously, hence increasing the competitive forces during the procurement process.
6. In a combinatorial auction, bidders can submit package bids. Please see Chapters 7 and 5 for further discussion of combinatorial auctions.
7. It is important to note that a growing area of multi-unit auction literature that has been left out of the discussion below is the design and use of combinatorial auctions. These auctions, where bidders can submit package or combinational bids, are often desirable when bidders realize synergies across objects in a multi-object auction. While extremely useful in helping to capture synergies, combinatorial auctions can be quite difficult to solve for the allocation that maximizes the seller's revenue (known as the winner determination problem).
8. Presentation on March 18,2002. Data based on analysis of US Manufacturing Companies in report by CAPS, The Watch Group, A.T. Kearney.
9. Based on discussions with auction makers at FreeMarkets.
10. Krishna and Rosenthal Krishna and Rosenthal, 1996 and Elmaghraby Elmaghraby, 2003a address the issue of asymmetric sized suppliers competing against each other in an auction, and its impact on the performance of an auction. Both study auctions where there are two types of bidders, global and local. Global bidders desire more than one object and their valuation for multiple objects is greater than the sum of each individual object's valuation, while local bidders desire at most one object. Krishna and Rosenthal, 1996 identify equilibrium bidding strategies when individual objects are auctioned individually and simultaneously. Elmaghraby, 2003a examines under what market settings it is profitable for a buyer to exclude local suppliers from a sequential 2nd price auction.
11. An alternative to the buyer explicitly bundling together products into lots and having bidders submit all-or-nothing bids for the entire lot, is to allow the bidders the flexibility to create their own bundles or packages. Such an auction format is referred to as a combinatorial or package auction, and is discussed further in Chapters 7 and 5. FreeMarkets has begun to develop such combinatorial auctions but uses them in limited market settings. Their combinatorial auction format allows bidders realtime market feedback, i.e., bidders can continuously update their bids, as opposed to discrete rounds. In addition, FreeMarkets is in the development stage with a *take-the-market* myopic best response feedback option: Based on the current bids this option tells the bidder what is the minimal bid necessary to win a particular package.
12. The issue of ordering in auction design is addressed in Benoit and Krishna Benoit and Krishna, 1998, Elmaghraby Elmaghraby, 2003b, Krishna Krishna, 1993 and Pitchik Pitchik, 1998. Under a complete information setting, Benoit and Krishna, 1998 examine the revenue generated by the sale of two objects that possess common values to all bidders under sequential and simultaneous auctions. Elmaghraby, 2003b studies the critical interplay between the sequencing of auctions and the performance of the auction in an incomplete information setting when the buyer wishes to procure two heterogenous objects. Krishna, 1993 examines the efficiency properties of a sequential auction of capacity to an incumbent and several potential new entrants. Pitchik, 1998 finds that, under certain equilibria (satisfying various assumptions about bidding behavior), the revenue from a sequential auction of two heterogenous objects to two bidders with private valuations only depends upon the sequence in which the goods are sold.
13. The identity of bidders is kept anonymous; rather bidders identity other bidders by some auction-given identifier, such as a number assigned to them at the start of the auction.
14. A supplier who participates in an on-line auction can be viewed as defector by a collusive supplier market and possibly subject to punishment, such retaliatory price competition.
15. In order to decrease the attractiveness of collusion, FreeMarkets has recently begun to institute a *penalty box*. If a supplier is suspected of colluding in an auction, then it is 'placed' in the penalty box and barred from participating in future auctions for some predetermined amount of time.

16. A larger bid decrement will lead to faster price changes, but at the possibly expense of excluding some suppliers' from bidding their true cost, when it falls between two bid steps.

17. In fact, the software does not incorporate any constraints, with the possible exception of restrictions imposed on the winning prices. For example, the supplier may incorporate a constraint requiring the price(s) or win probability to be above/below a certain threshold level. While supply constraints are not explicitly represented in NTP, the effects of the prospective supply constraints can be incorporated as opportunity costs and factored into the optimal price determination process.

18. Klemperer Klemperer, 1999, Krishna Krishna, 2002, McAfee and McMillan McAfee and McMillan, 1987 and chapters 7 and 19 in this handbook provide an excellent survey of the auction literature. In addition, Elmaghraby Elmaghraby, 2001 provides a in-depth survey of the use of procurement auctions.

19. The issue of bundling (or lot sizing) and its impact on the performance of an auction is studied in Palfrey Palfrey, 1983, Seshadri et al. Seshadri et al., 1991, and Levin Levin, 1997. Palfrey, 1983 explores the optimal bundling decision for a multiproduct monopolist selling via posted prices under incomplete information about buyers' valuations. Seshadri et al., 1991 consider a model where a buyer wishes to procure a divisible object from a subset of risk-neutral suppliers via an auction. The buyer must decide into how many (equal) parts to divide her demand, thereby determining the number of suppliers from whom she wishes to procure (multi-source). Drawing from the FCC spectrum auction experience, Levin, 1997 searches for the optimal auction design, i.e., what is the optimal manner to auction goods when there exist complementarities between two goods.

20. For further discussion of the role of multi-attribute decision making with auctions, please see Chapter 5.

21. It is worth noting that, while Manugistics' NTP does take into account the presence of competitors when formulating a bid, it does not take into consideration the repeated interactions amongst suppliers, and the possible effects one auction's outcome may have on another procurement auctions. This type of interdependence across auctions has been omitted from Manugistics' NTP, under the belief that NTP is used for relatively 'small' purchases that would not impact a supplier's ability to participate in future auctions.

22. The archetypal example of this is different prices for students, adults and senior citizens at the movie theater.

23. While P2 is used to help establish and maintain spot prices, these spot prices are sometimes used as "reference prices" in contract negotiations, particularly in contract renewal cases where current spot prices differ significantly from the spot prices in effect when the contract was initially negotiated.

24. For example, a very common form of price discrimination is coupons (or rebates). Ault et al. Ault et al., 2000 and Gerstner and Hess Gerstner and Hess, 1991 examine the rationale behind the use of 100% redeemable coupons, when a manufacturer sells through a retailer to his end customers. Ault et al., 2000 find that a universal rebate can increase manufacturer profits by reducing the incentive of downstream retailers to hoard inventories when optimal wholesale price vary over time in a predictable manner. Gerstner and Hess, 1991 assume that the end market can be segmented into high and low valuation customers. They find that 100% redeemable rebates are less costly for the manufacturer (then reducing the wholesale selling price to the retailer) when the manufacturer wishes to encourage the retailer to price so as to sell to both end market segments.

25. A very interesting website that tallies customers' responses to price discrimination by Amazon can be found at $http://www.kdnuggets.com/polls/2000/amazon\_prices.htm$

26. Such a prisoner's dilemma was shown to occur in Holmes Holmes, 1989. Holmes, 1989 studies a model where two firms sell differentiated products (partial substitutes) into the end market, which consists of two customer types. Both firms have the same constant marginal cost of production, sell to both markets and behave as Bertrand competitors, i.e., they compete in price, in each market. He finds examples in which the duopolists are made strictly worse off by practicing price discrimination.

27. The issue of switching cost and 'behavioral' price discrimination, whereby customers are differentiated according to their purchasing history, is a burgeoning area of research. Recent papers by Chen Chen, 1997, Villas-Boas Villas-Boas, 1999 Villas-Boas, 2001 and Fudenberg and Tirole Fudenberg and Tirole, 2000 address the issue of behavioral price discrimination, and how price discrimination across customer types effect equilibrium prices and seller profitability. Stole Stole, 2001 provides a nice summary of this work in Chapter 6. All of these papers assume that customers differ according to (i) whether or not they made a purchase previously, (ii) whether they are new or returning customers and (iii) their valuations for the product. However, in practice, customers will differ along multiple attributes.

References

AMR (2001). B2b marketplaces report, 2000-2005. *AMR Research*. August 1.

Anthes, G. (1999). The price had better be right. *Computerworld*, December 21. http://www.computerworld.com/cwi/story/0,1199,NAV47_STO33288,00.html.

Athey, S. and Levin, J. (1998). Information and competition in u.s. forest service timber auctions. Working Paper, Massachusetts Institute of Technology.

Ault, R., Beard, T., Laband, D., and Saba, R. (2000). Rebates, inventories, and intertemporal price discrimination. *Economic Inquiry*, 38(4):570–578.

Ausubel, L. and Cramton, P. (2002). Demand reduction and inefficiency in multi-unit auctions. Working Paper No. 96-07 University of Maryland.

Back, K. and Zender, J. (1993). Auctions of divisible goods: On the rationale for the treasury experiment. *The Review of Financial Studies*, 4:733–764.

Barling, B. (2001). Gxs brings hosted auctions to europe. *AMR Research*. October 22.

Benoit, J. and Krishna, V. (1998). Multiple-object auctions with budget constrained bidders. *Review of Economic Studies*. forthcoming.

Bichler, M. (2000). An experimental analysis of multi-attribute auctions. *Decision Support Systems*, 29:249–268. 3.

Bikhchandani, S. (1999). Auctions of heterogeneous objects. *Games and Economic Behavior*, 26:193–220.

Bitran, G. and Caldentey, R. (2002). An overview of pricing models for revenue management. *Manufacturing & Service Operations Management*. to appear.

Bushnell, J. and Oren, S. (1994). Bidder cost revelation in electric power auctions. *Journal of Regulatory Economics*, 6(1):5–26.

Chao, H. and Wilson, R. (2001). Multi-dimensional procurement auctions for power reserves: Robust incentive-compatible scoring and settlement rules. Working Paper, Stanford University.

Che, Y. (1993). Design competition through multidimensional auctions. *RAND Journal of Economics*, 24:668–681.

Chen, Y. (1997). Paying customers to switch. *Journal of Economics and Management Strategy*, 6:887–897.

DeGraba, P. (1990). Input market price discrimination and the choice of technology. *American Economic Review*, 80(5):1246–1253.

Economist (2001). B2b exchanges : The container case. *The Economist*. October 21.

Elmaghraby, W. J. (2001). Supply contract competition and sourcing policies. *Manufacturing & Service Operations Management*, 2(4):350–371.

Elmaghraby, W. J. (2003a). The impact of asymmetric bidder size on auction performance: Is more bidders always better? Working Paper, Georgia Institute of Technology.

Elmaghraby, W. J. (2003b). The importance of ordering in sequential auctions. *Management Science*, 49(5):673–682.

Elmaghraby, W. J. and Keskinocak, P. (2003). Dynamic pricing in the presence of inventory considerations: Research overview, current practices and future directions. *Management Science*. Special Issue on E-Business, forthcoming.

Engelbrecht-Wiggans, R. and Kahn, C. (1998). Multi-unit auctions with uniform prices. *Economic Theory*, 12:227–258.

Friedman, D. and Rust, J. (1993). *The Double Auction Market: Institutions, Theories and Evidence.* Addison Wesley Longman. SPI Studies in the Sciences of Complexity.

Fudenberg, D. and Tirole, J. (2000). Customer poaching and brand switching. *RAND Journal of Economics*, 31(4):634–657.

Gerstner, E. and Hess, J. (1991). A theory of channel price promotions. *American Economic Review*, 81(4):872–886.

Holmes, T. (1989). The effects of third-degree price discrimination in oligopoly. *American Economic Review*, 79(1):244–250.

Katz, M. (1987). The welfare effects of third-degree price discrimination in intermediate good markets. *American Economic Review*, 77(1):154–167.

Katzman, B. (1999). A two stage sequential auction with multi-unit demands. *Journal of Economic Theory*, 86:77–99.

Klemperer, P. (1999). Auction theory: A guide to the literature. *Journal of Economic Surveys*, 13(3):227–286.

Klotz, D. and Chatterjee, K. (1995). Dual sourcing in repeated procurement competitions. *Management Science*, 41(8):1317–1327.

Koppius, O. and van Heck, E. (2002). Information architecture and electronic market performance in multidimensional auctions. Working Paper Erasmus University.

Krishna, K. (1993). Auctions with endogenous valuations: The persistence of monopoly revisited. *American Economic Review*, 83(1):147–160.

Krishna, V. (2002). *Auction Theory*. Academic Press, San Diego.

Krishna, V. and Rosenthal, R. (1996). Simultaneous auctions with synergies. *Games and Economic Behavior*, 17:1–31.

Latham, S. (2000). Independent trading exchanges–the next wave of b2b e-commerce. *AMR Research*. March 1.

Levin, J. (1997). An optimal auction for complements. *Games and Economic Behavior*, 18:176–192.

McAfee, P. and McMillan, J. (1987). Auctions and bidding. *Journal of Economic Literature*, 15:699–738.

McGill, J. and van Ryzin, G. (1999). Revenue management: Research overview and prospects. *Transportation Science*, 33:233–256.

Milgrom, P. (2000). *A Theory of Auctions and Competitive Bidding II*. Edward Elgar Publishing, Inc. in The Economic Theory of Auctions P. Klemperer Ed. Cheltenham, UK.

Palfrey, T. (1983). Bundling decisions by a multiproduct monopolist with incomplete information. *Econometrica*, 51(2):463–483.

Phillips, R. (2003). *Pricing and Revenue Optimization*. Stanford University Press, forthcoming, Palo Alto, CA. Chap. 4.

Pigou, A. (1920). *The Economics of Welfare*. Macmillan, London.

Pitchik, C. (1998). Budget-constrained sequential auctions with incomplete information. Working Paper University of Toronto.

Prouty, K. (2001). Automotive and heavy equipment outlook: Gm tries to move covisint beyond auctions. *AMR Research*. October 31.

Rosenblum, P. (2002). Major tidal wave of auctions coming to apparel, soft home, and footwear retail. *AMR Research*. August 5.

Seshadri, S., Chatterjee, K., and Lilien, G. (1991). Multiple source procurement competition. *Management Science*, 10(3):246–263.

Stole, L. (2001). Price discrimination in competitive environments. Working Paper, Graduate School of Business, University of Chicago.

Villas-Boas, J. (1999). Dynamic competition with customer recognition. *RAND Journal of Economics*, 30(4):604–631.

Villas-Boas, J. M. (2001). Price cycles in markets with customer recognition. Working Paper, Haas School of Business, University of California at Berkeley.

Yoshida, Y. (2000). Third-degree price discrimination in input markets: Output and welfare. *American Economic Review*, 90(1):240–246.

David Simchi-Levi, S. David Wu, and Z. Max Shen (Eds.)
Handbook of Quantitative Supply Chain Analysis:
Modeling in the E-Business Era
©2004 Kluwer Academic Publishers

Chapter 7

DESIGN OF COMBINATORIAL AUCTIONS

Sven de Vries
Zentrum Mathematik
TU München
80290 München
Germany
devries@ma.tum.de

Rakesh V. Vohra
Department of Managerial Economics and Decision Sciences
Kellogg School of Management
Northwestern University
Evanston, IL 60208
USA
r-vohra@nwu.edu

Keywords: package bidding, mathematical programming

1. Introduction

Suppose you must auction off a dining room set consisting of four chairs and a table. Should you auction off the entire set or run five separate auctions for each piece? It depends, of course, on what bidders care about. If every bidder is interested in the dining room set and nothing less, the first option is preferable. If some bidders are interested in the set but others are interested only in a chair or two it is not obvious what to do. If you believe that you can raise more by selling off the chairs separately than the set, the second option is preferable. Notice, deciding requires a knowledge of just how much bidders value different parts of the ensemble. For this reason, economic efficiency is enhanced if bidders are allowed to bid directly on **combinations** or **bundles** of different assets instead of bidding only on individual items. Auctions where bidders are allowed to submit bids on combinations of items are usually called **combinatorial auctions.** 'Combinational auctions' is more accurate, but we will follow convention.

Auctions where bidders submit bids on combinations have recently received much attention. See for example Caplice [15], Rothkopf et al. [61], Fujishima et al. [31], and Sandholm [64]. However, such auctions were proposed as early as 1976 [38] for radio spectrum rights. Rassenti et al. [63], a little later, propose such auctions to allocate airport time slots. Increases in computing power have made combinatorial auctions more attractive to implement.

A number of large firms have actively embraced combinatorial auctions to procure logistics services (see section 7.2). Ledyard et al. [48] describe the design and use of a combinatorial auction that was employed by Sears in 1993 to select carriers. The objects bid upon were delivery routes (called **lanes**). Since a truck must return to its depot, it was more profitable for bidders to have their trucks full on the return journey. Being allowed to bid on bundles gave bidders the opportunity to construct routes that utilized their trucks as efficiently as possible. In fact, a number of logistics consulting firms tout software to implement combinatorial auctions. SAITECH-INC, for example, offers a software product called SBIDS that allows trucking companies to bid on bundles of lanes. Logistics.com's system is called OptiBid^{TM}. Logistics.com (acquired in 2002 by Manhattan Associates, Inc.) claims that more than $8 billion in transportation contracts have been bid by January 2001 using OptiBid^{TM} by Ford Motor Company, Wal-Mart, and K-Mart.

Since about 1995, London Transport has been auctioning off bus routes using a combinatorial auction. About once a month, existing contracts to service some routes expire and these are offered for auction. Bidders can submit bids on subsets of routes, winning bidders are awarded a five-year contract to service the routes they win. In this way, about 20 percent of London Transport's 800 bus routes are auctioned off every year, see Cantillon and Pesendorfer [17].

Epstein et al. [27] report of the successfull application of a combinatorial auction to procure different school meals by the Chilean government saving about US$40 million on a procurement of US$180 million.

Procurement auctions where bidders are asked to submit a collection of price-quantity pairs, for example, $4 a unit for 100 units; $3.95 a unit for 200 units, etc., are also examples of combinatorial auctions. Here each price-quantity pair corresponds to a bundle of homogenous goods and a bid. If the goods are endowed with attributes like payment terms, delivery, and quality guarantees, they become bundles of heterogenous objects. Davenport and Kalagnanam [21] describe a combinatorial auction for such a context that is used by a large food manufacturer. Davenport et al. [20] describe the procurement of raw materials by Mars, Inc. Ausubel and Cramton [2] and Bikhchandani and Huang [9] describe the auction for Treasury Securities that is actually used by the US Department of Treasury and compare it with other mechanisms. More details on some aspects of these business-to-business applications can be found in Chapter 5 by Kalagnanam and Parkes.

Usually an auction has one of two objectives. The first is revenue maximization for the auctioneer. The second is to promote economic efficiency. Here **efficiency** is to be interpreted in an allocative sense. The goal is to get the assets into the hands of those who will use them in the most productive manner. On occassion the two goals coincide. One instance noted by Ausubel and Cramton [3] occurs when there is an active secondary market for the assets. Frequently they diverge. In practice it is not a matter of choosing one over the other. For example, a goverment regulator who auctions off public assets is judged by the revenue raised even though the goal of the auction is to promote efficiency.

We ignore these issues and focus separately on revenue maximization and efficiency. We first identify what the theory has to say about the design of auctions to achieve each of these objectives. It will be seen that these auctions are impractical to implement as is and adjustments must be made. These adjustments come at a price and we will identify the tradeoffs that must be made.

2. Mechanism Design Perspective

There are a multitude of auction design parameters to play with. Once one specifies the objectives of the auction how is one to search through all these parameters to find the setting that best achieves them? This is where the revelation principle of mechanism design comes in.

In a **direct** mechanism, the auctioneer announces how the goods will be allocated and the payments set as a function of the private information reported by the bidders. It remains then for the bidders to report (not necessarily truthfully) their private information. Once the reports are made, the auctioneer 'applies' the announced function and the resulting allocations made and payments executed. One can think of a direct mechanism as being a sealed bid auction, but with the bids being complex enough to communicate the private information of the agents. If the direct mechanism has the property that it is in the bidders interests, suitably defined, to report their private information truthfully, call it a **direct revelation** mechanism.

One can imagine a mechanism that is not 'direct'. Consider one that involves multiple rounds where participation in one round is conditional on bids or other information revealed in previous rounds. These indirect mechanisms can be quite baroque. It is here that the 'revelation principle' comes to our aid.

Given any mechanism, there is a corresponding direct revelation mechanism which will produce the same outcome as the given mechanism. Thus, for the purposes of analysis, one can restrict attention to direct revelation mechanisms. The revelation principle does pose a problem. If a direct mechanism suffices why do we observe indirect mechanisms in practice? We defer a discussion of this to sections 5.3 and 6.

We now put the revelation principle to work to identify auctions that meet two obvious and popular goals. The first is revenue maximization, the second is economic efficiency. For a more detailed treatment of the mechanism design perspective see 2.2 of Chapter 5 by Kalagnanam and Parkes.

3. Optimal Auctions

Our examination of optimal auctions will be limited to the following simple setting, called the **independent private value model**.

1. Agents and auctioneer are risk neutral.

2. Let Γ be the set of feasible allocations of the resources amongst the agents and the auctioneer.

3. The parameter (possibly multidimensional) that encodes how an agent values different allocations is called her type. Let $T = \{t_1, t_2, \ldots, t_m\}$ be a finite set of possible types.[1] A collection of types one for each (of n agents) will be called a profile. Let T^n be the set of all profiles. A profile involving only $n - 1$ agents will be denoted T^{n-1}.

4. If $a \in \Gamma$ and a bidder has type t_i she assigns monetary value $v(a|t_i)$ to the allocation $a \in \Gamma$.

5. Uncertainty about the valuations of the bidders is captured assuming that types are independent draws from a common distribution that is commonly known. This is the **common prior assumption**. Let f_i be the probability that a bidder has type t_i.

6. The probability of a profile $t \in T^{n-1}$ being realized is $\pi(t)$

One could easily include costs for the seller into the setup, but it changes nothing essential in the arguments and clutters the notation.

The unsatisfactory feature of this setup is the common prior assumption. It is difficult to believe that the auctioneer and the bidders can express their beliefs about others' valuations probabilistically, let alone agree upon that belief. It (or something like it) is unavoidable. If the values that bidders have for the object are unknown to the auctioneer, then that uncertainty must be modeled. If the auction is such that how a bidder behaves depends on how she thinks

Design of Combinatorial Auctions

other bidders will behave, then again her uncertainty about other bidders must be modeled. This is not the forum for a full discussion of such issues. Suffice it to say that the common prior assumption is unpopular but there are no good alternatives. The reader interested in these matters should consult Morris [53] for a general discussion and Goldberg, Hartline, Karlin and Wright [33] and Baliga and Vohra [14] for a discussion focused on auctions.

By the revelation principle we can restrict attention to direct revelation mechanisms. Each agent is asked to announce her type and the auctioneer, as a function of the announcements, decides what element of Γ to pick and what payments each agent will make.

To describe a mechanism, let P_i be the expected payment that an agent who announces type t_i must make. An allocation rule assigns to each member of T^n an element of Γ.[2] If α is an allocation rule, the expected utility of agent with type t_i under this rule will be (under the assumption that agents announce truthfully)

$$\sum_{t^{n-1} \in T^{n-1}} v(\alpha[t_i, t^{n-1}]|t_i) \pi(t^{n-1})$$

which we will denote $E_{t^{n-1}}[v(\alpha[t_i, t^{n-1}]|t_i)]$. The '$E$' denotes expectation. The expected utility of agent with type t_i announcing t_j (perhaps $\neq t_i$) under allocation rule α is

$$\sum_{t^{n-1} \in T^{n-1}} v(\alpha[t_j, t^{n-1}]|t_i) \pi(t^{n-1}),$$

which will be abbreviated to $E_{t^{n-1}}[v(\alpha[t_j, t^{n-1}]|t_i)]$.

To ensure that an agent will report truthfully we impose **incentive compatibility** (IC): for each agent with type t_i:

$$E_{t^{n-1}}[v(\alpha[t_i, t^{n-1}]|t_i)] - P_i \geq E_{t^{n-1}}[v(\alpha[t_j, t^{n-1}]|t_i)] - P_j \ \forall t_i, t_j \in T.$$

In words, the IC constraint says that assuming other agents truthfully report their types, ones best response is to truthfully report ones type. Phrased differently, truth telling is a (Bayesian) equilibrium of the mechanism. It is sometimes common to require a stronger condition, dominant strategy incentive compatibility, which we discuss later in section ??.

To ensure that each agent has the incentive to participate in the mechanism we impose the following:

$$E_{t^{n-1}}[v(\alpha[t_i, t^{n-1}]|t_i)] - P_i \geq 0 \ \forall t_i \in T.$$

This constraint is called **individual rationality** (IR). If we add a dummy type, t_0 which assigns utility 0 to all allocations, and set $P_0 = 0$, we can fold the IR constraint into the IC constraint.[3] So, from now on T contains the dummy type t_0.

The auctioneer's problem is to maximize expected revenue subject to IC. Expected revenue (normalized for population) is $\sum_{i=0}^{m} f_i P_i$. Fix an allocation rule α and let

$$R(\alpha) = \max_{P_i} \sum_{i=0}^{m} f_i P_i$$

subject to

$$E_{t^{n-1}}[v(\alpha[t_i, t^{n-1}]|t_i)] - P_i \geq E_{t^{n-1}}[v(\alpha[t_j, t^{n-1}]|t_i)] - P_j \ \forall t_i, t_j \in T.$$

Call this program LP_α. If LP_α is infeasible set $R(\alpha) = -\infty$. Thus the auctioneer's problem is to find the allocation rule α that maximizes $R(\alpha)$.

One way to solve this optimization problem is to fix the allocation rule α. Then we can rewrite the IC constraint as follows:

$$P_i - P_j \leq E_{t^{n-1}}[v(\alpha[t_i, t^{n-1}]|t_i)] - E_{t^{n-1}}[v(\alpha[t_j, t^{n-1}]|t_i)] \ \forall i \neq j$$

If there is a feasible solution to this LP then we can find payments that implement α in an incentive compatible way.

To understand this LP it helps to flip to its dual. The dual is a network flow problem that can be described in the following way. Introduce one node for each type (the node corresponding to the dummy type will be the source) and to each directed edge (j, i), assign a length of $E_{t^{n-1}}[v(\alpha[t_i, t^{n-1}]|t_i)] - E_{t^{n-1}}[v(\alpha[t_j, t^{n-1}]|t_i)]$. Each P_i corresponds to the length of the shortest path from the source to vertex i. For fixed α, the optimization problem reduces to determining the shortest path tree in this network (union of all shortest paths from source to all nodes).

If the network has a negative cycle, the dual problem is unbounded, which means the original primal problem is infeasible. Thus there is no set of payments to make the allocation α incentive compatible. To summarize, given an allocation rule we have a way of checking whether it can be implemented in an incentive compatible way.

Naively, one could fix an allocation rule α, then solve the LP and repeat. The problem is that the set of allocation rules need not be finite or even countable! Thus, unless one is prepared to make additional assumptions about the structure of the problem, the optimal auction will be difficult to identify. With few exceptions, the solvable cases involve types that are one dimensional. See Rochet and Stole [62] for a discussion.

EXAMPLE 7.1 *Consider the allocation of one good among two agents with types* $\{0, 1, 2\}$, *here 0 is the dummy type. Let* $f_1 = f_2 = 1/2$ *and* $\pi(1) = \pi(2) = 1/2$ *be the probabilities. Choose as the allocation rule the following: assign the object to the agent with highest type, in case of ties randomize the allocation '50-50'. The possible allocations are: agent 1 wins, agent 2 wins,*

Design of Combinatorial Auctions

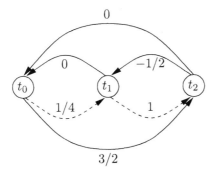

Figure 7.1. Shortest path problem to Example 7.1; shortest path tree is dashed.

agent 1 and 2 get '1/2' of the item, seller keeps it. For the valuations we have: an agent of type t_i who wins the item values it at t_i; if he tied with his competitor and gets only a half he derives value $t_i/2$ if finally he loses or the seller keeps it, he gets value 0. Now to the computation of expected utility when honest:

$$E_{t^{n-1}}[v(\alpha[t_i, t^{n-1}]|t_i)] = \sum_{t^1 \in \{0,1,2\}} v(\alpha[t_i, t^1]|t_i)\pi(t^1)$$

$$= \begin{cases} \frac{1}{4} & \text{if } t_i = 1 \\ \frac{3}{2} & \text{if } t_i = 2 \end{cases}$$

Similarly we obtain for $E_{t^{n-1}}[v(\alpha[t_j, t^{n-1}]|t_i)]$:

$t_i \setminus t_j$	0	1	2
0	0	0	0
1	0	$\frac{1}{4}$	$\frac{3}{4}$
2	0	$\frac{1}{2}$	$\frac{3}{2}$

So we obtain for the optimization problem

$$\begin{aligned}
\max \quad & \tfrac{1}{2}P_1 + \tfrac{1}{2}P_2 \\
\text{s.t.} \quad & P_1 - P_0 \leq \tfrac{1}{4} \quad P_0 - P_1 \leq 0 \quad P_0 - P_2 \leq 0 \\
& P_2 - P_0 \leq \tfrac{3}{2} \quad P_2 - P_1 \leq 1 \quad P_1 - P_2 \leq -\tfrac{1}{2} \\
& P_0 = 0
\end{aligned}$$

The resulting shortest path problem is depicted in Figure 7.1. Reading off the shortest path distances to t_1, t_2 yields $P_1 = 1/4$ and $P_2 = 5/4$. So the auctioneer can realize revenue of $3/4$ with this allocation rule.

Although the procedure specifies only expected payments there is a way to derive from them a payment rule that depends on the announcements. The reader is referred to Myerson [55] for the details.

Perhaps the main insight from the results (e.g. [55]) on optimal auctions, is the importance of a seller's reserve price. This has two effects. The first, is that a seller's reserve is analogous to adding another bidder, so this stimulates competition. The second is that if the reserve is not met, the seller winds up not selling anything. Thus, the higher expected revenue comes from taking on additional risk. When bidders are heterogenous, that is the types are draws from different distribution, the theory mandates varying reserve prices. Thus different bidders may face different reserve prices. In section 2.2.3 of Chapter 5 by Kalagnanam and Parkes a full analysis of the one-good case is carried out.

4. Efficient Auctions

One way to simplify the design problem is to fix the allocation rule and then ask how to choose the payments so as to maximize expected revenue. One popular choice of allocation rule is the efficient one. The efficient allocation rule is the one that for every profile $t \in T^n$ of types chooses an allocation $a^*(t) \in \arg\max_{a \in \Gamma} \sum_{i=1}^{n} v_i(a|t_i)$. Can we implement a^* in an incentive compatible way? To do so we must find a feasible solution to the IC constraints:

$$P_j - P_i \geq E_{t^{-i}}[v(a^*[t_j, t^{-i}]|t_i)] - E_{t^{-i}}[v(a^*[t_i, t^{-i}]|t_i)] \quad \forall i \neq j.$$

Here is where the IC constraints are useful. Fix a type t_i and call a type $s \in T \setminus t_i$ **equivalent** to t_i if

$$E_{t^{-i}}[v(a^*[t_i, t^{-i}]|t_i)] = E_{t^{-i}}[v(a^*[s, t^{-i}]|t_i)].$$

Notice that if type t_i announces type s instead, her expected utility under the mechanism is the same as announcing the truth. IC implies that type t_i cannot be charged more than type s, i.e. $P_i \leq P_s$. If not, type t_i has an incentive to say her type is s instead. This observation imposes an upper bound on the amount the auctioneer can extract from type t_i:

$$\min_{s \in T} E_{t^{-i}}[v(a^*[s, t^{-i}]|s)]$$

$$\text{s.t. } E_{t^{-i}}[v(a^*[s, t^{-i}]|t_i)] = E_{t^{-i}}[v(a^*[t_i, t^{-i}]|t_i)].$$

By solving this problem for each type, the auctioneer can identify upper bounds on the payments that maximize his expected revenue while still preserving IC and IR.

The approach is rather messy and if one is prepared to enlarge the domain of types then a very nice expression can be obtained. The associated auction is called the AGV-Arrow mechanism. We refer the reader to pg. 273 of Fudenberg and Tirole [32] for details. We omit the description since we intend to discuss something similar below.

4.1 The VCG Auction

The analysis above imposed (Bayesian) incentive compatibility upon the design. This is problematic because it requires bidders and auctioneer to know or believe in a *common* prior. For this reason, one may ask for a design that is independent of such considerations. One way of circumventing the common prior assumption is to require a design in which bidding truthfully is a dominant strategy as well as ex-post individually rational. That is, no matter what other bidders bid, one is (weakly) better off bidding truthfully. Compare this with IC which assumes other bidders bid truthfully. Requiring truthful bidding as opposed to IC imposes more constraints. Specifically:

$$v(a^*[t_i, t^{-i}]|t_i)] - P(t_i, t^{-i}) \geq v(a^*[s, t^{-i}]|t_i)] - P(s, t^{-i}) \ \forall (s, t_{-i}) \in T^n$$

The IC constraints are obtained by averaging over these inequalities and thus are weaker.

Now, amongst all auctions that produce the efficient allocation, are ex-post individually rational, in which bidding truthfully is a dominant strategy, which one maximizes the revenue of the auctioneer? The answer to this question was provided by Groves [35] and Clarke [16] with the impetus coming from Vickrey's seminal 1961 paper [66]. For this reason the auction is called the Vickrey-Clarke-Groves mechanism or VCG for short.[4] We show how to derive it using the ideas described above.

To economize on notation, we absorb the type information into the value function itself. That is agents types will be their value functions and they are free to report any non-negative value function they like.[5] We continue to let Γ be the space of feasible allocations.

Let v_i be the value function of agent i. Assuming the agents truthfully report their value functions, the efficient allocation is

$$a^* \in \arg\max_{a \in \Gamma} \sum_{i \in N} v_i(a).$$

For simplicity assume that arg max is single valued. Consider now agent j. We ask what other value function, say u, she could report without changing the efficient allocation. Specifically, we want a u such that

$a^* \in \arg\max_{a \in \Gamma} \left(u(a) + \sum_{i \neq j} v_i(a) \right).$

If $u(a^*) \leq v_j(a^*)$, we cannot charge agent j more than $u(a^*)$. If we do, agent j has the incentive to report u rather v_j. Doing so leaves the efficient allocation unchanged, lowers her payment and this makes her better off. Thus the most we can charge agent j is the objective function value of the following program:

$$\min_u u(a^*) \text{ s.t. } a^* \in \arg\max_{a \in \Gamma} \left(u(a) + \sum_{i \neq j} v_i(a) \right).$$

This can be rewritten to read:

$$\min_u u(a^*) \text{ s.t. } u(a^*) + \sum_{i \neq j} v_i(a^*) \geq u(a) + \sum_{i \neq j} v_i(a) \ \forall a \in \Gamma.$$

To make the objective function value as small as possible we should choose a u to make the righthand side of the constraint as small as possible. This can be done by choosing a u such that $u(a) = 0$ for all $a \neq a^*$. Restricting ourselves to such a function, the problem becomes

$$\min_u u(a^*) \text{ s.t. } u(a^*) \geq \sum_{i \neq j} v_i(a) - \sum_{i \neq j} v_i(a^*) \ \forall a \in \Gamma.$$

Observe that the largest right hand side is $\max_{a \in \Gamma} \sum_{i \neq j} v_i(a) - \sum_{i \neq j} v_i(a^*)$. In other words we should choose u so that

$$u(a^*) = \max_{a \in \Gamma} \sum_{i \neq j} v_i(a) - \sum_{i \neq j} v_i(a^*).$$

We can now describe the auction as follows:

1 Agents report their value functions. That is they report their value for each bundle of goods.

2 Using the reported value functions compute the efficient allocation a^*.

3 Charge each agent j, $(\max_{a \in \Gamma} \sum_{i \neq j} v_i(a)) - \sum_{i \neq j} v_i(a^*)$.

Reporting truthfully is a dominant strategy by construction. We simply asked what is the maximum rent agent j could gain from misreporting without changing the allocation and gave her that rent, so removing the incentive to misreport. Furthermore, each agent is charged the upper bound of what they can be charged and preserve the incentive to bid truthfully. Thus the auction amongst all that implement the efficient allocation, maximizes the revenue to the seller.

It is instructive to calculate the profits of agent j under VCG. They are

$$v_j(a^*) - \left[\max_{a \in \Gamma} \sum_{i \neq j} v_i(a) - \sum_{i \neq j} v_i(a^*) \right] = \sum_{i \in N} v_i(a^*) - \max_{a \in \Gamma} \sum_{i \neq j} v_i(a).$$

It is the difference between the value of the efficient allocation when all agents are included and when agent j is excluded. This quantity is called the **marginal product** of agent j and represents the economic value she brings to the table.

5. Implementing an Efficient Auction

It seems then, if efficiency is the goal, we have an answer as to what kind of auction to run: VCG. In the sections that follow we discuss some of the issues that arise in implementing the VCG auction. The reader interested in revenue maximization will still benefit for two reasons. First, these issues are not unique to the VCG. Second, for reasons not formally modeled, one may be forced to use an efficient auction rather than a revenue maximizing one. These have to do with resale and long term participation by bidders and are discussed in Ausubel and Cramton [3] as well as Milgrom [51].

Before continuing it is important to remember that the analysis above is conducted in a 'vacuum'. It is presumed that the auction is the sole interaction between auctioneer and bidder. In practice this need not be true. The potential for future interactions between bidders and the auctioneer will influence their behavior in the auction

5.1 Winner Determination

The most obvious hurdle to implementing the VCG auction is the computational one of identifying the efficient allocation. In the context of combinatorial auctions this amounts to solving an integer program, called the **winner-determination problem.** Here we focus on the formulation described in Rothkopf et al. [61] and by Sandholm [64]. To distinguish it from other possible formulations we call it the **combinatorial auction problem** (CAP).[6] To formulate CAP as an integer program, let N be the set of bidders and M the set of m distinct objects. For every subset S of M let $b^j(S)$ be the bid that agent $j \in N$ has announced he is willing to pay for S.[7] From the formulation it will be clear that bids with $b^j(S) < 0$ will never be selected. So, without loss of generality, we can assume that $b^j(S) \geq 0$. The reader will notice that we use bids rather than value functions. We do so to highlight the fact that the computational issues are not unique to the VCG auction.

Let $y(S, j) = 1$ if the bundle $S \subseteq M$ is allocated to $j \in N$ and zero otherwise.

$$\max \sum_{j \in N} \sum_{S \subseteq M} b^j(S) y(S, j)$$

$$\text{s.t.} \sum_{S \ni i} \sum_{j \in N} y(S, j) \leq 1 \quad \forall i \in M$$

$$\sum_{S \subseteq M} y(S, j) \leq 1 \quad \forall j \in N$$

$$y(S, j) = 0, 1 \quad \forall S \subseteq M, \ j \in N$$

The first constraint ensures that overlapping sets of goods are never assigned. The second ensures that no bidder receives more than one subset. Call this formulation CAP1. Problem CAP as formulated here is an instance of what is known as the **set-packing problem**.[8]

We have formulated CAP1 under the assumption that there is at most one copy of each object. It is easy to extend the formulation to the case when there are multiple copies of the same object and bidders may want more than one copy of the same object. Such extensions, called **multi-unit combinatorial auctions,** are investigated by Leyton-Brown et al. [47] as well as by Gonen and Lehmann [34].

The formulation for winner determination just given is not flexible enough to encompass some of the variations that have been considered in the literature. For example technical constraints may prevent certain combinations of assets being awarded. Also, the auctioneer may impose side constraints that limit amounts that any one bidder can obtain. For a more comprehensive formulation we refer the reader to de Vries and Vohra [26].

Had we cast the auction problem in procurement rather than selling terms the underlying integer program would be an instance of the **set partitioning** problem. The auctions used in the transport industry are of this set-partitioning type. In that setting, objects are origin-destination pairs, called **lanes.** Bidders submit bids on bundles of lanes that represent how much they must be offered to undertake the deliveries on the specified lanes. The auctioneer wishes to choose a collection of bids of lowest cost such that all lanes are served.[9]

It is natural to think that since CAP is an instance of the set packing problem it must be NP-hard. One must be careful in drawing such a conclusion, since it depends on how the input to an instance of CAP is encoded. Suppose one takes the number of bids as a measure of the size of the input and this number is exponential in $|M|$. Any algorithm for CAP that is polynomial in the number of bids but exponential in the number of items would, formally, be polynomial in the input size but impractical for large $|M|$.[10] In particular, an instance of CAP in which bids were submitted on every possible combination of items would be polynomial in the number of bids. On the other hand, if input size is measured by $|M|$, the number of bids is polynomial in $|M|$, then CAP is indeed NP-hard.

In any case, the upshot is that effective solution procedures for the CAP must rely on two things. The first is that the number of distinct bids is not large. The second is that the underlying set packing problem can be solved reasonably quickly. For a list of solvable instances of the set packing problem and their relevance to CAP see de Vries and Vohra [26].

Solvability of these instances arise from severe restrictions on the set of combinations that can be bid upon. Also, solvability of these cases typically evaporates when one appends constraints that limit a bidder to having only one of his bids accepted.

Leaving aside solvability, one may still be interested in restricting the bundles that can be bid upon so as to make communicating bids easier. Nisan [56] discusses how various bid restrictions prevent certain kinds of preferences being expressed. The problem of communicating bids will be discussed again later in section 5.6.

The designer thus faces a tradeoff. Restricting the bids could make the underlying winner determination problem easy. However, by restricting the bids one limits the preferences that can be expressed by bidders. This will produce allocations that are economically inefficient.

The most general class of preferences for which a positive computational result is available is the gross substitutes class (Kelso and Crawford [42]). To describe it let the value that bidder j assigns to the set $S \subseteq M$ of objects be $v_j(S)$. Given a vector of prices p on objects, let the collection of subsets of objects that maximize bidder j's utility be denoted $D_j(p)$, and defined as

$$D_j(p) = \{S \subseteq M : v_j(S) - \sum_{i \in S} p_i \geq v_j(T) - \sum_{i \in T} p_i \quad \forall T \subseteq M\}.$$

The **gross-substitutes condition** requires that for all price vectors p, p' such that $p' \geq p$, and all $A \in D_j(p)$, there exists $B \in D_j(p')$ such that $\{i \in A : p_i = p'_i\} \subseteq B$. Gross-substitutes formalizes the following intuition. If prices of some goods rise, an agent may switch his consumption from these goods to some other goods. However, on goods that did not see a price change, it is always utility maximizing to continue consuming them. A special case of the gross-substitutes condition is when bidders are interested in multiple units of the same item and have diminishing marginal utilities.

In the case when each of the $v_j(\cdot)$ have the gross-substitutes property, the linear-programming relaxation of CAP1 (with $b^j(S)$ replaced by $v_j(S)$ in the objective function) has an optimal integer solution. This is proved in Kelso and Crawford [42] as well as Gul and Stacchetti [36]. Murota and Tamura [54] point out the connection between gross substitutes and $M^\natural$-concavity. From this connection it follows from results about $M^{\natural}$-concave functions that CAP1 can be solved in time polynomial in the number of objects under the assumption of gross substitutes by using a proper oracle.

Exact methods for solving the SPP and the CAP come in three varieties: branch and bound, cutting planes, and a hybrid called **branch and cut.** Fast exact approaches to solving the CAP require algorithms that generate both good lower and upper bounds on the maximum objective-function value of the instance. Because even small instances of the CAP1 may involve a huge number of columns (bids) the techniques described above need to be augmented with **column generation.** An overview of such methods can be found in Barnhart et al. [10].

Exact methods for CAP have been proposed by Rothkopf et al. [61], Fujishima et al. [31], Sandholm [64], and Andersson et al. [6]. The first uses straight dynamic programming, while the second and third use refinements by substantially pruning the search tree and introducing additional bounding heuristics. Andersson et al. use integer programming utilizing the commercial CPLEX-solver.

Andersson et al. [6] point out that CPLEX dominates (in terms of run times) the algorithms of Sandholm [64] and appears competitive with Fujishima et al. [31]. They remark that solution times can be sensitive to problem structure. For this reason Leyton-Brown et al. [46] are compiling a suite of test problems.

On the commercial side, Logistics.com's OptiBid^{TM} software has been used in situations where the number of bidders is between 12 and 350 with the average being around 120. The number of lanes (objects) has ranged between 500 and 10,000. Additionally, each lane bid can contain a capacity constraint as well as a budget capacity constraint covering multiple lanes. The typical number of lanes is 3,000. OptiBid^{TM} does not limit the number of distinct subsets that bidders bid on or the number of items allowed within a package. OptiBid^{TM} is based on an integer-program solver with a series of proprietary formulations and starting heuristic algorithms.

SAITECH-INC's bidding software, SBID, is also based on integer programming. They report being able to handle problems of similar size as OptiBid^{TM}.

Methods that find feasible and possibly near optimal solutions to the CAP have also been proposed, see for example Akcoglu et al. [1]. Also any of the approximation algorithms for SPP would apply as well.

It is natural and important to ask about the effect of approximation on incentives. For example, suppose in the VCG scheme one were to use an allocation found by an approximation algorithm rather than an optimal one. Such a modification in the scheme does not in general preserve incentive compatibility (see Nisan and Ronen [57]). The reason is that for each allocation rule α, we must solve LP_α to find a payment rule that would implement α in an incentive compatible way. One cannot take a payment rule designed to implement a given allocation rule in an incentive compatible way and hope that will implement some entirely different rule in an incentive compatible way. There will of course be unusual instances where this will work. See for example Lehmann et al. [49]. These instances are too specialized to be generally relevant.

Even if one is willing to relax incentive compatibility, an approximate solution to the underlying optimization problems in the VCG can lead to other problems. There can be many different solutions to an optimization problem whose objective function values are within a specified tolerance of the optimal objective function value. The payments specified by the VCG scheme are very sensitive to the choice of solution. Thus the choice of approximate solution can have a significant impact on the payments made by bidders. This issue is discussed by Johnson et al. [41] in the context of an electricity auction used to decide the scheduling of short term electricity needs. Through simulations they show that variations in near-optimal schedules that have negligible effect on total system cost can have significant consequences on the total payments by bidders.

5.2 Supposed Problems with VCG

The VCG auction is sometimes traduced for being counter-intuitive (see for example Johnson et al. [41]). One can construct examples where the revenue from the VCG auction can fall as the number of bidders increase. The second is that the VCG auction (in a procurement setting) can end up paying many times more for an asset than it is worth. The third is that it is vulnerable to false name bidding (see Yokoo et al. [69]), that is a single bidder submitting bids under many different identities. We want to take some time to counter the view that these are counter-intuitive or odd. First the examples. All will be in a procurement context.

Consider two problems defined on graphs. Fix a connected graph G with edge set E and vertex set V. Every edge in the graph is owned by a single agent and no agent owns more than one edge.[11] The length/cost of an edge is private information to the agent who owns the edge. The first problem is the minimum cost spanning tree problem. If A is any subset of agents, denote by $M(A)$ the length of the minimum spanning tree using edges in A alone.[12]

The second problem is the shortest path problem. Designate a source, s and sink t and let $L(A)$ be the length of the shortest $s - t$ path using edges only in A.[13] Also to make things interesting, assume that the minimum $s - t$ cut is at least two.

In the first case the auctioneer would like to buy the edges necessary to build the minimum spanning tree. If the auctioneer uses a VCG auction, what does she end up paying for an edge, e, say, in the MST?

Recall, that in a VCG auction each agent must obtain their marginal product. So, what is the marginal product of agent e? To answer this we must determine what the length of the minimum spanning tree without edge e would be. Let T be the MST and f the cheapest edge that forms a cycle with e in $T \cup f$. The increase in cost from excluding agent e would be the difference in cost between f and e. Thus, if the number of edges in G were to increase, the price that the auctioneer would have to pay for an edge can only go down.

Second, if the ratio of the most expensive edge to the cheapest were β, say, each agent in T would receive at most β times their cost.

Third, could the agent e benefit by introducing dummy bidder? This can happen only if the dummy bidder serves to increase agent e's marginal product. If this dummy bidder does not form a cycle through e it cannot make a difference to the outcome. If it does, it must have a length smaller than edge f, so reducing agent e's marginal product.

In short everything about the VCG auction in this case conforms to our intuition about how auctions should behave.

Now turn to the shortest path problem. Consider a graph G with two disjoint $s - t$ paths each involving K edges. The edges in the 'top' path have cost 1 each while in the 'bottom' path they have cost $1 + r$ each. The VCG auction applied to this case would choose the top path. Each agent on the top path has a marginal product of rK and so receives a payment of $1 + rK$. The total payment by the auctioneer is $K + rK^2$, which, for appropriate r and K is many times larger than the length of the shortest path. A clear case of the 'generosity' of the VCG auction.

Furthermore, if we increase the number of agents by subdividing a top edge and assigning each segment a cost of 1/2, the payments increase without an increase in the length of the path! In fact an individual agent could increase her return by pretending to be two separate edges rather than one. So, we have two environments, one where the VCG confirms with 'intuition' and the other where it does not. What causes these anomalies to occur? To answer this question it is useful to associate with the auction a co-operative game defined as follows. As before let Γ be the space of feasible allocations, and v_j be the value function of bidder j. Let N denote the set of buyers, s the seller and $N' = N \cup \{s\}$. For any coalition $S \subseteq N'$ we denote the value of the coalition S to be $V(S)$. If $s \notin S$ then $V(S) = 0$, otherwise $V(S) = \max_{a \in \Gamma} \sum_{i \in S \setminus s} v_i(a)$. Define $V(s) = 0$. An **imputation** is a vector x such that $\sum_{i \in N'} x_i = V(N')$ and $x_i \geq V(i)$ for all $i \in N'$. An imputation is thus a way of dividing up the total value, $V(N')$ so that each agent gets at least as much as he would on his own. The **core** is the set of imputations that satisfy $\sum_{i \in S} x_i \geq V(S)$ for all $S \subseteq N'$. Thus an imputation is in the core if each subset of agents gets at least as much value as they generate. More vividly, if one were to divide $V(N')$ up amongst the agents in such a way that the division was in the core, no subset of agents, has an incentive to break away as they could only generate less value by themselves.

The outcome of the VCG auction is an imputation x where

$$x_i = V(N') - V(N' \setminus i) \qquad \forall i \in N$$
$$x_s = V(N') - \sum_{i \in N}[V(N') - V(N' \setminus i)].$$

If the imputation implied by the VCG auction is in the core, then the anomalous behavior described above cannot occur. For the imputation implied by the VCG outcome to be in the core we require that

$$\sum_{i \in S \setminus s}[V(N') - V(N' \setminus i)] + V(N') - \sum_{i \in N}[V(N') - V(N' \setminus i)] \geq V(S) \; \forall S \subseteq N'.$$

To describe this condition more succinctly, let $MP(S) = V(N') - V(N' \setminus S)$ for all $S \subseteq M$. Then the condition we need is:

$$MP(S) \geq \sum_{i \in S} MP(i) \; \forall S \subseteq N.$$

This is known as the **agents are substitutes condition** (Shapley [65]). A subset of agents together have greater economic value than all of them separately. It is implied by submodularity of $V(\cdot)$.

In the spanning tree setup the agents are substitutes condition holds (see Bikhchandani et al. [8]). Intuitively, an agent is included in the MST if it has low cost and there is no other agent with a lower cost that will perform the same 'function'. That function is to 'cover a cycle'.[14] Thus no agent depends on another to make it into the MST, but another agent could crowd it out of the MST. This pits the agents against each other and benefits the auctioneer. Alternatively, the agents profit by banding together, this in effect is what the substitutes property means.

In the shortest path case, the agents are substitutes condition is violated. Agents on the same path complement each other, in that no agent on the top path can be selected unless all other agents on the top path are selected. The marginal product of the agents on the top path is less than the sum of their individual marginal products.

To summarize, when the agents are substitutes condition holds, the VCG auction works without problem. When the agents are substitutes condition is violated, then economic power is in the hands of the bidders, unless, one is prepared to use an auction that does not share some of the properties of the VCG auction. Here we shall argue that the prospects in this regard are bleak.

Suppose you face severe complementarities amongst the bidders and your goal is to reduce the payments made by you to them. Is there another auction that is incentive compatible and individually rational that one might use?

If yes, it must give up on efficiency. This is a line taken by Archer and Tardos [5] who suggests that one relax the efficiency requirement. In the shortest path context this would mean accepting a path that is perhaps only slightly longer than the shortest path but the auctioneer ends up paying far less for it. Archer and Tardos [5] replace efficiency by a collection of 'monotonicity' type conditions and show that the same 'odd' behavior persists.[15] In light of the complementary nature of agents this is not surprising. In fact we think it is the complementary nature of agents coupled with incentive compatibility that raises the costs of the VCG auction rather than the desire for efficiency.

To see why, return to the simple graph with two disjoint $s-t$ paths. Now that we understand the complementary nature of the agents, we recognize that the auctioneer wants $s-t$ paths to compete with each other rather than edges. How can this be arranged? One thing the auctioneer could do is tell the agents on the top path to band together and submit a single bid and likewise for the ones on the bottom. This forces competition between the two paths and should lower payments. Now think about the winning group of bidders. They receive a certain amount of money to divide between them. How do they do it? Each agent, by threatening to withdraw from the coalition can claim a lion's share of the money. This makes it very difficult for the bidders to band together.

5.3 Ascending Implementations of VCG

One of the difficulties with implementing a VCG auction is the requirement that a bidder submits a bid for *every* combination of assets. Evaluating the worth of some combination of assets can be both time consuming and expensive. Especially when almost all bids will not be part of the efficient allocation. In the FCC spectrum auction, to prepare a bid for some combination of spectrum licenses, the bidder must develop a business plan for their use in order to estimate their worth. In logistics auctions an execution of some vehicle routing algorithm would be required to ascertain the costs of some collection of lanes. Since the cost of computing as well as communicating a bid in a combinatorial auction can be large, there has been an interest in developing indirect versions of the VCG auction. Such auctions might proceed in rounds, for example, where bidders are asked if some bundle's value exceeds some threshold at that round. This may be easier to do than evaluating the acutual value of the bundle. Also bidders could focus on bundles that 'matter' in some sense.

Here we focus on the class of indirect mechanisms that are called **ascending auctions.**

In the auction of a single object, it has long been known that an open (ascending) auction exists that duplicates the outcomes of the Vickrey auction. This is the English auction, in which the auctioneer continuously raises the price for the object. An agent is to keep his hand raised until the price exceeds the amount the agent is willing to pay for the object. The price continues to rise only as long as at least two agents have hands raised. When the price stops, the remaining agent wins the object at that price.

In such an auction, no agent can benefit by lowering his hand prematurely, or by keeping his hand raised beyond the price he is willing to pay, as long as every other agent is behaving in a way consistent with some valuation for the object. The ascending auction therefore has desirable incentive properties comparable to its sealed bid counterpart.

One reason why an ascending version of the Vickrey auction is preferable to its sealed bid counterpart is experiments (Kagel and Levin [43]) which suggest that bidders do not recognize that bidding truthfully is a dominant strategy in the sealed bid version but do behave according to the equilibrium prescription of the ascending counterpart.

Ascending auctions also have problems in that they make collusion amongst the bidders easier. This we discuss in section 5.4.

Ascending auctions come in two varieties (with hybrids possible). In the first, bidders submit, in each round, prices on various bundles. The auctioneer makes a provisional allocation of the items that depends on the submitted prices. Bidders are allowed to adjust their price offers from the previous rounds and the auction continues. Such auctions come equipped with rules to ensure rapid progress and encourage competition. Ascending auctions of this type seem to be most prevalent in practice.

In the second type, the auctioneer sets the price and bidders announce which bundles they want at the posted prices. The auctioneer observes the requests and adjusts the prices. The price adjustment is usually governed by the need to balance demand with supply.

Call auctions of the first type **quantity setting**, because the auctioneer sets the allocation or quantity in response to the prices/bids set by bidders. Call the second **price setting** because the auctioneer sets the price. Quantity setting auctions are generally harder to analyze because of the freedom they give to bidders. Each bidder determines the list of bundles as well as prices on them to announce. In price setting auctions, each bidder is limited to announcing which bundles meet their needs at the announced prices.

The simplest example is the auction of a single object. The popular English ascending auction is an example of a quantity setting auction. Bidders submit prices in succession, with the object tentatively assigned to the current highest bidder. The auction terminates when no one is prepared to top the current high bid. The 'dual' version to this auction has the auctioneer continuously raising the price. Bidders signal their willingness to buy at the current price by keeping their hands raised. The auction terminates the instant a single bidder remains with his hand raised. In fact this dual version of the English auction is used as a stylized model of the English auction itself for the purposes of analysis (see for example Klemperer [44]).

Our discussion of ascending auctions is motivated by the following observation: an ascending auction that produces the efficient allocation is an algorithm for finding the efficient allocation. Therefore, by studying appropriate algorithms for finding the efficient allocation we will identify ascending auctions. The idea that certain optimization algorithms can be interpreted as auctions is not new and goes back at least as far as Dantzig [19]. In particular many of the ascending auctions proposed are either primal-dual or subgradient algorithms for the underlying optimization problem for determining the efficient allocation.

In order to understand how to design ascending auctions it is important to identify what properties prices must have in order to produce an allocation that solves CAP1. Such an understanding can be derived from the duality theory of integer programs.

5.3.1 Duality in Integer Programming.
Consider the following set packing problem (SPP):

$$\max vx$$
$$\text{s.t.} \sum_{j=1}^{n} a_{ij} x_j \leq 1 \quad \forall i = 1, \ldots, m$$
$$x_j = 0, 1 \quad \forall j = 1, \ldots, n$$

Here $\{a_{ij}\}$ is a m-by-n 0-1 matrix and $v \in \mathbb{R}^n$ a non-negative vector.

To describe the dual to SPP let $\mathbf{1}$ denote the m-vector of all 1's and a^j the jth column of the constraint matrix A. The (superadditive) dual to SPP is the problem of finding a superadditive, non-decreasing function $F : \mathbb{R}^m \to \mathbb{R}^1$ that solves

$$\min F(\mathbf{1})$$
$$\text{s.t.} F(a^j) \geq v_j \quad \forall j \in V$$
$$F(\mathbf{0}) = 0$$

We can think of F as being a non-linear price function that assigns a price to each bundle of goods (see Wolsey [67]).

If the primal integer program has the integrality property, there is an optimal integer solution to its linear programming relaxation, the dual function F will be linear i.e. $F(u) = \sum_{i \in M} y_i u_i$ for some y and all $u \in \mathbb{R}^m$. The dual becomes:

$$\min \sum_{i \in M} y_i$$
$$\text{s.t.} \sum_{i \in M} a_{ij} y_i \geq v_j \quad \forall j \in V$$
$$y_i \geq 0 \quad \forall i \in M$$

That is, the superadditive dual reduces to the dual of the linear programming relaxation of SPP. In this case we can interpret each y_i to be the price of object i. Thus an optimal allocation given by a solution to the CAP can be supported by prices on individual objects.

Optimal objective function values of SPP and its dual coincide (when both are well defined). There is also a complementary slackness condition:

THEOREM 7.2 (NEMHAUSER AND WOLSEY [59]) *If x is an optimal solution to SPP and F an optimal solution to the superadditive dual then*

$$(F(a^j) - v_j)x_j = 0 \ \forall j \in V.$$

Solving the superadditive dual problem is as hard as solving the original primal problem. It is possible to reformulate the superadditive dual problem as a linear program (the number of variables in the formulation is exponential in the size of the original problem). For small or specially structured problems this can provide some insight. The interested reader is referred to Nemhauser and Wolsey [59] for more details. An indirect way is through the use of an extended formulation.

5.3.2   Extended Formulations.   Recall formulation CAP1 (with b^j replaced by v_j in the objective function):

$$V(N) = \max \sum_{j \in N} \sum_{S \subseteq M} v_j(S) y(S, j)$$
$$\text{s.t.} \sum_{S \ni i} \sum_{j \in N} y(S, j) \leq 1 \quad \forall i \in M$$
$$\sum_{S \subseteq M} y(S, j) \leq 1 \quad \forall j \in N$$
$$y(S, j) = 0, 1 \quad \forall S \subseteq M, \forall j \in N$$

Even though the linear relaxation of CAP1 admits fractional solutions let us write down its linear programming dual. To each constraint

$$\sum_{S \ni i} \sum_{j \in N} y(S, j) \leq 1 \quad \forall i \in M$$

associate the variable p_i, which will be interpreted as the price of object i. To each constraint

$$\sum_{S \subseteq M} y(S, j) \leq 1 \quad \forall j \in N$$

associate the variable λ_j which will be interpreted as the profit of bidder j. The dual is

$$\min \sum_{i \in M} p_i + \sum_{j \in N} \lambda_j$$
$$\text{s.t.} \; \lambda_j + \sum_{i \in S} p_i \geq v_j(S) \quad \forall S \subseteq M$$
$$\lambda_j \geq 0 \quad \forall j \in N$$
$$p_i \geq 0 \quad \forall i \in M$$

Let (λ^*, p^*) be an optimal solution to this linear program. It is easy to see that

$$\lambda_j^* = \left[\max_{S \subseteq M} \left[v_j(S) - \sum_{j \in M} p_j^* \right] \right]^+$$

which has a natural interpretation. Announce the price vector p^* and then let each agent choose the bundle that maximizes his payoff. Would the resulting choices be compatible and form an efficient allocation? No. Since CAP1 is not a linear program there is a duality gap, so that it would not be possible to engineer the efficient allocation by announcing prices on individual items alone. Unless CAP1 has the integrality property, no auction that works by setting prices on individual items alone, can be guaranteed to produce the efficient allocation. One instance of when CAP1 has an optimal integral solution is when all v_j's satisfy the gross substitutes property.

Since the linear relaxation of the above integer program admits fractional solutions, it is natural to look for a stronger formulation. Bikhchandani and Ostroy [12] offer one using an extended formulation.

To describe the first extended formulation, let Π be the set of all possible partitions of the objects in the set M. If π is an element of Π, we write $S \in \pi$ to mean that the set $S \subseteq M$ is a part of the partition π. Let $z_\pi = 1$ if the partition π is selected and zero otherwise. These are the auxiliary variables.

Using them we can reformulate CAP1 as follows:

$$\max \sum_{j \in N} \sum_{S \subseteq M} v_j(S) y(S,j)$$

$$\text{s.t.} \sum_{S \subseteq M} y(S,j) \leq 1 \quad \forall j \in N$$

$$\sum_{j \in N} y(S,j) \leq \sum_{\pi \ni S} z_\pi \quad \forall S \subset M$$

$$\sum_{\pi \in \Pi} z_\pi \leq 1$$

$$y(S,j) = 0,1 \quad \forall S \subseteq M, \ j \in N$$

$$z_\pi = 0,1 \quad \forall \pi \in \Pi$$

Call this formulation CAP2. In words, CAP2 chooses a partition of M and then assigns the sets of the partition to bidders in such a way as to maximize revenue. CAP2 is stronger than CAP1 in the sense that there are feasible fractional solutions to CAP1 that infeasible in CAP2. Nevertheless, CAP2 still admits fractional extreme points.

The dual of the linear relaxation of CAP2 involves one variable, λ_j, for each constraint that an agent gets only one bundle, a variable $p(S)$ for each constraint that only bundles are permissible that are part of the chosen partition, and one variable μ for the third constraint. μ can be interpreted as the revenue of the seller, while $p(S)$ as the price of the subset S. In fact the dual will be:

$$\min \sum_{j \in N} \lambda_j + \mu$$

$$\text{s.t.} \ \lambda_j \geq v_j(S) - p(S) \quad \forall j \in N \ \forall S \subseteq M$$

$$\mu \geq \sum_{S \in \pi} p(S) \quad \forall \pi \in \Pi$$

$$\lambda_j, p(S), \mu \geq 0 \quad \forall j \in N \ \forall S \subseteq M$$

and has the obvious interpretation: minimizing the bidders' surplus plus μ. Thus one can obtain bundle prices from the extended formulation. These prices do not support the optimal allocation since CAP2 is not integral. Unless CAP2 has the integrality property, no auction that works by setting prices on bundles alone, can be guaranteed to produce the efficient allocation. We know of no interesting class of value functions for which CAP2 has an optimal integral solution. In fact even if all the v_j's are submodular, there are still instances where CAP2 will not have optimal integral solution. This rules out the possibility of supporting an efficient allocation using non-linear prices for preferences that violate the gross substitutes condition.

To obtain an even stronger formulation, let μ denote both a partition of the set of objects *and* an assignment of the elements of the partition to bidders. Thus μ and μ' can give rise to the same partition, but to different assignments of the parts to bidders. Let Γ_P denote the set of all such partition-assignment pairs. We will write μ_j for the part assigned to agent j. Let $\delta_\mu = 1$ if the partition-assignment pair $\mu \in \Gamma_P$ is selected, and zero otherwise. Using these new variables the efficient allocation can be found by solving the formulation we will call **CAP3**.

$$V(N) = \max \sum_{j \in N} \sum_{S \subseteq M} v_j(S) y(S, j)$$

$$\text{s.t. } y(S, j) \leq \sum_{\mu \in \Gamma_P : \mu_j = S} \delta_\mu \quad \forall j \in N, \forall S \subseteq M$$

$$\sum_{S \subseteq M} y(S, j) \leq 1 \quad \forall j \in N$$

$$\sum_{\mu \in \Gamma_P} \delta_\mu \leq 1$$

$$y(S, j) = 0, 1 \quad \forall S \subseteq M, \forall j \in N$$

Bikhchandani and Ostroy [12] showed that this formulation's linear programming relaxation has optimal extreme points that are integral.

To write down the dual, we associate with each constraint $y(S, j) \leq \sum_{\mu \in \Gamma_P : \mu_j = S} \delta_\mu$ a variable $p_j(S) \geq 0$ which can be interpreted as the price that agent j pays for the set S. To each constraint $\sum_{S \subseteq M} y(S, j) \leq 1$ we associate a variable $\lambda_j \geq 0$ which can be interpreted as agent j's surplus. To the constraint $\sum_{\mu \in \Gamma_P} \delta_\mu \leq 1$ we associate the variable λ^s which can be interpreted as the seller's surplus. The dual **(DP3)** becomes

$$\min \sum_{j \in N} \lambda_j + \lambda^s$$

$$\text{s.t. } p_j(S) + \lambda_j \geq v_j(S) \quad \forall j \in N, \forall S \subseteq M$$

$$- \sum_{S \subseteq M : S = \mu_j} p_j(S) + \lambda^s \geq 0 \quad \forall \mu \in \Gamma_P$$

$$p_j(S) \geq 0 \quad \forall j \in N, \forall S \subseteq M$$

Because CAP3 has the integrality property, the dual above is exact. Hence, an efficient allocation can always be supported by prices that are non-linear and non-anonymous, meaning different agents see different prices for the same bundle. More significantly, the dual variable associated with the second constraint, $\sum_{S \subseteq M} y(S, j) \leq 1$, can be interpreted as agent j's marginal product.

To see why, for an agent $j \in N$, reduce the right hand side of the corresponding constraint (holding all other right hand sides fixed) to zero. This has the effect of removing agent j from the problem. The resulting change in optimal objective function value will be agent j's marginal product. This argument shows only that amongst all optimal dual solutions there is one such that λ_j is equal to agent j's marginal product. It does not prove that this dual solution gives at the same time the marginal product for all agents $i \neq j$.

The difficulty is to ensure that a dual solution exists such that, simultaneously, each λ_j variable takes on the value of the corresponding agent's marginal product. Bikhchandani and Ostroy [12] characterized precisely when this would occur. The proof follows from the coincidence of the co-operative game defined in section 5.2 and the set of optimal solutions to (DP3).

THEOREM 7.3 (BIKHCHANDANI AND OSTROY [12]) *The agents are substitutes condition holds if and only if there exists an optimal solution to (DP3), $((\lambda_j)_{j \in N}, \lambda^s)$, such that $\lambda_j = V(N) - V(N \setminus j)$ for all $j \in N$, i.e., the dual provides each agent's marginal product.*

The preceding discussion implies a tradeoff for the auction designer. The simpler the price structure of the auction, the less likely one is to achieve efficiency. On the other hand the more complex the prices, the closer to efficiency.

5.3.3 Primal-Dual Methods. Theorem 7.3 (of Bikhchandani and Ostroy [12]) provides a systematic way of deriving an ascending implementation of the sealed bid VCG auction in the case of private values and no externalities. It is described in de Vries, Schummer and Vohra [25] and we reproduce it here:

1 For the domain of preferences being considered, verify that the agents are substitutes condition holds.

2 Formulate as a linear program the problem of finding the efficient allocation. Appropriate variables in the dual can be interpreted as the prices paid by buyers as well the surplus they receive. There can be many different formulations of the same problem but not any formulation will do. The formulation must be sufficiently rich that the Vickrey prices are contained in the set of feasible dual solutions. We believe that with the exception of diminishing marginal utilities, the only formulation rich enough for the job is CAP3.

3 We have a primal problem whose optimal solutions correspond to efficient allocations. Given (1) and (2) we know that amongst the set of optimal dual solutions that support the efficient allocation is one that corresponds to the Vickrey prices. A corollary of Theorem 7.3 shows that

these will be the prices that minimize the surplus to the seller. Call this the **buyer optimal** dual solution. Choose a primal-dual algorithm for this linear program that terminates in the buyer optimal dual solution.

4. Interpret the primal-dual algorithm as an auction. This part is easy given the structure of such algorithms. They operate by choosing at each iteration a feasible dual solution (i.e. prices) and then forcing a primal solution that is complementary to the current dual solution. This corresponds to bidders reporting their demand correspondences at current prices. If a primal solution complementary to the current dual solution exists, the algorithm stops. This corresponds to all demand being satisfied at current prices. If no such primal solution exists, the Farkas lemma yields a direction along which dual variables must be adjusted. Non-existence of a complementary primal solution corresponds to an imbalance between supply and demand. The price change dictated by the Farkas lemma attempts to adjust the imbalance. The magnitude of the price increase is determined by the smallest difference between any two valuations. The Farkas lemma actually offers more than one direction in which prices may be adjusted. If all one cares about is convergence to the efficient allocation it does not matter which of these directions is chosen. However, if one desires the algorithm to produce increasing prices and terminate with the buyer optimal dual solution care must be exercised in choosing the direction.

5. Given the algorithm terminates with the efficient allocation and gives each buyer their marginal product, it follows immediately that bidding truthfully is an ex-post Nash equilibrium of the auction.

What fails if the agents are substitutes condition does not hold? First, there is no formulation which contains the Vickrey outcome in its dual. Second, one cannot guarantee that bidding sincerely will be an ex-post equilibrium of the auction (obtained from this primal-dual approach). In fact, when the "aggregate valuations are submodular condition", which is only slightly stronger than agents are substitutes, fails it is shown in [25] that no ascending auction (suitably defined) can implement the Vickrey outcome. However, assuming sincere bidding it is still the case that the auction terminates in the efficient allocation with buyer optimal prices. A reader interested in a fuller account should consult [25].

5.3.4 Lagrangean Relaxation. Another class of algorithms for linear programs that admit an auction interpretation are based on Lagrangean relaxation. We describe the bare bones of the method first and then give an 'auction' interpretation of it.

Recall formulation CAP1

$$V(N) = \max \sum_{j \in N} \sum_{S \subseteq M} v_j(S) y(S, j)$$

$$\text{s.t.} \sum_{S \ni i} \sum_{j \in N} y(S, j) \leq 1 \quad \forall i \in M$$

$$\sum_{S \subseteq M} y(S, j) \leq 1 \quad \forall j \in N$$

$$y(S, j) = 0, 1 \quad \forall S \subseteq M, \forall j \in N$$

Let $V_{LP}(N)$ denote the optimal objective function value to the linear programming relaxation of CAP1. Note that $V(N) \leq V_{LP}(N)$. Consider now the following relaxed problem:

$$V_p(N) = \max \sum_{j \in N} \sum_{S \subseteq M} v_j(S) y(S, j) + \sum_{i \in M} p_i [1 - \sum_{S \ni i} \sum_{j \in N} y(S, j)]$$

$$\text{s.t.} \sum_{S \subseteq M} y(S, j) \leq 1 \quad \forall j \in N$$

$$0 \leq y(S, j) \leq 1 \quad \forall S \subseteq M, \forall j \in N$$

The objective function can be rewritten to read:

$$\sum_{j \in N} \sum_{S \subseteq M} [v_j(s) - \sum_{i \in S} p_i] y(S, j).$$

Evaluating $V_p(N)$ is easy. For each agent j we must solve

$$\max \sum_{S \subseteq M} [v_j(S) - \sum_{i \in S} p_i] y(S, j)$$

$$\text{s.t.} \sum_{S \subseteq M} y(S, j) \leq 1$$

$$y(S, j) \geq 0 \quad \forall S \subseteq M$$

Thus, for each j, identify the set S that maximizes $v_j(S) - \sum_{i \in S}$. A basic result of the duality theorem of Linear Programming is that $V_{LP}(N) = \min_{p \geq 0} V_p(N)$.

If one can determine a p that solves $\min_{p \geq 0} V_p(N)$, one would have a procedure for computing $V_{LP}(N)$. The resulting solution (values of the y variables) while integral need not be feasible for CAP1. To produce a feasible solution one can round (heuristically) the final solution into a feasible one.

Finding the p that solves $\min_{p \geq 0} V_p(N)$ can be accomplished using the subgradient algorithm. Suppose the value of the Lagrange multiplier p at iteration

t is p^t. Choose any subgradient of $V_{p^t}(N)$ and call it s^t. Choose the Lagrange multiplier for iteration $t+1$ to be $p^t + \theta^t s^t$, where θ^t is a positive number called the step size. In fact if y^t is the optimal solution associated with $V_{p^t}(N)$, we can choose $s_i^t = \sum_{S \ni i} \sum_{j \in N} y^t(S,j) - 1$ for all $i \in M$. Hence

$$p_i^{t+1} = p_i^t + \theta^t s_i^t \quad \forall i \in M.$$

Notice that $p_i^{t+1} > p_i^t$ for any i such that $\sum_{S \ni i} \sum_{j \in N} y^t(S,j) > 1$. The penalty term is increased on any item for which demand exceeds supply.

Here is the auction interpretation. The auctioneer chooses a price vector p for the individual objects and bidders submit bids. If the bid, $v_j(S)$, for the bundle S exceeds $\sum_{i \in S} p_i$, this bundle is tentatively assigned to that bidder. Notice that the auctioneer need not know what $v_j(S)$ is ahead of time. This is supplied by the bidders after p is announced. In fact, the bidders need not announce bids, they could simply state which individual objects are acceptable to them at the announced prices. In effect bidders simply declare which individual objects are acceptable at current prices. No direct information about bundles is being exchanged. The auctioneer can randomly assign objects to bidders in case of ties. If there is a conflict in the assignments, the auctioneer uses the subgradient algorithm to adjust prices and repeats the process.

Now let us compare this auction interpretation of Lagrangean relaxation with the simultaneous ascending auction (SAA) proposed by P. Milgrom, R. Wilson and P. McAfee (see Milgrom [52]). In the SAA, bidders bid on individual items simultaneously in rounds. To stay in the auction for an item, bids must be increased by a specified minimum from one round to the next just like the step size. Winning bidders pay their bids. The only difference between this and Lagrangean relaxation, is that the bidders through their bids adjust their prices rather than the auctioneer. The adjustment is along a subgradient.[16] Bids increase on those items for which there are two or more bidders competing.

One byproduct of the SAA is called the **exposure problem**. Bidders pay too much for individual items or bidders with preferences for certain bundles drop out early to limit losses. As an illustration consider an extreme example of a bidder who values the item i and j at \$100 but each separately at \$0. In the SAA, this bidder may have to submit high bids on i and j to be able to secure them. Suppose that it loses the bidding on i. Then it is left standing with a high bid j which it values at zero. The presence of such a problem is easily seen within the Lagrangean relaxation framework. While Lagrangean relaxation will yield the optimal objective function value for the *linear relaxation* of the underlying integer program, it is not guaranteed to produce a feasible solution. Thus the solution generated may not satisfy the complementary slackness conditions. The violation of complementary slackness is the *exposure problem* associated with this auction scheme. To see why, notice that a violation of complementary slackness means

$$\sum_{i \in S} p_i > v_j(S) \text{ and } y(S, j) = 1.$$

Hence the sum of prices exceeds the value of the bundle that the agent receives. Notice that any auction scheme that relies on prices for individual items alone will face this problem.

To avoid this exposure problem we could, as is described by de Vries et al. [25], apply the subgradient algorithm to a stronger formulation, say, CAP3. We relax the constraints $y(S, j) \leq \sum_{\mu \ni S^j} \delta_\mu$. Let $p_j(S) \geq 0$ be the corresponding multipliers. Then

$$Z_p(N) = \max \sum_{j \in N} \sum_{S \subseteq M} [v_j(S) - p_j(S)] y(S, j) + \sum_{\mu \in \gamma} \delta_\mu [\sum_{\mu \ni S^j} p_j(S^j)]$$

$$\text{s.t.} \sum_{S \subseteq M} y(S, j) \leq 1 \quad \forall j \in N$$

$$\sum_{\mu \in \Gamma_P} \delta_\mu \leq 1$$

$$y(S, j) = 0, 1 \quad \forall S \subseteq M, \forall j \in N$$

Given $p_j(S)$, the solution to the relaxed problem can be found as follows.

1. Choose $\delta_\mu = 1$ to maximize $\sum_{\mu \in \gamma} \delta_\mu [\sum_{\mu \ni S^j} p_j(S^j)]$. In words, choose the solution that maximizes revenue at current prices. Call that solution the seller optimal allocation and label it μ^*.

2. For each $j \in N$ choose a $B^j \in \arg\max_{S \subseteq G} [v_j(S) - p_j(S)]$ and set $y(B^j, j) = 1$. In words, each agent chooses a utility maximizing bundle at current prices.

Now apply the subgradient algorithm. If B^j is assigned to agent j under μ^*, the corresponding constraint: $y(B^j, j) \leq \sum_{\mu \ni B^j} \delta_\mu$, is satisfied at equality. So, by the subgradient iteration there is no change in the value of the multiplier $p_j(B^j)$. In words, if an agent's utility maximizing bundle is in μ^*, there is no price change on this bundle.

If B^j is not assigned to agent j under μ^*, the corresponding constraint: $y(B^j, j) \leq \sum_{\mu \ni B^j} \delta_\mu$, is violated. So, by the subgradient iteration value of the multiplier $p_j(B^j)$ is increased by $\theta > 0$. In words, if an agent's utility maximizing bundle is not in μ^* that bundle sees a price increase.

If μ^* assigns to agent j a bundle $S \neq B^j$, then the left hand side of $y(S, j) \leq \sum_{\mu \ni S^j} \delta_\mu$ is zero while the right hand side is 1. In this case the value of the multiplier $p_j(S)$ is decreased by θ. In words, a bundle that is a part of μ^* but not demanded sees a price decrease. If one wishes to avoid a price decrease, this step can be dropped without affecting convergence to optimality.[17]

We have interpreted both the multipliers and the change in their value as the auctioneer setting and changing prices. We could just as well interpret the prices and price changes as bids and bid changes placed by the buyers. Thus, if a utility maximizing bundle is part of a seller optimal allocation, the bid on it does not have to be increased. If a utility maximizing bundle is not part of a seller optimal allocation, the bidder must increase her bid on it by exactly θ. This is precisely the i-bundle auction of Parkes and Ungar [60] as well as the package bidding auction of Ausubel and Milgrom [4]. The step size, θ of the subgradient algorithm becomes the bid increment. From the properties of the subgradient algorithm, for this auction to terminate in the efficient allocation, the bid increments must be very small and go to zero sufficiently slowly.

5.4 Threshold and Collusion Problems

The exposure problem is not the only problem that can occur with an ascending auction. Another is the **threshold problem** (see Bykowsky et al. [7]) which is an instance of the violation of the agents are substitutes condition.

To illustrate, consider three bidders, 1, 2 and 3, and two objects $\{x, y\}$. Suppose:

$$v_1(\{a\}) = v_1(\{b\}) = 0, \quad v_1(\{a,b\}) = 100,$$
$$v_2(\{a\}) = v_2(\{b\}) = 75, \quad v_2(\{a,b\}) = 0,$$
$$v_3(\{a\}) = v_3(\{b\}) = 40, \quad v_3(\{a,b\}) = 0.$$

Here $v_i(\cdot)$ represents the value to bidder i of a particular subset. Notice that the bid that i submits on the set S, $b^i(S)$, need not equal $v_i(S)$.

If the bidders bid truthfully, the auctioneer should award x to 2 and y to 3, say, to maximize his revenue. Notice however that bidder 2 say, under the assumption that bidder 3 continues to bid truthfully, has an incentive to shade his bid down on x and y to, say, 65. Notice that bidders 2 and 3 still win but bidder 2 pays less. This argument applies to bidder 3 as well. However, if they both shade their bids downwards they can end up losing the auction. A collection of bidders whose combined valuation for distinct portions of a subset of items exceeds the bid submitted on that subset by some other bidder may find it hard to coordinate their bids so as to outbid the large bidder on that subset. The basic problem is that the bidders 2 and 3 must decide how to divide $75 + 40 - 100$ between them. Every split can be rationalized as the equilibrium of an appropriate bargaining game.

Ascending auctions, because of their open nature, allow bidders to signal to one another and so coordinate their bidding. To illustrate suppose two distinct bidders each wanting at most one of the objects. Suppose one were to run SAA to sell off the two objects. If bidder 1 bids only on object A and bidder 2 only on object B, each gets an object for zero. How can the bidders coordinate like this? They could signal through the trailing digits of their bids (which would coincide with numerical label given each object) their intention to bid on one and not the other. The example is not far fetched and did occur in some of the well known auctions for FCC spectrum, see Cramton and Schwartz [18].

5.5 Other Ascending Auctions

Other ascending auctions that do not fit neatly into our division between primal-dual and lagrangean have also been proposed. Wurman and Wellman [68] propose an iterative auction that allows bids on subsets but uses anonymous, non-linear prices to 'direct' the auction. Bidders submit bids on bundles and using these bids, an instance of CAP1 is formulated and solved. Then, another program is solved to impute prices to the bundles allocated that will satisfy a complementary slackness condition. In the next round, bidders must submit a bid that is at least as large as the imputed price of the bundles.

The Adaptive User Selection Mechanism (AUSM) proposed by Banks et al. [11] is motivated by a desire to mitigate the threshold problem. AUSM is asynchronous in that bids on subsets can be submitted at any time. An important feature of AUSM is an arena which allows bidders to aggregate bids to exploit synergies. DeMartini et al. [22] propose an iterative auction scheme that is a hybrid of the SAA and AUSM that is easier to connect to the Lagrangean framework. In this scheme, bidders submit bids on packages rather than on individual items. Like the SAA, bids on packages must be increased by a specified amount from one round to the next. This minimum increment is a function of the bids submitted in the previous round. In addition, the number of items that a bidder may bid on in each round is limited by the number of items s/he bid on in previous rounds. The particular implementation of this scheme advanced by DeMartini et al. [22] can also be given a Lagrangean interpretation. They choose the multipliers (which can be interpreted as prices on individual items) so as to try to satisfy the complementary slackness conditions of linear programming. Given the bids in each round, they allocate the objects so as to maximize revenue. Then they solve a linear program (that is essentially the dual to CAP1) that finds a set of prices/multipliers that approximately satisfy the complementary slackness conditions associated with the allocation.

Kelly and Steinberg [45] also propose an iterative scheme for combinatorial auctions.[18] The auction has two phases. The first phase is an SAA where bidders bid on individual items. In the second phase an AUSM-like mechanism is used. The important difference is that each bidder submits a (temporary) suggestion for an assignment of all the items in the auction. Here a temporary assignment is comprised of previous bids of other players plus new bids of his own.

5.6 Complexity of Communication

The motivation for investigating ascending auctions was that they might reduce the burden of communicating preference information to the auctioneer. Recent work by Nisan and Segal [58] suggests otherwise. Nisan and Segal argue that any indirect mechanism for implementing the efficient allocation must produce enough information to certify that the resulting allocation is efficient. This certification can always take the form of prices. Fix a preference domain, for example, all value functions that are submodular. Nisan and Segal ask how 'rich' the space of prices must be to certify an allocation as being efficient for any profile of preferences from the domain. They show that there are profiles which would, at minimum, require a price be specified for every bundle. Thus, roughly speaking, in the worst case, any indirect mechanism must communicate (in order of magnitude terms) an amount of information exponential in the number of items.

Even if an indirect mechanism were to reduce the communication burden, the complexity might be loaded into some other part of the mechanism. As an illustration consider the auction of a single good amongst n bidders. The sealed bid Vickrey auction must collect n numbers and identify the largest and second largest of them. The total complexity is $O(n)$. The ascending counterpart on the other hand increases prices by some given increment, Δ. It terminates only when the prices hit the second highest valuation (or thereabouts). Thus the complexity of the procedure is $O(v/\Delta)$ where v is the second highest valuation. Thus the complexity is exponential in the size of the values.

6. Interdependent Values

Our discussion of auctions thus far has been restricted to the private values model. It is the simplest case and one for which there are the most results. It is restrictive and there are many important environments that simply cannot be modeled using it. One such case, which we discuss briefly, is that of **interdependent values**. For brevity we focus on the case of two bidders and one object. This is enough to highlight all the difficulties.

The model involves each bidder i receiving a signal, s_i, that is assumed to be an independent draw from a commonly known distribution. The value of the asset to bidder 1 is $v_1(s_1, s_2)$ and the value to bidder 2 is $v_2(s_1, s_2)$. What is important to note is that the value of the asset to a bidder depends not just on her private information, but the information of the other bidder as well. The classic example of such a situation is bidding on oil leases. One bidder, for example, may have carried out a geological survey of the tract in question, while the other owns and extracts oil on an adjacent parcel. Each bidder has information that would affect the other's estimate of the value of the lease.

To focus on how the interdependence affects the allocation, it is usual to suppose that the functional form of the valuations, $v_i(\cdot)$, are commonly known. Additional technical conditions are placed on the v_i to ensure existence of various properties. The reader is directed to Dasgupta and Maskin [23] for a discussion. Again, because this is where there are results, our interest will be in efficient allocations, i.e., the goal is to allocate the object to the person who values it the most.

Suppose one runs a sealed bid, first price auction in which bidders are asked to submit their *values*. The problem for each bidder is that her value is unknown at the start of the auction. It depends on the private signal of the rival bidder. A bidder could use her knowledge of the distribution from which the signals are drawn to estimate this value and bid the estimate or something below it. This is not guaranteed to produce the efficient outcome. Furthermore, it can, if bidders are bidding conservatively, depress revenues to the auctioneer.

Alternatively, one could ask bidders to submit their *signals*. Using the logic of the Vickrey auction one can construct (under appropriate conditions on the v_i's) an auction where truthful reporting is an equilibrium. Auctions where bidders report signals rather than monetary values are odd.

One could use an ascending auction. If the other bidder is still in, it allows one to infer something about their signal. To see how, suppose you are bidder 1 and the current price is p. Now ask the following: if bidder 2 knew your signal and were to drop out at this price, what would that say about their signal? One can answer this question by solving the following equation: $v_2(s_1, x) = p$. From the perspective of bidder 1, x is the only unknown. Given x, we can compute $v_1(s_1, x)$. If this exceeds p we stay in the bidding, if not, drop out. Indeed, this is one of the arguments in favor of ascending auctions. They permit private information about the value of the asset to be diffused amongst the bidders. This does not mean that it is impossible to accomplish the same with a sealed bid auction. It can, but it requires the bidders to submit more complex bids. One proposal (see Dasgupta and Maskin [23]) is for bidders to submit contingent bid functions. They are, as the term suggests, a series of contingencies of the form: if my rival bids x I bid y, if they bid x', I bid y' etc.

Much of what is known about auctions with interdependent values assumes that the signals are one dimensional. When signals are multidimensional the story is more complicated. First, without severe restrictions, no auction that relies on reports of signals alone can be efficient and incentive compatible (Jehiel and Moldovanu [40]). It is useful to see why this must be the case. Suppose there exists a mechanism which produces the efficient allocation and that it is an equilibrium for bidders to report their signals truthfully.

Let $s^1 = (s_{11}, s_{12})$ be the signal of agent 1 and $s^2 = (s_{21}, s_{22})$ be the signal of agent 2. The value that bidder i assigns to the object is $a^{i1} \cdot s^1 + a^{i2} \cdot s^2$ where a^{ij} are vectors and '·' is the vector dot product. The efficient allocation is found by solving:

$$\max (a^{11} \cdot s^1 + a^{12} \cdot s^2)x_1 + (a^{21} \cdot s^1 + a^{22} \cdot s^2)x_2$$
$$\text{s.t. } x_1 + x_2 = 1$$
$$x_1, x_2 \geq 0$$

Suppose the optimal solution is to give the object to agent 1. Following the Vickrey logic, we ask what is the 'smallest' valuation function that agent 1 can report (holding agent 2's report fixed) and it still be efficient to give the object to agent 1. This will represent an upper bound on what any auctioneer can charge agent 1. Thus we must solve:

$$\min a^{11} \cdot t + a^{12} \cdot s^2$$
$$\text{s.t. } (a^{11} \cdot t + a^{12} \cdot s^2) \geq (a^{21} \cdot t + a^{22} \cdot s^2)$$

Simplifying:

$$\min a^{11} \cdot t$$
$$\text{s.t. } (a^{11} - a^{21}) \cdot t \geq (a^{22} - a^{12}) \cdot s^2$$

Let t^* be the optimal solution to this program. The Vickrey logic would dictate that agent 1 be charged $a^{11} \cdot t^* + a^{12} \cdot s^2$. What is important to observe is that the objective function has a slope different from the one constraint. Thus there exists t', infeasible, but $a^{11} \cdot t' > a^{11} \cdot t^*$. Now suppose agent 1's signal was t'. Since t' is not feasible for the above program, efficiency dictates that the object be awarded to agent 2 instead. However, if agent 1 reported t^* as their signal instead, they would be assigned the object and be strictly better off.

On the other hand, if there is a second stage where bidders report their realized valuations then one can achieve efficiency and incentive compatibility. An example will be helpful to make the distinction clear. Consider the oil lease example. Pick any auction with the constraint that all payments and allocations must be made *before* the winner is allowed to drill for oil and determine the actual value of the lease. Then, no such auction can be guaranteed to be both efficient and incentive compatible. Now add a second stage where the winner reports the actual value of the lease and further monetary transfers are made contingent on the reported actual value. In this case it is possible to design an auction that is both incentive compatible and efficient, see Mezzetti [50].

7. Two Examples

7.1 The German UMTS-Auction

The German UMTS-Auction (Universal Mobile Telecommunications System) in August 2000 attracted, like similar auctions in Great Britain and the Netherlands earlier in the year 2000, a deal of public attention and money. At this auction 4 to 6 licenses to offer third generation cell-phone service in Germany were sold, raising a total revenue of 98.8072 billion DM. It was conducted by the *Regulatory Authority for Telecommunications and Posts* (RegTP)[19].

7.1.1 Design of the UMTS-Auction.

The auction involved two stages, the second of which was to sell unsold items from the first stage plus auxiliary unpaired spectrum.[20] Here we focus on the first stage. In the first stage twenty-four items of 5 MHz spectrum were to be auctioned. They were bundled together into 12 blocks of paired spectrum.[21] In order to obtain a licence a bidder had to win at least two blocks and no bidder was allowed more than 3 blocks. The actual location of the block in the spectrum was not to be specified until after the auction. This was to ensure that a winning bidder received adjacent blocks of spectrum. Thus, bidders face a multiple unit auction with identical items, 12, in this case. Each block had a reserve price of 100 million DM. All blocks were sold simultaneously but each separately in its own ascending auction. Bidders could bid in as many of these separate auctions subject to eligibility requirements. The main features of the auction are summarized below:

1. Each bidder had to provide bank-guarantees, that limited the number of blocks (eligibility) and the highest bid they could place.

2. In the first round all bidders submit simultaneously their secret bids for the different blocks; they can submit as many bids as their eligibility permits.

3. Bids to be made in whole multiples of 100 000 DM.

4. For a bid in the first round to be deemed valid it had to exceed 100 million DM.

5. In subsequent rounds a bid for a block was valid if it exceeded the current highest bid by the current minimum bid increment. If there was no previous valid bid for this block, the minimum bid qualified as a valid bid.

6. The minimum increment, announced before the start of each round, was a percentage of the current highest bid on that block rounded up to the nearest 100 000 DM.

7 Newly submitted bids in a round and standing highest bids from the previous round are deemed active.

8 The number of active bids of a bidder must not exceed the number of active bids in the previous round.

9 A bidder is eliminated from the auction, if he has at most one active bids in any round.

10 The auction ends when there are no new valid bids submitted for any of the 12 blocks.

Item (1) prevents a winner reneging on his bid. Item (3) guards against signalling through the trailing digits of the bids. Items (4)–(5) ensure that prices ascend in the auction and do so at something other than a snail's pace. Items (8)–(10) encourage bidders to bid on two blocks and discourage them from 'sitting' on a block just to drive its price up. Unsold blocks and additional blocks of auxiliary spectrum were auctioned off in the second stage to bidders who had won a licence in the first stage.

Should it be a revenue maximizing auction? For a private entity perhaps, but for a public one, not. In the case of the UMTS-auction, the government was primarily concerned with efficiency.[22] That is, does the auction allocate the licences to the ones that will make the most efficient use of them.[23] There is also the subsequent competition to provide communication services amongst the winners to be considered as well. To provide sufficient competition, the government must ensure enough winners at the auction.

Why an open auction? Favoritism is harder to disguise. Another is the interdependent values aspect of the spectrum auction alluded to above.

Why a simultaneous ascending auction? We suspect precedent. There are other (open) auction forms one could have suggested. For example, allow bidders to submit simultaneous bids on packets of 2 or 3 blocks (at most one of which would be honored). Alas they have not yet been implemented.[24] Our guess is that the RegTP had no wish to be pioneers. The Simultaneous Ascending Auction (SAA) had been used in the US to much praise and so was a natural choice to consider.

Given the choice of SAA, bundling the licences into blocks of two served to limit the exposure problem. If each licence had been auctioned off separately a bidder would have needed two of them to be able to provide service (most companies commented in the pre-auction-hearings that at least two—if not even three—blocks are necessary for providing high data-rates to customers). Limiting the number of blocks ensures sufficient competition in the delivery of services that consume the portion of spectrum being auctioned. There is still a threshold problem in that a bidder with bids on three blocks can crowd out two bidders with an interest in these three blocks. It is mitigated by the fact that bidders can switch between bidding for different blocks. One other feature of the SAA that appealed to RegTP was that it would result in all blocks being sold for about the same price. This happens because bidders can always shift their subsequent bids to the block with the current lowest price. There is a heterogenous private values aspect to the auction in that incumbents, because of prior investments in technology and existing markets, assign a higher value to the licences than new entrants. This would suggest higher reserve prices for them (or subsidies for the entrants). The UMTS auction treated all bidders the same. This heterogenity coupled with the 3 block limit may have given incumbent bidders an incentive to bid aggressively so as to prevent new entrants from winning (see Jehiel and Moldovanu [39]).

We describe the course of the auction by **phases**. A phase is a consecutive set of rounds, during which no bidder decreased his activity and the minimum increment was constant. The activity of a bidder was the number of items he was currently active on. Since dropping out of the auction requires reducing one's activity, the number of active bidders is constant during each phase. Our description is necessarily incomplete because RegTP provides for each round only two pieces of data. One is the highest bid and bidder in that round. The other item is a table that shows for each bidder the number of licenses he submitted the highest bids on.[25] The auction began with seven bidders: Debitel Multimedia GmbH, DeTeMobil Deutsche Telekom Mobilnet GmbH, E-Plus Mobilfunk GmbH, Group 3G, Mannesmann Mobilfunk GmbH, MobilCom Multimedia GmbH, and VIAG INTERKOM GmbH & Co.

The first phase began with each bidder placing bids on three blocks. It continued with a minimum increment of 10% in round 1 and ended between round 109 (the last time that Viag Interkom was the highest bidder on three blocks) and round 127 (Debitel dropped out). With few exceptions bids were raised only by roughly the necessary minimum increments. The major exception, MobilCom, put in bids, that were high enough to remain standing high bids for many rounds. On the first round they bid 1.002 billion DM for a pair. This bid was topped for the first time in round 45. In round 46 MobilCom bid 1.602 billion DM for two blocks; in round 61 they topped their own high bid with 2.5601 billion DM.

The last rounds in which Group 3G, MobilCom and E-Plus were observed to be highest bidders on 3 blocks were 129, 136, and 136, respectively. So possibly during that time some additional phase-transitions happened. The next observable transition happened with round 139, since beginning with that round, the minimum increment percentage was reduced to 5%. The minimum increment was reduced again during round 168, this time down to 2%. The auction ran over fourteen days. The winners of the auction were DeTeMobil Deutsche Telekom Mobilnet GmbH, E-Plus Mobilfunk GmbH, Group 3 G, Mannesmann Mobilfunk GmbH, MobilCom Multimedia GmbH, and VIAG

INTERKOM GmbH & Co. Each won two blocks at roughly 16 billion DM. Total revenue raised by the auction's first stage was 98.8072 billion DM.

Did the auction design promote efficiency? It depends on how bidders value the blocks. Suppose that values are independent and that the value of a third block is at most half the value of a pair of blocks, i.e., diminishing marginal benefits. In this case, a bidder should place bids on three licences as long as the total price of the three licences is less than the value he assigns to owning three licences. In this case bidders can express their preferences and efficiency is promoted. If the value of the third licence exceeds half the value of a pair, then the exposure problem of the SAA makes bidders reluctant to place a bid for three blocks that exceeds 1.5 times the value for two blocks. This is an obstacle to efficiency.

There is reason to believe that the incumbents (DeTeMobil Deutsche Telekom Mobilnet GmbH, E-Plus Mobilfunk GmbH, Mannesmann Mobilfunk GmbH, and VIAG INTERKOM GmbH & Co) assigned a higher value to the licences than the entrants. The incumbents had existing networks through which they offered second generation services. Thus the incremental cost of building a third generation network was smaller for them than the other three bidders. Also, a winner of the auction was permitted to offer second generation services. This made losing for incumbents more costly.[26] Given this heterogenity in bidders, an auction in which all bidders face the same reserve price can never be revenue maximizing. Efficiency also dictates that when bidders are heterogenous, different bidders will usually pay different prices. This heterogenity in valuations has other implications. Since a winner of two blocks has the right to offer second generation services, incumbent firms have an incentive to bid for three blocks so as to exclude an entrant. If this is the case, the auction limits competition in the provision of communication services. On the other hand, bidding for three licences opens the incumbent up to the exposure problem. Which of these two effects is dominant is difficult to see.

7.2 Logistics Auctions

A number of large firms have run combinatorial auctions to procure delivery services. The items on offer are origin-destination pairs (called lanes), along with volume and deadline specifications. For bidders there are at least three sources of synergy.

1. A lane at the right time might be a return trip for lanes which a carrier already serves.

2. Pool volume from many buyers along a lane, ensuring full truck loads.

3. A collection of lanes can be patched together into a single route.

Design of Combinatorial Auctions

Allowing bidders to submit bids on bundles of lanes would help in exploiting these synergies.

Auctions to procure delivery services are not new, but they have typically been run lane by lane. The Home Depot, for example, used to publish the zip codes and aggregated demand forecasts for all their needed OD-pairs but not information on demand or growth patterns. Based on this limited information, carriers had to submit individual bids for OD-pairs they were willing to serve. Not knowing the demand distribution for such a given lane, or whether they would win lanes that would match with the lane under consideration, bidders would bid conservatively.[27]

In this subsection we want to describe the logistics auctions run by Sears and The Home Depot, to procure shipping services from carriers. Both companies have a myriad of shops, suppliers and warehouses between which they need transportation services.

7.2.1   Sears Logistics Services Auction (SLS).   SLS[28] teamed up with Net Exchange and Jos. Swanson&Co. to implement a more efficient and money-saving auction. Key-features of the contracts were 3-year contracts (with surge and slack demand contingencies) and simultaneous letting of contracts for many lanes. There were concerns that carriers (even the better, reliable ones) might frown upon the prospect of participating in a complex auction and getting squeezed. So SLS decided to invite only a small number of desirable 'elite' carriers capable of covering all lanes to participate in the auction.[29]

On the basis of lab-experiments conducted at Caltech a multi-round auction was selected since it did not require participants to evaluate all contingencies exactly. Bidders were allowed to submit bids on packages and winning bidders paid their bids. Additionally, bidders could specify capacity constraints to enforce business rules (see Sect. 3.3 of Chapter 5). In each round a provisional allocation was chosen so as to minimize cost to the seller. Bids could be lowered or withdrawn. However a winning bid from the preceding round had to be resubmitted in the subsequent round. If the total cost did not drop by a specified percentage from one round to the next, the auction terminated. Even though bidders were permitted to submit bids for any bundle, the number of submitted bids did not explode. In fact the bids were multiattributed (see Sect. 3.2.4 of Chapter 5) to permit offers with different equipment for example.

[48] report, that their implemented auction saved SLS an estimated amount of $15 million per year, that is about 13%.

7.2.2   The Home Depot's Auction.   According to [28] The Home Depot's (THD) bidding process was, until 1996 almost fully manual. Subsequently they switched to a process where shippers submitted bids for various lanes on an EXCEL spreadsheets, thereby running a sort of simultaneous first-price auction. Bidders where provided with OD-pairs (as zip-codes) with aggregated demand forecasts.

In 2000, THD asked i2 Technologies[30] for a more efficient procedure. As THD was worried about ruinous bidding wars between bidders that might compromise quality of shipping, they decided on a combinatorial sealed bid first-price auction. Sealed bundle bids with XOR-restrictions (sometimes wrongly called OR-bids), capacity-constraints (so that bidders could limit the equipment capacity) and budget constraints (so that bidders could limit how much business they were willing to do with THD) were permitted. THD considered all submitted bids binding and the penalty for any shipper's attempt to renege would be the loss of all contracts for that shipper. In addition to the previously supplied information, THD provided for each OD-pair daily/weekly demand-forecasts.

Before the auction, THD asked for carrier profile information from the 192 applicants. Of these, 111 were invited to participate in the auction. They were selected on the basis of acceptable quality, promise of a favorable business relationship, and sufficient size.

The resulting winner determination problem was a set-partitioning problem. Cost was not the only criterion. Carrier-reliability, load balancing and incumbency concerns mattered to THD. These nonmonetary preferences were not revealed to the bidders.

After running the auction most of the lanes were awarded. Lanes which received too few bids, were offered again for auction. Bidding on these lanes was open only to a small number of desirable bidders determined by what-if-scenario analyses. We think that one reason for this reauctioning lays in the fact, that some soft constraints of THD were not incorporated fully into the model; therefore not every best-value solution was optimal for them. Another reason, why THD was able to change the rules of the game on the fly is that they, as a private company has substantially more leeway in deciding with whom to make business under which conditions than, say, the government.

Notes

1. It is more common to assume a continuous type space, but we believe that the main ideas are more transparent without a continuous type space.
2. Strictly speaking one should allow an allocation rule to be randomized. Given risk neutrality, we omit this possibility. To account for it requires introducing additional notation that is subsequently not used.
3. It is sometimes common to require that the *actual* payoff be non-negative rather than the expected payoff as we have done here. This stronger condition is called ex-post individual rationality.
4. It is also called the generalized Vickrey auction.
5. This amounts to dropping the assumption that there are finite number of types.
6. We assume that the auctioneer is a seller and bidders are buyers.
7. Implicit is the assumption that bidders care only about the combinations they receive and not on what other bidders receive.
8. For this reason the extensive literature on the set packing problem and its relatives is relevant for computational purposes.
9. In fact, one must specify not only lanes but volume as well, so this problem constitutes an instance of a multi-unit combinatorial auction.
10. It is not hard to construct such algorithms.
11. For this reason we use the words edge and agent interchangeably.
12. If A does not span the graph, take $M(A) = \infty$.
13. If A does not contain a $s - t$ path set $L(A) = \infty$.
14. Recall that one way to phrase to the minimum spanning tree problem is as follows: find the least weight collection of edges that intersect every cycle of the graph at least once.
15. One can think of these additional conditions as being properties of a path that would be generated by optimizing some 'nice' function on paths other than just length.
16. Not always. Movement along a subgradient might require that a price be decreased. The price decrease step is sometimes omitted or appears indirectly when bidders are allowed to withdraw bids.
17. This is not true in general, but is so for this class of optimization problems. The rate of convergence is slowed.
18. The description is tailored to the auction for assigning carrier of last resort rights in telecommunications.
19. Regulierungsbehörde für Telekommunikation und Post http://www.regtp.de
20. A complete description may be found at http://www.regtp.de/imperia/md/content/reg_tele/umts/9.pdf.
21. Spectrum was paired so that potential users would have one segment to send information and another to receive. While one portion of spectrum could do both this would require some kind of traffic control which is difficult to manage.
22. See §11 of the German Telekommunikationsgesetz, available at http://www.regtp.de/gesetze/start/in_04-01-00-00-00_m/index.html
23. Public opinion, however, judges the success of an auction by the revenue generated.
24. However the US is about to do so. See announcement DA00-1486 (from http://www.fcc.gov/wtb/auctions/700/da001486.pdf) of the FCC to implement auction #31 as a combinatorial auction.
25. RegTP's counterpart in the US, the FCC, discloses all bids.
26. These observations are made in Jehiel and Moldovanu [39].
27. Our description is based on i2 Technologies [37], Elmaghraby and Keskinocak [28–30], Bowman [13] and Ledyard, Olson, Porter, Swanson and Torma [48].
28. https://www.slslogistics.com/
29. The description and other details are from Ledyard, Olson, Porter, Swanson and Torma [48].
30. http://www.i2.com

References

[1] Karhand Akcoglu, James Aspnes, Bhaskar DasGupta, and Ming-Yang Kao. An opportunity-cost algorithm for combinatorial auctions. In Erricos John Kontoghiorghes, Berç Rustem, and Stavros Siokos, editors, *Computational Methods in Decision-Making, Economics and Finance*, volume 74 of *Applied Optimization*, chapter 23. Kluwer Academic Publishers, 2002.

[2] Lawrence M. Ausubel and Peter C. Cramton. Demand reduction and inefficiency in multi-unit auctions. Manuscript, University of Maryland, 1998.

[3] Lawrence M. Ausubel and Peter C. Cramton. The optimality of being efficient. Working Paper, Department of Economics, University of Maryland, MD, 1999.

[4] Lawrence M. Ausubel and Paul R. Milgrom. Ascending auctions with package bidding. *Frontiers of Theoretical Economics*, 1(1):Article 1, 1–42, 2002.

[5] Aaron Archer and Éva Tardos. Frugal path mechanisms. In *Proceedings of the 13th Annual ACM-SIAM Symposium on Discrete Algorithms*, pages 991–999, 2002.

[6] Arne Andersson, Mattias Tenhunen, and Fredrik Ygge. Integer programming for combinatorial auction winner determination. In *Proceedings of the Fourth International Conference on Multi-Agent Systems (ICMAS00), Boston, MA*, pages 39–46, 2000.

[7] Mark M. Bykowsky, Robert J. Cull, and John O. Ledyard. Mutually destructive bidding: The FCC auction design problem. *Journal of Regulatory Economics*, 17:205–228, 2000.

[8] Sushil Bikhchandani, Sven de Vries, James Schummer, and Rakesh V. Vohra. Linear programming and Vickrey auctions. In Dietrich and Vohra [24], pages 75–116.

[9] Sushil Bikhchandani and Chi-fu Huang. The economics of treasury securities markets. *Journal of Economic Perspectives*, 7:117–134, 1993.

[10] Cynthia Barnhart, Ellis L. Johnson, George L. Nemhauser, Martin W. P. Savelsbergh, and Pamela H. Vance. Branch and price: Column generation for solving huge integer programs. *Operations Research*, 46(3):316–329, 1998.

[11] Jeffrey S. Banks, John O. Ledyard, and David P. Porter. Allocating uncertain and unresponsive resources: An experimental approach. *RAND Journal of Economics*, 20(1):1–25, 1989.

[12] Sushil Bikhchandani and Joseph M. Ostroy. The package assignment model. *Journal of Economic Theory*, 107(2):377–406, 2002.

[13] Robert J. Bowman. Home Depot: An experiment in LTL carrier management. Manuscript, July 2001.

[14] Sandeep Baliga and Rakesh V. Vohra. Market research and market design. Working paper, Kellogg School of Management, Northwestern University, Evanston, IL, 2002.

[15] Christopher George Caplice. *An Optimization Based Bidding Process: A New Framework for Shipper-Carrier Relationship.* PhD thesis, Department of Civil and Environmental Engineering, MIT, 1996.

[16] Edward H. Clarke. Multipart pricing of public goods. *Public Choice*, 8:19–33, 1971.

[17] Estelle Cantillon and Martin Pesendorfer. Combination bidding in multi-unit auctions. Manuscript, Economics Department, Yale University, 2002.

[18] Peter Cramton and Jesse Schwartz. Collusive bidding: Lessons from the FCC spectrum auctions. *Journal of Regulatory Economics*, 17(3):229–252, 2000.

[19] George B. Dantzig. *Linear Programming and Extensions.* Princeton University Press, Princeton, NJ, 1963.

[20] Andrew J. Davenport, Chae An, Ed Ng, Gail Hohner, Grant Reid, Ho Soo, Jayant R. Kalagnanam, and John Rich. Combinatorial and quantity-discount procurement auctions benefit mars, incorporated and its suppliers. *Interfaces*, 33(1):23–35, 2003.

[21] Andrew J. Davenport and Jayant R. Kalagnanam. Price negotiations for procurement of direct inputs. In Dietrich and Vohra [24], pages 27–44.

[22] Christine DeMartini, Anthony M. Kwasnica, John O. Ledyard, and David P. Porter. A new and improved design for multi-object iterative auctions. Social Science Working Paper No. 1054, Division of the Humanities and Social Sciences, California Institute of Technology, Pasadena, CA, 1999.

[23] Partha Dasgupta and Eric Maskin. Efficient auctions. *The Quarterly Journal of Economics*, CXV:341–388, 2000.

[24] Brenda Dietrich and Rakesh V. Vohra, editors. *Mathematics of the Internet: E-Auction and Markets*, volume 127 of *The IMA Volumes in Mathematics and its Applications*. Springer-Verlag, New York, NY, 2002.

[25] Sven de Vries, James Schummer, and Rakesh V. Vohra. An ascending Vickrey auction for heterogeneous objects. Manuscript, 2003.

[26] Sven de Vries and Rakesh V. Vohra. Combinatorial auctions: A survey. *Informs Journal on Computing*, 15:26, 2003. to appear.

[27] Rafael Epstein, Lysette Henríquez, Jaime Catalán, Gabriel Y. Weintraub, and Cristián Martínez. A combinational auction improves school meals in Chile. *Interfaces,*, 32(6):1–14, 2002.

[28] Wedad Elmaghraby and Pinar Keskinocak. Technology for transportation bidding at the home depot. Case Study, 2000.

[29] Wedad Elmaghraby and Pinar Keskinocak. Combinatorial auctions in procurement. Technical report, The Logistics Institute - Asia Pacific, National University of Singapore, January 2002.

[30] Wedad Elmaghraby and Pinar Keskinocak. Technology for transportation bidding at the home depot. In Terry P. Harrison, Hau L. Lee, and John J. Neale, editors, *The Practice of Supply Chain Management: The Practice of Supply Chain Management*, volume 62 of *International Series in Operations Research and Management Science*, chapter 15. Kluwer, 2003.

[31] Yuzo Fujishima, Kevin Leyton-Brown, and Yoav Shoham. Taming the computational complexity of combinatorial auctions: Optimal and approximate approaches. In *Proceedings of IJCAI-99, Stockholm, Sweden*, pages 548–553, Stockholm, 1999.

[32] Drew Fudenberg and Jean Tirole. *Game Theory*. MIT Press, Cambridge, MA, 1992.

[33] Andrew V. Goldberg, Jason D. Hartline, Anna R. Karlin, and Andrew Wright. Competitive auctions. Manuscript, 2002.

[34] Ricca Gonen and Daniel Lehmann. Optimal solutions for multi-unit combinatorial auctions: Branch-and-bound heuristics. In *ACM Conference on Electronic Commerce (EC-00), Minneapolis, MN.*, pages 13–20, 2000.

[35] Theodore Groves. Incentives in teams. *Econometrica*, 41(4):617–631, 1973.

[36] Faruk Gul and Ennio Stacchetti. The English auction with differentiated commodities. *Journal of Economic Theory*, 92(1):66–95, 2000.

[37] i2 Technologies. Helping The Home Depot keep its promise to customers. i2 Success Story #131, 2002.

[38] Charles Lee Jackson. *Technology for Spectrum Markets*. PhD thesis, Department of Electrical Engineering, MIT, Cambridge, MA, 1976.

[39] Philippe Jehiel and Benny Moldovanu. A critique of the planned rules for the German UMTS/IMT-2000 license auction. Manuscript, March 2000.

[40] Philippe Jehiel and Benny Moldovanu. Efficient design with interdependent valuations. *Econometrica*, 69(5):1237–1259, 2001.

[41] Raymond B. Johnson, Shmuel S. Oren, and Alva J. Svoboda. Equity and efficiency of unit commitment in competitive electricity markets. *Utilities Policy*, 6(1):9–20, 1997.

[42] Alexander S. Kelso, Jr. and Vincent P. Crawford. Job matching, coalition formation and gross substitutes. *Econometrica*, 50(6):1483–1504, 1982.

[43] John H. Kagel and Dan Levin. Behavior in multi-unit demand auctions: Experiments with uniform price and dynamic Vickrey auctions. *Econometrica*, 69(2):413–454, 2001.

[44] Paul Klemperer. Auction theory: A guide to the literature. *Journal of Economic Surveys*, 13(3):227–286, July 1999.

[45] Frank Kelly and Richard Steinberg. A combinatorial auction with multiple winners for universal service. *Management Science*, 46(4):586–596, 2000.

[46] Kevin Leyton-Brown, Mark Pearson, and Yoav Shoham. Towards a universal test suite for combinatorial auction algorithms. In *Proceedings ACM Conference on Electronic Commerce (EC-00), Minneapolis, MN*, pages 66–76, 2000.

[47] Kevin Leyton-Brown, Yoav Shoham, and Moshe Tennenholtz. An algorithm for multi-unit combinatorial auctions. In *Proceedings of the Seventeenth National Conference on Artificial Intelligence and Twelfth Conference on Innovative Applications of Artificial Intelligence, Austin, TX*, pages 56–61, 2000.

[48] John O. Ledyard, Mark Olson, David Porter, Joseph A. Swanson, and David P. Torma. The first use of a combined-value auction for transportation services. *Interfaces*, 32(5):4–12, 2002.

[49] Daniel Lehmann, Liadan O'Callaghan, and Yoav Shoham. Truth revelation in rapid, approximately efficient combinatorial auctions. In *Proceedings ACM Conference on Electronic Commerce (EC-99), Denver, CO*, pages 96–102, November 1999.

[50] Claudio Mezzetti. Mechanism design with interdependent valuations: Efficiency and full surplus extraction. Manuscript, Department of Economics, University of North Carolina at Chapel Hill, February 2003.

[51] Paul Milgrom. An economist's vision of the B2B marketplace. White paper, perfect.com.

[52] Paul R. Milgrom. *Auction Theory for Privatization*, chapter Auctioning the Radio Spectrum. Cambridge University Press, 1995.

[53] Stephen Morris. The common prior assumption in economic theory. *Economics and Philosophy*, 11:227–253, 1995.

[54] Kazuo Murota and Akihisa Tamura. New characterizations of M-convex functions and connections to mathematical economics. RIMS Preprint 1307, Research Institute for Mathematical Sciences, Kyoto University, Kyoto, Japan, 2000.

[55] Roger B. Myerson. Optimal auction design. *Mathematics of Operations Research*, 6:58–73, 1981.

[56] Noam Nisan. Bidding and allocation in combinatorial auctions. In *Proceedings ACM Conference on Electronic Commerce (EC-00), Minneapolis, MN*, pages 1–12, 2000.

[57] Noam Nisan and Amir Ronen. Computationally feasible VCG mechanisms. In *Proceedings ACM Conference on Electronic Commerce (EC-00), Minneapolis, MN*, pages 242–252, 2000.

[58] Noam Nisan and Ilya R. Segal. The communication requirements of efficient allocations and supporting Lindahl prices. Working Paper, Institute of Computer Science, Hebrew University, Jerusalem, Israel, 2003.

[59] George L. Nemhauser and Laurence A. Wolsey. *Integer and Combinatorial Optimization*. John Wiley, New York, 1988.

[60] David C. Parkes and Lyle H. Ungar. Iterative combinatorial auctions: Theory and practice. In *Proceedings 17th National Conference on Artificial Intelligence, Austin, TX*, pages 74–81, 2000.

[61] Michael H. Rothkopf, Aleksandar Pekec, and Ronald M. Harstad. Computationally manageable combinatorial auctions. *Management Science*, 44(8):1131–1147, 1998.

[62] Jean-Charles Rochet and Lars Stole. The economics of multidimensional screening. Manuscript, Graduate School of Business, University of Chicago, Chicago, IL, 2002.

[63] Stephen J. Rassenti, Vernon L. Smith, and Robert L. Bulfin. A combinatorial auction mechanism for airport time slot allocation. *Bell Journal of Economics*, 13(2):402–417, 1982.

[64] Tuomas W. Sandholm. Algorithm for optimal winner determination in combinatorial auctions. *Artificial Intelligence*, 135(1–2):1–54, 2002.

[65] Lloyd Shapley. Complements and substitutes in the optimal assignment problem. *Naval Research Logistics Quarterly*, 9:45–48, 1962.

[66] William Vickrey. Counterspeculation, auctions, and competitive sealed tenders. *Journal of Finance*, 16:8–37, 1961.

[67] Laurence A. Wolsey. Integer programming duality: Price functions and sensitivity analysis. *Mathematical Programming*, 20:173–195, 1981.

[68] Peter R. Wurman and Michel P. Wellman. AkBA: A progressive, anonymous-price combinatorial auction. In *Proceedings Second ACM Conference on Electronic Commerce (EC-00), Minneapolis, MN*, pages 21–29, October 2000.

[69] Makoto Yokoo, Yuko Sakurai, and Shigeo Matsubara. Robust combinatorial auction protocol against false-name bids. *Artificial Intelligence*, 130(2):167–181, 2001.

Part III

SUPPLY CHAIN COORDINATIONS IN E-BUSINESS

David Simchi-Levi, S. David Wu, and Z. Max Shen (Eds.)
Handbook of Quantitative Supply Chain Analysis:
Modeling in the E-Business Era
©2004 Kluwer Academic Publishers

Chapter 8

THE MARKETING-OPERATIONS INTERFACE

Sergio Chayet
Graduate School of Business
University of Chicago
Chicago, IL 60637
sergio.chayet@gsb.uchicago.edu

Wallace J. Hopp
Department of Industrial Engineering
Northwestern University
Evanston, IL 60208
hopp@northwestern.edu

Xiaowei Xu
Department of Industrial Engineering
Northwestern University
Evanston, IL 60208
xuxiaowe@northwestern.edu

Keywords: Marketing, operations, e-commerce

Introduction

The American Marketing Association defines marketing as "the process of planning and executing the conception, pricing, promotion, and distribution of ideas, goods, and services to create exchanges that satisfy individual and organizational goals." Operations management is typically defined in texts as "management of an organization's production system, which converts inputs into products and services." Clearly the two are closely related. Marketing requires the operations function to deliver goods and services, while operations requires the marketing function to generate demand.

In practice, however, marketing and operations management often find themselves in fundamental conflict with one another (see e.g., Porteus and Whang, 1991). Marketing people want to offer customers a high degree of product variety and delivery flexibility to generate sales. Operations management people, want to limit variety and volume/mix changes to reduce costs. As a result, the interface between marketing and operations management is a perennial management challenge and a stimulating source of research topics.

The importance of the interface between marketing and operations management has long been recognized by researchers in both fields. For example, the *Journal of Operations Management* has devoted two entire issues to this topic (Vol. 10 (Aug.) 1991 and Vol. 20 (Jun.) 2002) and several authors have offered overviews (see, Eliashberg and Steinberg, 1993, Hausman and Montgomery, 1993, Karmarkar, 1996, Shapiro, 1977). In general, the focus has been on opportunities for better integration of the operations and marketing perspectives with respect to forecasting, capacity planning, production scheduling, delivery and distribution, quality assurance, breadth of product line, innovation and cost control.

The management and research challenges posed by the marketing-operations interface are being heightened by the rise of the Internet. In addition to providing a new channel for selling goods and services, the web is greatly enhancing the types and flow of information throughout the process. As such, it is having at least three profound effects on the process through which firms interact with customers, and hence on the interface between marketing and operations:

1 *Speed:* The Internet makes it possible to shorten the time required for many operations, both within the production system and at the interface with the customer. This raises a number of researchable questions, such as: How much do customers value faster response? How will firms compete on the basis of time? What kind of production system is required to underpin a quick response marketing system?

2 *Communication:* The web can be used to provide more information to customers (e.g., real-time stock levels or order status). This suggest research into questions such as: How much information should a firm reveal to customers? What impact does such information have on customers' propensity to change orders? How much does it cost to provide customers with flexibility to make changes in their orders. What impact does customer-to-customer communication have on the process?

3 *Data Collection:* E-commerce transactions offer the potential to collect and share information about customers. This raises research questions

such as: How can customer preference profiles be extracted (mined) from transactional data? How can demand curves be estimated? How can this information be used to improve the product development and delivery process? What is the impact of information sharing on customer behavior (e.g., lying).

The range of issues that arise in conjunction with the marketing-operations interface is so broad that a comprehensive analysis of research opportunities is not feasible. Moreover, we do not wish to repeat the excellent surveys cited above. Instead, we will focus specifically on impacts of the Internet on the marketing-operations interface. To organize these we consider the customer contact chain, shown in Figure 8.1, which breaks the process of providing goods and services to customers into:

1. *Product Development:* including long-term evolution of goods and services to meet customer needs (i.e., what to introduce and when to introduce it), as well as short-term configuration of products, such as a customer might do by designing his/her own product on-line.

2. *Sales:* in which the actual purchase transaction occurs. This presents the firm with decisions about positioning of products (what to offer and how to display them), pricing, and lead time (where both pricing and lead time could be dynamic, based on work backlog, inventory, competitor behavior, forecasts of customer demand, etc.)

3. *Delivery:* where the product is produced and shipped to the customer. The Internet raises the possibility of giving the customer visibility to this process and the option of making changes in his/her order. Such changes could, of course, come at a cost, which would then present the problems of valuation and pricing.

4. *Service:* includes the post-sale contact with the customer in support of the product. A long-term contract, in which updates or repair service is offered, is one option. User groups in which customers can get information from the firm or each other is another. Information sharing (including promotional contacts) is a third.

All of these stages present opportunities to enhance the overall customer experience, as well as to gain additional insight into customer wants and needs. Learning what information to collect at each stage and how to make use of it is an important research subject.

The remainder of this chapter is organized according to the customer contact chain. In Section 2 we examine the new product development process. In Section 3 we discuss the sales function. In Section 4 we review delivery. And in

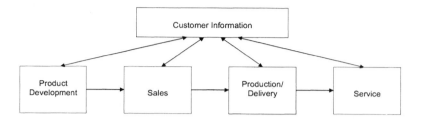

Figure 8.1. The customer contact chain

Section 5 we look at post-sales service. In Section 6 we conclude with general observations about research opportunities related to the marketing-operations interface in the world of e-commerce.

1. Product Development

Product development can be defined as "the transformation of a market opportunity into a product available for sale" (Krishnan and Ulrich, 2001). The actual process that underlies this transformation is often decomposed into five phases: concept development, system-level design, detailed design, testing and refinement, and production ramp-up (Ulrich and Eppinger, 1995). Each of these involves a series of dynamic decisions and can evolve over time in an iterative fashion. For recent literature reviews of research on product development see Krishnan and Ulrich, 2001, Ozer, 1999, and Wind and Mahajan, 1997.

At the highest strategic level, an organization can adopt either a proactive or a reactive approach toward product development. Which is appropriate depends on growth opportunities, protection for innovation, market scale, competition, and position in production and distribution system (Urban and Hauser, 1993). Since the Internet offers the potential to change these, one possible research question is how the advent of e-commerce may alter decisions about product development strategy (e.g., does it become more or less important to offer a stream of innovative new products as the web increases in importance as a marketing channel?)

Since product development is an interdisciplinary activity that involves almost every function of a firm, organizational structure is vital to the process. As customer tastes have become more diversified in recent years, organizations and strategies have had to evolve to provide greater levels of speed, variety and flexibility (e.g., through use of techniques such as concurrent engineering, design for manufacture, and quality function deployment (QFD), Fitzsimmons et al., 1991). By offering a speedy and flexible vehicle for interacting with customers, the Internet is likely to amplify these trends. Therefore, techniques for facilitating better understanding of customer needs, such as QFD, (Griffin and Hauser, 1993), will remain and increase in relevance. However, a customer focus is not a panacea, in traditional commerce or in e-commerce. Some authors (e.g., Bower and Christensen, 1995) have pointed out instances of leading companies who focused on customers and introduced new products only to find that their products eventually "overshot" the needs of the market. For instance, Digital continually improved the mini-computer to meet the needs of its best customers at the same time that the "disruptive technology" of PC's was capturing the low-end market and preparing to supplant mini-computers. This phenomenon occurs not only in high-tech industries, but also in retailing, where e-commerce can be regarded as a disruptive technology (Christensen and Tedlow, 2000). Hence, another research question that arises at the top strategic level is to what extent adoption of the Internet as a channel for designing and delivering products could undermine successful companies.

Having noted that the Internet may influence the strategy that guides the new product development process, we now turn to the process itself.

1.1 Conjoint Analysis For Concept Development

A key step in the concept development phase is identification of customer preferences. The needed data is often collected through one-on-one interviews, focus groups, observation studies, and written surveys. Conjoint analysis is a technique for measuring consumer trade-offs across two or more attributes of a product (often including price as an attribute) and seeks to explain why consumers choose one product among a group of competing products (Green et al., 2001). Usually a part-worth model is fitted for respondents' evaluative judgments and then a simulator is built to predict how consumers might react to changes in attributes of a new or existing product. For details on conjoint analysis and its applications see Green et al., 2001 Green and Srinivasan, 1990 and Wittink et al., 1994.

Conjoint analysis and related techniques are directly relevant to the marketing-operations interface because: (1) they address the trade-off between customer satisfaction, which determines an upper-bound on product price, and the production/distribution cost of the product, and (2) they facilitate acquisition of empirical evidence to evaluate operational strategies. For instance, Lindsley et al., 1991 used conjoint analysis to quantify the tradeoffs that retailers are willing to make among price discount, speed of delivery, delivery reliability, number of titles carried in inventory, offered by book distributors. The customer satisfaction equation, estimated by the authors, suggests that a decrease in the discount rate of 0.38% offsets a one-day reduction in delivery time and an increase of 1% title filling rate saves 0.16% discount rate. The equation could be used by managers to improve the operational efficiency further. In new product development, decisions regarding how many features to include and what pricing strategy to pursue depend on both the market information extracted via conjoint analysis and the operational capabilities of the system. To negotiate these tradeoffs, optimization models have been developed to determine how many product varieties within a product group should be introduced into the market and how they should be positioned in an attribute space (see Green and Krieger, 1989 and Kaul and Rao, 1995 for literature reviews).

Although conjoint analysis is very useful, it is time-consuming and expensive. For instance, a traditional conjoint analysis project may take six weeks or more and cost $20,000 to $200,000 (Paustian, 2001). Because it provides a fast and inexpensive means for communicating with potential customers, the Internet offers an attractive prospect for improving this situation. Dahan, Hauser and their group have been developing "The Virtual Customer" at MIT's center for innovation(mitsloan.mit.edu/vc, Dahan and Hauser, 2002), which includes several approaches for performing conjoint analysis on the Internet. Based on their experience they have noted three advantages of web-based methods:

1 rapid interaction and communication among the product development team and the respondents,

2 conceptualization via the multimedia capability of the Internet, and

3 real-time computation.

In the MIT system, web-based conjoint analysis, full-profile evaluation, and fast polyhedral adaptive conjoint estimation are similar to their non-Internet counterparts. But, when implemented on the web, the three methods provide respondents with animated and interactive stimuli, flexibility to participate at their convenience, and ease and speed of expressing their preferences through simple mouse clicks. Still, there are challenges in performing on-line conjoint analysis, presented by limited screen space, possible difficulties in understanding instructions and tasks, and easy escape of respondents (Dahan and Hauser, 2002). Furthermore, due to the easy impatience and wear-out of online respondents, the alleviation of respondent burden becomes critical. While adaptive conjoint analysis could dramatically reduce the number of questions required by customizing respondents' current questions based on their earlier answers, it might not achieve the online survey requirement. Fast polyhedral adaptive conjoint estimation is a way to further reduce the number of questions by using an interior-point algorithm (for details, see Toubia et al., 2002).

Typically, at the beginning of a product development project, the product development team faces difficulties in understanding the vocabulary used by customers for both existing products and new concepts. The information pump is a web-based discussion game that encourages customers to think hard and find new ways to verbalize the product features that are important to them. It is a guessing game involving an encoder, one or more decoders, and a dummy. In a typical implementation, the encoder and decoders are shown a series of photographs/renderings of the same concept (e.g., a car). The dummy does not see these. After seeing each picture, the encoder provides a statement about the concept (e.g., good for city driving). Given their knowledge of previous statements the decoders and the dummy guess the answer (e.g., true or false) and give confidence levels on their answers. To encourage truth telling, the decoders and dummy are rewarded on the accuracy of their answers. The encoder is rewarded on the accuracy of the decoder's answer relative to the accuracy of the dummy's answer. Since the decoders have more information than the dummy, the encoder is encouraged to provide new and accurate descriptions in each round so that the decoders have a better chance than the dummy of guessing correctly. Otherwise, the dummy may guess as well as the decoders do (see Dahan and Hauser, 2002 for more details on how to play the game). The result is a more accurate and understandable description of desired product features.

In addition to collecting information in advance of a product launch, via conjoint analysis or other survey methods, it is possible to make use of post-launch demand data to further understand customer preferences. The Internet, by allowing many options to be displayed and choices to be recorded, greatly facilitates this process. Since allowing customers to choose some features of the product to fit their needs best requires dramatic changes of the traditional design process, we will discuss this topic in the next section.

1.2 System Level and Detailed Design

The majority of questions that arise on the marketing-operations interface in conjunction with new product development relate to identifying what customers want in a product and how to ensure that the process provides it (see, e.g., Balakrishnan et al., 1997). Viewed as a one-time process, firms (i) collect information from customers, (ii) guess what they want, (iii) design, develop and launch the product, and (iv) see how good their guess was (Slywotzky, 2000). But in practice, the process is on-going. So, it is important to establish a process that systematically manages the risk of new product introductions and provides feedback so that future iterations can take advantage of the insights gained from previous ones.

In traditional production systems, the risk associated with introducing new products is substantial. In some industries, the cost of launching a product is so high that a failure can threaten a company's viability (e.g., the failure of the Edsel did serious financial damage to Ford in the 1950's). In other industries risk is associated more with the distribution channel (e.g., Barnes and Noble cannot stock every book it offers in every store and so must decide what to offer where). Over time, product proliferation can drive up costs and cause an organization to lose focus (Quelch and Kenny, 1994). As global competition places increasing pressure on profit margins, all of these types of mistakes in matching products to customer preferences are becoming more serious than ever.

One strategy for mitigating risk in the new product development process is mass customization (see Gilmore and Pine II, 1997 for some applications of mass customization). Mass customization can be defined as "a system that uses information technology, flexible processes, and organizational structures to deliver a wide range of products and services that meet specific needs of individual customers (often defined by a series of options), at a cost near that of mass-produced items" (Silveira et al., 2001). By giving customers a high degree of variety, mass customization strategies reduce the likelihood that a firm's offerings will fail to satisfy customer preferences. Moreover, if the mass customized product line is offered on-line, then the issue of matching products to physical sites also disappears (e.g., Amazon.com can offer an extremely broad product line because it does not maintain inventory in local retail sites).

Mass customization presents a host of challenges from both an operations and a marketing point of view. On the operations side, the problem is to make and distribute a high variety of products at a competitive cost. In many products, modular design is almost mandatory, so that customized products can be built quickly from standard components (Baldwin and Clark, 1997). In addition, postponement of product differentiation until as close as possible to the point of sale is used to reduce the cost (by keeping inventory as generic as possible) and speed delivery (through assemble-to-order processes) in mass customization systems (Feitzinger and Lee, 1997).

On the marketing side, the problem is how to identify and display an array of product choices that effectively meet customer preferences. The Internet can play an important role by supporting choice-boards. Choice-boards are "interactive, on-line systems that allow individual customers to design their own products by choosing from a menu of attributes, components, prices, and delivery options" (Slywotzky, 2000). Made famous by Dell (Magretta, 1998) and other Internet retailers, choice-boards provide an intuitive way to present customers with many (usually modular) products. But they also disclose what customers want, along with when and where they want it (see, e.g., Liechty et al., 2001). This is potentially valuable marketing information. Liechty et al., 2001, proposed a multivariate probit model for estimating the utility of each menu scenario as a linear function of features and price. This Bayesian model allows for correlation of the utilities across scenarios, accommodates customer heterogeneity, and incorporates constraints in the menu scenarios. As choice-boards become more common as components of mass customization strategies, research into better ways to extract customer preferences and methods to use these to adjust the production and supply chain process in a dynamic fashion will become increasingly important.

Although choice-boards allow customers to fulfill their individual needs independently, all significant product features must be included in a menu ex-ante to make them work. Hence, the task of understanding customer needs is not completely resolved. Indeed, some researchers question the ability of customers to understand their needs before trying out prototypes (i.e., they "learn by doing") and have therefore proposed the use of user toolkits to facilitate coordination between the design team and customer (see Thomke and von Hippel, 2002, von Hippel, 2001, and von Hippel and Katz, 2002). A user toolkit partitions the firm's solution information from the customer's need information and effectively outsources product development to the customers themselves, which saves the cost of transferring sticky information. Although a solution space must still be established, this innovative methodology goes one step beyond choice-boards. Toolkits include various module libraries and are therefore attractive for on-line implementation through use of functionality-based modular designs (Wang and Nnaji, 2001). An interesting research topic would be comparison of the costs of this method, choice-boards, and traditional methods under various scenarios. Such an analysis should also consider the strategic implications of outsourcing either design to customers or design of modularity to suppliers, especially on the Internet where the cost of entry and customer switching is low (Porter, 2001, von Hippel, 2001).

Beyond external communication, the success of a new product development effort depends strongly on the performance of the design team itself. Various structures have been proposed for interdisciplinary teams in different environments (see e.g., Wheelwright and Clark, 1992). Regardless of the structure of the team, the Internet (or intranet) can be used to improve communication and coordination. For example, Iansiti and MacCormack, 1997, has reported on successful internal use of the Internet/intranet to develop and test products at Netscape, Microsoft, and Yahoo! Hameri and Nihtilä, 1997, studied management and dissemination of project data via the Internet and concluded that "file-transfer activities typically come in bursts around project milestones." Although it has not yet been researched in depth, the introduction of the Internet into design team coordination will likely produce the three standard effects of new technology: the intended technical effects, the organizational adjustments, and the unintended social effects (Kiesler, 1986). Understanding these presents an interesting research challenge and a means for identifying how best to organize design teams in an Internet context.

1.3 Prototyping and Testing

A phase of the product development process where the Internet may ultimately play a very major role is prototyping/testing. Traditionally, this phase involved making physical prototypes and testing them, both from an engineering performance perspective and from a customer acceptance perspective. However, because physical prototypes are expensive and slow to use, the recent trend among designers has been toward increased use of computer simulation. For example, the Boeing 777 was developed entirely via virtual prototypes.

The Internet makes it possible to extend the use of virtual prototypes beyond the design process and into the marketing process, by presenting animated prototypes to potential customers to obtain their reactions. Dahan and Srinivasan, 2000, compared static and animated virtual prototypes of a bicycle pump with physical prototypes and attribute-only full-profile conjoint analysis and concluded that the market shares predicted by virtual prototypes are close to those obtained using the physical products and that both virtual and physical prototypes produced better predictions than conjoint analysis. Although a pump is a relatively simple product, their conclusions suggest that virtual prototypes may be important substitutes or complements for physical prototypes in assessing customer preferences in the future. Indeed, such beta-prototyping on the intranet is already widely used in the software industry (Iansiti and MacCormack, 1997).

The speed of an Internet process also makes it possible to set up a "market," in which customers trade conceptual products like buying and selling securities in a free market. This enhances interaction among customers, reveals their preferences, and identifies winning concepts. This method is very useful for measuring customers' interdependent preferences for fashion products. However, it is vulnerable to cheating by experienced participants (see Dahan and Hauser, 2002, for details).

By increasing the speed and reducing the cost of prototyping/testing, the use of virtual prototypes on the web could impact a number of decisions related to the new product development process, including:

- *Number of prototypes:* If they are cheaper, it may make sense to develop more prototypes. How many and how to best make use of the increased flow of customer preference information are researchable questions (see Dahan and Mendelson, 2001, for an analysis of how the optimal number of prototypes and the expected profit are affected by the cost of building and testing product concepts).

- *Number of respondents:* If customers can easily access new conceptual designs in the form of virtual prototypes on the Internet, their preferences for existing products may be affected (e.g., they may be induced to wait for a new product). The magnitude of this impact and the implications for how broadly to distribute virtual prototypes are open research questions.

- *Accuracy of information:* Virtual prototypes cannot fully replicate the experience of a physical product (i.e., because one cannot touch them). So, it is possible that customer preferences may be biased. For this reason, it may make sense to continue to use some classical product development approaches to avoid unexpected bias (Kiesler, 1986) generated by Web-based methods. Furthermore, unless the respondents are chosen systematically (instead of randomly based on who visits the firm's web site), there could be further bias due to the population sampled. Hence, effective use of virtual prototyping will require research into understanding the factors that lead to bias and developing methods to manage it.

- *Side effects:* There are some side effects of concept testing on the Internet. First, the Internet is more transparent to competitors and hence makes it more difficult for firms to maintain privacy than in traditional systems. For instance, Iansiti and MacCormack, 1997, has reported that "Netscape continually monitored the latest beta versions of Microsoft's competing product, Explorer, to compare features and formats." Second, it is possible that the users who try the new prototype and have unsatisfactory experience are unlikely to return. To solve this problem, Yahoo! puts its new service on Yahoo!'s Web sites without being linked to frequently visited parts of the site so that only leading-edge users are likely to find the new service (Iansiti and MacCormack, 1997).

1.4 Macro-Level Research

Beyond research on setting a new product strategy and managing the innovation process, there also exists the possibility of research into macro scale impacts of the Internet on new product development. For example:

- *Pace of Innovation:* Will the cost reductions and speed enhancements of the Internet promote more rapid introduction of new products and technologies? The pace of innovation has been a subject of research attention in the economics literature since the classic work of Schumpeter, 1942. Presumably this style of research can be aimed more specifically toward understanding the impacts of the Internet. With regard to the marketing-operations interface the question is how should firms react to any pace effects? Should they prepare to speed their own innovation processes? Or can they counter with other strategies (e.g., quality, service, speed)? The answers would have significant implications for long-term strategic planning and could enrich our understanding of policies aimed at influencing time-to-market (e.g., Datar et al., 1997).

- *Revolutionary versus Evolutionary Designs:* Will the Internet, and the virtual prototypes, choice-boards and toolkits it spawns, make firms more likely to emphasize rapid evolutionary designs (e.g., to present an ever changing web experience)? If so, how might this impact more fundamental revolutionary innovation? Could reliance on an e-commerce strategy actually make firms more vulnerable to attack by disruptive technologies?

- *Product Life Cycles:* Will the flow of information within firms, between firms and customers, and among customers affect product life cycles? Researchers have long tried to provide quantitative characterization of the rise and fall of products in the marketplace (e.g., the Bass diffusion model, see Mahajan et al., 1991 and Mahajan et al., 1995). If the Internet changes the underlying dynamics of product life cycles, then this will have significant impact on modeling and practice.

In summary, we can view the Internet as weakening the importance of customer location (Porter, 2001), lowering the cost of communicating with the customer, speeding the process of analyzing, prototyping and testing products, and enhancing options for customers to participate in the design process itself (see Figure 8.2 for a schematic illustrating the effect of the Internet on the product development process). The overall impact of these effects is not yet known. However, it is clear that competition, along with ever increasing demands by customers for speed, variety and flexibility, will continue to drive firms to seek innovative ways to exploit the power of the Internet in the new product development process. While research initiatives aimed at specific techniques, such as those suggested in the previous subsections, are clearly important, there is also a need to step back and evaluate some of these "big picture" issues in order to maintain a sense of perspective with which to view specific trends.

2. Sales

Sales is where products are presented to customers and the purchasing transaction occurs. As such, it is the purest meeting point of the marketing and operations functions. Marketing determines what products to offer and how to price and present them; operations determines how to make, stock and deliver them. In order to provide effective service at a competitive price, it is critical that the capabilities provided by operations are consistent with marketing strategy. For example, if the marketing strategy is to compete on extreme speed, then operations may have to use a make-to-stock approach. But if the marketing strategy is to provide high levels of variety and customization, then a make-to-order

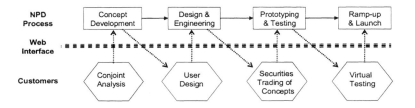

Figure 8.2. The effect of the Internet on the new product development (NPD) process

or assemble-to-order approach may be appropriate. Achieving strategic fit between marketing and operations is the core challenge in managing the interface between these two functions.

The primary competitive dimensions of a marketing strategy (i.e., the factors that motivate a sale) are: *price, speed, quality,* and *variety.* Therefore, the strategic fit problem amounts to finding an operations structure that supports a particular mix of targets for each of these factors. While myriad possibilities exist, we can identify most of the major issues and their relationship to the Internet by examining each of these issues in turn.

2.1 Pricing

Pricing is a classical problem in the marketing literature (Kotler, 2000, Chap. 15). Questions related to pricing strategy and competition, promotions, auctions, and characterization of demand functions (reservation prices, elasticities and price effects) have been studied for many years.

Because it allows changing of prices in real time, the rise of the Internet has intensified interest in dynamic pricing. It is not that these subjects are new. Dynamic pricing (i.e., offering different prices depending on variables such as demand, buyer type, and sales channel) has always been practiced via haggling between buyers and sales reps. Manufacturers have always scanned competitors' catalogs and used this information to negotiate deals with customers. And pricing often discounts items with excess inventory and charges a premium for scarce items. Hence, some research did address dynamic pricing prior to the Internet (see e.g., Lieber and Barnea, 1977).

But the Internet provides both a mechanism for sales reps to obtain much more information for making dynamic pricing decisions and a vehicle for offering dynamic prices to customers. As a result dynamic pricing is being practiced more widely and research interest in algorithms for setting prices has risen dramatically since the advent of the Internet. Recent reviews of dynamic pricing are given in Bitran and Caldentey, 2002 and by Elmaghraby and Keskinocak,

2002, and specific examples of models include Chatwin, 2000, Gallego and van Ryzin, 1994, Gallego and van Ryzin, 1997, Gurnani and Karlapalem, 2001, Masuda and Whang, 1999 and Zhao and Zheng, 2000). Because the appropriate dynamic pricing approach depends on product, market and system characteristics, we can expect research along these lines to continue as more specific settings are studied.

By itself, research into pricing does not necessarily address the marketing-operations interface. However, because prices cannot be set in a vacuum, there are ample opportunities to study the relationship of prices to other dimensions affected by operations management practices, primarily leadtime and quality.

Research into markets where customers are sensitive to price and delivery is particularly relevant to e-commerce systems because it is possible to price on the basis of work backlog. That is, when backlogs are high, and hence leadtimes are long, the firm may decide to price differently than when backlogs are low and leadtimes are short. Examples of research into pricing and leadtimes includes Chayet and Hopp, 2002, Dewan and Mendelson, 1990, Kalai et al., 1992, Lederer and Li, 1997, Li, 1992, Li and Lee, 1994, Loch, 1992, Mendelson and Whang, 1990, So, 2000, Tsay and Agrawal, 2000, van Mieghem and Dada, 1999. Unfortunately, finding market equilibria in systems where demand is a function of both price and leadtime is very difficult. As a result, virtually no leadtime quoting models take into consideration the dynamic pricing or leadtimes of competitors. The static game theory models that do try to account for competitive issues generally rely on extremely simple (e.g., M/M/1) models of the production process. Hence, none of the available models fully capture the pricing/leadtime decisions faced by retailers (e-tailers). Finding a practical method to set leadtime sensitive prices is a significant research opportunity that could have a major impact on e-commerce.

Analogous to price/leadtime models are price/quality models, in which customer demand is a function of both price and product quality. Examples of research that addresses this interface include Banker et al., 1998, Curry, 1985, Moorthy, 1988, van Mieghem and Dada, 1999. The difficulty with such models is that quality is a multi-dimensional attribute. Quality is measured in different ways in different settings. So, to be of use, such research must address specific aspects of quality that are subject to influence by operating policies.

The most pressing research challenge related to pricing posed by the Internet is development of models to support real-time pricing in an e-commerce environment. In addition to considering available capacity and inventory, such price quotes should also be sensitive to impacts on customers. For instance, a customer who makes a purchase only to find that the price was reduced shortly thereafter may be unhappy and less likely to return in the future. For this reason, some e-tailers provide some measure of price guarantee (e.g., if a lower price is offered within 30 days of purchase the difference will be refunded,

The Marketing-Operations Interface

provided the customer asks for the refund, of course). Presumably such price guarantees reduce the effectiveness of real-time pricing. Moreover, if they are to be offered, then pricing models should consider this in determining when and how to change price.

2.2 Lead Time Quoting

Competing on response time became a popular management practice in the 1990's (see Blackburn, 1990, Charney, 1991, Schmenner, 1988, Stalk and Hout, 1990, Thomas, 1990 for discussions of the virtues and practices of time based competition). This trend motivated researchers in operations management to develop models to understand leadtimes and how to shorten them (see Karmarkar, 1993, for a review).

A significant amount of the research into leadtimes has focused on how to reduce manufacturing cycle times in order to support shorter leadtime quotes to customers (see e.g., Hill and Khosla, 1992, Hopp et al., 1990, Karmarkar, 1987, Karmarkar et al., 1985, Lambrecht et al., 1998). Related research has addressed problems of predicting manufacturing lead times (e.g., Kim and Tang, 1997, Morton and Vepsalainen, 1987, Ruben and Mahmoodi, 2000, Shantikumar and Sumita, 1988), determining leadtimes within the production system (i.e., manufacturing or purchasing lead times) that will improve system performance (e.g., Hopp and Spearman, 1993, Yano, 1987a, Yano, 1987b), and characterizing the relationship between leadtimes and other parameters in production systems, such as lot sizes, inventories, dispatching rules, capacity and customer priorities (e.g., Ben-Daya and Raouf, 1994, Chaouch and Buzacott, 1994, Elwany and Baddan, 1998, Eppen and Martin, 1998, Glasserman and Wang, 1998, Karmarkar, 1987, Karmarkar et al., 1985, Song, 1994, Vepsalainen and Morton, 1988).

Most directly relevant to the marketing-operations interface is research directly into methods for setting due dates and quoting leadtimes (e.g., Baker, 1984, Baker and Bertrand, 1981, Bertrand, 1983a, Bertrand, 1983b, Duenyas, 1995, Duenyas and Hopp, 1995, Hopp and Roof, 2000, Hopp and Roof, 2001, Hopp and Spearman, 1993, Karmarkar 1994, Plambeck, 2003, Shantikumar and Sumita, 1988, Spearman and Zhang, 1999, Wein, 1991). In general, these models use well-defined traditional objective functions (e.g., minimize leadtime subject to a service constraint). However, in systems where it is possible to influence the customer experience during the leadtime (e.g., by hanging mirrors to distract customers waiting for elevators or serving wine to entertain customers waiting for a restaurant), research into how customers perceive waiting is also relevant (see Katz, et al., 1991).

In recent years, leadtime quoting has been addressed in commercial applications under the heading of *available to promise (ATP)* and *capable to promise*

(CTP). Typically, ATP systems determine whether a particular due date is feasible given current finished goods and WIP inventory levels, while CTP checks component availability in the supply chain, as well as FGI and WIP, to determine due date feasibility. These systems are gaining popularity as a means to offer more competitive leadtimes in traditional production systems and as a result are starting to get attention by modeling researchers (e.g., Chen et al., 2000 and Chen et al., 2001).

Automated ATP/CTP systems are even more relevant in e-commerce settings because these systems do not have a salesperson to make leadtime quotes. Hence, for e-commerce systems to implement dynamic leadtime quoting (to enable shorter average leadtimes for the same service level), they must make use of an automated ATP/CTP-type system. For example, Dell makes use of such an algorithmic approach to quote leadtimes on new computers ordered over their website. The concept of announcing leadtimes to customers to improve their experience is particularly attractive in service systems (see Whitt, 1999). For example, Disney World posts expected waiting times for rides, some call centers estimate waiting time, and many on-line help systems show the number of people in queue. It seems very likely that announcing leadtimes, both for delivery of products ordered on-line and for services obtained directly through the web, will become a standard feature of e-commerce systems as the industry matures.

However, while commercial ATP/CTP systems and tools for quoting leadtimes in queueing systems currently are available, the due date and leadtime quoting problem still presents research challenges. For example, while many of the research papers cited above consider stochastic effects in order to quote leadtimes with a predictable service level, many commercial systems do not consider uncertainty (e.g., they simply check for material availability and use fixed, MRP-type, leadtimes to consider timing). The reason is likely that the research models have considered simplified systems with elementary production structures, bills of material and supply chain configurations. So, stochastic optimization models that compute real-time leadtime quotes for realistic systems are a clear research need. Since these would be implemented inside the computer systems that support e-commerce sites, sophisticated computational solutions are entirely practical.

But an even more practical research goal would be techniques for using historical data to "tune" the leadtime quoting model over time. This is the approach used in Hopp and Roof, 2000, where a control chart methodology was adapted to dynamically update parameters in a due date quoting model for a single-product, single-stage production system. The advantages of such an approach over the traditional optimization approach are:

1 It does not require assumptions about the form of the distributions of demand or supply, since these are estimated statistically,

2 It enables models to adapt to a changing system, and

3 It has the potential to model complex systems that may resist analytic solution by capturing the behavior of complicated processes in empirical distributions.

By using the Internet to both collect the demand and delivery data needed to evolve the model and to present the resulting dynamic leadtime quotes, statistical leadtime quoting models could play a key role in the operation of e-commerce systems.

2.3 Quality Management

Quality has always been a concern of both the marketing and operations functions. From a marketing perspective, quality can be a differentiating factor that sells a product. From an operations factor, quality is a key driver of cost and productivity. The Total Quality Management (TQM) movement of the 1980's heightened awareness for an integrated view of quality and stimulated a great deal of research into quality methodologies. Most of this is only tangentially related to the marketing-operations interface. However, two threads have direct relevance:

1 Research into the linkage between quality and operations policies (see e.g., Baiman et al., 2000, Ben-Daya and Makhdoum, 1998, Goyal et al., 1993, Mefford, 1991, Miltenburg, 1993, Porteus, 1986, Soteriou and Zenios, 1999, Tapiero, 1987), and

2 Research on the role of quality in marketing and competition (see e.g., Gans, 1999, Grapentine, 1998, Hall and Porteus, 2000, Lele and Karmarkar, 1983, Parasuraman et al., 1985, Zeithaml at al., 1990).

The Internet offers a potentially significant impact on quality of service and hence on the marketing-operations interface. The availability or nature of human contact in the sales process, the accuracy of leadtime quotes, and the availability of updates on order status, are just a few of the issues where the Internet may change how a customer perceives the quality of service. Hence, several of the above cited references that deal with service quality offer (e.g., Gans, 1999, Grapentine, 1998, Hall and Porteus, 2000, Keeney, 1999, Lele and Karmarkar, 1983, Parasuraman et al., 1985, Zeithaml at al., 1990) provide valuable starting points for understanding how to provide strategically valuable kinds of quality experiences on the Internet. Specific research on this subject (e.g., Crowcroft, 2000) is just beginning; detailed investigation into what kind of operating structure is needed to support effective marketing strategies based on quality of service remains a promising research opportunity.

2.4 Product Variety

Within the sales arena, there is no issue that is more influenced by the Internet than that of product variety. By pooling inventory and streamlining distribution, on-line retailers can offer a much larger variety of offerings than can traditional storefronts (Brynjolfsson and Smith, 2000). Because of this, it is not surprising that a number of e-businesses have chosen to make variety a strong part of their strategy (e.g., the value proposition offered by Amazon.com is based primarily on an extremely high level of product variety, combined with service to support a convenient on-line experience). Research into how much variety is appropriate to offer (see e.g., Bawa, 1990, Yeh, 1991, van Ryzin and Mahajan, 1999) is certainly relevant, since variety is not an unambiguous benefit to the customer. For example, as Quelch and Kenny, 1994, points out "People do not eat more, drink more, or wash their hair more just because they have more products from which to choose." Moreover, Iyengar and Lepper, 2000, point out that more choices can actually be perceived as worse by customers than fewer choices. But the real issue at the heart of the operations/marketing interface is how to support whatever variety the e-tailer decides to offer.

E-commerce systems can be backed up by make-to-stock, make-to-order and hybrid (e.g., assemble-to-order) systems. While such systems have been studied for a long time, interest in them remains strong, partly due to the pressure of the Internet to offer high levels of service despite an extensive product catalog. Examples of recent research addressing the choice, structure and operation of production/inventory systems include Fegergruen and Katalan, 1999, Ha, 1997, Hillier, 1999, Johnson and Scudder, 1999, Kingsman et al., 1996, Lovejoy and Whang, 1995, Moon and Choi, 1997, Perez and Zipkin, 1997. These and other models offer a start to answering key questions of which products to stock, which to produce to order, and how to organize the system for doing this.

Beyond the question of how to meet orders within the firm's production system is how to structure the entire supply chain to support the process. For example, Amazon.com only stocks some of the items it sells. The others are stocked at and shipped from suppliers. Research into how to structure relationships with suppliers to support a high-variety marketing strategy is therefore extremely relevant to management of e-commerce systems. An excellent survey of the types of contracts available and the modeling frameworks in which they have been analyzed is given by Cachon, 2002. Examples of research into understanding the factors behind supplier contracts and the contracts themselves include Barnes-Schuster et al, 2002, Bassok and Anupindi, 1997, Cachon and Lariviere, 2000, Fisher et al., 1999, Chen, 1999, Corbett and DeCroix, 2001, Lee, 2000, Tsay and Lovejoy, 1999, Tsay et al., 1999. Because the provider of goods and/or services in most supply chain structures rely on various parties to transact the various operations decisions, the principle agent framework has been proposed as a mechanism for studying the coordination of the overall system (see Plambeck and Zenios, 2000 and Zenios, 2003 for examples of this framework applied to operations decisions).

Although there has been a flurry of research into supply chain contracts in recent years, there remains a gap between the contracting practices of firms and the models in the research. For example, most models examine the behavior of contracts in single period (e.g., news vendor) situations or in steady state (e.g., queueing) environments, while realistic systems evolve dynamically with the terms of the contracts being adjusted continually. Furthermore, although many contracts are used to mitigate some kind of risk (e.g., getting caught with excess inventory), few models explicitly consider risk attitude. Finally, real world contracts are often concerned with the details of cash flow (e.g., whether payment terms are net 60 days or net 30 days makes a tremendous difference to a firm's bottom line), but these are rarely modeled in anything beyond a highly simplified way. Hence, it would appear that there is still room for research into practical contracts that can be used to structure supply chains that support e-commerce sales channels.

Beyond the question of how to structure the production/inventory system and the supply chain is the issue of how to evolve the system over time. Specifically, how can an e-business make use of transactional data to determine what products to offer and how to treat them in the operations system? For instance, if an e-commerce system makes use of a choice board mechanism to allow customers to configure products, then their choices reveal a great deal about preferences. Making use of this data in forecasting, procurement and operating system reconfiguration (e.g., switching a product or component from make-to-stock to make-to-order, or vice versa) could enable the system to adapt over time to a changing environment.

3. Production/Delivery

The sales step is only part of the overall customer experience with an e-tailer. Providing convenient on-line ordering is certainly important, but if the firm doesn't back it up with prompt, reliable delivery of the product or service, the customer is unlikely to return. For this reason, customer leadtimes are a critical issue in e-commerce systems.

Unfortunately, it is easier to promise rapid delivery than to provide it. Laseter et al., 2000 pointed out that unless the density of online shoppers is very high, the cost of "the last mile" in the delivery process is extremely expensive. They also note that there is a "speed versus variety" tradeoff inherent in on-line retailing. Extremely rapid (i.e., same day) delivery requires local stocking and so generally precludes offering a great deal of variety. For example, Kozmo.com, a same-day deliverer, offered only about 15,000 items, while Amazon.com, a more traditional deliverer, offers over 10 million items. While it may be possible to tailor an attractive value proposition around fast delivery in some markets, the cost and variety obstacles are daunting. Hence it is not surprising that many of the early pioneers of web-based rapid delivery systems (e.g., Kozmo, Urbanfetch, Webvan) are out of business.

The delivery challenge presents a number of research opportunities at the interface between marketing and operations. To begin with, it raises the question of how much speed customers really want. The failure of some early e-tailers may have been partly due to their mistaken assumption that impatient on-line shoppers all wanted quick (and costly) delivery. But receiving a book in days or a car in weeks may be perfectly acceptable responsiveness to many customers, particularly if this reduces cost or enhances variety. Conventional market research methods can be applied to determining how customers value speed in particular on-line markets. Once we have gained a better understanding of customer priorities for different types of products, research can be directed at developing better operating systems with which to deliver products appropriately.

Given any speed target, a main challenge is to attain it at minimum cost. Indeed, a big reason for the survival (and near profitability) of Amazon.com is that they have made significant increases in efficiency of the order fulfillment process in recent years. By making a host of improvements in design and control of their warehousing and delivery systems Amazon has transformed delivery from a significant source of cost into a money making operation. Because of the importance of efficiency, traditional research into tactical OR models will remain and increase in importance. For example, warehousing (see Cormier and Gunn, 1992 and Rouwenhorst et al., 2000 for reviews) and vehicle routing (see Bertsimas and Smichi-Levi, 1996 and Fisher, 1995 for overviews) are key areas that present opportunities for efficiency upgrades.

A newer line of research stimulated by the rise of the Internet deals with the strategic question of how to structure the delivery process to achieve the desired performance. One option that has been adopted primarily by small e-tailers is drop-shipping, in which suppliers hold the inventory and ship directly to customers. Netessine and Rudi, 2000 examined drop-shipping under various power arrangements between a wholesaler and retailer and develop a coordination scheme that involves the wholesaler subsidizing part of the customer acquisition (marketing) process of the retailer. Netessine and Rudi, 2001 characterized situations in which drop-shipping is preferred to the traditional approach of stocking inventory at the retailer and also devised dual strategies in which local inventory is stocked and drop-shipping is used as a backup. Randall et al., 2002 performed an empirical analysis of Internet retailers to determine what factors are important in determining their inventory/delivery structure.

For retailers (e.g., Amazon), leadtime performance is purely determined by the structure and execution of inventory and distribution policies. But for firms that also manufacture or assemble products (e.g., Dell), the production function is also a driver of performance. Of course, all of the usual methods for achieving manufacturing efficiency (lean, agile, cellular manufacturing, etc.) are relevant to e-commerce settings. As we have noted above, modular product architectures and assemble-to-order production systems are particularly well-adapted to supporting quick-response manufacturing with which to support an e-commerce system. Since the issues of speed, variety, quality and flexibility were clearly priorities prior to the advent of e-commerce, research that addresses these remains relevant but has not been radically affected by the Internet.

An issue that has been affected by the push to enhance production and delivery systems, however, is that of integration. As an illustration of a failure to integrate properly Fisher et al., 1995, observed a mismatch between the manufacturing and distribution capability in the automobile industry. On the manufacturing side, investment in flexible equipment has made the production system capable of providing considerable variety to customers. However, because the market is finely segmented (into dealerships or regions) there is little pooling of demand. As a result local retailers/dealers/distributors face a highly volatile demand, which is not economical to buffer with stock. So, customers are faced with very limited variety in the showroom (or a long wait for a customized car). As Fisher et al., 1995 pointed out, this puts the assembly plant in the worst of two worlds; it must be configured to deliver an huge amount of variety, but is seldom called on to do so. In such an environment, the Internet offers the potential for pooling stock across regions and hence facilitating a better match between manufacturing and distribution.

In addition to cost, speed and fit with production, the Internet introduces another issue at the marketing-operations interface of the delivery function–visibility. The Internet makes it possible for customers to monitor the progress of their order as it moves through the production and delivery steps. Such systems have been in use for several years at delivery companies like Federal Express and UPS as well as at many Internet retailers. But while order tracking is becoming increasingly common in practice, relatively little research has considered it. The lead time quoting and available-to-promise models cited earlier may be relevant to the mechanics of order tracking systems, since they offer means for making estimates of when orders will complete the various stages of production and delivery. But we are not aware of research that addresses the value of such visibility to customers or its effect on their behavior. For example, is a customer less dissatisfied with a late delivery given advance warning of it through a tracking system? Are customers more likely to change their orders based on feedback about their status? How detailed should the information provided to customers be? How can a manufacturer/distributor/retailer structure its production/inventory/distribution system to support customer expectations in an environment of real time communication? Because it has received fairly little attention to date, research on understanding the implications of visibility may be as important in the long-run as research on speed and efficiency.

4. Service

Post-sales service, also referred to as after-sales or customer service, and product or customer support, can be defined as "assistance to help [customers] obtain maximum value from their purchases" (Goffin, 1999). The key post-sales service elements identified by Goffin, 1999, are: installation, user training, documentation, maintenance and repair, telephone support, warranty, and upgrades. Other forms of support include product returns, complaints, and account management. Some of the main reasons of why firms offer post-sales services are: to attain customer satisfaction and encourage customer loyalty (Armistead and Clark, 1992), to gain a competitive advantage (Hull and Cox, 1994, Cohen and Whang, 1997, Goffin, 1999), to improve knowledge of customers (Burns, 1995), and because they can represent an important source of revenue (Hull and Cox, 1994, Akşin and Harker, 1999).

4.1 Previous Research

A substantial portion of research on post-sales service is strategy oriented and empirical (survey and case-method based), and has focused on identifying effective service channels (Armistead and Clark, 1991, Burns, 1995, Loomba, 1996, Loomba, 1998, Goffin, 1999), and key drivers (Hull and Cox, 1994). For manufacturing firms in the early 1980s, fast and reliable repair was considered the most important aspect of support (Lele and Karmarkar, 1983) and received most of the research attention. Many relevant issues in this area lie on the marketing and operations interface.

For instance, as part of a service support strategy, Lele, 1986, proposed addressing support at the design level (e.g., improved component reliability, redundancy, and modular design), reducing customer risk (e.g., via warranties or service contracts), and improving support system operations. Loomba, 1996, suggested designing the support distribution structure based on the sales system. Lele, 1986, argued that rapid technological change makes it difficult to maintain a competitive advantage based on product features or design alone, and suggested a firm might differentiate itself in terms of superior post-sales service. Goffin, 1994, provided empirical evidence that firms which address service at the design level and have access to a superior support system command a significant market share. Mentzer, Gomes and Krapfel, 1989, formulated a model (subsequently tested empirically by Emerson and Grimm, 1996) to predict customer purchase decisions based on an extensive list of marketing and logistics factors. In particular, spare parts management plays a key role in post-sales support (e.g., Hull and Cox, 1994, Cohen, Zheng, and Agrawal, 1997). Cohen, Cull, Lee and Willen, 2000, illustrate this point with the case of automotive service at Saturn, where superior off-the-shelf parts availability by dealers dramatically improved customer satisfaction and loyalty.

Analytic modeling research in this area is limited. Examples include two competitive models involving products that require after-sales maintenance repair. Cohen and Whang, 1997, used product, price, and quality of after-sales service as strategic decisions and concluded that after-sales service can be used as competitive differentiator. Murthy and Kumar, 2000, developed a game-theoretic model representing the interaction between a manufacturer and a retailer, and studied the relationship between quality of service, quality of performance, and retail price.

During the last two decades, the manufacturing sector has experienced technological improvements resulting in more reliable but more complex products, while the service sector has experienced unprecedented growth. This has led to a shift in customer service emphasis from maintenance and repair to installation and telephone support. The proliferation of call centers has presented practitioners with challenges in the interface between marketing and operations as they attempt to provide quality service while trying to control costs. An outstanding overview and literature survey of call centers is Gans, et al., 2002. Specific examples of research on service systems include Akşin and Harker, 1999, Frei and Harker, 1999, Garnett et al., 2002 and Zhou and Gans, 1999. Attempts to combine support and sales functions have inspired additional research (e.g., Loomba, 1996). Akşin and Harker, 1999, studied this issue in the context of retail banking. But much remains to be done with regard to analytical modeling of post-sales services to evaluate relevant tradeoffs from an operations and marketing interface perspective. In the remainder of this section we identify areas of post-sales service affected by the Internet, and potential research opportunities at the empirical, strategic, and analytic modeling levels.

4.2 Research Opportunities

The Internet has brought several improvements in call center operations. For example, effective capacity can be increased by enabling workers at different locations to access, from any Web browser, a single database of customer requests. But more relevant to the marketing and operations interface is the emergence of new ways to interact with the customer. Both asynchronous and synchronous communication modes like e-mail, chat forums, discussion groups, and interactive form-based queries, reduce customer waiting times, and allow the firm to demonstrate responsiveness while reducing costs by easing capacity constraints. Some examples of the Internet's impact on user training, documentation, and technical support functions are: the possibility of offering self-service support (e.g., via FAQs—frequently asked questions—and searchable knowledge databases), dynamic information updates, and remote computer control (e.g., providing training, diagnostics, and troubleshooting with products such as Timbuktu, NetOp, and Apple Remote Desktop).

While shifting operations to the Web can lead to dramatic savings, around the clock availability, and reduced customer waiting, many questions remain about adopting these new customer contact modes. For instance: Which are the most effective contact modes for a particular segment-task combination? How does service variety affect operational requirements? What incentives should be used to improve customer satisfaction and operational efficiency across contact modes? What are the competitive implications of various types of post-sale support?

The notion of combining sales with support services lies at the core of the marketing and operations interface, but it is not exclusive to the Internet (see e.g., Akşin and Harker, 1999). However, the Internet's customer-driven approach to product information, mass-customization capabilities, and asynchronous communication modes make it a better suited channel for that practice since it allows more accurate targeting and is less intrusive than other contact modes (e.g., Anderson, 2000 reported that among customers purchasing a new computer at IBM, 80% sign up voluntarily for newer product information emails as part of product support).

Some examples of the Internet's impact on maintenance, repair, and upgrade services are electronic deployment of software upgrades and patches, as well as faster, more accurate and cost effective distribution of spare parts. The combination of the Internet's universal access with emerging appliance networking technologies (e.g., Bluetooth) opens the possibility for remote diagnostics and detailed product quality feedback. Although it remains to be seen if these services will be widely adopted in the future, they offer the prospect that customer feedback can become a key driver of post-sales service improvements. Besides its repercussions at the product design level, an immediate challenge is identifying the most relevant information to drive the process.

Internet-based account management services are now widely spread across the service sector (e.g., magazine or newspaper subscription services, residential telephone features selection, account statements in commercial banking, etc.). These offer added value to customers (e.g., convenient access, fewer errors, reduced paper clutter, etc.), and savings as well as additional sales opportunities to service providers. An interesting development in this area has been the emergence of account aggregation services by companies such as Yodlee, 724 Solutions, Kinexus, ByAllAccounts, and others, allowing customers one-click secure access to account statements from several institutions. This has led to the formation of conglomerates that, unlike airline frequent flyer alliances, are non-exclusive. Attempts to understand the economic, competitive, customer loyalty and operational implications of such conglomerates is likely to stimulate future research on the marketing and operations interface.

The promise of no-hassle returns is widely considered an essential component of customer service which may lead to increased sales by easing customer fears. Yet, product exchange and returns have represented a significant logistical challenge to retailers since the early days of e-commerce. This has led click-and-mortar retailers (e.g., Barnes and Noble and The Right Start) to offer store returns to online customers, and other retailers to outsource those services with companies such as The Return Exchange. These challenges are likely to generate research interest in "reverse logistics" issues, originally considered in the context of recycling. Related issues arise from the notion of a manufacturer moving from product selling to service leasing (e.g., instead of selling software, cars, or carpet, the company sells a software subscription, transportation services, or floor covering). In this framework, the products remain the property of the manufacturer, who now has an incentive to recycle them.

Customer complaints are a valuable form of customer feedback (Plymire, 1991). The Internet is a convenient way to submit customer complaints, and gives the firm an opportunity to positively influence the customer's perceptions by being responsive. Analyzing aggregate complaints is a valuable driver for product or service improvement (Schibrowsky and Lapidus, 1994). The integration of complaint management with product development and post-sales services is likely to inspire new research. The Internet also opens the possibility of allowing communication between customers in a company-sanctioned bulletin board. The company is then faced with the challenge of deciding whether to censor negative comments on their products in order to protect sales, or allowing them so as not to appear too defensive. Studying the relationship between those risks and the value of customer feedback, as well as the competitive implications of offering such a forum, represent research opportunities.

5. Conclusions

The marketing-operations interface has long been a challenge in practice and thus a source of research opportunities. In recent years, intense global competition in most industries has elevated customer expectations. Understanding these expectations and developing the capabilities to meet them will require increasingly close coordination of the marketing and operations functions. Hence, we can expect interest in the marketing-operations interface to intensify over time.

The rise of the Internet presents special challenges in coordinating marketing and operations. In this paper we have noted that the web offers significant opportunities with regard to speed, communication and data collection. Practitioners are rapidly developing e-commerce systems that exploit the Internet to provide rapid customer response, greater sharing of information and visibility to the production/delivery process, and methods for extracting customer information from transactional data. Concurrently, researchers are providing analytic support aimed at many steps in the value chain, from product development to post-sale service.

After reviewing the broad range of research that addresses the marketing-operations interface in the context of e-commerce, we feel that the studies can be roughly categorized into the following three stages:

1. *Traditional:* Many pre-existing research areas, ranging from conjoint analysis to inventory management to vehicle routing, are directly relevant to the marketing-operations interface in an e-commerce environment. Existing and new research in these traditional areas will continue to find application in the design and management of effective systems.

2. *Direct:* Use of the Internet as a marketing channel has posed some new research problems that either did not exist or were of much less importance under traditional channel structures. For example, the web is an essential prerequisite to use of animated prototypes in the market research process. Distribution systems involving drop-shipping become much more relevant in an e-commerce setting. And real-time dynamic pricing is a legitimate option in an on-line system. Direct research into these and other specific issues have provided academics with a new and interesting set of modeling and analysis challenges.

3. *Dynamic:* Using past experiences and results from traditional and direct research to set up effective e-commerce systems will be an important priority for the near term. However, in the longer term, as the web becomes a standard channel, companies will have to adapt to their customers more rapidly. This will necessitate using the web to collect information about customers and making adjustments to the system based on this information. For instance, a firm may track customer selections in a choice-board environment to update its estimates of customer preferences. Armed with this updated information it may adjust the production and inventory system used to support the choice-board interface. In the long run, it may be the ability to collect the right information and use it

quickly to dynamically evolve effective systems that will determine who is successful in the world of e-commerce.

Our review suggests that, while many opportunities remain, there has been impressive progress on the first two stages. A broad range of traditional and new modeling tools now exist to support practitioners in designing an e-commerce interface and the operating system with which to support it. However, work is only beginning on the third stage. As Internet businesses strive to cope with intensifying customer demands for personalized service, it will become increasingly important to evolve e-commerce systems to keep pace. In this environment, the data collection and feedback functions will loom large. Therefore, research into these issues has the potential to be both novel and influential.

References

Akşin, O. Z., P. T. Harker 1999. To Sell or Not to Sell: Determining the Tradeoffs Between Service and Sales in Retail Banking Phone Centers, Journal of Service Research 2(1), 19-33.

Anderson, D. 2000. Creating and Nurturing a Premier e-Business, Journal of Interactive Marketing 14(3), 67-73.

Armistead, C., G. Clark 1991. A Framework for Formulating After-sales Support Strategy, International Journal of Physical Distribution & Logistics Management 21(9), 111-124.

Armistead, C., G. Clark 1992. Customer Service and Support, Pitman, London.

Baiman, S., P.E. Fischer, M.V. Rajan. 2000. Information, Contracting, and Quality Costs. Management Science 46(6), 776-789.

Baker, K. 1984. Sequencing Rules and Due-Date Assignments in a Job Shop, Management Science 30, 1093-2004.

Baker, K., and J. Bertrand. 1981. A Comparison of Due-Date Selection Rules, AIIE Transactions 13, 123-131.

Baker, K., and J. Bertrand. 1982. An Investigation of Due-Date Assignment Rules with Constrained Tightness, Journal of Operations Management 3, 109-120.

Balakrishnan, N., A.K. Chakravarty, S. Ghose. 1997. Role of Design-Philosophics in Interfacing Manufacturing with Marketing. European Journal of Operational Research 103, 453-469.

Baldwin, C.Y., K.B. Clark. 1997. Managing in an Age of Modularity. Harvard Business Review 75(5), 84-93.

Banker, R.D., I. Khosla, K.K. Sinha. 1998. Quality and Competition. Management Science 44(9), 1179-1192.

Barnes-Schuster, D., Y. Bassok, R. Anupindi. 2002. Coordination and Flexibility in Supply Contracts with Options. Manufacturing & Service Operations 4(3), 171-207.

Bassok, Y., R. Anupindi. 1997. Analysis of Supply Contracts with Total Minimum Commitment. IIE Transactions 29, 373-381

Bawa, K. 1990. Modeling and Variety Seeking Tendencies in Brand Choice Behavior. Marketing Science 9(3), 263-278.

Ben-Daya, M., A. Raouf. 1994. Inventory Models Involving Lead Time as a Decision Variable. Journal of the Operational Research Society 45(5), 579-582.

Ben-Daya, M., M. Makhdoum. 1998. Integrated Production and Quality Model under Various Preventive Maintenance Policies. Journal of the Operational Research Society 49(8), 840-853.

Bertrand, J. 1983. The Effect of Workload Dependent Due-Dates on Job Shop Performance, Management Science 29, 799-816.

Bertrand, J. 1983. The Use of Workload Information to Control Job Lateness in Controlled and Uncontrolled Release Production Systems, Journal of Operations Management 3, 67-78.

Bertsimas, D., D. Simchi-Levi. 1996. A New Generation of Vehicle Routing Research: Robust Algorithms, Addressing Uncertainty. Operations Research 44(2), 286-304.

Bitran, G. and R. Caldentey. 2002. An Overview of Pricing Models for Revenue Management. To appear in Manufacturing & Service Operations Management.

Blackburn, Joseph. 1990. Time Based Competition. Irwin, Burr-Ridge, Illinois.

Bollo, D., M. Stumm. 1998. Possible Changes in Logistic Chain Relationships Due to Internet Developments. ITOR 5(6), 427-445.

Bower, J.L., C.M. Christensen. 1995. Disruptive Technologies: Catching the Wave. Harvard Business Review (Jan./Feb.), 43-53.

Bregman, R.L. 1995. Integrating Marketing, Operations, and Purchasing to Create Value. OMEGA 23(2), 159-172.

Brynjolfsson, E., M.D. Smith. 2000. Frictionless Commerce? A Comparison of Internet and Conventional Retailers. Management Science 46(4), 563-585.

Burns, J. 1995. Developing and Implementing a Customer Contact Strategy, Managing Service Quality 5(4), 44-48.

Cachon, G. 2002. Supply Chain Coordination with Contracts. to appear in Graves, S., T. de Kok (Eds.), Handbooks of Operations Research and Management Science: Supply Chain Management. Elsevier, New York.

Cachon, G., M. Lariviere. 2000. Supply Chain Coordination with Revenue Sharing: Strengths and Weaknesses. Working paper, University of Pennsylvania, Philadephia, PA.

Chaouch, B.A., J.A. Buzacott. 1994. The Effects of Lead Time on Plant Timing and Size. Production and Operations Management 3(1), 38-54.

Charney, C. 1991. Time to Market: Reducing Product Lead Time, Society of Manufacturing Engineers, Dearborn, MI.

Chatwin, R.E. 2000. Optimal Dynamic Pricing of Perishable Products with Stochastic Demand and a Finite Set of Prices. European Journal of Operational Research 125(1), 149-174.

Chayet, S., W.J. Hopp. 2002. Lead Time Competition under Uncertainty. Working Paper, Graduate School of Business, University of Chicago, Chicago, IL.

Chen, F. 1999. Decentralized Supply Chains Subject to Information Delays. Management Science 45(8), 1076-1090.

Chen, I.J., R.J. Calantone, C.-H. Chung. 1992. The Marketing-Manufacturing Interface and Manufacturing Flexibility. OMEGA 20(4), 431-443.

Chen, C.Y., Z.Y. Zhao, M. Ball. 2000. A Model for Batch Advanced Available-To-Promise. Working Paper, Smith School of Business, University of Maryland, College Park, MD 20742.

Chen, C.Y., Z.Y. Zhao, M. Ball. 2001. Quantity-and-Due-Date-Quoting Available-To-Promise. Working Paper, Smith School of Business, University of Maryland, College Park, MD 20742.

Choi, S. Chan, W.S. Desarbo, P.T. Harker. 1990. Product Positioning under Price Competition. Management Science 36(2), 175-199.

Christensen, C.M., R.S. Tedlow. 2000. Patterns of Disruption in Retailing. Harvard Business Review 78(1) 42-45.

Cohen, M. A., C. Cull, H. L. Lee, D. Willen 2000. Saturn's Supply-Chain Innovation: High Value in After-Sales Service, Sloan Management Review (Summer), 93-101.

Cohen, M. A., S. Whang 1997. Competing in Product and Service: A Product Life-Cycle Model, Management Science 43(4), 535-545.

Cohen, M. A., Y.-S. Zheng, V. Agrawal 1997. Service Parts Logistics: A Benchmark Analysis, IIE Transactions 29(8), 627-639.

Corbett, C., G. DeCroix. 2001. Shared Savings Contracts in Supply Chains. Management Science 47(7), 881-893.

Cormier, G., E. Gunn. 1992. A Review of Warehouse Models. European Journal of Operational Research 58(1), 3-13.

Crowcroft, J. 2000. Herding Cats: Modelling Quality of Service for Internet Applications. Philisophical Transactions of the Royal Society, London Series A (Physical Sciences and Engineering) 358(1773), 2209-2215.

Curry, D.J. 1985. Measuring Price and Quality Competition. Journal of Marketing, Spring 1985, 106-117.

Dahan, E., J.R. Hauser. 2002. The virtual customer. Journal of Product Innovation Management 19 332-353.

Dahan, E., H. Mendelson. 2001. An Extreme-Value Model of Concept Testing. Management Science 47(1), 102-116.

Dahan, E., V. Srinivasan. 2000. The predictive power of Internet-based product concept testing using visual depiction and animation. Journal of Product Innovation Management 17(2), 99-109.

Datar, S., C. Jordan, S. Kekre, S. Rajiv, K, Srinivasan. 1997. New product development structures and time-to-market. Management Science 43(4), 452-464.

Dewan, R., M. Freimer, A. Seidmann. 2000. Organizing Distribution Channels for Information Goods on the Internet. Management Science 46(4), 483-495.

Dewan, S., H. Mendelson 1990. User Delay Costs and International Pricing for a Service Facility, Management Science 36, 1502-1517.

Duenyas, I. 1995. Single Facility Due Date Setting with Multiple Customer Classes. Management Science 41(4), 608-619.

Duenyas, I. and W. Hopp. 1995. Quoting Customer Lead Times. Management Science 41(1), 43-57.

Eliashberg, J., J. Steinberg. 1993. Marketing-Production Joint Decision-Making. In: Eliashberg, J., G.L. Lillien (Eds.), Handbook in OR and MS, Vol.5. Elsevier, New York.

Elmaghraby, W., P. Keskinocak. 2002. Dynamic Pricing: Research Overview, Current Practices and Future Directions. Working paper, Georgia Institute of Technology, Atlanta, GA.

Elwany, M.H., A. El Baddan. 1998. The Sensitivity of Jobshop Due Date Lead Time to Changes in the Processing Time. Computers and Industrial Engineering 35(3), 411-414.

Emerson, C. J., C. M. Grimm 1996. Logistics and Marketing Components of Customer Service: An Empirical Test of the Mentzer, Gomes and Krapfel Model. International Journal of Physical Distribution & Logistics Management 26(8), 29-42.

Eppen, G.D., A.V. Iyer. 1997. Backup Agreements in Fashion Buying: The Value of Upstream Flexibility. Management Science 43, 1469-1484.

Eppen G., and R. Martin. 1988. Determining Safety Stock in the Presence of Stochastic Lead Time and Demand, Management Science 34, 1380-1390.

Federgruen, A., Z. Katalan. 1999. The Impact of Adding a Make-to-Order Item to a Make-to-Stock Production System. Management Science 45(7), 980-994.

Feitzinger, E., H.L. Lee. 1997. Mass Customization at Hewlett-Packard: The Pwer of Postponement. Harvard Business Review. 75(Jan./Feb.), 116-121.

Fisher, M.L. 1995. Vehicle Routing. In Handbooks in Operations Research and Management Science, the volume on Network routing, M. Ball, T. Magnanti, C. Monma and G. Nemhauser (Eds.) 1-33.

Fisher, M.L. 1997. What is the Right Supply Chain for Your Product? Harvard Business Review March-April, 105-116.

Fisher, M.L., A. Jain, J.P. MacDuffie. 1995. Strategies for product variety: Lessons from the Auto industry. In: Bowman, E., B. Kogut (Eds.), Redesigning the Firm, Oxford University Press, New York, 116-154.

Fisher, M.L., K. Ramdas, K.T. Ulrich. 1999. Component Sharing in the Management of Product Variety: A Study of Automotive Braking Systems. Management Science 45(3), 297-315.

Fitzsimmons, J.A., P. Kouvelis, D.N. Mallick. 1991. Design Strategy and Its Interface with Manufacturing and Marketing: A Conceptual Framework. Journal of Operations Management 10(3), 398-415.

Frei, F. X., P. T. Harker 1999. Value Creation and Process Management: Evidence from Retail Banking. Working Paper. The Wharton School, The University of Pennsylvania. Philadelphia, PA, 19104.

Gallego, G., and G. van Ryzin. 1994. Optimal Dynamic Pricing of Inventories with Stochastic Demand over Finite Horizons. Management Science 40(8), 999-1020.

Gallego, G., G. van Ryzin. 1997. A Multiproduct Dynamic Pricing Problem and its Applications to Network Yield Management. Operations Research 45(1), 24-41.

Gans, N. 1999. Customer Loyalty and Supplier Quality Competition. Working Paper, Wharton School, University of Pennsylvania, Philadelphia, PA 19104.

Gans, N., G. Koole, A. Mandelbaum. 2002. Telephone Call Centers: A Tutorial and Literature Review. to appear in Manufacturing & Service Operations Management.

Garnett, O., A. Mandelbaum, M. Reiman. 2002. Designing a Call Center with Impatient Customers. Manufacturing & Service Operations Management 4(3), 208-227.

Glasserman, P., J. Wang. 1998. Leadtime-Inventory Trade-Offs in Assemble-to-Order Systems. Operations Research 46(6), 858-871.

Goffin, K. 1994. Gaining a Competitive Advantage from Support: Five Case Studies, European Services Industry 1(4), 1-7.

Goffin, K. 1999. Customer Support: A Cross-Industry Study of Distribution Channels and Strategies, International Journal of Physical Distribution & Logistics Management 29(6), 374-397.

Goyal, S.K., A. Gunasekaran, T. Martikainen, P. Yli-Olli. 1993. Integrating Production and Quality Control Policies: A Survey. European Journal of Operational Research 69(1), 1-13

Grapentine, T. 1998. The History and Future of Service Quality Assessment. Marketing Research 10(4), 4-20.

Green, P.E., A.M. Krieger. 1989. Recent Contributions to Optimal Product Positioning and Buyer Segmentation. European Journal of Operational Research 41(2), 127-141.

Green, P.E., A.M. Krieger, Y. Wind. 2001. Thirty Years of Conjoint Analysis: Reflections and Prospects. Interfaces 31(3) 56-73.

Green, P.E., V. Srinivasan. 1990. Conjoint Analysis in Marketing: New developments with Implications for Research and Practice. Journal of Marketing 54(4), 3-19.

Griffin, A., J.R. Hauser. 1993. The Voice of the Customer. Marketing Science 12(1), 1-27.

Gilmore, J.H., B.J. Pine II. 1997. The Four Faces of Mass Customization. Harvard Business Review 75(Jan./Feb.) 91-101.

Gurnani, H., K. Karlapalem. 2001. Optimal Pricing Strategies for Internet-Based Software Dissemination. Journal of the Operational Research Society 52(1), 64-70.

Ha, A.Y. 1997. Inventory Rationing in a Make-to-Stock Production System with Several Demand Classes and Lost Sales. Management Science 43(8), 1093-1103.

Hall, J., and E.L. Porteus. 2000. Customer Service Competition in Capacitated Systems. Manufacturing & Service Operations Management 2(2), 144-165.

Hameri, A., J. Nihtilä. 1997. Distributed New Product Development Project Based on Internet and World-Wide Web: A Case Study. Journal of Production Innovation and Management 14(1), 77-87.

Hausman, W.H., D.B. Montgomery. 1993. The Manufacturing/Marketing Interface: Critical Strategic and Tactical Linkages. In: Buffa, S., R. Sarin (Eds.), Perspectives in Operations Management: Essays in Honor of Elwood. Kluwer Academic Publishers, Boston.

Hill, A.V., I.S. Khosla. 1992. Models for Optimal Lead Time Reduction. Production and Operations Management 1(2), 185-197.

Hillier, M.S. 1999. Product Commonality in Multiple-Period, Make-to-Stock Systems. Naval Research Logistics 46(6), 737-751.

Hopp, W.J., M.L. Roof. 2000. Quoting Manufacturing Due Dates Subject to a Service Level Constraint. IIE Transactions 32(9), 771-784.

Hopp, W.J., M.L. Roof. 2001. A Simple Robust Leadtime Quoting Policy. Manufacturing & Service Operations Management 3, 321-336.

Hopp, W., M. Spearman. 1993. Setting Safety Leadtimes for Purchased Components in Assembly Systems. IIE Transactions 25, 2-11.

Hopp, W., M. Spearman, D. Woodruff. 1990. Practical Strategies for Lead Time Reduction. Manufacturing Review 3, 78-84.

Huff, L.C., W.T. Robinson. 1994. The Impact of Leadtime and Years of Competitive Rivalry on Pioneer Market Share Advantages. Management Science 40(10), 1370-1377.

Hull, D. L., J. F. Cox 1994. The Field Service Function in the Electronics Industry: Providing a Link between Customers and Production/Marketing", International Journal of Production Economics 37(1), 115-126.

Iansiti, M., A. MacCormack. 1997. Developing Products on Internet Time. Harvard Business Review 75(Sept./Oct.), 108-117.

Iyengar, S. S., M. Lepper. 2000. When Choice is Demotivating: Can One Desire Too Much of a Good Thing? Journal of Personality and Social Psychology 76, 995-1006.

Johnson, M.E., and B. Scudder. 1999. Supporting Quick Response Through Scheduling of Make-to-Stock Production/Inventory Systems. Decision Sciences 30(2), 441-467.

Kalai, E., M.I. Kamien, and M. Rubinovitch. 1992. Optimal Service Speeds in a Competitive Environment. Management Science 38(8), 1154-1163.

Karmarkar, U. 1987. Lot Sizes, Lead Times, and In-Process Inventories. Management Science 33, 409-418.

Karmarkar, U.S. 1993. Manufacturing Lead Times, Order Release and Capacity Loading. Chapter 6 in Handbook in Operations Research and Management Science: Logistics of Production and Inventory, Graves S., et al. (eds.), North-Holland.

Karmarkar, U.S. 1994. A Robust Forecasting Technique for Inventory and Leadtime Management. Journal of Operations Management 12(1), 45-54.

Karmarkar, U.S. 1996. Integrative Research in Marketing and Operations Management. Journal of Marketing Research 33, 125-133.

Karmarkar, U., S. Kekre, S. Kekre, S. Freeman. 1985. Lotsizing and Lead Time Performance in a Manufacturing Cell. Interfaces 15, 1-9.

Karmarkar, U.S., M. Lele. 1989. The Manufacturing-Marketing Interface: Strategic and Operational Issues. Working Paper, William E. Simon Graduate School of Business Administration, Rochester, NY 14626.

Karmarkar, U.S. and R.V. Pitbladdo. 1997. Quality, Class, and Competition, Management Science 43(1), 27-39.

Katz, K.L., B.M. Larson, R.C. Larson. 1991. Prescription for the Waiting-in-Line Blues: Entertain, Enlighten, and Engage. Sloan Management Review 32(2), 44-53.

Kaul, A., V.R. Rao. 1995. Research for Product Positioning and Design Decisions: An Integrative Review. International Journal of Research in Marketing 12(4), 293-320.

Keeney, R.L. 1999. The Value of Internet Commerce to the Customer. Management Science 45(4), 533-542.

Kekre, S., and V. Udayabhanu. 1988. Customer Priorities and Lead Times in Long Term Supply Contracts, Journal of Manufacturing and Operations Management 1, 44-66.

Kiesler, S. 1986. The Hidden Message in Computer Networks. Harvard Business Review Jan./Feb., 46-60.

Kim, I., and Tang, C.S. 1997. Lead Time and Response Time in a Pull Production Control System. European Journal of Operational Research 101(3), 474-485.

Kotler, P. 2000. Marketing Management: The Millenium Edition. Prentice Hall: Upper Saddle River, NJ.

Kingsman, B., L. Hendry, A., Mercer, A. de Souza. 1996. Responding to Customer Enquiries to Make-to-Order Companies: Problems and Solutions. International Journal of Production Economics 46/47, 219-231.

Krishnan, V., K.T. Ulrich. 2001. Product Development Decisions: A review of the Literature. Management Science 47(1) 1-21.

Lee, H.L., K.C. So, C.S. Tang. 2000. The Value of Information Sharing in a Two-Level Supply Chain. Management Science 46(5), 626-643.

Lambrecht, M.R., P.L. Ivens, and N.J. Vandaele. 1998. ACLIPS: A Capacity and Lead Time Integrated Procedure for Scheduling. Management Science 44(11), I548-I561.

Laseter, T., P. Houston, A. Chung, S. Byrne, M. Turner, A. Devendran. 2000. The Last Mile to Nowhere. Strategy+Business 20(3), 40-48.

Lederer, P.J. and L. Li. 1997. Pricing, Production, Scheduling, and Delivery-Time Competition. Operations Research 45(3), 407-420.

Lele, M.M, U.S. Karmarkar. 1983. Good Product Support is Smart Marketing. Harvard Business Review, Nov-Dec, 124-132.

Lele, M. M. 1986. How Service Needs Influence Product Strategy, Sloan Management Review 28(1), 63-70.

Lele, M. M. 1997. After-Sales Service - Necessary Evil or Strategic Opportunity? Managing Service Quality 7(3), 141-145.

Li, L. 1992. The Role of Inventory in Delivery-Time Competition. Management Science 38(2), 182-197.

Li, L. and Y. S. Lee. 1994. Pricing and Delivery-time Performance in a Competitive Environment. Management Science 40(5), 633-646.

Lieber, Z., A. Barnea. 1977. Dynamic Optimal Pricing to Deter Entry under Constrained Supply. Operations Research 25(4), 696-705.

Liechty, J., V. Ramaswamy, S. Cohen. 2001. Choice-Menus for Mass Customization: An Experimental Approach for Analyzing Customer Demand with an Application to a Web-Based Information Service. Journal of Marketing Research 38(2), 183-196.

Lindsley, W.B., J.D. Blackburn, T. Elrod. 1991. Time and Product Variety Competition in the Book Distribution Industry. Journal of Operations Management 10(3), 344-362.

Loch, C.H. 1992. Pricing in Markets Sensitive to Delay. Ph.D. Thesis, Stanford University, Stanford, CA 94304.

Loomba, A. P. S. 1996. Linkages Between Product Distribution and Service Support Functions, International Journal of Physical Distribution & Logistics Management 26(4), 4-22.

Loomba, A. P. S. 1998. Product Distribution and Service Support Strategy Functions: An Empirical Validation, International Journal of Physical Distribution & Logistics Management 28(2), 143-161.

Lovejoy, W., S. Whang. 1995. Response Time Design in Integrated Order Processing/Production Systems. Operations Research 43, 851-861.

Magretta, J. 1998. The Power of Virtual Integration: An Interview with Dell Computer's Michael Dell. Harvard Business Review 76(Mar./Apr.), 73-84.

Mahajan, V., E. Muller, F. Bass. 1991. New Product Diffusion Models in Marketing: A Review and Directions for Research. Journal of Marketing 54(1) 1-26.

Mahajan, V., E. Muller, F. Bass. 1995. Diffusion of New Products: Empirical Generalizations and Managerial Uses. Marketing Science 14(3) G79-88.

Masuda, Y., S. Whang. 1999. Dynamic Pricing for Network Service: Equilibrium and Stability. Management Science 45(6), 857-869.

Mendelson, H. and S. Whang. 1990. Optimal Incentive-Compatible Priority Pricing for the M/M/1 Queue. Operations Research 38(5), 870-883.

Mefford, R.N. 1991. Quality and Productivity: The Linkage. International Journal of Production Economics 24(1/2), 137-145.

Mentzer, J. T., R. Gomes, and R. E. Krapfel Jr. 1989. Physical Distribution Service: A Fundamental Marketing Concept? Journal of the Academy of Marketing Science 17(1), 53-62.

Miltenburg, J. 1993. A Theoretical Framework for Understanding Why JIT Reduces Cost and Cycle Time and Improves Quality. International Journal of Production Economics 30/31, 195-204.

Moon, I., and S. Choi. 1997. Distribution Free Procedures for Make-to-Order (MTO), Make-in-Advance (MIA), and Composite Policies. International Journal of Production Economics 48(1), 21-28.

Moorthy, K.S. 1988. Product and Price Competition in a Duopoly. Marketing Science 7(2), 141-168.

Morton, T., and A. Vepsalainen. 1987. Priority Rules and Leadtime Estimation for Job Shop Scheduling with Weighted Tardiness Costs, Management Science 33, 1036-1047.

Murthy, D. N. P., K. R. Kumar 2000. Total Product Quality, International Journal of Production Economics 67(3), 253-267.

Netessine, S., N. Rudi. 2000. Supply Chain Structures on the Internet: Marketing-Operations Coordination Under Drop-Shipping. Working paper, Wharton School, University of Pennsylvania, Philadelphia, PA.

Netessine, S., N. Rudi. 2001. Supply Chain Choice on the Internet. Working paper, Wharton School, University of Pennsylvania, Philadelphia, PA.

Ottosson, S. 2002. Virtual Reality in the Product Development Process. Journal of Engineering Design 13(2), 159-172.

Ozer, M. 1999. A Survey of New Product Evaluation Models. Journal of Product Innovation Management 16(1), 77-94.

Parasuraman, A., V.A. Zeithaml, L.L. Berry. 1985. A Conceptual Model of Service Quality and Its Implications for Future Research. Journal of Marketing 49(4), 41-50.

Paustian, C. 2001. Better Products Through Virtual Customers. MIT Sloan Management Review 42(3), 14-16.

Perez, A.P., and P.H. Zipkin. 1997. Dynamic Scheduling Rules for a Multi-product Make-to-Stock Queue. Operations Research 45(6), 919-930.

Plambeck, E.L. 2003. Optimal Leadtime Differentiation via Diffusion Approximations. to appear in Operations Research.

Plambeck, E.L. and S.A. Zenios. 2000. Performance-Based Incentives in a Dynamic Principal-Agent Model. Manufacturing & Service Operations Management 2(3), 240-263.

Plymire, J. 1991. Complaints as Opportunities, Journal of Consumer Marketing 8(2), 39-43.

Porter, M.E. 2001. Strategy and the Internet. Harvard Business Review 79(3), 62-78. With a discussion on Harvard Business Review 79(6) 137-142.

Porteus, E.L. 1986. Optimal Lot Sizing, Process Quality Improvement and Setup Cost Reduction. Operations Research 34(1), 137-144.

Porteus, E.L., S.J. Whang. 1991. On Manufacturing/Marketing Incentives. Management Science 37(9), 1166-1181.

Quelch, J.A., D. Kenny. 1994. Extend Profits, Not Product Lines. Harvard Business Review 72(Sept./Oct.), 154-160.

Randal, T., S. Netessine, N. Rudi. 2002. Inventory Structure and Internet Retailing: An Empirical Examination of the Role of Inventory Ownership. Working paper, Wharton School, University of Pennsylvania, Philadelphia, PA.

Rouwenhorst, B., B. Reuter, V. Stockrahm, G. van Houtum, R. Mantel, W. Zijm. 2000. Warehouse Design and Control: Framework and Literature Review. European Journal of Operational Research 122(3), 515-533.

Ruben, R.A., F. Mahmoodi. 2000. Lead Time Prediction in Unbalanced Systems. International Journal of Production Research 38(7), 1711-1729.

Schibrowsky, J. A., R. S. Lapidus 1994. Gaining a Competitive Advantage by Analyzing Aggregate Complaints, Journal of Consumer Marketing 11(1), 15-26.

Schmenner, Roger W. 1988. The Merit of Making Things Fast. Sloan Management Review, Fall Quarter, 11-17.

Schmidt, G.M, E.L. Porteus. 2000. The Impact of an Integrated Marketing and Manufacturing Innovation. Manufacturing & Service Operations Management 2(4), 317-336.

Schumpeter, J.A. 1942. Capitalism, Socialism and Democracy. New York: Harper & Brothers.

Seidmann, A., and M. Smith. 1981. Due Date Assignment for Production Systems, Management Science 27, 401-413.

Shanthikumar, J., and U. Sumita. 1988. Approximations for the Time Spent in a Dynamic Job Shop with Applications to Due Date Assignment. International Journal of Production Research 26, 1329-1352.

Shapiro, B.P. 1977. Can Marketing and Manufacturing Co-exist? Harvard Business Review 55, 104-114.

Silveira, G.D., D. Borenstein, F.S. Fogliatto. 2001. Mass Customization: Literature Review and Research Directions. International Journal of Production Economics 72(1), 1-13.

Slywotzky, A.J. 2000. The Age of the Choiceboard. Harvard Business Review 78(1), 40-41.

Smith, M.D., E. Brynjolfsson. 2001. Consumer Decision-Making at an Internet Shopbot. The Journal of Industrial Economics 49(4), 541-558.

So, K.C. 2000. Price and Time Competition for Service Delivery. Manufacturing & Service Operations Management 2(4), 392-409.

Song, J-S. 1994. The Effect of Leadtime Uncertainty in a Simple Stochastic Inventory Model. Management Science 40(5), 603-613.

Soteriou, A., and S.A. Zenios. 1999. Operations, Quality, and Profitability in the Provision of Banking Services. Management Science 45(9), 1221-1238.

Spearman, M.L., and R.Q. Zhang. 1999. Optimal Lead Time Policies. Management Science 45(2), 290-295.

Stalk, G. and T.M. Hout. 1990. Competing Against Time: How Time-Based Competition is Reshaping Global Markets. The Free Press, New York.

Tapiero, C.S. 1987. Production Learning and Quality Control. IIE Transactions 19(4), 362-370.

Thomas, P.R. 1990. Competitiveness Through Total Cycle Time: An Overview for CEO's. McGraw-Hill, New York.

Thomke, S., E. von Hippel. 2002. Customers as Innovators: A New Way to Create Value. Harvard Business Review 80(4), 74-81.

Tielemans, P.F.J., R. Kuik. 1996. An Exploration of Models that Minimize Leadtime Through Batching of Arrived Orders. European Journal of Operational Research 95(2), 374-389.

Toubia, O., D. Simester, J.R. Hauser. 2002. Fast Polyhedral Adaptive Conjoint Estimation. Working paper, Center for Innovation in Product Development, MIT.

Tsay, A., N. Agrawal. 2000. Channel Dynamics under Price and Service Competition. Manufacturing & Service Operations Management 2(4), 372-391.

Tsay, A.A., W.S. Lovejoy. 1999. Quantity Flexibility Contracts and Supply Chain Performance. Manufacturing and Service Operations Management 1(2), 89-111.

Tsay. A.A., S. Nahmias, N. Agrawal. 1999. Modeling Supply Chain Contracts: A Review. In Tayur, S., R. Ganeshan, M. Magazine, (Eds.) Quantitative Methods for Supply Chain Management. Kluwer Academic Publishers, Norwell, MA.

Ulrich, K.T., S.D. Eppinger. 1995. Product Design and Development. McGraw-Hill, New York.

Urban, G.L., J.R. Hauser. 1993. Design and Marketing of New Products. Prentice-Hall, Englewood Cliffs, NJ.

van Mieghem, J. 1999. Differentiated Quality of Service: Joint Pricing and Scheduling in Queueing Systems. Working Paper, Kellogg Graduate School of Management, Northwestern University, Evanston, IL 60208.

van Mieghem, J.A., M. Dada. 1999. Price Versus Production Postponement: Capacity and Competition. Management Science 45(12), 1631-1649.

van Ryzin, G., and S. Mahajan. 1999. On the Relationship Between Inventory Costs and Variety Benefits in Retail Assortments. Management Science 45(11), 1496-1509.

Vepsalainen, A., and T. Morton. 1988. Improving Local Priority Rules with Global Lead Time Estimates" Journal of Manufacturing and Operations Management 1, 102-118.

von Hippel, E. 2001. Perspective: User Toolkits for Ennovation. Journal of Production Innovation and Management 18 247-257.

von Hippel, E., R. Katz. 2002. Shifting Innovation to Users Via Toolkits. Management Science 48(7) 821-833.

Wang, Y., B.O. Nnaji. 2001. Functionality-Based Modular Design for Mechanical Product Customization over the Internet. Journal of Design and Manufacturing Automiation 1(1&2), 107-121.

Wein, L.M. 1991. Due-Date Setting and Priority Sequencing in a Multiclass M/G/1 Queue, Management Science 37, 834-850.

Wheelwright, S., K.B. Clark, Revolutionizing Product Development: Quantum Leaps in Speed, Efficiency, and Quality, New York: Free Press, 1992.

Whitt, W. 1999. Improving Service By Informing Customers About Anticipated Delays. Management Science 45(2), 192-207.

Wind, J., V. Mahajan. 1997. Issues and Opportunities in New Product Development: An Introduction to the Special Issue. Journal of Marketing Research 34(1), 1-12.

Wittink, D.R., M. Vriens, W. Burhenne. 1994. Commericial Use of Conjoint Analysis in Europe: Results and Critical Reflections. International Journal of Research in Marketing 11(Jan.), 41-52.

Yano, C. 1987. Setting Planned Lead Times in Serial Production Systems with Tardiness Costs. Management Science 33, 96-106.

Yano, C. 1987. Stochastic Leadtimes in Two-Level Assembly Systems. IIE Transactions 19, 371-378.

Yeh, K.H., C.H. Chu. 1991. Adaptive strategies for Coping with Product Variety Decisions. International Journal of Operations and Production Management 11(8), 35-47.

Zhao, W., Y.S. Zheng. 2000. Optimal Dynamic Pricing for Perishable Assets with Nonhomogeneous Demand. Management Science 46(3), 375-388.

Zeithaml, V.A., A. Parasuraman, L.L. Berry. 1990. Delivering Quality Service: Balancing Customer Perceptions and Expectations. New York: Free Press.

Zenios, S. 2003. Decentralized Decision Making in Dynamic Technological Systems. In Simchi-Levi, D, D. Wu, and M. Shen (Eds.), Handbook of Supply Chain Analysis in the eBusiness Era. Kluwer Academic Publishers, Boston.

Zhou, Y. P., N. Gans 1999. A Single-Server Queue with Markov Modulated Service Times. Working Paper. The Wharton School, The University of Pennsylvania. Philadelphia, PA, 19104.

David Simchi-Levi, S. David Wu, and Z. Max Shen (Eds.)
Handbook of Quantitative Supply Chain Analysis:
Modeling in the E-Business Era
©2004 Kluwer Academic Publishers

Chapter 9

COORDINATION OF PRICING AND INVENTORY DECISIONS:

A Survey and Classification

L. M. A. Chan
School of Management
University of Toronto
achan@rotman.utoronto.ca

Z. J. Max Shen
Dept. of Industrial and Systems Engineering
University of Florida
shen@ise.ufl.edu

David Simchi-Levi
Dept. of Civil and Environmental Engineering and the Engineering Systems Division
Massachusetts Institute of Technology
dslevi@mit.edu

Julie L. Swann
School of Industrial and Systems Engineering
Georgia Institute of Technology
jswann@isye.gatech.edu

1. Introduction
1.1 Motivation

Recent years have seen scores of retail and manufacturing companies exploring innovative pricing strategies in an effort to improve their operations and ultimately the bottom line. Firms are employing such varied tools as dynamic pricing over time, target pricing to different classes of customers, or pricing to learn about customer demand. The benefits can be significant, including not only potential increases in profit, but also improvements such as reduction in demand or production variability, resulting in more efficient supply chains.

Pricing strategies such as dynamically changing the pricing of products are an important revolution in retail and manufacturing industries driven in large part by the Internet and the Direct-to-Customer (DTC) model. This business model, used by industry giants such as Dell Computers and Amazon.com, allows companies to quickly and easily change prices based on parameters such as demand variation, inventory levels, or production schedules. Further, the model enables manufacturers to collect demand data more easily and accurately [142].

Employing a DTC model over the Internet means that price changes are easy to implement, and the cost to make changes is very low without catalogs or price stickers to produce. Further, the Internet offers an unprecedented source of information about customer preferences and demand. Many examples of dynamic and target pricing can be found on the Internet; for example, online auctions allow buyers to bid on everything from spare parts to final goods or companies use questionnaires to determine market segmentation for product prices and offerings.

Initially, many firms and researchers focus on pricing alone as a tool to improve profits. However for manufacturing industries, the coordination of price decisions with other aspects of the supply chain such as production and distribution is not only useful, but is essential. The coordination of these decisions mean an approach that optimizes the system rather than individual elements, improving both the efficiency of the supply chain and of the firm. This *integration* of pricing, production and distribution decisions in retail or manufacturing environments is still in its early stages in many companies, but it has the potential to radically improve supply chain efficiencies in much the same way as revenue management has changed airline, hotel and car rental companies. Thus we are motivated in this paper to consider strategies which integrate pricing decisions with other aspects of the supply chain, particularly those related to production or inventory.

1.2 Scope

Pricing is a significant and important topic, and there are many aspects of it that offer challenges to companies and researchers alike. In this chapter, we will focus on one aspect of pricing–when it is integrated with inventory and production decisions. We review research in this area because it is an important one to manufacturing and retail industries, and because consideration of this aspect alone is a significant undertaking.

In all of the papers we consider, price is a decision variable, where customer demand depends on the price chosen. Thus, we do not review papers where demand is completely exogenous to the process (see [27] for instance, which examines papers in the area of contracting). The firm must also have some control over the inventory or production decisions, thus enabling the coordination with pricing.

Within this context, we will consider a number of different pricing strategies, including dynamic pricing (where price changes over time) as well as constant pricing (price remains fixed over time), even the pricing of multiple products simultaneously or across different customer segments. We also consider relatively new topics in the area of pricing and production control, such as when competition among firms is considered, or when learning of demand is incorporated in the decision making process. Additional topics include the coordination of price across multiple channels within the supply chain, and coordination of price with production leadtime decisions. Some problems incorporate additional elements such as the decision of capacity investment. Finally, we consider industry examples and applications of price and production control.

For each paper, we review the important elements of it, and we provide a classification to try to give more insight as to the differences between problems. This classification is described in the following section.

Ours is certainly not the first paper to review pricing strategies, or even the first to review strategies that cross both marketing and production areas. Elmaghraby and Keskinocak [50] provide a recent review of dynamic pricing strategies, in particular noting the distinction of different strategies based on whether they address short, medium, or long term problems. The overlap between the reviews is on papers that consider dynamic pricing with inventory decisions, although we also consider additional topics in this chapter (e.g., across customer segments or constant pricing).

Eliashberg and Steinberg [48] provide a notable review of problems in the intersection of marketing and production. We also review some of the papers contained in their article, particularly if they are important to the context of later research topics. However, significant research as been done since their review appeared, so we focus on the newer topics as much as possible. We note other review papers as they apply to a particular topic.

We do not claim to have an exhaustive list of papers in the area of pricing and production coordination; we apologize if we have missed significant work in this area. In several cases we include working papers, as they address the most recent topics, but we try to make the references available for these. Due to the scope of the problem, we have not verified accuracy of results in all of the cases. The ordering of papers can be difficult; in general, we try to associate papers most closely linked, so we do not necessarily follow a strict chronological order in each section.

1.3 Classification and Outline

We find it useful to group and describe papers in the area of price and production according to a number of characteristics of the problem or assumptions made by the researchers. In many cases the kind of assumption (deterministic versus stochastic parameters for instance), dictates the types of tools that are available to solve the problems or the kinds of solutions that will be possible. We describe many of the distinctions in this section, and provide a classification table at the end of Section 3 that includes a number of key papers we review.

- *Length of Horizon*
  A number of problems consider a single period problem, similar to a newsvendor setting. Other possibilities include a multiple period horizon, or an infinite horizon. Occasionally papers assume a two period problem, this is also noted when appropriate.

- *Prices*
  For the multiple period problems that we describe, they generally fit into one of two classes. As is common in many traditional business models, price may remain fixed or constant over the time horizon, even if the demand is non-stationary over time. The second classification allows for price to dynamically change over time (as a function of demand, inventory, or other parameters of the problem). The second case is particularly important for firms that sell product through electronic channels.

- *Demand Type*
  The first distinction we make about demand is whether it is deterministic, with a known function according to parameters like price. An example of a problem with deterministic demand is the Economic Order Quantity (EOQ) model, which a number of researchers have considered with the addition of price as a decision. Demand may also be assumed to be stochastic or random. Generally in this case, it is assumed that there is some known portion that is based on price (e.g., linear demand curve), with an additional stochastic element.

- *Demand Functional Form*
  Researchers assume a variety of functional forms for the relationship between demand and price (or other parameters like inventory). One common one is where demand is a linear function of price, i.e., $D =$

$aP+b$. If demand is also stochastic, the stochastic term may be additive ($D = \alpha P + \beta + \epsilon$) or multiplicative ($D = \epsilon \alpha P + \beta$). Demand may also be exponential or a Poisson process as well. In many papers, demand is assumed to be a Poisson arrival process, where arrivals in different units of time are independent of each other. Other forms are also possible and are noted when applicable.

- *Demand Input Parameters*
  Since we limit the papers under consideration to be those where demand is endogeneous to the decisions, at the least demand is generally a function of price. However, there are a number of other parameters which may impact the demand function, such as time (in which case demand may be referred to as non-stationary). For many problems it may also be appropriate to assume that demand is a function of inventory; for example, if demand decreases due to lower stock in the stores, as may be common in some retail applications. Other potential parameters include ads, promotions or sales, or product characteristics.

- *Sales*
  In the case of stochastic demand, or when production capacity limits exist, researchers make assumptions about how to treat excess demand. The primary assumptions are that demand is either backlogged to satisfy in a later period, or that excess demand is lost. In some cases neither assumption is needed, for example under deterministic demand with no limits or when price can be set to match demand exactly.

- *Restocking*
  For many papers focused on applications in the retail industry, it may not be possible to make ordering decisions in every period. Thus we note whether research assumes that reordering is allowed (yes or no).

- *Production Set-up Cost*
  For some manufacturing problems the addition of a fixed production set-up cost may be appropriate. In general, the addition of this fixed charge complicates the structure of the objective function and makes the problems more difficult to solve (also noted in [48] as a distinction between convex or non-convex cost functions). We will characterize problems by this element according to whether or not they include the fixed charge (yes or no).

- *Capacity Limits*
  For manufacturing problems, researchers sometimes include the fact that production may be limited by the capacity of the system. We will classify papers as yes or no according to this element, although the default is that most papers do not consider capacity limits.

- *Products*
  Most papers that we review consider a single product or multiple products that do not share resources, in which case the problems are separable into the single product case. A few papers consider multiple products that share common production resources or components, or share demand from customers.

In the chapter, we first consider single period models in Section 2. Next we address multiple period models in Section 3, including separate sections for topics like those that apply to the retail industry. A classification of many multiple period papers is provided at the end of Section 3.

In Section 4, we address a number of emerging areas in price and production coordination, such as demand learning or pricing under competition. We also consider a few papers in the area of Supply Chain Coordination, where there are multiple agents or channels to consider. We consider papers and applications to industry in Section 5. Finally, we suggest additional areas for research in the area of pricing and production control, of which there are many.

2. Single Period Models

In this section, we review pricing models with only a single selling period. In general, single period problems with stochastic demand are called newsvendor problems, where a decision maker has to decide how much to stock for a perishable product before the random demand realizes, and this is a well studied problem in inventory literature. The original newsvendor problem assumes that pricing is an exogeneous decision, but there are also applications that incorporate pricing as a decision variable.

The newsvendor problem assumes that unsatisfied demand is lost. The information available to the decision maker includes the demand D, which follows a known distribution with continuous cdf $F(\cdot)$, the unit production (order) cost c, the selling price p, and the salvage value per unit, v. The objective is to minimize the expected cost. It is well known that the optimal production (order) quantity, S, can be decided easily, which should satisfy the following condition:

$$\Pr\{D \leq S\} = \tfrac{p-c}{p-v}$$

See Porteus [124] for an excellent review of research on general newsvendor problems.

Whitin [159] was the first to add pricing decisions to inventory problem, where the selling price and order quantity are set simultaneously. In his model for staple merchandise, there is a fixed cost for each order, and the per unit order cost and inventory holding cost are also given. The firm faces deterministic demand that is a linear function of price, and the objective is to maximize expected profit. He shows how to optimize the price and order quantity decisions.

Mills [107] concentrates on showing the effect of uncertainty on a monopolist's short-run pricing policy under the assumption of additive demand. In his model, the demand is specified as $D(p, \varepsilon) = y(p) + \varepsilon$, where $y(p)$ is a decreasing function of price p and ε is a random variable defined within some range. In particular, he shows that the optimal price under stochastic demand is always no greater than the optimal price under the assumption of deterministic demand, called the riskless price. Lau and Lau [90] and Polatoglu [122] both study different cases of demand process for linear demand case where $y(p) = a - bp$, where $a, b > 0$.

On the other hand, Karlin and Carr [76] used the following multiplicative case where the demand $D(p, \varepsilon) = y(p)\varepsilon$. They show that the optimal price under stochastic demand is always no smaller than the riskless price, which is opposite of the corresponding relationship found to be true by Mills [107] for the additive demand case.

Petruzzi and Dada [116] provide a unified framework to reconcile this apparent contradiction by introducing the notion of a base price and demonstrating that the optimal price can be interpreted as the base price plus a premium.

Much of the research considers demand to be a function of price alone, but Urban and Baker [152] study a model in which the demand is deterministic, multivariate function of price, time, and inventory level. They also study the case with a price markdown during the season and show how to decide the optimal order quantity.

Dana and Petruzzi [41] consider a model where the firm faces uncertain demand which depends on both price and inventory level. They assume that there is also an exogenous outside option for the customer. The outside option represents the utility the consumer forgoes when she chooses to visit the firm before knowing whether or not the product will be available. For the case in which a firm's price is exogenous, they show that by internalizing the effect of its inventory on demand, a firm holds more inventory, provides a higher service level, attracts more customers, and earns higher expected profits than a firm that ignores this effect. For the case in which the price is endogenous, they show the problem is as easy to solve as the exogenous-price case by reducing a two dimensional decision problem to two sequential single variable optimization problems.

Raz and Porteus [128] study the newsvendor problem with pricing from a discrete service levels perspective. In stead of making specific distributional assumptions about the random component of demand (for example, additive, multiplicative, or some mixture of the two), they assume that demand is a discrete random variable that depends on the retail price. Specifically, they assume there are a finite number of market states, each arising with a given probability, and with a demand function for each. They also assume that each of these outcomes is a piecewise linear function of the price. They show their model can be specialized to a discretized version of the additive and multiplicative models. They call these market demand functions "service level functions". They decompose the problem into two subproblems: first, finding the best price and quantity when the service level is given as one of the finite number possible, and then choosing the service level to offer. They find that both the service level constrained problem and the unconstrained problem with specialized demand (additive or multiplicative or a mixture of the two) are well behaved, while the unconstrained general problem produces some counterintuitive results. For example, when the marginal cost of the product increases, the optimal price for the service level constrained problem increases, but the optimal price for the overall problem can decrease. This is caused by the fact that when faced with changes in its environment, in addition to changing its optimal price and quantity, the firm might also change the service level for the product. They examine the effect of changes on the optimal behavior of the firm and consider two main changes: changes in the cost structure (the unit cost or the penalty cost decrease, or the salvage value increases) and changes in the demand environment. They show that when the service level remains unchanged, the optimal price increases when the firm faces changes in the cost structure that makes it worse off, but the price can increase or decrease when the firm is faced with changes in its demand environment. However, when the firm can change its service level, they show that in the general case (without additive or multiplicative assumptions), the optimal price will be non-monotonic both when the cost structure parameters change or when the demand environment worsens.

Usually in the newsvendor setting, it is assumed that if any inventory remains at the end of the period, one discount is used to sell it or it is disposed of. Khouja [79] extends the newsvendor problem to the case in which multiple discounts with prices under the control of the newsvendor are used to sell excess inventory. They develop two algorithms, under two different assumptions about the behavior of the decision maker, for determining the optimal number of discounts under fixed discounting cost for a given order quantity and realization of demand. Then, they identify the optimal order quantity before any demand is realized for Normal and Uniform demand distributions. They also show how to get the optimal order quantity and price for the Uniform demand case when the initial price is also a decision variable.

Weng and Parlar [158] study the benefits of encouraging customers to commit themselves to purchasing the seasonal product before it becomes available by providing price incentives to customers. They model the price incentive and one-time stocking decisions as a two-stage stochastic dynamic program. The materialized demand commitments based on price incentives may reveal useful information about the distribution of the future demand before the one-time stocking decision is made. Some customers may react to early promotion because they are either price-sensitive or shortage-sensitive. They show that price-incentives can increase the expected total demand while reducing its variability. They also explicitly obtain the price incentive and stocking decisions that maximize expected profit.

Bernstein and Federgruen [14] study a model with many newsvendors. Specifically, they consider a two-echelon supply chain with one supplier servicing a network of (competing) retailers. The retailers face random demands, the distribution of which may depend on its own price as well as those of the other competing retailers according to general stochastic demand functions. They identify the equilibrium behavior of the retailers and design contractual arrangements between the parties that allow the decentralized chain to perform as well as a centralized one.

3. Multiple Period Models
3.1 Models to Explain Price Realizations

A number of researchers, mostly from economics backgrounds, have been interested in developing models that incorporate decisions such as price and production with the purpose of understanding why price behaves as it does. In particular, researchers have been interested in explaining price smoothing or "stickiness", where price is fairly rigid over time despite system changes. Price stickiness could be a result of market exploitation or price-fixing to deter entry into a market, or it could be the result of profit-maximizing production and sales strategies. In this stream, for early examples see [22], [68], [119], and [129], researchers often develop an optimization model to derive the optimal sales, price and production policy of firms, and use the model to explain realized price policies.

Amihud and Mendelson [5] build a model that maximizes the discounted expected profit and show that the resulting policy may have relatively smooth prices as a policy. They assume that demand is stochastic with a concave revenue function, and excess demand results in negative inventory with shortage costs. Production is also stochastic, and decisions in each period are made before the realization of stochastic components. The authors show that the existence of inventory (positive or negative), acts to absorb demand shocks, thereby leading to smoother prices. Furthermore, the optimal pricing policy is in an assymetric interval around the classic monopoly price, where the assymetry depends on the ratio of backlogging cost to inventory holding cost. When backlogging poses a lesser penalty on the firm, the lower bound on price reduces, and price can be used to increase demand (the result is vice-versa when the backlogging has a heavier penalty).

Ashley and Orr [8] disallow negative inventories or backlogging, but still find that production smoothing can bring about price smoothing. Under the assumption of deterministic demand and concavity of revenue, they find that price stickiness is realized when demand is foreseen to decrease but not when demand is predicted to increase.

One of the first papers to incorporate production capacity constraints in a pricing and inventory model is Lai [88]. Under linear deterministic demand and quadratic production costs, Lai shows that price is more responsive to demand shocks when a stockout is realized or when capacity is soon to be exhausted, and that considering production capacity constraints reduces the asymmetry in price behavior.

Several economists have addressed pricing specific to the fashion retail industry. For instance, Lazear [91] develops a model to address a number of questions in this industry, including differences in prices based on the type of good, the volume of sales in a particular industry, or the time a good has been on a shelf. Lazear considers a two-period pricing model where the probability of making a sale is constant over time the exact value function of the consumer is not known to the retailer. The retailer makes a price decision in the first period, and based on whether or not items are sold has an opportunity to make a second price decision. Using Bayesian probability methods, Lazear's model predicts decreasing prices based on time on a shelf, and decreased prices associated with obsolescence. Further, he demonstrates that the uniqueness of the good does not affect the price path, and that perishable goods are more likely to have an initial lower price than non-perishable goods.

Building on Lazear's work, Pashigian [114] allows for industry equilibriua, and particularly focuses on explaining the fact that retail mark-ups and markdowns increased in percentage in the 1970s and 1980s. Using a model similar to Lazear's as well as empirical data, Pashigan indicates the price differences can be attributes to larger price uncertainty, which is a function of the greater product variety in the fashion industry.

3.2 General Pricing and Production Models

3.2.1 Variable Production Cost.
Among the earliest to consider pricing and inventory in a multi-period setting are [76], [81], [111], [156], and [157]. The initial focus of research in this area is on solution methods or showing properties of the problem ultimately leading to efficient solution methods. In particular, different assumptions on the structure of the demand distribution or costs sometimes leads to concavity of profit or expected profit, making a problem easier to solve to optimality.

For an example of an early paper, Zabel [161] focuses on showing the existence and uniqueness of a solution under certain assumptions. The objective is to maximize discounted expected profit, and excess demand is lost. Zabel considers both multiplicative and additive demand with a stochastic component, but finds that the latter has properties that make the problem easier to solve. The author shows that when there is additive demand, $d = \mu(p) + \eta$, with $\mu(p)$ concave and where the random variable η has a probability function that is uniform or exponential, then a unique solution may be found.

Similarly, Thowsen [151] considers the determination of price and production in a nonstationary model where demand is a general function of price and has an additive stochastic component. Thowsen assumes proportional ordering costs (without set-up) and convex holding and stockout costs, and he considers the conditions of backlogging, partial backlogging, and lost sales as well as partial spoilage of inventory under a number of common-sense assumptions. A critical number inventory policy is considered, denoted by later researchers as (y, p), which is similar to critical number inventory policies, and conditions for optimality are shown.

The (y, p) policy outlined in [151], states that, for each period t, if inventory is above y_t no production should be made, and the price should be set to the point $p_t(I)$ on the optimal price trajectory based on a current inventory level of I. If inventory is below y_t, production should be made to bring the inventory level to y_t, and the price should be set to $p_t(y_t)$.

Since the conditions in general are difficult to verify, Thowsen discusses special cases under which the (y, p) policy is optimal for the problem considered. For instance, under a linear expected demand curve, linear stockout costs, convex holding costs, and a demand distribution that is a PF_2 distribution[1], the (y, p) policy is optimal. Furthermore, if excess demand is backlogged, the demand curve is concave and the revenue is collected a fixed number of periods after the time orders are placed, then no assumptions are needed on the cost and demand distribution for optimality of the critical number policy, and for this case the decision on price and quantity decisions can be made separately. Thowsen also shows that if negative demand is disallowed, the optimal price will be a decreasing function of increasing initial inventory.

Similarly, Federgruen and Heching [51] study a problem where price and inventory decisions are coordinated under linear production cost, stochastic demand and backlogging of excess demand. They assume revenue is concave (e.g., as under a linear demand curve with an additive stochastic component), and inventory holding cost is convex. All parameters are allowed to vary over time, and price changes may be bi-directional over a finite or infinite horizon.

Using the structure of the problem, Federgruen and Heching show that expected profit-to-go is concave, and that the optimal price is nonincreasing as a function of initial inventory. Further, similar to Thowsen, they characterize the optimal policy in each period according to a base-stock level and list price pair, (y_t^*, p_t^*), where the inventory level is increased to y_t^* if it falls below that value, and a price of $p_t^*(I)$ is charged as a function of the initial inventory in period t if $I > y_t^*$. The authors provide an efficient algorithm to solve for the optimal policy. Using a numerical study, they indicate dynamic pricing may result in a 2% increase in profit over fixed pricing, and the base stock level and list price increase with demand uncertainty. Their results also extend to the case with production capacity limits.

Very few integrated pricing and production problems have explicitly considered production capacity limits, Chan et al [30] is another exception to this. They assume that sales are lost if product is not available, and further that the firm may use *discretionary sales*, where some product may be set aside to satisfy future demand even if this results in lost demand in the current period. The authors consider *Delayed Strategies*, where a decision such as price (resp., production) is made at the beginning of the horizon, and the delayed production decision (resp., price) is made at the beginning of each period. This strategy is called delayed production (resp., delayed pricing). In all cases, customer demand arrives in each period after the pricing and production decisions. Demand is a general stochastic function with linear production and inventory holding costs; all parameters are non-stationary. The problem they study is also related to [154], which examines postponed decisions in a single period when the postponed decisions is made after customer demand is realized.

Chan et al show that given an initial price vector, Delayed Production has a concave expected profit function, and thus the optimal policy is characterized by two parameters: a modified base stock level and a target inventory to save (y_t^*, s_t^*), both of which are independent of the initial inventory level. In both cases, the policy is to order or save as much of the target as possible, as determined by the available inventory and production capacity. However, for the Delayed Pricing strategy with an initial production vector, the expected profit is not concave, even for linear expected demand curves. The result is that the optimal price and the target save inventory both depend on initial inventory, and for this strategy, the discretionary sales decision can be improved with additional information on realized customer demand. The problem of setting a constant price under stochastic demand and capacity limits is a special case of Delayed Production.

The authors provide heuristics to determine a complete pricing and production policy for both strategies, using a deterministic approximation as a starting point for the initial decision. In computational analysis, they show that Delayed Pricing often outperforms Delayed Production when uncertainty and production costs are low, and Delayed Production is often a better strategy in other situations, except under some cases with high production cost or differing product (similar to single period results in [154]).

Additional dynamic pricing papers that incorporate capacity limits include Biller et al [18] and Kachani and Perakis [73]; these are reviewed with papers that consider multiple products in Section 4.1.1.

Control Theory under Continuous Time. A large stream of research on pricing and production problems with control theory techniques was begun in Pekelman [115], who considered choosing intertemporal price and production rates as the control variable. Demand is a deterministic linear function of price, and a constraint forced inventory to be non-negative. This work was extended in Feichtinger and Hartl [53], who allowed demand to be non-linear with marginal revenue an increasing function of price and included shortage costs for negative demand. Thompson et al [150] also extended [115] by including a fixed production charge. Other related papers include [5] above and additional ones in the section on retail papers.

There are numerous papers that assume that demand is a Poisson process. One important early one is Li [97], who studied price and production problems with capacity investment at the beginning of the horizon. Given an initial production capacity, the demand and production rates are Poisson counting processes; if there is demand in excess of production, sales are lost. Li shows that the optimal production policy under a single fixed price is a barrier or threshold policy, where production is optimal if inventory is below a certain value. Further, he characterizes the optimal policy when price is dynamic over time, and he shows that the stochastic price is always higher than the deterministic price. In extensions, he considers the case with production learning effects; that is, the production rate over time becomes closer to the ideal capacity.

3.2.2 Fixed Production Set-Up Cost.

Early research also considered the addition of production set-up costs to the problem of simultaneously determining price and production in a multiple period horizon. As described in [48] as well as others, with the addition of production set-up cost the objective becomes a concave cost function and becomes much more difficult to solve. For example, Kunreuther and Richard [85] consider a problem with deterministic demand that is stationary over time, with and without production set-up costs.

Many research problems that address pricing and production decisions with fixed production set-up cost fall within the area of the Economic Order Quantity (EOQ) model. The general EOQ model inventory model has been studied frequently in inventory literature (see [164] for a review). The problem consists of multiple periods in a fixed time horizon, with a stationary deterministic function in each period; ordering or production costs have a fixed and variable component. Since demand is deterministic, the optimal policy will leave zero inventory at the end of each time cycle, so each period may be considered independently. With the addition of price as a decision, demand is generally assumed to be a deterministic function that decreases with price. Usually the focus of research under these assumptions is to demonstrate an algorithm for finding the optimal price and production quantity (using calculus, derivatives, or iterative methods), while adding additional considerations to the basic EOQ model. Many papers with the EOQ based model are described in the next paragraph, although some also appear in other sections of this article.

For example, early efforts at including price in the EOQ model are [87] and [157]. Arcelus and Srinivasan [7] expanded the work to include three different kinds of profit functions (profits, return on investment, or residual income) and used a derivative based solution technique. Cheng [37] added storage space considerations and inventory investment limitations to the pricing-EOQ model. Using linear based demand curves, Chen and Min [31] consider objective functions similar to [7] but provides a solution that guarantees optimality without an iterative method. The addition of an extended credit period to the pricing-EOQ model is considered in Hwang and Shinn [70], where inventory deteriorates both with demand and time. Marketing expenditures were added to the problem in Lee and Kim [94], and Kim and Lee [80] included the effect of capacity limits (fixed and variable) on the pricing-EOQ model. Arcelus [6] allowed for a one-time discount in the model, and Abad [3] considered a problem with partial backlogging based on the impatience of customers in relationship to the waiting time. Other papers in this area include [1], [24], [93], [123], [131], and [133].

In a set of papers [148], [149], Thomas considers the multi-period pricing problem with production set-up costs. The first paper addresses a problem with deterministic non-stationary demand of a general form, where backorders are not allowed. The objective is to determine production periods and levels as well as prices in each period, and a forward algorithm is provided to solve the problem.

In the second paper, Thomas considers a related problem but incorporates a general stochastic demand function and backlogging of excess demand. Specifically, Thomas considers a periodic review, finite horizon model with a fixed ordering cost and stochastic, price-dependent demand. The paper postulates a simple policy, referred to by Thomas as (s, S, p), which can be described as follows. The inventory strategy is an (s, S) policy: If the inventory level at the beginning of period t is below the reorder point, s_t, an order is placed to raise the inventory level to the order-up-to level, S_t. Otherwise, no order is placed. Price depends on the initial inventory level at the beginning of the period. Thomas provides a counterexample which shows that when price is restricted to a discrete set this policy may fail to be optimal. Thomas goes on to say:

"If all prices in an interval are under consideration, it is conjectured that an (s, S, p) policy is optimal under fairly general conditions."

Chen and Simchi-Levi [33] consider a model identical to the one by Thomas and assume a general demand function of the form

$$D_t(p, \epsilon_t) := \alpha_t D_t(p_t) + \beta_t,$$

where the random perturbation over time, ϵ_t, is composed of a pair of random variables (α_t, β_t). Without loss of generality, they assume that $E\{\alpha_t\} = 1$ and $E\{\beta_t\} = 0$. Both additive demand functions $(D_t(p) + \beta_t)$ and multiplicative demand functions $(\alpha_t D_t(p))$ are special cases. Special cases of the function $D_t(p)$ include $D_t(p) = a_t - b_t p$ $(a_t > 0, b_t > 0)$ in the additive case and $D_t(p) = b_t p^{-a_t}$ $(a_t > 1, b_t > 0)$ in the multiplicative case; both are common in the economics literature (see [116]).

In case of the additive demand model, Chen and Simchi-Levi prove that the (s, S, p) policy identified by Thomas is indeed optimal. On the other hand, for the general demand model, they show that a different policy, referred to as an (s, S, A, p) policy is optimal policy. In such a policy, the optimal inventory strategy is characterized by two parameters s_t and S_t and a set $A_t \subset [s_t, (s_t + S_t)/2]$, possibly empty. When the inventory level, x_t, at the beginning of period t is less than s_t or if $x_t \in A_t$ an order of size $S_t - x_t$ is made. Otherwise, no order is placed. Thus, it is possible that an order will be placed when the inventory level $x_t \in [s_t, (s_t + S_t)/2]$, depending on the problem instance. On the other hand, if $x_t \geq (s_t + S_t)/2$ no order is placed. Price depends on the initial inventory level at the beginning of the period.

Clearly, the results obtained by Chen and Simchi-Levi also apply to the special case in which the ordering cost function includes only variable but no fixed cost, i.e., the model analyzed by Federgruen and Heching [51]. Indeed, Chen and Simchi-Levi pointed out that their results generalize the results in [51] to more general demand processes, such as the multiplicative demand model.

In [34], Chen and Simchi-Levi consider the infinite horizon model with stationary parameters and general demand processes. They show that in this case, the (s, S, p) policy identified by Thomas is optimal under both the average and discounted expected profit criteria. They further consider the problem with continuous inventory review in [35], and show that a stationary (s, S, p) policy is optimal for both the discounted and average proft models with general demand functions and general inter-arrival time distribution.

A related paper is the one by Polatoglu and Sahin [121]. They consider a model similar to the one by Thomas but assume that the backorder cost is a non-decreasing function of price. The authors give sufficient conditions for the optimal policy to be of the (s, S) type.

While many researchers assume that demand is a function of price, Datta and Paul [42] consider the case where demand depends on both price and the existing level of stock (and thus the shelf space). This situation might occur in the fashion industry for perishable grocery items, and in these cases the demand would tend to increase with a higher stock level. Other researchers who consider demand as a function of price and inventory include Rajan et al [125] (with shrinkage and a different functional demand form) and Smith and Achabal [136] (clearance pricing of initial inventory with different cost assumptions).

In the case of [42], demand (D) is a function of the inventory level at time t, $i(t)$, and the price mark-up rate, k, which is the ratio of the selling price to the unit cost of the product. The demand function is written as

$$D(i(t), k) = f(k)(i(t))^\beta$$

where $f(k)$ is a positive non-increasing function of the mark-up rate k, and $0 < \beta < 1$. The demand model is similar to that in some inventory literature but with the addition of the price term. [42] formulate an EOQ-type model with deterministic demand and production set-up cost and provide an algorithm that gives the optimal price and ordering policy.

See Bernstein and Federgruen [15] in Section 4.3 for an example of pricing coordination with fixed ordering costs and multiple retailers.

Exponential Decay with Production Set-up Cost. Although many papers have considered demand to be a function of price and possibly time-based seasonalities, others have considered that demand for some products may be affected by conditions such as obsolescence or deterioration, e.g., computers or grocery items.

One of the first to consider pricing and inventory decisions for products experiencing deterioration in demand is Cohen [39]. The author considers a modified Economic Order Quantity (EOQ) model with production set-up costs and zero inventory at the beginning of each cycle, where the objective is to determine a price and order quantity for each cycle. Cohen assumes demand is a deterministic linear function of price with exponential decay that is proportional to the on-hand inventory, and he derives the profit maximizing solution. Sensitivity analysis indicates that for a fixed price, the optimal cycle length decreases as the decay rate increases. Further, the optimal order rate increases with an increase in the decay rate and decreases with increasing price if price is an external parameter.

Chakravarty and Martin [29] consider a related problem that allows for demand to decay exponentially with time, where the function has a known and constant rate of decay. Production by the seller and ordering by the buyer incur a fixed set-up charge, but holding cost is not considered in the model; the leadtime is deterministic, and shortage is not allowed. The authors provide a method of determining both price and the replenishment order frequency for the buyers. Further, they consider the case where multiple buyers may be grouped based on common ordering schedules, and they demonstrate a procedure to optimal assign buyers to groups.

Rajan et al [125] considers a modified EOQ model similar to [39], and extends the work by allowing price to change over the cycle. In addition, [125] allow demand to decay according to a general function of time elapsed since the beginning of the selling season (wastage) and the on-hand inventory (value drop). They show that the optimal dynamic price may be determined independently of the cycle length when maximizing average profit, and that this price is unique under conditions of demand such as linear, exponential, and constant price elasticity. They demonstrate conditions under which the cycle time is also unique, among these is the case of linear demand. They also derive the optimal fixed price solution under deterministic demand and compare the profit to the dynamic price case.

Abad [2] generalizes the model in [125] to include partial backordering and derives the optimal policy.

3.3 Retail, Clearance, and Promotion

Perhaps the first uses of dynamic pricing were in retail goods, exhibited by sales or promotions over time. The retail industry, and more specifically the fashion industry, is typically characterized by a large variety of goods, a high level of demand uncertainty, and a short selling season. Further, many fashion goods are ordered well in advance, and retailers may not have the opportunity to restock during the selling season. Thus, price offers an important tool to manage demand and inventory over time.

Marketers and other analysts have been addressing price in this context for some time, see for example Rao [126] for a review. However, we are particularly interested in the situation when the pricing decisions are incorporated with inventory decisions. In this context, there are a number of researchers to consider pricing and inventory problems that are specific to retail industries, where production decisions are usually not incorporated. Although these problems are contained within the scope of a manufacturing price and inventory problem, focusing on the characteristics of the retail industry can lead to more robust models and solutions specific to the situation. Therefore in this section we address research specific to the area of retail.[2]

Many of the papers in this section also have applications to *yield* or *revenue management* problems such as those faced by the airlines. For example, in the airlines there is usually a fixed supply of a product (seats on a plane), with a specific deadline for the selling season (time the plane takes off). Price is a tool to manage demand and may be used jointly with inventory control to maximize revenues, see [102] for a review. This problem also arises in service industries such as hotels or rental car agencies; recent applications include varied products such as entertainment tickets, restaurant reservations, broadcast advertising. Only the papers most pertinent to the retail applications will be reviewed.

3.3.1 Clearance Pricing and Markdown Pricing.

Gallego and van Ryzin [62] focus on dynamic pricing with stochastic demand where selling must stop after a deadline and restocking may not be allowed. They use control theory methods to formulate the problem and determine price as function of the inventory level and time left in the horizon. They assume that demand is a function of the price i and is a Poisson process with intensity rate $\lambda_i = \lambda(p_i)$ (similar to [39] and [76]). Since it is a common economic assumption that marginal revenue is decreasing in output, the authors assume revenue is concave and increasing in the demand intensity. The demand curve is stationary over the time horizon, but realized demand depends on price, which is a function of time. Sales are neither backlogged or lost, as price is set to infinity when inventory is zero (a "null-price condition").

For the specific case where demand is Poisson process, Gallego and van Ryzin find the optimal solution to the pricing problem, where price changes continuously over time. For the general case, they formulate and solve a deterministic problem, which provides bounds on the expected profit. Further, they show that a fixed price is asymptotically optimal with increasing expected sales or as the time-to-go approaches zero. When price is chosen from a discrete set of choices, using a small set of neighboring prices is close to optimal. For the discrete price case, they describe an heuristic where a number of items and amount of time are specified apriori, and a price is switched to a second price when either the time elapses or the items are sold, whichever occurs first. In extensions they consider the cases where demand is a compound Poisson process (similar to [97]), or where restocking is allowed.

In [63], the same authors extend their work to focus on a multi-market problem, with multiple products charing common resources. They model demand as a stochastic point process function of time and the prices of all products: the vector of demand for n products, $\lambda = (\lambda^1, \lambda^2, \ldots, \lambda^n)$, is determined by time and the vector of prices, $p = (p^1, p^2, \ldots, p^n)$. As before, revenue is assumed to be concave, and they apply the null price condition.

Gallego and van Ryzin formulate a deterministic problem, which they show gives a bound on the expected revenue. This problem also motivates the creation of a Make-to-Stock (MTS) and a Make-to-Order (MTO) heuristic. The MTS heuristic requires that all products be pre-assembled, and the price path is determined from the deterministic solution. The MTO heuristic also uses the prices from the deterministic solution but produces and sells products as they are requested. An order is rejected if the components are not available to assemble it. The authors show that each of these heuristics is asymptotically optimal as the expected sales increases.

Feng and Gallego [54] focus on a very specific question: what is the optimal time to switch between two pre-determined prices in a fixed selling season? They consider both the typical retail situation of switching from an initial high price to a lower price as well as the case more common in the airlines of switching from an initial low price to a higher one later in the season. They particularly focus on the case where revenue depends on both the time-to-go and the current inventory level. They assume that demand is a Poisson process that is a function of price. Feng and Gallego show that the optimal policy for this problem is a threshold policy, whereby a price is changed (decreased or increased) when the time left in the horizon passes a threshold (resp., below or above) that depends on the unsold inventory. For the problem where the direction of price change is not specified, they show a dual policy, with two sequences of monotone time thresholds. Although they do not explicitly consider the choice of the two starting prices for the problem, a firm could use the policy they develop to determine the expected revenue for each pair of prices and choose the pair that maximizes expected revenue.

Some firms may be sensitive to the variability of sales or over time, so Feng and Xiao [56] extend the work of [54] to consider the addition of a risk factor. They add a penalty (or premium) to the objective function, which is a constant multiplied by the net monetary loss or gain per unit of revenue variance and obtain an analytical solution for this problem.

The same authors further extend the area of research from [54] in [58], where they consider the retail pricing and inventory problem with multiple predetermined prices. They assume that price changes are either decreasing or increasing, i.e., monotone and non-reversible. As before, initial inventory is fixed and demand at each price is a Poisson process with constant intensity rate. Under these assumptions, the authors develop an exact solution for the case with continuous time and show that the objective function of maximizing revenue is piecewise concave with respect to time and inventory.

Noting that there may be scenarios where reversible price changes are allowed, the authors consider this extension in [57]. Unlike [62], the prices are a predetermined set. Many assumptions are as before, for instance demand as a Poisson process and a fixed level of initial inventory. Feng and Xiao show that at any point in time a subset of the possible prices form the maximum concave envelope, and prices which cannot be optimal at that time are not in the envelope. The authors derive the optimal solution for this problem, where each inventory level has a set of time thresholds that lead to a price change, and they show the time thresholds are monotone decreasing in price and inventory.

Further, they consider the relationship between time, inventory, and price, and show that the time thresholds are monotone. For instance, the time thresholds are monotone decreasing in price for a given inventory level of n; $z_n^i \leq z_n^{i-1}$, where $p^i > p^{i-1}$ and z is measured from the beginning of the horizon. The time thresholds are also monotone decreasing in inventory for a fixed price: $z_n^i \leq z_{n-1}^i$. Intuitively, if inventory is higher, the price should be switched to a lower price sooner.

Feng and Gallego [55] further contribute to this area of research by considering pricing of perishable goods when demand is non-homogeneous over time. Specifically the demand intensity rates may be time-varying and depend on the sales-to-date (e.g., word of mouth effect). Rather than allowing for time dependent demand intensities, they model time dependent fares (prices), where the set of allowed fares at time t is $P_t = p_1(t,n), p_2(t,n), \ldots, p_k(t,n)$, where n is the inventory level. They define k as the action at state (t,n) associated with fare path $p_k(t,n)$, and the objective is to select the switching times from fare path k to $k+1$ in a way that maximizes expected revenue. They provide an efficient algorithm that comuptes the optimal pricing policy and objective function.

A similar topic to that of [62] is considered by Bitran and Mondschein [21], except that they model customers with hetergeneous valuations of products. Each customer has a maximum price they are willing to pay for a product (*reservation price*); the probability distribution of reservation prices is known to the retailer and may change over time. Customer arrivals are a non-homogeneous Poisson process with rate that is a function of time, and the seller may increase or decrease the price over time. They examine the optimal pricing policy when reservation prices are time stationary, and they show that price is a nonincreasing function of inventory for a given period of time, and price is a nonincreasing function of time for a given inventory level. However, the overall optimal policy has price decreasing in time with sharp increases when a unit is sold, thus it is not a monotonic policy (nor is the expected price policy). The authors consider continuous and periodic review practices and demonstrate that periodic review policies can be quite close to optimal.

Zhao and Zheng [162] further examine the problem considered in [21] and focus on deriving structural properties of the solution. While [21] showed that if reservation prices are non-stationary that price may not decrease over time, Zhao and Zheng show a necessary condition for the property to hold, namely that the probability a customer is willing to pay a premium must not increase over time. If prices are limited to a discrete set, the authors show a set of critical numbers determine the optimal policy.

In many cases in the retail industry, pricing policies may be coordinated across multiple stores of one retail chain. Thus, Bitran et al [20] extend the research in [21] to allow for prices to be coordinated across multiple stores with different arrival patterns. As before, customers have a non-homogeneous arrival rate (now to each store in the chain), and the seller knows the probability distribution of the reservation price of a customer. They consider the cases where inventory transfers are and are not allowed between stores. The authors develop heuristics to approximate the optimal solution to the dynamic program since the state space is quite large. These heuristics are based on pricing policies which allow 0 or 1 price changes over some or all of the time horizon, and they use a Fibonacci algorithm to solve a nonlinear problem with price as a variable. Applications of the heuristic to sample problems indicated they achieved 97 - 99 % of the optimal pricing policy, and applications to a real set of data showed an expected improvement over the implemented policy for that data (e.g., 16% in one case).

Liang [99] also considers a problem related to Gallego and van Ryzin [62], except that the author allows for multiple fare classes with different demand intensity functions, which are inhomgeneous Poisson processes with no specific order arrival. As before, the demand depends on both the price and the inventory that is available. Liang solves the problem using a dynamic programming approach, and shows that the optimal solution is a threshold type bid price policy that may be determined at the beginning of the horizon if re-forecasting is not needed. The difficulty in applying this to a case in the retail industry is determining what the application of the multiple classes should be or determining the demand curves for these classes.

Whereas some research on clearance pricing assumes time-stationary demand (e.g., [54] or [62]), and some assumes that demand is dependent on the inventory level (e.g., [149]), Smith and Achabal [136] consider pricing when demand is both seasonal and dependent on initial inventory. These assumptions are similar to Rajan et al [125], but instead they have a fixed selling season and no shrinkage due to the short selling season. Based on discussions with retail buyers, they assume the dependence on initial inventory is one-sided–that is, if there is low inventory then demand is slow, but high inventory does not have an effect on the demand level. Assuming deterministic demand that is a continuous function of price, inventory and time ($x(p, I, t)$), the authors formulate a profit maximizing problem where inventory commitments may be changed. When demand has an exponential sensitivity to price, and is both multiplicative and separable, a closed form solution can be obtained for the inventory and price trajectory policy. Smith and Achabal implemented their inventory and pricing policies at several retailers, one of the more successful of these resulted in a 1% revenue increase for the products studied and an increase in the clearing of markdown merchandise.

Generally, research on clearance pricing and inventory decisions assumes that customers are myopic, i.e., they do not exhibit strategic behavior; Elmaghraby et al [49] is an exception to this. The authors focus on markdown pricing when prices follow a pre-announced schedule, similar as seen in some Internet applications (e.g., Sam's Club). The initial supply of inventory is limited, and customers purchase one or more units according to their reservation prices and the available supply. They consider both full information and partial information cases; in both cases the buyers know the valuations of other customers, in the latter case the seller does not. Elmaghraby et al demonstrate a number of analytical properties about their problem. In particular, a 2-step markdown is preferred to policies with more cuts (similar to results found in [54], [62], and [136]). They also compare markdown policies to fixed price policies. For the full information case, they found that markdown policies can dominate only if the supply is determined exogenously, but under partial information markdown policies can also dominate when the seller determines the initial supply.

3.3.2 Promotion. While many retailers employ clearance pricing to manage inventory over a short selling season, in some cases promotional pricing is also appropriate. Promotion pricing, advertising, or marketing efforts may be directed to increase the knowledge about new products, increase traffic to a store, or lure customers away from competitors. Firms often have flexibility about the timing and length of promotions as well as the amount of any discounts. There are many examples of promotion pricing in the marketing literature but relatively few that coordinate promotion decisions with inventory or production decisions.

Sogomonian and Tang [140] develop a model that integrates production and promotion decisions as well as ones that considers the decisions separately so that the benefit of the joint decision-making can be assessed. With the initial retail prices given, the decisions in the models are the promotion periods and the level of the promotion (chosen from a discrete set). Demand is assumed to be deterministic and a function of the time since and level of the last promotion as well as the list price in the current period. Production costs include set-up plus a variable cost per unit, and no backlogging is allowed. The resulting integrated model to maximize revenue is a mixed-integer program, and the authors develop a Wagner-Whitin type algorithm and solve the problem as a nested longest path problem. Using an example, the authors show that the integrated model can result in decreased inventory levels as well as increased profit.

In Smith et al [137], the authors develop a model to plan promotions and advertising, and use scenarios to represent market conditions. The decisions of the firm are the promotion price and the ad size or cost, where the ads are limited by a budget. The deterministic demand depends on the price as well as the ad type, and demand scenarios occur with some probability specified by the user. With additional constraints such as a limit on the number of markdowns, the authors develop a optimization problem that maximizes expected profit over multiple products.

A different approach is taken by Cheng and Sethi [36], where demand is a Poisson process that depends on the current promotion level. In each period, the firm determines if there is a promotion or not and how much inventory to order. Promotions incur a fixed cost (e.g., advertising cost), and backlogging of demand includes a per unit cost. With a goal of maximizing expected profit, the authors show that the optimal inventory policy has two base-stock levels (S^0 and S^1), where each are a function of the demand state, and the choice between the base-stock levels is contingent on the promotion decision (0 or 1). In general, the authors show the promotion decision is of the multi-threshold type, where each inventory threshold indicates a change in the promotion decision. However, under certain conditions of stochastic dominance of demands, a single threshold policy is optimal for the promotion decision, where the product is promoted if the inventory is above the threshold (P). The optimal policy in this case is thus of the form (S^0, S^1, P).

Sethi and Zhang [134] are interested in the joint production and marketing problem where both demand and capacity are stochastic. Given the complexity of this problem, the general case is intractable, so they develop hierarchical models with two levels of decisions, including a production plan and an advertising portfolio. They assume that the demand and production capacity are a joint Markov process of the production and advertising decisions, and the objective is to maximize expected profit. They include an inventory and shortage cost function as well as constraints on the production and advertising decisions.

They are particularly interested in the case where either the rate of change in capacity states is an order of magnitude different than the rate of change in demand states (or vice versa). For these cases, they develop a model where one decision (i.e., capacity) is approximated using average capacity rather than stochastic capacity, and the solution is used to determine average production and advertising plans. The high level decision is implemented (i.e., advertising,), and the lower level decision (i.e., production plan), uses the average plan along with stochastic demand to develop a feasible schedule. The authors show that this hierarchical approximation (and the corresponding reverse problem where demand is considered the high level decision), is asymptotically optimal as the rates of change of the top level stochastic variable become large with respect to the second.

The focus in Neslin et al [109] is on the manufacturer's advertising and promotion decisions, while taking into account the actions of retailers selling the products and consumers buying them. The authors assume that demand is deterministic and non-seasonal, and the functional form is dependent on several factors such as the advertising rate and the promotion level, and the demand may lag behind advertising. Retailers determine whether to offer a given promotion as well inventory replenishment for current and future orders, but they do not observe the manufacturer's advertising expenditures. The behavior of the retailers and consumers are built into a model that optimizes over the manufacturer's advertising and promotion decisions; the result is a discrete time non-linear problem. The authors develop reasonable parameters for the model based on their experiences with companies, and they run test cases to show the trade-offs between many of the variables.

3.4 Fixed Pricing

Although much of the research in pricing and inventory control has centered around dynamic prices, some has also considered the problem of choosing a fixed or constant price over the lifetime of a product. Typically, this is less motivated in an e-commerce environment, but there may be cases where it is desired, and thus the topic bears mentioning here.

The earliest known example of a problem that integrate a fixed price decision with inventory policies is that of Kunreuther and Schrage [86]. They consider a problem with demand that is deterministic, a linear function of price, and varying over a season, and they include production set-up costs. Their model does not have lost sales or backlogging, since demand is exactly that predicted by price and time, and there are no production capacity limits. The objective is to determine price, production order period, and production quantities so as to maximize profit. The authors provide a "hill-climbing" algorithm that provides upper and lower bounds on the price decision.

Gilbert [64] focuses on a similar problem but assumes that demand is a multiplicative function of seasonality, i.e., $d_t(p) = \beta_t D(p)$, where $D(p)$ represents the intensity of demand, thus demand the ratio of demands in any two periods does not depend on price. Gilbert also assumes that holding costs and production set-up costs are invariant over time, and the total revenue is concave. He develops a solution approach that guarantees optimality for the problem, employing a Wagner-Whitin time approach for determining production periods.

In Gilbert [65], the author extends his previous research to the problem of determining a single price for each of a number of goods when the goods share production capacity. He again assumes that revenue is concave, but does not consider production set-up costs in the multiple product model. Demand is a function of time and product characteristics, and is multiplicative with seasonality, i.e., $D_{jt} = \beta_{jt} D_J$. By formulating a deterministic optimization problem and using the dual, Gilbert develops an iterative algorithm that solves the problem to optimality. By applying the procedure to a numerical example, Gilbert also demonstrates that a firm that is pricing multiple products may want to be more aggressive in pricing products that have high demand early in the season.

Further sensitivity shows that if a product has greater seasonality than another product, the price may be higher.

Although most of the work on constant pricing assumes determininistic demand, fixed pricing under stochastic demand is a special case of the Delayed Production problem in Chan et al [30]. In this research, the authors consider a general stochastic demand function over multiple periods, where production capacity is limited but set-up costs are not incurred. Excess demand is lost, and sales are discretionary, i.e., inventory may be set aside to satisfy future demand even at the expense of lost sales in the current period. The authors develop a dynamic program that solves the problem to optimality for discrete possibilities of fixed prices and compare the results to a deterministic approach. In a numerical study, making the production and price decisions under stochasticity is more important when there is limited deterministic seasonality but high levels of uncertainty about the actual demand realization.

4. Extension Areas

4.1 Multiple Products, Classes, and Service levels

4.1.1 Multiple Products. The product line design problem (see Yano and Dabson [160] for a review) is concerned with the selection of a mix of products to offer in the market. As Morgan et al. [108] point out, this problem has typically been considered from a marketing perspective, while the operational aspects of product line decisions have been largely ignored. Morgan et al. consider individual product costs and relevant cost interactions among products in their product line design model, but the prices of products are given as inputs.

Some papers consider a multiproduct inventory problem with substitution. For example, Bassok et al. [13] consider a model with N products and N demand classes with full downward substitution, i.e., excess demand for class i can be satisfied using product j for $i \geq j$. They show that a greedy allocation policy is optimal. However, there are no pricing decisions in their model. Meyer [106] considers a multiproduct monopoly pricing model under risk. The firm must make decisions for prices, productions, and capacities before actual demand is known. However, his model does not consider explicitly the inventory related costs.

Alternatively, papers in Section 2 consider pricing and inventory model where the demand is a function of the pricing decision. Most of this work is extendable to multiple products, but they do not share a common resource.

A single period paper that considers the pricing multiple products with substitution is Birge et al [19]. The authors also set the capacity levels for production, thus this paper is described below in the section on capacity decisions. A

Table 9.1. Legend for Classification System

Elements	Code	Descriptions	Codes
Prices	($)	Dynamic, Fixed	D, F
Demand Type	(DT)	Deterministic, Stochastic	D, S
Demand Form	(DF)	Concave demand or revenue, Exponential, General, Linear Multiplicative, Non-linear, Poisson	C, E, G, L, M, NL, P
Demand Input	(DI)	Price, Time, Inventory Sales, Ads, Products, Market Production, Reservation Price	P, T, I, (words)
Sales	(Sa)	Backlogged, Lost, Neither	B, L, n
Restocking	(R)	Yes/No	Y, N
Set-up costs	(Se)	Yes/No	Y, N
Capacity limits	(CL)	Yes/No	Y, N
Products	(#P)	Single, Multiple	S, M
All Categories		Other	O

Table 9.2. Classification of Multiple Period Papers

Reference	$	DT	DF	DI	Sa	R	Se	CL	#P
Whitin (1955)	D	D	L	T,P	B	Y	Y	N	S
Thomas (1970)	D	D	G	T,P	n	Y	Y	N	S
Zabel (1972)	D	S	L, M	P	L	Y	N	N	S
Pekelman (1974)	D	D	L	T,P	n	Y	N	N	S
Thomas (1974)	D	S	G	T,P	B	Y	Y	N	S
Thowsen (1975)	D	S	G	T,P	o	Y	N	N	S
Cohen (1977)	F	D	G,L	P,I	B	Y	Y	N	S
Amihud & Mendelson (1983)	D	S	G,L	T,P	B	Y	N	N	S
Feichtinger & Hartl (1985)	D	D	non-L	T,P	n	Y	N	N	S
Li (1988)	D	S	P	P	L	Y	N	Y	S
Lai (1990)	D	D	L	T,P	n	Y	N	Y	S
Lazear (1990)	D	S	O	P	L	N	N	N	S
Rajan et al (1992)	D	D	G,L	T,P,I	B	Y	Y	N	S
Sogomanian & Tang (1993)	D	D	G	P,T,Sales	n	Y	Y	N	S
Gallego & van Ryzin (1994)	D	S	P,R	P	n	N	N	N	S

Table 9.3. Classification of Multiple Period Papers (cont.)

Reference	$	DT	DF	DI	Sa	R	Se	CL	#P
Feng & Gallego (1995)	D	S	P	T,I	n	N	N	N	S
Neslin et al (1995)	D	D	G	Ads,T, Sales	n	both	N	N	S
Sethi & Zhang (1995)	D	S	P	Ads,T, Prodn	B	Y	N	Y	S
Abad (1997)	D	D	L	T,P	PB	Y	Y	N	S
Bitran & Mondschein (1997)	D	S	P	T, Resv P	L	N	N	N	S
Gallego & van Ryzin (1997)	D	S	P,R	P, Product	n	N	N	N	M
Bitran et al (1998)	D	S	P	T, Resv P	L	N	N	N	S
Smith & Achabal (1998)	D	D	G,E	T,P,I	n	Y	N	N	S
Smith et al (1998)	D	D	G	P,Ad, Market	n	N	N	N	M
Cheng & Sethi (1999)	D	S	P	Sales	B	Y	N	N	S
Federgruen & Heching (1999)	D	S	C	T,P	B	Y	N	Y	S
Feng & Xiao (1999)	D	S	P	P	n	N	N	N	S
Gilbert (1999)	F	D	M,C	T,P	n	Y	Y	N	S
Feng & Gallego (2000)	D	S	P	T,I	n	N	N	N	S
Feng & Xiao (2000a,b)	D	S	P	T,I	n	N	N	N	S
Gilbert (2000)	F	D	M,C	T,P, Product	n	Y	N	Y	M
Datta & Paul (2001)	D	D	O	P,I	n	Y	Y	N	S
Biller et al (2002)	D	D	C	P,I, Product	L	Y	N	Y	M
Chan et al (2002)	D,F	S	G	T,P	L	Y	N	Y	S
Chen & Simchi-Levi (2002a,b,2003)	D	S	G	T,P	B	Y	Y	N	S
Kachani & Perakis (2002)	D	D	O,L	P,I	n	Y	N	Y	M
Elmaghraby et al (2002)	D	D	O	all P	L	N	N	N	S
Bernstein & Federgruen (2002)	F	D	L	all P	N	Y	Y	N	S

few papers with multiple time periods have also considered multiple products. Gallego and van Ryzin [63] is described above in the section on Clearance Pricing, as is Smith et al [137].

In Biller et al [18], the authors analyze a pricing and production problem where (in extensions), multiple products may share limited production capacity. When the demand for products is independent and revenue curves are concave, the authors show that an application of the greedy algorithm provides the optimal pricing and production decisions.

Kachani and Perakis [73] study a pricing and inventory problem with multiple products sharing production resources, and they apply fluid methodology to make their pricing, production, and inventory decisions. In their case, they consider the *sojourn* or delay time of a product in inventory, where the delay is a deterministic function of initial inventory and price (including competitor's prices). For the continuous time formulation, they establish when the general model has a solution, and for the discretized case they provide an algorithm for computing pricing policies.

Recently, Zhu and Thonemann [163] study a pricing and inventory-control problem for a retailer who sells two products. The demand of each product depends on the prices of both products, that is, the cross-price effects exist between two products. They show that the base-stock list price policy, which is optimal for the single product problem, is no longer optimal for their problem. They derive the optimal pricing and inventory-control policy and show the retailer can greatly improve profits if she manages the two products jointly and consider the cross-price effects. Specifically, they prove that the optimal pricing policy depends on the starting inventory levels of both products and the optimal expected profit is submodular in the inventory levels.

Karakul and Chan [74] consider a company that produces a well-established product is introducing a new product. The existing product is priced at its optimum equilibrium value and has a stable pool of major customers. The new product targets for a more demanding market segment and has all the functionality of the existing product. As a result, the company has the option to offer the new product as a substitute at a cut price in case the existing product runs out. Both products are seasonal and have fairly long production lead-times. The company faces the single period problem of pricing the new product and having optimal quantities of both products on hand till replenishment is possible. Demand of the existing product during the period is represented by a discrete distribution generated by the pool of major customers. However, demand of the new product has to be estimated and is represented by a set of price dependent continuous distributions. The objective is to find the optimal price for the new product and inventory level for both products so as to maximize the single period expected profit.

The authors show that the problem can be transformed to a finite number of single variable optimization problems. Moreover, for some general new product demand distributions, including the set of log-concave distribution, example of which are Normal, Log-Normal, Uniform, Exponential, Gamma, etc., the single variable functions to be optimized have only two possible roots each. They also show that besides the expected profit, both the price and production quantity of new products are higher when it is offered as a substitute.

4.1.2 Multiple Classes.

There are many cases when firms may have multiple classes of customers competing for scarce resources. These type of problems arise in service industries (e.g., computer processing or Internet providers), travel industries (e.g., airlines or hotels), as well as production environments. A firm may desire to set different prices for different classes, especially if the classes differ in their service requirements, and inventory or production may be determined based on the requirements of the customers.

Much of the initial work on pricing problems with multiple classes of customers has been done in the context of a queueing network, where hetergeneous customers arrive with different sensitivites to delay, and pricing may be determined to match the supply (server ability) with the demand (customer's needs). Although in many of these problems inventory is not an explicit decision, the ability to accept or reject orders arriving to the queue implies a corresponding production decision, so we will provide a brief review of this area. For additional references, see [153] for instance.

Work on problems with multiple classes owes much to a few seminal papers that studied optimal pricing of homogeneous customers arriving to a service facility, see for example [44], [104], and [144]. The focus is generally on determining a price that balances the delay cost to the customer and the utilization of the server. The long-term service rate is also considered in many of these papers, which is effectively a capacity decision for the facility.

Building on this, Mendelson and Whang [105] extend [104] to include multiple classes of users, the authors derive incentive-compatible pricing controls for an M/M/1 queue with constant delay cost per unit job. As defined by Mendelson and Whang, optimal incentive-compatible means that "the arrival rate and execution priorities jointly maximize the expected net value of the system while being determined on a decentralized basis", i.e., individual customers determine their service level and whether to join the queue.

In [66], Ha initially extends [104] to include a service rate decision by homogeneous customers in a GI/GI/1 queue where FIFO (first-in first-out) scheduling is used. For example, the service rate decision may be applicable when a customer has the ability to choose an amount of preprocessing on a task that would affect the server effort required to complete the task. Ha shows that a pricing scheme with a variable fee and fixed rebate based on delay cost maximize the net-value of the system, and the facility profit maximization can also be induced with a fixed and variable fee.

In [67], Ha extends this work to consider multiple customer classes that differ in their demand, delay costs, and service costs. For an M/G/s processor sharing queue, he shows that a single fee dependent on the time in the system can coordinate the system, and for an M/G/1 FIFO server, coordination can be effected through a pricing scheme that is quadratic in time of service.

Van Mieghem [153] focuses on using price to coordinate a queueing system where customer delay costs are nonlinear curves. Since the optimal scheduling rule may be quite complex under general delay cost functions, the author approximates the optimal scheduling rule with the generalized $c\mu$ ($Gc\mu$) rule, which is optimal in congested systems.

A different approach is taken by Feng and Xiao [59], who build on their work in [58], described in Section 3.3, to consider multiple classes. In particular, they focus on perishable product industries that have a short selling season, and they assume that demand arrives as a Poisson process. The classes are served simultaneously at the beginning of the time horizon, and inventory is set aside to satisfy the higher priority classes to ensure a service level target. They focus on the decision of when to close the lower price (and priority) segment. By showing the concavity of the value function, they are able to provide time thresholds that depend on the inventory and time left in the selling season.

Karakul and Chan [75] consider a single period, single product model with two customer categories. Under a monopoly situation, demand from the first customer category follows a price dependent demand continuous distribution. Customers from the second category are willing till the end of the period to buy the product by bulk at a fixed discounted price. Demand from the second customer category is represented by a discrete distribution. The objective is to find the optimal product price and inventory level so as to maximize expected profit.

The objective function assumes several different functional forms according to the product inventory level. For some general first customer category demand distributions, including the set of log-concave distribution, Karakul and Chan show that all except one of these functions are concave, and at most one of them has a feasible local maximum. They also show that both the price and production quantity of are higher in the presence of the second customer category.

Some of the papers that consider multiple classes of customer differentiate them by service, and more particularly by leadtime within the production system. In the next section we consider papers with an explicit differentiation of leadtime, some of which could also clearly fall in the multiple classes section of this paper; there is clearly overlap between these two sections.

4.1.3 Leadtime Differentiation.

Leadtime differentiation is becoming an increasingly important aspect of service in a manufacturing environment. It offers additional flexibility to a manufacturer in scheduling his plant, and if used in conjunction with pricing, it may offer additional revenue choices to customers. Thus a problem that considers leadtime differentiation may combine pricing decisions with leadtime quotation and production scheduling, although few have considered all of these simultaneously. For a thorough review of papers in leadtime quotation (including some with a pricing decision), see Chapter 12 in this book.

One of the earliest examples of a paper which considers simultaneous price and leadtime decisions is Easton and Moodie [45]. For their demand model, the authors use a logit based model, where customer's utility is based on the work required for a job, the price, and the leadtime. The manufacturer gives a bid to a customer for a job, and the customer decides whether to accept or reject the job; thus the available shop time is a random variable. The authors develop a procedure for quoting a price and leadtime for one job at a time based on the current status of the system.

Elhafsi considers price and delivery time quotation in [46] and [47]. In particular, the focus is on the leadtime aspect, with price calculated as a function of the cost of the job.

In [77], Keskinocak et al focus on scheduling and leadtime quotation, particularly when customers exhibit sensitivity towards leadtime. Although price is not an explicit decision, the revenue of a joib depends on the leadtime. They develop algorithms to solve this problem and provide bounds on their performances.

Plambeck [120] considers two classes of customers that differ by price and delay sensitivity. An initial static price and production rate are fixed by the manufacturer, and the firm dynamically quotes leadtimes and determines the processing of customer orders. The author shows her solution is asymptotically optimal in heavy traffic in the system.

Chen [32] considers multiple classes of customers that are differentiated in their service level as well, in particular a firm may give a discount to customers who are willing to wait longer for a delayed shipment. Customer arrivals are Poisson, and they receive a menu of price and leadtime pairs from which to choose. The cost of waiting is assumed to be concave (i.e., with marginal cost decreasing), and excess demand is backlogged. The goal is to use advance demand orders to provide information to improve the performance of the supply chain. Chen shows that a base stock inventory policy is optimal for a given price vector, and he provides an heuristic for computing prices.

A number of firms may also compete on different leadtime capabilities, see papers [92], [98], [112], [138], and [139] reviewed in Section 4.4.

4.2 Capacity as a Decision

A few papers that consider pricing and production decisions have also incorporated capacity investment decisions as well. For instance, Li [97] studies the short-term strategies (price) as well as long-term decisions (capacity) where cumulative production and demand are both Poisson processes.

In Maccini [101], the author examines the long and short run pricing and inventory decisions where capacity investments can be made. Maccini assumes that expected sales have the following functional form,

$$N(t) = k[P(t/V(t), H(t)/Q(t)]Q(t), k_1 < 0, K_2 > 0,$$

where $V(t)$ and $Q(t)$ are the expected market price and demand, respectively, and $H(t)$ is the on-hand inventory. The goal is to maximize expected profit, consisting of revenue minus labor and capital costs. Maccini derives optimality conditions and long-run decisions for the system, and develops short run decision rules from approximations of the differential equations in the optimality conditions. He focuses on the gap between the long and short term decisions, and the interaction between the various decisions. For instance, excess capital implies that prices will decrease and output will increase compared to the long run levels, but excess inventory results in decreases in both price and output.

Gaimon [61] studies the problem of determining price, production, inventory, and capacity investment over time. The research is also related to that of [53] and [115]. The author assumes that demand is a deterministic function of the time in the horizon, and it is linearly related to price. Gaimon models the problem as one of optimal control and provides a solution method based on a numerical solution algorithm. Using numerical examples, she also shows that capacity investment is often delayed until later in the horizon; this result may also be dependent on her assumption that new capacity reduces costs more as time increases. Investing in additional capacity can also lower average production cost, thus leading to lower prices and higher demand.

There is limited literature on pricing and capacity-setting decisions for substitutable products. Birge et al. [19] is one of the first to address this problem. They consider a single-period model in which a firm produces two products with price-dependent demands. The firm has the ability to make pricing or capacity decisions for one or both of its products. By assuming the demands to be uniformly distributed, they are able to show that the pricing and capacity decisions are affected greatly by the system parameters that the decision makers can control. They consider two different cases. In the first case, the capacity is fixed for both products, but the firm can set prices. In the second case, each product is managed by a product manager trying to maximize individual product profits rather than overall firm profits and analyze how optimal price and capacity decisions are affected.

Van Mieghem and Dada [154] focus on a single-period, two-stage process with an initial decision, e.g. production decision, followed by a realization of demand, followed by another decision, e.g. pricing decision; they also consider the capacity investment decision. After the capacity is determined, production is limited, so excess sales are lost. They show how to solve this problem, and they also consider the impact of competition. They find that conditions dictate whether price postponement or production postponement is more valuable to a firm. Specifically they show that the former is likely to be more valuable if demand variability, marginal production, and holding costs are low.

In Netessine [110], the author studies the determination of prices and timing of price changes for a piece-wise constant pricing policy. Although for most of the results, the author assumes that production is make-to-order with unlimited capacity or the initial inventory is given exogenously (and thus, those results are outside of the considerations of this review), in some cases the capacity decision is considered explicitly. Specifically, Netessine models demand as a deterministic time-varying function of price. Under some assumptions of complementary of the objective function, he shows that when prices (timing) are exogenously given, that the timing (pricing) is monotone in capacity. Furthermore, under a linear pricing model, when making all three decisions (price, timing and capacity) then the optimal capacity decision is independent of the price discrimination policy and a closed form solution is obtained.

4.3 Supply Chain Coordination

Issues about coordination of business activities have been studied extensively in recent years. In a typical distribution channel, coordination can be characterized as *Vertical* or *Horizontal*. Vertical coordination focuses on the issues of why and how the different channel members (for example, supplier and retailer) should consolidate their operations to achieve better system performance. It is also known as *Supply Chain Coordination*. For a thorough review of vertical channel coordination, please refer to Cachon [27] and several chapters in Tayur et al. [147].

Horizontal coordination refers to the coordination of different functional areas (for example, marketing and production) of the same firm to achieve better performance of the whole firm. Recently, many models have been developed to identify the benefits of coordination between the marketing and the production departments. We only review papers on horizontal coordination in a particular channel of distribution; general papers on horizontal coordination are included in earlier sections.

Jeuland and Shugan [71] study a simple system where one manufacturer sells items to a single retailer and the retailer resell it to consumers. The consumer demand is a function of retail price and has a downward slope. They show that quantity discounts, which are a method of profit sharing, can provide an optimal means for achieving coordination. They also claim that by horizontal coordination, a manufacturer can realize higher profits and at the same time increase profits for the entire channel.

For a general system with a single seller and a single buyer, Lal and Staelin [89] show it is possible for the seller to design a pricing policy where the average price of the seller's product is a monotonically decreasing function of the amount ordered by the buyer and the buyer is no worse off while the seller's profits are increased.

Kumar et al. [84] study a channel of distribution consisting of a manufacturer and a value-adding distributor. The distributor can decentralize decision-making between his marketing and production functions. By analyzing the model in a non-repeated perfect information game setting, they show that this functional decentralization increases the price sensitivity of the distributor concerning the price he faces from the manufacturer, causes a shift in the distributor's demand curve. The decentralized distributor carries uniformly more inventory than his centralized counterpart and under certain conditions, the decentralized manufacturer also carries more inventory than his counter part in the centralized case. Their numerical experiments show that functional decentralization can improve the overall channel profitability. They claim that this counter-intuitive result comes from the Pareto-dominant equilibrium outcome in their non-repeated perfect information game setting.

Boyaci and Gallego [23] analyze coordination issues in a supply chain consisting of one wholesaler and one or more retailers under deterministic price-sensitive customer demand. The costs components considered in their model include purchasing, setup, order processing, and inventory costs. They show that an optimal coordinated policy can be implemented as a nested replenishment policy where the retailer pays the wholesaler the unit wholesale price as the items are sold, and the retail price is set jointly with the manufacturer. Since this inventory consignment policy adjusts the wholesale price, it is also capable of distributing the gains of channel coordination without requiring side-payments. The authors also show that pricing and lot sizing decisions can be done sequentially without a significant loss of optimality when the demand rate is high. This justifies the practice of making marketing decisions that ignore operations costs, followed by operations decisions that take the retail price as given, for high sales volumes.

Federgruen and Heching [52] consider the problem of joint pricing and inventory-control problem in distribution systems with geographically dispersed retailers. They assume the distribution of demand in each period, at a given retailer, depends on the item's price according to a stochastic demand function. These stochastic demand functions may vary by retailer and by period. A distribution center place orders with an outside supplier at the beginning of some periods and then allocate the products to the retailers. The order and allocations both require a certain leadtime. They develop an approximate model that is tractable and in which an optimal policy of simple structure exists. In addition, they develop combined pricing, ordering, and allocation strategies and show that the system's performance under these strategies is well gauged by the above approximations. They analyze how different system parameters impact the system performance and show the benefits of coordinated replenishments under dynamic pricing. A comprehensive numerical study based on data obtained from a nationwide department store chain is reported at the end of the paper.

A rare example of supply coordination under fixed ordering costs is found in Bernstein and Federgruen [15]. The authors consider a supply chain with 1-supplier and N competing retailers. Both the supplier and retailers incur a fixed ordering cost, with an additional variable cost differing by facility. Bernstein and Federgruen study a Bertrand competition where each retailer chooses a price and replenishment strategy as well as a Cournot competition where each retailer selects a sales target and replenishment strategy. Demand is a deterministic function of all retailer prices in the Bertrand and all sales targets in the Cournot. The authors provide a centralized inventory and pricing policy based on the Power of Two heuristic (see Roundy [132]) with upper and lower bounds on the performance compared to the optimal policy. For the decentralized system they demonstrate that a unique equilibrium exists under some conditions, and they derive a discount scheme that coordinates the supply chain perfectly.

4.4 Competition

There is a rich literature on price and quantity competition in economics literature. In this chapter, we only review the ones that include competition under price and quantity coordination.

The basic models of oligopoly assume that there is competition over price (Bertrand, [16]) or quantity (Cournot, [40]) and price and quantity are related through a demand function. Kirman and Sobel [82] develop a multi-period model of oligopoly where a set of competing firms need to decide during each period the price and the amount of goods it will produce to satisfy random demand. They show the existence of equilibrium price-quantity strategies for the firm and show that an equilibrium strategy may be found by solving an appropriate static game. Further, the quantity decision is often a time-invariant constant. In each period, each of the retailers uses a randomized strategy to choose from all feasible price levels according to a given distribution and his optimal order policy is an order-up-to policy.

Loch [100] studies a Bertrand duopoly competing for homogeneous customers in a make-to-order environment. Demand arises in a Poisson fashion with an average rate that depends on price and delay. Two firms have independently distributed processing times for a unit of work, and they compete by specifying prices. The author shows that the game has a unique Nash equilibrium when firms are symmetric, that is, their processing time distributions have the same mean and variance.

Eliashberg and Steinberg [48] study the problem of determining production and marketing equilibrium strategies for two competing firms under a seasonal demand condition. The firms' production cost structures are asymmetric. The first firm faces a convex production cost and a linear inventory holding cost, while the second firm faces linear production cost and holds no inventory. The demand each firm faces is a function of the prices each firm charges. The objective of each firm is to maximize his respective profit over the known time horizon, and all the decisions are made simultaneously and non-cooperatively. They characterize and compare the equilibrium strategies of the two firms. If the first firm finds it optimal to hold inventory, then he begins the season by building up inventory, then lets the demand draw down inventory until it reaches zero; no production will be scheduled after that. For the pricing decisions, they show that the prices charged by each of the two firms first increase and then decrease. They also characterize the condition under which the second firm's price over the entire period is strictly lower than the first firm's price.

Reitman [130] studies a competition model in which firms sell a product with congestion externalities to a heterogeneous population of customers. Each firm chooses its price and service capacity, and the capacity chosen by the firm affects the quality level received by customers. They show that the general oligopoly equilibrium, if exists, is asymmetric. They also numerically compute a Nash equillibrium in prices and capacity for several examples.

In an n-firm market game in which firms compete for orders by early delivery, Li [95] considers the effect of inventory in such competition. He shows that firms are more likely to keep inventory in such a market with competition.

Some papers consider the effects of other factors, such as the delivery time, in market competition models. For example, So and Song [139] study the impact of using delivery time guarantees as a competitive strategy in service industries. The demand in their model is a function of both price and delivery time. They focus on a single-firm model and develop an optimization model for the firm to understand the interrelationships among pricing, delivery time guarantees, demand, and the overall profitability of offering the services. By adjusting the different parameters of the system, they offer insights on how managers should adjust the optimal decisions under changing operating environments. For example, if the system capacity increases, the manager should always promise a shorter delivery time guarantee. The corresponding price changes depends on the demand. That is, the manager should raise the price if the demand is elastic and reduce the price if the demand is inelastic. They also show that if a higher service reliability is desired, the manager should quote a longer delivery time and reduce the price. Palaka et al. [112] study a similar system to that of So and Song [139]. They add the lateness penalty costs and congestion related costs in their model while assume the demand is linear in price and quoted lead-time.

Li and Lee [98] propose a model of market competition in which a customer values cost and quality as well as speed of delivery. In their model, there are two firms in the market providing substitutable goods or services in a make-to-order fashion. When demand arises, a customer chooses the firm that maximizes its expected utility of price, quality and response time. They study the relationships between price and processing speed (capacity) and show that the rapid response does give a competitor the advantage of price premium and market share. They assume that customers can observe the congestion levels of the firms, and their choices are dynamic.

In a multiple firm setting, So [138] assumes the overall market size is constant and show that a unique Nash equilibrium solution exists and is easy to compute. Several managerial insights are offered in the paper. For example, he shows that with all other factors being equal, a firm with high capacity should compete with shorter time guarantee, while a firm with lower unit operating cost should compete with lower price.

Lederer and Li [92] study perfect competition between firms that produce goods or services for customers sensitive to delay time. Firms compete by setting prices and production rates for each type of customer and by choosing scheduling policies. They prove that there exists a competitive equilibrium. Since each customer has incentive to truthfully reveal his/her delay cost, the equilibrium is well defined whether or not a firm can differentiate between customers based upon physical characteristics. Two special cases are studied further and a unique equilibrium exists for each of the cases. In the first case, firms are differentiated by cost, mean processing time, and processing time variability, but customers are homogeneous. The conclusions include that a faster, lower variability and lower cost firm always has a larger market share, higher capacity utilization, and higher profits. However, this firm may have higher prices and faster delivery time, or lower prices and longer delivery time. In the second case, firms are differentiated by cost and mean processing time, but customers are differentiated by demand function and delay sensitivity. The results include that customers with higher waiting costs pay higher full prices, and that each firm charges a higher price and delivers faster to more impatient customers. Competing firms that jointly serve several types of customers tend to match prices and delivery times.

Carr et al. [28] study a single-stage game in which firms with random production capacity engage in oligopolistic price competition. The uncertain demand is price dependent. They identify situations under which improvements in a firm's production process can actually lead to reduced profits for the firm. For example, one surprising result is that an oligopolist's profits can increase in its capacity variance. This is derived from standard non-cooperative Nash equilibrium solutions to a single-stage game with complete information and normally distributed random variables, such as demand and capacity. The managerial implication is that process improvement efforts in a competitive environment may end up transferring wealth from firms to society. The consumers at large can benefit through lower prices and higher volumes, as firms end up worse off by competing away their potential gains from capacity improvement. Van Mieghem and Dada [154] analyze a similar model in which the retailer's capacity limits are selected endogenously and the products offered by the retailers are homogeneous. The retailers compete in terms of their sales targets. They show that the relative value of operational postponement (production postponement, price postponement) techniques seems to increase as the industry becomes more competitive.

Li and Whang [96] consider a make-to-order environment where both customers and firms are heterogeneous. Firms are differentiated by processing time distribution and production cost, and customers are differentiated by sensitivity to price and delay. Customers demand identical goods or services, each of which requires one unit of work. Firms compete by specifying prices and production rates for each type of customer as well as scheduling policies. They show the existence and uniqueness of a competitive equilibrium and analyze some important operations strategy analysis, especially on how firms differentiate their competitive strategies, which include price and/or delivery service, according to their own competencies, which include cost and/or processing capability.

To study the impacts of choosing pricing strategy or quantity strategy in different business competition environment, Tanaka [146] considers a two stage game of an oligopoly in which $n(n \geq 2)$ firms produce substitutable goods. The two sequential stages of the game are as follows:

1. The firms choose their strategic variable, price or quantity.
2. The firms choose the levels of their strategic variables.

The firms are identical except for their strategy choices, and the demand functions are symmetric. The author considers four equilibrium configurations in the second stage of the game corresponding to the strategy choice of the firms in the first stage as follows.

- *The Bertrand equilibrium:* All firms choose a price strategy in the first stage.

- *The Quasi Bertrand equilibrium:* Only one of the firms chooses a quantity strategy, and other $n-1$ firms choose a price strategy in the first stage.

- *The Cournot equilibrium:* All firms choose a quantity strategy in the first stage.

- *The Quasi Cournot equilibrium:* Only one of the firms chooses a price strategy, and other n-1 firms choose a quantity strategy in the first stage.

The author shows the following results. (1) The profit of the firm who chooses a quantity strategy in the quasi Bertrand equilibrium is larger than its profit in the Bertrand equilibrium. Thus, a quantity strategy is the best response for each firm when all other firms choose a price strategy. (2) The profit of the firm who chooses a price strategy in the quasi Cournot equilibrium is smaller than its profit in the Cournot equilibrium. Thus, a quantity strategy is also the best response for each firm when all other firms choose a quantity strategy.

Parlar and Weng [113] is perhaps the first paper to jointly considers the pricing decision faced by the marketing department and the production quantity decision faced by the production department while allowing the two departments to coordinate their decisions to compete against another firm with a similar organizational structure for price-sensitive random demand. They predict that coordinating the pricing and production quantity decisions in a firm may lead to a stronger competitive position in terms of offering lower prices, increasing production quantity, and higher expected profit. Although this is intuitive, their models provide a means for formalizing and quantifying the differences between coordination and non-coordination policies.

For a review and discussion on the effect of retail competition on supply chain coordination, we refer the readers to Cachon [27].

4.5 Demand Learning and Information

Most of the existing models including both stocking and pricing as decisions assume that a firm has knowledge about the relationship between demand and price. In reality, of course, there are many situations where a firm does not have full knowledge of the demand function, when new products are introduced for example, or the demand function may be changing in ways that are not necessarily predictable. In the literature, the process of incorporating information about customer demand is often called "demand learning".

Alpern and Snower [4] analyze a dynamic pricing and inventory control model with learning, where they assume upper and lower bounds on the initial uncertainty in the price-dependent demand function. They assume that the firm chooses a sequence of price-quantity combinations through time. The firm infers the position of its demand curve by observing its inventory level.

Jorgensen et al. [72] develop a dynamic model of pricing, production and inventory management which allows for learning effects on both the demand and the production side. They refer 'demand learning' to the situation where current demand is influenced by past demand. That is, the demand rate at the current period depends not only on price, but also on cumulative sales by now. Production learning suggests that when a firm introduces a new product, the costs of production will decline as the accumulated output increases. They combine the essential elements from three streams of research. The first of these streams deals with optimal production and inventory policy when the demand is exogenously given but time-varying. The second stream is an area in marketing research which investigates the impact of demand learning on dynamic pricing policy. The third stream is in microeconomics and industrial organization and the research focus is on the effects of production learning in a firm's production process. They formulate the problem as an optimal control model and find that the optimal price and production rate are both increasing on any interval of time during which inventory is positive.

Subrahmanyan and Shoemaker [145] develop a model for use by retailers to determine the optimal pricing policy and stocking policy for new fashion items that incorporates the updating of demand. In this multi-period decision problem where pricing and stocking decisions made in earlier weeks affect options and decisions in future weeks, the retailer has little or no solid information to use in forecasting the level of demand at various prices. In their model, the probability distribution of sales is updated every period and depends upon the initial priors and the history of prices, sales results and inventory levels from the first period to the current period. They show by examples that expected retailer profits can be substantially higher when the model includes learning. The increase in expected profits with learning is a function of the decision maker's degree of uncertainty about the distribution of demand.

Burnetas and Smith [26] consider the combined problem of pricing and ordering for a perishable product with unknown demand distribution and censored demand observations resulting from lost sales, faced by a monopolistic retailer. They develop an adaptive pricing and ordering policy with the asymptotic property that the average realized profit per period converges with probability one to the optimal value under complete information on the distribution. The pricing mechanism is modeled as a multiarmed bandit problem, while the order quantity decision, made after the price level is established, is based on a stochastic approximation procedure with updates. The stochastic approximation can be seen as a stochastic generalization of Newton's method for estimating the root of an unknown regression function.

Petruzzi and Dada [117] analyze the problem of determining inventory and pricing decisions in a two-period retail setting when an opportunity to refine information about uncertain demand is available. The model extends the newsvendor problem with pricing by allowing for multiple suppliers, the pooling of procurement resources, and more general informational dynamics. The sequence of events and decisions are as follows: Before the beginning of Period 1, determine how much to buy from a supplier for delivery in Period 1 and in Period 2, respectively. The price at which to sell the product is also determined here. After observing the sales in Period 1, refine the characterization of the demand uncertainty prescribed for the second period. Given the refined estimate for demand in Period 2, then finally determine the stocking quantity and selling price for Period 2.

They show that all decisions, including recourse decisions, can be determined uniquely as a function of the first-period stocking factor. The 'stocking factor' z is defined as follows:

$$z = \begin{cases} q - y(p) & \text{for the additive demand case;} \\ q/y(p) & \text{for the multiplicative demand case;} \end{cases}$$

They also find that the cost of learning is a consequence of censored information and shared with the consumer in the form of a higher selling price when demand uncertainty is additive. They also apply the results to three motivating examples: a market research problem in which a product is introduced in a test market prior to a widespread launch; a global newsvendor problem in which a seasonal product is sold in two different countries with nonoverlapping selling seasons; and a minimum quantity commitment problem in which procurement resources for multiple purchases may be pooled.

The same authors consider a related problem in Petruzzi and Dada [118], in which they assume that demand in each of the T periods is a deterministic function of price. They present a dynamic model that simultaneously links price, learning from sales, and inventory. Instead of specifying upper and lower bounds on the initial uncertainty in the price-dependent demand function, as Alpern and Snower [4] did, they use a subjective probability distribution to characterize the function. They determine optimal stocking and pricing policies over time when a given market parameter of the demand process, though fixed, initially is unknown. Learning occurs as the firm monitors the market's response to its decisions and then updates its characterization of the demand function. For the case of a perishable product, they show that the structure of the optimal policy depends crucially on whether uncertainty has an additive or a multiplicative effect on demand. For example, in the additive case, the optimal price rises until uncertainty is resolved, while in the multiplicative case, the optimal price likely falls until uncertainty is resolved.

They also provide some interesting managerial issues. When managers are uncertain of the market potential, but can learn more by observing the reaction of the market to their decisions, it is advisable to set a higher z. This increases the likelihood of having leftovers, so this decision can be viewed as an investment today to reap higher benefit in the future. Since a higher z results in a higher price, this interpretation suggests that the investment is at least partially subsidized by the customer.

5. Industry

5.1 Dynamic Pricing Practice

The proliferation of Internet and related technology is transforming the business world. In 1998, 11,000 US companies engaged in e-business, generating revenues of $101.89 billion. This number is expected to grow to $4 trillion by 2004 [69].

This increase in traffic and interest in the Internet has influenced many companies to try new pricing models. In this section, we survey some research that relates to the industrial application of pricing strategies on the internet, and provide examples of companies using different kinds of pricing practices, especially those driven by the shift to e-commerce. We give particular emphasis to pricing strategies that are coordinated with inventory or production control. We also note several possible pitfalls associated with innovative pricing strategies.

5.2 Related Research

Several papers study the impacts of the Internet on firms' marketing and product strategies [17], [127]. Smith et al. [135] provide a review and assessment of the digital markets. They identify four dimensions of efficiency in Internet markets when compared to brick and mortar markets: price levels, price elasticity, menu costs, and price dispersion. They review evidence that Internet markets are more efficient than conventional markets with respect to the first three dimensions. They note that there are several studies (Bailey [9], [10]; Clemons et al [38]; and Brynjolfsson and Smith [25]) that find substantial and persistent dispersion in prices on the Internet.

Baker et al. [11] identify the opportunities for companies to improve their financial performance by pricing smarter on the internet. They first point out that two dominant approaches to pricing on the internet, that is, offering untenably low prices, or simply transferring the off-line prices onto the Internet, are not optimal. The transparency and efficiency of doing online pricing gives companies greater precision in setting prices, more flexibility in changing prices, and improved customer segmentation. They also suggest three important steps to make an on-line pricing strategy work: 1) identify degrees of freedom consistent with strategy and brand, 2) build appropriate technological capabilities, and 3) create an entrepreneurial pricing group.

Bakos [12] focuses on the impacts of one important aspect of e-business, that is, the reduced search cost for the buyers, on price and the efficiency of allocation of the electronic markets. He shows that electronic marketplaces will increase price competition and reduce seller monopoly power. Socially optimal allocation may be possible if search costs become low enough.

The Internet commerce technologies and advanced manufacturing technologies allow an online seller to offer customized products to different customers. Dewan et al. [43] examine the impact of reduced costs of collecting buyer information and menu costs on a firm's product and pricing strategies. Using a game-theoretic model, they show analytically that a seller who sell customized products at discriminatory prices will gain market share and profits at the expense of the conventional seller.

5.3 Price Discrimination in Practice

One popular pricing model for e-commerce applications is price differentiation, where different customers may pay different prices for the same good. Differential pricing is not a brand-new concept, it is just a new version of the old practice of price discrimination [83]. Department stores often charge more for goods in well-off neighborhoods than in stores in poorer areas; the movie industry charge different prices for different customers, such as adults versus students and senior citizens; insurance companies charge different premiums to different customers based on the characteristics of the customers, such as location and service history, etc. Successful price discrimination requires that a firm know its customers' utility functions, which has been almost impossible to do because of the very high cost associated with determining these functions.

A second pricing model that is increasingly popular for e-commerce applications is dynamic pricing, where the price of an item may change over time. This may used as a tool to manage variability in supply of product or components, or variability of customer demand, or as a reaction to competitors.

Online dynamic pricing is possible today mainly because of the growth of Internet and new technology. First, compared to the conventional markets, it is very easy for online companies to change their prices and the cost of doing it is negligible. Second, it is easier for online companies to segment the market. Computer "cookies" and other similar technological tools can help companies trace the customer's buying history and identify the customer's buying habits, so different customer can be charged with different prices.

Research indicates that several companies may dynamically adjust the price of their products over time. Indeed, dynamic pricing is a popular strategy in computer industry. Dell Computer Corp., which is the No. 1 PC maker in the world, is one of the pioneers in this area. By working with the company main suppliers, Dell can forecast prices for each component in its computers several months into the future. This enables Dell to adjust PC prices based on quantity, delivery dates and even the degrees of competition among computer buyers. As a result, Dell now accounts for nearly a quarter of PC sales, compared with 6.8% in 1996, according to International Data Corp. [103].

Other major computer makers, including IBM, Compaq, and Hewlett-Packard, are also actively investigating dynamic pricing models for their online retail sites. IBM automatically adjusts pricing on its server line in real time based on metrics such as customer demand and product life cycle. As a result, customers will find that pricing will dynamically change when they visit IBM's Web site on any given day. Compaq has been testing dynamic pricing on its website which sells excess and refurbished products, and they may extend dynamic pricing to its main web site if all goes well [155]. HP is also investigating the benefits of doing dynamic pricing, they hope to use price as a strategic lever to gain more market share and increase their profits.

Many other industries practice "dynamic pricing". For example, Amazon.com charged different customers different prices for the same DVD movie. [141]. Coca-Cola was reported to have been adjusting the prices of soft drinks in vending machines according to the surrounding temperature [60]. Airlines may charge different prices for different seats in the same row depending on a lot of different factors; the movie the magazine industry charge different subscription prices for different method of solicitation; grocery stores change the prices of seasonal fruits and some perishable products from time to tome.; even restaurants charge different prices for weekday lunch meals and weekend lunch meals.

5.4 Potential problems with Dynamic Pricing

1) Is dynamic pricing suitable for your company?

Whether customers deem the price discrimination is fair depends largely on the products you are offering. Many customers may be willing to tolerate a certain amount of price discrimination if doing so can assure the provision of particular goods and services. Many people get used to the variation of air tickets, but not quite used to the different prices of CDs or DVDs [141].

2) Competition may mis-direct the direction of dynamic pricing.

A March 2002 survey by the Professional Pricing Society shows that the real issues pricers face in maintaining profitability with their pricing and promotions strategy comes from competitive pricing moves. According to Eric Mitchell, the President of the Professional Pricing Society, "The competition issue means that pricers are expending too much energy trying to match competition rather than pricing to customer demand and value expectations. This is a shift in the trends we were seeing in past. While it may increase the perception of competitiveness in the short run, in the long run these practices can damage the enterprises ability to capture profit and value". He also suggests that in order to avoid this kind of unhealthy competition, pricers need to better understand customer demand by collecting and analyzing past customer buying behavior before setting new prices, rather than just looking up recent competitor pricing and promotion activity to arrive at a new price, and to begin managing pricing more strategically. In many cases that means using better technology to price smarter. That is real competitive differentiation.

3) Some customers may use technology to give the pricer wrong information. They may mask them as lower profit potential customers to your company.

Currently, many companies are using sophisticated pricing technologies to optimize prices and manage revenue. Also from the survey, they found 28 percent (of the pricers) are using their ERP-based software or custom or in house solutions to price and "a surprising 9 percent of the pricers have already adopted and are using new pricing optimization and revenue management technologies". According to Mitchell, pricing is taking on an expanded role and importance in the enterprise. The next logical step is to integrate pricing as a strategic capability into the value chain. This will require firms to develop flexible pricing strategies and scenarios that can only be accomplished with sophisticated pricing technologies and trained pricers who understand how to use these systems (for details of the survey, see [143]).

6. Conclusions and Future Research

We have observed that there has been increased interest by manufacturers in coordinating pricing with inventory and production decisions. This is particularly true for manufacturers who may be operating in an e-commerce environment, since price may be changed easily and demand data is more readily available.

The increased interest by industry has been matched with a corresponding increase in research in this area. We have a reviewed a number of papers in this area, and there are a number of significant results. For example, researchers have studied problems with stochastic demand, production set-up cost, multiple classes of customers, capacity as a decision variable, demand learning, and competition, all of which offer significant challenges. Researchers have examined ways of solving these problems as well as developing general insights for manufacturers in many cases.

However, significant challenges still exist, and there is clearly still a gap between the research that has been done and the problems that exist in the real world. We mention a few ideas based on our experiences and reviews of current literature, and we encourage further exploration of additional topics.

Very few papers to date have considered multiple products that share a common resource such as production capacity. This problem becomes even more difficult when there are interactions among products, such as demand diversions or cross-elasticities, but many companies are quite interested in these interaction effects. It may also be possible to bundle multiple products and determine pricing and production policies in a way that make the system more efficient. Also, determining pricing and production strategies when common components of limited supply make up numerous products is an interesting topic.

Strategic behavior is another area that holds great promise for future research. In the real world, customers are likely to buy according to expectations of future prices, and this can have important implications for determining pricing decisions.

Some work has considered multiple classes, which may be differentiated by service level (leadtime for instance). The initial work in this area is promising, although significant challenges still exist, such as having general arrival or processing distributions and intertemporal parameters. Additionally, many of the current models take a queueing approach, but examining multiple period models may also be of use for manufacturing contexts.

Although some work has been done in the area of demand learning, this continues to remain an area of fruitful progress. In particular, many companies may not have good estimates of demand distributions, so it can be quite important to establish how to set price and use the resulting information in an effective manner.

Likewise, we reviewed the research on vertical coordination between supply chain members with the purpose of maximizing their profits and horizontal competition among similar firms to gain larger market shares. It is not clear what the joint effect of competition and coordination is on the firm's strategy. We believe this is an interesting and important future research direction to go, although the analysis would not be easy. The reality is that most companies are not monopolies, and many specifically compete on product prices (or service differentiation), so these effects need to be considered when establishing coordinated pricing and production policies.

In addition to incorporating competititor's strategies, additional research can be performed that includes customers acting rationally or strategically. For instance, some promotional work accounts for "buy-ahead" (where a low price in the current period may divert demand that would have appeared later). Including consumer gaming is an interesting and challenging area.

Other research areas of interest are some that include different kinds of buying policies; for instance, demand that arrives or is purchased in batches. In addition, some companies may be interested in incorporating customer returns or cancellations, and may employ techniques such as overbooking of cancellations are significant.

Finally, we note that other topics may be particularly pertinent to firms operating on the Internet as a sales channel. For example, this channel may compete with a given firm's traditional retail channel, so considering these simultaneously for pricing and production strategies may be important.

We look forward to continued research in the intersection of marketing and production, and we anticipate that the gap between research and industry in this area will continue to close.

Notes

1. PF_2 includes many common distributions such as exponential, uniform, normal and truncated normal.
2. Research in [91] and [114] is also considered to be retail; however, they are included in Section 3.1 since their models are focused on explaining the current state of pricing rather than developing models to change the current state.

References

[1] P. L. Abad. Determining optimal selling price and lot size when the supplier offers all-unit quantity discounts. *Decision Sciences*, 19(3):622–634, 1988.

[2] P. L. Abad. Optimal pricing and lot-sizing under conditions of perishability and partial backordering. *Management Science*, 42(8):1093–1104, 1996.

[3] P. L. Abad. Optimal pricing and lot-sizing under conditions of perishability, finite production and partial backordering and lost sale. *Eur J of Operational Research*, in preparation, 2002.

[4] S. Alpern and D. J. Snower. A search model of pricing and production. *Engineering Costs and Production Economics*, 15:279–285, 1989.

[5] Y. H. Amihud and H. Mendelson. Price smoothing and inventory. *Rev. Econom. Studies*, 50:87–98, 1983.

[6] F.J. Arcelus. Ordering policies under one time only discount and price sensitive demand. *Iie Transactions*, 30(11):1057–1064, 1998.

[7] F.J. Arcelus and G. Srinivasan. Inventory policies under various optimizing criteria and variable markup rates. *Management Science*, 33:756–762, 1987.

[8] R. A. Ashley and B. Orr. Further results on inventories and price stickiness. *American Economic Review*, 75(5):964–75, 1985.

[9] J. P. Bailey. Intermediation and electronic markets: Aggregation and pricing in internet commerce. Working Paper, MIT, 1998a.

[10] J. P. Bailey. Electronic commerce: Prices and consumer issues for three products: Books, compact discs, and software. *Organisation for Economic Co-Operation and Development*, 98(4), 1998b.

[11] W. Baker, M. Marn, and C. Zawada. Price smarter on the net. *Harvard Business Review*, 79(2):122–126, 2001.

[12] J. Y. Bakos. Reducing buyer search costs: Implications for electronic markets. *Management Science*, 43:1676–1692, 1997.

[13] Y. Bassok, R. Anupindi, and R. Akella. Multi-product inventory models with substitution. Working Paper, New York University, 1994.

[14] F. Bernstein and A. Federgruen. Decentralized supply chains with competing retailers under demand uncertainty. *Management Science*, forthcoming, 2003.

[15] F. Bernstein and A. Federgruen. Pricing and replenishment strategies in a distribution system with competing retailers. *Operations Research*, 51(3):409–426, 2003.

[16] J. Bertrand. Review of walrass theorie mathematique de la richesse social and cournots recherches sur les principes mathematiques de la theorie des richesses. *Journal des Savants*, September:499–508, 1883.

[17] J. Bessen. Riding the marketing information wave. *Harvard Business Review*, 71(5):150–160, 1993.

[18] S. Biller, L.M.A. Chan, D. Simchi-Levi, and J. Swann. Dynamic pricing and the direct-to-customer model in the automotive industry. *To appear in E-Commerce Journal, Special Issue on Dynamic Pricing*, 2002.

[19] J. R. Birge, J. Drogosz, and I. Duenyas. Setting single-period optimal capacity levels and prices for substitutable products. *International Journal of Flexible Manufacturing Systems*, 10(4):407–430, 1998.

[20] G. Bitran, R. Caldentey, and S. Mondschein. Coordinating clearance markdown sales of seasonal products in retail chains. *Operations Research*, 46(5):609–624, 1998.

[21] G. R. Bitran and S. V. Mondschein. Periodic pricing of seasonal products in retailing. *Management Science*, 43(1):64–79, 1997.

[22] A. S. Blinder. Inventories and sticky prices: More on the microfoundations of macroeconomics. *American Economic Review*, 72(3):334–348, 1982.

[23] T. Boyaci and G. Gallego. Coordinating pricing and inventory replenishment policies for one wholesaler and one or more geographically dispersed retailers. *Interntional Journal of Production Economics*, 77:95–111, 2002.

[24] A.C. Brahmbhatt and M.C. Jaiswal. An order-level-lot-size model with uniform replenishment rate and variable mark-up of prices. *International Journal of Production Research*, 18(5):655–664, 1980.

[25] E. Brynjolfsson and M. Smith. Frictionless commerce? a comparison of internet and conventional retailers. Working Paper, MIT, 1999.

[26] A. N. Burnetas and C. E. Smith. Adaptive ordering and pricing for perishable products. *Operations Research*, 48(3):436–443, 2000.

[27] G. P. Cachon. Supply chain coordination with contracts. In S. Graves and T. de Kok, editors, *forthcoming in Handbooks in Operations Research and Management Science: Supply Chain Management*. North-Holland, 2002.

[28] S. C. Carr, I. Duenyas, and W. Lovejoy. The harmful effects of capacity improvements. Working Paper, UCLA, 1999.

[29] A. Chakravarty and G. E. Martin. Discount pricing policies for inventories subject to declining demand. *Naval Research Logistics*, 36:89–102, 1989.

[30] L. M. A. Chan, D. Simchi-Levi, and J. L. Swann. Dynamic pricing models for manufacturing with stochastic demand and discretionary sales. *Working Paper*, 2002.

[31] C. K. Chen and K. J. Min. An analysis of optimal inventory and pricing policies under linear demand. *Asia-Pacific Journal of Operational Research*, 11(2):117–129, 1994.

[32] F. Chen. Market segmentation, advanced demand information, and supply chain performance. *Manufacturing and Service Operations Management*, 3(1):53–67, 2001.

[33] X. Chen and D. Simchi-Levi. Coordinating inventory control and pricing strategies with random demand and fixed ordering cost: The finite horizon case. Working Paper, MIT, 2002a.

[34] X. Chen and D. Simchi-Levi. Coordinating inventory control and pricing strategies with random demand and fixed ordering cost: The infinite horizon case. Working Paper, MIT, 2002b.

[35] X. Chen and D. Simchi-Levi. Coordinating inventory control and pricing strategies with random demand and fixed ordering cost: the continuous review model. Working Paper, MIT, 2003.

[36] F. Cheng and S. P. Sethi. A periodic review inventory model with demand influenced by promotion decisions. *Management Science*, 45(11):1510–1523, 1999.

[37] T. C. E. Cheng. An eoq model with pricing consideration. *Computers and Industrial Engineering*, 18(4):529–534, 1990.

[38] E. K. Clemons, I. Hann, and L. M. Hitt. The nature of competition in electronic markets: An empirical investigation of online travel agent offerings. Working Paper, The Wharton School of the University of Pennsylvania, 1998.

[39] M. A. Cohen. Joint pricing and ordering policy for exponentially decaying inventory w/ known demand. *Naval Research Logistics Quarterly*, 24:257–268, 1977.

[40] A. A. Cournot. *Researches into the Mathematical Principles of the Theory of Wealth*. Irwin, 1963 edition, 1883.

[41] J. D. Dana and N. C. Petruzzi. Note: The newsvendor model with endogenous demand. *Management Science*, 47(11):1488–1497, 2001.

[42] T. K. Datta and K. Paul. An inventory system with stock-dependent, price-sensitive demand rate. *Production Planning and Control*, 12(1):13–20, 2001.

[43] R. Dewan, B. Jing, and A. Seidmann. Adoption of internet-based product customization and pricing strategies. *Journal of Management Information Systems*, 17(2):9–28, 2000.

[44] S. Dewan and H. Mendelson. User delay costs and internal pricing for a service facility. *Management Science*, 36:1502–1517, 1990.

[45] F. F. Easton and D. R. Moodie. Pricing and lead time decisions for make-to-order firms with contingent orders. *European Journal of Operational Research*, 116(2):305–318, 1999.

[46] M. ElHafsi. An operational decision model for lead-time and price quotation in congested manufacturing systems. *European Journal of Operational Research*, 126(2):355–370, 2000.

[47] M. Elhafsi and E. Rolland. Negotiating price delivery date in a stochastic manufacturing environment. *Iie Transactions*, 31(3):255–270, 1999.

[48] J. Eliashberg and R. Steinberg. Marketing-production joint decision making. In J. Eliashberg and J.D. Lilien, editors, *Management Science in Marketing, Handbooks in Operations Research and Management Science*. Elsevier Science, North Holland, 1991.

[49] W. Elmaghraby, A. Gulcu, and P. Keskinocak. Optimal markdown pricing in the presence of rational customers. Working Paper, Georgia Institute of Technology, 2002.

[50] W. Elmaghraby and P. Keskinocak. Dynamic pricing: Research overview, current practices and future directions. Working Paper, Georgia Tech, 2002.

[51] A. Federgruen and A. Heching. Combined pricing and inventory control under uncertainty. *Operations Research*, 47(3):454–475, 1999.

[52] A. Federgruen and A. Heching. Multilocation combined pricing and inventory control. *Manufacturing and Service Operations Management*, 4(4):275–295, 2002.

[53] G Feichtinger and R. Hartl. Optimal pricing and production in an inventory model. *Eur. J of Operational Research*, 19(1):45–56, 1985.

[54] Y. Y. Feng and G. Gallego. Optimal starting times for end-of-season sales and optimal stopping-times for promotional fares. *Management Science*, 41(8):1371–1391, 1995.

[55] Y. Y. Feng and G. Gallego. Perishable asset revenue management with markovian time dependent demand intensities. *Management Science*, 46(7):941–956, 2000.

[56] Y. Y. Feng and B. C. Xiao. Maximizing revenues of perishable assets with a risk factor. *Operations Research*, 47(2):337–341, 1999.

[57] Y. Y. Feng and B. C. Xiao. A continuous-time yield management model with multiple prices and reversible price changes. *Management Science*, 46(5):644–657, 2000.

[58] Y. Y. Feng and B. C. Xiao. Optimal policies of yield management with multiple predetermined prices. *Operations Research*, 48(2):332–343, 2000.

[59] Y. Y. Feng and B. C. Xiao. Revenue management with two market segments and reserved capacity for priority customers. *Advances in Applied Probability*, 32(3):800–823, 2000.

[60] C. Fracassini. Amazon 'hikes prices' for loyal shoppers. *Scotland on Sunday*, Oct 1 2000.

[61] C. Gaimon. Simultaneous and dynamic price, production, inventory and capacity decisions. *Eur J of Operations Research*, 35(3):426–441, 1988.

[62] G. Gallego and G. Van Ryzin. Optimal dynamic pricing of inventories with stochastic demand over finite horizons. *Management Science*, 40(8):999–1020, 1994.

[63] G. Gallego and G. Van Ryzin. A multiproduct dynamic pricing problem and its applications to network yield management. *Operations Research*, 45(1):24–41, 1997.

[64] S. M. Gilbert. Coordination of pricing and multi-period production for constant priced goods. *European Journal of Operational Research*, 114(2):330–337, 1999.

[65] S. M. Gilbert. Coordination of pricing and multiple-period production across multiple constant priced goods. *Management Science*, 46(12):1602–1616, 2000.

[66] A. Y. Ha. Incentive-compatible pricing for a service facility with joint production and congestion externalities. *Management Science*, 44(12):1623–1636, 1998.

[67] A. Y. Ha. Optimal pricing that coordinates queues with customer-chosen service requirements. *Management Science*, 47(7):915–930, 2001.

[68] G. A. Hay. Production, price and inventory theory. *The American Economic Review*, 60(4):531–536, 1970.

[69] http://www.emarket.com.

[70] H. Hwang and S. W. Shinn. Retailer's pricing and lot sizing policy for exponentially deteriorating products under the condition of permissible delay in payments. *Computers and Operations Research*, 24(6):539–547, 1997.

[71] A. Jeuland and S. Shugan. Managing channel profits. *Marketing Science*, 2:239–272, 1983.

[72] S. Jorgensen, P. M. Kort, and G. Zaccour. Production, inventory, and pricing under cost and demand learning effects. *Eur J of Operational Research*, 117:382–395, 1999.

[73] S. Kachani and G. Perakis. A fluid model of dynamic pricing and inventory management for make-to-stock manufacturing systems. Working Paper, MIT, 2002.

[74] M. Karakul and L.M.A. Chan. Optimal introduction of substitutable products: Combined pricing and ordering decisions. Working Paper, University of Toronto, 2003a.

[75] M. Karakul and L.M.A. Chan. Newsboy problem with pricing of a monopolist with two customer categories. Working Paper, University of Toronto, 2003b.

[76] S. Karlin and C.R. Carr. Prices and optimal inventory policies. In S. Karlin K. J. Arrow and H. Scarf, editors, *Studies in Applied Probability and Management Science*. Stanford University Press, Stanford, CA, 1962.

[77] P. Keskinocak, R. Ravi, and S. Tayur. Scheduling and reliable lead-time quotation for orders with availability intervals and lead-time sensitive revenues. *Management Science*, 47(2):264–279, 2001.

[78] P. Keskinocak and S. Tayur. Due date management policies. In *Supply Chain Analysis in the eBusiness Era*. Kluwer, 2002.

[79] M. J. Khouja. Optimal ordering, discounting, and pricing in the single-period problem. *International Journal of Production Economics*, 65(2):201–216, 2000.

[80] D. Kim and W. J. Lee. Optimal coordination strategies for production and marketing decisions. *Operations Research Letters*, 22:41–47, 1998.

[81] W.M. Kincaid and D. Darling. An inventory pricing problem. *J of Math. Analysis and Applications*, 7:183–208, 1963.

[82] A. Kirman and M. Sobel. Dynamic oligopoly with inventories. *Econometrica*, 42(2):279–287, 1974.

[83] P. Krugman. What price fairness? *New York Times*, Oct 4 2000.

[84] K. R. Kumar, A. P. S. Loomba, and G. C. Hadjinicola. Marketing-production coordination in channels of distribution. *European Journal of Operational Research*, 126(1):189–217, 2000.

[85] H. Kunreuther and J.F. Richard. Optimal pricing and inventory decisions for non-seasonal items. *Econometrica*, 39(1):173–175, 1971.

[86] H. Kunreuther and Schrage. Joint pricing and inventory decisions for constant priced items. *Management Science*, 19(7):732–738, 1973.

[87] S. Ladany and A. Sternlieb. The interaction of economic ordering quantities and marketing policies. *AIIE Transactions*, 6:35–40, 1974.

[88] K. S. Lai. Price smoothing under capacity constraints. *Southern Economic Journal*, 57(1):150–159, 1990.

[89] R. Lal and R. Staelin. An approach for developing an optimal discount pricing policy. *Management Science*, 30:1524–1539, 1984.

[90] A. H. L. Lau and H. S. Lau. The newsboy problem with price-dependent demand distribution. *Iie Transactions*, 20(2):168–175, 1988.

[91] E. Lazear. Retail pricing and clearance sales. *The American Economic Review*, 76(1):14–32, 1986.

[92] P. J. Lederer and L. D. Li. Pricing, production, scheduling, and delivery-time competition. *Operations Research*, 45(3):407–420, 1997.

[93] W. J. Lee. Determining order quantity and selling price by geometric programming: Optimal solution, bounds and sensitivity. *Decision Sciences*, 24(76-87), 1993.

[94] W. J. Lee and D. Kim. Optimal and heuristic decision strategies for integrated production and marketing planning. *Decision Sciences*, 24(6):1203–1213, 1993.

[95] L. Li. The role of inventory in delivery-time competition. *Management Science*, 38(2):182–197, 1992.

[96] L. Li and S. Whang. Game theory models in operations management and information systems. In K. Chatterjee and W. Samuelson, editors, *Game Theory and Business Applications*. Kluwer Academic Publishers, Boston, 2001.

[97] L. D. Li. A stochastic-theory of the firm. *Mathematics of Operations Research*, 13(3):447–466, 1988.

[98] L. D. Li and Y. S. Lee. Pricing and delivery-time performance in a competitive environment. *Management Science*, 40(5):633–646, 1994.

[99] Y. G. Liang. Solution to the continuous time dynamic yield management model. *Transportation Science*, 33:117–123, 1999.

[100] C. Loch. Pricing in markets sensitive to delay. Working Paper, Stanford University, 1991.

[101] L. J. Maccini. The interrelationship between price and output decisions and investments. *Journal of Monetary Economics*, pages 41–65, 1984.

[102] J. I. McGill and G. J. van Ryzin. Revenue management: Research overview and prospects. *Transportation Science*, 33(2):233–256, 1999.

[103] G. McWilliams. Lean machine: How dell fine-tunes its pc pricing. *Wall Street Journal*, June 8 2001.

[104] H. Mendelson. Pricing computer services: queueing effects. *Comm. ACM*, 28:312–321, 1985.

[105] H. Mendelson and S. Whang. Optimal incentive-compatible priority pricing for the m/m/1 queue. *Operations Research*, 38:870–883, 1990.

[106] R. A. Meyer. Risk-efficient monopoly pricing for multiproduct firms. *Quarterly Journal of Economics*, 90(3):461–474, 1976.

[107] E. S. Mills. Uncertainty and price theory. *Quarterly J. of Economics*, 73:117–130, 1959.

[108] L. O. Morgan, R. L. Daniels, and P. Kouvelis. Marketing/manufacturing trade-offs in product line management. *Iie Transactions*, 33(11):949–962, 2001.

[109] S. A. Neslin, S. G. Powell, and L. S. Stone. The effects of retailer and consumer response on optimal manufacturer advertising and trade promotion strategies. *Management Science*, 41(5):749–766, 1995.

[110] S. Netessine. Dynamic pricing of inventory/capacity when price changes are costly. Working Paper, The Wharton School, University of Pennsylvania, 2002.

[111] A. J. Nevins. Some effects of uncertainty: Simulation of a model of price. *Quarterly J. of Economics*, 80(1):73–87, 1966.

[112] K. Palaka, S. Erlebacher, and D. H. Kropp. Lead-time setting, capacity utilization, and pricing decisions under lead-time dependent demand. *Iie Transactions*, 30(2):151–163, 1998.

[113] M. Parlar and Z. K. Weng. Vertical coordination of pricing and production decisions to hedge against horizontal price competition. Working Paper, University of Wisconsin, 2000.

[114] B. P. Pashigian. Demand uncertainty and sales: A study of fashion and markdown pricing. *The American Economic Review*, 78(5):936–953, 1988.

[115] D. Pekelman. Simultaneous price-production decisions. *Operations Research*, 22:788–794, 1974.

[116] N. C. Petruzzi and M. Dada. Pricing and the newsvendor problem: A review with extensions. *Operations Research*, 47(2):183–194, 1999.

[117] N. C. Petruzzi and M. Dada. Information and inventory recourse for a two-market, price-setting retailer. *Manufacturing and Service Operations Management*, 3(3):242–263, 2001.

[118] N. C. Petruzzi and M. Dada. Dynamic pricing and inventory control with learning. *Naval Research Logistics*, 49(3):303–325, 2002.

[119] L. Phlips. Intertemporal price discrimination and sticky prices. *Quarterly J. of Economics*, 94(3):525–542, 1980.

[120] E. Plambeck. Optimal incentive compatible control of a queue with some patient customers. Working Paper, Stanford University, 2002.

[121] H. Polatoglu and I. Sahin. Optimal procurement policies under price-dependent demand. *International Journal of Production Economics*, 65(2):141–171, 2000.

[122] L. H. Polatoglu. Optimal order quantity and pricing decisions in single-period inventory systems. *International Journal of Production Economics*, 23(1-3):175–185, 1991.

[123] E. L. Porteus. Investing in reduced setups in the eoq model. *Management Science*, 31:998–1010, 1985.

[124] E. L. Porteus. Stochastic inventory theory. In D. P. Heyman and M. J. Sobel, editors, *Handbooks in Operations Research and Management Science*, volume 2, pages 605–652. Elsevier, North Holland, 1990.

[125] A. Rajan, Rakesh, and R. Steinberg. Dynamic pricing and ordering decisions by a monopolist. *Management Science*, 38(2):240–262, 1992.

[126] V. R. Rao. Pricing research in marketing: The state of the art. *J of Business*, 57(1):S39–S64, 1984.

[127] J. Rayport and J. Sviokla. Managing in the marketspace. *Harvard Business Review*, 72(6):141–150, 1994.

[128] G. Raz and E. Porteus. A discrete service levels perspective to the newsvendor model with simultaneous pricing. *Working Paper, Stanford University*, 2001.

[129] P. B. Reagan. Inventory and price behavior. *Review of Economic Studies*, 49(1):137–142, 1982.

[130] D. Reitman. Endogenous quality differentiation in congested markets. *Journal of Industrial Economics*, 39:621–647, 1991.

[131] D. Rosenberg. Optimal price-inventory decisions - profit vs roii. *Iie Transactions*, 23(1):17–22, 1991.

[132] R. Roundy. 98% effective integer-ratio lot-sizing for one warehouse multi-retailer systems. *Management Science*, 31:1416–1430, 1985.

[133] E. Sankarasubramanyam and S. Kumaraswamy. Eoq formula under varying marketing policies and conditions. *AIIE Transactions*, 13(4):312–314, 1981.

[134] S. P. Sethi and Q. Zhang. Multilevel hierarchical decision making in stochastic marketing-production systems. *SIAM J. Control Optim.*, 33(1):528–553, 1995.

[135] M. D. Smith, J. Bailey, and E. Brynjolfsson. Understanding digital markets. In E. Brynjolfsson and B. Kahin, editors, *Understanding the Digital Economy*. MIT Press, 1999.

[136] S. A. Smith and D. D. Achabal. Clearance pricing and inventory policies for retail chains. *Management Science*, 44(3):285–300, 1998.

[137] S. A. Smith, N. Agrawal, and S. H. McIntyre. A discrete optimization model for seasonal merchandise planning. *Journal of Retailing*, 74(2):193–221, 1998.

[138] K. C. So. Price and time competition for service delivery. *Manufacturing and Service Operations Management*, 2(4):392–409, 2000.

[139] K. C. So and J. S. Song. Price, delivery time guarantees and capacity selection. *European Journal of Operational Research*, 111(1):28–49, 1998.

[140] A. G. Sogomonian and C. S. Tang. A modeling framework for coordinating promotion and production decisions within a firm. *Management Science*, 39(2):191–203, 1993.

[141] staff. Amazon.com varies prices of identical items for test. *The Wall Street Journal*, Sept 7 2000.

[142] staff. On the web, price tags blur; what you pay could depend on who you are. *The Washington Post*, Sept 27 2000.

[143] staff. Keeping up with competitors and increased customer haggling driving price setting in today's economy, report nation's pricers. Technical report, Professional Pricing Society, 2002.

[144] S. Stidham. Optimal control of admission to a queueing system. *IEEE Trans. Automatic Control*, AC-30:705–713, 1985.

[145] S. Subrahmanyan and R. Shoemaker. Developing optimal pricing and inventory policies for retailers who face uncertain demand. *Journal of Retailing*, 72(1):7–30, 1996.

[146] Y. Tanaka. Profitability of price and quantity strategies in an oligopoly. *Journal of Mathematical Economics*, 35(3):409–418, 2001.

[147] S. Tayur, R. Ganeshan, and M. Magazine, editors. *Quantitative Models for Supply Chain Management*. Kluwer Academic Publishers, 1998.

[148] L. J. Thomas. Price-production decisions with deterministic demand. *Management Science*, 16(11):747–750, 1970.

[149] L. J. Thomas. Price and production decisions with random demand. *Operations Research*, 22:513–518, 1974.

[150] G. L. Thompson, S. P. Sethi, and Teng. J. T. Strong planning and forecast horizons for a model with simultaneous price and production decisions. *Eur J of Operational Research*, 16:378–388, 1984.

[151] G. T. Thowsen. A dynamic, nonstationary inventory problem for a price/quantity setting firm. *Naval Research Logistics*, 22:461–476, 1975.

[152] T. L. Urban and R. C. Baker. Optimal ordering and pricing policies in a single-period environment with multivariate demand and markdowns. *European Journal of Operational Research*, 103(3):573–583, 1997.

[153] J. A. Van Mieghem. Price and service discrimination in queuing systems: Incentive compatibility of gc mu scheduling. *Management Science*, 46(9):1249–1267, 2000.

[154] J. A. Van Mieghem and M. Dada. Price versus production postponement: Capacity and competition. *Management Science*, 45(12):1631–1649, 1999.

[155] M. Vizard, E Scannell, and D. Neel. Suppliers toy with dynamic pricing. *InfoWorld*, May 11, 2001.

[156] H. M. Wagner. A postscript to "dynamic problems in the theory of the firm". *Naval Research Logistics*, 7(1):7–13, 1965.

[157] H. M. Wagner and T. M. Whitin. Dynamic problems in the theory of the firm. *Naval Research Logistics*, 5(1):53–74, 1958.

[158] Z. K. Weng and M. Parlar. Price-incentive and stocking decisions for seasonal products: A dynamic programming model. Working Paper, University of Wisconsin, 2000.

[159] T. M. Whitin. Inventory control and price theory. *Management Science*, 2(1):61–68, 1955.

[160] C. Yano and G. Dobson. Profit-optimizing product line design, selection and pricing with manufacturing cost considerations. In T.H. Ho and C. S. Tang, editors, *Product Variety Management: Research Advances*, pages 145–176. Kluwer, Norwell, MA, 1998.

[161] E. Zabel. Multi-period monopoly under uncertainty. *Economic Theory*, 5:524–536, 1972.

[162] W. Zhao and Y. S. Zheng. Optimal dynamic pricing for perishable assets with nonhomogeneous demand. *Management Science*, 46(3):375–388, 2000.

[163] K. Zhu and U. W. Thonemann. Coordination of pricing and inventory control across products. Working Paper, Stanford University, 2002.

[164] P. Zipkin. *Foundations of Inventory Management*. McGraw-Hill Companies, Inc., 2000.

David Simchi-Levi, S. David Wu, and Z. Max Shen (Eds.)
Handbook of Quantitative Supply Chain Analysis:
Modeling in the E-Business Era
©2004 Kluwer Academic Publishers

Chapter 10

COLLABORATIVE FORECASTING AND ITS IMPACT ON SUPPLY CHAIN PERFORMANCE

Yossi Aviv
Washington University
St. Louis, MO 63141
aviv@olin.wustl.edu

Introduction

In recent years, electronic data sharing has become an industry standard. Starting as a tool to cut costs through elimination of expensive paperwork and reduction in human errors, *electronic data interchange* (EDI) evolved during the last two decades into an enabler for the creation of new supply chain business models. For example, managers realized that by cutting lead times they can bring their products to the market faster and thus save much in inventory costs and at the same time reduce lost sales. This gave rise to important practices such as *quick response (QR)* (Hammond 1990, Frazier 1986) and *Efficient Consumer Response (ECR)* (Kurt Salomon Associates 1993). Visibility turned quickly into a popular word in the supply chain manager's vocabulary: For example, by providing suppliers with access to inventory and *Point-of-Sale (POS)* data, companies such as Wal-Mart gained a substantial increase in the level of product availability due to the ability of their vendors to better plan and execute their logistics processes – hence, replacing costly inventory by information. In recent years, rapid evolution in electronic commerce and information technology has paved the way for the implementation of a variety of coordination mechanisms in supply chains, and in many cases it has provided the necessary means for supply chain partners to restructure the way they conduct business (Andersen Consulting 1998, Magretta 1998). Visibility through EDI is, for instance, at the backbone of advanced collaborative manufacturing and e-trade concepts (i2 TradeMatrix 2001).

Consider *Vendor-Managed Inventory (VMI)*, a supply chain partnership that has gained popularity in industry since early 1980's. In a VMI program, the supplier takes the responsibility for replenishing the customer's inventory. In some cases, VMI was adopted when retailers (the customers) lacked the technological infrastructure for managing complex forecasting-integrated inventory systems. Vendors that had good decision support systems in place, and were intimately familiar with the characteristics of the demand for their product could bring substantial value to their customers, by better matching supply and demand. But VMI was also adopted to enable coordination among some key decision processes in the supply chain – e.g., inventory replenishment, transportation scheduling and production capacity allocation. A vendor that proactively manages the inventory downstream the supply chain, can coordinate his actions with other logistics decisions, taking into account his production, transportation, and warehousing constraints. In order to enable the supplier to perform the inventory management task under a VMI arrangement, visibility into the retailer's system is necessary and is usually provided through the share of demand forecasts and inventory status data.

But how successful have VMI programs been? Recall that the underlying logic of their value proposition, is that VMI enables the decision-maker (the vendor) to take a system-wide view of the entire chain's inventory process. As a result, the retailrs gain an improvement in stock availability without the need to hold excessive amount of inventory. The vendors benefit due to improved certainty about their overall resource requirements, since they no longer have to passively respond to retailers' orders. Nevertheless, in the statement of this value proposition, one can easily see some of the potential problems VMI may encounter, which in fact were reasons for many failures in practice. For instance, what happens if vendors take advantage of retailers by over-stocking their distribution centers (DC's) with goods? How could the trading partners ensure that the overall potential benefits that VMI brings to their supply chain, are shared appropriately among them? Clearly, the vendor and the retailer have to develop a process with shared goals and design the right metrics to track the performance of VMI. For instance, it is common in certain settings to establish a consignment agreement in which the retailer pays the vendor for the product only upon usage, in order to reduce the incentive for the supplier to unnecessarily push inventory downstream (this works particularly well if the vendor is committed to remove obsolete or stale products from the DC at his own expense). Another way to better align incentives in the supply chain is through constraints in the contractual agreements. For example, in some VMI programs I was invloved in, lower and upper bounds were set for the safety stocks held at the DC. Hence, if a retailer anticipates the demand to be about 100 units during a given period of time, the vendor has to ensure that this number of units, plus an additional safety stock will be available at the DC to serve the actual demand during that period. The vendor promises that the level of safety stock will be somewhere in a pre-specified range that depends on the level of forecast accuracy.

A second possible problem with VMI programs exists when they are limited to replenishments at the DC level only, with the retailers managing the inventory allocation from their DCs to their retail outlets. Indeed, VMI was generally structured this way during the early stages of its evolution. Letting the vendor manage only the vendor-to-DC part of the inventory process bears the risk of creating a broken link in the supply chain, that might prevent the partners from taking a full advantage of the potential benefits of VMI. The extent to which such problem is serious depends on how well the vendor-to-DC and DC-to-store replenishment processes are synchronized. Without supporting technology in place, product availability at the DC does not translate well to sales to end consumers.

A third problem with VMI, which is more relevant to this chapter, has to do with the ability of the decision-makers to forecast demand and integrate it into their inventory planning and control processes. Sometimes, the data provided by the retailer to the vendor is only partial and hence does not transfer all relevant information about future demand. For example, the retailer may have important information about an upcoming promotion, which is not conveyed via the history of sales or through inventory status data. After all, in the food industry, knowledge about an upcoming Salsa promotion should be taken into account by the corn chip supplier. If this kind of information is absent, it results in a vendor making replenishment decisions based on partial information. The lack of complete information may explain why a large number of VMI programs have failed because of poor supplier forecasting performance (see examples in Aviv 2002a). It is instructive to note that there are various possible reasons for transferring only partial information to the vendor. Naively ignoring the value of sharing full information is one reason, but there is also the inability of retailers to share information due to non-disclosure agreements (if data may reveal, e.g., promotion or product introduction plans of a competing firm). Additionally, lack of IT support for information sharing is yet another reason for not sharing full information between trading partners (see additional comments below).

The Voluntary Interindustry Commerce Standard (VICS) Association established the *Collaborative Planning, Forecasting and Replenishment (CPFR)* subcommittee in 1993, with the aim of developing standards for collaborative processes for the industry (Koloczyc 1998, Aviv 2001). Various types of CPFR partnerships have been experimented with in industry. In mid 1990's, Wal-Mart and Warner-Lambert embarked on the first CPFR pilot, involving Listerine products. In their pilot, Wal-Mart and Warner-Lambert used special CPFR software to exchange forecasts. Supportive data, such as past sales trends, promotion plans, and even the weather forecasts, were often transferred in an iterative fashion to allow them to converge on a single forecast in case their original forecasts differed. This pilot was very successful, resulting in an increase in Listerine sales and better fill rates, accompanied by a reduction of inventory investment. On a pilot involving 22 products, Nabisco and Wegmans reported benefits of 54% growth in sales, and 18% reduction in days of supply. Other examples of CPFR pilots include Sara Lee's Hanes and Wal-Mart, involving 50 stock-keeping units of underwear products supplied to almost 2,500 Wal-Mart's stores (see Hye 1998). Heineken USA employed CPFR to cut its order-cycle time in half in 1996, and provided Collaborative Planning and Replenishment software to its top 100 distributors (Beverage Industry 1998). Procter and Gamble experimented with several CPFR pilots, and Levi Strauss & Co. incorporated certain aspects of the CPFR business process into their retail replenishment service (e.g., by creating joint business plans, and identifying exceptions to the sales and/or order forecasts).

CPFR improves VMI in several fronts. First, it calls for standardization of processes, so that relationships between vendors and their customers do not over-rely on specific, highly-proprietary retailer's applications. For example, in the early 1990's, the Bose company established a type of VMI relationship, entitled *Just-in-Time II* (JIT-II), having representatives of their vendors stationed at the Bose facility, and placing orders for Bose (see HBS Case 9-694-001). More than 10 vendors were selected for this program, in categories such as plastics, trucking, corrugated packaging, office supplies, and printed materials. Each vendor representative acted as a "VMI agent," although JIT-II benefited also from the representative's assistance in resolving problems and participation in design and engineering processes (i.e., representatives' duties were not limited to inventory replenishment). Bose's JIT-II model was proven successful when the volume of transactions between the companies were sufficiently high, and when involvement of a person in collaborative design, resource planning, and problem resolution was vital to the supply chain. Today, the business-to-business standards and technology has been matured substantially, with CPFR offering business process guidelines that are scalable across multiple trading partners (allowing the shift to multiple-tier supply channels). This progress in technology and process definition enables companies to participate in multiple CPFR programs without incurring high customer-specific setup costs (which, as a consequence, alleviates concerns about high switching costs).

On a second front, CPFR calls for a much larger degree of information sharing than traditional VMI programs. The primary source for forecast development in a VMI framework is warehouse withdrawals from the retail distribution center. In some VMI programs the vendor receives in addition POS data, to enable a clearer understanding of the underlying consumer demand process. But this is not enough! The collaborative forecasting (CF) part of CPFR is a process through which members of the supply chain jointly develop forecasts of future demand, so that they all have the same reference to base their decisions on. By sharing the knowledge and utilizing the unique expertise of the trading partners, CF has the potential of producing forecasts that are substantially more accurate than those each party can develop individually. By recognizing that a good forecasting system is perhaps the key enabler in matching supply with demand, partners in the supply chain need to understand what factors affect the demand process, and how the demand correlates with various types of market information. After identifying the relevant information sources, they need to address the problem of making them available to the decision-makers, in order to assist them in anticipating future demand as perfectly as possible. For example, if weather conditions affect demand for a particular type of medicine, this information has to be calculated in the manufacturer's production plans, but also transferred to the managers responsible for determining the inventory at the store-levels. Similarly, if a competitor that sells a similar type of medicine has offered discounts for his product, a drugstore chain may want to provide this information to their vendor, so that the vendor can take this into account when setting his manufacturing resource plans. Clearly, one has to be aware of the fact that sharing proprietary information is generally not simple, and in fact is a big concern in CPFR implementation. Yet, in order to extract benefits from such restricted information, it is not necessarily true that the retailer (or the vendor) has to pass very detailed data to the vendor (or retailer, respectively). In other words, if the retailer knows about an upcoming product introduction that may have a negative impact on the demand for the product sold by the CPFR vendor, the retailer may simply reduce the forecasts of future demands, without making a specific reference to the new product arrival.

Today, after many pilot programs have resulted in promising success stories, CPFR has undoubtedly engendered a great deal of interest in the supply chain management world. This is clearly evidenced through the increasing number of references to this subject at professional conferences and in publications targeting practitioners. Anderson and Lee (2001) discuss the rapid emergence of supply chain collaboration, and its role in creating new business models – *the organization's core logic for creating value in a sustainable way.* Variety of collaborative cross-company processes come to mind, including customer-driven design, collaborative design, collaborative manufacturing, as well as our topic of discussion – collaborative forecasting and replenishment.

The objectives of this chapter

After following the evolution of CPFR in recent years and conducting academic research on the forecasting aspects of CPFR, we identified a need to review a set of modeling approaches available for the study of collaborative forecasting programs. We have chosen to take an academic, management-science approach, by summarizing stylized frameworks designed to capture essential parameters of collaborative forecasting processes. Therefore, our targeted audience are researchers as well as pratitioners involved in system design and analyses. Naturally, this chapter is not designed to discuss the very specific details of forecasting or inventory models. This is impossible to do in a single chapter due to the vast amount of literature on these subjects. Rather, we attempt to provide a flavor of some analytical methods that exist in the literature.

We note that this chapter does not attempt to provide an overall coverage of CPFR. Important issues such as technology choice and change management are excluded from our discussion.

Preliminary literature review

Aviv (2001) was the first to treat the topic of collaborative forecasting in the management science literature. Its purpose was threefold: (i) to propose a supply chain model that can potentially serve as a building block for quantifying the inventory and service performance of supply chains in settings where forecasts are dynamically updated at more than one location in the system; (ii) to obtain insights as to how key system parameters affect the performance and the benefits of collaborative forecasting; and (iii) to further the area of CPFR for academic discussion and consideration. This paper argues that collaborative forecasting initiatives can provide substantial benefits to the supply chain, but the magnitude of these benefits depends on the specific setting: Collaborative forecasting practices are particularly beneficial when forecasting capabilities are diversified, or in other words, when the trading partners can bring something unique to the table. Collaborative forecasting programs and two other efficient consumer response initiatives appear to be complementary, in the sense that the benefits of collaborative forecasting increase when both or either of them are implemented. These types of initiatives are Quick Response (QR), and advanced demand information. Aviv (2001) also discusses a so-called *composite uncertainty measure* that enables a quick and tractable assessment of the supply chain's inventory performance. This single measure reflects the accuracies of the forecasts of lead-time demands, generated at *both* levels of the supply chain. Nevertheless, §6 in this chapter which was developed later by Aviv (2002ab) refines the treatment of cost assessment, and hence we refer the reader to this more recent analysis.

In Aviv (2002a), we extend the model of Aviv (2001) by specifically considering questions related to VMI and CPFR, and by using an auto-correlated time series framework for the underlying demand process (we shall provide some motivation for this later in this chapter). We examine three types of two-stage supply chain configurations, each consisting of a single retailer and a single supplier. In the first setting, the retailer and the supplier coordinate their policy parameters in an attempt to minimize system-wide costs, but they do not implement a collaborative forecasting process. In the second setting, that resembles many original VMI programs, the supplier takes the full responsibility of managing the supply chain's inventory, but the retailer does not share all of his knowledge about future demand on an on-going basis. The third setting, reminiscent of CPFR partnerships, is one in which inventory is managed centrally, and all demand related information is shared. Through numerical examples, the paper compares between VMI and CFAR programs in settings with different levels of inter-temporal correlation in the demand process. We argue that the best choice of business relationship (i.e., VMI, CPFR, etc.) greatly depends on the interaction between the explanatory power of the supply chain members. In other words, the benefits of CPFR depend on the information resources available to the supply chain partners, and how unique they are in enabling the explanation of future demand. In this paper, we utilized the time-series framework for supply chain inventory management developed in Aviv (2002b). The latter is in fact a focal point of this chapter.

The main objective of Aviv (2002b) is to propose a *methodology* for studying inventory systems in which demands evolve according to a relatively general class of statistical time-series patterns, and to propose building blocks for the construction of adaptive inventory policies for such settings. In addition, the paper provides a framework for assessing the impact of various information sharing mechanisms on supply chain performance. Our work in Aviv (2002b) is hence motivated by the following requirements faced by many supply chain inventory managers in various industrial contexts. First, inventory planning tools should handle a sufficiently *general* family of demand models. This is because inventory systems need to manage a wide array of stock keeping units (SKUs), sometimes numbering in the thousands, with each SKU having its own type of time series demand pattern. Particularly, we need to consider demand processes that are auto-correlated (Kahn 1987, Lee et al. 2000), avoiding the traditional assumption that demands over consecutive time periods are statistically independent. Second, demand models should capture dependencies between demand realizations and other types of explanatory variables that can also evolve as correlated time series. This need is particularly important in systems where data such as advanced customer orders, promotion plans, and expected weather conditions, are collected and revised frequently. Finally, practitioners and researchers are faced today with the need to design and analyze information sharing mechanisms in supply chains. As a consequence, a third requirement of analysis and planning tools is that they can accommodate cases in which information may not be commonly known across the supply chain. In addition, they need to provide the ability to assess the potential benefits that can be gained by the supply chain members if they share part or all of the data available to them with each other.

In recent years, researchers have suggested several stylized models to study the problem of inventory control when demand information is not commonly known throughout the supply chain. Chen et al. (2000) propose a two-stage supply chain model in which neither the retailer nor the supplier know the exact form of the demand process, which is modeled as an auto-regressive process of order 1. They assume that at the beginning of each ordering period, the retailer uses a simple moving average procedure to generate two values (mimicking a mean and standard deviation), based on which orders are determined. Chen et al. then investigate the magnitude of the so-called *"bullwhip phenomenon"*, or the level by which the variability of order quantities is higher than the variability of the actual demand. Watson and Zheng (2000) analyze a supply chain facing a serially correlated demand model. Here, again, the supply chain partners do not know the exact characteristics of the demand process. The authors claim that inventory managers tend to over-react to demand changes, and show that this in turn leads to sub-optimal replenishment policies internally and causes an undue bullwhip effect externally. Croson and Donohue (2001) conduct an experimental study, based on the popular Beer Distribution Game, to uncover behavioral factors that impact a company's ability to manage its supply chain. They measure how the institutional designs of supply chains impact the magnitude and timing of the bullwhip effect in a multi-echelon system. For other works investigating the impact of forecasting in cases where the underlying characteristics of the demand distribution are unknown to the decision makers, see, e.g., Miyaoka and Kopczak (2000) and Chen, Ryan and Simchi-Levi (2000).

One of the key challenges of CPFR programs is to overcome the obstacle of trust concerns, and incentive misalignment. Consider, for instance, a retailer that would like to ensure supply availability for a product he sells. Then, by inflating the demand forecasts he passes to the vendor (this is often referred to as *phantom ordering*), the retailer may mislead the vendor to hold a virtually unlimited supply. Companies in the high-tech industry, such as IBM and HP, felt the negative impact of such retailer behavior that led them to over-invest in production capacity (see, e.g., Zarley and Damore 1996). But vendors are not always naive! Since vendors may suspect that customers do not truthfully share forecasts, they may take the demand estimates provided to them with a grain of salt. The understanding of self-interest behavior in uncertain environments characterized by assymetric information, is of fundamental importance. Cachon and Lariviere (2001) study a single-time procurement setting in which a manufacturer (the buyer) plans to purchase a critical component from a single supplier. The manufacturer offers a non-negotiable contract to the supplier, specifying a commitment to purchase a specific amount of the component, plus another quantity reserved as an option. Additionally, the manufacturer passes along his forecast of the demand. If accepting the contract, the supplier starts building up production capacity. Later, upon realization of demand, the manufacturer places a firm order with the supplier. The paper demonstrates that compliance terms significantly impact the outcome of this supply chain interaction. In other words, we would expect the contract terms and expected profits to be different in settings where the supplier has to build up enough capacity to cover all of the quantity the manufacturer specified in the contract (forced compliance), than in settings where the supplier can build capacity at any level he desires (voluntary compliance). The paper also discusses the circumstances under which it is beneficial to truthfully share forecasts with the supplier. Miyaoka (2003) studies incentive issues in a collaborative forecasting environment. In her paper, CF is not modeled as an ongoing process, but rather as a single time procurement transaction. Specifically, at some point in advance of the purchase period, the initial forecast of demand and the terms of the purchasing agreement are negotiated. Later, the retailer places his order to serve demand. Miyaoka introduces the concept of *Collaborative Forecasting Alignment* (CFA), as the case under which the trading partners have the right incentives to credibly share forecasts. But how can we achieve CFA? First, Miyaoka argues that CFA exists if and only if two conditions are satisfied: (1) neither party has an incentive to unilaterally distort information, and (2) given true information, both parties' expected profits increase as the forecast of the demand process becomes more accurate. Under price-only agreement, CFA is not generally achieved. Nevertheless, if a buyback contract can be used, its parameters can be chosen so as not only to achieve CFA, but also to bring channel coordination (i.e., the parties act in a way that results in the highest total profit in the channel).

Mishra et al. (2001) discuss the benefits related to information sharing in a supply chain, where prices are key decision variables (both the manufacturer and the retailer set prices based on their forecasts of an unknown demand). Demand is a linear function of retail price. They analyze a make-to-order setting, in which the production takes place after realizing the demand, and a make-to-stock setting, in which the production is completed before realizing the demand. In both settings prices are set prior to demand realization. The paper analyzes how prices are set and forecasts are shared in this complex environment.

The organization of this chapter

Models for collaborative forecasting need to consider all of the following components simultaneously: (i) The demand process; (ii) The forecasting process; (iii) The inventory control process; and (iv) The collaboration process. As you will see, this mix of processes is fully integrated in our main set of models, referred to as the *linear state-space framework*. In addition, this chapter suggests other possible types of frameworks that can be used to model collaborative forecasting.

This chapter is organized as follows: In §1 we specify our main model assumptions. These assumptions may be limiting in certain settings, but we make them in order to keep the presentation at a reasonable length. In §2 we briefly survey several types of approaches for modeling demand and forecast evolution. This will provide the interested reader with a flavor of common types of demand models that can be used in an integrative collaborative forecasting model. Particular emphasize is given to the linear state space model (§2.3), the model used in our main framework. We continue with §3 which deals with single-location inventory systems and discusses the dynamic inventory control problem for each of the demand processes described in §2. We pay attention to the similarity between all of the models, in terms of the structure of the optimal policies. We end §3 with a description of an adaptive inventory model that is used in our main framework, but the discussion within this section is structured in a way that may be useful for developing heuristic inventory control policies for the other demand models that one may wish to analyze. Section 4 presents our integrative model, in which we describe the evolution of information and replenishment policies in a two-stage supply chain. We end this section with some remarks about extensions of the framework to more complex supply chain structures, and discuss how the value of information sharing can be analyzed using this framework. As an illustration, we present in §5 the model of Aviv (2002a) which was used to study the benefits of collaborative forecasting processes in a certain auto-regressive demand environment. We continue in (§6) with a detailed discussion about the assessment of inventory cost performance. Section 7 concludes with some ideas for further research.

1. Notation and Preliminaries

We consider a supply chain that has a *distribution network* structure, in which each inventory stocking location has a unique supplier. Each inventory location, also known as a *member* of the supply chain, is identified by an index m, for which the precise definition depends on the actual network structure. As a building block for analyzing supply chain inventory networks, we shall mainly discuss a simple two-level supply *chain* consisting of a single retailer (denoted by $m = 1$ or $m = r$) and a single supplier ($m = 2$ or $m = s$). External demand placed (at the retail point) by the customers of the supply chain during period t is denoted by D_t. Below, we present some of the notation and common assumptions that are relevant to this chapter.

Periodic-review settings: We discuss supply chains that work in a periodic-review fashion. In other words, inventories are monitored at the beginning of each fixed time interval (a *period*). This assumption has no significant limitation in practice, as the length of a period selected for the control policy can vary arbitrarily. Usually, the type of inventory that we will be interested in is the so-called *inventory position* – i.e., the inventory that was ordered but not yet received at the member's location ("outstanding orders"), plus the inventory on hand at the location, minus any accumulated backlogs. We assume full backlogging.

Placement of replenishment orders and sequence of events: In parallel to monitoring the inventory status, each member updates its demand forecast at the beginning of a given period. The next step, still at the beginning of the period, is to make a decision as to whether or not to place an order with the immediate supplier, and if yes, to determine the exact quantity. For this to be done, a forecasting-integrated inventory process has to be in place, and we elaborate on this process in §2. We shall assume, without significant limitation, that the placement of replenishment orders in a given period is sequenced in a way that corresponds with the hierarchical level in the supply chain: For instance, in the simple two-level chain structure, the retailer orders first, then the supplier, so that the latter can update its own inventory position and demand forecast *after* observing the retailer's order. This allows the supply chain to avoid unnecessary information delays (Lee et al. 1997ab). We use the notation A_t to denote the order placed by the retailer at the beginning of period t.

Lead times: Normally, orders placed by member m in period t are received at the end of period $t + L^m$, and are available to fill the demand and any accumulated backlogs at the beginning of period $t + L^m + 1$. The effective *nominal lead time* is therefore $L^m + 1$ periods. Of course, when shortages occur in the supply chain, it is possible for a member to experience a delay that is longer than the nominal lead time.

Cost structure: Each member of the supply chain is charged a holding cost at a rate of h^m per unit of inventory on hand at the end of a period. A shortage penalty cost at a rate of p per period is charged for each unit of external customer demand that is back ordered.

The information structure: In the basic form of the supply chain, day-to-day exchange of information is not possible, except for order placement. On the other hand, in a collaborative forecasting setting, sharing of forecast data and inventory status may be possible. Some of the models we shall present later are quite flexible in terms of their capability of handling a variety of information structures. We shall generally assume that a fair amount of *static* information is commonly known to the members of the supply chain. This includes the cost structure of the supply chain, the characteristics of the demand process, the types of forecasting processes used at each member's location, and the statistical relationship between them (the latter point will become clear later).

2. Common Approaches for Modeling Demand Uncertainty and Forecast Evolution in the Inventory Management Literature

In this section we shall begin our presentation by examining common ways in which demand uncertainty is modeled in the inventory management literature. We intend to be brief in our presentation, as our purpose here is only to provide some basic terminology to which we can refer back later in this chapter. The most basic model is one that pertains to products that face uncertain demand, with the demand during each period having the same (known) statistical distribution, and in which demands in different periods are statistically independent. In short, this demand process is well-known as the "i.i.d. (independent and identically distributed) demand model." In this model, the demand realized during a single period is characterized by a statistical distribution (say) $D \sim G$. It follows trivially from this model that the forecasting process of any future demand is fixed with respect to the time the forecast is generated, and is simply given by a sufficient characterization of G. A simple extension to the stationary model would be to characterize the demand using a period-specific statistical distribution. In other words, we would say that the demand during period t has the distribution $D_t \sim G_t$. The set of distributions $\{G_t\}$ can have a periodic pattern (i.e., repeat the same pattern every fixed and known number of periods) to represent seasonality. But the pattern $\{G_t\}$ can be specified arbitrarily. Here, as in the basic model above, forecasts can be generated upfront without the need to model their evolution, since the statistical characterizations $\{G_t\}$ are known in advance and do not change over time. The following alternative models are also common in the literature.

2.1 Demand models with unknown parameters

Sometimes, even when one has the reason to believe that a product experiences a demand that is stationary with independent realizations, one is still uncertain about the exact parametric specification of the demand distribution. A stream of inventory management research that considers the case in which certain parameters of the demand distributions are unknown, started with the papers Dvoretzky et al. (1952), Scarf (1959,1960) and Iglehart (1964). Azoury (1985) is cited very often in this context, and the following model follows closely the presentation there with slight changes of notation: Suppose that the demands $\{D_t\}$ are continuously distributed, with a probability density function that has a known statistical *form*, but with an unknown parameter (or vector of parameters) w. Specifically, let $\phi(d|w)$ represent the probability density function of each D_t if w was known. Let $\theta(w)$ be the known prior density function of w. Azoury restricts her attention to the case in which: (i) the demand distribution has a sufficient statistic of fixed dimension for a sample of independent observations; and (ii) $\theta(w)$ belongs to a conjugate family of distributions (e.g., normal-normal, gamma-Poisson, or beta-binomial), so that the posterior distribution of w given a sample of n observations $d_1, \ldots, d_n$ belongs to the same family. At the beginning of period t, after observing the demand realizations $d_1, \ldots, d_{t-1}$, a sufficient statistic denoted by S can be calculated. Let $\theta(w|S, t-1)$ be the posterior distribution of w given S and the number of demand observations $t-1$. Then, the updated distribution density function of the future demands can be presented in terms of $(S, t-1)$ as follows:

$$\phi(d|S, t-1) = \int_{-\infty}^{\infty} \phi(d|w)\theta(w|S, t-1)dw.$$

The latter equation defines the forecast evolution in the system. For example, Karlin (1960) considers densities of the exponential family

$$\phi(d|w) = \beta(w)e^{wd}r(d), \quad \frac{1}{\beta(w)} = \int_0^{\infty} e^{wx}r(x)dx$$

For which it is easy to demonstrate that a sufficient statistic based on $t-1$ observations $d_1, \ldots, d_{t-1}$ is $S = \sum_{i=1}^{t-1} d_i$. Karlin shows that for a given priori distribution $\theta(w)$, the conditional density $\theta(w|S, t-1)$ is given by

$$\theta(w|S, t-1) = \beta^n(w)e^{wS}f(w)/[\int_{-\infty}^{\infty} \beta^n(x)e^{xS}f(x)dx].$$

We refer the reader to Karlin (1960), Scarf (1959) and Iglehart (1964) for further statistical results related to the exponential family.

An additional interesting example arises from the normal distribution. Suppose that the demand for a product behaves as a sequence of independent samples from a normal distribution with a mean of μ and standard deviation σ. However, even though the parameter σ is known, the exact value of μ can only be characterized by a priori normal density $N(\mu_0, \sigma_0)$. It can be shown that for a given sequence of demand observations $d_1, \ldots, d_{t-1}$, a sufficient statistic is $S = \sum_{i=1}^{t-1} d_i$, and that the posterior distribution of the mean demand μ is given by

$$\mu|(S, t-1) \sim N\left(\mu_{t-1} = \frac{\frac{S}{\sigma^2} + \frac{\mu_0}{\sigma_0^2}}{\frac{t-1}{\sigma^2} + \frac{1}{\sigma_0^2}}, \sigma_{t-1} = \sqrt{\frac{1}{\frac{t-1}{\sigma^2} + \frac{1}{\sigma_0^2}}}\right)$$

Additionally, the posterior distribution of each of the future demand volumes is

$$D_{t+i}|(S, t-1) \sim N\left(\mu_{t-1}, \sqrt{\sigma_{t-1}^2 + \sigma^2}\right); \quad i \geq 0.$$

This model was recently employed in a periodic-review inventory management model by Aviv and Federgruen (2001).

2.2   A Markov-modulated demand process

When the demand for a product fluctuates in accordance to some identifiable, yet unpredictable *environmental factors*, non-stationary models with fixed patterns, such as $\{G_t\}$, may be inadequate. Song and Zipkin (1993) provide three examples to such types of settings: (i) Products that are sensitive to economic conditions, and their demand can be statistically explained through a set of certain economic variables; (ii) Products that are subject to obsolescence, where both the timing and the magnitude of the reduction in demand are uncertain; and (iii) New products for which the timing and the level of increase in the demand is uncertain. Song and Zipkin propose an inventory-management model that aims to deal with such types of environments, and in their setting the demand follows a Poisson process with a rate that fluctuates according to a continuous-time Markov process. Beyer and Sethi (1997) discuss a periodic-review inventory model, using the following type of demand evolution (with some changes of notation): Let the demand *state* during a specific period t be denoted by I_t, and suppose that it can take one of N values $\{1, 2, \ldots, N\}$. The state I_t evolves according to a Markov chain with a known transition-probability matrix $P = \{p_{xy}\}$. Let $\phi(d|i)$ denote the conditional density function of the demand during period t, given that $I_t = i$. When the state of the demand (i.e., the process $\{I_t\}$) is fully observable, the demand evolution and the forecasts are given by the transition matrix P and the conditional distributions ϕ.

The potential use of the Markovian framework in the context of collaborative forecasting seems to be promising. Recall that a collaborative forecasting process derives its benefits from sharing and using information that is not commonly observable by all members of the supply chain. Such situation can be modeled as a case in which the demand depends on some *state* (using the terminology of Beyer and Sethi) that is only partially observable by some or all of the supply chain members. Consider a Markovian demand model in which there is a single forecaster that only partially observes the actual state. Treharne and Sox (2002) propose a model in which the distribution of the demand in each period is determined by the state of a Markov chain (named the *core process*). They consider the special case in which the state of this core process is not directly observed but only the *actual* demand is observed by the decision maker. This situation gives rise to a so-called partially observed Markov decision process (POMDP), and we shall discuss some related inventory policies briefly in §3. For the moment, we will only present the demand and forecast evolution in this type of a model, again with a slight change of notation. Recall the model above, and suppose that the distribution $\phi(d|i)$ is discrete. Since the exact state I_t is not directly observed during period t, we need to keep some sufficient information of the history $\{I_1, \ldots, I_t\}$. To do so, let π_t be a vector of probabilities $(\pi_{1t}, \ldots, \pi_{Nt})$ that represent our knowledge about the current state of the demand. Given the vector π_t, the demand in period t has the following distribution: $\Pr\{D_t = d|\pi_t\} = \sum_{i=1}^{N} \phi(d|i)\pi_{it}$. Treharne and Sox also provide the calculation method of the conditional probability distribution function of the lead time demand (i.e., $\Pr\{\sum_{i=0}^{l} D_{t+i} = d|\pi_t\}$); see there for more details. Given a realization of the demand d_t, the vector π_t is updated as follows:

$$\pi_{i,t+1}|(\pi_{it}, d_t) = \frac{\sum_{j=1}^{N} \pi_{jt}\phi(d_t|j)p_{ji}}{\sum_{j=1}^{N} \pi_{jt}\phi(d_t|j)}.$$

Clearly, this updating scheme is an essential component of the forecast evolution process.

2.3 A linear state-space model

In this section we briefly describe a linear state space model that serves as a building block for the main models of collaborative forecasting processes we present in this chapter. We then present a well-known forecasting technique associated with this model; namely, the Kalman filter. Let $\{X_t\}$ be a finite, n-dimensional vector process called the *state of the system*. In the context of inventory management, this vector may consist of early indicators of future demand in the channel, actual demand realizations at various points of the channel, and so forth. Suppose that the state vector evolves according to:

State space dynamics: $\quad X_t = FX_{t-1} + V_t \quad$ (10.1)

with F being a known, time invariant, $n \times n$ matrix. The vectors $\{V_t \in \mathbf{R}^n : t \geq 1\}$ represent a white noise process, with each random vector V_t having a multivariate normal distribution with mean $0_{n \times 1}$ and covariance matrix Σ_V. Next, assume that the decision-maker partially or fully observes the state vector X_t. In other words, during period t, only the vector

The observation equation: $\quad \Psi_t = HX_t \in \mathbf{R}^m \quad$ (10.2)

is observed where H is a known $m \times n$ matrix. Hence, one can think of Ψ_t as the information "collected" by the decision-maker during period t. In addition, let D_t represent the demand realized during period t, and suppose that the demand is a deterministic, linear function of the observed state vector:

The demand equation: $\quad D_t = \mu + R\Psi_t \quad$ (10.3)

where μ is a known scalar, and R is a $1 \times m$ vector of known parameters. To shorten the notation, we define the $1 \times n$ matrix $G \doteq RH$, so that $D_t = \mu + GX_t$[1]. The following is an elementary example of such model:

EXAMPLE 1 (AR(1)) *Consider the auto-regressive process of order 1, described by $D_t = \rho D_{t-1} + c + \epsilon_t$ (with $|\rho| < 1$), for all $t \geq 1$. An equivalent representation (10.1)-(10.3) for this case is: (i) $X_t = \rho X_{t-1} + \epsilon_t$, and (ii) $D_t = X_t + c/(1 - \rho)$. Using our notation, we have: $F = \rho$, $R = H = 1$, $\mu = c/(1 - \rho)$, and $\Sigma_V = \sigma^2$. (The choice of R and H is made arbitrarily from all combinations of parameters that satisfy $RH = 1$.)*

We refer to the AR(1) model as an elementary example, because in this case one can think of the process $\{X_t\}$ as the actual demand realizations. But X_t may include more information than the demand history. The need for such extension arises in various cases, such as: (i) when customers provide early information about their prospective orders; (ii) when planned promotions are known to one or more of the decision-makers in the supply chain; or (iii) when a detailed point-of-sale data (e.g., at a store-level) are provided directly to suppliers.

2.3.1 The Kalman Filter.

As in the partially observable Markov decision process we presented above, in order to forecast future demand the forecaster needs to estimate the actual state of the system. To this end, suppose that we would like to compute the minimum mean-square errors (MMSE) estimate of the state X_t, given the history of observations $\{\Psi_{t-1}, \Psi_{t-2}, \ldots, \Psi_1\}$, and denote this estimate by $\hat{X}_t = \mathsf{E}[X_t | \Psi_{t-1}, \Psi_{t-2}, \ldots, \Psi_1]$. Also, let $\Omega_t^X \doteq \mathsf{E}[(X_t - \hat{X}_t)(X_t - \hat{X}_t)']$ be the error covariance matrix associated with the estimate $\hat{X}_t$. A well known method for computing the values of $\hat{X}_t$ and Ω_t^X for linear state space models of the type (10.1)-(10.2), is the *Kalman filter*, and in this section we provide a brief summary of this tool.

Generally speaking, the Kalman filter is a set of recursive equations used to regenerate estimates of the system state after every transition the system makes (i.e., at the beginning of each period, in our terms). Assume an initial state vector estimate $\hat{X}_1$ and an initial error covariance matrix Ω_1^X. Then, at the beginning of every period t, the estimate $\hat{X}_t$ is produced by:

$$\hat{X}_t = F\hat{X}_{t-1} + FK_{t-1}(\Psi_{t-1} - H\hat{X}_{t-1}) \quad (10.4)$$

and its associated error covariance matrix is computed by

$$\Omega_t^X = F\Omega_{t-1}^X F' - FK_{t-1}H\Omega_{t-1}^X F' + \Sigma_V \quad (10.5)$$

where $K_{t-1} = \Omega_{t-1}^X H'(H\Omega_{t-1}^X H')^{-1}$ is an $n \times m$ matrix, called the *Kalman gain* matrix. The matrix $(H\Omega_{t-1}^X H')^{-1}$ represents any *generalized inverse* of $H\Omega_{t-1}^X H'$ (a generalized inverse of an $m \times n$ matrix G always exists, and it is an $n \times m$ matrix G^{-1}, such that $GG^{-1}G = G$). Recall that the vector Ψ_t is observed by the decision-maker *during* period t, *after* revising his forecasts and making his replenishment decisions at the beginning of that period, and just prior to the beginning of period $t+1$. This is the reason for including Ψ_{t-1}, and not Ψ_t in (10.4). The first term in (10.4) is the best forecast of X_t that can be generated using only the previous state estimate $\hat{X}_{t-1}$. The second term in (10.4) represents the adjustment to the estimate of X_t, based on the most recent observation, Ψ_{t-1}. Particularly, this term stands for the conditional expectation $\mathsf{E}\{X_t|\Psi_{t-1} - \mathsf{E}[\Psi_{t-1}|\Psi_{t-2},\ldots,\Psi_1]\}$.

As mentioned in the introduction, we confine ourselves to cases in which the level of uncertainty surrounding future demand volumes is constant across time, or at least becomes constant when t is large. This property is clearly satisfied if the mean square error matrices Ω_t^X converge to a constant matrix (say) Ω^X, as t grows to infinity. When the Kalman filter applied to the model (10.1)-(10.3) has (or converges to) a time-invariant error covariance matrix Ω^X, we say that it is in (or converges to) a *steady state* (for related convergence and steady state conditions, see Anderson and Moore 1979, and Chan et al. 1984). For Kalman filters in steady state, we shall also use the time-invariant notation K, instead of K_{t-1}. If the matrix F is stable (i.e., has all eigenvalues inside the unit circle), this property is satisfied as long as the matrix Ω_1^X is *positive semi-definite (p.s.d.)*. But one can assume, for instance, that $X_0 = 0_{n\times 1}$ and $\Omega_1^X = \Sigma_V$ are used, with the latter clearly being a p.s.d. matrix. When F is *not* stable, Chan et al. (1984) show that $\lim_{t\to\infty} \Omega_t^X = \Omega^X$ if the system is *observable*, and if $\Omega_1^X - \Omega^X$ is positive definite or $\Omega_1^X = \Omega^X$. The system (10.1)-(10.3) is observable if the state space X_t can be determined exactly given the observations $\Psi_t, \ldots, \Psi_{t+n-1}$, where n is the dimension of X_t. When F is unstable, it is necessary to characterize the distribution of the state vector X at a particular reference point (without loss of generality we refer to such point as point zero). This can be done, for instance, by providing a mean vector $\mathsf{E}[X_0]$ and a covariance matrix Σ_0 for the initial state X_0. Then, a recursive application of (10.1) yields $X_t = F^t X_0 + \sum_{j=0}^{t-1} F^j V_{t-j}$, and hence $D_t = \mu + G \sum_{j=0}^{t-1} F^j V_{t-j} + G F^t X_0$. Consider, for instance the ARIMA(0,1,1) model:

EXAMPLE 2 (ARIMA(0,1,1)) *Consider the auto-regressive, integrated moving average process:* $D_t = D_{t-1} - (1-\alpha)\epsilon_{t-1} + \epsilon_t$ *for all* $t \geq 2$. *In addition, we assume that* $D_1 = \mu + \epsilon_1$ *(or, alternatively,* $D_0 = 0$ *and* $\epsilon_0 = 0$*). An equivalent representation (10.1)-(10.3) for this case is:*

$$X_t = \begin{pmatrix} 1 & \alpha \\ 0 & 0 \end{pmatrix} X_{t-1} + \begin{pmatrix} 0 \\ \epsilon_t \end{pmatrix}; \quad D_t = \begin{pmatrix} 1 & 1 \end{pmatrix} X_t + \mu \quad (10.6)$$

with the interpretation $X_t = (\alpha \sum_{i=1}^{t-1} \epsilon_{t-i}, \epsilon_t)'$. *Here,* $H = (1,1)$, $R = 1$, *and* Σ_V *is a* 2×2 *matrix with* $\Sigma_{V(2,2)} = \sigma^2$, *and all of its other components are zeros.*

In this example, the matrix F is unstable (eigenvalues=$\{0,1\}$), but $\Omega_1^X = \Omega^X = \Sigma_V$. Thus, the Kalman filter for this model converges to a steady state, and so the framework in this paper can still be used to devise and characterize appropriate classes of *stationary* inventory policies for the ARIMA(0,1,1) case. In fact, a detailed treatment of this model is provided in Graves (1999).

The construction and applications of Kalman filters are extensively documented in the literature, and we refer the reader to Harvey (1990), Brockwell and Davis (1996; Chapter 12), and Hamilton (1994; Chapter 13) for a thorough description of this tool. Furthermore, the Kalman filter can be applied to state-space forms that are more general than the one we assume in this paper, and it is widely used in control theory applications.

3. Common Types of Single-Location Inventory Control Policies

In the previous section we presented a set of models of demand/forecast evolution that are fairly common in the inventory management literature. Clearly, this collection of models was not intended to be exhaustive. Nevertheless, these models share an important advantage: They are sufficiently descriptive of demand processes in a large variety of settings, and at the same time, they are simple enough to be embedded into inventory decision models without making them virtually intractable. Indeed, the purpose of this section is to explain some of the complexities associated with the control of inventory for products that face the above types of demand processes, and to direct the reader to the relevant literature.

In order to construct our framework for inventory management in the supply chain, we first consider the inventory management problem at the consumption points; i.e., at the locations where customer demands are realized. We call these locations the *front-end points* of the supply chain, or simply the *end points*. An end point will usually refer to a combination of place (geographic) and product type (stock-keeping unit). Initially, we shall assume that the end point does not face any supply delays for products it orders from its supplier.

3.1 Dynamic models for inventory management

Ideally, given a particular characterization of the demand process, and given the set of assumptions specified in §1, inventory should be managed in a way that provides the best balance between holding too much inventory versus running into shortages. In most stylized models, this balance is usually captured in a cost structure that charges inventory carrying cost as well as shortage penalty costs. The objective of the decision-maker is usually to minimize either the average cost per period, or the total discounted costs charged over the entire planning horizon. To keep the presentation in this chapter at a reasonable length, we shall confine our discussion to the long-run average-cost measure. Let $c(x)$ be the single-period cost function, when the net inventory at the end of the period is x. When x is positive it means that inventory is carried over to the next period, and when x is negative it means that there is a shortage. Specifically, we charge:

$$c(x) = h \cdot x^+ + p \cdot x^-, \tag{10.7}$$

where $x^+ = \max(x, 0)$ and $x^- = \max(-x, 0)$.

3.1.1 The demand model with statistically independent realizations.
We begin with the simple and well-known i.i.d. demand model. To find the best inventory policy, a stochastic dynamic program can be formulated as follows. Let:

$z_{t-L}, \ldots, z_{t-1}$ = deliveries in transit (pipeline inventory), where z_{t-i} being the order placed in period $t - i$.

q_t = the inventory on hand at the beginning of period t, just after receiving the delivery of z_{t-L-1} units, placed in period $t - L - 1$.

$C_n(q_t, z_{t-L}, \ldots, z_{t-1})$ = the minimum expected cost over a planning horizon of the n periods $t, \ldots, t+n-1$. We set $C_0 \equiv 0$.

The functions C_n satisfy the dynamic-programming scheme:

$$C_n(q_t, z_{t-L}, \ldots, z_{t-1}) = \min_{z_t \geq 0} \Big\{ \mathsf{E}_{D_t}[c(q_t - D_t)]$$
$$+ \int_x C_{n-1}(q_t - x + z_{t-L}, z_{t-L+1}, \ldots, z_t) dG(x) \Big\}. \tag{10.8}$$

When the planning horizon is infinite, and if the demand D_t does not take negative values, the solution to the dynamic program is given by a solution to the following problem $\min_\beta \mathsf{E}_{D_t^L}[c(\beta - D_t^L)]$, where $D_t^L = \sum_{i=0}^{L} D_{t+i}$ is the lead time demand. In other words, at a time when a replenishment decision is made, one only has to take into account the consequences of this decision on the expected costs at the end of lead-time periods into the future. The interpretation of the resulting policy is as follows: Let β^* be the solution to the latter minimization problem. Then, at the beginning of period t place an order $z_t \geq 0$ so as to bring the inventory position $q_t + z_{t-1} + \ldots + z_{t-L}$ up to the level β^*. This is named a base-stock policy, and β^* is named the base-stock (target) level. Furthermore, β^* is the solution to the well known Newsvendor formula

$$G^L(\beta^*) = \frac{p}{p+h}, \tag{10.9}$$

where G^L is the (continuous) distribution function of the lead time demand D_t^L. The results above are very elementary and are known for decades; see e.g., Karlin and Scarf (1958) and a later survey of Lee and Nahmias (1993).

3.1.2   Non-stationary demand patterns with independent realizations.   When the statistical distribution of the demand in different periods varies, it is no longer true that a newsvendor solution of the type (10.9) is optimal, even if G^L is replaced with G_t^L – the period-dependent distribution function of the lead time demand D_t^L. The main reason for this is that in addition to the impact that the decision variable z_t has on the expected cost at the end of period t, it may also have a negative consequence on the expected costs in later periods. Specifically, it may be the case that after placing the order z_t and after the demand D_t is realized, the inventory position at the beginning of period $t+1$ will be higher than desired. It is easy to see that in the i.i.d. case, in which the target level β^* is fixed, this can never happen.

Karlin (1960) has treated the dynamic decision model for this problem under a discounted cost criterion, while confining himself to stationary cost parameters. Zipkin (1989) extended Karlin's results to the case in which cost parameters are not necessarily fixed throughout the planning horizon. Essentially, even under non-stationary demand patterns, the optimal policy is of a base-stock type; i.e., for each period t there exists a target level β_t^* so that it is optimal to place an order z_t that brings the current inventory position to the level β_t^*. If, however, the inventory position is already higher than the target level, no order is placed. Under certain conditions, it may be possible that β_t^* would be equal to the newsvendor quantity described above (with G_t^L replacing G^L). But in general it may be necessary to solve the dynamic programming model to obtain the optimal values $\{\beta_t^*\}$.

3.1.3   The Bayesian framework.   When some of the characteristics of the demand process are not known with certainty, the dynamic control of inventory becomes significantly harder. Not only that the decision maker has to consider the cost impact of his decisions in the regular way described in Equation (10.8), but he also has to simultaneously consider how the information about future demands evolves based upon the ongoing revelation of the demand realizations. Fortunately, the Bayesian models described in §2.1 above allows one to write a dynamic program associated with the optimal management of the inventory system: Let S be the sufficient statistic based on $t-1$ demand observation $D_1, \ldots, D_{t-1}$. Then, the resulting dynamic program is given by

$$C_n(q_t, z_{t-L}, \ldots, z_{t-1}, S, t-1) = \min_{z_t \geq 0} \left\{ \int_x c(q_t - x)\phi(x|S, t-1)dx \right.$$
$$+ \int_x C_{n-1}(q_t - x + z_{t-L}, z_{t-L+1}, \ldots, z_t, S_{t+1}(S, x, t-1), t)$$
$$\left. \cdot \phi(x|S, t-1)dx \right\}. \tag{10.10}$$

But Bayesian dynamic programs such as (10.10) have a significant disadvantage. Due to their higher dimensionality (compared to non-Bayesian optimization schemes) they are usually very hard to solve, even if the statistic S is of a single dimension. As common in the inventory management literature (see, e.g., Veinott 1965), a reduction of the state of problem (10.10) is possible. This is done through a collapse of the sub-vector $(q_t, z_{t-L}, \ldots, z_{t-1})$ into the sum of its components $\tilde{q}_t = q_t + z_{t-L} + \ldots + z_{t-1}$, which is the *inventory position* at the beginning of period t. Then, the trick is to use a phased cost accounting scheme that charges in period t the expected costs at the end of period $t+L+1$ (this is fine since in the long-run, this phased accounting scheme does not have any effect on average cost calculations). Let

$$\tilde{c}(\tilde{q}_t + z_t, S, t-1) = \int_x c(\tilde{q} + z_t - x)\phi^{*(L+1)}(x|S, t-1)dx$$

where $\phi^{*(L+1)}(x|S, t-1)$ is the density function of the lead time demand D_t^L, given the information $(S, t-1)$. The reduced dynamic program is as follows

$$C_n(q_t, z_{t-L}, \ldots, z_{t-1}, S, t-1) = \tilde{C}_n(\tilde{q}_t, S, t-1)$$
$$= \min_{z_t \geq 0} \left\{ \tilde{c}(\tilde{q}_t + z_t, S, t-1) \right.$$
$$\left. + \int_x \tilde{C}_{n-1}(\tilde{q}_t + z_t - x, S_{t+1}(S, x, t-1), t) \cdot \phi(x|S, t-1)dx \right\}.$$

Early works in the literature (e.g., Scarf 1959, Karlin 1960 and Iglehart 1964) show that a state-dependent base stock policy structure is optimal. In other words, there exists a target level $\beta_n^*(S, t-1)$ so that it is optimal to order $z_t = \max(\beta_n^*(S, t-1) - \tilde{q}_t, 0)$ during that period. We refer the reader to Azoury (1985), Azoury and Miller (1984) and Lovejoy (1990) for more detailed discussion of inventory control policies (both optimal and heuristic) for the case of demand models with uncertain parameters.

3.1.4 The Markovian model.
We find the Markovian framework appealing due to its potential applications in the analysis of collaborative forecasting settings. While we refer the reader to Song and Zipkin (1993, 1996)

for detailed discussions of the fully-observable Markovian decision model, we have chosen to elaborate more on the partially observable Markov decision problem (POMDP). As we explained above, the applicability to collaborative forecasting processes comes from the fact that POMDP's can describe situations in which demand-related information is not commonly known in the supply chain. But let us first describe a dynamic program for inventory management in a single facility. To keep this part short, we plan to provide a similar presentation as for the case of the Bayesian model. Lets define first the phased cost function

$$\tilde{c}(\tilde{q}_t + z_t, \pi_t) = \mathsf{E}_{D_t^L | \pi_t}[c(\tilde{q} + z_t - D_t^L)]$$

where π_t is the vector of information, with π_{it} representing the likelihood that the actual state of the underlying Markov chain is i. A dynamic program follows immediately:

$$\tilde{C}_n(\tilde{q}_t, \pi_t) = \min_{z_t \geq 0} \left\{ \tilde{c}(\tilde{q}_t + z_t, \pi_t) \right.$$
$$\left. + \int_x \tilde{C}_{n-1}(\tilde{q}_t + z_t - x, \pi_{t+1}(\pi_t, x)) \cdot \sum_{i=1}^{N} \phi(x|i) \pi_{it} dx \right\}.$$

Monahan (1982) surveys the theory associated with POMDP's. A useful survey of algorithmic methods for POMDP's can be found in Lovejoy (1991). Recently, Treharne and Sox (2002) examined different types of inventory policies for the above type of POMDP. Particularly, they compared the performance of their suggested policies with that of a so-called *certainty equivalent control* policies. In the latter type of policy, the decision-maker maintains the forecasting updating scheme $\pi_{t+1}(\pi_t, D_t)$. However, instead of solving the tedious dynamic program above, the decision-maker replaces the uncertain future demand with their expected values, hence reducing the problem from one with a stochastic nature to a deterministic and solvable one. Nonetheless, Treharne and Sox showed that in some instances, their policies significantly outperform the commonly practiced certainty equivalent control policies.

3.1.5 The linear state-space model.

The linear state space model which stands as the main focal point of this chapter includes, as special cases, many of the "classical" time series demand models. Time series models such as auto-regressive or moving-average processes gain particular interest in the literature; see, e.g., empirical observations in Erkip et al. (1990) and Lee et al. (2000). Among the references mentioned in this book, the reader can find examples for the use of the auto-regressive process (Kahn 1987, Lee et al. 2000), the auto-regressive moving-average process (Veinott 1965, Johnson and Thompson 1975), the ARIMA process (Graves 1999), and the Martingale model for forecast evolution (Graves et al. 1986, Heath and Jackson 1994). As shall be demonstrated, the linear state space model is not only a unified model that embeds the above time-series processes, but it provides an important modeling flexibility.

Suppose that the demand evolves according to the linear state space model (10.1)-(10.3). Aviv (2002b) presents a dynamic programming model for this case by extending the state-space as follows. Let

$C_n(q_t, z_{t-L}, \ldots, z_{t-1}, \Psi_{t-1}, \Psi_{t-2}, \ldots)$ = the minimum expected cost over a planning horizon of the n periods $t, \ldots, t+n-1$. Again, we set $C_0 \equiv 0$.

Then,

PROPOSITION 7 *Consider the Kalman filter mechanism (10.4)-(10.5) in steady state, and suppose that the matrix $F - FKH$ has all its eigenvalues inside the unit circle. Then, the functions C_n satisfy the dynamic-programming scheme:*

$$C_n(q_t, z_{t-L}, \ldots, z_{t-1}, \Psi_{t-1}, \Psi_{t-2}, \ldots) = \min_{z_t \geq 0} \left\{ \mathsf{E}_{D_t}\left[c(q_t - D_t) \right.\right.$$
$$+ \mathsf{E}_{\xi_t}[C_{n-1}(q_t - D_t + z_{t-L}, z_{t-L+1}, \ldots, z_t,$$
$$\left.\left. \xi_t + H\sum_{i=1}^{\infty}(F - FKH)^i FK\Psi_{t-i}, \Psi_{t-1}, \Psi_{t-2}, \ldots)]|\Psi_{t-1}, \Psi_{t-2}, \ldots\right]\right\}$$

where $\xi_t \in \mathbf{R}^p$ has a multivariate normal distribution with mean zero and covariance matrix $H\Omega^X H'$, and it is independent of $\Psi_{t-1}, \Psi_{t-2}, \ldots$.

Aviv simplified the dynamic program of Proposition 7 as follows:

PROPOSITION 8 *Define*

$$\tilde{c}(\tilde{q}_t, \hat{X}_t, \Psi_{t-1}, \Psi_{t-2}\ldots) \doteq \mathsf{E}[c(\tilde{q}_t - \sum_{l=0}^{L} D_{t+l})|\hat{X}_t, \Psi_{t-1}, \Psi_{t-2}\ldots].$$

The function $\tilde{c}$ depends on the history $\{\Psi_{t-1}, \Psi_{t-2}\ldots\}$ through the most recent estimate $\hat{X}_t$ only, and hence can be written as $\tilde{c}(\tilde{q}_t, \hat{X}_t)$. The decision-making problem can be defined as follows:

$$\tilde{C}_n(\tilde{q}_t, \hat{X}_t) = \min_{z_t \geq 0} \left\{ \tilde{c}(\tilde{q}_t + z_t, \hat{X}_t) \right.$$
$$\left. + \mathsf{E}_{\tilde{\xi}_t}[\tilde{C}_{n-1}(\tilde{q}_t + z_t - RH\tilde{\xi}_t - RH\hat{X}_t - \mu, F\hat{X}_t + FKH\tilde{\xi}_t)]\right\}$$

where $\tilde{\xi}_t \doteq X_t - \hat{X}_t$. The vectors $\{\tilde{\xi}_t\}$ are i.i.d., each having a multivariate normal distribution with mean zero and covariance matrix Ω^X.

The dynamic program in Proposition 8 represents the decision-maker's problem as an attempt to find the best inventory position level, given a multidimensional information state $\hat{X}_t$. For this type of problem, a state-dependent order-up-to policy is optimal. In other words, there exists a sequence of target level functions $\{\beta_n^*(\hat{X}_t) : n = 1, 2, \ldots\}$, such that if $\tilde{q}_t < \beta_n^*(\hat{X}_t)$, an order for the difference $\beta_n^*(\hat{X}_t) - \tilde{q}_t$ is placed. Otherwise, no order is placed. The optimality of order-up-to policies in the finite-horizon case follows from the convexity of $\tilde{c}$, the linearity of the Kalman filter ($\hat{X}_{t+1} = F\hat{X}_t + FKH\tilde{\xi}_t$), and from the fact that all moments of the lead-time demands $\sum_{l=0}^{L} D_{t+l}$ and error terms $\tilde{\xi}_t$ are finite (due to the properties of the normal distribution). A formal proof can be easily written, using an induction on n. Furthermore, the extension to the case $n \to \infty$ only requires some standard adaptation from the inventory management literature; see, e.g., Karlin (1960). In the long-run problem, when $n \to \infty$, the target level is given by a function $\beta^*(\hat{X}_t)$.

3.2 Heuristic policies

In the previous section, we saw that under the objective of minimizing the long-run average inventory carrying costs and shortage penalty costs, it is optimal to use a base-stock policy that prescribes order quantities that bring the inventory position up to a periodically-calculated target level whenever possible (orders usually have to be non-negative). Nevertheless, despite this deceivingly-simple structure of the optimal policy, one still faces the hard challenge of calculating the target level for each possible value of the state space. For example, observe that in the linear state space model it is necessary to calculate the value of β^* for each possible value of the vector $\hat{X}_t$. But, usually, dynamic programs become impossible to solve when the dimensionality of their state space increases. It is for this reason that researchers have been looking at alternative policies that may be sub-optimal on one hand, but with parameters that are more easily calculable on the other hand. Of course, the tricky part is to identify policies that are simple enough, but not significantly inferior to the optimal policy. One such approach is to adopt a *myopic* policy, as explained below.

3.2.1 Myopic policies.
Consider a dynamic program of the type presented, e.g., in Proposition 8, and define the following *myopic* policy: At the beginning of period t, given a state space value $(\tilde{q}_t, \hat{X}_t)$, solve the problem

$$\min_{z_t \geq 0} \left\{ \tilde{c}(\tilde{q}_t + z_t, \hat{X}_t) \right\}.$$

The solution to this problem will be the order quantity for that period. In other words, in a myopic policy, we ignore the future consequences (beyond a lead time) of the actions we take today. Recall that in the elementary i.i.d. demand case this is indeed optimal to do. But, as expected, myopic policies may perform poorly if demand patterns fluctuate significantly from one period to another. To illustrate this, suppose that during a peak period the demand did not turn to be as high as expected, and so excess inventory is carried to the following period during which demand is expected to be very low. Moreover, the expected demand is so low that it is the best for the system if it could get rid of some of the inventory in the pipeline. This represents a situation in which the decision-maker may regret he had not considered in the past all of the future cost-implications of the orders he placed.

Let $\hat{D}_t^{(L)}$ be an estimate of the lead time demand D_t^L, and suppose, as common in the literature on inventory management, that $\tilde{c}(\tilde{q}_t, \hat{X}_t)$ can be written as a convex function $\tilde{c}'(\tilde{q}_t - \hat{D}_t^{(L)})$. Let γ be the value of e that minimizes $\tilde{c}'(e)$. Suppose further that γ is unique (this assumption holds in our main set of models). Using similar arguments as in Veinott (1965), we can easily show that if $\tilde{q}_t - \hat{D}_t^{(L)}$ is smaller than γ for every t, then it is optimal to order in each period t the amount

$$z_t = \hat{D}_t^{(L)} + \gamma - \tilde{q}_t. \qquad (10.11)$$

This is a myopic policy, since replenishment decisions are based on the best way to handle the single-step cost function $\tilde{c}'$. Johnson and Thompson (1975) prove the optimality of a *myopic* order-up-to policy for a system that faces a demand that follows an auto-regressive moving-average (ARMA) process. To do so, they truncate the normal distributions that characterize the components of the ARMA process, by imposing a set of conditions on the parameters of the underlying demand process (see §3 there). Consider also the inventory model with zero set-up costs discussed in §5 of Gallego and Özer (2001). In Theorem 5 of their paper, they prove that a myopic order-up-to policy is optimal for an inventory system that faces a demand with advanced orders. This is shown by observing that the "myopic" target level can be reached in each period (see specifically the observation "$x < y_{n+1}^m$" in the proof of Theorem 5 of their paper). We note that in Gallego and Özer's paper, orders for a specific periods are placed up to N periods in advance; if these orders can be characterized as normally distributed, their model in §5 would turn to be a special case of our linear state space model. As another illustration, in the context of Markov-modulated demand, take for instance the zero order cost model discussed in §4.2 of Song and Zipkin (1993). They show (Theorem 9) that when the demand rate does not decrease over time, the myopic order-up-to level is alway attainable, and thus it is optimal. Nevertheless, they demonstrate later, in Song and Zipkin (1996), that a myopic policy can be substantially sub-optimal when demand rate can drop over time, as is often experienced when products become obsolete. We refer the reader to Veinott (1965), Azoury (1985), and to the later paper of Lovejoy (1990) for further detailed discussion on the optimality (or near-optimality) of myopic policies for inventory models with parameter-adaptive (Bayesian) demand process. Specifically, the models treated by Lovejoy consider situations in which a parameter of the demand process is unknown and estimates of it are regenerated progressively over time, as demands are realized.

A slightly different approach to the justification of myopic order-up-to policies was taken by Lee et al. (2000) and most recently by Dong and Lee (2001): Consider a base-stock policy and let $\beta_t = \hat{D}_t^{(L)} + \gamma$ be the myopic target level calculated for period t. If one can show that β_t can be met *most* of the time, instead of at the beginning of each and every period, then a myopic order-up-to policy will be close to optimal. Note that in this case, the replenishment order z_t set by the end-point at the beginning of period t, can be approximately given by

$$\begin{aligned} z_t &= \gamma + \hat{D}_t^{(L)} - \tilde{q}_t = (\gamma + \hat{D}_t^{(L)}) - (\gamma + \hat{D}_{t-1}^{(L)} - D_{t-1}) \\ &= \hat{D}_t^{(L)} - \hat{D}_{t-1}^{(L)} + D_{t-1}, \end{aligned} \quad (10.12)$$

or, in other words, $z_t = \gamma + \hat{D}_t^{(L)} - \tilde{q}_t$, instead of $z_t = \max(\gamma + \hat{D}_t^{(L)} - \tilde{q}_t, 0)$. But how appropriate is this assumption? Clearly, equation (10.11) may be viewed as a good approximation for policies that do *not* permit negative orders, as long as $\Pr\{\hat{D}_t^{(L)} - \hat{D}_{t-1}^{(L)} + D_{t-1} \geq 0\} \approx 1$. In the linear state-space model, when the matrix F is stable, the latter probability is given by $1 - \Phi(-\mathsf{E}[z_t]/\sqrt{\mathsf{Var}(z_t)})$. We provide later, in Proposition 11, a formula for calculating the long-run variance of the orders $\{z_t\}$. For instance, in the AR(1) model of Example 1, Proposition 11 yields:

$$\begin{aligned} &\Pr\{\hat{D}_t^{(L)} - \hat{D}_{t-1}^{(L)} + D_{t-1} \geq 0\} \\ &= 1 - \Phi\left(-\frac{c}{\sigma} \cdot \frac{1}{\sqrt{\rho^{2L+4} \cdot (1-\rho)/(1+\rho) + (1-\rho^{L+2})^2}}\right) \end{aligned}$$

When $\rho = 0$ the last equation is reduced to $1 - \Phi(-c/\sigma)$ which is close to one when σ/c is relatively small. Furthermore, it can be easily seen that the probability above becomes even closer to one as ρ increases in the range $[0, 1)$. This specific result is also shown in Lemma 1 of Lee et al. 2000.

In our integrated model for collaborative forecasting (§4 and thereafter) we shall assume that (10.11) *always* holds, or alternatively, we shall permit negative values of z_t. This assumption is essential when we expand our discussion to more complex information-rich settings.

3.3 An adaptive replenishment policy for the linear state-space model

In the linear state-space demand model described in §2.3, the estimate of the lead-time demand D_t^L is given by

$$\hat{D}_t^{(L)} \doteq \mathsf{E}[D_t^L | \Psi_{t-1}, \Psi_{t-1}, \ldots] = (L+1)\mu + RH\left(\sum_{l=0}^{L} F^l\right)\hat{X}_t. \quad (10.13)$$

When a myopic base-stock policy is applied to this case, the target level

$$\beta_t = \hat{D}_t^{(L)} + \gamma, \quad (10.14)$$

is used. Furthermore, it is easily seen that given the linear cost rates h and p, the value $\gamma = \sqrt{\Omega^D} \cdot \Phi^{-1}(p/(p+h))$, minimizes the (phased) single-period expected costs, where Ω^D is equal to $\Omega_t^D \doteq \mathsf{E}[(\hat{D}_t^{(L)} - D_t^L)^2]$ when t is large. The value of Ω^D is shown (see Aviv 2002b) to be equal to

$$\begin{aligned}\Omega^D &= \sum_{l=1}^{L}\left[G\left(\sum_{j=0}^{L-l} F^j\right)\Sigma_V\left(\sum_{j=0}^{L-l} F^j\right)'G'\right] \\ &+ G\left(\sum_{j=0}^{L} F^j\right)\Omega^X\left(\sum_{j=0}^{L} F^j\right)'G'. \end{aligned} \quad (10.15)$$

The value of γ is interpreted as a fixed safety-stock level set for the end-point. Thus, although the long-run optimal base-stock level $\beta^*(\hat{X}_t)$ can take any functional form, we restrict ourselves to a target level β which is a very specific linear function of the vector-estimate $\hat{X}_t$. The reason we use a fixed safety-stock in (10.14) is that the level of uncertainty (Ω^D) surrounding the lead-time demands is constant over time.

The restriction to the above class of policies is obviously convenient from a practical point of view: The Kalman filter technique is readily available to generate the estimates $\hat{X}_t$, and (10.13) is used to compute the corresponding values of $\hat{D}_t^{(L)}$. In fact, Lee et al. (2000) and Graves (1999) use the above class of inventory replenishment policies in their analysis of the AR(1) and ARIMA(0,1,1) models, respectively. Both of these papers use the same MMSE forecasts that can be obtained by applying the Kalman filter tool, although they do not refer to this technique explicitly. Another example is the Martingale Model for Forecast Evolution (MMFE) developed by Graves et al. (1986). Here, as well, the evolution of forecasts is equivalent to what would be generated if the Kalman filter technique is used. Other examples for the use of this policy class can be found in Lee et al. (1997a) and Chen et al. (1999), to mention but a few.

In fact, in many papers, the policy (10.14) is specifically confined to *myopic* order-up-to policies in which the safety-stock parameter γ is set in a way that minimizes the expected cost at the end of a lead-time. However, this should not necessarily be the correct thing to do. Recall that a myopic order-up-to policy may fail to be optimal if the order-up-to level cannot be reached in almost every period. In such cases, myopic policies may perform poorly, and it would be reasonable to assume that their performance vis-a-vis optimal policies would worsen as $\Pr\{\hat{D}_t^{(L)} - \hat{D}_{t-1}^{(L)} + D_{t-1} < 0\}$ grows. Thus, it may be beneficial to set the safety stock γ to a level that is different than that suggested by a myopic policy, while still maintaining the base stock structure (10.14). We start with an illustration of this point, using the following special case:

PROPOSITION 9 *Recall Proposition 8. If $RHF^{L+1}\hat{X}_t = 0$ for all $\hat{X}_t \in \mathbf{R}^n$, and $\tilde{c}(\tilde{q}_t, \hat{X}_t)$ can be written as a convex function $\tilde{c}'(\tilde{q}_t - \hat{D}_t^{(L)})$, then the policy structure (10.14) is optimal.*

In his proof, Aviv (2002b) defines $e_t = \tilde{q}_t - \hat{D}_t^{(L)}$ as the "effective" inventory position, and proves by induction that $\tilde{C}_n$ can be written as a function $\tilde{C}_n'$ of e_t. In other words,

$$\tilde{C}_{n+1}(\tilde{q}_t, \hat{X}_t) = \min_{z_t \geq 0} \left\{ \tilde{c}'(e_t + z_t) \right.$$
$$\left. + \mathsf{E}_{\tilde{\xi}_t}[\tilde{C}_n(\tilde{q}_t + z_t - RH\tilde{\xi}_t - RH\hat{X}_t - \mu, F\hat{X}_t + FKH\tilde{\xi}_t)] \right\}$$
$$= \min_{z_t \geq 0} \left\{ \tilde{c}'(e_t + z_t) + \mathsf{E}_{\tilde{\xi}_t'}[\tilde{C}_n'(e_t + z_t - \tilde{\xi}_t')] \right\}$$
$$= \tilde{C}_{n+1}'(e_t),$$

where $\tilde{\xi}_t' = \mu + RH(I + \sum_{l=1}^{L} F^l KH)\tilde{\xi}_t$. For the simple single-dimensional dynamic program above, a static order-up-to level policy is optimal, which in turn implies that the policy structure (10.14) is optimal. Despite the fact that Proposition 9 only considers a special case, it raises an important point: On one hand, the myopic policy (i.e., $\gamma = e^*$) may fail to be optimal if $\hat{D}_t^{(L)} - \hat{D}_{t-1}^{(L)} + D_{t-1}$ is sometimes negative, or in other words, if $\tilde{\xi}_t'$ is a random variable that can often take negative values. On the other hand, the structure of the optimal policy does not change – policy (10.14) can still be used, albeit with γ being set to γ^* – the target-level solution of the dynamic-program $\tilde{C}_{n+1}'(e_t) = \min_{z_t \geq 0} \left\{ \tilde{c}'(e_t + z_t) + \mathsf{E}_{\tilde{\xi}_t'}[\tilde{C}_n'(e_t + z_t - \tilde{\xi}_t')] \right\}$.

The second possible reason for why a myopic order-up-to policy may not work well is because of potential supply delays beyond the nominal lead-time L, that may occur due to shortages at the supplier. In order to design our inventory policies to cope with potential supply delays, we propose an upward adjustment in the parameter γ. The numerical analyses in Aviv (2001,2002a) suggest that substantial benefits can be gained by such an adjustment. Some technical details about the determination of γ^* will be described in §6.

To summarize, we propose a so-called *MMSE forecast adaptive base-stock policy*. This policy employs the *Kalman filter* technique to calculate minimum mean square error (MMSE) forecasts of future demands at the beginning of each period. A fixed safety stock γ^* set at the beginning of the planning horizon, is then added to the MMSE forecast $\hat{D}_t^{(L)}$ to form the target level β_t for this period. Then, the following rule is applied: if the current inventory position is lower than the target level, an order is placed to fill this gap; otherwise, no order is placed. The advantage of our policy is that it is intuitive and easily implementable. But, not less importantly, it can be tailored for use in information-rich supply chains, for which the characterization of optimal policies is virtually impossible.

Remark. A different type of approach to study forecasting issues in inventory management is that taken, e.g., by Chen et al. (1999). In their models they assume that the decision-makers are not aware of the exact characteristics of the demand process, and hence they resort to popular forecasting mechanisms such as the moving-average technique when making replenishment decisions. Chen et al. also propose heuristic policies that are very similar in their nature to those proposed in this paper.

4. Models for Decentralized Forecasting Processes

From this section and on, we restrict ourselves to the linear state-space demand model. We first extend our discussion from the single-location setting to a simple supply chain that consists of *two* members: A retailer (end-point), and its supplier. These members are indexed 1 and 2, respectively. In order to develop inventory policies for the supply chain members, one needs to keep in mind the interaction between these members. On one hand, the retailer needs to take into account the delivery performance of the supplier, or in other words, possible delays of shipments of goods beyond the nominal lead-time $L^1 = L$. On the other hand, the supplier needs to forecast the orders that will be placed by the retailer, using all available information the supplier has. As we shall see, this is not a simple task, and in fact, the main purpose of this section is to discuss the supplier's forecasting process. First, for the retailer, we still propose the same type of forecasting and adaptive inventory policy described above. The way the retailer deals with possible supply shortages is by setting the safety-stock level γ^1 appropriately (see §3.3). As for the supplier's policy, we proceed as follows: In §4.1 we study the behavior of the process $\{A_t\}$ – the sequence of orders placed by the retailer (note that A_t was previously denoted by z_t). We show that the retailer's orders evolve according to a linear state-space. A myopic inventory policy similar in structure to that of the retailer then follows. In §4.2, we extend the discussion to consider cases in which the supplier, and not only the retailer, observes part of the system state vector X_t. While the analysis of this model extension becomes more complex, we are still able to show that the supplier's demand and information processes evolve according to a linear state space model of type (10.1)-(10.3). This is quite pleasing, since we can adopt for the supplier the same *type* of forecasting and replenishment policy used by the retailer.

The system still works in a periodic review fashion, and the order of events is such that the retailer places his order for period t *before* the supplier needs to determine his replenishment order (dealing with the opposite order of events requires some straightforward alterations to our models). Therefore, the supplier can make use of the information A_t when making his replenishment decision in period t. The supplier faces a fixed delivery lead-time of $L^2 = \tau$ periods. We shall assume, again, that the retailer's inventory and forecasting process is in steady state.

4.1 The orders generated by the retailer

Understanding the way by which the order quantities A_t evolves is essential for developing an inventory policy for the supplier. We shall assume initially that the only source of demand information for the supplier is the history of orders placed by the retailer. We begin with the following observation.

THEOREM 7 *Suppose that the order quantities $\{A_t : t \geq 1\}$ are determined according to (10.11), and that the Kalman filter (10.4)-(10.5) is in steady state. Then,*

$$A_t = RHF^{L+1}\hat{X}_{t-1} + \Theta(\Psi_{t-1} - H\hat{X}_{t-1}) + \mu \qquad (10.16)$$

where $\Theta \doteq R[H(\sum_{l=1}^{L+1} F^l)K + I_{m \times m}]$ *is a* $1 \times m$ *vector.*

Essentially, the representation of A_t in the theorem is equivalent to the expression

$$A_t = \mathsf{E}[D_{t+L}|\Psi_{t-2}, \Psi_{t-3}, \ldots] + \mathsf{E}[\sum_{l=0}^{L+1} D_{t-1+l}|\Psi_{t-1} - H\hat{X}_{t-1}],$$

where the first component is the estimate of A_t based on the information available at the retailer at the beginning of period t, and is given by $RHF^{L+1}\hat{X}_{t-1} + \mu$. The second term is based on the *forecast error* $\Psi_{t-1} - H\hat{X}_{t-1}$, observed by the retailer during period $t - 1$, and it corresponds to the second term in (10.16). The interpretation of Equation (10.16) is quite interesting. Suppose that at the beginning of period t, the retailer wishes to forecast the next demand volume D_t, and that the supplier wants to forecast his next demand volume A_{t+1}. Furthermore, suppose that the supplier can retrieve the information $\{\Psi_{t-1}, \Psi_{t-2}, \ldots\}$ from the history of orders $\{A_t, A_{t-1}, \ldots\}$. Then, the standard errors of these forecasts are given by:

$$\text{Std}(A_{t+1}|\Psi_{t-1}, \Psi_{t-2}, \ldots) = \sqrt{\Theta H \Omega^X H' \Theta'},$$

and

$$\text{Std}(D_t|\Psi_{t-1}, \Psi_{t-2}, \ldots) = \sqrt{R H \Omega^X H' R'}.$$

Thus, we observe that even if both supply chain members observe the data Ψ, the forecast errors of the demands at both levels may be different. Consider, for example, the ARIMA(0,1,1) model of Graves (1999), and recall that in his setting the supplier fully observes the information $\{\Psi_{t-1}, \Psi_{t-2}, \ldots\}$ through the history of orders $\{A_t, A_{t-1}, \ldots\}$. It may be easily shown that for the ARIMA model, $\Theta = 1 + (L + 1)\alpha$ and that $H\Omega^X H' = \sigma^2$. Thus, $\text{Std}(A_{t+1}|\Psi_{t-1}, \Psi_{t-2}, \ldots)/\text{Std}(D_t|\Psi_{t-1}, \Psi_{t-2}, \ldots) = 1 + (L+1)\alpha$. The last term describes the amplification of demand uncertainty as one goes upstream the supply chain, moving from the retailer to its supplier. Similar result can be shown with respect to the AR(1) model, described in Example 1. In this case, $\Theta = (1 - \rho^{L+2})/(1 - \rho)$, and $H\Omega^X H' = \sigma^2$. Therefore, the amplification of demand uncertainty in this case is given by $(1 - \rho^{L+2})/(1 - \rho)$; see also Observation 2 in Raghunathan (2001).

In general, we observe that the retailer's orders follow the evolution:

PROPOSITION 10 *Let $\xi_t \doteq \Psi_t - H\hat{X}_t$, and define the $(n+m) \times 1$ state vector $Y_t \doteq (\hat{X}'_t \ \xi'_t)'$. The following state-space form represents the evolution of the process $\{A_t\}$: $Y_t = F_A Y_{t-1} + \Xi_t$, and $A_{t+1} = G_A Y_t + \mu$, where*

$$F_A = \begin{pmatrix} F & FK \\ 0_{m \times n} & 0_{m \times m} \end{pmatrix}, \quad \Xi_t = \begin{pmatrix} 0_{n \times 1} \\ \xi_t \end{pmatrix},$$

and

$$G_A = (\ RHF^{L+1} \quad \Theta\ )$$

The matrix F_A is stable if and only if F is stable.

Like the process $\{V_t\}$, the process $\{\Xi_t\}$ is a white noise, with vector-mean zero, and covariance matrix

$$\Sigma_\Xi = \begin{pmatrix} 0_{n \times n} & 0_{n \times m} \\ 0_{m \times n} & H\Omega^X H' \end{pmatrix}$$

The significance of Proposition 10, is that it demonstrates that the orders placed with the supplier evolve according to the same type of dynamics that governs the underlying demand process. From the supplier's point-of-view, the system state vector consists of the most recent retailer's estimate of the underlying system state X_t and the retailer's *observed* forecast error $\Psi_t - H\hat{X}_t$. Hence, in an analogy to the retailer's forecasting and inventory process, we offer the following policy for the supplier: At the beginning of period t, update the estimate of Y_t (denoted by $\hat{Y}_t$) using the Kalman filter equations (10.4)-(10.5), with $F_A, K_A, A_t, G_A, \Omega^Y$, and Σ_Ξ, instead of $F, K, \Psi_{t-1}, H, \Omega^X$ and Σ_V. Next, the forecast of the supplier's lead-time demand $\sum_{l=0}^{\tau-1} A_{t+1+l}$, denoted by $\hat{A}_t^{(\tau)} \doteq \mathsf{E}[\sum_{l=0}^{\tau-1} A_{t+1+l}|A_t,\ldots,A_1]$, is computed via $\hat{A}_t^{(\tau)} = \tau\mu + G_A(\sum_{l=0}^{\tau-1} F_A^l)\hat{Y}_t$; see (10.13). Finally, a replenishment order is placed for a number of units that would bring the supplier's inventory position as close as possible to the target level

$$\beta_t^2 = \hat{A}_t^{(\tau)} + \gamma^2 \qquad (10.17)$$

for some fixed safety-stock γ^2. We describe in §6 possible ways to determine the value of γ^2.

The analysis above also allows us to provide a simple equation for calculating the long-run variance of the orders $\{A_t\}$, denoted below by $\mathsf{Var}(A_t)$. This measure is useful, for instance, when one wants to assess the long-run portion of periods in which negative orders are prescribed; see our discussion in §3.2.1. For the measure $\mathsf{Var}(A_t)$ to exist, we require that the matrix F is stable.

PROPOSITION 11 *Suppose that the matrix F of (10.1) is stable. Then,*

$$\mathsf{Var}(A_t) = RHF^{L+1}\Big[\sum_{j=1}^{\infty} F^j KH\Omega^X H'K'(F^j)'\Big](RHF^{L+1})'$$
$$+ \Theta H\Omega^X H'\Theta'. \qquad (10.18)$$

The variance of the orders $\{A_t\}$ are also of significant interest when the supplier faces capacity constraints. Thus, the term in (11) has a potential use in developing replenishment strategies for the supplier in such applications.

4.1.1 Illustration: demand processes with early signals.

We consider here an example suggested (without treatment) by Raghunathan (2001): Suppose that the demand process evolves according to the dynamics:

$$D_t - \mu = \rho(D_{t-1} - \mu) + \kappa U_{t-1} + \epsilon_t, \quad \epsilon_t \sim N(0,\sigma) \qquad (10.19)$$

where U_t is a retailer's observation during period t (e.g., promotion, price reduction, etc.), that directly affects the demand during the next period. Each observation U_t is seen by the retailer only, and has a normal distribution with mean $\mu_u = 0$ and standard deviation σ_u (dealing with $\mu_u \neq 0$ does not complicate the treatment). Also, suppose for simplicity that the process $\{U_t\}$ is independent of the random noise process $\{\epsilon_t\}$. A linear state-space model that represents this case is given as follows: let $X'_t \doteq (D_t - \mu, U_t)$. Then,

$$X_t = \begin{pmatrix} \rho & \kappa \\ 0 & 0 \end{pmatrix} X_{t-1} + \begin{pmatrix} \epsilon_t \\ U_t \end{pmatrix}, \quad \Psi_t = X_t, \quad D_t = \mu + \begin{pmatrix} 1 & 0 \end{pmatrix} X_t$$

In this case, $K = H = I_2, \Omega^X = \Sigma_V, R = (1,0)$, and

$$\Theta = R \sum_{l=0}^{L+1} F^l = \frac{1}{1-\rho} \begin{pmatrix} 1 - \rho^{L+2}, & \kappa(1 - \rho^{L+1}) \end{pmatrix}$$

Even under the relatively simple type of model presented above, the process $\{A_t\}$ does *not* follow an ARMA(1,1) evolution as under the case in which the demand evolves according to the elementary auto-regressive process of Example 1. In fact, under (10.19), the orders $\{A_t\}$ maintain the following recursive scheme:

$$A_t - \mu = \rho(A_{t-1} - \mu) + \Theta \xi_{t-1} + \Theta_1 \xi_{t-2} \quad (10.20)$$

where $\Theta_1 = \rho/(1-\rho) \ (1 - \rho^{L+1}, \ \kappa(1 - \rho^L) \ )$, and the *vector* process $\{\xi_t\}$ consists of i.i.d. components, each having a binormal distribution. We shall continue with this illustration in §4.2.1.

4.2 Enriched information structures

One of the strongest advantages of the framework proposed by Aviv (2002b), is that it deals with cases in which the supplier may have additional streams of information (on top of the observation of the retailer's orders) through which he can learn about future retailer's orders. For example, suppose that the supplier plans a promotion for his product, which in turn will cause future retailer orders to be larger (statistically). Below, we explain how to accommodate the supplier's desire to use all information in his possession when forecasting demand and placing replenishment orders. Aviv (2002b) suggests the following approach: We keep the demand evolution structure (10.1) and (10.3) the same. However, we replace the single observation equation (10.2) by two linear observation equations. The retailer observes $\Psi_{r,t} = H_r X_t$, and for him, we offer the same forecasting process and policy class described in §3.3. The supplier's observation during period t is given by the linear filtration $\Psi_{s,t} = H_s X_t$, where H_s is a known $q \times n$ matrix. The supplier's policy remains of the same type discussed in §4.1, but we modify his forecasting process to:

$$\hat{A}_t^{(\tau)} \doteq \mathsf{E}[\sum_{l=0}^{\tau-1} A_{t+1+l} | \{H_s X_{t-1}, H_s X_{t-2}, \ldots\}, \{A_t, A_{t-1} \ldots\}]$$

Luckily, it turns out that the supplier's forecasting problem needs to deal with producing estimates of future retailer's orders based on a partially observed state vector that evolves according to the linear model of the type discussed above. The following proposition shows a sufficient representation of such linear model.

PROPOSITION 12 *Let $\hat{X}_t, K$ and Θ be the values associated with the retailer's steady state forecasting process, and define the extended state vector: $Z_t \doteq \begin{pmatrix} \hat{X}'_t & X'_t \end{pmatrix}'$. This state vector follows the evolution*

$$\underbrace{\begin{pmatrix} \hat{X}_t \\ X_t \end{pmatrix}}_{Z_t} = \underbrace{\begin{pmatrix} F - FKH_r & FKH_r \\ 0_{n \times n} & F \end{pmatrix}}_{\tilde{F}} \cdot \underbrace{\begin{pmatrix} \hat{X}_{t-1} \\ X_{t-1} \end{pmatrix}}_{Z_{t-1}} + \underbrace{\begin{pmatrix} 0_{n \times 1} \\ V_t \end{pmatrix}}_{\tilde{V}_t}$$

The complete supplier's observation structure is given by

$$\Psi_{s,t} = \begin{pmatrix} A_{t+1} \\ H_s X_t \end{pmatrix} = \underbrace{\begin{pmatrix} RH_r F^{L+1} - \Theta H_r & \Theta H_r \\ 0_{q \times n} & H_s \end{pmatrix}}_{\tilde{H}} \cdot Z_t$$

Finally, $A_{t+1} = \tilde{R} \Psi_{s,t} = \tilde{R} \tilde{H} Z_t$, where $\tilde{R} = (1, 0, \ldots, 0)$.

The representation in Proposition 12 is sufficient in the sense that a reduction in its dimension may be possible. But more importantly, the proposition enables us to suggest again the Kalman filter algorithm for estimating future order quantities, based on their history and on the supplier's observations of the underlying state vector process $\{X_t\}$. The estimates $\hat{Z}_t$ and their associated error covariance matrices Ω_t^Z, can be generated through the Kalman filter algorithm (10.4)-(10.5), replacing F, H, Ψ, and Σ_V, with $\tilde{F}, \tilde{H}, \Psi_s$, and $\Sigma_{\tilde{V}}$, respectively. Hence,

$$\hat{A}_t^{(\tau)} = \tau \mu + \tilde{G} \Big(\sum_{l=0}^{\tau-1} \tilde{F}^l \Big) \hat{Z}_t \tag{10.21}$$

and Aviv (2002b) showed that the mean square error of the supplier's estimates $\hat{A}_t^{(\tau)}$ satisfies

$$\begin{aligned} \Omega_t^A &= \sum_{l=1}^{\tau-1} \Big[\tilde{G} \Big(\sum_{j=0}^{\tau-1-l} \tilde{F}^j \Big) \Sigma_{\tilde{V}} \Big(\sum_{j=0}^{\tau-1-l} \tilde{F}^j \Big)' \tilde{G}' \Big] \\ &+ \tilde{G} \Big(\sum_{j=0}^{\tau-1} \tilde{F}^j \Big) \Omega_t^Z \Big(\sum_{j=0}^{\tau-1} \tilde{F}^j \Big)' \tilde{G}' \end{aligned} \tag{10.22}$$

In a steady state analysis, one may suppress the subscript t in the value of Ω^A.

4.2.1 Illustration: demand process with early signals (cont.).

Consider the following extension to the demand model (10.19) of §4.1.1:

$$D_t - \mu = \rho(D_{t-1} - \mu) + \kappa_r U_{r,t-1} + \kappa_s U_{s,t-1} + \epsilon_t, \quad (10.23)$$

where $\epsilon_t \sim N(0,\sigma)$ and $(U_{r,t-1}, U_{s,t-1})$ are "market signals" observed by the retailer and the supplier, respectively. We assume that the bivariate process $\{(U_{r,t}, U_{s,t}) : t \geq 1\}$ consists of i.i.d. components, each having a binormal distribution with mean $\mu_u = 0_{2\times 1}$ and covariance matrix Σ_u. The motivation for studying such types of demand processes was provided recently by Lee et al. (2000), Raghunathan (2001), and Aviv (2002a). Under the demand structure (10.23), the model representation is: let $X'_t \doteq (D_t - \mu, U_{r,t}, U_{s,t})$. Thus,

$$X_t = \begin{pmatrix} \rho & \kappa_r & \kappa_s \\ 0 & 0 & 0 \\ 0 & 0 & 0 \end{pmatrix} X_{t-1} + \begin{pmatrix} \epsilon_t \\ U_{r,t} \\ U_{s,t} \end{pmatrix}, \quad \Sigma_V = \begin{pmatrix} \sigma^2 & 0_{1\times 2} \\ 0_{2\times 1} & \Sigma_u \end{pmatrix},$$

$$H_r = \begin{pmatrix} I_2 & 0_{2\times 1} \end{pmatrix}, \quad H_s = \begin{pmatrix} 0 & 0 & 1 \end{pmatrix}, \text{ and } R = \begin{pmatrix} 1 & 0 \end{pmatrix}.$$

Aviv (2002a) used the framework described here to devise inventory control policies for various types of two-stage supply chain structures that face demand processes with early market signals. The demand process considered by Aviv (2002a) is more complex than (10.23), allowing both the retailer and the supplier observe signals (i.e., U-values) about future demands, starting from several periods in advance. We shall provide more details on Aviv's model in §5.

In addition to specifying operating policies for the supply chain, one may find the linear state-space framework useful in studying important issues in supply chain management, such as the assessment of the potential benefits of information sharing between the retailer and the supplier (see more details in §4.3). In fact, it has been shown that in the elementary AR(1) and ARIMA(0,1,1) cases, there is no value for the retailer sharing information about the actual demand realizations with the supplier. This is because the supplier can observe the values of $\{D_{t-1}, D_{t-2}, \ldots\}$ from the history of retailer's orders $\{A_t, A_{t-1}, \ldots\}$ (see Raghunathan 2001 and Graves 1999, respectively). When the demand depends on "signals," we would expect that information sharing will enhance the ability of *both* the retailer and the supplier to more accurately forecast demand at their levels. As can be expected, when the information structure in the supply chain is complex, the derivation of exact analytical formulae for forecasting and inventory target levels is tedious. Therefore, in addition to the elegance of matrix representation, the models above offer an easily implementable method for analysis and operation of information-rich inventory systems.

4.3 Assessment of the benefits of information sharing

The purpose of this section is to demonstrate that in addition to the development of forecasting and control policies for supply chain inventory systems, the framework above provides the means for assessing the benefits of information usage and information sharing practices, one of which is *collaborative forecasting*. For example, in the model of Aviv (2002a) that will be described in §5 below, the difference between various supply chain settings is determined in terms of the information available to the retailer and the supplier at the times they need to forecast future demand.

The topic of information sharing in supply chains has received a great deal of attention in the academic literature as well as in the business press (see the survey chapter of Chen in this handbook). This is particularly due to the fact that today, supply chains are very rich in the amount and types of information they possess. Very often, inventory managers need to choose between the uses of different sets of data streams when determining replenishment orders. This need arises, for example, when the cost of collecting a specific type of data is high, or when a particular type of information exists in a different database than that used by the inventory planner, and so an investment in hardware and software is required in order to make this data available. Aviv (2002b) describes examples for possible studies in supply chain inventory management that can exploit this framework. For instance, three extensions to the two-stage supply chain model are discussed there: The case of a serial supply chain, the case of multiple products and retail locations, and *distribution networks*, in which each inventory stocking location has a unique supplier.

We provide below an illustration to demonstrate the applicability of the linear state space framework in assessing the benefits of information usage and sharing. Consider the demand model (10.19), and suppose that the retailer contemplates whether or not to use the data $\{U_t\}$. In such case, one can examine the performance of the retailer's inventory policy under the case $H_r = I_{2\times 2}$ (i.e., the retailer observes and uses $\{U_t\}$), and under the case $H_r = (1,0), R = 1$, which represents the situation in which the retailer does not use the values of $\{U_t\}$. A potential application of such analysis is in supply channels for spare parts. One way by which inventory managers can reduce their uncertainty about future demand for spare parts, is to collect usage data from a representative subset of their customers. While this can certainly improve forecasting performance, the main question is whether the *magnitude* of such benefits will surpass the investment in data collection and analysis. With regards to the assessment of the value of information sharing, our methodology follows quite intuitively: Since both H_r and H_s are given as linear functions of the state vector X_t, information sharing can be described by the exact same model, but with an appropriate modification of the matrices H_r, H_s, and R. Consider, for example, the demand model (10.23), and suppose that the retailer and the supplier wish to gauge the benefits of "complete" information sharing. In that case, the original system (i.e., no information sharing) can be compared with that with $H_r = H_s = I_{3\times 3}$, and $R = (1, 0, 0)$. Clearly, various levels of information sharing can be examined, and compared with each other in order to identify the most promising (cost-beneficial) data sharing arrangement.

5. An inventory model of collaborative forecasting

In this section, we summarize an inventory model for collaborative forecasting studied in Aviv (2002a). In Aviv's model, the demand follows an autoregressive statistical time-series of order 1, as follows:

$$d_t - \mu = \alpha(d_{t-1} - \mu) + q_t \tag{10.24}$$

where μ represents the long-run average demand per period, and α is a known constant ($0 \leq \alpha < 1$). The sequence $\{q_t\}$ consists of independent and identically distributed (i.i.d.) random components, each of which is normally distributed with a mean of zero. The random components $\{q_t\}$ represent a sequence of "shocks," where q_t has a full impact on the demand during period t, and a partial, diminishing impact on future demands $(d_{t+1}, d_{t+2}, \ldots)$. Model (10.24) is common for describing demands in settings in which inter-temporal correlation between demands in consecutive periods exist (see, e.g., Chen et al. 2000a, Lee et al. 2000 and Kahn 1987). Aviv (2002a) proposes an extension to this model that enables one to consider situations in which supply chain members are able to learn about future demand through a variety of early *signals*, as described above.

Suppose that the random error term q_t can be *explained* statistically by the general linear regression model

$$q_t = \theta_r \sum_{i=1}^{T} \delta_{t,i}^r + \theta_s \sum_{i=1}^{T} \delta_{t,i}^s + \epsilon_t \tag{10.25}$$

where the $\delta_{t,i}^r$ and $\delta_{t,i}^s$, can be fully determined by the retailer and the supplier, respectively, during period $t - i$. For example, $\delta_{t,i}^r$ by itself can be a function of new information about the expected weather condition in period t, early order placement by customers, etc. Hence, the δ-variables in (10.25) represent a set of locally observable *explanatory variables*, and ϵ_t is an independent, unobservable error component. We refer to $\delta_{t,i}^m$ as the *market signal* about q_t observed during period $t - i$, by member m of the supply chain ($m \in \{r, s\}$ represents the retailer and the supplier, respectively). Specifically, we assume that the market signals are observed by the supply chain members *progressively* over time, beginning from a maximal horizon of T periods in advance. We further assume that $\delta_{t,i}^m \sim N(0, \sigma_{m,i})$, and $\epsilon_t \sim N(0, \sigma_0)$. Clearly, the information that the retailer and the supplier have about future demand are expected, in general, to be correlated. Therefore, while Aviv (2002a) assumes that the δ and ϵ-components are independent, he makes the important exception that for every combination of $t \geq 1$ and $i = 1, \ldots, T$, the couple $(\delta_{t,i}^r, \delta_{t,i}^s)$ can be correlated. For simplicity, Aviv assumes a correlation $\rho = \mathrm{corr}(\delta_{t,i}^r, \delta_{t,i}^s)$, that is fixed across t and i. It hence follows that the variance of q_t is equal to: In the linear regression model (10.25), the parameters θ_r and θ_s represent, to a certain degree, the explanatory strengths of the retailer and the supplier. For instance, if $\rho = 0$, and $\sigma_{r,i} = \sigma_{s,i}$, the meaning of $\theta_r > \theta_s$ is that the retailer's market signals are more informative than the supplier's market signals. But in order to sharpen this interpretation, it is useful to examine the possible reduction in the uncertainty about q_t that can be achieved by observing the sequence $\{\delta_{t,i}^m\}$. Because of the dependency between $\delta_{t,i}^r$ and $\delta_{t,i}^s$, the possible *reduction* in the unexplained variability of q_t during period $t - i$ is: $(\theta_r \sigma_{r,i} + \rho \theta_s \sigma_{s,i})^2$ for the retailer, and $(\rho \theta_r \sigma_{r,i} + \theta_s \sigma_{s,i})^2$ for the supplier. Aviv defines the ratio between the latter two values as the *relative explanation power* of the retailer.

5.1 Installation-based inventory systems

In his first model, named *locally-managed inventory (LMI) system*, Aviv (2002a) studies a supply chain structure in which the members of the supply chain do not share their observations of explanatory market signals (the δ-variables). Inventory is managed according to the installation-based MMSE adaptive base-stock policy, with target levels

$$\begin{aligned} \text{Retailer's base-stock level:} & \quad \beta_t^r = \hat{D}_t^{(L)} + \gamma^r \\ \text{Supplier's base-stock level:} & \quad \beta_t^s = \hat{A}_t^{(\tau)} + \gamma^s \end{aligned} \quad (10.26)$$

(To read more about installation-based policies, see Axsäter and Rosling 1993.) One can easily verify that the linear state space form applies as follows: Let

$$X_t' = (d_t - \mu, \Delta_{t+1,1}^r, \Delta_{t+2,2}^r, \ldots, \Delta_{t+T,T}^r, \Delta_{t+1,1}^s, \Delta_{t+2,2}^s, \ldots, \Delta_{t+T,T}^s),$$

with X_t' denoting X_t-transposed. The variable $\Delta_{t+l,l}^m$ represents the accumulated information about q_{t+l} acquired by member m up until the end of period t. In other words, $\Delta_{t,l}^m \doteq \sum_{i=l}^T \delta_{t,i}^m$ ($l = 1, \ldots, T$, $m = r, s$). Then, the state space follows the evolution $X_t = FX_{t-1} + V_t$ where

$$F = \begin{pmatrix} \alpha & \theta_r & 0 & \cdots & 0 & \theta_s & 0 & \cdots & 0 \\ 0 & 0 & 1 & 0 & 0 & 0 & \cdots & \cdots & 0 \\ \vdots & 0 & \ddots & \ddots & 0 & \vdots & & & \vdots \\ \vdots & \vdots & & \ddots & 1 & \vdots & & & \vdots \\ 0 & 0 & \cdots & 0 & 0 & 0 & \cdots & \cdots & 0 \\ 0 & 0 & \cdots & \cdots & 0 & 0 & 1 & 0 & 0 \\ \vdots & \vdots & & & & \vdots & 0 & \ddots & 0 \\ \vdots & \vdots & & & & \vdots & \vdots & \ddots & 1 \\ 0 & 0 & \cdots & \cdots & 0 & 0 & \cdots & 0 & 0 \end{pmatrix},$$

and

$$V_t = \begin{pmatrix} \epsilon_t \\ \delta^r_{t+1,1} \\ \vdots \\ \delta^r_{t+T,T} \\ \delta^s_{t+1,1} \\ \vdots \\ \delta^s_{t+T,T} \end{pmatrix},$$

Next, $\Psi^r_t = H_r X_t$ is the $(T+1)$-dimensional vector representing the information about X_t actually observed by the retailer. Clearly, Ψ^r_t includes the demand realization observed during period t, and all Δ^r-data, and so $H_r = (\ I_{(T+1)\times(T+1)} \quad \mathbf{0}_{(T+1)\times T}\ )$. Furthermore, the demand equation is given by $D_t = GX_t + \mu$ where $G = (1,0,0,\ldots,0) \in \mathbf{R}^{2T+1}$. The supplier's observation equation is given by $\Psi^s_t = H_s X_t$, where $H_s = (\ \mathbf{0}_{T\times(T+1)} \quad I_{T\times T}\ )$.

Aviv (2002a) shows that from the supplier's perspective, the retailer's order in period $t+1$ is a linear function of a certain state vector Z_t, and that the supplier observes a linear filtration of this state during period t. Furthermore, the proposition demonstrates that A_{t+1}, Z_t, and the observed state follow the linear state space described in §2.3.

PROPOSITION 13 *Suppose that $\sigma_{r,i} = \sigma_{s,i}$ for all $i \geq 1$, and that the retailer's forecasting system is in steady state. Define the state vector $Z_t \doteq (\ H_r X'_{t-1} \quad X'_t\ )'$. Then, the following system represents the co-evolution of the retailer's order quantities and the supplier's observations:*

$$\underbrace{\begin{pmatrix} H_r X_{t-1} \\ X_t \end{pmatrix}}_{Z_t} = \underbrace{\begin{pmatrix} \mathbf{0}_{(T+1)\times(T+1)} & H_r \\ \mathbf{0}_{(2T+1)\times(T+1)} & F \end{pmatrix}}_{\tilde{F}} \cdot \underbrace{\begin{pmatrix} H_r X_{t-2} \\ X_{t-1} \end{pmatrix}}_{Z_{t-1}} + \underbrace{\begin{pmatrix} \mathbf{0}_{(T+1)\times 1} \\ V_t \end{pmatrix}}_{\tilde{V}_t}$$

$$\tilde{\Psi}^s_t = \begin{pmatrix} A_{t+1} \\ H_s X_t \end{pmatrix} = \underbrace{\begin{pmatrix} -M & G + MH_r \\ 0_{T\times(T+1)} & H_s \end{pmatrix}}_{\tilde{H}} \cdot Z_t$$

where the vector $M \in \mathbf{R}^{T+1}$ is given by:

$$M_i = \begin{cases} \sum_{l=1}^{L+1} \alpha^l & i = 0 \\ (\theta_r + \rho\theta_s)\sum_{l=0}^{L+1-j} \alpha^l & 1 \le j \le \min(L+1, T) \\ 0 & \text{otherwise.} \end{cases}$$

with the index j ranging between 0 and T. Finally, note that the process $\{\tilde{V}_t\}$ consists of independent components, and that

$$A_{t+1} = \underbrace{\tilde{R}\tilde{H}}_{\tilde{G}} Z_t ; \quad \tilde{R} = (1, 0, \dots, 0).$$

It is instructive to note that the supplier's observable vector $\tilde{\Psi}^s_t$ includes, in addition to the market signals observed by the supplier (i.e., the Ψ^s_t component), the retailer's order A_{t+1}. Using the above linear state space representation, the supplier can calculate the estimates of future lead-time demands using the exact same mechanism as used in the retailer's forecasting process. Here, the values of $\hat{Z}_{t|t-1}$ and the error covariance matrices $\Omega^Z_{t|t-1}$, are generated through the Kalman filter algorithm, merely making the appropriate matrix substitutions. At the beginning of period t, immediately after observing A_t, the supplier's estimate of the aggregate order quantity $\sum_{l=1}^{\tau} A_{t+l}$ is given by $\hat{A}^{(\tau)}_t = \tau\mu + \tilde{G}(\sum_{l=0}^{\tau-1} \tilde{F}^l) \cdot \hat{Z}_{t|t-1}$. In the long-run, the mean-square error of this estimate is given by

$$\Omega^{\text{LMI}}_A = \sum_{l=0}^{\tau-2} \left[\tilde{G}\left(\sum_{j=0}^{l} \tilde{F}^j\right) \Sigma_{\tilde{V}} \left(\sum_{j=0}^{l} \tilde{F}^j\right)' \tilde{G}' \right] + \tilde{G}\left(\sum_{j=0}^{\tau-1} \tilde{F}^j\right) \Omega^Z \left(\sum_{j=0}^{\tau-1} \tilde{F}^j\right)' \tilde{G}'$$

where Ω^Z is the steady state error covariance matrix of the supplier's estimation procedure.

5.2 Echelon-based inventory systems

The inventory management literature (Clark and Scarf 1960) show that an *echelon-based* policy is optimal for a centrally-managed supply chain. However, no optimality results where demonstrated for a *general* linear state space demand model, in which different echelons observe different information about the demand, and the demand itself follows a correlated time-series pattern. Nevertheless, there are some interesting and relevant works related to supply chains that face auto-correlated demands: Erkip et al. (1990) treat a distribution network including a depot and several retail outlets. The demand is placed at the retail level with the demands being correlated not only across retailers but also across time. Dong and Lee (2001) study the optimality of inventory policies in serial multi-echelon supply chains, when the underlying demand follows an auto-regressive time-series. They also provide an approximation procedure that yields an attractive set of operating policies.

Aviv (2002a) studies two echelon-based settings. The first, named *Supplier-Managed Inventory (SMI)* is a system in which the supplier takes the full responsibility of managing the supply chain's inventory, but the retailer's observations of market signals are not transfered to him. Therefore, this model resembles many original VMI programs, in which forecasting data available to the retailer is not fully shared with the vendor. The SMI setting is typical to cases in which the retailer and the supplier do not have the required infrastructure, such as communication links between inventory planning systems and statistical software packages, to support a smart exchange of forecast information (hence, SMI should not be viewed as a VMI setting in which the supplier naively "throws out" information provided by the retailer). In the second setting, reminiscent of CPFR partnerships, inventory is managed centrally, and all demand related information is shared. This setting is named *Collaborative Forecasting and Replenishment (CFAR)*. In both of these settings there is a single decision-maker (the *supplier*, for the purpose of our discussion) that has the responsibility of replenishing the inventories both at the retailer's site and at the supplier's site. In order to make such a task practical, the supplier would need to gain access to inventory status information at the retailer's level. With this information available, the supplier is able to determine the replenishment quantities for the retailer by using the same type of adaptive MMSE base-stock policy that the retailer uses in the locally-managed settings. However, here the supplier would be in charge of forecasting the demand $d_t + \ldots + d_{t+L}$, to be used in the determination of the base-stock level for the retailer's inventory position. For the supplier, an echelon-based policy is used. Under such policy, the supplier looks at the inventory position of the *entire* supply chain, instead of his local inventory position. This echelon inventory position consists of all goods within the supply chain (on hand, plus in-transit from the supplier to the retailer), minus backlogs at the retailer level (only), plus all outstanding orders at the supplier's level. The echelon position is then compared to the supplier's best estimate of the total demand during a lead time of $L + \tau + 1$ periods ($d_t + \ldots + d_{t+L+\tau}$), plus a fixed safety stock. To summarize, order quantities that are based on the two base-stock levels:

$$\begin{aligned} \text{Retailer's echelon base-stock level:} \quad B_t^r &= \hat{D}_t^{(L)} + \Gamma^r \\ \text{Supplier's echelon base-stock level:} \quad B_t^s &= \hat{D}_t^{(L+\tau)} + \Gamma^s \end{aligned} \quad (10.27)$$

and the overall inventory replenishment process is hence fully defined by the choice of the safety-stock parameters Γ^r and Γ^s.

The treatment of the SMI and CFAR settings is easier than that of the LMI setting, because decisions are based on centralized information (even if the information is incomplete, such as in the SMI case). First, the linear model (10.25) can be written as:

$$q_t = \sum_{i=1}^{T} \hat{\delta}_{t,i}^E + \hat{\epsilon}_t$$

where $\hat{\delta}_{t,i}^E$ represents the best update to the forecast of q_t that can be generated using the information available to the decision-maker. The superscript 'E' is used to denote echelon-based settings. In the SMI setting, when $\hat{\delta}_{t,i}^E$ is solely based on the signal $\delta_{t,i}^s$, we have $\hat{\delta}_{t,i}^E = (\theta_s + \theta_r \rho \sigma_{r,i}/\sigma_{s,i})\delta_{t,i}^s$. In the CFAR setting, $\hat{\delta}_{t,i}^E = \theta_r \delta_{t,i}^r + \theta_s \delta_{t,i}^s$, since both market signals $\delta_{t,i}^r$ and $\delta_{t,i}^s$ are known to the decision-maker.

Aviv shows the following: Let $\hat{\Delta}_{t,i}^E \doteq \sum_{l=i}^{T} \hat{\delta}_{t,l}^E$, and define the $(T+1)$-dimensional state vector X_t as follows:

$$X_t \doteq (d_t - \mu, \hat{\Delta}_{t+1,1}^E, \hat{\Delta}_{t+2,2}^E, \cdots, \hat{\Delta}_{t+T,T}^E)'$$
$$= F_* X_{t-1} + W_t, \qquad (10.28)$$

where

$$F_* = \begin{pmatrix} \alpha & 1 & 0 & \cdots & 0 \\ 0 & 0 & 1 & 0 & 0 \\ \vdots & 0 & \ddots & \ddots & 0 \\ \vdots & \vdots & & \ddots & 1 \\ 0 & 0 & \cdots & 0 & 0 \end{pmatrix}$$

Next, note that $D_t = G_* X_t + \mu$, with $G_* = (1, 0, \ldots, 0)$. Each vector W_t is equal to $(\hat{\epsilon}_t, \hat{\delta}_{t+1,1}^E, \ldots, \hat{\delta}_{t+T,T}^E)$, and the vectors $\{W_t\}$ are statistically independent across t. Finally, the decision-maker fully observes the vector X_t during period t (i.e., $\Psi_t = X_t$ for all t). It follows that:

PROPOSITION 14 *The estimates of the lead time demands, $\hat{D}_t^{(L)}$ and $\hat{D}_t^{(L+\tau)}$, are given by*

$$\hat{D}_t^{(L)} = (L+1)\mu + \left(\sum_{l=1}^{L+1} \alpha^l\right) \cdot (d_{t-1} - \mu)$$
$$+ \sum_{j=1}^{\min(T,L+1)} \left(\sum_{l=0}^{L+1-j} \alpha^l\right) \hat{\Delta}_{t+j-1,j}^E \qquad (10.29)$$

with $\hat{D}_t^{(L+\tau)}$ being the same, replacing L with $L+\tau$ in (10.29). Furthermore, for all $t \geq T$, the uncertainty measure Ω_D associated with the MMSE estimate $\hat{D}_t^{(L)}$ is given by:

$$\Omega_D^{(L)} \doteq \sum_{l=0}^{L} \left[G_* \left(\sum_{j=0}^{L-l} F_*^j \right) \Sigma_W \left(\sum_{j=0}^{L-l} F_*^j \right)' G_*' \right]$$

$$= \text{Var}(\hat{\epsilon}_t) \sum_{l=0}^{L} \left(\sum_{j=0}^{L-l} \alpha^j \right)^2$$

$$+ \sum_{i=1}^{\min(L,T)} \text{Var}(\hat{\delta}_{t,i}^E) \cdot \left\{ \sum_{l=0}^{L-i} \left(\sum_{j=0}^{L-i-l} \alpha^j \right)^2 \right\}. \quad (10.30)$$

Thus far, we provided a characterization of the forecast evolution and the forecasting errors associated with the estimates of the lead-time demands. As we shall see in the next section, this lays the foundation for devising an inventory policy for the channel, and for assessing the overall supply chain's cost performance. Using this tool, one can easily compare between the three settings – LMI, SMI, and CFAR (see §6.4 below).

6. Cost analysis

In this section we describe a method developed in Aviv (2002ab) for assessing the inventory and shortage penalty costs for a two-stage supply chain consisting of a retailer and a supplier. The analysis here is different than that in the existing literature primarily due to the intricate relationship between the retailer's and the supplier's forecasting processes. We shall use the location index $i = 1, 2$ to denote the retailer and the supplier, respectively. Let y_t^i be the net inventory of location i at the end of period t. Then, the following costs are charged in period t: $c^i(y_t^i) = h^i \cdot \max(0, y_t^i) + p^i \cdot \max(0, -y_t^i)$, where q^i is the cost function for location i. The non-negative parameters h^i and p^i represent the per-unit inventory holding cost and shortage penalty cost, respectively, at location i. We shall focus on the long-run average cost functions

$$g^i \doteq \lim_{N \to \infty} \sum_{t=1}^{N} c^i(y_t^i), \quad i = 1, 2.$$

The long-run average total supply chain cost is hence given by $g \doteq g^1 + g^2$.

6.1 Cost assessment

This part considers the assessment of long-run average inventory costs for a given set of policy parameters.

6.1.1 Installation-based inventory systems.

Consider the installation-based inventory system with the base stock levels (10.14) and (10.17). Aviv (2002b) provides the following expression for the long-run average costs as functions of the safety-stock parameters γ^1 and γ^2:

PROPOSITION 15 *Let* $E_t^1 = \sum_{l=0}^{L} D_{t+l} - \hat{D}_t^{(L)}$, *and* $E_{t-\tau}^2 = \sum_{l=1}^{\tau} A_{t-\tau+l} - \hat{A}_{t-\tau}^{(\tau)}$. *Then,*

$$g(\gamma^1, \gamma^2) = \lim_{N \to \infty} \frac{1}{N} \sum_{t=\tau+1}^{N+\tau} \left\{ c^1(\gamma^1 - E_t^1 - \max(E_{t-\tau}^2 - \gamma^2, 0)) \right.$$
$$\left. + c^2(\gamma^2 - E_{t-\tau}^2) \right\} \quad (10.31)$$

The terms E_t^1 and $E_{t-\tau}^2$ stand for the forecast errors of the retailer's and the supplier's estimates of their lead-time demand. We are particularly interested in the *lagged* forecast errors; i.e., that made by the supplier during a particular period, and that made by the retailer τ periods later. Proposition 15 tells us that the cost performance of the supply chain is uniquely determined by the safety stock parameters γ^1 and γ^2, and by the characteristics of the joint distribution of the lagged forecasts E_t^1 and $E_{t-\tau}^2$, which is described in the next theorem.

THEOREM 8 *Consider the error terms E_t^1, and $E_{t-\tau}^2$ of Proposition 15. Under the assumption of Kalman filters in steady state, their joint distribution is bivariate normal:*

$$\begin{pmatrix} E_t^1 \\ E_{t-\tau}^2 \end{pmatrix} \sim N\left(\mathbf{0}_{2 \times 1}, \begin{pmatrix} \Omega^D & C \\ C & \Omega^A \end{pmatrix} \right) \quad (10.32)$$

where Ω^D and Ω^A are the mean-square errors associated with the retailer's and the supplier's estimates of their lead-time demands: (10.15) and (10.22). In addition,

$$C \doteq \mathsf{Cov}(E_t^1, E_{t-\tau}^2) =$$
$$\tilde{G}\left(\sum_{j=0}^{\tau-1} \tilde{F}^j\right)\left\{\sum_{l=\tau}^{\infty}(\tilde{F} - \tilde{F}\tilde{K}\tilde{H})^{l-\tau}\mathsf{E}[\tilde{V}_t V_t']([F - FKH]^l)'\right\}\left(\sum_{j=0}^{L} F^j\right)' G'$$
$$+ \tilde{G}\left\{\sum_{l=1}^{\tau-1}\left(\sum_{j=0}^{l-1}\tilde{F}^j\right)\mathsf{E}[\tilde{V}_t V_t']([F - FKH]^l)'\right\}\left(\sum_{j=0}^{L} F^j\right)' G'$$

It follows from Proposition 15 and Theorem 8 that for a generic pair of random variables R^1 and R^2 that maintain the time-invariant distribution described in (10.32), the cost functions g^1 and g^2 are given by $g^1(\gamma^1, \gamma^2) = \mathsf{E}[c^1(\gamma^1 - R^1 - \max(R^2 - \gamma^2, 0))]$, and $g^2(\gamma^2) = \mathsf{E}[c^2(\gamma^2 - R^2)]$. This result gives rise to a straightforward simulation-based method for estimating the supply chain's long-run average inventory costs for any choice of γ^1 and γ^2: A large set of random values $\{(u_n^1, u_n^2), n = 1, \ldots, N\}$ is generated from the joint distribution (10.32). Then, the expected cost is evaluated by

$$\frac{1}{N} \sum_{n=1}^{N} [c^2(\gamma^2 - u_n^2) + c^1(\gamma^1 - \max(u_n^2 - \gamma^2, 0) - u_n^1)]$$

This final cost estimation procedure can be repeated enough times, with different streams of $\{(u_n^1, u_n^2), n = 1, \ldots, N\}$, so as to generate a desirable statistical confidence interval. For an illustration, see the LMI model of Aviv (2002a).

6.1.2 Echelon-based inventory systems.

Consider next an echelon-based inventory system with a single decision-maker, that applies the base stock levels (10.27). For this setting Aviv (2002a) shows the following result: Let $E_t^1 = \sum_{l=0}^{L} D_{t+l} - \hat{D}_t^{(L)}$, and $E_{t-\tau}^2 = \sum_{l=0}^{L+\tau} D_{t-\tau+l} - \hat{D}_{t-\tau}^{(L+\tau)}$. Also, let $\gamma = \Gamma^2 - \Gamma^1$ and $\tilde{E}_{t-\tau}^2 = E_{t-\tau}^2 - E_t^1$. Then, the cost functions g^1 and g^2 are given by $g^1(\Gamma^1, \Gamma^2) = \mathsf{E}[c^1(\Gamma^1 - R^1 - \max(\tilde{R}^2 - \gamma, 0))]$, and $g^2(\Gamma^1, \Gamma^2) = \mathsf{E}[c^2(\gamma - \tilde{R}^2)]$. The generic pair of random variables $(R^1, \tilde{R}^2)$ represents the joint distribution of $(E_t^1, \tilde{E}_{t-\tau}^2)$ for every t. It can be easily seen that the cost functions g^1 and g^2 are given by the exact same type of expressions as for the installation-based policies, but with a slight transformation of variables. We also refer the interested reader to Axsäter and Rosling (1993) in which the equivalence between installation-based and echelon-based is studied.

6.2 Policy coordination in the supply chain

In this section we describe a way for finding the pair of safety-stock parameters γ^{*1} and γ^{*2}, that minimize the long-run average total supply chain cost. We assume that installation-based policies are used. The extension of the discussion to echelon-based policies is relatively easy, given the observations made in §6.1.2. We refer to a supply chain in which these optimal safety stocks are used, as a *coordinated two-level inventory system*[2]. We assume that $h^2 < h^1$ and set $p^2 = 0$. Aviv (2002b) shows the following:

THEOREM 9 *Suppose that $\Omega^A, \Omega^D > 0$ and consider the cost function $g(\gamma^1, \gamma^2) = g^1(\gamma^1, \gamma^2) + g^2(\gamma^2) = \mathsf{E}_{(R^1, R^2)}[c^1(\gamma^1 - R^1 - \max(R^2 - \gamma^2, 0)) + c^2(\gamma^2 - R^2)]$. Also, let $f(\gamma^2) \doteq \min_{\gamma^1} g(\gamma^1, \gamma^2)$. Then:*

(i) The functional solution $\gamma^{*1}(\gamma^2)$ to $f(\gamma^2)$ is unique, and is given by

$$\gamma^{*1}(\gamma^2) = F_{\gamma^2}^{-1}\left(\frac{p^1}{h^1 + p^1}\right) \qquad (10.33)$$

where F_{γ^2} is the continuous, cumulative distribution function of $R^1 + \max(R^2 - \gamma^2, 0)$, with support on the entire real line.

(ii) For all $\gamma^2 \in \mathbf{R}$: $-1 < \frac{d\gamma^{*1}(\gamma^2)}{d\gamma^2} < 0$.

(iii) The first order condition $\frac{df(\gamma^2)}{d\gamma^2} = 0$ is satisfied for any point γ^2, that solves the equation

$$\Pr\{R^1 \leq \gamma^{*1}(\gamma^{*2}) | R^2 \leq \gamma^{*2}\} = \frac{h^2 + p^1}{h^1 + p^1} \tag{10.34}$$

From (10.34), we see that when γ^2 is set to its best value γ^{*2}, the retailer sets his own policy parameter γ^1 by adjusting his cost parameters accordingly ($\tilde{h} = h^1 - h^2$ and $\tilde{p} = p^1 + h^2$), and plans only for the cases in which there are no supply delays (i.e., $E^2_{t-\tau} \leq \gamma^{*2}$). This is done by setting γ^1 so that the probability of a stockout occurring at the retailer's location after a lead-time (i.e., $E^1_t \leq \gamma^{*1}$), is given by the cost ratio $\tilde{p}/(\tilde{p} + \tilde{h}) = (h^2 + p^1)/(h^1 + p^1)$. This result is reminiscent of that of Clark and Scarf (1960): In the case of $\mathsf{Cov}(R^1, R^2) = 0$, (10.34) reduces to $\Pr\{R^1 \leq \gamma^{*1}\} = \tilde{p}/(\tilde{p} + \tilde{h})$, or $\gamma^{*1} = \sqrt{\Omega^D} \cdot \Phi^{-1}(\tilde{p}/(\tilde{p} + \tilde{h}))$. Part (ii) of the theorem shows that regardless of the correlation between the lagged forecast error terms, an increase in the supplier's inventory will enable the retailer to decrease his safety stock, when the latter is concerned with local cost performance (i.e., the function $f(\gamma^2)$). Part (ii) also supports the intuition that in order to achieve a certain reduction in the retailer's safety stock inventory, at least the same amount would need to be added to the supplier's safety stock.

Since the correlation between the lagged forecast errors E^1_t and $E^2_{t-\tau}$ is a determining factor in the cost performance of the supply chain, the framework described above provides both the control policy and cost estimation procedures that may add important new insights into practical settings. Some of the previous models in the literature did not have to take this factor into account since $\mathsf{Cov}(E^1_t, E^2_{t-\tau}) = 0$. Aviv (2002b) shows, for instance, that if $\hat{X}_t = FX_{t-1}$, then the covariance factor is equal to zero. This condition applies in cases in which the retailer fully observes the value of X_t during period t; see, e.g., Lee et al. 2000, and Graves 1999. In contrast, in the model of Aviv (2002a) described in §5 above, the covariance factor plays an important role in the understanding of the relationship between a supplier's and a retailer's forecasting correlation pattern, and the benefits the supply chain can gain from collaborative forecasting programs.

The results in Theorem 9 lead to a simple method for optimizing the policy parameters. Using part (i), one can write a simple routine to calculate the optimal value of $\gamma^{*1}(\gamma^2)$ for any given value of γ^2. Then, a simple and efficient line search procedure can be written to calculate the best value of γ^{*2}. In fact, when $\text{Cov}(R^1, R^2) \geq 0$, it is easy to show that γ^{*2} is unique, and that the function $f(\gamma^2)$ is quasi-convex. It is possible that this result extends to $\text{Cov}(R^1, R^2) < 0$ as well. Aviv (2002b) suggests a simple (but possibly crude) upper bound on the cost $g(\gamma^{*1}, \gamma^{*2})$:

$$g(\gamma^{*1}, \gamma^{*2}) \leq (h^1 + p^1) \cdot \phi\left[\Phi^{-1}\left(\frac{p^1}{h^1+p^1}\right)\right]$$
$$\cdot \sqrt{\Omega^D + 2 \cdot \text{Cov}(E_t^1, E_{t-\tau}^2) + \Omega^A} \quad (10.35)$$

The upper bound (10.35) is based on a setting in which the supplier uses a base-stock policy, but with a very large negative target level ($\gamma^2 \to -\infty$). See Aviv (2002b) for more details and discussion. A different upper bound on $g(\gamma^{*1}, \gamma^{*2})$ is based on a setting in which the retailer is responsible for forecasting the demand during a lead-time of $L + \tau$ periods, and the supplier's only role is to pass on the retailer's replenishment orders:

$$(h^1 + p^1) \cdot \phi\left[\Phi^{-1}\left(\frac{p^1}{h^1+p^1}\right)\right] \cdot \sqrt{\mathsf{E}[(\hat{D}_t^{(L+\tau)} - \sum_{l=0}^{L+\tau} D_{t+l})^2]}$$

A comparison between the upper bounds is often important. For example, if the latter upper bound is the smaller between the two, then it may be appropriate for the supply chain to use the supplier as a cross-docking point, and let the retailer take responsibility of the forecasting process.

6.3 Decoupled two-level inventory systems

In this section we discuss a so-called *decoupled inventory system*, in which the supplier maintains a high level of product availability to the retailer. Consequently, the retailer will rarely face supply delays, and hence he does not have to be concerned about supply shortages when setting safety-stocks at his level. Unlike the models we described earlier in this section, the calculation of the policy parameters and cost estimation is much easier. Let p^2 be the per-period penalty cost charged to the supplier for any unit delayed more than the nominal lead time L. This cost rate will be set to be high enough so that it is almost guaranteed that no delay will occur. Yet, p^2 is *not* actually considered as a net cost charged to the supply chain, but rather as an internal payment mechanism that is set to provide the incentives for the supplier to hold a satisfactory amount of inventory. The retailer will set $\gamma_D^1 = \Phi^{-1}\left(\frac{p^1}{h^1+p^1}\right) \cdot \sqrt{\Omega^D}$, and his long-run average costs will be (approximately) $(h^1 + p^1) \cdot \phi\left[\Phi^{-1}\left(\frac{p^1}{h^1+p^1}\right)\right] \cdot \sqrt{\Omega^D}$. The supplier will set

$\gamma_D^2(p^2) = \Phi^{-1}\left(\frac{p^2}{h^2+p^2}\right) \cdot \sqrt{\Omega^A}$, and the inventory *holding* costs at the supplier's location is hence given $h^2\left[\frac{p^2}{p^2+h^2} \cdot \Phi^{-1}(\frac{p^2}{p^2+h^2}) + \phi(\Phi^{-1}\frac{p^2}{p^2+h^2}))\right] \cdot \sqrt{\Omega^A}$. We use the subscript "D" in the safety-stock parameters, since they relate to a decoupled system. As can be seen, decoupled policies are attractive because of their simplicity and the ease of their performance evaluation. Nevertheless, they may be significantly outperformed by better non-decoupled policies. Although the use of a decoupled inventory control policies may sometimes be well-justified (if, for instance, inventory holding costs at the supplier's facility are not significant, or if the supplier almost accurately forecasts the demand he faces), we advise some caution when selecting this type of policy structure merely due to its technical convenience.

During the study reported in Aviv (2002a), the author observed that from the overall supply chain perspective, a decoupled chain can perform very poorly compared to a coordinated one. Aviv (2002b) suggests a simple formula that specifies a range of values of p^2 for which the associated decoupled system performs even poorer than a non-decoupled setting in which $\gamma^2 = -\infty$: Consider the supply chain's long-run average cost $g(\gamma_D^1, \gamma_D^2(p^2))$, and let

$$u_t \doteq \frac{h^1 + p^1}{h^2} \cdot \phi\left[\Phi^{-1}\left(\frac{p^1}{h^1 + p^1}\right)\right]$$
$$\times \left[\sqrt{\frac{\Omega^D + 2 \cdot \text{Cov}(E_t^1, E_{t-\tau}^2) + \Omega^A}{\Omega^A}} - \sqrt{\frac{\Omega^D}{\Omega^A}}\right]$$

and let u^* be the solution to the equation $u\Phi(u) + \phi(u) = u_t$. Then, for all $p^2 \geq h^2\frac{\Phi(u^*)}{1-\Phi(u^*)}$, $g(\gamma_D^1, \gamma_D^2(p^2)) \geq g(\gamma^{*1}(\gamma^2 = -\infty), \gamma^2 = -\infty)$. The upper bound on p^2 is rather crude, and thus Aviv (2002b) suggests the use of the above-mentioned numerical simulation method to find the best operating policy (within our policy class) for the supply chain, and compare that to the performance $g(\gamma_D^1, \gamma_D^2(p^2))$.

6.4 Results from the study of CF

In Aviv (2002a) we conducted a numerical study to analyze the three settings (LMI, SMI, and CFAR; see §5), over a large set of system parameters combinations. For each specific vector of parameters, we used the cost assessment and optimization method to find the best average cost for the supply chain, under LMI, SMI, and CFAR. We found that when information about future demand is not collected periodically in the system (i.e., forecasts only rely on demand estimates that are known well in advance), the supply chain gains no benefits by providing the supplier with direct access to demand and inventory information at the retailer's level. We thus argue that the classic AR(1) model is insufficient to explain the benefits of information sharing practices (e.g., in the grocery industry; see Lee et al. 2000), since it does not capture the existence of advanced demand information. Therefore, the model in Equations (10.24)-(10.25) offers a valuable extension to the literature. When demand forecasts can be revised periodically by the supply chain members, a performance gap exists between the three settings, even in the case $\alpha = 0$ (the case of i.i.d. demand).

Additionally, we observe that costs are declining as the explainable portion of the error terms q_t increases. The magnitude of the performance gap between CFAR and SMI becomes larger when more advanced demand information prevails in the system. This suggests that as companies improve their ability to explain future demands on the basis of early market signals, the potential of CFAR vis-a-vis SMI increases. Hence, a standard VMI program (i.e., no collaborative forecasting) that has worked well up until a certain point of time, may need to be replaced by CFAR as the trading partners improve their forecasting performance. As a conjecture, we expect that as technology advances in industry (e.g., making data mining solutions more affordable and powerful), we shall see CFAR replacing most standard VMI programs. An exception to that would be in settings in which the supplier's forecasting capabilities are substantially better than those of the retailer.

In agreement with earlier results by Lee et al. (2000) and Graves (1999), we observed that the costs increase considerably with α, i.e., as demands in consecutive periods are more correlated. For the case $\alpha = 0$, Aviv (2001a) has illustrated that significant benefits can be achieved by taking into account early demand signals. It seems that when $\alpha > 0$ the sizes of these benefits are even larger. But why should a company be concerned about the value of taking into account early demand information? Shouldn't they do it anyway? The answer to this has to do with the ability of the company to make use of market signals. Consider, for example, Longs Drug Stores, a major U.S. drug chain with $3.7 billion in annual sales. Lee and Whang (2001) report that although the pharmacists and buyers of this chain were intimately familiar with seasonal patterns of demand for major drugs, they lacked the data and decision-support systems that would enable them to effectively manage the enormous number of stock keeping units sold by the company. It is therefore to the advantage of such companies to be able to use models for gauging the potential benefits that decision-support technology can provide them with. Longs Drug Stores partnered with a third party demand management solution provider, Nonstop Solutions, that offers state-of-the-art methodologies to optimize their demand management process. Among the strengths provided by Nonstop Solutions is the capability of analyzing rich and timely data to generate forecasts.

Both Aviv (2001) and Aviv (2002a) study how the relationship between the supply chain members' forecasting capabilities impact the difference between various supply chain structures. We refer the interested readers to these papers for a detailed discussion.

7. Summary

We provided in this chapter an overview of existing and potential modeling approaches available for the study of collaborative forecasting programs. A main part of the chapter was occupied by a presentation of the linear state space framework, which simultaneously models the demand process, the forecasting process evolution, the inventory control processes, and the collaboration process. Nevertheless, two sections of the chapter provide the reader with an overview of alternative forecast evolution models, and how they can be integrated in inventory planning models. We hope that researchers and practitioners find this chapter useful for a continuing research on CPFR.

We conclude this chapter with several ideas for future research. One research path, which is currently in progress, has to do with the study of the benefits of CPFR in settings limited by resource constraints. Several research papers have demonstrated that the value of information sharing is limited in capacity constrained environments because decision-makers are limited in their ability to respond to information (see, e.g., Cachon and Fisher 2000, and Gavirneni et al. 1999). Nevertheless, what happens when capacity is not a constraint, but is also a decision variable? We await a completion of the analysis of this model, and hence we do not present the results yet.

Other key issues in the implementation of CPFR programs include the following: (1) Incentive alignment and truthful share of information. We have seen some good research in this area (e.g., Cachon and Lariviere 2001, Mishra et al. 2001, and Miyaoka 2003), but there is still a lot of room for future research, modeling CF as an ongoing process, rather than a single-procurement transaction. (2) Modeling Multi-lateral CPFR agreements (e.g., N-tier chain structures). Extensions of the linear state space models are certainly possible in this direction. (3) Analyses that fall in the interface between marketing and supply chain operations. For example, how promotion plans or prices can be integrated into the forecasting and inventory management processes. The list of additional topics for future research is long, and interested readers should refer to the material published through the CPFR web page (currently http://www.cpfr.org/) for updates on current trends in CPFR and the managerial concerns and issues most relevant to industry.

Notes

1. Aviv (2002b) shows how the model can easily accommodate the cases in which demand equation is of the form $D_t = \mu + GX_t + W_t$, with $\{W_t\}$ being a white noise process with $W_t \sim N(\mu_W, \sigma_W)$ for all $t \geq 1$, as long as the processes $\{V_t\}$ and $\{W_t\}$ are uncorrelated at all non-zero lags (i.e., $E[W_s \cdot V_t] = 0_{n \times 1}$ for all $s \neq t$).

2. In fact, in the simple i.i.d. case, Lee and Whang (1999) discuss ways in which installation-based policies can be set, through an establishment of an appropriate performance measure, to yield the same performance as a centralized serial supply chain.

References

Andersen Consulting Report. 1998. *Unlocking hidden value in the personal computer supply chain.*

Anderson, B.D.O., J.B. Moore. 1979. *Optimal Filtering.* Prentice-Hall, Englewood Cliffs.

Anderson, D.L., H. Lee. 2001. New supply chain business models – the opportunities and challenges. *Achieving Supply Chain Excellence through Technology.* Montgomery Research, Inc.

Aviv, Y. 2001. The effect of collaborative forecasting on supply chain performance. *Management Sci.* **47** (10) 1326-1343.

—. 2002a. Gaining benefits from joint forecasting and replenishment processes: The case of auto-correlated demand. *Manufacturing and Service Operations Management* **4** (1) 55-74.

—. 2002b. A Time Series Framework for Supply Chain Inventory Management. To appear in *Operations Research.*

—, A. Federgruen. 2001. Design for postponement: A comprehensive characterization of its benefits under unknown demand distributions. *Operations Research* **49** (4), 578-598.

Axsäter, S., K. Rosling. 1993. Notes: Installation vs. echelon stock policies for multilevel inventory control. *Management Sci.* **39** (10) 1274-1280.

Azoury, K.S. 1985. Bayes solution to dynamic inventory models under unknown demand distribution. *Management Sci.* **31** 1150-1160.

—, B. L. Miller. 1984. A comparison of the optimal ordering levels of Bayesian and non-Bayesian inventory models. *Management Science* **30** 993-1003.

"Heineken redefines collaborative planning through the Internet," *Beverage Industry,* Sept 1, 1998 p47(1).

Beyer, D., S.P. Sethi. 1997. Average cost optimality in inventory models with Markovian demands. *J. Optim. Theor. Appl.* **92** (3) 497-526.

Bourland, K., S. Powell, D. Pyke. 1996. Exploring timely demand information to reduce inventories. *European Journal of Operational Research* **92** 239-253.

Brockwell, P.J., R.A. Davis. 1996. *Time Series: Theory and Methods (2ed).* Springer-Verlag, New York.

Cachon, G.P., M. Fisher. 2000. Supply chain inventory management and the value of shared information. *Management Sci.* **46** (8) 1032-1048.

Cachon, G.P., M.A. Lariviere 2001. Contracting to assure supply: How to share demand forecasts in a supply chain. *Management Sci.* **47** (5) 629-646.

Chan, S.W., G.C. Goodwin, K.S. Sin. 1984. Convergence properties of the Riccati difference equation in optimal filtering of nonstabilizable systems. *IEEE Transactions on Automatic Control* **AC-29** 10-18.

Chen F., Z. Drezner, J.K. Ryan, D. Simchi-Levi. 1999. The bullwhip effect: Managerial insights on the impact of forecasting on variability in a supply chain. Tayur, S., R. Ganeshan, M. Magazine, eds. *Quantitative Models for Supply Chain Management.* Kluwer Academic Publishers Group. 417-440.

Chen, F., Z. Drezner, J.K. Ryan, D. Simchi-Levi. 2000. Quantifying the bullwhip effect in a simple supply chain: The impact of forecasting, lead times, and information. *Management Science* **46** (3) 436-443.

Chen, F., J.K. Ryan, D. Simchi-Levi. 2000. The impact of exponential smoothing forecasts on the bullwhip effect. *Naval Research Logistics* **47**, 269-286.

Clark, A.J., H. Scarf. 1960. Optimal policies for a multi-echelon inventory problem. *Management Sci.* **6** 475-490.

Croson R., Donohue, K.L. 2001. Experimental economics and supply chain management. To appear in *Interfaces*.

Dong, L., H.L. Lee. 2001. Optimal policies and approximations for a serial multi-echelon inventory system with time correlated demand. Working Paper. *Washington University and Stanford University.*

Dvoretzky, A., J. Kiefer, J. Wolfowitz. 1952. The inventory problem I. Case of unknown distributions of demand. *Econometrica* (20) 450-466.

Erkip, N., W.H. Hausman, S. Nahmias. 1990. Optimal centralized ordering policies in multi-echelon inventory systems with correlated demands. *Management Sci.* **36** (3) 381-392.

Frazier, R.M. 1986. Quick response in soft lines. *Discount Merchandiser.* January. 86-89.

Gallego, G., Ö. Özer. 2001. Integrating replenishment decisions with advanced demand information. *Management Sci.* **47** (10) 1344-1360.

Gavirneni, S., R. Kapuscinski, S. Tayur. 1999. Value of information in capacitated supply chains. *Management Sci.* **45** (1) 16-24.

Graves, S.C. 1999. A single-item inventory model for a nonstationary demand process. *Manufacturing and Service Operations Management* **1** (1) 50-61.

—, H.C. Meal, S. Dasu, Y. Qiu. 1986. Two-stage production planning in a dynamic environment. Axsäter, S., C. Schneeweiss, E. Silver, eds. *Multi-Stage Production Planning and Control. Lecture Notes in Economics and Mathematical Systems.* Springer-Verlag, Berlin.

Hamilton, J.D. 1994. *Time Series Analysis.* Princeton University Press.

Hammond, J. 1990. Quick response in the apparel industry. *Harvard Business School Note 9-690-038.*

HBS. 1990. Bose corporation: The JIT-II program (A). *Harvard Business School Case 9-694-001.*

Harvey, S.C. 1990. *Forecasting, Structural Time Series Models and The Kalman Filter.* Cambridge University Press.

Hye, J., "Wal-Mart, Sara Lee near completion of CPFR pilot project," *Daily News Record,* April 22, 1998 n48 p12(1).

Heath, D. P., P. L. Jackson. 1994. Modeling the evolution of demand forecasts with application to safety stock analysis in production/distribution systems. *IIE Transactions* **26** 17-30.

i2 TradeMatrix. 2001. *TradeMatrix Solutions Overview (TMSO).* Presentation notes. i2 Field Education Program.

Iglehart, D. 1964. The dynamic inventory model with unknown demand distribution. *Management Sci.* **10** 429-440.

Johnson, G., H. Thompson. 1975. Optimality of myopic inventory policies for certain dependent demand processes. *Management Sci.* **21** 1303-1307.

Kahn, J. 1987. Inventories and the volatility of production. *Amer. Econ. Rev.* 667-679.

Karlin, S. 1960. Dynamic inventory policy with varying stochastic demands. *Management Sci.* **6** 231-258.

Karlin, S., H.E. Scarf. 1958. Optimal inventory policy for the Arrow Harris Marschak dynamic model with a time lag. Chapter 14 in: Arrow, Karlin and Scarf (eds.), *Studies in the mathematical theory of inventory and production.* Stanford University Press. Stanford, CA.

Koloszyc, G. 1998. Retailers, suppliers push joint sales forecasting. *Stores,* June.

Kurt Salomon Associates Inc. 1993. *Efficient Consumer Response: Enhancing consumer value in the grocery industry.* Food Marketing Institute, Washington, D.C.

Lee, H.L., P. Padmanabhan, S. Whang. 1997a. Information distortion in a supply chain: The bullwhip effect. *Management Sci.* **43** 546-558.

—, —, —. 1997b. Bullwhip effect in a supply chain. *Sloan Management Review* **38** (Spring) 93-102.

Lee, H.L., S. Nahmias. 1993. Single-product, single-location models. Chapter 1 in: Graves, Rinnooy Kan and Zipkin (eds.), *Logistics of production and inventory. Handbooks in Operations Research and Management Science,* vol. 4. North Holland, Amsterdam.

—, K.C. So, C.S. Tang. 2000. The value of information sharing in a two-level supply chain. *Management Sci.* **46** 626-643.

—, S. Whang. 1999. Decentralized multi-echelon inventory control systems: Incentives and information. *Management Sci.* **45** (5) 633-640.

—, —. 2001. Demand chain excellence: A tale of two retailers. *Supply Chain Management Rev.* **5** 40-46.

Lovejoy, W.S. 1990. Myopic policies for some inventory models with uncertain demand distributions. *Management Sci.* **36** (6) 724-738.

Margetta, J. 1998. The power of virtual integration: An interview with dell computer's Michael Dell. *Harvard Business Review*, March-April.

Miyaoka, J. 2003. Implementing collaborative forecasting through buyback contracts. Working paper. *Department of Management Science and Engineering, Stanford University, Stanford, CA.*

Miyaoka, J., L.R. Kopczak. 2000. Using forecast synchronization to reduce the bullwhip effect of the forecast error. Working paper. *Department of Management Science and Engineering, Stanford University, Stanford, CA.*

Mishra, B.K., S. Raghunathan, X. Yue. 2001. Demand forecast sharing in supply chains. Working paper. *School of Management, The University of Texas at Dallas.*

Monahan, G.E. 1982. A survey of partially observable Markov decision processes: Theory, models, and algorithms. *Management Sci.* **28** (1) 1-16.

S. Raghunathan. 2001. Information sharing in a supply chain: A note on its value when demand is nonstationary. *Management Sci.* **47** 605-610.

Scarf, H. 1959. Bayes solutions of the statistical inventory problem. *Ann. Math. Statist.* **30** 490-508.

—. 1960. Some remarks on Bayes solutions to the inventory problem. *Naval Logistics Research Quarterly* **7** 591-596.

Song, J.S., Zipkin, P. 1993. Inventory control in a fluctuating demand environment. *Operations Research* **41** 351-370.

—, —. 1996. Managing inventory with the prospect of obsolescence. *Operations Research* **44** 215-222.

Treharne, J.T., Sox, C.R. 2002. Adaptive inventory control for nonstationary demand and partial information. *Management Sci.* **48** (5) 607624.

Watson, N., Y.S. Zheng. 2000. Adverse effects of over-reaction to demand changes and improper demand forecasting. Working Paper. *University of Pennsylvania.*

Veinott, A. 1965. Optimal policy for a multi-product, dynamic, non-stationary inventory problem. *Management Sci.* **12** 206-222.

Zarley, C., K. Damore 1996. Backlogs plague HP – Resellers place phantom orders to get more products. *Computer Reseller News.* May 6.

David Simchi-Levi, S. David Wu, and Z. Max Shen (Eds.)
Handbook of Quantitative Supply Chain Analysis:
Modeling in the E-Business Era
©2004 Kluwer Academic Publishers

Chapter 11

AVAILABLE TO PROMISE

Michael O. Ball
Robert H. Smith School of Business & Institute for Systems Research,
University of Maryland, College Park, MD 20742

Chien-Yu Chen
School of Management,
George Mason University, Fairfax, VA 22030

Zhen-Ying Zhao
Robert H. Smith School of Business & Institute for Systems Research,
University of Maryland, College Park, MD 20742

Keywords: Available to Promise (ATP), Push-Pull Framework, Order Promising and Fulfillment, Mixed Integer Programming

1. Introduction

Advanced information technologies and expanded logistics infrastructure are reshaping the global business environment. Buyers and sellers can now share information and exchange decisions in real time while products can be moved from place to place globally within days. In order to gain competitive advantage in the new business landscape, firms are redefining their business models not only to enhance back-end logistics efficiency but also to improve front-end customer satisfaction. Available to Promise (ATP) is a business function at the forefront of this trend since it plays the prominent role of directly linking customer orders with enterprise resources and must evaluate the tradeoff between front-end and back-end performance.

1.1 Push-Pull Framework

Increasingly, companies are viewing supply chain management as a competitive weapon, and as such are shifting their focus from cost minimization to profit maximization (see Lee and Billington, 1995). Along with this shift, there is a growing trend for changing business processes and corporate organization from company-centric (or product-centric) toward customer-centric. Characteristics of this trend include offering a huge variety of customizable products and adopting Make-to-Order (MTO) and Assemble-to-order (ATO) production strategies. Traditional planning functions alone are no longer capable of matching supply with demand under this high product mix and low order volume environment since it is not possible to accurately foresee the short-term demand and efficiently prepare each customized finished product by implementing pure Make-to-Stock (MTS) strategy. Advanced ATP, therefore, can play a vital role in mapping customer demands to enterprise resources.

Before we discuss ATP strategies in more detail, we would like to introduce a Push-Pull framework. *There are two fundamental control mechanisms along a supply chain: push control and pull control. Push control consists of those mechanisms for production planning and product distribution that are based on a forecast of future demand. Pull control consists of those mechanisms that are executed in response to customer orders.* For detailed discussion, see Simchi-Levi, et al. (2000) and Hopp and Spearman (2001). In general, pull control is inherently better at reducing inventory, increasing flexibility and improving responsiveness to customer product requirements. Push control is associated with increasing manufacturing and production efficiency, improving stability and reducing the response time to customer orders. Closer observation shows that most practical production systems employ both push controls and pull controls to gain the advantages of each and avoid disadvantages of both. Depending on the degree of product customization, product lead-time can be divided into two components. The first component involves those production resources made available by the push control activities. The second component involves those activities that are executed after a customer order is received. It becomes obvious that any production framework must define a push-pull boundary. Any inconsistencies and contradictions between push and pull controls can degrade the balance between supply and demand on the boundary. This is especially true in high mix low volume production environments. ATP is the business function that coordinates demand and supply across the push-pull boundary. As the location and nature of the push-pull boundary has changed in recent years so has the nature of required ATP functionality.

1.2 Available to Promise (ATP)

The fundamental business role of the ATP function is to provide a response to customer order requests based on resource availability. In order to make a reliable response to a customer order, an ATP system must insure that the quantity promised can be delivered on the date promised. Thus, an ATP system must include both order promising and order fulfillment capabilities. In addition, an ATP system should be able to dynamically adapt resource utilization and to prioritize the customer orders so as to coordinate supply and demand in a way that maximizes profit. *By its very nature, the ATP system should operate within a short-term operational environment where most resource availability is considered fixed because of procurement lead-time limitations. This distinguishes ATP systems from traditional planning, scheduling and inventory management processes.*

As shown in Figure 11.1, any production framework actually defines a production process push-pull boundary across the order fulfillment cycle. On the left of the boundary, we have the production resources made available by the push-based planning processes. Customer orders arise, on the right of the boundary, to compete for these limited short-term production resources. The ATP system executes at the interface of the push-driven flow and pull-based flow systems. It plays a critical role in making best use of available resources, including raw materials, work-in-process, finished goods, and even production and distribution capacities, to commit customers' order requests over a period of time across a supply chain. By doing so, it coordinates activities across the push-pull boundary of a supply chain.

2. Business Examples

This section provides several examples of actual ATP business practices. Each example has its own special features in terms of customer order format, customer response strategy, problem complexity, and resource allocation approach. We start with a generic description of the so-called "conventional" ATP, which represents the traditional view of ATP and which is probably the approach still most broadly used today.

2.1 Overview of Conventional ATP

Conventional ATP is associated with a traditional make-to-stock (MTS) production environment associated with long process lead times, relatively standard products and stable demand. In the Materials Requirements Planning (MRP) II framework, production decisions are based on the embedded Master Production Schedule (MPS), which takes into account a demand forecast, committed customer orders, existing inventory and production capacity. Hence,

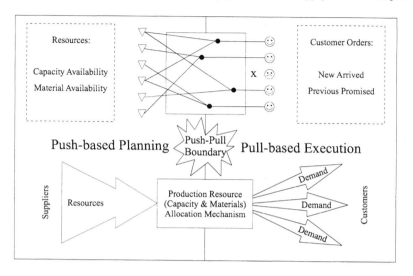

Figure 11.1. Push-pull boundary and role of ATP

APICS defines ATP as *"The uncommitted portion of a company's inventory and planned production, maintained in the master schedule to support customer order promising."* Traditionally, the ATP scope includes the on-hand inventory and the planned production at a designated location. The MPS becomes "moderately firm" or even "frozen" once a designated time window is reached. This implies that the planned production quantity becomes static as the planned production time approaches.

The ATP quantity could be calculated in a few ways: discrete ATP, cumulative ATP without look-ahead, and cumulative ATP with look-ahead. The difference between the discrete and cumulative calculations is that the discrete calculation only computes the ATP quantities for those periods with non-zero planned production but the cumulative calculation also computes the ATP quantities for other periods by subtracting the new orders from the ATP in the previous time period. For the look-ahead case, the system does not allow negative ATP quantities and reduces ATP quantities in earlier time periods to cover any possible shortage.

Figure 11.2 shows that conventional ATP can be viewed as a bookkeeping function in MPS, which keeps track of uncommitted availability, including planned production and existing inventory, in a database for each finished product. Typical conventional ATP practice formulates a response to an order based on a simple check or straightforward search algorithm. However, for cases involving multiple production and inventory locations, a more complex pull-based mechanism may be needed to choose the sourcing location for each order.

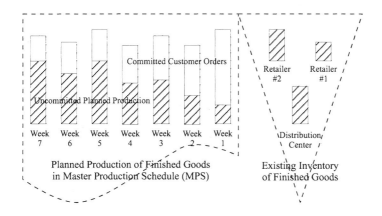

Figure 11.2. An Availability Database in a Conventional ATP

2.2 Toshiba Electronic Product ATP System

We now describe the ATP system for a particular electronic product (denoted by EP) manufactured by Toshiba Corporation. The EP supply chain consists of multiple final assembly and testing (FAT) factories all located in Japan, which provide EPs delivered directly to domestic business customers. An ATO production framework is employed. The order promising and fulfillment process involves several thousand product models. Order sizes range from a very small number of units to a few hundred. Orders are generated by one of several sales units and are processed by a single central order processing system. The ATP system collects orders over a $1/2$ hour time interval and returns commitments to the sales offices at the end of each ATP run ($1/2$ hour interval). order commitments are booked up to ten weeks in advance of delivery.

For EP, Toshiba employs the business practice of never denying an order. If an order cannot be fulfilled before its requested due date, then a promise date beyond the requested date is given, i.e. it is backordered, or the order is split with a portion given an early promise date, e.g. before the due date and a portion given one or more later promise dates. In order to emphasize customer satisfaction for EP, Toshiba weights due date violation higher than any holding costs in its order fulfillment decision models. Occasionally, the sales staff will book "pseudo orders" to reserve critical resources for anticipated future high priority demands.

The order-promising process employed partitions the due date time horizon into three intervals: fixed product, flexible product and flexible resource. For the fixed product interval, which spans from approximately the present time to two weeks into the future, resources, in the form of manufacturing orders (MO) are fixed. An MO specifies the production quantity for each product at each assembly line in each factory. That is, a fixed production schedule is set, which takes into account both production capacity availability and critical material availability. Having a fixed schedule stabilizes production dynamics in the near term and allows for the required materials to be set up and put in place. Any order commitments made for this time interval must fit within the fixed production schedule. In the flexible product interval, two kinds of resources, capacity and material, are considered. The capacity consists of both production capacity in different factories and transportation capacity from factories to sales subsidiaries. The production capacity is given daily at factory level in terms of machine-hour and manpower availability, while the transportation capacity is specified as weekly maximum quantity from factories to sales subsidiaries. The weekly availability of individual critical materials is aggregated into finished good level availability grouped based on the bill of material (BoM). Any order commitments made for this time interval must satisfy the capacity and material availability constraints. The flexible product interval spans from approximately two weeks to two months into the future. For the flexible resource interval, which covers due dates more than two months into the future, the only constraint considered is production capacity. This interval starts beyond the resource lead times so any resource commitments can be met.

2.3 Dell Two-stage Order Promising Practice

The Dell Computer Corporation is famous of its "direct" business model (see Dell and Fredman, 1999) that assembles customized computer systems based on customer orders, and ships directly from the factory to the customers. Based on Kraemer, et. al. (2000), Dell basically segments its customers into Transaction, Relationship, and Public/International customer segments, in which more refined sales channels, including Home & Home Office, Small Business, Medium & Large Business, State & Local Government, Federal Government, Education, Healthcare, are offered at Dell. By doing so, Dell is able to satisfy special needs in each customer channel.

In most transaction-based consumer channels, Dell uses a two-stage order promising practice that is widely adopted by e-commerce companies (see Bayles, 2001). The U.S. patented on-line configuration service (see Henson, 2000) at "dell.com" allows customers to configure their computer system by choosing options over CPUs, memory, operating systems, etc. As a result, Dell has to handle potentially thousands of possible configurations for each product family in its assembly process. Closely collaborating with its component suppliers, Dell is able to assemble each customer order in just-in-time (JIT) mode according to the customer's specification with only hours of inventory of components in the factory. When a customer finishes configuring a computer and placing the order, Dell offers a near real-time order confirmation message via e-mail stating that the order has been successfully placed and providing a standard shipment date usually about 14 days after receiving the order. Then, a few days later, more precise order promising information on the exact shipment date is provided to the customers via e-mails and/or Dell's on-line order status monitoring system. This two-stage order promising strategy is very popular in e-commerce business practices; the company gives initial "soft" order confirmation immediately then generates "hard" order delivery date after checking resource availability (based on some type of "batch ATP" functionality).

It is not unusual for the actual shipment date to be a few days different from the original promised shipment date due to uncertainty in supply chain processes. However, this unreliability might not be acceptable to relationship-based customers. Therefore, Dell offers a special service through Dell Premier accounts at "premier.dell.com" for business and public sector customers. The service not only provides reports and tools to assist purchasing, asset management, and product support, but also allows customers to hook up their ERP/EDI systems with Dell's to perform real-time computer systems procurement. Through Premier accounts, Dell commits to more responsive and reliable order promising and fulfillment solutions for its relationship-based customers with support from advanced IT systems (see Kraemer and Dedrick 2001).

2.4 Maxtor ATP Execution for Hard Disk Drive

In one of its final assembly factories, Maxtor Corporation offers a component-level product, Hard Disk Drive (HDD), to its customers for further assembly of desktop PCs, high-end servers, consumer electronics products, network attached storage (NAS) server appliances, etc. (see www.maxtor.com). Maxtor has partnership-based B2B relationships with its customers. In this setting, customer orders are not given as specific order quantities and order due dates. Instead, only the total order quantity in each week is specified with permitted minimum and maximum quantity limits. In its order promising and fulfillment processes, Maxtor does not postpone customer orders to later weeks (i.e. no backorders), but it may deny customer orders subject to liability and/or penalties. While customer orders are promised weekly, the order fulfillment process is executed daily to provide accurate resource utilization and production schedules (see Ali, et al. 1998).

Since the HDD is a subassembly component for finished goods, customers are sensitive to the suppliers of the critical components used to produce HDD. Therefore, when a customer places an order, he/she is required to configure the order by specifying not only the individual preferred component but also preferred suppliers of each component. The customer is encouraged to include multiple preferred supplier options in order to provide Maxtor with substitution flexibility within the assembly processes.

However, customer preferences are subject to material incompatibility constraints among alternative suppliers of different material types. Material incompatibility can be characterized by the supplier groups whose components cannot be assembled together due to the differences in design and technology. Material incompatibility issues can become very complex because of the large number of suppliers.

Production capacity is managed from day to day in terms of aggregate manpower and machine hour availability. Day-to-day production smoothness is an important consideration so day-to-day capacity variations are controlled to be within a certain plus and minus percentage window.

2.5 ATP Functionality in Commercial Software

The conventional ATP quantity in the MPS has been widely implemented in many commercial Enterprise Resource Planning (ERP) and Supply Chain Management (SCM) software suites. Advanced ATP functionality has recently grown to be an important component of ERP and SCM solutions, and is becoming a tool for improving competitive advantage. Many Advanced Planning and Scheduling (APS) products, such as i2's Rhythm Module, QAD's Supply Chain Optimizer, Oracle's APS module and SAP's Advanced Planning and Optimizing (APO) module, have enhanced ATP functionality using techniques such as heuristic search, rule-based search or constraint-based models. In these APS Modules, the ATP function is integrated with other modules like transportation scheduling, production scheduling, etc.

i2's Rapid Optimized Integration for R/3 and Demand Fulfillment? (ROI for R/3-DF) has added the advanced Allocated ATP (AATP) function that allows users gain more control over the R/3 order management process through interaction with the Demand Fulfillment module. The allocated ATP module employs a heuristic search algorithm to obtain global allocated availability over a complete supply chain. Accordingly, the allocated ATP module in ROI for R/3-DF supports a promising capability for both simple catalog products and highly configurable products (see www.i2.com/Home/ SolutionAreas/ SupplyChainManagement/).

The global ATP capability of SAP's APO suite serves as a key component for supporting integrated supply chain decision support. Global ATP is a part of the Order Fulfillment solution for SAP-APO. It employs rule-based algorithms to enable effective management of, and commitment processing for, thousands of orders in a real-time mode. User-defined ATP rules are used for determining the preferred distributor for each customer while searching for end-product availability. Moreover, global ATP provides multilevel availability checks considering material substitution and location selection (see www.sap.com/solutions/scm/planning.asp).

Oracle®Global ATP Server extends the order promising capabilities of other Oracle applications. The extended features include distributed global order promising and multi-level supply-chain ATP quantity checking. The global ATP server integrates with other planning modules and front-end order management systems in call centers. It can be triggered by an individual sales order or a set of orders from different sales locations (see www.oracle.com/applications/).

Further examples of systems that provide advanced ATP functionality include: Manugistics Supply Chain Planning and Optimization (SCPO) (see www.manu.com/ solutions/scm.asp), Paragon Applications Suite 4.0 (see www.adexa.com/getstarted.asp) and QAD's Supply Chain Optimizer (see www.qad.com/solutions/ mfgpro/). In all cases, the emphasis is on providing fast search for product availability across a global supply chain.

3. ATP Modelling Issues

Using ATP systems to accomplish order promising and fulfillment involves complicated modeling issues under different business environments. In this section, we systematically summarize the implementation dimensions, discuss factors affecting ATP implementations, and define push-pull ATP strategies.

3.1 ATP Implementation Dimensions

We now summarize the major dimensions that distinguish various ATP models.

ATP Decision Support: The ATP function could provide a simple query capability or provide a specific response to the order(s) under consideration. If only a query capability is provided then the user would be expected to formulate a specific response based on the information provided.

ATP Decision Space: The primary purpose of the ATP function is to provide a response to customer orders. There are two levels of response, one involves the direct response to the customer and the other involves the underlying activities required of the production and distribution systems to carry out the customer commitment.

- *Customer Response Space:* the most fundamental decision related to an order is whether or not to accept the order. Assuming it is accepted, other decision choices could involve delivery date, order quantity and a decision on whether to split the order into multiple pieces. Which of these components must be specified by the ATP function depends on the policies of the organization.

- *Product Processing Space:* the source location for the product must be specified, e.g. the factory that will produce it or the inventory location from which it will be shipped. In addition the shipping mode may need to be specified. For MTO or ATO product a variety of information may be needed regarding how to produce the finished product. In some cases, detailed production schedules are required.

ATP Execution Scope: A supply chain network consists of supply network, manufacturing network and distribution network. ATP execution scope refers to the extent of the supply chain elements included within an ATP model or system's domain of interest. Distribution ATP (d-ATP) considers a distribution network including distribution centers and retailers across a supply chain; Make ATP (m-ATP) considers both the distribution network and the manufacturing network; and Supply Chain ATP (sc-ATP) considers the entire supply chain including distribution network, manufacturing network and supplier network.

ATP Execution Mode: We distinguish between two fundamental modes of ATP execution: batch mode ATP and real-time mode ATP. Under batch mode ATP orders are collected over a batching interval and the ATP function is executed at the end of that interval to generate a response to each customer order, e.g. depending on the application requirements batching intervals could be an hour in length, a day, a week, etc. Under real-time mode ATP, execution of the ATP function is triggered by each individual customer order. When a customer requests an order, a sales representative can promise the customer a delivery date and quantity based on a real-time interaction with the ATP system.

3.2 Factors Affecting ATP Implementations

The ATP implementation dimensions described above clearly impact the nature of ATP implementation. However, the appropriateness of a potential ATP model depends on a number of other characteristics related to the fundamentals of the underlying business. We attempt to summarize the key factors into categories and explain each of them in detail below.

Front-End Factors: Many front-end factors residing near the customer interface play an important role in designing ATP functionality. There are many ways categorizing customer relationships. We use the categories between *partnership-based* and *transaction-based* to distinguish between long-term strategic alliances or supply contracts relationships and the occasional purchasing relationship. We also distinguish between business-to-business ($B2B$) and business-to-consumer (*B2C*) relationships as these can have substantial impact on the nature of ATP. Moreover, we use the terms *e-commerce* vs. *traditional customer* relationships to emphasize the different expectations of order-promising speeds. Each of the categorizations impact important factors related to ATP strategy and model choice. We summarize these front-end factors as follows:

- *Order Specification Flexibility* – The flexibility in configuration, due date, delay, quantity, and order-splitting is usually higher under a partnership-based or B2B relationship. Therefore, sophisticated ATP models that can take full advantage of the order specification flexibility should be employed in these settings.

- *Customer response time* – Customer response time measures the time lapse between order placement and the timing of the corresponding promise. Under an e-commerce relationship (especially associated with B2C), customers generally expect to receive an order promise (or "response") at "web-speed". An ATP system should be implemented in real time mode to satisfy this expectation. However, few ATP systems provide a "hard" promise in a real-time mode today, due to the challenges both in processing large-scale real-time information and in carrying out real-time optimization. Instead, most order promising systems that provide an immediate response use a two-stage approach by giving an immediate "soft" promise and then producing a "hard" promise based on a batch ATP module executed later (see Section 2.3).

- *Profitability and Priority* – In general, a manufacturer prefers higher profit orders in terms of quantity, unit price, or both. However, in the short term other priority considerations may sometimes supersede purely profit-based criteria, particularly under B2B and partnership-based relationships. A well-designed customer response space, e.g. employing differentiated sales channels, can be used to effectively balance the trade-offs between profitability and priority criteria.

- *Order Promising Reliability* – Customers under B2B and B2C relationships often have different tolerances for the order-promising reliability. Generally, in a B2B setting adherence to promised due dates and quantities is more important as production schedules and downstream processes can be adversely impacted by poor order fulfillment.

Back-End Factors: Similarly, there are important back-end factors that determine ATP strategy and model choice. Back-end factors are related to material suppliers and production processes.

- *Production Framework* – The fundamental production strategy ranging from make-to-stock (MTS), make-to-order (MTO), assemble-to-order (ATO), to Configure-to-Order (CTO) has perhaps the most significant impact on the nature of ATP functionality. For example, available finished products make up the base ATP resource in an MTS environment, while available materials and production capacity are allocated by an ATP system in ATO and MTO environments. We can classify resource availability into four levels: 1) local finished-product inventory, 2) remote finished-product inventory, 3) planned finished-product production, and 4) unallocated production capabilities (including available materials and capacity). An appropriate product processing space must be selected in an ATP system to be aligned with different production strategies based on the level of resource availability considered.

- *Product variety & resource commonality* – A company often offers multiple products or a variety of similar models that share common materials or production capacity. The commitment to deliver a specific finished product represents an allocation of these common resources. The degree of product variety and the complexity of the relationship between products and resources directly impacts the complexity of the ATP resource allocation problem.

- *System scope* – An order fulfillment process may involve different locations across a supply chain. Retailers, warehouses, manufacturers and even suppliers can be associated with alternate locations. In the more complex cases, determining how to satisfy a set of orders can represent a complex distribution management problem.

- *Resource Type* – ATP allocation models must potentially deal with three resource types: *material resources, factory resources* , and *distribution resources*. Figure 11.3 illustrates the variety of combinations that could exist in different applications. These directly impact the nature of models required. For example, in some cases the factory resource must be modeled in detail and the ATP problem can be modeled as a type of factory scheduling problem, while in other cases, its distribution planning models are appropriate.

Information Technology Infrastructure Factors: ATP execution requires information like customer orders, resource availability, production processes, inventory, etc. from supply chain business systems. The ATP execution mode is directly affected by the information retrieval speed. Hence, IT infrastructure and production information systems like enterprise resource planning (ERP), manufacturing execution system (MES) and legacy systems are essential to the successful implementation of an ATP strategy. Moreover, ATP outputs also

Available to Promise 459

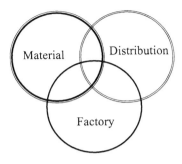

Figure 11.3. Resource types and model complexity

need advanced enterprise systems to effectively execute and implement their results.

3.3 Push vs. Pull ATP Models

There are potentially many ways to classify different ATP models. For example, from a logistics perspective, we can classify ATP models along the execution scope dimension into transportation/distribution, material requirement, job-shop scheduling, supply chain coordination models, etc. From the operations research perspective, we may name an ATP model, based on decision space and problem structure, as an assignment/transportation, network, location, knapsack, resource allocation, or yield management model. However, it is important to differentiate ATP models based on their functional role in the order promising business process. We therefore adopt the push-pull framework and classify ATP models into two fundamental categories: 1) *Push-based ATP models*, and 2) *pull-based ATP models*. The key differences involve not only model implementation timing but also resource allocation granularity. Push-based ATP models are designed to proactively allocate available resources into demand classes and to prepare searchable availability databases for promising future customer demands. In contrast, pull-based ATP models are used to directly react to actual customer orders and to determine detailed commitment parameters for each order. We describe both categories of ATP models in more detail below.

Push-based ATP Models: In anticipation of future potential customer orders, a push-based ATP model pre-allocates lower-level availabilities (usually resources, including material, production, distribution and so on) into higher-level availabilities (e.g., finished products at a certain location in a certain time period) based on forecasted demand. The resulting "allocated" availability (or

ATP quantities) will be used to support future order promising upon actual order placements.

The main advantage of using push-based ATP models is that the up-to-date pre-calculated ATP quantities over multiple time periods are readily available to support real-time and reliable customer order promising. Another significant advantage is the ability to insert longer-term profitability consideration into daily order-promising decisions by emphasizing more profitable demand categories in the pre-allocation. However, as there is more dependence on pre-allocation and less dependence on the reactive pull-based models, the fundamental limitations of basing decisions on (inaccurate) demand forecasts become significant. The resulting inefficiencies worsen as the complexity of product structure, production processes, or the distribution network grows.

Pull-based ATP Models: Pull-based ATP models perform dynamic resource allocation in direct response to actual customer orders. Models of this type can range from a simple database lookup to sophisticated optimization. The purpose of pull-based ATP models is to make best use of available resources (including raw materials, work-in-process, finished goods, and even production and distribution capacities) to commit customer order requests over a period of time across a supply chain. The time horizon in pull-based ATP models is usually so short that a company can assume full knowledge about the availability of production resources. Pull-based ATP models are responsible for matching complicated customer requests with diversified resource availability. The specific decisions usually involve which orders to accept and, for each order, what quantity and which due date to promise.

The main advantage of pull-based ATP models is their ability to mitigate the discrepancy between forecast-driven resource replenishments and order-driven resource utilization. Furthermore, by implementing pull-based ATP models, a production system can retain flexibility and quickly react to special customer preferences. However, the repetitive execution of a pull-based ATP model over time represents a type of "greedy" algorithm and, as such, can at times be overly myopic. The extreme case would be a real-time mode ATP system where each order is committed and resources are allocated as each order arrives. The myopic nature of such an approach can be mitigated by going to batch mode ATP, where orders are collected over a batching interval and commitments are made for the batch based on a batch ATP model. Of course, as the length of the batching interval increases, the response time to customer orders increases as well, which represents a degradation in customer service.

Depending on where the push-pull boundary is positioned and how many echelons are considered in a supply chain network, different ATP strategies can be used by structuring push-based and pull-based ATP models properly. In general there will be a push component and a pull component, but the importance and complexity of the underlying models will vary by context.

4. Push-Based ATP Models

From a production point of view, ATP quantities are often categorized in terms of end-item or product family. In order to meet customers' expectations and improve customer service level, sales and marketing departments usually group and configure end-items along multiple *demand classes*, which can represent customer groups, sales channels, distribution channels, specific physical retail locations, etc. The different demand classes typically have very different demand characteristics, customer priority, sensitivity to delivery dates, as well as unit profitability and overall demand quantity. Push-based ATP models allocate resources to demand classes. In an MTS environment the principal resource allocated is finished product whereas in an MTO environment raw materials/components and production capacity are the resource. Push-based ATP models must consider constraints and characteristics of order promising and fulfillment, demand categories and configurations, customer satisfaction, etc. Vollmann et al. explain the conventional ATP models in their book *Manufacturing Planning and Control System* (Vollmann 1997).

Consider the simple case of a single product and two demand classes A and B. In general, the product available would be partitioned into three quantities Y_A: available to A only, Y_B: available to B only and Y_{AB}: available to A or B. There are two possible cases for the allocation:

1 The allocation constitutes a logical designation in the sense that there is no physical repositioning or packaging of the products, specific to the demand class;

2 The allocation implies the product is moved to a location specific to the demand class or is in some way prepared or packaged for that demand class.

Under case 1, the purpose of the allocation is to reserve a certain amount of product for each demand class to insure that the demand class receives an appropriate level of service. Thus, if over the demand time horizon the demand for A exceeded $Y_A + Y_{AB}$, but the demand for B was less that Y_B, demand for A would go unmet even though product designated for B remained available. The motivation for this approach is to insure that there is product available to insure future demand for B is met. This might be done to insure a certain service level to B or in the case where B had a much higher profit margin than A. Under case 2, once the allocation is carried out, there is a cost associated with transferring product from one demand class to another. For example, Y_A and Y_B could represent amounts shipped to particular retail outlets and Y_{AB} could represent product kept in a warehouse that supplied both A and B. In this case, as demand is realized it could be that product is moved from one category to another but the associated cost must be incurred, e.g. a transshipment from retail outlet A to retail outlet B.

Although push-based ATP models are similar to traditional inventory control and planning models, there are three major differences:

1. Push-based ATP models span a relatively short time horizon, usually several days or a few weeks into the future. Within this time period decisions are based principally on a mix of confirmed customer orders and short term forecasts. In contrast, traditional inventory control and planning models cover longer time periods and decisions are based completely on (longer-term) demand forecasts.

2. Because of the short planning horizon, the bulk of the resource allocated by push-based ATP models is fixed. Inventory control and planning models generally assume that resources are open and infinite.

3. Push-based ATP models must take into account very specific customer-related constraints such as customer priorities, sales channel characteristics, variations in customer service levels and so on. Inventory control and planning models usually operate at a more aggregate level.

4.1 Push ATP Rules and Policy Analysis

The conventional approach to performing push-based ATP allocation is based on allocation rules. Kilger and Schneeweiss (2000) summarize major rule classes as follows:

- Rank based rules: ATP quantity allocation is based on the rank or priority of the demand dimension. This kind of rule is particularly helpful in supporting sales to specific markets, e.g. developing markets in their early stages.

- Per committed rule: Allocation of a limited overall ATP quantity is done in proportion to the forecasted demand for each demand class. This kind of rule is applied in the business scenarios where the share of the forecasted demand among different demand classes reflects a fare share of the actual customer orders. The goal would be that the expected percentage orders that could not be fulfilled would be approximately the same for all demand classes.

- Fixed splitting rule: The ATP quantity is allocated according to pre-defined split factor for each demand class. The resulting ATP quantities would be based on historical business information or other business performance requirements. This kind of rule is usually used in the case when resources are limited or the company has predefined objectives among different demand classes, e.g. education channels, government channels, public consumer channels, etc.

A company might apply a combination of the first two rules as follows. Three emerging markets (demand classes) would be designated as high priority and all others as medium priority. Each of the three high priority demand classes would be allocated a quantity equal their expected maximum demand. Assuming there was not enough remaining quantity to allocate the maximum demand estimate to all other classes, the amount allocated to each of the medium priority demand classes would be an equal proportion of the estimated maximum demand, e.g. each would be allocated 80% of their maximum estimated demand. This rule is meant to insure that all orders in high priority demand classes are satisfied and that the percentage of unfulfilled orders in the other demand classes is approximately the same.

A variety of ATP allocation rules have been evaluated in conjunction with inventory control of multi-echelon distribution systems, which has received considerable research attention in recent years. Eppen and Schrage (1981) introduced a *fair share allocation rule* for a two-echelon system without intermediate stocks. The allocation rule ensures that at the end-stockpoints stockout probabilities are equalized. Extensions of the results of Eppen and Schrage are given by Federgruen and Zipkin (1984). The focus of this line of research is to determine allocation policies that minimize holding and shortage costs. In order to explicitly take into account demand uncertainty, some results from the inventory control literature, which derive *rationing policies* based on models of uncertainty, can also be borrowed for ATP allocation along different demand dimensions. For example, Topkis (1968) uses a dynamic inventory model to study an optimal ordering and rationing policies for a single product in inventory model with multiple demand classes. Ha (1997) uses a queuing model to derive a stock rationing policies of a single item, make-to-stock production system with several demand classifications.

Rule-based ATP allocation mechanisms are simple and straightforward, which leads to relative simplicity in implementation and integration with other business processes. Thus, they can be easily modified and combined with other business rules to achieve broader objectives. In the next section we discuss the use of optimization-based allocation models, which bring a higher level of complexity but allow for consideration of more problem details.

4.2 Deterministic Optimization-Based Push ATP Models

The alternative to rule-based models and allocation policy is the use of optimization-based resource allocation models. Such models can explicitly take into account variations in profitability at the customer order level, complex resource constraints, interactions between similar products, etc. We now describe a general optimization framework that we will use to describe several models that have been described in the literature. We note that push-based ATP models are structurally similar to traditional production planning models and, in fact, that literature provides useful background. However, the push-based ATP models have a number of subtle but important differences including: shorter time horizons and time granularity, different objective functions, fixed resource availability, allocation to general demand classes rather than demand locations or customers as well as others.

We start with a simple model that rations available raw materials and production capacity among demand classes. The model operates over a set of time periods, T; has product set, P; raw materials set M; and set of demand classes, K. The indices for product model, raw material type, demand dimension and time period are respectively: $i \in P, j \in M, k \in K$ and $t \in T$. The basic input data are:

$d_{i,k,t}$:forecasted demand for product model i from demand class k in period t

$v_{i,k}$:per unit net revenue for demand for product model i from demand class k

$b_{i,j}$:quantity of material type j needed to produce product model i, (from BoM)

$s_{j,t}$:amount of material type j that becomes available in period t

q_t :production capacity available in period t

$c_{i,t}$:cost of producing a unit product i in period t.

$h_{i,t}, h'_{j,t}$:cost of holding unit product i and unit material j in period t, respectively.

The decision variables associated with the push-based ATP model are as follows:

$Y_{i,k,t}$:quantity of product model i allocated to demand class k in period t

$X_{i,t}$:quantity of product model i produced in period t

$I_{i,t}$:quantity of product model i held in inventory in period t

$R_{j,t}$:quantity of material type j held in inventory in period t

We can now state the push-based ATP model:

Maximize

$$\sum_{i,k,t} v_{i,k} \cdot Y_{i,k,t} - \sum_{i,t} h_{i,t} \cdot I_{i,t} - \sum_{j,t} h'_{j,t} \cdot R_{j,t} - \sum_{i,t} c_{i,t} \cdot X_{i,t} \qquad (11.1)$$

Subject to:
Demand limitation:

$$\sum_{t'=1}^{t} Y_{i,k,t'} \leq \sum_{t'=1}^{t} d_{i,k,t'} \quad \forall i \in P, k \in K, t \in T \qquad (11.2)$$

Product inventory balance:

$$I_{i,t-1} + X_{i,t} = I_{i,t} + \sum_{k \in K} Y_{i,k,t} \quad \forall i \in P, t \in T \qquad (11.3)$$

Material inventory balance:

$$R_{j,t-1} + s_{j,t} = R_{j,t} + \sum_{i \in P} b_{i,j} \cdot X_{i,t} \quad \forall j \in M, t \in T \qquad (11.4)$$

Capacity:

$$\sum_{i \in P} X_{i,t} \leq q_t \quad \forall t \in T \qquad (11.5)$$

Initialization and non-negativity:

$$I_{i,0} = p_i^0, \quad R_{j,0} = s_j^0, \quad X_{i,t} \geq 0, \quad Y_{i,k,t} \geq 0, R_{j,t} \geq 0, \quad I_{i,t} \geq 0 \qquad (11.6)$$

This model allocates product to the demand classes in order to maximize profit. It is clear that solutions can result where certain demand classes receive a small or zero allocation. It is often the case that such solutions are considered unacceptable from a customer service point of view. To address this issue a *fill rate* is very often computed and incorporated into the model. The fill rate for demand class k, is defined:

$$W_k = \frac{\sum_{i,t} Y_{i,k,t}}{\sum_{i,t} d_{i,k,t}} \qquad (11.7)$$

This fill rate variables can be either included in the objective function with an appropriate weight, or constrained as follows:

$$W_k \geq w_k \qquad (11.8)$$

where w_k is the minimum fill rate for the demand class k.

The simple production capacity constraint (11.5) could be enhanced in a number of ways using standard techniques depending on the complexity of the production environment. For example, allocation among multiple factories or production lines within a factory could be made. The possibility of overtime could be included. Also, production scheduling factors, such as lot sizing, could be modeled (e.g., see Constantino (1996) and Wolsey (1997)). Özdamar and Yazgac (1997) provide an example of using this kind of model to determine the optimal production quantity for each product family in an MPS.

It is very often the case that certain material types are not available in sufficient quantities to satisfy anticipated demand. To address this material scarcity it is common to consider a variety of material substitution strategies. The following material inventory balance constraint (11.9) can be used to replace constraint (11.4) to take into account material substitution.

$$R_{j,t-1} + s_{j,t} = R_{j,t} + \sum_{i \in P} b_{i,j} \cdot X_{j,t} - \sum_{j' \in SU(j)} S_{j,j',t} + \sum_{j' \in SU(j)} S_{j',j,t}$$

$$\forall j \in M, t \in T \quad (11.9)$$

where $S_{j,j',t}$ is the quantity of material type j substituted by material type j' in period t and $SU(j)$ is the set of material types that can be substituted for material type j. The objective function could also include a material substitution cost given by:

$$\sum_{j,j',t} c_{j,j'} \cdot S_{j,j',t} \quad (11.10)$$

Balakrishnan and Geunes (2000) have addressed a dynamic requirement planning problem for two stage multiple product manufacturing systems using substitutions of this type.

The general modeling approach just outlined, is appropriate for an MTS environment where finished products are being allocated. In an ATO, CTO or MTO setting, the resources allocated are raw materials or components, not finished products. Forecasts are commonly made on a product model or product family basis. However, the product model or product family does not have a predefined BoM. Rather the specific materials used are determined by the specific configuration chosen by the customer. An approach based on feature sets is described by Ervolina and Dietrich (2001). A feature set, which corresponds to a particular customer product configuration option, maps to a set of required materials. Associated with the product forecasts, are a set of ratios that relate product family quantities to features. Specifically, let F be a set of features and $r_{f,i}$ the fraction of product family i that demands feature f. In this setting, since features map to material requirements, we replace $b_{i,j}$ with $b'_{f,j}$ which is defined as the material j quantity associated with feature f. Constraint (11.4) can now be replaced with:

$$R_{j,t-1} + s_{j,t} = R_{j,t} + \sum_{i \in P} \sum_{f \in F} r_{f,i} b'_{f,j} \cdot X_{i,t} \quad \forall j \in M, t \in T \quad (11.11)$$

More refined version of this basic approach can be developed, e.g. where feature/product family combinations are satisfied at different levels within a demand class. Ervolina and Dietrich (2001) provide details on approaches of this type.

The solution to the model we have described effectively commits resources to certain demand classes. In general there could be left over capacity. In such cases, it is desirable to make this excess capacity available for order promising either outside the normal demand classes or for unanticipated demand. Taylor and Plenert (1999) propose a Capacity Available to Promise (CATP) concept to support customer order promising in a CTO environment. The CATP reflects the machine slack time for a corresponding time period. Its units are measured in time available on a machine. A general CATP,$X_{i,l,t}$, for demand class i on production line (or machine) l, in period t is given by

$$X_{i,l,t} = LS_{il} - t \quad \text{for} \quad EF_{i-1,l} \leq t < ES_{il} \quad (11.12)$$

$$X_{i,l,t} = LS_{il} - ES_{il} \quad \text{for} \quad ES_{i,l} \leq t < EF_{il} \quad (11.13)$$

where $ES_{i,l}, EF_{i,l}, LS_{i,l}$ are earliest start time, earliest finish time and latest start time of demand class i on production line (or machine) l. The CATP availability for all production lines can be used for marketing personnel to establish realistic order promising dates and quantities.

4.3 Stochastic Push ATP Models

Since the previous models depend significantly on demand forecasts, more explicit modeling of stochastic elements would seem appropriate. De Kok (1999) studied a continuous processing plant with single uniform product, which is packed in barrels of different sizes before shipping to company-owned geographically dispersed stockpoints. He proposed a so-called Balanced Stock (BS) rationing rule for capacity allocation to different package sizes under random demands from external customers. Balakrishnan et al. (1999) describe a capacity rationing model by which MTO firms facing expected total demand in excess of installed capacity can reserve a portion of the available capacity for products yielding higher profit. This is a kind of capacity reservation model that is based on a known demand distribution. They determine the reservation quantity for period, t,R_t based on the following inequality,

$$\text{Probability}(X_{Ht} \geq R_t) \geq P_L/[P_H - (P_H - P_L)\varpi] \quad (11.14)$$

where P_H, P_L is the profit contribution per unit of capacity used to fill orders of the higher profit demand classification and lower profitable demand classification, respectively; X_{Ht} is the total demand (random variable) for the higher profit class of demands between time t and T; $\omega = \theta_{Lt}/(\theta_{Ht} + \theta_{Lt})$ and θ_{Ht}, θ_{Lt} are parameters of the probability density function for higher profit demands and lower profit demand.

In fact, the stochastic push ATP problem can be viewed as a type of revenue (or yield) management problem and so the literature on this topic is quite relevant. Most revenue management models assume fixed (or "almost fixed") resource availability (e.g., airline seats) and balance resource allocation among multiple demand classes (e.g., fair segments of price-sensitive customers). Yield management was initiated by the airline industry to maximize revenue through pre-allocating seats on a flight into different fair classes and determining the closing time for each fair class. Later, many other industries also applied these techniques to control their perishable or even non-perishable assets. Weatherford and Bodily (1992) not only propose the new name, Perishable-Asset Revenue Management (PARM), but also provide a comprehensive taxonomy and research overview of the field. They identify fourteen important elements for defining revenue management problems. Although most of these elements are airline-oriented, many push-based ATP problems share similar characteristics, particularly the last three modeling-related elements: bumping procedure (for handling "overbooking"), asset control mechanism (for resource reservation), and decision rule (for resource allocation). More recently, McGill and van Ryzin (1999) classify over 190 revenue management research papers into four groups: 1) forecasting, 2) overbooking research, 3) seat inventory control, and 4) pricing. The papers in the third group are more relevant to the push-based ATP models discussed here.

For example, Bodily and Weatherford (1995) present a generic multiple-class PARM allocation problem. They first study a simplified two-class problem without diversion. The problem assumes that there are two demand classes, full-price and discount, sharing a fixed available capacity of q_0 units, and that no full-price customers would pay less than their willingness to pay (i.e. the full price R_0). The purpose is to determine how many units (denoted by a decision variable q_1) should be allocated to discount-price customers, paying discount price R_1, before the number of full-price customers (denoted by a random number X) is realized. Through a newsboy-type marginal analysis, the authors derive the decision rule:
increase discount share from q_1 to $q_1 + 1$ if:

$$\Pr(q_1 + X < q_0) > \frac{R_0 - R_1}{R_0}. \tag{11.15}$$

The authors further extend the problem to allow diversion in a multiple-class setting. Available resources q_0 are offered in N "nested buckets" of size q_0, q_1, q_2, ..., and q_N at full price R_0 and discount prices R_i (for $i = 1, 2, ..., N$), respectively (assuming $R_0 > R_1 > R_2 > ... > R_N$). Customers from

discount class i paying R_i can access the allocated resource q_i, which actually also contains all the resources that can be used for deeper discount classes (i.e., $i+1, i+2, \ldots, N$). Hence, the bucket q_N will sell out first because all demand classes have access to it at the price of R_N. After closing this deepest discount class, only customers who are willing to pay at least R_{N-1} will be served. Similarly, each succeeding discount class will be closed one by one after the corresponding bucket is filled. This implies that only the difference $q_i - q_{i+1}$ is actually protected for class i (at price R_i) from classes $i+1, i+2, \ldots$, and N. Following a similar decision-tree analysis by Pfeifer (1989), the authors compare the expected marginal gain by changing the bucket size from q_i to $q_i + 1$. They obtain the decision rules for q_i, $i = 1, 2, \ldots, N$, as:
accept an additional class i customer if

$$\left(\frac{\beta_i}{\beta_i + \beta_{i-1} + \ldots + \beta_0}\right) \cdot p_{i-1} > \frac{R_{i-1} - R_i}{R_{i-1}}, \qquad (11.16)$$

where β_i = the probability that the next arriving customer is in class i, and p_i = the conditional probability that all customers willing to pay as much as R_i can be satisfied, given that an additional reservation in class $i + 1$ is accepted.

Sen and Zhang (1999) work on a more complicated similar problem by treating the initial availability as a decision variable and model the problem as a newsboy problem with multiple demand classes. While Pfeifer (1989) employs a marginal gain approach, they use a total expected profit function approach. Complex optimality conditions are given for both increasing-price (e.g., airline seats) and decreasing-price (e.g., computer chips) scenarios in a product life cycle. Clearly, "no reservation" policy should be implemented for decreasing prices, while "protection levels" should be in place to safeguard higher fare classes.

We note that there is a growing body of literature on revenue management models that include price as a variable. We will not cover such approaches here as they are treated by other chapters in this volume (see Chapters 9 & 12).

5. Pull-Based ATP Models

Pull-based ATP models are executed in response to one or more customer orders. In an MTS production environment addressed in Section 5.1, the problem to be solved is one of assigning customer orders to some product availability pool, which could be a location of existing inventory or a planned production batch. The simplest optimization models of this category can be viewed as an assignment/transportation problem. In an ATO/MTO/CTO environment addressed in Sections 5.2 and 5.3, the problem to be solved is more complex. It involves assigning customer orders to appropriate manufacturing and material resources.

5.1 Pull-Based ATP Models for an MTS Production Environment

The most basic pull-based ATP model is to use the allocated ATP quantities to commit customer orders within each demand class. The simple approach, which is probably the one most widely used in practice, is to use order-promising rules to generate "real-time" order promises. The widely accepted *first-come-first-served policy* does not differentiate customer orders within a demand class and assign priority purely based on order arrival time. Among other characteristics, this policy has the advantage of preserving a widely accepted notion of fairness. There would typically be a preferred product availability pool, e.g. local inventory, associated with each demand class, which each order would be given access to in turn. However, when this ATP quantity becomes zero, an alternate product pool would have to be determined. For such cases, a search process would have to be executed. We note that a major factor in the ability to locate product is the capability of the organization's IT infrastructure. Companies have invested substantially in supply chain IT, e.g. ERP and supply chain management software (see Section 2), in order to provide fast order promising response. Capabilities in this area are viewed as providing a significant competitive advantage (see Simchi-Levi, et al., 2000).

Given an underlying IT infrastructure, search procedures must locate one or more appropriate product sources to satisfy the order. Kilger and Schneeweiss (2000) have proposed an ATP search method along three dimensions: the time dimension, the customer dimension and the product dimension. The process proceeds through three levels: 1) the product dimension is searched over product that is pre-allocated to one specific customer group and that is available by the requested date. 2) If the ATP quantity found in 1) is not sufficient, then the time dimension, either backward or forward up to certain limits, is searched for additional ATP quantity. 3) If the ATP quantity is still not sufficient in 2), steps 1) and 2) will be repeated along the customer dimension; i.e., the product pre-allocated to other customer groups will be considered. If there is a priority or hierarchy among customer groups, the search along the customer dimension may be ordered from lower priority to higher priority.

In a complex supply chain network, multiple business units including retailers, distribution centers and factories may be considered in an ATP search. These business units can be dispersed geographically. Hence, an intuitive method for ATP search is based on distance (or lead-time) between inventory location and customer order location. The search process may also be based on a pre-assigned location priority scheme, or on the echelon-level of the business units, e.g. retailer level, distribution center level, manufacturer level. Considering a complex supply chain with multiple distribution centers and factories, Jeong et al. (2002) proposed a global search model to find available resources across the supply chain in order to promise customer orders. The global search model employs the following distribution center priority measure to ration available inventory among customer orders with different priorities:

$$P_{jl} = Q_{jl}/T_{jl}, \qquad (11.17)$$

where Q_{jl} is the quantity from order j that can be supplied from distribution center location l and T_{jl} is the associated lead time. The priorities P_{jl} essentially form the basis for a "greedy algorithm" that commits available product to orders. At each step the highest P_{jl} value is found and the quantity Q_{jl} from location l is committed to order j. Note that this rule favors higher quantity orders and shipment sizes over smaller ones and shorter lead times over longer ones

5.2 Real Time Order Promising and Scheduling

In order to provide a real-time response to an order request in an ATO/MTO/CTO environment, a fast procedure is required to allocate material and production resources to a single order in real time. The type of approach required depends on the complexity in production processes and material requirements. For example, at the simplest level, the required procedure can be viewed as one of incrementally "grabbing" the best available materials and production resources as each order arrives. The problem becomes more complex in the case of a job shop production environment. Here the literature of on-line algorithms for job shop scheduling becomes relevant.

Due-date assignment is one important research area, which closely relates to real-time order promising and scheduling. The major focus of this line of research is on scheduling machine production or resource utilization to meet pre-specified order due date. Hegedus and Hopp (1999) have addressed due-date assignment problem for material-intensive environments. Their work is using cost-minimization approach in an environment with certain vendor lead-times. Due dates are set to minimize the cost of quoting a date later than requested, completing an order later than promised, and holding inventory. Many other due date assignment models using parametric approaches, queuing theory and deterministic optimization have been widely studied. Most of these works assumes that due dates of individual jobs for these orders are exogenously determined. In case of negotiable order due dates, Kaminsky and Lee (2001) proposed a novel model for due date quotation. Most of the results are for single machine systems and a survey of work in this area recently appeared in Gordon et al. (2002). For multiple machines (or resource systems), statistical estimation of the distribution of flowtime to set due date for future arrivals is an important approach (Kaplan and Unal 1993). Lawrence (1995) used demand forecasting of order flowtime by one of six different estimators and then added some function of the forecast error distribution to this estimation so as to achieve a certain performance objective. Hopp and Sturgis (2000) proposed leadtime estimation approaches from a control perspective. Their work combines factory physics, statistical estimation, and control charting to create a leadtime estimator that is generally applicable and also adaptive to changes in the system over time.

Moses et al. (2002) proposed a method for real-time promising (RTP) of order due dates in a discrete make-to-order environments facing dynamic order arrivals. Different from conventional scheduling to meet due dates, this method considers how to assign due dates without actually scheduling the orders. When computing a due date, this method considers: 1) dynamic time-phased availability of resources that are required for each operation of the order, 2) individual order-specific characteristics, and 3) existing commitments to orders that arrived previously. In the model, orders are assumed to arrive dynamically and one-at-a-time. In this case, to assign a due date D_i to an order i arriving at time T_i with a time buffer B, they estimate the order flowtime based on conditions at the time of arrival as:

$$D_i = T_i + \hat{F}_i + B \qquad (11.18)$$

where $\hat{F}_i$ is the estimated flowtime. The authors further proposed a real-time approach to estimate the flowtime by determining the time when each individual operation $j (j = 1, 2, \cdots \theta_i)$ required to fulfill order i can be completed, as

$$t_{i,1}^{\min} \leftarrow T_i$$

for all operations j do

$$t_{i,j}^* \leftarrow RO(t_{i,j}^{\min}, p_{i,j})$$

$$t_{i,j+1}^{\min} \leftarrow t_{i,j}^*$$

$$\hat{F}_i \leftarrow t_{i,\theta_i}^* - t_{i,j}^{\min}$$

where $t_{i,1}^{\min}$ is the earliest feasible start time of operation j on order i, $t_{i,j}^*$ is the estimated completion time of operation j on order i, and $p_{i,j}$ is the total set-up and processing time required for operation j on order i. Function $RO(\cdot)$ determines the earliest time when a resource will be available to complete the operation. It does this by reviewing the tasks currently planned on the resources to compute a resource load profile. Real-time estimation of resource availability poses a unique challenge: the algorithm must dynamically determine when the

Available to Promise   473

system has the requisite amount of capacity available, but time is not available to perform detailed scheduling which is an NP-hard problem. The author has run experiments to demonstrate the effectiveness of the RTP method.

5.3 Optimization-Based Batch ATP Models

In order to model the batch ATP problem as a decision problem, we first must characterize the decision space. The specific decisions directly related to the order response are: 1) whether to accept or reject the order, 2) determining the order quantity and 3) determining the order due date. Based on the hard-disk drive assembly of Maxtor, Chen et al. (2002) introduced a pull-based ATP model to determine the committed order quantity for customer orders whose due dates are given (i.e., decisions 1) and 2)). Chen et al. (2001) further extend the quantity-quoting ATP model to a more general problem of determining both quantity quoting and due date quoting based on Toshiba Corporation's notebook computer production. This model carries out order quantity and due date quoting within a certain range of flexibility. Both models not only use unallocated material availability and production capacity as production resources, but also consider a flexible BOM with material incompatibility and material substitution.

More specifically, the models assume multiple products are offered. The materials needed in assembling end-items are grouped into different *component types*. The manufacturer may have multiple suppliers to provide materials of the same type, which differ in features such as quality, price and technology. The authors denote the combination of a component type and a supplier as a *component instance*, which represents the very basic material element in the models. Each customer order has an associated BOM that specifies the quantity of each component type required to build the customized product. Furthermore, for each selected component type, a customer can specify a set of preferred suppliers. The manufacturer is allowed to take advantage of this customer preference flexibility relative to components. This implies that the models only allow component substitution at component-instance level. However, the manufacturer has to take into consideration the incompatibility between certain pairs of component instances . This further complicates the formulation of these advanced pull-based ATP models.

The two other kinds of flexibility associated with customer orders in the models are order quantity range and delivery time window. As mentioned earlier, the general order specification regime allows an order to include an allowable range on the desired order quantity and an acceptable range of delivery due dates. If the manufacturer cannot promise a quantity and a delivery time within the respective windows, the order is denied and assumed lost.

The models employ an objective function of maximizing *overall profit*, which is calculated by subtracting certain *intangible penalties* from *tangible profit*. Tangible profit is defined as the difference between revenues and *tangible costs,* which include component costs and inventory holding costs for finished-product, work-in-process (WIP), and component-instance. Intangible penalties reflect the consequences resulting from order denial, capacity under utilization, and insufficient material reserve. The purpose of intangible penalties is to balance short-term tangible profit with long-term profit and customer service. We will discuss a material reservation policy in detail later.

As an order promising and order fulfillment engine, a pull-based ATP model is responsible for quoting a committed quantity and a due date for each order, for scheduling production to fulfill promised orders, and for configuring finished products at the component instance level. As mentioned earlier, customer orders are collected over a *batching interval,* the time lapse between successive ATP executions. The major decision variables in the advanced pull-based ATP models include:

Z^i :order fulfillment indicator (1, if a new order i is promised and will be fulfilled; 0, otherwise);

$D^i(t)$:delivery time period indicator (1, if order i is promised a delivery in time period t; 0, otherwise);

$C^i(t)$:quantity delivered for order i in time period t;

$P^i(t)$:quantity produced for order i in time period t (note that the entire production process actually starts in time period $t - \varpi$);

$X^i_{j,k}(t)$:quantity of component instance (j, k) used to produce order i in time period t.

The mixed integer programming models of Chen et al. (2000, 2001) includes constraints for order promising and fulfillment, finished product inventory, material/component inventory, material requirements, capacity utilization, and material compatibility. We highlight the major constraints as follows.

$$C^i(t) \leq u^i \cdot D^i(t), \text{ for all } i \in O, \text{ and } t_e + 1 \leq t \leq t_e + T, \quad (11.19)$$

$$C^i(t) \geq l^i \cdot D^i(t), \text{ for all } i \in O, \text{ and } t_e + 1 \leq t \leq t_e + T, \quad (11.20)$$

$$\sum_{d^i_e \leq t \leq d^i_l} D^i(t) = Z^i, \text{ for all } i \in O. \quad (11.21)$$

The first two constraints above enforce the acceptable range $[l^i, u^i]$ of delivery quantity for all new customer order i (denoted by the set O) on a possible due date t within the planning horizon T from the current time epoch t_e. The third constraint defines the feasible delivery time window $[d_e^i, d_l^i]$. If the manufacturer decides to accept an order (i.e., $Z^i = 1$), the delivery quantity has to satisfy the allowable quantity range on a promised due date within the acceptable due date window. Otherwise, the order should be denied (i.e., $Z^i = 0$).

The flexible BOM is modeled through the following two constraints.

$$\sum_{k \in M_j^i} X_{j,k}^i(t) = b_j^i \cdot P^i(t+\varpi) \tag{11.22}$$
$$\forall\, j \in M,\, i \in O \cup \hat{O},\, \text{and}\, t_e+1 \le t \le t_e+T$$

$$\sum_{k \in M_j \setminus M_j^i} X_{j,k}^i(t) = 0 \tag{11.23}$$
$$\text{for all}\, j \in M,\, i \in O \cup \hat{O},\, \text{and}\, t_e+1 \le t \le t_e+T$$

The first constraint defines the BOM relationship between the total material requirements from the preferred set of component instances M_j^i, and product quantity $P^i(t+\varpi)$ with production lead-time ϖ for both new and old orders (with the set $\hat{O}$ denoting all old orders). Non-preferred component instances should not be used as shown in the second constraint.

The presence of component instances from multiple suppliers, can lead to incompatibility among certain components, e.g. a disk drive from manufacturer XX cannot be used in a product with a power supply from manufacturer YY. Such incompatibilities can be characterized by a set of incompatible component-instance pairs. Within the math programming model, there is a need to insure that product configurations that avoid such incompatibilities can from the resourced (component instances) allocated to an order. This constraint can be modeled using a class of constraints derived from similar constraints related to non-bipartite graph matching:

$$\sum_{k \in N} X_{j,k}^i(t) \le \left(\frac{b_j}{b_j'}\right) \cdot \sum_{k' \in \Gamma_{j,j'}(N)} X_{j',k'}^i(t)$$
$$\forall\, N \subset M_j^i,\, j \in M,\, j' \in M,\, i \in O \cup \hat{O},\, t_e+1 \le t \le t_e+T \tag{11.24}$$

where the set $\Gamma_{j,j'}(N)$ is defined to include all type-j' component instants that are compatible with at least one component instant in any subset N of type j. Consider the example in Figure 11.4. The dashed links represent incompatible pairs of component instances between types j and j'. By definition, when $N = \{(j,2), (j,3), (j,4)\}$, we have $\Gamma_{j,j'}(N) = \{(j',2), (j',3), (j',4), (j',5)\}$. Thus, the associated constraint is

$$X^i_{j,2} + X^i_{j,3} + X^i_{j,4} \leq \left(\frac{b_j}{b'_j}\right) \cdot \left(X^i_{j',2} + X^i_{j',3} + X^i_{j',4} + X^i_{j',5}\right). \quad (11.25)$$

Note that only component instance $(j', 1)$ is excluded from $\Gamma_{j,j'}(N)$. This is due to the fact that $(j', 1)$ is incompatible with every component instance in N.

By using this class of constraints, it becomes unnecessary to explicitly include in the model (a potentially huge number of) produce configuration variables. Ball et al. (2001) analyzed this constraint set based on its relationship to non-bipartite matching and proved that the constraint set represents a necessary for any class of products and a necessary and sufficient condition for classes of products where material types can be ordered so that incompatibility situations only exist between successive material types.

In the quantity and due date quoting model, a material reservation policy can be adopted to reserve components for future important customers (see Chen et al 2001). There are two control parameters associated with each component instance (j, k) in the sth time period of the planning horizon: 1) reserve level $\gamma_{j,k,s}$ and 2) shortfall penalty $\eta_{j,k,s}$. If the corresponding ending inventory is above the reserve level, no penalty occurs. However, when the ending inventory is below the reserve level, each shortfall unit will be charged at the shortfall penalty. The constraint can be formulated as:

$$\begin{array}{c} R_{j,k}(t) + S_{j,k}(t) \geq \gamma_{j,k,t-t_e} \\ \forall\, k \in M_j,\ j \in M,\ t_e + 1 \leq t \leq t_e + T \end{array} \quad (11.26)$$

where the decision variables $R_{j,k}(t)$ and $S_{j,k}(t)$ denote the ending inventory and shortfall amount, respectively, of component instance (j, k) in time period t.

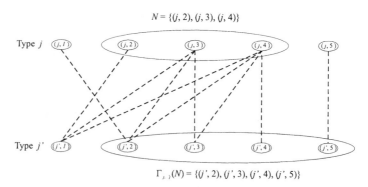

Figure 11.4. An Example of Material Compatibility Constraint

Available to Promise 477

5.4 Experimental Implementation

A pull-based batch ATP model would always be executed over a rolling time horizon so the model's objective function only reflects short-term profits for a single execution time horizon. To study the dynamic aspect Chen et al (2001, 2002) implemented and tested their models within a simulation that covered an extended time period. The discussions here summarize the results of those experiments.

For these experiments, the production activities were organized in terms of discrete *time periods* (a time period could be an 8 hour shift or a working day). Each execution of the MIP model planned activities over an *ATP Planning horizon* that consisted of an ensuing number of time periods. Order-promising decisions were made for a batch of orders collected over a batching interval. Figure 11.5 provides an example of this setting assuming a batching interval two time periods in length.

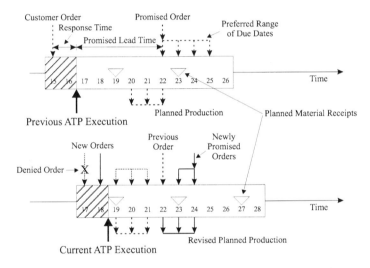

Figure 11.5. A Rolling ATP Execution

Orders are collected over the two-period batching interval and order promises are generated when the ATP model is executed at the end of the batching interval. This figure illustrates a single order arrival in the first batching interval and two order arrivals in the second. Each ATP execution generates a promised due date and order quantity together with a *tentative* production schedule for this order. After the planning horizon advances by two time periods, the ATP model is executed again. The second execution of the model treats the order quantities and due dates for all previous order promises as constraints, but has the flexibility to reallocate resources, e.g. production date, in order to accommodate new order requests.

Several experiments were carried out. One issue considered was the impact of batching interval size on overall performance. Both Figures 11.6 and 11.7 illustrate the relationship between batching interval size and tangible profit. While Figure 11.6 shows results of the quantity quoting model based on Maxtor data, Figure 11.7 illustrates results of the quantity and due date quoting model using business data from Toshiba. Both results do show a slight initial profit growth because the ATP models can simultaneously optimize over a larger order set as batching interval size increases. A subsequent decrease for larger batching intervals is observed in Figure 11.6. For very large batching intervals, the manufacturer does not have enough time to react to more urgent orders (i.e. orders whose due dates are within the batching interval). The growth of such "missed orders" causes the eventual decrease in tangible profit in Figure 11.6. We expect the same decreasing trend in Figure 11.7, if we keep increasing the batching interval size in the experiment.

Customer response time, defined as the time lag between a customer's order request and the corresponding promise confirmation, is an important component of customer service. On average, a customer needs to wait for one half of the batching interval (plus some constant communications and processing delay) to obtain the promised quantity and due date. Hence, shorter batching intervals contribute to better customer response time. However, a longer batching interval offers more order-pattern information and a larger order set to optimize over. A manufacturer would certainly gain by using a longer batching interval from a profit-maximization point of view. From a manager's perspective, the choice of a batching interval is a multi-objective decision that involves tradeoffs among three performance criteria: tangible profit, denied or missed orders, and customer response time.

One interesting phenomenon we observed is that the quantity and due date quoting ATP model always promises the earliest available due date to a customer order regardless the size of the acceptable time windows since early delivery saves overall inventory-holding costs (the order can be shipped as soon as it is produced in the flexible due date case). This aggressive behavior may jeopardize the ability to accommodate more profitable or urgent orders in future batching intervals. Longer batching intervals not only allow optimization over a larger set of orders but also maintain a higher degree of assembly-flexibility. Hence, the improvement in profit figures as a function of increasing batch size is more significant for the quantity and due date quoting ATP model than for the quantity quoting ATP model. Not surprisingly, we also found that most of the production activities happened just in time to meet the promised due date, if there was sufficient material availability and production capacity. This was mainly due to the higher holding cost of a finished product compared with the total holding costs of its raw-material components (in the fixed due date case, the order is not shipped until its due date).

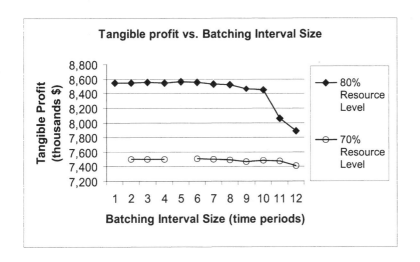

Figure 11.6. Tangible Profit as a Function of Batching Interval Size

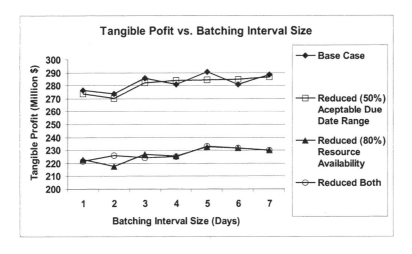

Figure 11.7. The effects of batching interval size on tangible profit

6. Conclusions

ATP is not yet well-established as a research area. Rather the name is more associated with a set of business practices and software products. However, recent research programs have started to define it as a research area and produce ATP models and results. We note in particular the work at IBM (Ervolina and Dietrich, 2001), the work at Oklahoma (Moses et al, 2002) and our own work (Chen et al, 2000, 2001; Ball et al, 2001). In reviewing this body of work as well as related research and business practice, we characterized ATP models as either push-based or pull-based. Further, it became clear that ATP models filled a "niche" downstream in time from inventory control models. Specifically, they carried out allocation over a very short time horizon of resources that were fixed by other business functions, most notably inventory control functions. When viewed in this light, two established bodies of research, namely, certain revenue management models and certain on-line scheduling models can be seen to fall within the ATP domain. We hope that by viewing these research streams as well as the evolving work specifically associated with ATP, that rich and useful new research directions will emerge.

Acknowledgments

This work was supported by the National Science Foundation (NSF) under grant numbers DMI9908137. Matching support was provided by the Robert H. Smith School of Business and the University of Maryland Institute for Advanced Computer Studies (UMIACS). Toshiba Corporation provided business scenarios and data. The opinions expressed in the paper are those of the authors and do not necessarily reflect the opinions of NSF or Toshiba.

References

Ali, SK Ahad, Zhao, Zhenying and R. de Souza (1998). "Product mix and material match based on genetic optimization," Proceedings of the Rensselaer's Sixth International Conference on Agile Intelligent and Computer Integrated Manufacturing Systems (CAICIM'98), Oct. 7-9, RPI, Troy, New York, USA.

Balakrishnan, A., and J. Geunes (2000). "Requirements Planning with Substitutions: Exploring Bill-of-Materials Flexibility in Production Planning," Manufacturing & Service Operations Management, 2, 166-185.

Balakrishnan, Nagraj, J. Wayne Patterson, and V. Sridharan (1999). "Robustness of capacity rationing policies", European Journal of Operational Research, 115, 328-338.

Ball, M. O., C.-Y. Chen, and Z.-Y. Zhao (2001). Material Compatibility Constraints for Make-to-Order Production Planning. to be published in Operations Research Letters.

Bayles, Deborah L. (2001). E-commerce logistics and fulfillment: delivering the goods, Upper Saddle River, NJ : Prentice Hall PRT.

Bodily, S. E. and L. R. Weatherford (1995). "Perishable-asset Revenue Management: Generic and Multiple-price Yield Management with Diversion," Omega, 23, 173-185.

Chu, Sydney C. K. (1995), "A mathematical programming approach towards optimized master production scheduling," International Journal of Production Economics, 38, 269-279.

Chen, C.-Y., Z.-Y. Zhao and M. O. Ball (2002). "A Model for Batch Advanced Available-to-Promise," Production and Operations Management, 11, 424-440.

Chen, C.-Y., Z.-Y. Zhao, and M.O. Ball (2001). "Quantity and Due Date Quoting Available to Promise," Information Systems Frontiers, 3, 477-488.

Constantino, M. (1996) "A Cutting Plane Approach to Capacitated Lotsizing with Start-up Costs," Mathematical Programming, 75, 353-376.

Dell, Michael, and Catherine Fredman (1999). Direct from Dell: Strategies That Revolutionized an Industry, New York: HarperBusiness.

De Kok, Ton G. (2000). "Capacity allocation and outsourcing in a process industry", International journal of production economics, 68, 229-239.

Eppen, G. and Schrage, L. (1981). "Centralized Ordering Policies in a Multi-warehouse System with Random Lead Times and Random Demand". Management Science, 16, 51-67.

Ervolina, Tom and Brenda Dietrich (2001). "Moving Toward Dynamic Available to Promise," in Saul I. Gass and Albert T. Jones, editors, Supply Chain Management Practice and Research: Status and Future Directions, R. H. School of Business, University of Maryland, Manufacturing Engineering

Laboratory, National Institute of Standards, and National Science Foundation, 15.1-19.

Federgruen, A. and P. Zipkin (1984). "Approximations of Dynamic, Multilation Production and Inventory Problems," Management Science, 30, 69-84.

Gordon, Valery, Jean-Marie Proth and Chengbin Chu (2002). "A Survey of the state-of-art common due date assignment and scheduling research", European Journal of Operational Research, 139, 1-25.

Ha, Albert Y., (1997) "Inventory rationing in a Make-to-Stock Production System with Several Demand Classes and Lost Sales", Management Science, 43, 1093 - 1103.

Hegedus, M. G. & W. J. Hopp (2001). "Due Date Setting with Supply Constraints in Systems Using MRP," Computers & Industrial Engineering, 39, 293-305.

Henson, Ken (2000). "Method and Apparatus for Providing Customer Configured Machines at an Internet Site," United States Patent No. 6,167,383.

Hopp, Wallace J. and Mark L. Spearman (2001). Factory physics: foundations of manufacturing management, Boston : Irwin/McGraw-Hill.

Hopp, W. J. & M. L. Sturgis (2000). "Quoting Manufacturing Due Dates Subject to a Service Level Constraint," IIE Transactions, 32, 771-784.

Jeong, Bongju, Seung-Bae Sim, So-Sang Jeong, & Si-Won Kim (2002). "An Available-to-Promise System for TLT LCD Manufacturing in Supply Chain," Computers and Industrial Engineering, 43, 191-212.

Kaminsky, P. and Z-H Lee (2001). "Analysis of On-Line Algorithms for Due Date Quotation, Working Paper, Industrial Engineering and Operations Research, University of California, Berkeley, CA 94720-1777.

Kaplan, A.C. and A.T. Unal (1993). "A Probabilistic Cost-based Due-date Assignment Model for Job Shops," International Journal of Production Research, 31, 2817-2834.

Kilger, C., & L. Schneeweiss (2000). "Demand Fulfillment and ATP," in: Stadtler, H., & Kilger, C., eds. Supply Chain Management and Advanced Planning: Concepts, Models, Software and Case Studies. Berlin, Germany: Springer, 135-148.

Kraemer, Kenneth L., Jason Dedrick, and Sandra Yamashiro (2000). "Refining and Extending the Business Model with Information Technology: Dell Computer Corporation," The Information Society, 16, 5-21.

Kraemer, Kenneth L. and Jason Dedrick (2001). "Dell Computer: Using E-commerce to Support the Virtual Company," Center for Research on Information Technology and Organizations, University of California, Irvine, CA.

Lawrence, S. R. (1995). "Estimating Flowtimes and Setting Due-dates in Complex Production Systems," IIE Transactions, 27,657-668.

Lee, H. L. and C. Billington (1995). "The Evolution of Supply-Chain-Management Models and Practice at Hewlett-Packard," Interfaces, 25, 43-63.

McGill, Jeffrey I. and Garrett J. van Ryzin (1999). "Revenue Management: Research Overview and Prospects," Transportation Science, 33, 233-256.

Moses, Scott, Hank Grant, Le Gruenwald, & Simin Pulat (2002). "Real-Time Due-Date Promising by Build-to-Order Environments," working paper, School of Industrial Engineering, University of Oklahoma, OK 73019.

Özdamar, Linet and Tülin, Yazgac (1997). "Capacity driven due date setting in make-to-order production systems," International Journal of Production Economics, 49, 29-44.

Pfeifer, Phillip E. (1989). "The Airline Discount Fare Allocation Problem," Decision Science, 20, 149-157.

Sen, Alper and Alex X. Zhang (1999). "The Newsboy Problem with Multiple Demand Classes," IIE Transactions, 31, 431-444.

Simchi-Levi, D., P. Kaminsky, and E. Simchi-Levi (2000). Designing and Managing the Supply Chain: Concepts, Strategies, and Case Studies, Irwin/McGraw-Hill, New York, NY.

Taylor, S. G., & G. J. Plenert (1999). "Finite Capacity Promising," Production and Inventory Management Journal, 40, 50-56.

Topkis, D. M. (1968). "Optimal Ordering and Rationing Policies in a Nonstationary Dynamic Inventory Model with n Demand Classes," Management Science, 15, 160-176.

Vollmann, Thomas E., William L. Berry, David C. Whybark (1997). Manufacturing Planning and Control Systems, 4th ed., New York : Irwin/McGraw-Hill.

Weatherford, Lawrence R. and Samuel E. Bodily (1992). "A Taxonomy and Research Overview of Perishable-Asset Revenue Management: Yield Management, Overbooking, and Pricing," Operations Research, 40, 831-844.

Wolsey, Laurence A. (1997). "MIP Modeling of Changeovers in Production Planning and Scheduling Problems," European Journal of Operational Research, 99, 154-165.

David Simchi-Levi, S. David Wu, and Z. Max Shen (Eds.)
Handbook of Quantitative Supply Chain Analysis:
Modeling in the E-Business Era
©2004 Kluwer Academic Publishers

Chapter 12

DUE DATE MANAGEMENT POLICIES

Pınar Keskinocak*
School of Industrial and Systems Engineering
Georgia Institute of Technology, Atlanta, GA 30332
pinar@isye.gatech.edu

Sridhar Tayur
Graduate School of Industrial Administration
Carnegie Mellon University, Pittsburgh, PA 15213
stayur@cyrus.andrew.cmu.edu

Introduction

To gain an edge over competitors in an increasingly global and competitive marketplace, companies today need to differentiate themselves not only in cost, but in the overall "value" of the products and the services they offer. As customers demand more and more variety of products, better, cheaper, and faster, an essential value feature for customer acquisition and retention is the ability to quote short and reliable lead times. Reliability is important for customers especially in a business-to-business setting, because it allows them to plan their own operations with more reliability and confidence [67].

We define the *lead time* as the difference between the promised due date of an order (or job) and its arrival time[1]. Hence, quoting a lead time is equivalent to quoting a due date. The importance of lead time quotation becomes even more prevalent as many companies move from mass production to mass customization, or from a make-to-stock (MTS) to a make-to-order (MTO) model to satisfy their customers' unique needs. Hence, companies need to determine in real time if and when an order can be fulfilled profitably. The Internet as a new sales channel further increases the importance of effective lead time

---

*Pınar Keskinocak is supported by NSF Career Award DMII-0093844.

quotation strategies, as customers who place orders online expect to receive reliable lead time as well as price quotes. For example, many customers were extremely dissatisfied with their online purchasing experience during the 1999 holiday season, mainly due to unreliable delivery date quotes, lack of order status updates, and significant order delays [48].

Quoting unreliable lead times not only leads to potential loss of future business, but may also result in monetary penalties. Seven e-tailers, including Toys R Us and Macy's had to pay a total of $1.5 million to settle a Federal Trade Commission (FTC) action over late deliveries made during the 1999 holiday season [40]. According to the FTC, the e-tailers promised delivery dates when fulfillment was not possible and failed to notify customers when shipments would be late. Sometimes a company may self-impose a penalty for missed due-dates. For example, due to the increasing insistence of many steel users on consistent reliable deliveries, Austin Trumanns Steel started to offer a program called Touchdown Guarantee in 1986. Under the program, if the company agrees to a requested delivery date at the time an order is placed, it has to deliver on time or pays the customer 10% of the invoice value of each item not delivered on time [52].

A common approach to lead time quotation is to promise a constant lead time to all customers, regardless of the characteristics of the order and the current status of the system [66] [112]. Despite its popularity, there are serious shortcomings of fixed lead times [63]. When the demand is high, these lead times will be understated leading to missed due dates and disappointed customers, or to higher costs due to expediting. When the demand is low, they will be overstated and some customers may choose to go elsewhere.

The fundamental tradeoff in lead time quotation is between quoting short lead times and attaining them. In case of multiple customer classes with different capacity requirements or margins, this tradeoff also includes capacity allocation decisions. In particular, one needs to decide whether to allocate future capacity to a low-margin order now, or whether to reserve capacity for potential future arrivals of high-margin orders.

Lead-time related research has developed in multiple directions, including lead time reduction [54] [100], predicting manufacturing lead times, the relationship between lead times and other elements of manufacturing such as lot sizes and inventory [42] [62] [69], and *due date management* (DDM) policies. Our focus in this survey is on DDM, where a DDM policy consists of a due date setting policy and a sequencing policy. In contrast to most of the scheduling literature [49] [79] [89], where due dates are either ignored or assumed to be set exogenously (e.g., by the sales department, without knowing the actual production schedule), we focus on the case where due dates are set endogenously. Most of the research reviewed here does not consider inventory decisions and hence is applicable to MTO systems. Previous surveys in this area include [5] [26].

Most of the research on DDM ignores the impact of the quoted due dates on the customers' decisions to place an order. Recently, a small but increasing number of researches studied DDM from a profit maximization rather than a cost minimization perspective, considering order acceptance decisions (or the effect of quoted lead times on demand) in addition to due date quotation and scheduling decisions. Although this is a step forward from the earlier DDM research, it still ignores another important factor that affects the demand: price. With the goal of moving towards integrated decision making, the latest advances in DDM research focus on simultaneous price and lead time quotation.

The paper is organized as follows. In Section 1, we discuss the characteristics of a DDM problem, including decisions, modeling dimensions, objectives and solution approaches. In Section 2, we discuss commonly used scheduling rules in DDM policies. Offline DDM models, which assume that the demand and other input about the problem are available at the beginning of the planning horizon, are discussed in Section 3. Online models, which consider dynamic arrivals of orders over time, are presented in Section 4. Models for DDM in the presence of service level constraints are discussed in Section 5. We review the DDM models with order acceptance and pricing decisions in Section 6. We conclude with future research directions in Section 7.

We reviewed a broad range of papers in this survey; however, we do not claim that we provide an all-inclusive coverage of all the papers published in the DDM literature and regret any omissions. We would be delighted to receive copies of or references to any work that was not included in this survey.

1. Characteristics of a Due Date Management Problem

In this section, we discuss the characteristics of DDM problems, including the decisions, modeling dimensions, and the objectives. The notation that is used throughout the paper is summarized in Table 12.1.

1.1 Due Date Management Decisions

The main decisions in DDM are order acceptance (or demand management), due date quotation, and scheduling. In determining a DDM policy, ideally one would consider due date setting and scheduling decisions simultaneously. There are a few papers in the literature that follow this approach. However, given their complexity, most papers consider these two decisions sequentially, where first the due dates are set and then the job schedules are determined. Commonly used scheduling policies in DDM are discussed in Section 2. We denote a due date policy by D-S, where D refers to the due-date setting rule

Table 12.1. Notation for DDM models. $(.)_{jo}$ denotes the value of parameter $(.)$ for the o-th operation of job j, if a job has multiple operations. $E[(.)]$ and $\sigma_{(.)}$ denote the expected value and the standard deviation of $(.)$, when $(.)$ is a random variable.

Indices:	
j:	order or job
k:	job class
o:	operation of a job
t:	time period

	Notation
r_j	: arrival (or ready) time of job j
p_j	: (total) processing time of job j
R_j	: revenue (price) of job j
w_j	: the weight of job j (e.g., may denote the importance of the job)
g_j (g_{jt})	: number of (remaining) operations on job j (at time t)
U_{jt}	: set of remaining operations of job j at time t
n_t (n_{kt})	: number of (class k) jobs in the system at time t
l_j	: quoted lead time (or quoted flow time) of job j
C_j	: completion time of job j
$W_j = C_j - r_j - p_j$	: wait time of job j
$F_j = C_j - r_j$	: flow time of job j
$T_j = (C_j - d_j)^+$	: tardiness of job j
$E_j = (d_j - C_j)^+$	: earliness of job j
$L_j = C_j - d_j$	: lateness of job j
τ_{max}	: maximum fraction of tardy jobs (or missed due dates)
T_{max}	: maximum average tardiness
ρ	: steady-state capacity utilization
α, β, γ	: parameters

and S refers to the scheduling rule. While most researchers are concerned in comparing the performance of alternative DDM policies, some focus on finding the optimal parameters for a given policy [19] [88].

Most of the papers in the literature assume that order acceptance decisions are exogenous to DDM, and all the orders that are currently in the system have to be quoted due dates and be processed. Equivalently, they ignore the impact of quoted due dates on demand and assume that once a customer arrives in the system, he will accept a due date no matter how late it is. In reality, the quoted lead times affect whether a customer places an order or not, and in turn, they affect the revenues. In many cases, there is a maximum limit on the lead time, either imposed by customers or by law. For example, under FTC's mail-and-telephone order rule, orders must be shipped within 30 days or customers must be notified and given the option of agreeing to a new shipping date or canceling the order. Even if the customers accepted any kind of lead time, due to the associated penalties for longer lead times or missed due dates, it might be more profitable for a manufacturer not to accept all customer orders. Hence, in contrast to most of the papers in the literature, one needs to consider DDM from a profit maximization perspective rather than a cost minimization perspective. Incorporating order acceptance decisions into DDM policies by considering the impact of quoted lead times and prices on demand is an important step towards integrated DDM policies. We discuss the literature on DDM with order acceptance (and price) decisions in Section 6.

1.2 Dimensions of a Due Date Management Problem

There are several dimensions that distinguish different due date management problems. Depending on the system (i.e., the manufacturing or service organization) under consideration, a combination of these dimensions will be present and this combination will affect the appropriate mathematical model for studying DDM in that setting.

Offline vs. online: In an offline setting, all the information about the problem, such as the job arrival and processing times, are available at the beginning of the scheduling horizon. In contrast, in an online setting future arrivals are not known with certainty and the information about a job becomes available only at the time of its arrival. In general, it is assumed that the arrivals follow a known probability distribution. An online setting is *stringent* if the decisions about a job, such as its due date, have to be made immediately at the job's arrival time. In contrast, in an online setting with *lookahead*, one can wait for some time after a job arrives before making a decision about that job, but there is usually some penalty for delaying the decision.

Single vs. multiple servers: In a single-server setting, only one resource needs to be considered for satisfying customer requests. In a multiple-server setting, multiple resources are available, which may be identical or different. In case of multiple non-identical servers, there are different possible system (or shop) characteristics, such as job shop and flow shop. In a job shop, each job has a sequence of operations in different machine groups. In a flow shop, the sequence of operations is the same for each job.

Preemptive vs. nonpreemptive: In a nonpreemptive setting, once the processing of a job starts, it must be completed without any interruption. In contrast, in a preemptive setting, interruption of the processing of a job is allowed. In the preempt-resume mode, after an interruption, the processing of a job can be resumed later on without the need to repeat any work. Thus, the total time the job spends in the system is p_j, although the difference between the completion and start times can be longer than p_j. In the preempt-repeat mode, once a job's processing is interrupted, it has to start again from the beginning. Hence, if job j is interrupted at least once, its total processing time is strictly larger than p_j.

Stochastic vs. deterministic processing times: When the processing times are not known with certainty, it is usually assumed that they follow a probability distribution with known mean and variance.

Setup times/costs: When changing from one order (or one customer class) to another, there may be a transition time. Most of the current literature on DDM ignores setup times.

Server reliability: The capacity of the resources may be known (deterministic) over the planning horizon, or there may be random fluctuations (e.g., machine breakdowns).

Single vs. multiple classes of customers: Customers (or jobs) can be divided into different classes based on the revenues (or margins) they generate, or based on their demand characteristics, such as (average) processing times, lead time related penalties, or maximum acceptable lead times.

Service level constraints: Commonly used service level constraints include an upper bound on (1) the fraction (or percentage) of orders which are completed after their due dates, (2) the average tardiness, and (3) the average order flow time.

Common vs. distinct due dates: In case of common due dates, all customers who are currently in the system are quoted the same due date. In case of distinct due dates, different due dates can be quoted to different customers.

1.3 Objectives of Due Date Management

When order acceptance decisions are assumed to be exogenous, the key decisions are due date setting and scheduling. Note that these decisions have conflicting objectives. One would try to set the due dates as tight as possible, since this would reflect better responsiveness to customer demand. However, this creates a conflict with the scheduling objectives because tight due dates are more difficult to meet than loose due dates. There are two approaches for resolving this conflict: (1) use an objective function that combines the objectives of both decisions, (2) consider the objective of one decision and impose a constraint on the other decision. A relatively small number of papers follow the first approach [11] [24] [68] [88] [87] using an objective function that is the weighted sum of earliness, tardiness and lead times, where the weights reflect the relevant importance of short lead times vs. meeeting them on time. Among the papers that follow the second approach, some try to achieve due date tightness subject to a service level constraint, while others consider a service objective (e.g., minimizing tardiness) subject to a constraint on minimum due date tightness. For example, in [6] the objective is to minimize the average due date subject to a 100% service guarantee (i.e., tardiness is not allowed). In contrast, in [7] the objective is to minimize average tardiness subject to a constraint on the average flow allowance (i.e., due date tightness).

In general, the objective in DDM is to minimize a cost function that measures the length and/or the reliability of quoted lead times. Common objectives include minimizing:

- Average/total (weighted) due date (subject to service level constraints) [1] [6] [61] [109]

- Average tardiness [7] [8] [9] [10] [27] [33] [37] [41] [55] [60] [71] [104] [105] [106]; number (or percentage) of tardy jobs [1] [9] [10] [14] [27] [34] [60] [71] [104] [105]; maximum tardiness [60] [71]; conditional mean tardiness [1][9] [60]

- Average lateness [1] [12] [14] [27] [33] [34] [60] [104] [105] [106] [107]; standard deviation of lateness [12] [27] [33] [60] [77] [83] [104] [105] [107]; sum of the squares of latenesses [19] [68]; mean absolute lateness [41] [68] [77]

- Average earliness [14] [55] [106]; number (or percentage) of early jobs [14] [33]

- Total (weighted) earliness and tardiness [33] [34] [55] [78] [106]

- Total (weighted) earliness, tardiness and lead times [11] [24] [68] [88] [87]

- Average queue length [14] [34]; waiting time [14] [34] or flow time [14] [27] [34] [41] [60] [104] [105] [107]; standard deviation of flow times

[60] [107]; number of (incomplete) jobs in system (WIP) [39]; total processing time of unfinished jobs in the system (CWIP) [39]; variance of queue length over all machines [39]

Note that the objectives in the last bullet focus primarily on "internal" measures of the shop floor, whereas the objectives in the previous bullets focus on "external" measures of customer service.

Minimizing the mean flow time, which is the average amount of time a job spends in the system, helps in responding to customer requests more quickly and in reducing (work-in-process) inventories. Minimizing earliness helps in lowering finished goods inventories leading to lower holding costs and other costs associated with completing a job before its due date. Earliness penalty may also capture a "lost opportunity" for better service to the customer, since a shorter lead time could be quoted to the customer if it were known that the job would complete early. Minimizing tardiness or conditional mean tardiness, which is the average tardiness measured over only the tardy jobs, helps in completing the jobs on or before their due dates, to avoid penalties due to delays, loss of goodwill, expedited shipment, etc. As noted by Baker and Kanet [9], average tardiness is equal to the product of the proportion of tardy jobs and the conditional mean tardiness, hence, looking at the two latter measures separately provides more information about the performance than just looking at the aggregate measure of average tardiness. Minimizing due dates helps in attracting and retaining customers. Average lateness measures whether the quoted due dates are reliable on average, i.e., the 'accuracy' of the due dates. Standard deviation of lateness measures the magnitude of digression from quoted due dates, i.e., the 'precision' of the due dates. WIP and CWIP measure the congestion in the system (through work-in-process inventory) and the average investment in work-in-process inventory, respectively, assuming that the investment in a partially finished job is proportional to the completed amount of processing.

The performance of a due date management policy depends both on due date setting and sequencing decisions. For certain objective functions, such as minimizing lateness or tardiness, due-date setting rules have direct and indirect effects on performance [107]. The direct effect results from the type of due-date rule employed and its parameters, which determine the due date tightness. The indirect effects result from the due-date being a component of some of the dispatching and labor assignment rules. The due-date setting rules have only indirect effects for some objectives, such as minimizing the average flow time, where sequencing rules have more direct impact.

An interesting question is which of the two decisions, due date setting and sequencing, has a bigger impact on performance. The answer depends on the measure of performance (objective function), the type of DDM policy and its parameters, service constraints, and system conditions. For example, Weeks and Fryer [107] find that due-date assignment rules are the most important decisions for mean lateness, variance of lateness and proportion of tardy jobs, whereas sequencing rules are the most important class of decision rules to impact mean flow time and variance of flow time. In addition, for some objective functions (the variance of flow time, variance of lateness and proportion of tardy jobs) the relative influence of sequencing rules depends on the tightness of the assigned due dates. Wein [109] finds that for the objective of minimizing average flow time, for the four due date setting rules he proposed, the impact of sequencing on performance is minimal. For other due date rules, however, the impact of sequencing on performance is significant.

One has to be very careful in choosing an appropriate objective function to satisfy business goals, as different objective functions may lead to remarkably different DDM policies. For example, a DDM policy that yields a zero lateness on average would result in approximately 50% of the jobs early and the remaining 50% tardy. If the cost of earliness is considerably lower for a firm than the cost of tardiness, then clearly minimizing lateness might not be an appropriate performance measure. In general, we can divide the objectives into two broad categories, depending on whether they penalize only for completing the jobs after their due dates, or penalize both for early and late completion. There are important differences among the objectives within the same category as well. For example, consider the three objective functions, minimizing average tardiness, the number of tardy jobs, and maximum tardiness. Consider two DDM policies, one resulting in one job that is 100 time units late, whereas the other one resulting in 10 jobs that are each 10 units late. While these two policies are equivalent with respect to the average tardiness objective, the first policy is better than the second one with respect to the number of tardy jobs, and the second policy is better than the first one with respect to maximum tardiness. Hence, depending on the choice of the objective function, one can obtain a result that is significantly poor with respect to another objective or service constraint. For example, Baker and Bertrand [7] note that when the due date tightness is below a threshold for the DDM policy SLK-EDD in an $M/M/1$ queuing system, even if the average tardiness is low, the proportion of tardy jobs is approximately equal to the utilization level in the system. While this characteristic might be quite acceptable for the objective of minimizing average tardiness, it might be undesirable for a secondary objective of minimizing the number of tardy jobs. Also, see [99] for a discussion on DDM policies which lead to unethical practice by quoting lead times where there is no hope of achieving them, while trying to minimize the number of tardy jobs.

The objective functions discussed above take a cost minimization perspective and are appropriate when the jobs to be served are exogenously determined. When order acceptance is also a decision (or equivalently, if the demand depends on quoted lead times), then the objective is to maximize profit;

that is, the revenues generated from the accepted orders minus the costs associated with not completing the orders on time.

1.4 Solution Approaches for Due Date Management Problems

To study the effectiveness of due date management policies, three approaches are used: simulation, analytical methods and competitive analysis. While they are quite interesting from a theoretical perspective, the application of analytical models to DDM has been mostly limited to simple settings, such as special cases of offline single machine problems. To study the effect of DDM policies in more realistic multi-machine environments where orders arrive over time, researchers have commonly turned to simulation.

As an alternative approach, a few researchers have used competitive analysis to study DDM policies, where an on-line algorithm A is compared to an optimal off-line algorithm [94]. Given an instance I, let $z_A(I)$ and $z^*(I)$ denote the objective function value obtained by algorithm A and by an optimum off-line algorithm, respectively. We call an algorithm A *c-competitive*, if there exists a constant a, such that

$$z_A(I) \leq c_A z^*(I) + a$$

for any instance I. (Here we assume that A is a deterministic on-line algorithm and we are solving a minimization problem.) The factor c_A is also called the *competitive ratio* of A. The performance of off-line algorithms can be measured in a similar way. An off-line algorithm A is called a $\rho-approximation$, if it delivers a solution of quality at most ρ times optimal (for minimization problems). Although the use of competitive analysis in studying DDM problems have been limited, it has received much attention recently in the scheduling literature [43] [44] [53].

2. Scheduling Policies in Due Date Management

Most DDM policies proposed in the literature propose a two-step approach: assign the due dates first, and then schedule the orders using a *priority dispatch policy*. While it would be desirable to consider due date and scheduling decisions simultaneously, the use of such a two-step approach is understandable given that scheduling problems even with preassigned due dates are quite difficult to solve. In a priority dispatching policy for scheduling, a numerical value (priority) is computed for each job that is currently in the system waiting to be processed. The job with the minimum value of priority is selected to be processed next. The priorities are usually updated at each new job arrival.

Commonly considered priority dispatch sequencing rules as part of DDM policies are summarized in Table 12.2. The simplest of the dispatching rules,

RANDOM and FCFS, consider no information about the jobs or the system, except the job arrival time. SPT, LPT, LROT/NOP, EFT, and WSPT consider the work content of the jobs. EDD, SLK, SLK', SLK/p_j, SLK/g_j, MDD, OP-NDD and P+S/OPN(a,b) consider the due dates as well as the work content. Some industry surveys suggest that due-date based rules, such as EDD and EDD$_o$ are the most commonly used rules, whereas other rules widely studied by researchers, e.g., SPT, are not preferred in practice [47] [50] [112].

We can categorize these rules as static and dynamic. The priority value assigned to a job by a static sequencing rule does not change while the job is in the system, e.g., SPT or EDD. In contrast, the priority value assigned to a job by a dynamic sequencing rules changes over time, e.g., SLK. It is interesting to note that sometimes a static rule and a dynamic rule can result in the same sequence, e.g., SLK and EDD^2.

Some of these rules are parametric, e.g., SLK and P+S/OPN have parameters α and β. A common approach is to use simulation to find the appropriate values for these parameters. The choice of the parameters depends on the system environment factors and can have a significant impact on the performance of the DDM policy.

As in the case of SPTT, some sequencing rules first divide the available jobs into two classes, and then apply a (possibly different) sequencing rule in each class. The division of jobs into classes can be random, or based on job characteristics. Three such rules are considered in [92]: 2C-SPT, 2C-NP and 2RANDOM-SPT. In the first two rules, jobs with processing time less than $\bar{p}$ belong to class 1 and others belong to class 2. In the third rule, jobs are assigned randomly into one of two classes. Higher priority is given to class 1, and within each class, jobs are sequenced based on SPT or FCFS.

The rules presented at the top portion of Table 12.2 are applied at the job level, using aggregate information about the operations of the job, whereas the rules presented in the lower portion of the table (with a subscript "o" denoting "operation") are applied at the operation level. For example, when scheduling the operations on a machine using the shortest processing time rule, one can sort the operations based on the corresponding jobs' total processing time ($\sum_o p_{jo}$), or based on the processing times of the operations to be performed on that machine (p_{jo}). For a rule like SPT, the application either at the job or at the operation level is straightforward, since the processing time information exists at both levels. On the other hand, it is not obvious how a rule like EDD can be applied at the operation level, since we usually tend to think of due dates at the job level. One needs to keep in mind that the due dates that are quoted to the customers do not have to be the same due dates as the ones used for scheduling. In other words, it is possible to have 'external' and 'internal' due dates, where the latter ones are dynamically updated based on shop conditions and used for scheduling purposes only. For example, the EDD_o rule is similar to EDD, but priorities are based on internally set operation due dates.

A number of researchers have focused on the performance of sequencing rules under different due-date setting rules. The performance of scheduling rules as part of a DDM policy depend on many factors, including the due date rule, the objective function under consideration, the tightness of the due dates, the workload/congestion level in the system. The scheduling rules have especially a significant impact on the performance of due date rules that consider shop congestion [104].

No single scheduling rule is found to perform 'best' for all performance measures. Elvers [37] and Eilon and Hodgson [34] find that (under the TWK rule) in general SPT performs best for minimizing job waiting times, flow times, machine idle times and queue lengths, and LPT performs worst. Several researchers have noticed that SPT also performs well in general for the objective of minimizing average tardiness (since it tends to reduce average flow time), but it usually results in higher variation. Bookbinder and Noor [14] find that earliness and tardiness have much higher standard deviation under SPT than under other sequencing rules. Hence, SPT might not be an appropriate choice for the objective functions that penalize both earliness and tardiness. For example, Ragatz and Mabert[83] find that for the objective of minimizing the standard deviation of lateness (σ_L), SPT performs worse than SLK and FCFS under eight different due date rules. Similar observations were made in [105] for SPTT, where the average flow time is significantly lower under SPTT, but the variance is higher. Weeks [106] finds that for the objectives of mean earliness, tardiness and missed due dates, sequencing rules that consider due date information perform better than the rules that do not.

The performance of scheduling rules also depends on the due date policy and how tight the due dates are. For the objective of minimizing tardiness, Elvers [37] finds that some scheduling rules work better under tight due dates while others work better under loose due dates. In particular, the scheduling rules which are based on processing times, such as LROT, LROT/NOP, SPT and SPT_o, perform best when the due dates are tight. (SPT's good performance under tight due dates is also observed in several other papers, e.g., [107].) On the other hand, scheduling rules which consider the operation due date or slack, such as EDD, SLK, SLK/OPN and SLK/P, perform better when the due dates are loose[3]. Intuitively, SPT tries to minimize average flow time without paying much attention to tardiness. This might hamper SPT's performance (with respect to minimizing tardiness) when the flow allowance is high where most due dates can be met using the EDD rule. But when the flow allowance is low, most jobs are late and SPT is advantageous because shorter mean flow times lead to less tardiness. Having noticed the complementary strengths of different rules depending on due date tightness, Baker and Bertrand [8] propose

Table 12.2. Dispatch sequencing policies

Policy	Priority value	References
RANDOM	$\beta_j \sim \text{Uniform}[0,1]$	[27] [34]
First come first serve (FCFS, FCFS$_o$)	r_j (or r_{jo})	[7] [8] [14] [27] [29] [33] [34] [37] [39] [68] [71] [83] [99] [104] [105] [107] [108], [37]
Shortest processing time (SPT, SPT$_o$)	p_j (or p_{jo})	[7] [8] [9] [27] [33] [34] [77] [78] [82] [83] [103] [106] [108] [109], [37] [107]
Least remaining operation time (LROT)	$\sum_{o \in U_{jt}} p_{jo}$	[37]
LROT/NOP	$\sum_{o \in U_{jt}} p_{jo}/|U_j|$	[37]
Truncated SPT (SPTT)	p_{jo}, if $W_{jo} < \alpha$ $\forall o$ in queue; W_{jo}, otherwise	[38] [39] [104] [105]
Weighted SPT (WSPT)	$w_j p_{jo}$	[109]
Longest processing time (LPT, LPT$_o$)	$1/p_j$ (or $1/p_{jo}$)	[27] [34], [37]
Truncated LPT (LPTT)	$1/p_{jo}$, if $W_{jo} < \alpha$ $\forall o$ in queue; W_{jo}, otherwise	[38]
Earliest finish time (EFT)	$r_j + p_j$	[7] [8]
Earliest due date (EDD, EDD$_o$)	d_j (or d_{jo})	[6] [7] [8] [27] [34] [37] [38] [39] [41] [60] [72] [82] [103] [109], [9] [39] [60]
Slack (SLK, SLK$_o$)	$d_j - t - \sum_{o \in U_{jt}} p_{jo} - \alpha$, $d_{jo} - t - p_{jo}$	[7] [8] [27] [37] [38] [83] [108] [109], [60]
SLK'	Put jobs with SLK≤ 0 into a priority queue and SLK> 0 into a normal queue. Apply SPT to each queue.	[33]
SLK/P	$(d_j - t - \sum_{o \in U_{jt}} p_{jo})/\sum_{o \in U_{jt}} p_{jo}$	[37] [38] [109]
SLK/OPN	$(d_j - t - \sum_{o \in U_{jt}} p_{jo})/g_j$	[9] [37] [71] [106] [107]
SLK/RAT	$(d_j - t - \sum_{o \in U_{jt}} p_{jo})/(d_j - t)$	[71]
COVERT	c_j/p_{jo}, where $c_j = 0$, if the job is ahead of schedule; $c_j = 1$, if the job has no slack; and $c_j = \bar{c}_j$, otherwise, where $\bar{c}_j$ is the proportion of the job's planned waiting time that has been consumed	[9]
Critical ratio (CR, CR$_o$)	$(d_j - t)/\sum_{o \in U_{jt}} p_{jo}$, $(d_{jo} - t)/p_{jo}$	[39] [60], [39]
R/OPN	$(d_j - t)/g_j$	[10] [71]
Modified Due Date (MDD, MDD$_o$)	$\max\{d_j, t + p_j\}$, $\max\{d_{jo}, t + p_{jo}\}$	[8] [9] [10] [82] [109] [104] [105]
Earliest operation due date (OPNDD)	$r_j + (d_j - r_j)(o/g_j)$	[27]
P+S/OPN	$\alpha p_{jo} + (1-\alpha)\dfrac{d_j - t - \sum_{l=o}^{g_j} p_{jl}}{(g_j - o + 1)^\beta}$	[27]

a modified due date (MDD) rule, which is a combination of EDD and SPT rules. It works like SPT when the due dates are very tight and like EDD when the due dates are very loose. They test the MDD rule on the same test data used in [6] and observe that MDD sequencing results in lower average tardiness compared to SPT or EDD, under both TWK and SLK due date rules, for a large range of allowance factor values. Note that in the single machine model with equal job release times, average tardiness is minimized by MDD under either TWK or SLK due date rules [8]. Wein [109] also finds that MDD performs better than other sequencing rules in general for the objective of minimizing the weighted average lead time subject to service constraints.

A interesting question is whether to use job due dates or individual operation due dates for sequencing. Kanet and Hayya [60] study the effect of considering job due dates versus operation due dates on the performance of scheduling rules in the context of DDM. Under the TWK due-date setting rule (with three different parameters), they compare three due-date based scheduling rules and their operation counterparts: EDD vs. EDD_o, SLK vs. SLK_o and CR vs. CR_o. They consider the objectives of minimizing: mean lateness, standard deviation of lateness, fraction of tardy jobs, conditional mean tardiness, mean flow time, standard deviation of flow time, and maximum job tardiness. They find that operation-due-date based scheduling rules outperform their job-due-date-based counterparts for almost all performance measures. In particular, operation-due-date based scheduling rules consistently result in lower flow times, and reduce the mean and the variance of the probability density function of lateness. Among these rules, EDD_o outperforms SLK_o for all performance measures under all conditions. In particular, EDD_o results in the minimum value of maximum tardiness in all cases. Intuitively, EDD_o extends Smith's rule [95], which states that in an offline single machine problem maximum tardiness is minimized by EDD sequencing. They also look at the impact of flow allowance on the performance. Note that when we sufficiently increase α in TWK, then the job with the largest processing time will always have the largest due date regardless of when it arrives. In that case, EDD sequencing becomes equivalent to SPT sequencing. Hence, for the rules EDD, EDD_o, SLK and SLK_o an increase in flow allowance results in a decrease in average flow time.

In a related study to [60], Baker and Kanet [9] compare the performance of MDD and MDD_o with a number of other well-studied scheduling rules under the TWK due date rule, focusing on the impact of due date tightness and shop utilization. Using industrial data, they note that the average manufacturing capacity utilization in the U.S. ranged from 80 to 90% between 1965 and 1980. Hence, they test low=80% (high=80%) utilization with α values 2.5, 5 and 7.5 (5, 10, 15, 20), where α is the parameter of the TWK rule indicating due date tightness. They consider three performance measures: proportion of tardy jobs, conditional mean tardiness, and average tardiness. A comparison of MDD and

MDD_o indicates the superior performance of MDD_o except in environments with high utilization and very loose due dates, in which case both rules perform quite well. MDD_o also outperforms SPT and EDD_o in all environments except under very loose due dates. Similar to [37], the observations of Baker and Kanet indicate that the tightness of due dates has an important effect on the performance of scheduling rules. For example, the performance of S/OPN and EDD_o is quite good with loose due dates but deteriorates as the due dates become tighter. The opposite is true for SPT. Therefore, a "hybrid" rule such as MDD tends to exhibit a more robust and superior performance over a wide range of settings.

Enns [38] [39] has experimented both with job- and operation-due-date dependent dispatch policies. He considers internal as well as external measures of the shop floor. He notes that the sequencing rule has a significant impact on internal measures such as the number of and the investment in work-in-process jobs, smoothness of work flow, and the balancing of queue lengths. He finds that EDD and SPT lead to less amount of work completed per job at any given time compared to FCFS, while jobs tend to move slower at first and then faster under EDD, whereas they progress at a more steady pace under EDD_o. Intuitively, operation due dates create artificial milestones along the way rather than a single deadline, and ensure a more steady progress of the jobs. While this might be desirable from a balanced workload point of view, it could increase work-in-process inventories and related costs.

Scheduling rules based on operation due dates have received limited attention in the literature, but the current results suggest that they are quite promising and in some cases perform better than the corresponding rules based on job due dates [8] [9] [12] [60]. In general, if the variability in the system is low, one can obtain better estimates of flow times and hence quote more accurate due dates. The scheduling policy impacts the variability of flow times. One possible reason for the observed superior performance of the operation due date based scheduling rules is because the operation due dates result in a more steady flow of the jobs through the system, reducing flow time variability.

3. Offline Models for Due Date Management

Offline models for DDM assume that the arrival times, and possibly the processing times, of the jobs are known at the beginning of the planning horizon. The arrival times can be equal ([3] [11] [19] [21] [24] [76] [81] [80] [87] [98], Section 3.1) or distinct ([6] [16] [45], Section 3.2). All of the papers discussed in this section consider a single machine.

3.1 Equal Order Arrival Times

Under the assumption of equal job arrival times, a number of authors have studied the *common due date problem* (CON), where all the jobs are assigned a common due date d and the goal is to jointly determine the common due date and a sequence [3] [11] [21] [24] [76] [81] [80]. The objective functions considered in these papers can be characterized by the following general function of earliness, tardiness and the due date: $\sum_j \beta_j E_j^a + \gamma_j T_j^b + \alpha_j d$. In most papers, $a = b = 1$, i.e., the first two summations correspond to weighted earliness and tardiness. In [76], $\beta_j = \beta$, $\gamma_j = \gamma$ and $\alpha_j = \alpha$; in [3], $\beta_j = \beta$, $\gamma_j = \gamma$ and $\alpha_j = 0$; in [21], $\beta_j = \gamma_j = w_j$ and $\alpha_j = 0$; and in [24], $\beta_j = \gamma_j = w_j$ and $\alpha_j = \alpha$. In this problem, it is easy to show that the optimal due date d is equal to the completion time of some job j^*, i.e., $d = C_{j^*}$. For the general function with $a = b = 1$, the following simple condition determines the optimal due date $d = C_{j^*}$ for a given sequence: j^* is the first position in the sequence such that $\sum_{k=1}^{j^*}(\beta_k + \gamma_k) \geq \sum_{k=1}^{n}(\gamma_k - \alpha_k)$ [11] [22] [80]. Note that all the jobs before j^* are early, and are sequenced in non-increasing order of the ratio C_j/β_j. The jobs after j^* are tardy and are sequenced in non-decreasing order of the ratio C_j/γ_j. Such a schedule is called a V-shaped sequence [3] [11]. The optimality of V-shaped schedules for common due date problems have also been observed in [84]. Optimality conditions for the case $a = b = s$ along with an iterative procedure for determining d (for a given sequence) are given in [24]. Enumeration algorithms for jointly determining the due date and the optimal sequence are presented in [3] and [21].

A related problem, where customers have a common preferred due date A, is studied in [87]. A lead time penalty is charged for lead time (or due date) delay, $A_j = \max\{0, d_j - A\}$, which is the amount of time the assigned due date of a job exceeds the preferred due date, A. The objective is to minimize the weighted sum of earliness, tardiness and lead time penalty. The authors propose a simple policy for setting the due dates and show that this policy is optimal when used in conjunction with the SPT rule.

The three papers discussed next [19] [25] [45] study the TWK and SLK due date rules presented in Table 12.3.

Cheng [19] studies the TWK due date rule under the objective of minimizing total squared lateness. Both for deterministic and random processing times (with known means and equal coefficient of variation), he finds the optimal value of the parameter α and shows that the optimal sequencing policy is SPT. Further extensions of this work are presented in [20] and [23].

Cheng [25] studies the SLK due date rule under the objective of minimizing the (weighted) flow allowance plus the maximum tardiness. He shows that the optimal sequence is EDD and derives a simple function to compute the optimal SLK due dates. Gordon [45] studies a generalized version of this problem with distinct job release times and precedence constraints on job completion times, allowing job preemption.

Soroush [98] considers the objective of minimizing the (expected) weighted earliness and tardiness under random processing times. First, he derives the optimal due dates for a given sequence, and shows that the due dates depend on the jobs' earliness and tardiness costs and the mean and the standard deviation of the jobs' completion times. Next, he addresses the problem of simultaneous scheduling and due date determination. Unfortunately, he is unable to provide an en efficient procedure for obtaining the optimal sequence, and hence provides lower and upper bounds on the expected total cost as well as two heuristics. Using examples, he shows that the heuristics perform well, and that treating the processing times as deterministic could lead to significantly inaccurate due dates and higher costs.

3.2 Distinct Order Arrival Times

Unlike the previous papers discussed in this section, the two papers we discuss next consider the case when the order arrival times are not necessarily equal and allow order preemption in an offline setting.

Baker and Bertrand [6] study DDM with job preemption under the objective of minimizing the average due date subject to a 100% service guarantee, i.e., all the jobs must be completed on time. This problem can be solved optimally by finding a schedule that minimizes the average completion time, and then by setting $d_j = C_j$ for each job j. Hence, using the three-field notation described in [46] it is equivalent to $1|r_j, pmtn| \sum_j C_j$. When the jobs have equal release times, the objective becomes equivalent to minimizing the mean flow time, and SPT sequencing gives the optimal solution. Unfortunately this approach requires that all the information about the jobs must be known in advance, which is rarely the case in practice. Hence, the authors consider three simple due date setting rules, CON, SLK, and TWK, where the due date decision depends only on the information about the job itself. They first consider these rules for the case of equal job arrival times and find the optimal value of the parameter α for each of these rules. Note that when the arrival times are equal, EDD sequencing minimizes maximum lateness, hence, given a set of due dates, one can use the EDD rule to find out if those due dates satisfy the service constraint of zero tardiness. Also, when the due dates are set using TWK or SLK due date rules, EDD and SPT sequences are equivalent. The authors show that in case of equal job arrival times, CON is dominated by TWK and SLK, but no dominance relationship exists between TWK and SLK. They also compare the worst case performance of these rules, by comparing them to the optimal solution. The main results are the following: (1) For the case of equal processing times, $c_{TWK} = c_{SLK} = c_{CON} = 2n/(n+1)$. That is, the worst case performance of the three rules is the same, approaching 2 as the number of jobs increases. (2) When the processing times are equal to 1 for all but one job j, for which $p_j > 1$, $c_{TWK} = n - 1$, $c_{SLK} = 1$, $c_{CON} = n$. In this case, TWK and CON have similar worst case performance, which can be arbitrarily large, whereas SLK produces the optimal solution. (3) When $p_j = j$ for all j, $c_{TWK} = 1.5$, $c_{SLK} = 3$, $c_{CON} = 3$. The competitive ratios provide information about the worst case performance of these rules, but they can be overly pessimistic. To gain an understanding about the average performance of CON, TWK and SLK, the authors perform simulation studies. In case of equal arrival times (with processing times drawn from exponential, normal or uniform distributions), TWK produces the best results in terms of due date tightness. Furthermore, compared to SLK and CON, TWK produced due dates which are less sensitive to problem size.

When the arrival times are distinct, a preemptive version of the EDD rule minimizes the maximum lateness. Under the CON rule, since all the jobs have the same flow allowance, EDD becomes equivalent to FCFS and preemption is not necessary. Under SLK, the waiting time allowance is the same for all jobs, and the EDD rule will sequence the jobs in the increasing order of $r_j + p_j$. Hence, preemption might be necessary under SLK. Note that in case of CON and SLK rules, one does not need the value of the parameter α to implement the EDD rule. However, in case of TWK, the value of α is needed for implementing EDD. Baker and Bertrand [6] provide simple procedures for computing the due dates under each of these rules. To test the performance of these rules, they run simulations for two different workload patterns. For the "random" workload pattern, they simulate an $M/M/1$ queue. For the "controlled" workload pattern, they use a scheme which releases a new job as soon as the workload in the system falls below a threshold. Experimental results suggest that TWK has the best average performance for the random workload pattern. In the case of controlled workload pattern, SLK produces the best results for light workloads; however, the due dates of TWK are less sensitive to workload, suggesting that TWK might be the preferred policy. Although TWK exhibits better performance on average than SLK, in unfavorable conditions it can perform considerably worse. (This is also indicated by the arbitrarily large competitive ratio of TWK when all but one job have $p_j = 1$.) The results suggest that when the variance of processing times is low, there is little difference among the performances of the three rules.

Charnsirisakskul et al. [16] take a profit-maximization rather than a cost-minimization perspective and consider order acceptance in addition to scheduling and due-date setting decisions. In their model, an order is specified by a unit revenue, arrival time, processing time, tardiness penalty, and preferred and latest acceptable due dates. While they allow preemption, they assume that the entire order has to be sent to a customer in one shipment, i.e., pieces of an order

Due Date Management Policies 503

that are processed at different times incur a holding cost until they are shipped. The shipment date of an accepted order has to be between the arrival time and the latest acceptable due date. Orders that are completed after the prefered due date incur a tardiness penalty. The processing and holding costs are allowed to vary from period to period. The goal is to decide how much of each order to produce in each period to maximize profit (revenue minus holding, tardiness and production costs).

Charnsirisakskul et al. consider both make-to-order (MTO) and make-to-stock (MTS) environments. In the first case, the processing of an order cannot start before the order's arrival (release) time while in the latter case it is possible to process an order and hold in inventory for later shipment. For both cases, they model the problem as a mixed integer linear program and study the benefits of lead time and partial fulfillment flexibility via a numerical study. Lead time flexibility refers to a longer lead time allowance (higher difference between the latest and preferred due dates) and partial fulfillment flexibility refers to the option of filling only a fraction of the ordered quantity. Their results show that lead time flexibility leads to higher profits, and the benefits of lead time flexibility outweigh the benefits of partial fulfillment flexibility in both systems. Lead time flexibility is more useful in MTO where early production is not an option. Numerical results also show that the benefits of lead time and partial fulfillment flexibility depend on the attributes of the demand (demand load, seasonality, and the order sizes).

4. Online Models for Due Date Management

Online models for DDM assume that the information about a job, such as its class or processing time, becomes available only at the job's arrival time. The arrival times are also not known in advance. Such models are sometimes referred to as *dynamic*.

Based on the 'dimensions' discussed in Section 1, the papers that study the DDM problem in an online setting can be further divided into two categories based on whether they study the problem in a single machine [7] [8] [14] [61] [82] [109] or a multi-machine setting [12] [27] [33] [37] [39] [41] [36] [38] [60] [71] [83] [103] [104] [105] [106] [107]. Although a single vs. multi-machine categorization seems natural at first, most of the research issues are common to both settings, and the resulting insights are often times similar. Therefore, we discuss the papers in this section by categorizing them based on their approach to due-date setting decisions and based on some of the other modeling dimensions and the related research questions.

For online DDM problems, very few researchers have proposed mathematical models (see Section 4.3). The most common approach for setting due dates is to use *dispatch* due date rules which follow the general form $d_j = r_j + f_j$

where f_j is the *flow allowance*. The tightness of the flow allowances (and the due dates) is usually controlled by some parameters. Alternative flow allowances are summarized in Table 12.3 and are discussed in Section 4.1.

As in the case of dispatch rules for scheduling (Section 2), most of the due date rules are parametric. These parameters may be constant (e.g., α in SLK) or dependent on the job or system conditions (e.g., γ_j in BERTRAND). The appropriate choices for these parameter values are not straightforward and usually determined via simulation experiments. A small number of papers in the literature focus on choosing the appropriate parameters for a given due date rule, and they are discussed in Section 4.2.

Most of the DDM policies proposed in the literature do not result in any service guarantees. A small number of researchers have proposed DDM policies with service guarantees, such as the maximum expected fraction of tardy jobs or maximum expected tardiness. DDM policies with service guarantees are discussed in Section 5.

Table 12.3: Due date setting rules with a flow allowance

Policy	Flow allowance	References
RND	$\alpha\beta_j$, $\beta_j \sim \text{Uniform}[0,1]$	[27]
CON	α	[6] [8] [27] [30] [72] [73] [88] [103] [109] [107]
SLK	$p_j + \alpha$	[6] [8] [103] [109]
TWK	αp_j	[6] [7] [8] [9] [10] [19] [27] [33] [34] [41] [60] [71] [72] [73] [77] [78] [103] [106] [107] [108] [109] [83]
TWK'	$\alpha(p_j)^\beta$	[33]
NOP	αg_j	[27] [77] [78] [83]
TWK+NOP	$\alpha p_j + \beta g_j$	[77] [78] [83] [104] [105]
BN	$\max\{d_{j-1} - r_j\} + \alpha_j p_j$	[14]
JIQ	$\alpha p_j + \beta Q_j$, where Q_j is the number of jobs waiting to be processed ahead of job j	[14] [33] [72] [73] [77] [78] [83] [104] [105] [109]
JIS	$\alpha p_j + \beta WIS_j$, where WIS_j is the total number of jobs waiting to be processed in the system at time r_j	[72] [83]
WIQ	$\alpha p_j + \beta WIQ_j$, where WIQ_j is the total processing time of the jobs waiting to be processed ahead of job j	[77] [78] [83]
WINS	$\alpha WINS$, where WINS is the sum of processing times of all the jobs currently in the system	[41]
TWKCP	$\alpha TWKCP$, where TWKCP is the sum of all operation times on the critical path of the BOM	[41]
FRY-ADD1	$\alpha TWKCP + \beta WINS$	[41]

Table 12.3 (continued)

FRY-ADD2	$\alpha p_j + \beta WINS$	[41]
FRY-MULT1	$\alpha p_j(WINS)$	[41]
FRY-MULT2	$\alpha(TWKCP)(WINS)$	[41]
RMR	$\alpha W_{SPT} + \sum_{i=1}^{k} \alpha_i WIQ_{ij} + \beta_1 g_j + \sum_{i=1}^{k} \gamma_i JIQ_{ij} + \beta_2 WIS_j + \beta_3 WIQ_j + \beta_4 JIS_j$ where WIQ_{ij} and JIQ_{ij} are the work and the number of jobs in queue on the i-th machine in the routing of job j	[77] [78][83]
FTDD	$E[F] + \alpha \sigma_F$	[10] [71]
TWK-RAGATZ	$p_j(1 + \alpha W'_j/E[W])$, where W'_j is the estimated workload in the system when j arrives	[82]
EC3	$\alpha p_j + \beta E[W]$	[33], [106]
WEEKS6	$\alpha p_j + \beta g_j W^o$ where W^o is the expected wait time per operation	[106]
WEEKS7	$\alpha p_j + \beta W'$, where W' is an estimated wait time based on shop congestion level	[106] [83]
OFS	$\alpha \bar{F} n_j + \beta n_j + \gamma P_j$ where $\bar{F}$ is the average operation time of the last three jobs that are completed	[73] [104] [105]
COFS	$\alpha \bar{F} n_j + \beta_1 Q_j + \beta_2 n_j + \gamma P_j$	[104] [105]
MFE	$(1-\alpha)f_{sj} + \alpha f_{dj}$ where f_{sj} and f_{dj} are static and dynamic flowtime estimates	[105]
CON-BB	α_j, where $\alpha_j = a(W'_j/E[W])$	[7]
SLK-BB	$p_j + \alpha_j$, where $\alpha_j = E[p](a-1)W'_j/E[W]$	[7]
TWK-BB	$\alpha_j p_j$, where $\alpha_j = aW'_j + E[p]/\bar{W}$	[7] [82]
BERTRAND	$p_j + \alpha E[p_{jo}]g_j + \gamma_j$, where γ_j is the additional flow allowance based on the congestion/workload in the shop	[12]
WEIN-PAR I	Equation (12.1)	[109]
WEIN-PAR II	Equation (12.2)	[109]
HRS	$E[F(n_j)] + \beta z_\alpha \sigma_{LT}(n_j)$ where α is the service level and z_α is the α-percentile of the standard normal distribution	[55]

Most of the papers that consider multiple machines in setting due dates use simulation as the method of study. Among these papers, a majority focus on job-shop environments, which are are summarized in Table 12.4. These simulation models assume that job preemption is not allowed and there are no service constraints. With the exception of a few papers, including [12] [35] [68], they assume that the machines are perfectly reliable, i.e., there is no machine downtime.

While generating routings for the jobs in the job-shop simulations, most of these papers assume that a job is equally likely start its first operation on each machine (or machine group), and upon completion in one machine a job moves to another machine or leaves the shop with some probability [27]. Let P_{ik} denote the probability that when a job completes operation on machine i it will move to machine k. Depending on the processing requirements and the associated routings of the jobs, a shop has different workload characteristics, including the following:

Balanced shop: In a balanced shop, the assignment of an operation to any work center has equal probablilities (i.e., $P_{ik} = 1/m$, $k = 1, \ldots, m$) and hence leads to approximately equal workload in each work center.

Unbalanced shop: In an unbalanced shop, some work centers might have higher loads than others.

- In a *bottleneck* shop, one work center has a significant higher load than others.

- In a *flow shop*, all jobs follow the same sequence of machines, that is $P_{i,i+1} = 1, i = 1, \ldots, m-1$.

- In a *central server job shop* all job transfers take place thorugh a central server, i.e., $P_{1k} > 0$ and $P_{k1} = 1, j = 2, \ldots, m$.

Most researchers base their experiments on balanced shops while a few consider both balanced and bottleneck shops [104] [105]. In general, two consecutive operations on the same machine are not permitted (exceptions include [83]), and there are two alternatives for repeated machines on a routing: (1) a job can return to a machine after being processed on another machine [33] [37] [39] [60] [104] [105], and (2) each job has at most one operation in each work center [106] [107].

Some papers explicitly consider multiple divisions and work centers with multiple machines as well as dual-constrained job-shops, where there are labor as well as machine constraints [106] [107] [108].

In the remaining part of this section, we discuss the papers with online DDM models in more detail. We use the term 'machine' to refer to any type of resource. We say that due date management policy A dominates B, if both policies satisfy the same service level constraints and policy A achieves a better objective function than policy B.

4.1 Due-Date Setting Rules

In quoting a lead time for a job, one can consider different types of information, such as the job content, general shop congestion (workload), information about the routing of a job and the congestion ahead of that job, and the sequencing policy.

Due Date Management Policies 507

Table 12.4. Simulation models for DDM in a job shop

	# of machs.	Interarrival times	Processing times	# of opns. per job	Shop utilization
[12]	5	Neg. exp.	Neg. exp.	Mean=4.463, Max=10	0.83, 0.9, 0.93
[27]	9	Exponential	Exponential	Mean=9	0.90
[33]	5	Constant	Normal$(20, 6^2)$	Uniform~[1,5]	0.9
[37]	8	weekly arrivals Uniform~ [8, 12]	Uniform~ [6, 15] Triang.~ [0.8, 1, 1.9]	P(1)=.2, P(2)=.16 P(3)=.128, P(4)=.512	
[41]	11	Exponential	Normal	depends on BOM	0.7, 0.8. 0.9
[38]	4	Exponential	Neg. exp.	Uniform~ [2, 6]	0.8, 0.9
[39]	4	Neg. exp.	Neg. exp.	Uniform~ [2, 6]	0.9
[9] [60]	8	Exponential	Exponential	Uniform~ [1, 8]	0.8, 0.9
[71]	5,7	Exponential	Exponential	Uniform~ $[1, 2m-1]$	0.8, 0.85, 0.9
[10]	4	Exponential	Exponential	2-6	0.8, 0.9
[77] [83]	9	Neg. exp.(1)	Neg. exp.(0.76)	Geometric~[1,15]	0.89
[103]	8	Exponential	Exponential	Uniform~ [4, 8]	0.9
[104]	5	Exponential	2-Erlang	Uniform~ [1, 10]	0.85 - 0.95
[105]	5	Exponential	Exponential	Uniform~ [1, 10]	0.85-0.95
[106] [107] [108]	24	Neg. exp.	Neg. exp.		0.9

The simplest due date rules are RND and CON, which do not consider job or system characteristics in setting due dates. Given the drawbacks of these simple rules, researchers have proposed other rules (e.g., SLK, TWK, TWK', NOP, TWK+NOP and BN), which ensure that the job content, such as the processing time or the number of operations, affects the due dates. Although these rules are one step up over the simple rules, they do not consider the state of the system, and hence, additional rules are proposed to consider the workload in the system (or the congestion) while setting due dates (e.g., EC3, JIQ, OFS, COFS). An important research question is whether the inclusion of detailed information about the job content and system congestion improves the performance of DDM policies. To answer this question, researchers have compared the performance of various due-date setting rules in different settings via simulation. In this section, we discuss the due date setting rules and their relative performance.

4.1.1 Due Date Rules with Job Information.
The due-date rules we discuss in this section incorporate only the information about a job in quoting due dates: the processing time (SLK, TWK, TWK', BN), the number of operations (NOP), or both (TWK+NOP).

One of the earliest papers that studies DDM is by Conway [27]. He considers four due date policies, CON, NOP, TWK and RND, and tests the performance of nine priority dispatching rules for each of these policies in a job shop. For the objective of minimizing average tardiness, the author finds that under FCFS sequencing, the four due date policies exhibit similar, mediocre performance. Under EDD, SPT and P+S/OPN, the due date rules rank in the following order of decreasing performance: TWK, NOP, CON, RND. This suggests that due date rules that take the work content into account perform better. Among sequencing rules, SPT's performance is less sensitive to the due date policy, mainly because it does not take the due dates into account. SPT is superior to other rules when the workload in the system is high, and has the best performance in general. EDD and P+S/OPN perform better with due date policies that take the work content into account.

Baker and Bertrand [7] study combinations of three due-date rules (CON, SLK, TWK) and five sequencing policies (FCFS, SPT, EDD, EFT, MST). Via simulation, they test due date management policies in $M/M/1$ and $M/G/1$ queues under low and high utilization. For the objective of minimizing average tardiness (subject to a maximum average flow allowance factor a, i.e., $E[F] = aE[p]$), they find that SPT and EDD always dominate the other sequencing rules. Furthermore, in combination with SPT or EDD, CON is always dominated by SLK which is dominated by TWK. Hence, they focus on the combinations of SPT and EDD with SLK and TWK. The dominance of SLK by TWK indicates the importance of using job information in determining due dates. The authors find that the performance difference between TWK, SLK and CON rules is very small when the variance of job processing times (σ_p^2) is low. However, the difference becomes significant when σ_p is high, and in that case TWK dominates SLK dominates CON.

The papers that are discussed in the remainder of this section consider dual-constrained job shops and investigate the performance of labor assignment policies in conjunction with due date quotation and sequencing.

DDM policies with labor assignment decisions

In case of dual-constrained job-shops, in addition to due date and sequencing decisions, one also needs to consider labor assignment decisions. Most papers assume that laborers may be transferred between work centers within a division, but not between divisions [106] [108]. Under this assumption, a laborer can be considered for transfer when he finishes processing a job (CENTLAB) or when he finishes processing a job and there are no jobs in the queue at the work center he is currently assigned to (DECENTLAB). The following reassignment rules are considered in the literature in assigning a laborer to a work center: assign to the work center (1) with the job in the queue which has been in the system for the longest period of time, (2) with the most jobs in the queue, and (3) whose queue has the job with the least slack time per operation remaining.

Weeks and Fryer [107] investigate the effects of sequencing, due-date setting and labor assignment rules on shop performance. They simulate a job show with three divisions, each containing four work centers of different machine types and four laborers. There are two identical machines in each work

Due Date Management Policies

center and each laborer can operate all machine types. For labor assignments, they consider CENTLAB and DECENTLAB policies and the three reassignment rules discussed above. They test FCFS, SPT_o and SLK/OPN sequencing rules with CON and TWK due date rules and compare their performance using multiple regression. Their experiments lead to the following three observations: (1) The impact of due-date assignment rules: Due-date assignment rules are the most important decisions for mean lateness, variance of lateness and proportion of tardy jobs. Tight due dates perform better than loose due dates in terms of mean flow time, variance of flow time, and variance of lateness, and the reverse is true for mean lateness and percent of jobs tardy. (2) The impact of sequencing rules: Sequencing rules are the most important class of decision rules to impact mean flow time and variance of flow time. Sequencing rules also have pronounced effects on mean lateness, variance of lateness and proportion of tardy jobs. The relative effects of sequencing rules are independent of the due date assignment rule for the mean flow time, mean lateness and total labor transfers. For the variance of flow time, variance of lateness and proportion of tardy jobs the relative influence of sequencing rules depends on the tightness of the assigned due dates, but their relative performance does not change. SPT_o has better performance than other rules for mean flow time, mean lateness, proportion of jobs late and total labor transfers. FCFS performs best in terms of variance of flow time and SLK/OPN performs best in terms of variance of lateness. SPT performs best in terms of the proportion of jobs late, regardless of the due-date assignment rule. (3) The impact of labor assignment rules: By increasing the flexibility of the shop, CENTLAB performs better than DECENTLAB in all measures except total labor transfers. The rule which assigns laborers to the work center with the job in the queue which has been in the system for the longest period of time performs best.

4.1.2 Due Date Rules with Job and System Information.

There are various ways of incorporating the system workload information in setting the due dates. Ultimately, the goal is to estimate the flow time of a job, and quote a due date using that estimate. Some researchers have proposed using simple measures about the system conditions for estimating flow times and quoting lead times:

- the number of jobs waiting to be processed in the system (JIS),
- the number of jobs waiting to be processed ahead of job j (JIQ),
- the total processing time of the jobs waiting to be processed ahead of job j (WIQ),
- the number and the total processing time of the jobs waiting to be processed on all machines on the routing of job j (RMR),

- the sum of processing times of all the jobs currently in the system (WINS), and

- the sum of all operation times on the critical path of the BOM (TWKCP).

The flow time has two components: the job processing time and the waiting (or delay) time. In general, finding analytical estimates of flow times in job shops is very difficult, even for very special cases [57] [93]. This is because the flow time depends on the scheduling policy, the shop structure, job characteristics (e.g., the number of operations, the variability of the processing times) and the current state (e.g., the congestion level) of the system. However, for relatively simple environments, such as exponential processing and inter-arrival times with one or two customer classes, finding analytical estimates for expected job flow (or waiting) times might be possible [7] [71] [109]. When the flow allowance is chosen to be a multiple of the average processing time, i.e., $f_j = aE[P]$, we refer to the multiple a as the *flow allowance factor*. For example, in the TWK rule, by setting $\alpha = a$, where $a = E[F]/E[P]$, the flow allowance becomes equal to the average flow time, which would lead to an average tardiness of zero. We denote by a^* the minimum flow allowance factor which would lead, on average, to zero tardiness for a given DDM policy.

The flow time estimates in a system can be static or dynamic, depending on whether they consider time-dependent information. The *static* estimates focus on steady-state, or average flow time, whereas the *dynamic* estimates consider the current job or shop conditions. Static estimates are usually based on queuing models or steady state simulation analysis. The due date rules bases on static flow time estimates include EC3, WEEKS6, FTDD. Several methods, ranging from simple rules to regression analysis on the information of recently completed jobs, are proposed for dynamic flow time estimates. Due date rules based on dynamic flow time estimates incorporate time-phased workload information: WEIN-PAR I-II, WEIN-NONPAR I-II, JIQ, WEEKS7, OFS, COFS, BERTRAND, and EFS. For example, the flow allowance in BERTRAND depends on the observed congestion in the shop at any time. The flow allowance in OFS and COFS depend on the average operation time of the last three jobs that are completed, as well as the number of operations in the queue ahead of a job.

There are two types of measures that can be used in evaluating the quality of a flow time estimate: *accuracy* and *precision*. Accuracy refers to the closeness of the estimated and true values, i.e., the expected value of the prediction errors. Precision refers to the variability of the prediction errors. Dynamic rules usually result in better precision (lower standard deviation of lateness) because they can adjust to changing shop conditions; however, even if the prediction errors are small, they may be biased in certain regions leading to poorer accuracy (deviations from the true mean) compared to static models. Note that if the quoted lead times are equal to the estimated flow times, accuracy and precision would measure the average and the standard deviation of lateness, respectively.

A small number of due date rules are based on the probability distribution of the flow time (or the workload in the system). For example, Baker and Bertrand [7] modify the TWK rule by setting $\alpha = 2aF(W'_j)$ where $F(x)$ denotes the cumulative distribution function (CDF) of the workload. Although it is hard to determine $F(x)$ for general systems, it is known to be exponential with mean $E[p]/(1-\lambda E[p])$, when the processing times are exponential. The authors find that this TWK modification used in conjunction with EDD or SPT produces the best results for some a values among all the policies they tested. The idea of using a CDF function to estimate workload information is further explored by Udo (1993) [103] and extended to multiple machines.

Given the difficulty of finding analytical estimates for flow times, several researchers have proposed approximations. For example, estimates for the mean and the variance of the flow time are proposed in [90] [91] and a method for approximating the flow time distribution is proposed in [92]. Shanthikumar and Sumita [92] develop approximations for estimating the distribution of flow time in a job shop under a given sequencing rule. Their approximation uses the concept of *service index* (SI) of a dynamic job shop, which is the squared coefficient of variation of the total service time of an arbitrary job. Depending on the value of the service index, they divide dynamic job shops into three classes: G ($SI \ll 1$, e.g, flow shop), E ($SI \approx 1$, e.g., balanced job shop) and H ($SI \gg 1$, e.g., central server job shop). They approximate the flow time distribution for the jobs in G, E and H by generalized Erlang, exponential, and hyper-exponential distributions, respectively. Numerical examples show that the approximations are good, especially when SI is close to $(\sigma_F/E[F])^2$. Recall that Seidmann and Smith [88] showed how to find the optimal CON policy (for the objective of minimizing the sum of weighted earliness, tardiness and deviations from a desired due date A) assuming that the distribution of the flow time is known. Using a numerical example, Shanthikumar and Sumita show how their estimate of the flow time distribution can be used to find the optimal CON due date policy following the method proposed in [88].

In the remainder of this section we discuss in more detail the DDM policies which take both job and system workload information into account. We first present workload-dependent rules which use simple formulas to incorporate the information about the state of the system. We then continue with the rules which are based on more sophisticated but steady-state estimates of the workload, such as the mean, standard deviation or distribution of flow time. Finally, we present the due date rules which are based on time- and job-dependent estimates of flow/waiting times.

Eilon and Chowdhury [33] are among the first researchers to consider workload-dependent due date rules. They propose EC3 and JIQ, which extend the TWK rule by incorporating the mean waiting time in the system and the number of operations ahead of an arriving job, respectively. Via simulation, they test DDM policies which result from combinations of three sequencing rules (FCFS, SPT and SLK′) and four due date policies (TWK, TWK′, EC3 and JIQ) for the objective of minimizing a (weighted) sum of earliness and tardiness. The simulation results suggest that JIQ is the best due date policy in general, and SLK′ performs better than FCFS and SPT. SPT seems to be the least effective, especially when the penalty for late jobs is very high.

Miyazaki [71] derives two formulae for estimating the mean and standard deviation of flow times, and proposes a due date rule (FTDD) based on these estimates. He tests this policy with three sequencing rules (SLK/OPN, R/OPN and SLK/RAT), which take into account the due dates. For comparison purposes, he also tests DDM policies TWK-FCFS and FTDD-FCFS. He considers three objectives: minimizing the number of tardy jobs, average tardiness and maximum tardiness. For all these objectives, he finds that (1) FTDD-FCFS performs better than TWK-FCFS, (2) sequencing rules SLK/OPN and R/OPN yield better results than FCFS when used in conjunction with FTDD, and (3) sequencing rule R/OPN is slightly better than SLK/OPN especially for tighter due dates. In a follow-up study, Baker and Kanet [10] compare FTDD and TWK due date rules with R/OPN and MDD scheduling rules in a setting similar to the one in [71]. They conclude that using TWK-MDD might be a better choice than FTDD-R/OPN or FTDD-SLK/RAT for the following three reasons. (1) SLK/RAT might perform in undesirable ways when the slack is negative. (2) Miyazaki claims that SLK/RAT is very effective in reducing the number of tardy jobs when the due dates are tight. However, a relatively small proportion of jobs was tardy in Miyazaki's experiments, and in such settings CR and SLK/RAT have similar performances (assuming that the slack is positive). In earlier studies as well as this one, MDD is found to outperform various rules, including CR and R/OPN, in a variety of settings. (3) The parameters of the FTDD rule are difficult to compute, especially under any priority rule other than FCFS.

Weeks [106] proposes two new workload-dependent due date rules, WEEKS6 and WEEKS7, and tests their performance against TWK and EC3 under SPT and SLK/g_j sequencing rules for the objectives of mean earliness, tardiness, and missed due dates. He finds that when employed with due-date based sequencing rules, due date rules that consider the current state of the system in addition to job characteristics perform better that the rules that only consider the job characteristics. Similarly, sequencing rules that consider due date information perform better than the rules that do not. He also finds that even though the shop size does not seem to affect the performance of due date management policies significantly, increased shop complexity degrades the performance.

Bertrand [12] proposes a due date policy (BERTRAND) that takes into account the work content of a job and the workload (or congestion) in the shop at the time of the job's arrival. He compares this policy against a simpler one (obtained by setting $\gamma = 0$ in BERTRAND) which ignores the workload in the shop. The mean and standard deviation of job lateness are used to measure the performance of DDM policies. The scheduling (or loading) policy ensures that during any time duration (1) the load factor of a machine (the ratio of the cumulative load on a machine to the cumulative machine capacity) does not exceed a specified cumulative load limit (CLL), and (2) there is a minimum waiting (or flow time) allowance, α, per operation. In a situation where the load factor would exceed CLL, the flow allowance of the job/operation is increased.

Using simulation experiments, Bertrand finds that an increase in α or a decrease in CLL increases the standard deviation of quoted flow times. Similarly, an increase in α increases the standard deviation of actual flow times. The covariance of actual and quoted flow times increase in α and decrease in CLL. He also finds that using time-phased workload information improves the standard deviation of lateness (σ_L); however, the improvement decreases for larger values of α. The parameter CLL does not seem to have much impact on σ_L; however, it has a significant impact on the mean lateness ($E[L]$) and its best value depends on α. For decreasing CLL values, the sensitivity of mean quoted flow time to mean flow time increases. These results indicate that for the objective of minimizing σ_L (1) DDM policies which use time-phased workload and capacity information perform quite well, and (2) the performance of BERTRAND seems to be very sensitive the choice of α and relatively insensitive to the type of job stream. Finally, the performance of the due date rule is sensitive to the sequencing rule, especially if changes in the sequencing rule shift mean flow time.

Ragatz and Mabert [83] test the performance of 24 DDM policies obtained by pairwise combination of eight due date rules (TWK, NOP, TWK+NOP, JIQ, WIQ, WEEKS7, JIS, and RMR) and three sequencing rules (SPT, FCFS and SLK). For the objective of minimizing the standard deviation of lateness (σ_L), simulation results suggest that the parameter values of the due date rules are sensitive to the sequencing policy, and due date rules that consider both job characteristic and shop information (RMR, WIQ and JIQ) perform significantly better than the rules that only consider job characteristics information (TWK, NOP, TWK+NOP). Among the rules that consider shop information, the ones that consider information about the shop status in the routing of a job (JIQ, WIQ) perform better than the rules that only consider general shop information (WEEKS7, JIS). It is also interesting to note that JIQ and WIQ exhibit similar performance to RMR, which incorporates the most detailed information about the jobs and the shop status. This suggests that the use of more detailed information in setting due dates, as in RMR, only brings marginal improvement over rules, such as WIQ and JIQ, which use more aggregate information.

The next three papers by Baker and Bertrand [7], Ragatz [82] and Udo [103] modify the CON, SLK and TWK due date rules to incorporate workload information. The modifications in [7] and [82] are based on simple estimates of the waiting time in the system, whereas the modifications in [103] are based on the estimates of the cumulative distribution function.

In a single-machine environment, Baker and Bertrand [7] modify the CON, SLK and TWK rules, such that the parameter α depends on the wait time (estimated analytically) in the system at the arrival time of job j, as well as the job's processing time and the average wait time. In particular, they set α equal to $aW'_j/E[W]$, $E[p](a-1)W'_j/E[W]$, and $a(W'_j + E[p])/\bar{W}$, for CON, SLK, and TWK, respectively, where W'_j is the estimated workload at the arrival of job j and $\bar{W}$ is a constant to meet the requirement $E[F] = aE[p]$. We refer to this as the proportional workload function (PWF). Surprisingly, simulation results show that the use of workload information does not always lead to better results. In particular, the inclusion of workload information frequently results in higher tardiness under the SPT rule when the due-dates are very tight. Similarly, in high utilization settings the DDM policy TWK-SPT without workload information produces the best results when the flow allowance factor a is between 2 and 4. On the positive side, the inclusion of workload information significantly improves the performance in some cases. For example, the DDM policy SLK-EDD, which is always dominated when workload information is not included, produces the best results when due dates are very loose. Note that under loose due dates, i.e., when $a \geq a^*$, SLK produces zero tardiness (on average). When the due-date tightness is medium, TWK-EDD produces the best results. In summary, for tight due dates (small a), the best policy is to use TWK or SLK with SPT, without using workload information. As the due date tightness increases, the TWK-EDD with workload information performs better. Finally, for really loose due dates ($a \geq a^*$), the best choice is SLK-EDD.

The dynamic version of the TWK rule proposed by Baker and Bertrand, TWK-BB, adjusts the flow allowance according to the congestion in the shop and quotes longer due dates in case of high congestion. Although this rule results in lower tardiness in many settings, it usually results in higher fraction of tardy jobs. Ragatz [82] proposes and alternative modification to the TWK rule, TWK-RAGATZ, which outperforms TWK-BB by reducing the number of tardy jobs, while achieving the same or lower average tardiness in most settings. Ragatz notices that when the shop is lightly loaded or if the flow allowance parameter α is small, then TWK-BB might result in a flow allowance smaller than the job's processing time. Hence, some jobs are a priori assigned 'bad' due dates which leads to a high number of tardy jobs. When the workload is not very high (less than 90%) the percentage of tardy jobs is quite high even for high allowance factors (relatively loose due dates). Ragatz's modification of the TWK rule ensures that the flow allowance is always larger than the average processing time. In almost all settings, TWK-BB leads to a significantly higher fraction of tardy jobs than TWK-RAGATZ which results in a higher fraction of tardy jobs than TWK.

Udo [103] investigates the impact of using different forms of workload information on the performance of DDM policies. In particular, as in [7], he considers two variations of the TWK and SLK rules, modified based on workload information in the form of proportional workload function (PWF) and cumulative distribution function (CDF). As in [7], Udo considers the objective of minimizing average tardiness (subject to a minimum average flow allowance a, i.e., $E[F] = aE[p]$). Based on simulations in a job-shop environment, he finds that the CDF model dominates the PWF model, which dominates the basic SLK and TWK models (under two sequencing rules, SPT and EDD). Except in TWK-SPT, the difference between PWF and CDF models is statistically significant. Furthermore, the difference between these models is larger for small allowances and becomes smaller as the allowance factor increases (as the due dates become looser). Among all the DDM policies tested, SLK-EDD results in the best performance, followed by SLK-SPT. These results suggest that (1) including workload information in DDM results in lower tardiness, (2) for loose due dates (low congestion), using detailed information about the workload in setting due dates might not be critical, and (3) the impact of workload information depends on the choice of the DDM policy; for certain policies (e.g., TWK-SPT) there is not much payoff in the additional information required by more complex workload models. Most of these observations hold for the single-machine environment as well, but interestingly, the SLK-EDD policy, which has the best performance in the job-shop setting, results in the highest tardiness in the single-machine environment. In addition, the mean tardiness performance of all the DDM policies tested was much poorer in a multi-machine job-shop than in a single-machine shop. These observations once again indicate that the performance of a DDM policy depends on the shop structure.

Lawrence [68] uses a forecasting based approach for estimating flow times. In particular, he focuses on approximating the flow time estimation error distribution (which is assumed to be stationary, and denoted by G) by using the method of moments [111]. He uses six methods for forecasting flow times. NUL sets flow times to zero, such that G becomes an estimator for the flow time. ESF uses exponential smoothing, such that the flow time estimate after the completion of k jobs is $\tilde{f}_k = \alpha F_k + (1 - \alpha)\tilde{f}_{k-1}$. The other four rules are WIS, which is WIQ with parameters $\alpha = \beta = 1$; EXP, which uses the (exponential) waiting time distribution of an $M/M/1$ queue; ERL, which uses Erlang-k with mean kp, if k jobs are currently in the system; and BNH, proposed in [14], which uses a hypoexponential distribution. As due date rules, Lawrence uses (1) $\tilde{f} + \tilde{G}^{-1}(0.5)$, where $\tilde{G}$ is the current estimate of the error distribution, (2) $\tilde{f}+$ the sample mean of G, (3) the due date rule proposed by Seidmann and Smith [88], and (4) $\tilde{f}+\tilde{G}^{-1}(1-MFT)$ where MFT is the mean fraction of tardy jobs. He notes that due date rules (1)-(4) minimize mean absolute lateness, mean squared lateness, the sum of earliness, tardiness and due date penalties, and due dates (subject to a service level constraint), respectively, in a single machine environment. He tests the quality of the proposed flow time estimators as well as due date rules via simulation both on single machine (in three environments, $M/M/1$, $M/G/1$, and $G/G/1$, as in [14]) and job-shop environments (as in [92]) under FCFS scheduling. For the single machine environment, simulation results indicate that flow time estimators that use current system information (WIS, ERL, BNH, and ESF) perform better than the estimators that use no or only long-run system information (NUL, EXP). For the multi-machine environment, in addition to using the flow time estimation methods discussed above by considering the system as a single aggregate work center ("aggregation" approach), he also tests variations of these six methods by estimating flow times (and forecast errors) for individual work centers and then combining them ("decomposition" approach). In addition, he tests TWK and JIQ rules to estimate flow times. Simulation results indicate that the decomposition approach works better for estimating flow times and setting due dates than the aggregate approach, and WIS and ESF perform at least as good as or better than other flow time estimators. These results are encouraging given the minimal data requirements for implementing these estimators. However, since the simulations consider only the FCFS scheduling rule, these results should be interpreted with some caution. The impact of the scheduling rules on flow times is critical, and these results might not generalize to environments using different scheduling policies.

Vig and Dooley [104] propose two due date rules, operation flow time sampling (OFS) and congestion and operation flow time sampling (COFS) which use shop congestion information for estimating flow times and setting due dates. In particular, these rules incorporate the average operation time of the last three jobs that are completed. They test the performance of due date rules OFS, COFS, JIQ and TWK+NOP in combination with three scheduling rules: FCFS, SPTT and MDD_o. These three rules are significantly different. FCFS contains no info about the job or shop status. SPTT is a dynamic rule considering job info. MDD_o is also a dynamic rule but it is based on internally set operation due dates. To test the impact of work center balance on the performance of DDM policies, they test both a balanced shop and a shop with a bottleneck. To estimate the best values for the parameters of the DDM policies, they use multiple linear regression techniques. The coefficients found for each DDM policy minimize the variations of the predicted flow times from actual flow times. In experiments, they find that the due-date rule, scheduling rule and shop balance levels all significantly affect the performance of DDM policies. For the objective of % tardy jobs, TWK+NOP performs best under all three scheduling rules and shop balance conditions. But in all other objectives, DDM policies that contain shop information outperform TWK+NOP. Among sequencing rules, for the objectives of average and standard deviation of lateness, MDD_o performs best in general. In terms of standard deviation of lateness (σ_L), i.e., due date precision, shop-information based rules OFS, COFS and JIQ perform 15-50% better than the job information rule TWK+NOP. In addition, when MDD_o rule is used, σ_L decreases by up to an additional 25%. In general, COFS, JIQ or OFS used in conjunction with MDD_o result in the best performance for all objectives other than % tardy jobs. COFS and JIQ perform very good in general. OFS, which is simpler, also performs comparably well for lateness, and provides improvements for % tardy jobs. No statistical significance is detected in average lateness among the three rules, hence more detailed shop info does not seem to help much. The differences between the performances of DDM policies are more pronounced in balanced shops than in unbalanced shops. Perhaps when the shop is imbalanced the performance of all policies degrade, making it more difficult to observe significant performance differences.

In a follow-up study, Vig and Dooley [105] propose a model which includes both static and dynamic flow time estimates into flow time prediction. Let f_{sj} and f_{dj} be the static and dynamic flow time estimates for job j. The mixed flow time estimate proposed by Vig and Dooley [105] is a convex combination of static and dynamic estimates, namely, $f_{mj} = (1 - \alpha)f_{sj} + \alpha f_{dj}$, where the dynamic flow time estimate f_{sj} is based on one of the following four due date rules, TWK+NOP, JIQ, OFS and COFS. The parameters for f_{sj} and f_{dj} are obtained from steady-state simulation experiments. Static and dynamic flow time estimates have complementary strengths: better accuracy and better precision, respectively. The mixed rule is designed with the goal of improving flow time accuracy (achieving a mean prediction error close to zero) without sacrificing precision (keeping the variability of flow time errors small). As in [104], they test three scheduling rules: FCFS, SPTT and MDD_o. Simulation results show that the performance of all due-date rules improved under the mixed strategy, for all scheduling rules, and the performance differences between the rules became less crucial. In particular, for the objective of average lateness and percent tardy jobs, the mixed due dates perform significantly better than their dynamic counterparts, especially when used with the MDD_o scheduling rule. There is no significant difference between dynamic and mixed rules in terms of the standard deviation of lateness. The results also show that the improved accuracy was not obtained at the cost of losses in the precision of flow time estimates.

As an alternative to the commonly studied rules (TWK, NOP, JIQ, WIQ, TWK+NOP, and RMR) for setting due dates, Philipoom et al. [77] propose a neural networks (NN) approach and nonlinear regression. Essentially, the neural network is "trained" on sample data and then used to predict flow times. Using a simulation environment similar to [83] under SPT sequencing, for "structured" shops (where all jobs follow the same sequence of work centers) they find that both NN and the nonlinear regression approaches outperform the conventional rules for the objectives of minimizing mean absolute lateness and standard deviation of lateness (SDL), and the neural network outperforms nonlinear regression. For unstructured shops, RMR slightly outperforms NN for standard deviation of lateness. In a follow-up study [78] they use a similar approach for minimizing objectives that are (linear, quadratic, and cubic) functions of earliness and tardiness. Again, they test NN against the six due date rules above, as well as nonlinear regression and linear programming (LP). Their results indicate that NN is fairly robust as it performs quite well regardless of the shape of the cost function, whereas some other approaches show significantly poor performance under certain cost functions (e.g., LP's performance with nonlinear cost functions or RMR's performance for linear cost functions). As in [68], one limitation of this study is that it only considers a single sequencing rule, SPT. Further testing is needed for assessing the performance and robustness of NN under different sequencing rules.

In estimating the flow time of a job and quoting a lead time, in addition to shop conditions and the operation processing times, the 'structure' of the job (or product) may also play a role. This is especially true in implementing a Materials Requirements Planning (MRP) system where one estimates the flow time at different levels and then offsets the net requirements to determine planned order releases. The total time a job spends in the system depends on the bill of materials (BOM) structure. For example, a 'flat' BOM, where there are multiple components or subassemblies but only a few levels, could permit concurrent processing of operations in multiple machines. In contrast, a 'tall' BOM with many levels would require most operations to be performed sequentially. Previous work on sequencing suggests that jobs with tall BOMs tend to have higher tardiness than jobs with flat BOMs [86].

Few researchers have studied the impact of product structures on flow times, scheduling and due date management policies [41] [56]. For assembly jobs, delays in the system occur not only due to waiting in queues (due to busy resources), but also due to "staging," i.e., waiting for the completion of other parts of the job before assembly. Fry et al. [41] study the impact of the following job parameters on the flow time: total processing time (sum of all operation times, denoted by p_j for job j), sum of all operations on the critical (longest) path of the BOM (TWKPC$_j$), total number of assembly points in the BOM (NAP), number of levels in the BOM (NL), number of components in the BOM (NB). In addition, they consider the impact of the total amount of work in the system (WINS) on flow times. WINS is computed by adding the processing times of all the jobs in the system, which are currently being processed or are in the queue. The authors find that NAP, NL and NB are not significant. On the other hand, p_j, TWKPC$_j$ and WINS are significant. In particular, WINS becomes more important than p_j or TWKCP in predicting flow times as the congestion level increases. The authors propose four models (see Table 12.3), two additive [74] and two multiplicative [4], for estimating flow times and setting due dates based on these parameters. They test these due date rules as well as three other simple rules via simulation under EDD sequencing. The results suggest that the performance of DDM policies depend both on the due date rule and the product structure. For the objective of minimizing mean flow time, TWKCP and TWK, which consider only job characteristics, perform best for flat or mixed product structures. For the objective of minimizing average tardiness, FRY-ADD1 and FRY-ADD2 perform best, except for high congestion levels, where WINS performs slightly better. These two rules consider both the job characteristics and the shop congestion. For the objective of minimizing mean absolute lateness, FRY-ADD2 performs the best under all BOM types and congestion levels. For all objectives, FRY-MULT1 and FRY-MULT2 are the worst performers.

One of the most interesting results of [41] is the relatively poor performance of TWKCP, which follows the traditional approach of setting lead times based on the critical path in the BOM. The best performers were those rules which considered both the critical path and the system congestion in an additive rather than a multiplicative form.

In a related study Adam et al. [1] propose dynamic coefficient estimates for CON, TWK and TWKCP for multi-level assembly job shops and test them under EDD and EDD$_o$ scheduling rules for the objectives of minimizing the average lead time, lateness, conditional tardiness, and fraction of tardy jobs. In contrast, most of the earlier studies concerning assembly jobs focus in fixed coefficients determined by a priori pilot simulation studies and regression. They also propose an extension to TWKCP, called critical path flow time (CPFT), where the flow allowance is based not only on the processing times (as in TWKCP) but also on expected waiting times on the critical path. The dynamic estimates of waiting times are based on the Little's law at an aggregate (micro) level, where the estimated waiting time (at a work center i) at time t is set to $\bar{W}_t = n_t/\bar{\lambda}_t$ ($\bar{W}_{ti} = n_{ti}/\bar{\lambda}_{ti}$), where n_t and $\bar{\lambda}_t$ (n_{ti} and $\bar{\lambda}_{ti}$) are the number of jobs in and the estimated arrival rate to the system (work center i) at time t. Through an extensive study that simulates various job structures they observe that the dynamic updates improve the overall performance of the due date rules, and reduce the performance differences among them. They also show that the performance of the job vs. operation-based scheduling rules (EDD vs. EDD_o) heavily depends on the job structure, the due date rule, and the objective function, and the performance differences are more prevalent under complex job structures.

4.2 Choosing the Parameters of Due Date Rules

Several researchers have investigated the impact of due date tightness or other parameter settings on the performance of DDM policies. For example, some suggested $\alpha = 10$ in the TWK rule claiming that a job's flow time in practice is 10% processing and 90% waiting [74]. The 'optimal' tightness is found to be a factor of overall shop utilization, relative importance of missed due dates, market and customer pressures, and relative stability of the system [104].

The choice of the parameters of a due date rule determines how tight the due dates are. The tightness of the due dates, in turn, affect the performance of the accompanying sequencing rule and the overall performance of the DDM policy. For example, in the study of Baker and Bertrand [7], when the flow allowance factor is low (i.e., due dates are tight), the choice of the sequencing rule impacts the performance of DDM policies significantly, whereas the choice of the due date rule doesn't seem to matter that much. In contrast, when the due dates are extremely loose, one can obtain no tardiness only with specific due date rules, but the sequencing rule does not need to be sophisticated. Finally, when the tightness of the due dates are medium, then both the due date policy and the sequencing policy impact the performance.

Analytical studies for choosing the parameters of due date rules include [88] and [108]. Seidmann and Smith [88] consider the CON due date policy in a multi-machine environment under the objective of minimizing the weighted sum of earliness, tardiness and lead time penalty (the same objective function is also considered in [87]). They assume that the shop is using a priority discipline for sequencing, such as FCFS or EDD, and the probability distribution of the flow time is known and is common to all jobs. They show that the optimal lead time is a unique minimum point of strictly convex functions can be found by simple numerical search.

Weeks and Fryer [108] consider the TWK rule in a dual-constrained job-shop along with sequencing policies FCFS, SPT and SLK. Their focus is on developing a methodology (based on regression analysis) for estimating the optimal parameter α for the TWK rule. They first run simulations using different values of α and use linear and nonlinear regression models to estimate the resulting costs of mean job flow time, earliness, tardiness, due date and labor transfer. They then combine these component costs into a total cost objective function and use regression to estimate the total cost as a function of α. Given the estimated total cost function, one can then do a numeric search to find the best value of α.

4.3 Mathematical Models for Setting Due Dates

All the papers discussed so far in this section propose heuristic "rules" for setting due dates. In contrast, one could possibly model the due date setting problem as a mathematical program. Given the difficulty of developing and solving such models for DDM, the research in this direction has been very limited.

Kapuscinski and Tayur [61] propose a finite-horizon discrete-time model for DDM with two customer classes. In their model, the processing times are deterministic, quoted lead times are 100% reliable, one unit of capacity is available per period, and the goal is to minimize the (expected) total weighted lead time where the weights for class 1 customers are higher than that of class 2 customers (i.e., class 1 customers are more sensitive to the quoted lead time). In this setting, it is easy to observe that quoting the shortest possible lead time for a class 1 customer is optimal. Hence, the authors focus on class 2 customers and first find a pseudo-optimal policy in which the due dates for all previously accepted orders are contiguous. That is, at time t, capacity at periods $t, t+1, \ldots, t+R-1$ is reserved and capacity at periods $t+R$ and later are free, where R denotes the outstanding orders (reservations) at time t after class-1 demand is scheduled. They then show that the optimal due date quote for a class 2 customer depends only on R and the latest due date of class 2 customers that has been quoted before time t. Combining these results with the convexity of the cost function with respect to the lead time of an individual order, the authors show that the pseudo-optimal quotation is optimal, if it does not conflict with the previously scheduled demand; otherwise, it can be converted to the optimal quotation by pushing (delaying) the new order(s) to the earliest feasible location. The authors present an exact algorithm for finding the pseudo-optimal policy. However, implementing the algorithm requires the computation of many curves and each curve requires a large number of parameters. Therefore, the authors propose an approximate computation of the pseudo-optimal policy (and hence, the optimal policy) and test their proposed method against three commonly used heuristics, namely, Greedy, CON, and Constant Slack (CS). Each of these heuristics gives priority to class 1 customers as in the optimal policy. In quoting due dates to class 2 customers, Greedy quotes each order the shortest feasible lead time. CON assigns a constant lead time to each class 2 order, whenever feasible, or the shortest feasible lead time. CS quotes the shortest feasible lead time plus a constant slack. The authors also consider smart versions of CON and CS, namely, CON-S and CS-S, which quote a zero lead time if the system is empty. Simulation results suggest that the approximate method is very close to the optimal policy, and performs significantly better than the best heuristic, in this case, CS-S. To the best of our knowledge, this is the only paper that identifies the optimal policy for multiple customer classes in an online setting.

In the remainder of this section we discuss two papers, by Elhafsi [35] and Elhafsi and Rolland [36], who propose a model for determining the delivery date and cost/price of an order. Elhafsi and Rolland [36] consider a manufacturing system with multiple processing centers. At the arrival of a new order, the manufacturer needs to decide where to process the order, given the current state of the system, i.e., given the orders currently being processed and the queues in the processing centers. The centers experience random breakdowns and repairs. The unit processing time and cost, the setup time and cost, and the inventory carrying cost differ across centers. The manufacturer can split an order across multiple processing centers, but the entire order has to be delivered at once. That is, the manufacturer incurs a holding cost for the part of the order that is completed before the delivery date. The completion time of the order has to be between the earliest and the latest acceptable times specified by the customer. The authors propose a mathematical model to assign a newly arrived order to the processing centers so as to minimize the total cost associated with that order. Note that this "greedy" approach only optimizes the assignment of one order at a time, without considering the impact of that assignment on future orders. Furthermore, it does not differentiate between multiple customer classes and assumes that all arriving orders must be accepted and processed.

Elhafsi and Rolland claim that one can estimate and quote a due date to the customer using the results of this model. In particular, they consider two types of customers: time-sensitive and cost-sensitive. Time-sensitive customers want to have their order completed as soon as possible, regardless of the cost, whereas cost-sensitive customers want to minimize the cost, regardless of the actual order completion time. The authors modify their original model according to the objectives of these two customer types, and propose solution methods for solving the corresponding models. While the models return an (expected) completion time C for the order, the authors note that quoting C as the due date might result in low service levels, depending on the variance of the completion time. They compute an upper bound σ_{max} on the standard deviation of the completion time, and propose a due date quote $C + \alpha \sigma_{max}$ which contains a safety factor $\alpha \sigma_{max}$. Using numerical experiments, they show that for $\alpha = 0.5$, the maximum average percentage tardiness (tardiness divided by the actual completion time) is 3.3% (2%) for time-sensitive (cost-sensitive) customers and the maximum safety factor is around 11% (6.5%). These results suggest that using $\alpha = 0.5$ in quoting due dates results in fairly high service levels for both customer types. In a follow-up paper, Elhafsi [35] extends this model by considering partial as well as full deliveries and proposes both exact algorithms and two heuristics. A major limitation of these papers is that FCFS scheduling rule is assumed and the results do not carry over to other scheduling rules.

5. Due Date Management with Service Constraints

Due date setting decisions, like many other decisions in a firm, need to balance competing objectives. While it is desirable to quote short lead times to attract customers, for the long-term profitability of the firm, it is also important that the quoted lead times are reliable. The papers discussed so far in this survey incorporate a lateness penalty into their models and analyses to ensure the attainability of the quoted lead times. Although this is a very reasonable modeling approach, in practice is very difficult to estimate the actual 'cost' of missing a due date. Furthermore, unlike the common assumption in the literature, lateness penalties in practice are usually not linear in the length of the delay. Therefore, the papers we review in this section impose service level constraints rather than lateness penalties to ensure lead time reliability. The two commonly used service constraints are upper bounds on the average fraction of tardy jobs (τ_{max}) and the average tardiness (T_{max}).

Bookbinder and Noor [14] study DDM policies in a single machine environment with batch arrivals (with constant interarrival times) subject to τ_{max}. For job sequencing, they test three rules: SPT, FCFS and SDD, where SDD uses SPT and FCFS for sequencing the jobs within a batch and the batches, respectively. They propose a due date rule (BN), which incorporates, through the parameter α_j, the job content and shop information at the arrival time of a job, and ensures that a due date will be met a given probability. For setting α_j, they propose a formula which uses information on the jobs in the queue or in process, their estimated processing times, and the sequencing rule. They test the BN rule for the objectives of mean lateness and earliness, percentage of jobs late and early, the mean queue length, waiting time and flow time, and compare against JIQ. They find that RM (when used with SDD) performs worse than JIQ in terms of earliness (i.e., quotes more conservative due dates), however, it performs better in the other objectives. The performance of BN under sequencing rules other than SDD degrades with respect to the lateness measure, since it quotes more conservative due dates to achieve service level constraints. This is also true for JIQ. Both due-date rules perform worse under FCFS.

Wein [109] studies DDM under the objective of minimizing the weighted average lead time subject to τ_{max} (Problem I) and T_{max} (Problem II). He considers a multiclass $M/G/1$ queuing system, where order arrivals are Poisson (with rate λ_k for class k), the processing times for each class are independent and identically distributed (iid) random variables with mean $1/\mu_k$, and preemption is not allowed.

The due date rules proposed by are based on the estimates of the *conditional sojourn times* of the jobs. The sojourn time (or flow time) of a job is the length of time the job spends in the system. The expected conditional sojourn time $E[S_{k,t}]$ of a class k job arriving at time t is defined as the total time the job spends in the system if the WSPT (or the $c\mu$) rule is used for sequencing. The standard deviation of the conditional sojourn time is denoted by $\sigma[S_{k,t}]$. Note that one needs to consider the state of the system as well as job information while estimating the conditional sojourn time.

Let $D_{k,t}$ denote the due date quoted to a class k job that arrives at time t. Wein (1991) proposes the following due date setting rules:

$$\text{WEIN-PAR I} \quad : \quad D_{k,t} = t + \alpha E[S_{k,t}] \quad (12.1)$$

$$\text{WEIN-PAR II} \quad : \quad D_{k,t} = t + \alpha E[S_{k,t}] + \beta \sigma[S_{k,t}] \quad (12.2)$$

$$\text{WEIN-NONPAR I} \quad : \quad D_{k,t} = t + f^I_{k,t} \quad (12.3)$$

$$\text{WEIN-NONPAR II} \quad : \quad D_{k,t} = t + f^{II}_{k,t} \quad (12.4)$$

where $\tau_{max} = P(S_{k,t} > f^I_{k,t})$, $T_{max} = \int_{f^{II}_{k,t}}^{\infty} (x - f^{II}_{k,t}) dG_{k,t}(x)$ and the random variable $S_{k,t}$ has distribution $G_{k,t}$.

The first two of these rules are parametric and the parameters α and β are chosen (based on simulation experiments) such that the service level constraints are satisfied. These rules apply to both problems I and II. The third and fourth rules set the shortest possible due date at any time to satisfy the service level constraints of the maximum fraction of tardy jobs and the maximum average tardiness, and apply to problems I and II, respectively. The proposed due date management policies use these rules with SPT sequencing between different classes, and FCFS sequencing within each class.

Note that the rules in Equations (12.1)-(12.4) do not consider the previously set due dates. Therefore, utilizing WEIN-NONPAR I and WEIN-NONPAR II, Wein proposes two other rules, WEIN-HOT-I and WEIN-HOT-II, for problems I and II, respectively. Under these rules, if there is enough slack in the system, a new job can be quoted a shorter lead time and scheduled ahead of some previous jobs of the same class. The sequencing policy is still SPT for different classes, but EDD (rather than FCFS) within each class. Although these rules may result in shorter due dates, they may violate the service level constraints.

The rules proposed by Wein are not easy to compute for general multiclass $M/G/1$ queues, and hence, their applicability in practice may be limited. Wein derives these rules for a special case with only two classes of jobs, exponential processing times and equal weights for each class. To test the performance of these rules, he conducts a simulation study on this example and compares his proposed rules against CON, SLK and TWK [6] and a modification of JIQ [33], under five different sequencing policies (i.e., a total of 40 due date management policies are tested for each problem). For the due date setting policies proposed by Wein simulation results indicate the following: (1) The performance is significantly better than the previous policies. For Problem I, the best policy proposed by Wein, (WEIN-HOT-I)-MDD, reduced the average lead time by 25.2%, compared to the best of the other policies tested, which was JIQ-EDD. JIQ-EDD's performance was 49.7% better than the worst policy tested, which was CON-SLK/P. (2) There is not a large difference in performance between parametric and nonparametric rules. (3) Due date setting has a bigger impact on performance than sequencing (however, this observation does not hold for other due date setting policies). For some due date policies, however, the impact of the sequencing rule was significant. For example, for Problem I, SLK/P worked well with SPT, and for Problem II, JIQ performed better under due-date based sequencing policies, especially under MDD. Wein also finds that JIQ, which uses information about the current state of the system, performs better than CON, SLK and TWK, which are independent of the state.

Wein's results indicate that the performance of a DDM policy also depends on the service constraint. For example, TWK-SPT, which performs well under τ_{max} (Problem I), does not perform well under T_{max} (Problem II).

Spearman and Zhang [99] study the same problems as in [109], in a multistage production system under the FCFS sequencing rule. They show that the optimal policy for Problem I is to quote a zero lead time if the congestion level is above a certain threshold, knowing that the possibility of meeting this quote is extremely low. Intuitively, since the service level is on the number of tardy jobs, then knowing that an arriving job will be tardy anyway (unless a very long lead time is quoted, which would affect the objective function negatively), it does make sense to quote the shortest possible lead time to reduce the objective function value. Clearly, such a policy would be quite unethical in practice and reminds us one more time that one has to be very careful in choosing the objective function or the service measures. Using numerical examples, the authors show that a customer is more likely to be quoted a zero lead time when the service level is low or moderate, rather than high, creating service expectations completely opposite of what the system can deliver. For Problem II, they show that (1) the optimal policy is to quote lead times which ensure the same service level to all customers, and (2) in cases where the mean and variance of the flow time increases with the number of jobs in the system, quoted lead times also increase in the number of jobs in the system. Note that the optimal policy for Problem II quotes longer lead times as the congestion level in the system increases. This is similar to some other rules proposed earlier in the literature, such as JIQ, JIS and WIQ, and is more reasonable in practice than the optimal policy found for Problem I. Also, observation (1) leads to an easy implementation of the optimal policy for problem II, which is to use the optimal policy for Problem I by choosing an appropriate Type I service level. The information required to implement the policy is an estimate on the conditional distribution of flow times, which may not be easy to determine in practice.

Hopp and Roof-Sturgis [55] study the problem of quoting shortest possible lead times subject to the service constraint τ_{max}. In their proposed due date policy (HRS), which is based on control chart methods, they break the lead time quote of job j into two components: the mean flow time, which is a function of the number of jobs in the system at the arrival time of job j, plus a safety lead time, which is a function of the standard deviation of flow time. They use the following quadratic functions (applicable to a flow shop) to estimate the mean and the standard deviation of the flow time

$$E[F(n)] = \begin{cases} T_0 + \beta_1(n-1) + \beta_2(n-1)^2 & \text{if } n \leq 2W_0 \\ (1/r_p)n & \text{if } n > 2W_0 \end{cases} \quad (12.5)$$

$$\sigma_{F(n)} = \begin{cases} \gamma_1 + \gamma_2 n & \text{if } n \leq 2W_0 \\ \sigma_p \sqrt{n} & \text{if } n > 2W_0 \end{cases} \quad (12.6)$$

where

T_0 = practical process time (the average time for one job to traverse the empty system)

$1/r_p$ = average interdeparture time from the system (inverse of the practical production rate)

W_0: critical WIP (the minimum number of jobs in a conveyor system that achieve the maximum throughput, $W_0 = r_p T_0$

σ_p: standard deviation of the interdeparture time from the system

The advantage of this approach is that it does not require any assumptions on probability distributions and it can be adjusted in response to changing environmental conditions.

Using simulation, the authors fine tune the parameters in the above equations as well as the parameter β in rule HRS, and show that these functions indeed provide very good estimates for $E[F(n)]$ and $\sigma_{F(n)}$. They also show that $z_\alpha \sigma_F(n)$ provides a good estimate of the tail of the flow time distribution. Next, they test the performance of the due date quotes which are based on these estimates, first in a simple $M/M/1$ system where the exact due date quotes can be computed analytically, and then in another system with nearly deterministic processing times and random machine breakdowns. They also compare their method to the JIS rule [106], after adjusting JIS to achieve the same service level as their rule. For the objectives of mean earliness, mean tardiness, and the sum of mean earliness and tardiness, they find that their method outperforms the modified JIS rule.

Hopp and Roof-Sturgis also propose a method inspired from statistical process control for dynamically setting and adjusting the parameters of the lead time quote. They define the model to be 'out of control' if the current service level (number of tardy jobs) differs from the target service level by more than three standard deviations of the service values. The service value of a job is one, if it is early or on time, and zero, if it is tardy. Then the process is out of control it indicates that the lead time quotes are either too short or significantly too long, and the model parameters need to be changed. They test their method in systems where two types of changes occur over time: increasing capacity and speeding up repairs. In both cases, their method reduces the lead times in response to system improvements. Finally, the authors discuss possible extensions of their methods to more complex multi-machine systems such as multi-product systems with intersecting or shared routing.

In most of the due date setting rules in Table 12.3, the flow allowance is based on simple estimates of the flow time plus a safety allowance for the estimated waiting time. Furthermore, these allowances are usually parametric, and choosing the appropriate parameters is not straightforward. It is especially not clear how one should set these parameters to satisfy service constraints, e.g., a maximum fraction of tardy jobs. As the congestion level increases in the shop, longer flow allowances are needed to achieve the same service objective, but how much longer? To address these issues, Enns [38] proposes a method for estimating the flow times and the variance of the forecast error, and proposes a due date rule based on these estimates. The goal is to set the tightest possible due dates (minimize average lead time) while satisfying a service constraint on the maximum fraction of tardy jobs.

The flow time estimate proposed by Enns has a similar flavor to the flow allowance in the TWK+NOP rule, where the parameter β is set equal to an estimate of the average waiting time per operation. More formally, the estimated flow time for job j is $\sum_o p_{jo} + \beta_j g_j$ where $\beta_j = \frac{n_t E[p]}{m\rho} - E[p]$. Note that here the parameter β depends on the job, hence, the rule is dynamic and we denote it by TWK+NOP$_d$. Via simulation, Enns tests the effectiveness of TWK+NOP$_d$ against TWK+NOP for estimating flow times. He uses these two rules for setting due dates, where the flow allowance is equal to the estimated flow time. Noting that the forecast error tends to have a normal distribution with the TWK+NOP$_d$ flow time estimates, one would expect to have approximately 50% of the jobs tardy, which turns out to be the case. Compared to TWK+NOP with the best constant multiplier, he finds that TWK+NOP$_d$ provides better estimates of the flow times. Deviations from estimated flow times are higher under non-due-date dependent sequencing rules and as utilization increases.

The next step is to estimate the variance of the forecast error. Enns proposes two estimates, based on the forecast error at the operation level ($\hat{\sigma}^2_{j,OLV}$) and at the job level ($\hat{\sigma}^2_{j,JLV}$). Based on these estimates, Enns proposes the following flow allowances for setting due dates:

$$ENNS-OLV = \sum_o p_{jo} + \beta_j + \gamma\sqrt{g_j}\hat{\sigma}_{j,OLV} \quad (12.7)$$

$$ENNS-JLV = \sum_o p_{jo} + \beta_j + \gamma\sqrt{g_j}\hat{\sigma}_{j,JLV} \quad (12.8)$$

The first two terms in equations (12.7) and (12.8) estimate the flow time, and the last term is a waiting time allowance based on the forecast error. Normal probability tables can be used to choose the γ value to satisfy the service level constraint on the maximum number of tardy jobs.

Enns tests the performance of this policy via simulation under various sequencing rules. As performance measures, he looks at the percentage of tardy jobs, and deviations from the desired service level (PTAE). He finds that (1) the performance of the proposed due date setting policy is affected more by the utilization level and the service level requirements than the sequencing policy, (2) due-date dependent sequencing rules in general outperform non-due-date dependent rules, especially for high service level requirements and as utilization increases.

While most research in DDM focuses on "external" measures on customer service, in a follow-up study Enns [39] focuses on internal measures such as the number of and the investment in work-in-process jobs, smoothness of work flow, and the balancing of queue lengths. To ensure a balanced workload in the shop, Enns enhances the scheduling rules with the queue balancing (QB) mechanism: among the top two candidate jobs to be chosen next, pick the one that has the smaller number of jobs in queue at the machine for the next operation. In addition to the TWK+NOP$_d$ rule proposed in [38], he proposes TWK$_d$, a dynamic version of the TWK rule, by setting $\alpha = \frac{n_t}{m\rho}$. Having these two flow time estimates, he then sets the due dates by adding to them a flow allowance $K_{SD}\sqrt{g_j}\hat{\sigma}_{j,OLV}$ or $K_{SD}\hat{\sigma}_{j,JLV}$, where K_SD is a parameter chosen to satisfy the service constraint on the maximum fraction of tardy jobs (0.05 in the experiments). Note that TWK$_d$ assigns tighter flow allowances to jobs with short processing times than TWK+NOP$_d$. Hence, when due-date dependent sequencing rules are used, short jobs get higher priority under TWK$_d$ (similar to the SPT rule), leading to shorter lead times. Enns' computational testing of these due dates under various sequencing rules (FCFS, SPTT, EDD, CR, EDD$_o$, and CR$_o$ and their QB versions) confirms this intuition and also indicates that the performance under internal and external measures is positively correlated. Comparing job vs. operation based due date rules, Enns finds that under TWK_d operation based rules perform better in general; this observation parallels the findings of Kanet and Hayya [60] for the job- and operation-based versions of the TWK rule. However, the superior performance of the operation-based due dates do not hold under $TWK+NOP_d$. About the impact of shop floor balancing, Enns finds that combining $TWK+NOP_d$ with the QB versions of the sequencing rules leads to reduced lead times, whereas this is not the case for TWK_d. These results once again suggest that there are strong interactions between due date and scheduling rules.

We say that a DDM policy has 100% service guarantee if an order's processing must be completed before its quoted due date. (Note that a 100% service guarantee is possible only if the processing times are deterministic and there are no machine breakdowns). Zijm and Buitenhek [113] propose a scheduling method, which is based on the shifting bottleneck procedure [2], that utilizes maximum lateness as a performance measure and finds the earliest time a new-coming job can be completed, without delaying any of the existing jobs in the system (assuming that the existing jobs already have due dates assigned and can be completed without tardiness). If the due date for each arriving job is set later than the earliest completion time for that job, then all jobs can be completed before their due dates. Hence, this procedure results in 100% service guarantee. This approach significantly differs from most of the existing approaches in the DDM literature: it "greedily" creates a tentative schedule for a new job to minimize its completion time (and due date). Note that most due date rules (e.g., TWK, NOP, SLK, etc.) do not consider a tentative schedule while assigning due dates.

Kaminsky and Lee [58] study a single-machine model (similar to Q-SLTQ of [65], discussed in Section 6.2) with 100% reliable due dates where the objective is to minimize the average due date. The authors first note that the SPT rule is asymptotically (as the number of jobs goes to infinity) optimal for the offline version of this problem, i.e., provides a lower bound for the online version. They propose three heuristics for this problem. First Come First Serve Quotation (FCFSQ) heuristic simply sequences the jobs in the order of their arrival, and quotes them their exact completion time, which can be computed easily. Sequence/Slack I (SSI) heuristic maintains a tentative sequence of the jobs in queue in an SPT-like order. A newly arriving job j is inserted at the end of this sequence and then moved forward one job at a time until the preceding job has a smaller processing time or moving job j would make at least one job late. After the position of the new job is established in the current sequence, the due date is quoted by adding a slack to the completion time of this job according to the current schedule. Note the following tradeoff in determining the slack: If the slack is too small, then in case other jobs with shorter processing times arrive later, they have to be scheduled after job j, violating the SPT rule. On the other hand, if the slack is too large, then the quoted due date of job

j might be unnecessarily large, degrading the value of the objective function. In SSI, the authors determine the slack by multiplying the average processing time of all the jobs which have processing time smaller than p_j (denoted by EPJ', and provides an estimate of the average processing time of a future job which would have to be scheduled ahead of job j in an SPT sequence) with the number of jobs ahead of job j in the current schedule which have smaller processing times than p_j (which provides an estimate on the number of jobs that might might arrive later and move ahead of job j). The SSII heuristic is similar to SSI, but it assumes that the total number of jobs is known. In SSII, the slack is computed by multiplying EPJ' with the expected number of yet to arrive jobs with processing time smaller than p_j. The main idea of the SSI and SSII heuristics is to build just enough slack into the schedule so as to obtain a final sequence that is as close to an SPT sequence as possible. The authors show that when the expected processing time is shorter than the expected interarrival time, then all three heuristics are asymptotically optimal. Furthermore, SSII is asymptotically optimal even if the expected processing time exceeds the expected interarrival time. A nice feature of the algorithms proposed in [58] is that they "learn" the environment over time by considering the processing times of all the jobs that arrived until the current time.

Another study that considers DDM with 100% service guarantee is [65]. In addition to due date setting and scheduling, the authors consider order acceptance decisions and take a profit maximization perspective; hence, their model and results are discussed in Section 6.1. Other papers that consider service level constraints along with pricing decisions in DDM include [75] and [97], which are discussed in Section 6.2.

6. Due Date Management with Price and Order Selection Decisions

Before a business transaction takes place, both the buyer and the seller evaluate the short and long term profitability of that transaction and make a decision on whether to commit or not based on that evaluation. Within the context of make-to-order environments, before placing an order for a product, the buyer evaluates various attributes of the product, such as its price, quality, and lead time. Similarly, before a supplier agrees to accept an order, she evaluates the 'profitability' of that order given the resources (e.g., manufacturing capacity) required to satisfy that order, currently available resources, and other potential orders that could demand those resources. For an order to become a firm order, the buyer and the seller have to agree on the terms of the transaction, in particular, on the price and the lead time. If the price or the lead time quoted by the seller is too high compared to what the buyer is willing to accept, the buyer may choose not to place the order. Alternatively, if the price the buyer is willing to pay is low or the lead time requested by the buyer is too short to make it a profitable transaction for the seller, the seller might decide not to accept the order. In either scenario, the final demand the seller sees is a function of the quoted price and lead time. In effect, by quoting prices and lead times, the seller makes an order selection/acceptance decision by influencing which orders finally end up the system.

Most of the existing literature on DDM ignores pricing decisions and the impact of the quoted prices and lead times on demand. Hence, it ignores the order selection problem and assumes that all the customers that arrive in the system place firm orders, which have to be accepted and scheduled. In this setting, the price and order selection decisions are exogenous to the model. This would be the case, for example, if the marketing department quotes prices and makes order acceptance decisions without consulting the manufacturing department (or without considering the current commitments and available resources), and taking these firm orders as input, the manufacturing department makes lead time quotation and production decisions.

Ideally, a manufacturer should take a global perspective and coordinate its decisions on price, lead time quotation and order acceptance for increased profitability. However, due to the nature of the business environment or company practices, this might not always be possible. For example, prices in certain industries are largely dictated by the market, and the manufacturer may not have much flexibility in setting the prices. Even if the manufacturer could set prices, due to the current organizational structure and practices, it might not be easy to integrate pricing decisions with lead time quotation and production decisions. But even in such environments where the price is exogenous to DDM, the manufacturer would still benefit from incorporating order selection decisions into DDM. For example, in periods of high demand or tight resources, it might be more profitable to reject some low-margin orders. Similarly, if there are multiple classes of customers/orders with different cost/revenue characteristics, it might be more profitable to reject some low-margin orders to "reserve" capacity for potential future arrivals of high-margin orders.

The papers we review in this section study DDM with order selection decisions by incorporating the impact of the quoted lead times on demand. In Section 6.1 we discuss papers which take price as given but consider the impact of quoted lead times on demand. In Section 6.2, we review the new but growing area of literature on combined price and lead time quotation decisions.

6.1 Due Date Management with Order Selection Decisions (DDM-OS)

In most businesses, the quoted lead times (or due dates) affect the customers' decisions on placing an order. In general, the longer the lead time, the less likely a customer will place an order. In most cases, a customer might have a firm deadline and would not consider placing an order if the quoted due date exceeds that deadline. Within acceptable limits, some customers might not care what the actual lead time is, while others might strongly prefer shorter lead times.

The models proposed in the literature for capturing the impact of quoted lead times on demand all follow these observations: demand (or the probability that an arriving customer places an order) decreases, as the quoted lead time increases. Let $P(l)$ denote the probability that a customer places an order given quoted lead time l and let l_{max} denote the maximum acceptable lead time to the customer. The proposed demand models include the following:

(D1) : $P(l) = 1 - l/l_{max}$

(D2) : $P(l) = \begin{cases} 1, & \text{if } l \leq l_{max} \\ 0, & \text{otherwise} \end{cases}$

(D3) : $P(l) = e^{-\lambda l}$, where λ is the arrival rate of the customers

(D4) : $P(l)$ is a decreasing concave function of l

DDM models which consider order selection decisions take a profit maximization perspective rather than a cost minimization perspective. In general, the revenue from an accepted order (in class j) is R (R_j) and there are earliness/tardiness penalties if the order is completed before/after its quoted due date.

Dellaert [29] studies DDM-OS using demand model (D1) in a single server queuing model. Unlike most other models in the literature, a setup is needed before the production of a lot can be started. When all the currently available demand has been produced, the machine goes down and a new setup is required before a new lot can be produced. Assuming FCFS sequencing policy, the goal is to find a combination of a due date setting rule and a production rule (deciding on when to start a setup) to maximize the expected profit (revenue minus earliness, tardiness and setup costs). Relying on earlier results in [28], the author focuses on the following production rule: perform a setup only if the number of jobs waiting is at least n. The value of n needs to be determined simultaneously with the optimal lead time.

Dellaert studies two lead time policies, CON and DEL, where DEL considers the probability distribution of the flow time in steady-state while quoting lead times. He models the problem as a continuous-time Markov chain, where the states are denoted by (n, s)=(number of jobs, state of the machine). Interarrival, service and setup times are assumed to follow the exponential distribution, although the results can also be generalized to other distributions, such as Erlang. For both policies, he derives the pdf of the flow time, and relying on the results in [88] (the optimal lead time is a unique minimum of strictly convex functions), he claims that the optimal solution can be found by binary search.

For the case of CON, computational results indicate that for small values of n, the lead time is larger than the average flow time, while the opposite is true for larger values of n. Intuitively, when n is large, batches and flow times are also large and in order to generate revenues, one needs to quote shorter lead times in comparison to flow times. The results also indicate that there is little loss in performance if the optimal CON lead time is replaced by the average flow time. In terms of implementation, quoting lead times equal to average flow times (which can be determined, for example, via simulation) might be much easier than computing the pdf of flow times, which are used to find the optimal CON policy. The comparison of the CON policy with DEL, in which jobs can be quoted different lead times, indicates that DEL has significantly better performance.

Duenyas and Hopp [31] consider demand models (D2)-(D4) and study DDM-OS using a queuing model. They first consider a system with infinite server capacity. For the special case of exponential processing times and model (D3), they find a closed form solution for the optimal lead time l^*, which is the same for all customers. The structure of l^* implies that: (1) For higher revenues per customer, the firm quotes shorter lead times. This allows the firm to accept more orders and the higher revenues offset the penalties in case due dates are missed. (2) For longer processing times, the firm quotes longer lead times and obtains lower profits.

Next, they consider the capacitated case studying a single server queue $GI/GI/1$ where processing times have a distribution in the form of increasing failure rate (IFR). That is, as the amount spent processing a job increases, the probability that it will be completed within a certain time duration increases as well. They use the FCFS rule for sequencing accepted orders. They first study the problem for model (D2), which might be appropriate if the lead times are fixed for all customers, e.g., 1 hour film processing. In this case, the firm's main decision is to decide which orders to accept (reject), by quoting a lead time less (greater) than l_{max}. They show that the optimal policy has a control-limit structure: for any n, the number of orders currently in the system, there exists a time $t(n)$ such that a new order is accepted if the first order has been in service for more than $t(n)$ time units. Next, they consider model (D4). Note that choosing the optimal lead time l implies an effective arrival rate of $\lambda(l)$. Hence, setting lead times is equivalent to choosing an arrival rate for the queue. Let l_{nt}^* denote the optimal lead time quote to a new customer when there are n customers in the system and the first customer has been in service for t time units. For an $M/M/1$ queue, they show that the optimal lead time to quote is increasing in n. Although they cannot find general structural results for the GI/GI/1 queue, they show that l_{nt}^* is bounded below by b_{nt}^*, which is the optimal lead time computed by disregarding the congestion effects (the customers currently in the system). When the service and interarrival times are exponentially distributed, the optimal lead times increase in the number of orders in the system.

The results of Duenyas and Hopp [31] assume the FCFS sequencing policy. When a different sequencing rule is used, the impact of accepting a new order on the current schedule has to be considered. Depending on the new sequence, the lead times and hence the penalties of existing orders might change. The authors show that when the penalty is proportional to tardiness, an optimal DDM policy processes all jobs using the EDD rule. This result does not hold when the penalty is fixed for each tardy job (e.g., as in the case of minimizing the number of tardy jobs).

The models studied in [29] and [31] have a single class of customers; the net revenue per customer is constant, customers have the same preferences for lead times, and the arrival and processing times follow the same distribution. Duenyas [30] extends these results to multiple customer classes, with different net revenues and lead time preferences. Customers/jobs from class j arrive to an $M/G/1$ queue according to a Poisson process with rate λ_j. A customer from class j accepts a quoted lead time l with probability $P_j(l)$, and generates net revenue R_j. If an accepted order is tardy for x time units, the firm pays a penalty wx. Duenyas first studies the problem assuming FCFS sequencing and shows that: (1) quoted lead times increase in the number of orders in the system, (2) customers who are willing to pay more and/or impatient are more sensitive to quoted lead times, and (3) ignoring future arrivals result in lead times that are lower than optimal. Next, he considers the simultaneous problem of due date setting and sequencing. Since the state and action spaces of the Semi-Markov decision process (SMDP) used for modeling this problem are uncountable, he is not able to find an exact solution, and therefore proposes a heuristic. He tests this heuristic on a set of problems with two customer classes against CON, JIQ, and two other rules obtained from modifying WEIN-PAR I and WEIN-PAR II. To achieve a fair comparison, the parameters of these rules are adjusted so that they all result in the same average tardiness. Simulation results indicate that the proposed heuristic outperforms the other rules, especially at high utilization levels. The two modified rules also performed very well, and since they require less information about customers' due date preferences, they may be a good alternative to the proposed heuristic if such information is not available. Duenyas also notes that it is possible to extend the heuristic to simultaneously decide on prices, due dates and sequencing, by assuming that a customer from class j accepts a (price, lead time) pair (R_j, l_j) with probability $P_j(R_j, l_j)$. The work of Duenyas [30] suggests that collecting information on customers' preferences about due dates and using this information in due date management can help manufacturers to achieve higher profitability.

Keskinocak et al. [65] study DDM for orders with availability intervals and lead time sensitive revenues in a single-server setting. In their model, revenues obtained from the customers are sensitive to the lead-time, there is a threshold of lead-time above which the customer does not place an order and the quoted lead times are 100% reliable. More formally, the revenue received from an order j with quoted lead time l_j is $R(d) = w_j(\hat{l}_j - l_j)$, where $\hat{l}_j$ is the maximum acceptable lead time for customer j. Quoting a lead time longer than $\hat{l}_j$ is equivalent to rejecting order j. Note that this revenue model is equivalent to a profit model where there is a fixed revenue (or price) per order $R_j = w_j\hat{l}_j$ and a penalty $w_j l_j$ for quoting a lead time l_j. The concept of lead-time-dependent price is also studied in [85], discussed in Section 6.2.

Keskinocak et al. [65] study the offline (F-SLTQ) and three online versions of this problem. In the offline models, all the orders are known in advance whereas in the online models orders arrive over time and decisions are made without any knowledge of future orders. F-SLTQ generalizes the well known scheduling problem of minimizing the sum of weighted completion times subject to release times [64], denoted by $1|r_j|\sum w_j C_j$. They study F-SLTQ by methods from mathematical programming and show several polynomially or pseudo-polynomially solvable instances. In the pure online model (O-SLTQ), decisions about accepting/rejecting an order and lead time quotation can be delayed, whereas in the quotation online model (Q-SLTQ) these decisions have to be made immediately when the order arrives. In the delayed quotation model (D-SLTQ), which is between the pure online and quotation models, decisions have to be made within a fixed time frame after the arrival of an order. To evaluate the performance of algorithms for the on-line models, they use competitive (worst case) analysis and show lower and upper bounds on the performance of their algorithms, relative to an optimum offline algorithm. Most of the previous research on the competitive analysis of on-line scheduling algorithms considers models similar to O-SLTQ, where the scheduling decisions about an order can be delayed. In many cases, this delay has no bound as the orders do not have latest acceptable start or completions times, which is not realistic for real-life situations. The authors show that the quotation version (Q-SLTQ), where order acceptance and due date quotation decisions have to be made immediately when an order arrives, can be much harder than the traditional on-line version (O-SLTQ). Partially delaying the quotation decision (Q-SLTQ) can improve performance significantly. So, the difficulty does not only lie in not knowing the demand, but in how soon one has to make a decision when an order arrives. They also observe that in order to obtain high revenues, it is important to reserve capacity for future orders, even if there is only a single type of orders (i.e., $w_j = w$ and $\hat{l}_j = \hat{l}$ in the revenue function).

Asymptotic [58] and worst case [65] analysis of online algorithms received very little attention so far within the DDM context. These approaches are criticized for being applicable only to very simple models and asymptotic analysis is further criticized for being valid only for a large number of jobs. However, these criticisms hold for steady-state analysis of most queuing models within the DDM context as well. Kaminsky and Lee [58] show via computational experiments that in a variety of settings the asymptotic results are obtained in practice for less than 1000-2500 jobs; most simulation studies are based on at least as many, and sometimes a larger number of jobs. An advantage of worst-case analysis is that it does not assume any information about the probability distribution of job processing or interarrival times, unlike most queuing studies which assume $M/M/1$. Asymptotic analysis usually assumes that the processing and interarrival times are independent draws from known distributions, but this information is used mainly for proving results, and is not necessarily used in the algorithms. For example, the heuristics proposed in [58] do not use any information about these distributions. In short, we believe that worst case and asymptotic analysis hold promise for analytically studying DDM problems.

Chatterjee et al. [18] pose DDM-OS in a decentralized marketing-operations framework where the marketing department quotes the due dates without having full information about the shop floor status. Upon the arrival of a new order, marketing learns the processing time p and selects a lead time $l(p)$, and the customer places an order with probability $e^{-\psi l(p)}$ (model (D3)). Let $\delta = 1$ if the customer stays and $\delta = 0$ otherwise. Manufacturing sees a filtered stream of arrivals $\Lambda = P(\delta = 1)\lambda$ where λ is the arrival rate observed by marketing. Marketing *assumes* that the delay time W in operations has PDF $F_W = 1 - \rho e^{-\gamma x}$, $x \geq 0$ (which is the waiting time distribution in an $M/M/1$ queue with FCFS scheduling), where $\rho = \Lambda E[p]$ and $\gamma = (1-\rho)/E[p]$. Note that γ denotes the difference between the service rate and the arrival rate at operations. The revenue and tardiness penalty for a job with processing time p are given by $R(p)$ and $w(p)$. The authors show that the optimal lead time has a log-linear structure and propose an iterative procedure for computing its parameters. For the special case of constant unit tardiness penalty $w(p) = w$, they show that the optimal lead time is zero for any job with processing time larger than a threshold $\bar{p}$, which is similar to the "unethical" DDM practice observed by Spearman and Zhang [99]. Through numerical examples, the authors illustrate that choosing lead times based on manufacturing objectives such as minimizing cycle time or maximizing throughput usually reduce profits.

6.2 Due Date Management with Price and Order Selection Decisions (DDM-P)

Most of the research in economics and marketing models the demand only as a function of price, assuming that firms compete mainly on price. In contrast, the operations literature usually takes the price (and demand) as given, and tries to minimize cost and/or maximize customer service. However, for most customers the purchasing decision involves trading off many factors in-

Due Date Management Policies

cluding price and quality, where delivery guarantees are considered among the top quality features. Hence, in most cases demand is a function of both price and lead time and therefore a firm's DDM policy is closely linked to its pricing policy. In this section, we provide an overview of the literature which considers DDM in conjunction with pricing decisions.

The first four papers we discuss, So and Song [97], Palaka et. al. [75], Ray and Jewkes [85], and Boyaci and Ray [15] consider capacity selection/expansion decisions in addition to price and lead time decisions. These papers study DDM-P using an M/M/1 queuing model with FCFS sequencing, where the expected demand is modeled by a linear function ($\Lambda(R,l) = a - b_1 R - b_2 l$) in [75] and [85], and a Cobb-Douglas function ($\Lambda(R,l) = \lambda R^{-a} l^{-b}$) in [97]. Note that the price elasticity of demand (the percentage change in demand corresponding to a 1% change in price) is constant in the second model whereas it increases both in price and quoted lead time in the first model. These papers consider a constant lead time and a service level constraint (the minimum probability, s, of meeting the quoted lead time).

Palaka et al. consider three types of costs: direct variable costs (proportional to production volume), congestion costs (proportional to the mean number of jobs waiting in the system), and tardiness costs. They first consider the case of fixed capacity (service rate μ), where the goal is to choose a price/lead-time pair (R, l) and a demand rate $\lambda \leq \Lambda(R, l)$ for maximizing the firm's expected profits. Noting that the demand rate λ will always be equal to $\Lambda(R, l)$ in the optimal solution, they focus on the two decision variables λ and l. They show that (i) the service constraint is binding in the optimal solution if and only if $s \geq s_c = 1 - b_2/b_1 w$, where w is the tardiness penalty per unit time, (ii) when s increases, the firm both increases its quoted lead time and decreases its demand rate (and expected lead time). In contrast, an increase in the tardiness penalty decreases the demand rate but the quoted lead time decreases (increases) if the service constraint is binding (non-binding). For small values of the tardiness penalty, the firm increases the price to reduce the demand rate, which does not decrease the probability of tardiness but reduces the tardiness penalty since there are fewer orders late overall. However, when the tardiness penalty is high, the firm needs to reduce the probability of tardiness and hence quotes higher lead times. Palaka et al. extend these results to the case where marginal capacity expansion is possible, that is, the firm can choose z, the fractional increase in the processing rate at a cost of c per job/unit time up to a limit of $\bar{z}$. Hence, the service rate becomes $\mu(1+z)$. The authors show that in the optimal solution, the firm uses both capacity expansion and a reduced arrival rate to achieve shorter lead times. Finally, they look at the sensitivity of the profits to the errors in estimating the lead time and conclude that guaranteeing a shorter than optimal lead time usually results in higher profit loss than guaranteeing a longer than optimal lead time.

The objective function in So and Song [97] is to maximize profit, which in their setting is the revenue minus direct variable costs and capacity expansion costs. Note that this is a special case of the objective function considered in Palaka et al., ignoring the congestion and tardiness costs. The qualitative results of So and Song are generally consistent with Palaka et al.

Ray and Jewkes [85] study a variant of the model in [75] by modeling the market price as a function of lead time, namely, $R = \bar{R} - el$ where $\bar{R}$ is the maximum price when the lead time is zero[4]. Hence, the demand function becomes $\lambda(R, l) = a - b_1 R - b_2 l = (a - b_1 \bar{R}) - (b_2 - b_1 e)l = a' - b'l$. This model naturally leads to a distinction of lead time and price sensitive customers: (i) when $b' > 0$, demand decreases in lead time, i.e., customers are lead time sensitive (LS) and are willing to pay higher prices for shorter lead times; (ii) when $b' < 0$, demand increases in lead time (as price decreases), i.e., customers are price sensitive (PS) and are willing to wait longer to get lower prices. The dependence of price on lead time reduces the number of variables, and the firm needs to choose a lead time (l) and capacity level (μ) subject to a service constraint, with the goal of maximizing profit (revenue minus operating and capacity costs). The authors consider both the cases of constant and decreasing convex operating costs (the latter models economies of scale). The authors show that the optimal lead time depends strongly on whether the customers are price or lead time sensitive, as well as operating costs. Comparing their model with the lead-time-independent price model, the authors show that the optimal solutions for these two models might look significantly different, and ignoring the dependency of price on lead time might lead to large profit losses.

Boyaci and Ray [15] extend the previous models to the case of two substitutable products served from two dedicated capacities. They assume that these two products are essentially the same, and are differentiated only by their price and lead time. As in [85], there is a marginal capacity cost c_i and a unit operating cost m, and as in [75] and [97], the price is not a direct function of lead time. The market demand for product $i = 1, 2$ is given by $\lambda_i = a - \beta_R R_i + \theta_R(R_j - R_i) - \beta_l l_i + \theta_l(l_j - l_i)$, $i \neq j$. In this demand model, (i) θ_R and θ_l denote the sensitivity of switchovers due to price and lead time, respectively, (ii) β_R and β_l denote the price and lead time sensitivity of demand, (iii) the total market demand $\lambda_1 + \lambda_2$ is independent of the switchover parameters (θ). The authors assume that the lead time (l_2) for product 2 ("regular" product) is given, and the firm needs to determine a shorter lead time ($l_1 < l_2$) for product 1. Express vs. regular photo processing or package delivery are some examples motivating this model. As in the previous papers, the goal is to maximize profits subject to service level constraints $1 - e^{-(\mu_i - \lambda_i)t_i} \geq s_i$, $i = 1, 2$. The authors first study the uncapacitated model (Model 0, $c_1 = c_2 = 0$) with a given l_1 (it turns out that in the uncapacitated case the optimal l_1 is zero), and differentiate between price and time sensitive customers as in [85]: (i) Price and time difference sensitive (PTD): When $\beta_R \theta_l > \theta_R \beta_l$, the market is more sensitive to price but switchovers are mainly due to the difference in lead times. Note that this condition is equivalent to $[\theta_R/(\beta_R+\theta_R)] < [\theta_l/(\beta_l+\theta_l)]$, i.e., the fraction of customers that switch due to price is smaller than the fraction of customers that switch due to lead time. (ii) Time and price difference sensitive (TPD): When $\beta_R \theta_l < \theta_R \beta_l$, the market is more sensitive to lead time but switchovers are mainly due to the difference in prices. When $\theta_R \approx \theta_l$, we are back to the standard price and lead time sensitive markets. Next, they study the case where only product 1 is capacitated (Model 1, $c_1 > 0$, $c_2 = 0$). Comparing Model 1 to Model 0, they show that the change in the optimal prices depend on the market type (TPD vs. PTD) as well as c_1. Interestingly, while price and time differentiation go hand in hand in Model 0, in Model 1 the firm price differentiates only if the marginal cost of capacity is sufficiently high. Finally, the authors consider the case where both products are capacitated (Model 2, $c_1 \geq c_2 > 0$). They find that for the same c_1, the optimal l_1 is shorter in Model 2, i.e., the firm offers a better delivery guarantee even though its cost is higher for the regular product. This is because a shorter l_1 lures customers away from the regular product and attracts them to the more profitable express product. The authors conclude their study by comparing their results to the case where substitution is ignored or not present.

So [96] studies a competitive multi-firm setting, where the firms have given capacities (μ_i) and unit operating costs (γ_i), i.e., they are differentiated by their size and efficiency, and choose prices (R_i) and delivery time guarantees (l_i). As in [75] and [97], each firm chooses a uniform delivery time guarantee. The market size (λ) is assumed to be fixed and the market share of each firm is given by

$$\lambda_i = \lambda \left(\frac{\alpha_i R_i^{-a} t_i^{-b}}{\sum_{j=1}^n \alpha_j R_j^{-a} t_j^{-b}} \right)$$

In this model, the parameter α_i denotes the overall "attractiveness" of firm i for reasons other than price and lead time, such as reputation and convenience of the service location. Parameters a and b denote the price and time sensitivity of the market, respectively. To ensure the reliability of the quoted lead times, the firms seek to satisfy a service constraint, which states that on average an s fraction of the orders will be on time. Using an $M/M/1$ queue approximation, this constraint is modeled as $1 - e^{-(\mu_i - \lambda_i)t_i} \geq s$, or equivalently, $(\mu_i - \lambda_i)t_i \geq -\log(1-s)$. Focusing only on the case where s is given and the same for each firm, So finds the "best response" of firm i in closed form given the other firms' price and lead time decisions. He shows that the price is decreasing in lead time. He also shows that (1) in the combined solution of price and lead time, the firm always charges the highest price $\bar{R}$ if $a \leq 1$, and

(2) the optimal price and lead time are increasing in α_i. These results suggests that firms compete only based on delivery time when the market is not price sensitive, and firms with lower attractiveness need to compete by offering better prices and lead times. So characterizes the Nash equilibrium in closed form for N identical firms, and proposes an iterative solution procedure for identifying the equilibrium in case of non-identical firms. Comparing the results to the single-firm case studied in [97], a capacity-increase in the multi-firm case leads to lower prices, whereas the reverse is true in the single-firm case. This is quite intuitive since an increased capacity in the multi-firm case leads to more intense competition in a fixed-size market. Numerical results with two firms, $s = 0.95$ and equal α_i's indicate that (1) the advantage of higher capacity increases as the market becomes more time sensitive, (2) low cost firm offers a lower price, longer lead time, and captures more of the market share and profits, (3) an increase in price sensitivity leads to lower prices overall, and shorter and longer lead times for the high capacity and low cost firms, respectively, (4) an increase in time sensitivity leads to an increase in prices, and shorter and longer lead times by the low cost and high capacity firms, respectively, reducing the difference between the lead times offered by the two firms, and (5) as the time (price) sensitivity increases, profits and market share of the high capacity (low cost) firm increase. He also conducts experiments in a three-firm setting, where one of the firms dominates the others both in terms of capacity and operating cost. Interestingly, in such a setting the dominant firm offers neither the lowest price, nor the shortest lead time, while the other two firms strive to differentiate themselves along those two dimensions by offering either the lowest price (low cost firm) or the shortest lead time (high capacity firm).

The papers discussed so far assume that once the firm quotes a customer a lead time (and price), the customer immediately makes a decision as to whether to accept or reject the offer. In reality, customers might "shop around", i.e., request quotes from multiple firms, and/or need some time to evaluate an offer. Hence, there might be a delay before a customer reaches a decision on whether or not to accept an offer. Easton and Moodie [32] study DDM-P in the face of such contingent orders, which add variability to the shop congestion and hence to the lead time of a new order. In their model, the probability that the customer will accept a quoted price/due-date pair (R, l) follows an S-shaped logit model

$$P(R, l) = \left[1 + \beta_0 \exp\left(\beta_1 \frac{l - p}{p} + \beta_2 \left(\frac{R}{cp} - 1\right)\right)\right]^{-1}$$

where p is the estimated work content of the order, c is the cost per unit work content, and β_0, β_1 and β_2 are parameters to be estimated from previous bidding results. The two terms in this expression refer to the lead time as a multiple of total work content, and the markup embedded in the quoted price. For choosing the price R and the lead time l for a new order (assuming FCFS sequencing), they propose a model that myopically maximizes the expected profit (revenue minus tardiness penalties) from that order considering both firm and contingent orders in the system but ignoring any future potential orders. The solution method they propose involves evaluating all possible accept/reject outcomes for contingent orders, i.e., 2^N scenarios if there are N contingent orders, which is clearly not efficient for a large number of contingent orders. Via a numerical example, the authors show that their model outperforms simple due-date setting rules based on estimates on the minimum, maximum or expected shop load.

All the papers we discussed in this survey so far assume that the due date (and price) decisions are made internally by the firm. This is in contrast to the scheduling literature where it is assumed that the due dates are set externally, e.g., by the customer. In practice, most business-to-business transactions involve a negotiation process between the customer and the firm on price and due date, i.e., neither the customer nor the firm is the sole decision maker on these two important issues. With this in mind, Moodie [72] and Moodie and Bobrowski [73] incorporate the negotiation process into price and due date decisions. In their model, both the customer and the firm have a reservation tradeoff curve between price and due date, which is private information. Customer arrivals depend on the delivery service reputation (SLR) of the firm, which depends on the firm's past delivery performance as follows: $SLR_{new} = (1 - \alpha)SLR_{old} + \alpha s$, where s is the fraction of jobs completed on time in the last period. If the firm's service level is below the customer's requirement, the customer does not place an order. Hence, by choosing its due dates, the firm indirectly impacts the demand through its service level. Given a new customer, the firm first establishes an earliest due date. The firm then chooses one of the four negotiation frameworks for price and lead time: negotiate on both, none, or only one of the two. Third, the firm chooses a price (a premium price for early delivery and a base price for later delivery) and due date to quote. And finally, if the order is accepted, it needs to be scheduled. Moodie [72] proposes and tests four finite-scheduling based due date methods, as well as four of the well-known rules from the literature: CON, TWK, JIQ, and JIS. Simulation results (under EDD scheduling) suggest that (1) due date methods based on the jobs' processing times (especially TWK and JIS) perform better than the proposed finite-scheduling based methods, (2) negotiating on both price and lead time provides higher revenues, and (3) it may be profitable to refuse some orders even if the capacity is not too tight. A more extensive study of this model is performed by Moodie and Bobrowski [73]. They find that full bargaining (both on price and lead time) is useful if there is a large gap between the quoted and the customers' preferred due dates. If this gap is small, then price-only bargaining seems more beneficial.

Charnsirisakskul et al. [17] study the benefits of lead time flexibility (the willingness of the customers to accept longer lead times) to the manufacturer in an offline setting with 100% service guarantees. They consider a discrete set of prices $\{R_j^1, \ldots, R_j^{n_j}\}$ the manufacturer can charge for order j. The demand quantity (expressed in units of capacity required, or processing time) corresponding to price R_j^k is p_j^k. Each customer has a preferred and acceptable lead time, denoted by f_j^k and l_j^k, respectively, if the manufacturer quotes price R_j^k. There is also an earliest start time for starting the production of and an earliest delivery time for each order. If an order's production (partially) completes before the earliest delivery time, the manufacturer incurs a holding cost. If the quoted lead time is between f_j^k and l_j^k, i.e., after the customer's preferred due date, the manufacturer incurs a lead time penalty. The authors model this problem as a linear mixed integer program where the decisions are: which orders to accept, which prices to quote to the accepted orders, and when to produce each order. Note that this model simultaneously incorporates order acceptance, pricing and scheduling decisions. The authors also consider a simpler model where the manufacturer must quote a single price (again, chosen from a discrete set of prices) to all customers. They propose heuristics for both models, since the solution time can be quite large for certain instances. To establish the link between prices and order quantities in their computational experiments, they consider the case where each customer is a retailer who adopts a newsvendor policy for ordering. They model the retailer's demand by a discrete distribution function and compute the retailer's optimal order quantity as a function of the manufacturer's quoted price. In an extensive computational study, they test the impact of price (single vs. multiple prices), lead time, and inventory holding (whether or not the manufacturer can produce early and hold inventory) flexibility on the manufacturer's profits. They find that lead time flexibility is useful in general both with and without price flexibility. The benefit of lead time flexibility is higher if there is no inventory flexibility, suggesting that lead time and inventory flexibilities are complementary. However, they also observe that in certain environments price, leadtime, and inventory flexibilities can be synergistic.

7. Conclusions and Future Research Directions

Lead-time/due-date guarantees have undoubtedly been among the most prominent factors determining the service quality of a firm. The importance of due date decisions, and their coordination with scheduling, pricing and other related order acceptance/fulfillment decisions has increased further in recent years with the increasing popularity of make-to-order (MTO) over make-to-stock (MTS) environments.

In this paper, we provided a survey of due date management (DDM) literature. The majority of the research in this area focuses solely on due-date setting and scheduling decisions, ignoring the impact of the quoted lead times on demand. Assuming that the orders arriving in the system are exogenously determined and must all be processed, these papers study various DDM policies with the goal of optimizing one or a combination of service objectives such as minimizing the quoted due dates, average tardiness/earliness, or the fraction of tardy jobs. Clearly, no single DDM policy performs well under all environments. Several factors influence the performance of DDM policies, such as the due date rule, the sequencing rule, job characteristics (e.g., product structures, variability of processing times), system utilization (the mean and variance of load levels), shop size and complexity, and service constraints. Due date policies that consider job and shop characteristics (e.g., JIQ) in general perform better than the rules that only consider job characteristics (e.g., TWK) which in turn perform better than the rules that ignore both job and shop characteristics (e.g., RND, CON) [7] [33] [83] [106]. However, the performance of a due-date rule also depends on the accompanying sequencing policy. For example, due-date based sequencing rules perform better than other rules, such as SPT, in conjunction with due date policies that consider work content and shop congestion [33] [106]. Scheduling rules that use operation due-dates may result in improvements over similar rules that use job-due-dates. The best values of the parameters for the due date rules also depend on the sequencing rule [33] [83]. While shop size does not seem to affect the performance of due date management policies significantly, increased shop complexity degrades the performance [103] [106]. Due date performance seems to be fairly robust to utilization levels of balanced shops but changes significantly in shops with unbalanced machine utilization [104].

Very few studies in the DDM literature use real-world data for testing purposes, or report implementations of DDM policies in practice. Most papers use hypothetical job-shop environments in simulations or use analytical models with very limiting assumptions. Exceptions include [68], which tests the proposed methods in a real-world example taken from [13]; [110], which discusses order acceptance and DDM practices at Lithonia Lighting, a lighting fixture manufacturer; [101], which discusses how the use of inventory management with cyclic scheduling, along with many concurrent improvements in supplier management, customer relations, product design, and quality, enabled the quotation of shorter and more accurate lead times at a laminate plant; and [102], which reports results from a survey of 24 make-to-order firms (subcontracting component manufacturers and manufacturers of capital equipment) in the U.K on these firms' pricing and lead time quotation practices.

There is a limited but growing body of literature on DDM that considers the impact of quoted due-dates on demand. By modeling the demand as a function of quoted lead times, these papers endogenize order selection decisions in addition to due-date setting and scheduling. Taking this approach one step further, a small number of papers model demand as a function of both price and lead time, and consider simultaneous pricing, order acceptance, lead time quotation and scheduling policies.

While there is a large body of literature on DDM, there are several directions for research that are still open for exploration. One area that received a lot of attention but is still open for research is the development and testing of new due date rules, which are easy to compute and implement based on readily available data. A related research topic is the selection of the parameters for the due-date rules. The due-date rules currently available in the literature use static parameters, that is, the parameters are estimated once at the beginning and remain constant throughout the implementation. One could possibly change the parameters of a due date or scheduling rule in response to the changes in the system, which is an area that received very little attention so far [55]. Alternatively, one could use different predictive models or due-date setting or scheduling rules depending on the state of the system, e.g., measured by the number of operations of an arriving job or shop congestion. Such mixed DDM strategies might result in better performance. We expect that some of the simple rules for DDM will be incorporated into enterprise strength software soon leading the way for more research once a large group from industry has had an opportunity to benefit from the first wave.

To ensure that the quoted lead times are short and reliable, some of the DDM models impose service constraints, such as the maximum fraction of tardy jobs or the maximum average tardiness. The service level is assumed to be fixed, regardless of the changes in the system. In practice, firms do not necessarily offer the same service guarantees all the time. For example, most e-tailers guarantee delivery times only for orders that are placed sufficiently in advance of Christmas. It would be interesting to study models with variable service guarantees, e.g., making the service level a function of the number of jobs in the system. In this case, one needs to simultaneously optimize the service levels and the due date quotes dynamically.

Another interesting question facing a firm is what kind of lead time guarantee to offer. Many companies offer constant lead times, and in case of lateness, they offer a discount or free service. For example, Bennigan's, a restaurant chain promised service within 15 minutes at lunch time, or the meal was free [51]. Since the choice of the service guarantee and the associated lateness penalties directly impact the delivery performance of the firm, it is important to understand what kind of service guarantees are appropriate in different business environment. In particular, a firm should not commit to a hard to meet service guarantee unless such a guarantee would positively impact its demand. For example, a restaurant's guarantee of less than 10 minutes wait might be valuable during lunch, but not necessarily at dinner time. In general, a firm needs to understand the impact of lead times or lead time guarantees on demand, especially on future demand. For example, the Touchdown guarantee program of Austin Trumanns Steel provided promotional value in approaching new customers, and it helped the company to win (and retain) many new customers [52].

The papers reviewed in this survey assume that all (accepted) orders must be completed even when the system is very congested and there are significant lateness penalties. Alternatively, one could consider the possibility of "outsourcing" an order if the system is highly congested and completing all the orders in-house would lead to very high tardiness penalties. In some industries, outsourcing might be necessary if the demand is time-sensitive. For example, in the air cargo industry if a carrier does not have enough capacity to ship all its contracted cargo within the promised lead time (within 3 days for most time-sensitive cargo), they purchase capacity from competing carriers in the spot market, usually at a high cost. While the option of outsourcing might be quite costly, it might actually help lower the overall costs by reducing the congestion in the system and lowering additional future delays.

Note that outsourcing is just one form of (short-term) capacity expansion. Very few papers in the DDM literature consider flexible capacity [75] [97], where a firm can increase its (short-term) capacity by incurring some additional cost. More research is needed to better understand under what conditions capacity flexibility is most useful and how it impacts the performance of DDM policies.

Most of the research in DDM considers MTO environments and hence, does not consider the option of producing early and holding inventory. However, it may be possible to keep inventory of partially finished products (vanilla boxes) or fast moving items in a MTO or hybrid MTO/MTS environment. The development of DDM policies for such hybrid systems is for the most part an untapped research area.

As we pointed out earlier, the incorporation of order selection and pricing decisions into DDM is an important area that needs further exploration. In addition to just reacting to demand, a firm can influence the demand via its choice of prices and lead times. Hence, rather than assuming that the demand and the associated revenues are input to the DDM policy and taking a cost minimization perspective, the firm can take a profit maximization perspective and coordinate its decisions on pricing, order selection, lead time quotation and scheduling. So far, very little research has been done on DDM policies with pricing decisions, and the available models have significant limitations, such as focusing on quoting a single price/lead time pair to all customers and/or ignoring multiple customer classes. Given that customers have different sensitivity to prices and lead time, it is important to incorporate multiple customer classes into the models. We believe that the development of DDM policies with pricing and order selection decisions is an important research direction with a significant potential impact in practice.

In their quest towards profitably balancing supply and demand, companies traditionally focused on the supply side. Among the decisions that influence supply availability over time are inventory policies, capacity selection and production scheduling. In contrast, price and lead time decisions help companies manage the demand side, that is, to shift or adjust the demand over time to better match supply. In essence, a company achieves different types of flexibility by being able to adjust "levers" such as price, lead time or inventory. An interesting research direction is to investigate which of these flexibility levers have more impact on profits in different types of business environments. Furthermore, are the benefits from having multiple types of flexibility subadditive or superadditive?

Most of the research in scheduling either ignores due dates or assumes that they are set exogenously (e.g., by the customer). Most of the research in DDM assumes that the due dates are set internally by the firm. The reality is in between these two extremes. In most business-to-business transactions, due dates are set through a negotiation process between the customer and the firm. DDM research that incorporates buyer-seller negotiation is still at its infancy [72] [73] and we believe this to be an interesting and fertile area for future research. Other interesting extensions to the existing literature include substitutable products and competition.

Notes

1. In this paper, we focus on lead times quoted to customers. Lead times can also be used for internal purposes, e.g., planned lead times are used for determining the release times of the orders to the shop floor [70]. We do not discuss planned lead times in this paper.
2. We are indebted to an anonymous referee for this observation.
3. The SLK/OPN rule can behave in undesirable ways when SLK is negative. See [59] for a discussion on such anomalous behavior and on modifications for correcting it.
4. Recall that a lead-time-dependent price model was also studied earlier in [65], discussed in Section 6.1. Since price is not an independent decision variable in these models, we could have discussed [85] in Section 6.1 as well. However, due to its relevance to some of the other papers in this section, we decided to discuss it here.

References

[1] N.R. Adam, J.W.M. Bertrand, D.C. Morehead, and J. Surkis. Due date assignment procedures with dynamically updated coefficients for multi-level assembly job shops. *European Journal of Operational Research*, 77:429–439, 1994.

[2] J. Adams, E. Balas, and D. Zawack. The shifting bottleneck procedure for job shop scheduling. *Management Science*, 34:391–401, 1988.

[3] U. Bagchi, Y.L. Chang, and R. Sullivan. Minimizing absolute and squared deviations of completion imes with different earliness and tardiness penalties and a common due date. *Naval Research Logistics Quarterly*, 34:739–751, 1987.

[4] K.R. Baker. The effects of input control in a simple scheduling model. *Journal of Operations Management*, 4:94–112, 1984.

[5] K.R. Baker. Sequencing rules and due-date assignments in a job shop. *Management Science*, 30(9):1093–1104, 1984.

[6] K.R. Baker and J.W.M. Bertrand. A comparison of due-date selection rules. *AIIE Transactions*, pages 123–131, June 1981a.

[7] K.R. Baker and J.W.M. Bertrand. An investigation of due date assignment rules with constrained tightness. *Journal of Operations Management*, 1(3):37–42, 1981b.

[8] K.R. Baker and J.W.M. Bertrand. A dynamic priority rule for scheduling against due-dates. *Journal of Operations Management*, 3(1):37–42, 1982.

[9] K.R. Baker and J.J. Kanet. Job shop scheduling with modified due dates. *Journal of Operations Management*, 4(1):11–23, 1983.

[10] K.R. Baker and J.J. Kanet. Improved decision rules in a combined system for minimizing job tardiness. *Journal of Operations Management*, 4(1):11–23, 1984.

[11] K.R. Baker and G.D. Scudder. On the assignment of optimal due dates. *Journal of the Operational Research Society*, 40(1):93–95, 1989.

[12] J.W.M. Bertrand. The effect of workload dependent due-dates on job shop performance. *Management Science*, 29(7):799–816, 1983.

[13] G.R. Bitran and D. Tirupati. Multiproduct queuing networks with deterministic routing: decomposition approach and the notion of interference. *Management Science*, 34(1):75–100, 1988.

[14] J.H. Bookbinder and A.I. Noor. Setting job-shop due-dates with service-level constraints. *Journal of the Operational Research Society*, 36(11):1017–1026, 1985.

[15] T. Boyaci and S. Ray. Product differentiation and capacity cost interaction in time and price sensitive markets. *Manufacturing & Service Operations Management*, 5(1):18–36, 2003.

[16] K. Charnsirisakskul, P. Griffin, and P. Keskinocak. Order selection and scheduling with lead-time flexibility. Working Paper, ISYE, Georgia Institute of Technology.

[17] K. Charnsirisakskul, P. Griffin, and P. Keskinocak. Pricing and scheduling decisions with lead-time flexibility. Technical report, School of Industrial and Systems Engineering, Georgia Institute of Technology, 2003.

[18] S. Chatterjee, S.A. Slotnick, and M.J. Sobel. Delivery guarantees and the interdependence of marketing and operations. *Production and Operations Management*, 11(3):393–409, 2002.

[19] T.C.E. Cheng. Optimal due-date determination and sequencing of n jobs on a single machine. *Journal of the Operational Research Society*, 35(5):433–437, 1984.

[20] T.C.E. Cheng. Optimal due-date assignment for a single machine sequencing problem with random processing times. *International Journal of Systems Science*, 17:1139–1144, 1986.

[21] T.C.E. Cheng. An algorithm for the con due-date determination and sequencing problem. *Computers and Operations Research*, 14:537–542, 1987.

[22] T.C.E. Cheng. An alternative proof of optimality for the common due-date assignment problem. *European Journal of Operational Research*, 1988.

[23] T.C.E. Cheng. Optimal total-work-content-power due-date determination and sequencing. *Computers and Mathematics with Applications*, 1988.

[24] T.C.E. Cheng. On a generalized optimal common due-date assignment problem. *Engineering Optimization*, 15:113–119, 1989.

[25] T.C.E. Cheng. Optimal assignment of slack due dates and sequencing in a single-machine shop. *Appl. Math. Lett.*, 2(4):333–335, 1989.

[26] T.C.E. Cheng and M.C. Gupta. Survey of scheduling research involving due date determination decisions. *European Journal of Operational Research*, 38:156–166, 1989.

[27] R.W. Conway. Priority dispatching and job lateness in a job shop. *The Journal of Industrial Engineering*, 16(4):228–237, 1965.

[28] N.P. Dellaert. Production to order. *Lecture Notes in Economics and Mathematical Systems*, 333, 1989. Springer, Berlin.

[29] N.P. Dellaert. Due-date setting and production control. *International Journal of Production Economics*, 23:59–67, 1991.

[30] I. Duenyas. Single facility due date setting with multiple customer classes. *Management Science*, 41(4):608–619, 1995.

[31] I. Duenyas and W.J. Hopp. Quoting customer lead times. *Management Science*, 41(1):43–57, 1995.

[32] F.F. Easton and D.R. Moodie. Pricing and lead time decisions for make-to-order firms with contingent orders. *European Journal of operational research*, 116:305–318, 1999.

[33] S. Eilon and I.G. Chowdhury. Due dates in job shop scheduling. *International Journal of Production Research*, 14(2):223–237, 1976.

[34] S. Eilon and R.M. Hodgson. Job shop scheduling with due dates. *International Journal of Production Research*, 6(1):1–13, 1967.

[35] M. Elhafsi. An operational decision model for lead-time and price quotation in congested manufacturing systems. *European Journal of Operational Research*, 126:355–370, 2000.

[36] M. Elhafsi and E. Rolland. Negotiating price/delivery date in a stochastic manufacturing environment. *IIE Transactions*, 31:225–270, 1999.

[37] D.A. Elvers. Job shop dispatching rules using various delivery date setting criteria. *Production and Inventory Management*, 4:62–70, 1973.

[38] S.T. Enns. Job shop lead time requirements under conditions of controlled delivery performance. *European Journal of Operational Research*, 77:429–439, 1994.

[39] S.T. Enns. Lead time selection and the behaviour of work flow in job shops. *European Journal of Operational Research*, 109:122–136, 1998.

[40] L. Enos. Report: Holiday e-sales to double. *E-Commerce Times*, September 6 2000. http://www.ecommercetimes.com/perl/story/4202.html.

[41] T.D Fry, P.R. Philipoom, and R.E. Markland. Due date assignment in a multistage job shop. *IIE Transactions*, pages 153–161, June 1989.

[42] P. Glasserman and Y. Wang. Leadtime-inventory trade-offs in assemble-to-order systems. *Operations Research*, 46(6):858–871, 1998.

[43] M.X. Goemans, M. Queyranne, A.S. Schulz, M. Skutella, and Y. Wang. Single machine scheduling with release dates. *SIAM Journal on Discrete Mathematics*, 15:165–192, 2002.

[44] M.X. Goemans, J. Wein, and D.P. Williamson. A 1.47-approximation algorithm for a preemptive single-machine scheduling problem. *Operations Research Letters*, 26:149–154, 2000.

[45] V.S. Gordon. A note on optimal assignment of slack due-dates in single-machine scheduling. *European Journal of Operational Research*, 70:311–315, 1993.

[46] R.L. Graham, E.L. Lawler, J.K. Lenstra, and A.H.G. Rinnooy Kan. Optimiztion and approximation in deterministic sequencing and scheduling: A survey. *Annals of Discrete Mathematics*, (5):287–326, 1979.

[47] G.I. Green and L.B. Appel. An empirical analysis of job shop dispatch rule selection. *Journal of Operations Management*, 1:197–203, 1981.

[48] A. Greene. The flow connection for e-business. *Manufacturing Systems Magazine*, February 2000.

[49] S.K. Gupta and J. Kyparisis. Single machine scheduling research. *Omega*, 15:207–227, 1987.

[50] D.N. Halsall, A.P. Muhlemann, and D.H.R. Price. A review of production planning and scheduling in smaller manufacturing companies in the uk. *Production Planning and Control*, 5:485–493, 1984.

[51] C.W.L. Hart. The power of unconditional service guarantees. *Harvard Business Review*, pages 54–62, July-August 1988.

[52] N. Hill. Delivery on the dot - or a refund! *Industrial Marketing Digest*, 14(2):43–50, 1989.

[53] D.S. Hochbaum, K. Jansen, J.D.P. Rolim, and A. Sinclair, editors. *Randomization, Approximation, and Combinatorial Algorithms and Techniques, Third International Workshop on Randomization and Approximation Techniques in Computer Science, and Second International Workshop on Approximation Algorithms for Combinatorial Optimization Problems RANDOM-APPROX'99, Berkeley, CA, USA, August 8-11, 1999, Proceedings*, volume 1671 of *Lecture Notes in Computer Science*. Springer, 1999.

[54] M. Hopp, W. Spearman and D. Woodruff. Practical strategies for lead time reduction. *Manufacturing Review*, 3:78–84, 1990.

[55] W.J. Hopp and M.L. Roof Sturgis. Quoting manufacturing due dates subject to a service level constraint. *IIE Transactions*, 32:771–784, 2000.

[56] P.Y. Huang. A comparative study of priority dispatching rules in a hybrid assembly/job shop. *International Journal of Production Research*, 22(3):375–387, 1984.

[57] J.R. Jackson. Job shop-lie queuing systems. *Management Science*, 10:131–142, 1963.

[58] P. Kaminsky and Z.-H. Lee. Asymptotically optimal algorithms for reliable due date scheduling. Technical report, Department of Industrial Engineering and Operations Research, University of California, Berkeley, CA, 2003.

[59] J.J. Kanet. On anomalies in dynamic ratio type scheduling rules: A clarifying analysis. *Management Science*, 28(11):1337–1341, 1982.

[60] J.J. Kanet and J.C. Hayya. Priority dispatching with operation due dates in a job shop. *Journal of Operations Management*, 2(3):167–175, 1982.

[61] R. Kapuscinski and S. Tayur. Reliable due date setting in a capacitated MTO system with two customer classes. Working Paper, GSIA, Carnegie Mellon University.

[62] U. Karmarker. Lot sizes, manufacturing lead times and throughput. *Management Science*, 33:409–418, 1987.

[63] R. Kaufman. End fixed lead times. *Manufacturing Systems*, pages 68–72, January 1996.

[64] P. Keskinocak. Satisfying customer due dates effectively. Ph.D. Thesis, GSIA, Carnegie Mellon University, 1997.

[65] P. Keskinocak, R. Ravi, and S. Tayur. Scheduling and reliable lead time quotation for orders with availability intervals and lead time sensitive revenues. *Management Science*, 47(2):264–279, 2001.

[66] B. Kingsman, L. Worden, L. Hendry, A. Mercer, and E. Wilson. Integrating marketing and production planning in make-to-order companies. *Interntional Journal of Production Economics*, 30:53–66, 1993.

[67] B.G. Kingsman, I.P. Tatsiopoulos, and L.C. Hendry. A structural methodology for managing manufacturing lead times in make-to-order companies. *Management Science*, 22(12):1362–1371, 1976.

[68] S.R. Lawrence. Estimating flowtimes and setting due-dates in complex production systems. *IIE Transactions*, 27:657–668, 1995.

[69] Y. Lu, J.-S. Song, and D.D. Yao. Order fill rate, lead time variability, and advance demand information in an assemble-to-order system. *Working Paper*, 2001. Columbia University, http://www.ieor.columbia.edu/ yao/lsy1rev1b.pdf.

[70] H. Matsuura, H. Tsubone, and M. Kanezashi. Setting planned lead times for multi-operation jobs. *European Journal of Operational Research*, 88:287–303, 1996.

[71] S. Miyazaki. Combined scheduling system for reducing job tardiness in a job shop. *International Journal of Production Research*, 19(2):201–211, 1981.

[72] D.R. Moodie. Demand management: The evaluation of price and due date negotiation strategies using simulation. *Production and Operations Management*, 8(2):151–162, 1999.

[73] D.R. Moodie and P.M. Bobrowski. Due date demand management: negotiating the trade-off between price and delivery. *International Journal of Production Research*, 37(5):997–1021, 1999.

[74] J. Orlicky. *Materials Requirements Planning*. McGraw Hill, Inc., New York, 1975.

[75] K. Palaka, S. Erlebacher, and D.H. Kropp. Lead-time setting, capacity utilization, and pricing decisions under lead-time dependent demand. *IIE Transactions*, 30:151–163, 1998.

[76] S. Panwalkar, M. Smith, and A. Seidmann. Common due date assignment to minimize total penalty for the one machine scheduling problem. *Operations Research*, 30(1):391–399, 1982.

[77] P.R. Philipoom, L.P. Rees, and L. Wiegman. Using neural networks to determine internally-set due-date asignments for shop scheduling. *Decision Sciences*, 25(5/6):825–851, 1994.

[78] P.R. Philipoom, L. Wiegman, and L.P. Rees. Cost-based due-date assignment with the use of classical and neural-network approaches. *Naval Research Logistics*, 44:21–46, 1997.

[79] M. Pinedo. *Scheduling: Theory, Algorithms and Systems*. Prentice Hall, New Jersey, 2002.

[80] M.A. Quaddus. A generalized model of optimal due date assignment by linear programming. *Journal of the Operational Research Society*, 38:353–359, 1987.

[81] M.A. Quaddus. On the duality approach to optimal due date determination and sequencing in a job shop. *Engineering Optimization*, 10:271–278, 1987.

[82] G.L. Ragatz. A note on workload-dependent due date assignment rules. *Journal of Operations Management*, 8(4):377–384, 1989.

[83] G.L. Ragatz and V.A. Mabert. A simulation analysis of due date assignment rules. *Journal of Operations Management*, 5(1):27–39, 1984.

[84] M. Ragavachari. A v-shape property of optimal schedule of jobs about a common due date. *European Journal of Operational Research*, 23:401–402, 1986.

[85] S. Ray and E.M. Jewkes. Customer lead time management when both demand and price are lead time sensitive. *European Journal of Operational Research*, 2003.

[86] R.S. Russell and B.W. Taylor. An evaluation of sequencing rules for an assembly shop. *Decision Sciences*, 16:196–212, 1985.

[87] A. Seidmann, S.S. Panwalkar, and M.L. Smith. Optimal assignment of due-dates for a single processor scheduling problem. *International Journal of Production Research*, 19(4):393–399, 1981.

[88] A. Seidmann and M.L. Smith. Due date assignment for production systems. *Management Science*, 27(5):571–581, 1981.

[89] T. Sen and S.K. Gupta. A state-of-art survey of static scheduling research involving due-dates. *Omega*, 12:63–76, 1984.

[90] J.G. Shanthikumar and J.A. Buzacott. Open queuing network models of dynamic job shops. *International Journal of Production research*, 19:255–266, 1981.

[91] J.G. Shanthikumar and J.A. Buzacott. The time spent in a dynamic job shop. *European Journal of Operational Research*, 17:215–226, 1984.

[92] J.G. Shanthikumar and U. Sumita. Approximations for the time spent in a dynamic job shop with applications to due-date assignment. *International Journal of Production Research*, 26(8):1329–1352, 1988.

[93] B. Simon and R. Foley. Some results on sojourn times in acyclic jackson networks. *Management Science*, 25:1027–1034, 1979.

[94] D.S. Sleator and R.S. Tarjan. Amortized efficiency of list update and paging rules. *Communications of the ACM*, 28:202–208, 1985.

[95] W.E. Smith. Various optimizers for single-stage production. *Naval Research Logistics Quarterly*, pages 59–66, March 1956.

[96] K.C. So. Price and time competition for service delivery. *Manufacturing & Service Operations Management*, 2(4):392–409, 2000.

[97] K.C. So and J.-S. Song. Price, delivery time guarantees and capacity selection. *European Journal of Operational Research*, 111:28–49, 1998.

[98] H.M. Soroush. Sequencing and due-date determination in the stochastic single machine problem with earliness and tardiness costs. *European Journal of Operational Research*, 113:450–468, 1999.

[99] M.L. Spearman and R.Q. Zhang. Optimal lead time policies. *Management Science*, 45(2):290–295, 1999.

[100] R. Suri. *Quick Response Manufacturing: A Companywide Approach to Reducing Lead Times*. Productivity Press, Portland, OR, 1998.

[101] S. Tayur. Improving operations and quoting accurate lead times in a laminate plant. *Interfaces*, 30(5):1–15, 2000.

[102] N.R. Tobin, A. Mercer, and B. Kingsman. A study of small subcontract and make-to-order firms in relation to quotation for orders. *International Journal of Operation and Production Management*, 8(6):46–59, 1987.

[103] G. Udo. An investigation of due-date assignment using workload information of a dynamic shop. *International Journal of Production Economics*, 29:89–101, 1993.

[104] M.M. Vig and K.J. Dooley. Dynamic rules for due date assignment. *International Journal of Production Research*, 29(7):1361–1377, 1991.

[105] M.M. Vig and K.J. Dooley. Mixing static and dynamic flowtime estimates for due-date assignment. *Journal of Operations Management*, 11:67–79, 1993.

[106] J.K. Weeks. A simulation study of predictable due-dates. *Management Science*, 25(4):363–373, 1979.

[107] J.K. Weeks and J.S. Fryer. A simulation study of operating policies in a hypothetical dual-constrained job shop. *Management Science*, 22(12):1362–1371, 1976.

[108] J.K. Weeks and J.S. Fryer. A methodology for assigning minimum cost due-dates. *Management Science*, 23(8):872–881, 1977.

[109] L.M. Wein. Due-date setting and priority sequencing in a multi class m/g/1 queue. *Management Science*, 37(7):834–850, 1991.

[110] Z.K. Weng. Strategies for integrating lead time and customer-order decisions. *IIE Transactions*, 31(2):161–171, 1999.

[111] R.L. Winkler and W.L. Hayes. *Statistics: Probability, Inference, and Decision*. Holt, Rinehart & Winston, New York, 1975.

[112] J.D. Wisner and S.P. Siferd. A survey of us manufacturing practices in make-to-order machine shops. *Production and Inventory Management Journal*, 36(1):1–7, 1995.

[113] W.H.M. Zijm and R. Buitenhek. Capacity planning and lead time management. *International Journal of Production Economics*, 46-47:165–179, 1996.

Part IV

MULTI-CHANNEL COORDINATION

David Simchi-Levi, S. David Wu, and Z. Max Shen (Eds.)
Handbook of Quantitative Supply Chain Analysis:
Modeling in the E-Business Era
©2004 Kluwer Academic Publishers

Chapter 13

MODELING CONFLICT AND COORDINATION IN MULTI-CHANNEL DISTRIBUTION SYSTEMS:

A Review

Andy A. Tsay
OMIS Department,
Leavey School of Business,
Santa Clara University, Santa Clara, CA 95053

Narendra Agrawal
OMIS Department,
Leavey School of Business,
Santa Clara University, Santa Clara, CA 95053

Keywords: channel conflict, competition, coordination, retail sales, direct sales, catalog sales, manufacturer outlets, intermediaries, resellers, channels of distribution, dual distribution, franchising, supply chain management, electronic commerce, Internet commerce, mathematical modeling, game theory

1. Introduction

1.1 Business Setting

For any company with a product to sell, how to make that product available to the intended customers can be as crucial a strategic issue as developing the product itself[1]. While distribution channel choice is a very traditional concern, for many companies it has recently come under intense scrutiny due to a number of major developments. The expanding role of the Internet in consumer and business procurement activity has created new opportunities for access to customers. Information and materials handling technologies have broadened the feasible set of sales and distribution activities that a producer might reasonably perform. The economics of materials delivery has been transformed by the pervasive logistical networks deployed by third-party shipping powerhouses such as Federal Express and United Parcel Services. As a result, many manufacturers are reconsidering their approaches to distribution, with particular attention to the role of intermediaries[2].

The prospect of reducing or even eliminating the reliance on resale intermediaries has always offered certain lures for manufacturers, including the following: (1) intermediaries carry only small assortments of a manufacturer's products, (2) direct control of distribution and pricing can lead to higher profit margins, (3) intermediaries can use their power to extract various concessions from the manufacturers, (4) manufacturers can provide a broader product selection in a better ambiance with higher service in direct outlets, (5) more flexibility in experimenting with product attributes, (6) closer contact with customers, and (7) protection from crises faced by intermediaries (Stern et al. 1996). Eliminating intermediaries ("disintermediation") can also improve supply chain efficiency by allowing upstream parties better visibility into market demand (cf. Lee et al. 1997). While these arguments have long supported the use of print catalog sales and company-owned stores, the explosion in possibility of electronic commerce has been particularly influential in drawing many manufacturers into the realm of direct sales[3].

Elimination of intermediaries is not without disadvantage. The role of intermediaries is to efficiently create and satisfy demand, through activities that include building brand and product awareness through advertising and customer education, providing market coverage, gathering market information, providing breadth of assortment, breaking bulk, processing orders, customer support, etc. If a manufacturer cannot otherwise attend to these functions efficiently, elimination of intermediaries may cause an erosion of profits, market share, or both (cf. Ghosh 1998). As noted by Stern et al. (1996, p.115), "It is an old axiom of marketing that it is possible to eliminate wholesalers (or any middlemen, for that matter) but impossible to eliminate their functions."

The "age of eBusiness" has now been underway for a few years, and evidence that has accumulated thus far indicates a trend towards a portfolio approach that includes both intermediated and manufacturer-owned channels (each type of which can take either bricks-and-mortar or online[4] form). This exploits the relative strengths of each and their appeal to different market segments. Indeed, leveraging multiple channel types may allow greater market penetration than using any one alone, and may enable innovative methods of value-delivery yet to be imagined (cf. Balasubramanian and Peterson 2000).

This vision faces a number of implementation challenges. An obvious obstacle that comes with increasing system complexity is the difficulty in maintaining coherence across channels with respect to strategy and execution. Perseverant manufacturers can presumably overcome this with management effort and appropriate information technologies. However, a different issue altogether is one that is frequently cited among the most significant barriers to multi-channel strategies, and is unlikely to be remedied purely internally since the root cause is interfirm conflict. Specifically, the existence of a manufacturer-owned channel may establish the manufacturer as a direct competitor to its intermediary. For example, Nike's opening of a Niketown store in downtown Chicago was considered a serious threat by retailers carrying Nike products (Collinger 1998). Estee Lauder's plans to sell its flagship Clinique brand directly over the Internet put the firm squarely in competition with the department stores whose cosmetics counters feature Clinique products so prominently (Machlis 1998(b)). Similar conflicts have been reported by Avon Products Inc. (Machlis 1998(c)), Bass Ale (Bucklin *et al.* 1997), IBM (Nasiretti 1998), the former Compaq (McWilliams 1998), Mattel (Bannon 2000), and others. Some trade groups such as the National Shoe Association and the National Sporting Goods Association have gone to the point of urging members to reduce or eliminate purchases from manufacturers establishing direct sales outlets (Stern *et al.* 1996). A well-publicized incident involved letters sent in May of 1999 by Home Depot to more than 1,000 of its suppliers, stating,

> "Dear Vendor, It is important for you to be aware of Home Depot's current position on its' [sic] vendors competing with the company via e-commerce direct to consumer distribution. We think it is short-sighted for vendors to ignore the added value that our retail stores contribute to the sales of their products.... We recognize that a vendor has the right to sell through whatever distribution channels it desires. However, we too have the right to be selective in regard to the vendors we select and we trust that you can understand that a company may be hesitant to do business with its competitors." (Brooker 1999)

In general, this type of channel conflict can undermine attempts to develop cooperative relations in the intermediated channel, possibly to the ultimate detriment of all involved parties.

To effectively assess the costs and benefits of multi-channel distribution, manufacturers and intermediaries alike must understand the cross-channel tensions that can arise. The desire to use multiple channel types may ultimately compel a manufacturer to redefine its relationship with its intermediaries, with careful attention to the division of labor and any associated financial terms. Indeed, the management of channel conflict is a key B2B concern that will profoundly influence supply chain success in the age of eBusiness.

1.2 Scope of Discussion

The title of this chapter declares our interest in multi-channel distribution, but we must still clarify what we mean by "multi-channel." Following the language of practitioners, we regard a particular method of accessing end customers as a single "channel" type even if the actual execution involves multiple outlets. For example, our framework views traditional retail as a single channel even if the manufacturer distributes through a retail firm with multiple physical stores, or through multiple, competing retail firms. Adding one or more physical outlets owned by the manufacturer would create a distinct channel. Yet another could arise from the various forms of manufacturer-managed mail-order (Internet or print catalog). Our terminology must also distinguish between control and materials flow. We will use the term "manufacturer-owned channel" to cover both of the latter two examples of channel types. The term "direct sales" will be reserved for the specific case in which the manufacturer controls the sales and marketing activities and a concrete purchase by the end customer is what triggers the product's flow from the manufacturer's warehousing/fulfillment operation (possibly co-located with the factory) to that customer.

Our primary interest is in how an intermediary reacts to creation of a new channel (that is assumed to enjoy some economic advantage and/or favorable treatment from the manufacturer), and if the resulting conflict might overwhelm any potential advantages. This is motivated by the recent attention of the business community to the scenario in which a manufacturer's direct online channel disrupts a status quo in which intermediaries are used heavily (although similar issues have long arisen in non-eBusiness settings as well, as examined by some of the papers in Section 3). This focus is intended to go beyond the manufacturer-intermediary conflict that can arise when a manufacturer sells exclusively through an intermediary (*vertical competition*). Similarly, it requires more than *horizontal competition* among sellers vying for the same customers. (These two and other related cases will be summarized in Section 2.) Described in these terms, our notion of channel conflict is associated with the case in which a manufacturer and its intermediary are engaged in horizontal and vertical competition simultaneously[5].

Two distinctions are central to understanding popular usage of the term "channel conflict." One is between actual harm and the perception of harm. Another is between effects on intermediate outcomes (e.g., sales volume or revenue) and on bottom-line objectives (e.g., profit). A strictly rational firm should care only about real impacts to its main objective (although conflict can still arise when individual employees do not share the firm's immediate goal, perhaps a consequence of the internal performance measurement and reward mechanisms). From that perspective, incumbent channels should not automatically be alarmed when additional channels are introduced. Consider that under certain circumstances one channel's efforts can drive traffic to another channel, especially when conducted with such an intent (Bucklin *et al.* 1997, McIntyre 1997, Schmid 1999). And losing sales need not hurt overall profitability. For instance, a new channel might be targeted at an existing channel's least profitable customers. Or a manufacturer opening a direct channel might at the same time sweeten the wholesale terms offered to existing intermediaries. However, real behavior is often driven by perceived impacts (either to intermediate impacts or the bottom line), perhaps due to difficulty in proving linkages to bottom-line consequences. As McIntyre (1997) notes, "This is the type of channel conflict we do hear a lot about, not because it is real, but because the fear is real." So, managing channel conflict might have a psychological component as well, which can give rise to practices whose primary or even sole purpose is to signal good intentions or reinforce a certain message. Our discussion will not pass judgment on the relative importance of managing perceptions. Instead, we will simply report the findings of existing research, which does occasionally take the perspective of demonstrating why popular perceptions may be inconsistent with rational economic conclusions.

1.3 Contribution

The intent of this chapter is to review quantitative approaches to modeling conflict in multi-channel distribution systems (as described previously) and policies that may coordinate the actions taken by channel partners, thereby improving system performance. To this end, attention will be focused on research that meets the following criteria:

- analytical modeling approach
- the distribution system is a design variable (this includes cases where the structure is set but the action strategies are not)
- the manufacturer and intermediary(ies) are independent
- the multiple channels interact in some way, primarily competing either for demand or supply.

An implication of these criteria is that research that takes a purely descriptive (e.g., empirical research that seeks to validate certain hypothesized relationships) rather than prescriptive approach will not be subject to our detailed review, although we will briefly mention some such works.

Because much of the research in this area is quite recent, many of the papers to be discussed have not yet appeared in the open literature, and are available as working papers only. Our including a paper in this review does not mean we believe it to be completely correct, or that it will eventually pass peer review and appear in print. We also seek to remain impartial about the significance of recent works, as many are very similar and only time will tell which will have the greatest impact. When we call attention to key assumptions, our intent is not to challenge the validity of those assumptions, but to identify the potential drivers of certain results and offer possible explanations for why models of similar settings might offer divergent conclusions.

While our primary focus has been very explicitly defined, there are a number of other bodies of literature that touch upon this setting. Section 2 delineates a framework to help organize the related work, and discusses each category briefly. Section 3 provides some commentary about modeling multi-channel settings, and detailed descriptions of the papers falling directly within the scope defined earlier. Section 4 concludes by summarizing key limitations of existing models and identifying areas open for future research.

2. Related Literature

Our general interest is in settings that offer end customers multiple ways of obtaining a given manufacturer's product, at least one of which involves intermediaries. A vast amount of research is relevant in some way, much of which we will not discuss in detail because of space limitations. Nevertheless, organizing this will sharpen the statement of our intended focus.

We make a primary distinction with respect to methodological approach. There is a vast amount of descriptive research about channel structure and conflict that performs analysis of empirical data or discusses evidence obtained anecdotally or by case study. We comment on this in Section 2.1. This chapter is primarily interested in analytical model-based research, which is introduced in Section 2.2. This literature will subsequently be partitioned with respect to assumptions concerning control structure and channel type.

2.1 Descriptive research

Marketing and economics researchers have conducted a substantial amount of empirical research that focuses on a variety of aspects of channel design and management, and on channel conflict and coordination in particular. Frazier (1999) and Balasubramanian and Peterson (2000) offer multidisciplinary reviews.

While this body of work is rich in the breadth of issues addressed, it is primarily descriptive in approach. The mechanism by which the various factors interact and ultimately affect system performance is generally not investigated mathematically. For that reason, and also due to the sheer amount of this type of research, here we will briefly summarize only the portion that is most relevant to our focus: dual-distribution in franchise systems. The interested reader is urged to consult Stern *et al.* (1996) and the references therein for further details.

Franchising is common when the establishing and promoting of a brand name are centralized, but production and/or distribution of the good or service are decentralized (Scott 1995). In other words, the franchisor (analogous to a *manufacturer* in our discussion) supplies a brand name and also a model of business for the franchisees (analogous to an *intermediary*, even though the franchisee may actually be manufacturing the product) to copy. Franchising typically involves an initial conveyance of industrial property rights leading to a common appearance, ongoing transfer of know-how, and regular technical assistance (Lafontaine 1992). See Dnes (1996) for a review of the economics literature on franchising.

A dual-distribution arrangement arises when the franchisor also owns and operates some of the stores. The mix of ownership types has been observed to correlate with the heterogeneity across stores, some salient dimensions of which are listed below:

- *Firm-specific investment in outlets* (Brickley and Dark 1987, Brickley *et al.* 1991(a)): Firm-specific investments in the franchise system by the franchisees will generate quasi-rents that can be expropriated by an opportunistic franchisor. This risk may deter the franchisees from making such investments when the required levels are high, which favors company ownership of stores.

- *Distance from the headquarters and monitoring costs* (Brickley and Dark 1987, Minkler 1992): The cost of monitoring the outlets increases with the distance from the franchisor's headquarters. Consequently, more distant outlets are more likely to be franchised than company owned.

- *Density or physical dispersion of outlets* (Caves and Murphy 1976, Brickley and Dark 1987, Brickley *et al.* 1991(b)): Geographic dispersion of the outlets increases the costs of monitoring the performance of company employees, which makes franchising more likely.

- *Repeat customers* (Caves and Murphy 1976, Klein 1980, Brickley and Dark 1987, Brickley *et al.* 1991(a,b)): When the frequency of repeat business is high, the impact of any debasement in quality on the operation's revenues will be high. Consequently, franchising makes the most sense when the costs of quality debasement are borne primarily by the franchisee. This is a reason why fast-food restaurants situated along freeways are more likely to be company owned.

- *Technological factors* (Caves and Murphy 1976, Klein and Saft 1985, Norton 1988): Monitoring the local production of goods or services becomes easier (less critical) if the process is more capital intensive and requires a greater proportion of machine effort than human effort. When monitoring costs are low, company ownership is a more likely outcome.

- *Age of the franchise* (Oxenfeldt and Kelley 1969, Caves and Murphy 1976, Martin 1988, Minkler 1990): The early stages of a franchise's life cycle are characterized by a greater degree of risk, and a greater need to raise capital for expansion, which may lead the franchisor to rely more heavily on franchising. As the franchisor learns about the local markets over time, the need for reliance on franchisees will diminish, which increases the likelihood of company ownership. Another rationale for franchising is based on the need to motivate both the franchisee and the franchisor to exert the effort needed for the business to succeed. Vertical integration suffers from the inherent difficulty in monitoring the actions of employees (whose compensation is not tied to the outlet's profits), while franchise arrangements align interests by giving each side a claim on a percentage of the outlet's profits. In contrast to the resource-based argument, this incentive-based argument implies that franchisors would want to move toward a fully franchised chain over time (see Lafontaine and Kaufmann 1994 for further discussion).

- *Signaling* (Gallini and Lutz 1992, Scott 1995): When there is uncertainty about the demand or profitability of the good or service being distributed, potential franchisees may be reluctant to make investments. Franchisors can send positive signals about demand or profitability by distributing their products through a high proportion of company-owned stores.

As an example of the methodology that is typical among descriptive research such as described above, consider Scott (1995), which tested hypothesized relationships between various factors and the proportion of stores franchised. Here, firm-level data from 47 different industry groups were used to estimate a regression equation of the following form:

Percent Franchised = $\beta_0 + \beta_1$ [Royalty Rate] + β_2 [Franchise Fee]
+ β_3 [Age] + β_4 [Area] + β_5 [Cash Investment] + β_6 [Training]
+ β_7 [Capital-Labor Ratio] + β_8 [Franchise Purchases].

The vast number of findings in the empirical dual-distribution franchising literature certainly can assist in corroborating the assumptions or assessing the predictions of analytical models of more general multi-channel settings, which are the topic of Section 2.2.

2.2 Analytical research

The analytical paradigm formulates a model of a decision problem and then recommends courses of actions based on rigorous mathematical justification. In contrast to the descriptive literature, for which numerous summaries and discussions already exist (as noted in Section 2.1), the analytical studies of our focal setting have not previously been reviewed to the extent attempted by this chapter.

Our specific interest is in the multi-echelon models that are necessary to explicitly represent channel intermediation, and therefore conflict between producers and their intermediaries. However, since one dimension of such conflict is due to a producer and its intermediary battling over the same market, models of horizontal competition provide a natural building block.

Horizontal competition arises when multiple sellers pursue the same pool of customers. The phenomenon is well studied in the economics literature and elsewhere, dating at least as far back as the classic models of oligopoly, and notions of Cournot, Bertrand, and Stackelberg competition. Variants of this are too numerous to review here, so we direct the reader to Shapiro (1989). In the inventory literature, most treatment of competition has been in a single echelon environment. For instance, Parlar (1988) and Lippman and McCardle (1997) study a pair of "newsvendor" firms who become competitors because their products are partially substitutable, while Bernstein and Federgruen (2002) examine an oligopoly in which sales are awarded based on the competitors' relative selling prices and fill rates. Cachon (2002, Section 5) summarizes this literature.

This area also includes studies of markets that are served either by multiple types of sellers and/or by individual sellers using multiple non-intermediated channel types (where the various modes of selling are differentiated by cost structure and/or market reach). Balasubramanian (1998) and Druehl and Porteus (2001) both model a horizontal competition between a direct marketer and conventional retailers. The topic of Cattani *et al.* (2002) is a competition between a traditional and an Internet channel that are either owned by the same retailer or by different entities. In Reinhardt and Levesque (2001), a seller decides how to allocate its product across two markets that are each reached a different way (online vs. offline sales channels), given a competitor in one of the markets. Huang and Swaminathan (2003) study the pricing strategies that might arise when a retailer with both traditional and Internet channels competes with a pure Internet retailer. Lal and Sarvary (1999) and Zettelmeyer (2000) examine a competition between two retailers that each sell both online and offline. In Chen *et al.* (2002), all sales occur at one of two competing bricks-and-mortar retailers, but customers can subscribe to an independent, online service that offers price quotes (possibly communicating discounts from affiliated retailers) and other product information. Balasubramanian *et al.* (2002) offer a vision of a "wireless" version of this marketing approach, in which retailers can compete on a customer-by-customer basis by broadcasting personalized discounts through mobile devices. Since these models do not explicitly represent intermediation, they do not investigate manufacturer-intermediary conflict or coordination.

Figure 13.1 provides a framework for organizing the types of multi-echelon distribution systems, and helps characterize what will and will not be highlighted in this review. This framework is motivated by our belief that at an aggregate level, control structure and choice of channel type are two of the most important determinants of the overall performance of any supply chain. Even this classification will not be perfect, as some papers study multiple settings. We will exercise judgment in classifying such papers by the system type that we believe to be their primary emphasis, or for which the most substantive results are obtained.

As indicated in the figure, the following three subsections will provide brief overviews of each class of literature related to our primary focus. As necessary we will emphasize the details of any interaction across channels, such as the decision variables available to each firm and the mechanism by which these have cross-channel effects. We do not claim our review to be comprehensive. Rather, our primary intent is to provide the reader some particularly recent or meaningful examples (and ideally to call attention to appropriate review papers) that might serve as a starting point for further investigation. This will create context for Section 3, which details the extant analytical research regarding the phenomenon of central interest: conflict and coordination in multi-channel systems.

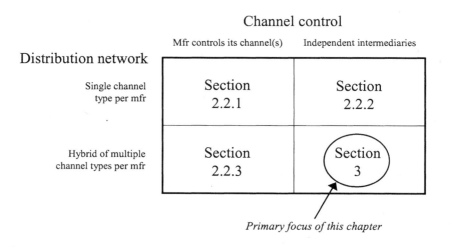

Figure 13.1. Classification of models of multi-echelon distribution systems

2.2.1 Single channel type systems under manufacturer control.
In this class of literature the product stops at intermediate points between a manufacturer and its end customers, for instance at distribution centers or depots. These locations add value in a variety of ways, including risk pooling of demand uncertainty across multiple retail locations, lower produc-

tion and distribution costs through coordination of shipments, etc. But because these locations are not independent, they are not intermediaries by our definition.

Many papers whose primary interest is in settings with independent agents begin with this scenario as a benchmark, as system performance is theoretically maximized under unified control (achievable perhaps by "forced compliance"). However, they quickly move on to studying mechanisms that might induce the independently managed parts to replicate the efficiency of central control.

The benchmark case is of greater stand-alone interest when the network exhibits complexities such as stochastic demand that is filtered through the inventory policies implemented at the successive echelons, intricate cost structures, lead times, and multiple time periods of activity. This is the theme of the bulk of multi-echelon inventory theory, dating back to the serial supply chain analysis of Clark and Scarf (1960). This literature also considers system architectures that move product from a manufacturer to end customers along multiple parallel paths, such as a 1-depot N-warehouse network.

We do not review this class of research since centralized control prevents internal conflict altogether, between manufacturers and intermediaries or otherwise. Instead, we refer the reader to reviews by Muckstadt and Roundy (1993) for deterministic demand models, Federgruen (1993) for discrete time models with stochastic demands, and Axsater (1993) for continuous time models with stochastic demands.

2.2.2 Single channel type systems with independent intermediaries. The most general form of this setting has multiple manufacturers selling exclusively through multiple independent retailers. While these cases do not qualify as multi-channel, understanding issues of conflict and coordination between manufacturers and intermediaries provides insights for managing the intermediated channel within a hybrid system. We highlight five major classes of research.

Bilateral monopoly

The vast majority of works assumes a single manufacturer and single retailer, which certainly is the most tractable case. This does not fall within our scope, since customers then have only one source for the product.

The typical analytical approach is to study the source of inefficiency (often related to *double marginalization*), and various mechanisms for achieving channel coordination and/or Pareto improvement. Mechanisms so studied recently include resale price maintenance, full-line forcing, quantity discounts, manufacturer returns, quantity flexibility, sales rebates, revenue sharing, price protection, and markdown allowances. New studies continue to be produced at a prodigious rate by generalizing the basic framework, perhaps by adding retailer decision variables (e.g., some form of effort, in addition to retail price or quantity decisions) or adding informational asymmetry. Examples of the former are Krishnan et al. (2001) which gives the retailer control over promotional effort (which is possibly unobservable), and Netessine & Rudi (2001(a)) which investigates the incentive effects of drop-shipping strategies (i.e., the retailer handles all sales, but avoids inventory ownership as fulfillment occurs directly from the supplier's warehouse to the end customer) when demand is dependent on the retailer's spending for customer acquisition. Examples of the latter are Corbett and de Groote (2000) and Ha (2001) in which the retailer's cost information is private, Bali et al. (2001) in which the retailer has private information about its own inventory level, and Kolay et al. (2002) in which a retailer has private information about demand. Cachon (1999), Lariviere (1999) and Tsay et al. (1999) review this literature in depth, and Cachon (2002) provides an updated perspective.

Single manufacturer with multiple retailers, where the retailers do not interact with each other
This generalization allows the product to flow through multiple independent retailers. However, the retailers do not interact because end customers are assumed to be captive to a particular retailer (i.e., exclusive territories) and the manufacturer produces enough to satisfy all retailer requests. Recent examples include Ingene and Parry (1995(a)), Chen et al. (2001), Fransoo et al. (2001), Netessine and Rudi (2001(b)). The first considers price-sensitive end demand while the other three include inventory effects.

Single manufacturer with multiple retailers, where the retailers interact with each other
Two types of retailer interaction that provide basis for conflict are (1) common interest in an item in scarce supply, and (2) competition for customers.

Recent works about the first phenomenon include Cachon and Lariviere (1999(a,b)) and Deshpande and Schwarz (2002), who investigate the forecast gaming behavior observed among retailers attempting to secure a disproportionate share of the supply. Such research typically investigates methods for mitigating dishonesty.

The second phenomenon has been studied much more extensively by building on insights from the literature of horizontal competition that was described earlier. The majority of papers assume a single dimension of retailer competition. For instance, Ingene and Parry (1995(b), 1998, 2000) examine issues of channel coordination faced by a manufacturer selling through two retailers that compete on price. Padmanabhan and Png (1997) investigate the role of manufacturer return policies in a similar setting. Marx and Shaffer (2001(b)) explore why a manufacturer might benefit from using a nondiscrimination ("most favored customer") clause when negotiating wholesale prices with multiple retailers sequentially. Lal (1990) considers the coordination of a franchise system in which the retailers engage in service competition. Models that broaden the breadth of decisions made by each retailer include the following. Mathewson and Winter (1984) include advertising as a decision, although not directly as a dimension of competition. Perry and Porter (1990) focus on a type of retailer service with positive externality effect across the retailers. In Winter (1993), Iyer (1998), and Tsay and Agrawal (2000) the retailers compete directly along both price and non-price dimensions.

In the majority of works the retailers are similar in their mode of sales. One exception is Purohit (1997), in which a durable goods manufacturer's two intermediaries differ significantly: one only sells the product new, while the other can rent out the new product for a period before subsequently selling the used good.

Some papers that are primarily focused on the bilateral monopoly will discuss applicability to multiple retailer settings. A typical question raised is whether a given manufacturer-retailer contract will retain its effectiveness if applied uniformly to multiple retailers that are asymmetric in some way. This may be motivated both by legal concerns (e.g., Robinson-Patman[6] considerations) and the desire to minimize the costs of negotiating and administering the contracts. For instance, the channel-coordinating properties of the manufacturer return contract of Pasternack (1985) are known to extend to the case of multiple retailers with identical costs but different market demand distributions (provided that the markets do not overlap). This is because the coordinating contract is independent of the retailer's market demand. Whether and when a common contract can coordinate a channel comprising asymmetrical, competing retailers remains a generally unresolved issue. Interestingly, O'Brien and Shaffer (1994) find that forcing a manufacturer to treat competing retailers equally can lead to substantial welfare loss.

Because of the multi-level setting and the tension across sellers, the formulations that arise when a single manufacturer sells through multiple, interacting retailers are very closely related to those of direct interest in this chapter. Indeed, similar solution methodologies are typically used.

Multiple manufacturers with one or more common retailers

This class of models moves towards greater network complexity and realism by introducing competitive dynamics at the manufacturer level, although still within a single channel type strategy. Choi (1991) examines two manufacturers whose partially substitutable products are sold through a common retailer, while in Choi (1996) and Trivedi (1998) the two manufacturers also use a second common retailer that price-competes with the first. These study the effect of decision structure (channel leadership), channel interaction, and product differentiation. O'Brien and Shaffer (1993) address the question of whether competing manufacturers should sell through a common retailer instead of through exclusive retailers. O'Brien and Shaffer (1997) allow for both nonlinear pricing and exclusive dealing arrangements when two manufacturers contract with a retail monopolist. Marx and Shaffer (1999, 2001(a,c)) examine sequential contracting with two manufacturers and their common retailer, focusing on the role of bargaining power and implications for rent-shifting. Shaffer and Zettelmeyer (2002) show how different types of information shocks can affect the allocation of profits among a retailer and the two competing manufacturers whose products it carries. Raju *et al.* (1995) investigate the profitability of introducing a store brand into a product category that consists of price-competing national brands sold by different manufacturers. In Corbett and Karmarkar (2001), the number of firms at each echelon is a function of an entry and exit model. Shaffer (2001) investigates how the balance of power between manufacturers and retailers influences not only the terms of trade, but also the bargaining process used to allocate channel profits.

Competition between two manufacturer-retailer dyads
This class of models considers two manufacturers selling through dedicated intermediaries that compete for end customers. One research stream asks whether the manufacturers are better off using the intermediaries instead of vertically integrating. Representative studies include McGuire and Staelin (1983, 1986), Coughlan (1985), Moorthy (1988), Coughlan and Wernerfelt (1989), Gupta and Loulou (1998), and Gupta (2001). Here a key conjecture is that intermediaries may be able to mitigate the competition between manufacturers. Choi (2002) examines the impact of channel choice on industry structure, specifically the degree of total market coverage by the incumbents and ease of entry by challengers when all manufacturers use the same channel type (direct, exclusive retailer, or common retailer). Other models explore implications of using various channel policies in this competitive setting (e.g., slotting allowances and resale price maintenance are examined in Shaffer 1991). We exclude this class of literature because while different modes of selling may be considered (e.g., a vertically integrated channel competing against an intermediated channel), each manufacturer uses only a single mode at a time.

2.2.3 Multiple channel type systems under manufacturer control. In this class of research, the term "coordination" takes the meaning of optimizing decisions across a complex system, rather than overcoming conflicts of interest. Consequently, such models tend to present the structural complexities (especially the differences across channel types) in greater detail. Central control also gives the manufacturer greater influence over the design of the system's physical topology.

Building upon initial work by Artle and Berglund (1959) and Balderston (1958), Baligh and Richartz (1964) consider the problem of designing the optimal distribution system to transfer materials from multiple manufacturers to multiple retailers for a single product. They determine the number of levels in this system (with zero levels indicating direct sales) as well as the number of firms within each level to minimize the communication and contact costs in the network. Blumenfield *et al.* (1985) present a cost-minimizing framework when end users can be served directly by the manufacturers or through a consolidating warehouse for deterministic end user demands. Jaikumar and Rangan (1990) and Rangan and Jaikumar (1991) study how price rebates offered to different intermediary levels affect the channel choice decision (buying direct or through intermediaries), and determine the optimal pricing and distribution strategy. Cohen *et al.* (1995) perform an industry-level analysis of distribution networks by focusing on the specific functions performed by intermediaries (redistributors), for which the intermediaries charge their customers (distributors) a premium relative to prices for direct purchases from the manufacturers, and derive profit maximizing channel management policies (pricing and rebates) for the manufacturers. Cohen *et al.* (1990) and Cohen *et al.* (1999) analyze service parts logistics systems where parts can reach customers through service depots as well as directly from the manufacturer's warehouses. Chiang and Monahan (2001) advise a manufacturer on how to set inventory levels when distribution occurs through one direct sales channel and one company-owned store, given that each customer has an initial preference for one of the channel types but there is some spillover across channels on stockout.

3. Analytical Research on Conflict and Coordination in Multi-Channel Systems With Both Manufacturer-Owned And Intermediated Channels

This section provides detailed examination of works that consider our primary focus, in which the manufacturer is simultaneously a supplier to and a competitor[7] of its retail partner(s)[8]. Figure 13.2 illustrates how goods and customers physically come together in the setting. Solid lines represent the basic structure, while dotted lines suggest generalizations pursued by some researchers.

The primary research objectives are to enlighten the following questions faced by the various parties:

- <u>manufacturer</u>: which channel or channel portfolio to use, and how to coordinate strategies and decisions across channels

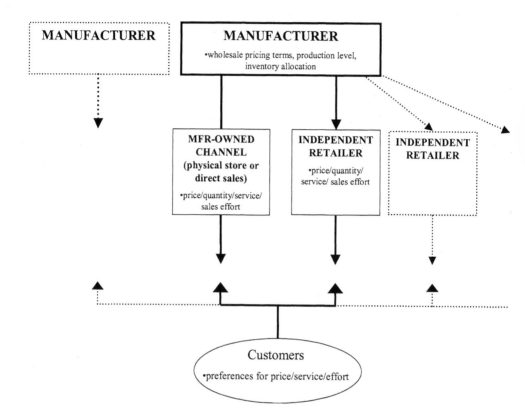

Figure 13.2. Multi-channel distribution system with independent retailer(s)

- independent retailer: whether to participate in a given distribution system, and if so, how to adjust behavior in response to the presence of the competing channel

- customers: which channel to use to serve which needs

The prevailing modeling approach is to formulate a game between a manufacturer and a retailer (or multiple retailers) transpiring during the single selling season of a single product (the same sold in all channels). The horizontal competition is modeled in a fairly standard way. All selling firms simultaneously compete for end customers via a primary decision variable such as price (or, less commonly, quantity). But there also needs to be a way for fundamentally different types of sellers to exist in the same marketplace without necessarily matching each other along that dimension[9]. This is typically done by one of two approaches: (1) including some Hotelling-style representation of customer heterogeneity (such as tastes, or aversion to travel or search); or (2) giving each seller a second control variable, such as some measure of service or sales effort. These assumptions dictate how seller decisions will affect aggregate buying behavior. Vertical competition is added by giving one seller (the one designated as manufacturer) the role of supplying product to other seller(s) (the retailer(s)). This entails the setting of a wholesale price, which the manufacturer typically does as Stackelberg leader prior to the horizontal competition for end customers. The manufacturer is typically assumed to enjoy infinite capacity. The costs incurred within the channels are typically summarized by assigning each channel its own per-unit cost parameter to cover production and logistics expenses. All firms maximize individual profit.

Given these assumptions, the analysis proceeds to solve (by reverse induction) for all equilibrium decisions and profit levels under each different distribution strategy, and characterize various sensitivities. Some studies also evaluate mechanisms for channel coordination and/or Pareto improvement. Unless highlighted, the reader should assume that this general framework applies.

As always, expectations regarding analytical conclusions should be tempered by an understanding of the modeling challenges. Most significantly, competition for customers is known to be difficult to model in a multi-echelon setting with independent parties. Moreover, the equations that arise in such models are easier to solve with symmetry across channels (e.g., all channels are independent, and identical intermediaries face the same types of demand), but hybrid distribution systems are inherently asymmetrical. For reasons such as these, the models we have encountered have all made some fairly restrictive assumptions. These issues and others will be discussed in more depth in Section 4.

Table 13.1 provides a quick comparison of the papers reviewed in this section, which are grouped according to whether the manufacturer-owned channel is direct sales or a manufacturer-owned retail store. This makes a distinction between the channel ownership/control and the physical supply chain structure. Selling through a manufacturer-owned physical store is clearly different from direct sales; the latter's logistics and marketing activities are dramatically dissimilar from a reseller channel's while the former's are not necessarily so. Also, the consumer's experience differs more markedly from conventional retail shopping when buying direct than from a company store.

There is no obvious sequence in which to present the papers in each subset. Indeed, with the exception of Rhee and Park (2000) and Rhee (2001), none of these papers can be considered an extension or generalization of any of the others since the formulations are so distinct. Overall, these works appear to have been developed largely in parallel by researchers from multiple disciplines.

Table 13.1. Classification of analytical research on conflict and coordination in multi-channel systems

	Basis of channel competition (seller decision variables)			Investigation of channel coordination	Strategic significance of mfr-owned channel	Dual channel preferred by both mfr and retailer?
	Price	Inventory	Other			
Manufacturer-owned channel is direct sales						
Chiang et al. (2003)	X				Implicit mechanism for keeping retail prices low	Maybe
Kumar & Ruan (2002)	X		Retailer service with no externality		Price discrimination between customer segments and influence over retailer service	Not addressed
Rhee & Park (2000)	X		Retailer service with no externality		Implicit mechanism for keeping retail prices low	Maybe
Rhee (2001)	X		Retailer service with no externality		To compete with a direct-selling challenger for price-sensitive customers	No
Hendershott & Zhang (2001)	X				Price discrimination through a higher direct price	No
Tsay & Agrawal (2001)	X		Marketing effort with positive externality	X	Drives traffic to the more efficient channel	Maybe
Peleg & Lee (2002)		X			Access to a price-sensitive customer base	Maybe
Manufacturer-owned channel contains physical stores						
Ahn et al. (2002)	X				Price discrimination and access to a distinct geographic market	Not addressed
Bell et al. (2002)	X		Marketing effort with positive externality		Implicit mechanism for keeping retail prices high	Maybe
Boyaci (2001)		X		X	Backup supply for retail stockout	Not addressed

3.1 Manufacturer-owned channel is direct sales

Chiang et al. (2003)

In Chiang et al. (2003), a manufacturer is considering whether to sell direct over the Internet, exclusively through a retailer, or through a hybrid of both approaches.

The manufacturer sells to the retailer at a unit wholesale price and incurs a unit cost which includes manufacturing and logistics, and potentially also sells through a direct channel at a different unit cost. The retailer decides on the retail price as Stackelberg follower.

Consumers have a willingness-to-pay parameter that is uniform on [0,1]. To impose an assumption that retail provides a superior shopping experience, the value to the consumer is scaled down by a multiplier $\theta < 1$ when the product is obtained through the direct channel. θ can be interpreted as consumers' willingness to tolerate the inconveniences of the direct channel. This structure produces deterministic demand curves that indicate how the market will be split between the channels as a linear function of the prices. Competition in this model is purely along the price dimension. Non-price differences between the channels are captured in θ, but this is a parameter of consumer preferences, not a decision variable that either channel can influence.

The key finding is that the manufacturer may use a direct channel as a way to combat double marginalization in the retail channel, opposing the retailer's tendency to price too high and sell too little. The effectiveness of this strategy depends on the viability of the manufacturer's threat to sell direct, which is driven by θ. When θ is sufficiently low (say, due to consumer perception of delays in direct delivery, potential mismatch of the catalog description and performance of the product, etc.), adding a direct channel creates no threat to the retailer. The retailer can effectively ignore the potential cannibalization of customers by the direct market, so the manufacturer does not profit by adding a direct channel. On the other hand, when θ is sufficiently high, the consumer sees little distinction between a traditional retailer and a direct marketer. Because the direct channel is a credible threat to cannibalize retail sales, the retailer will cut prices more aggressively, partially counteracting the double marginalization problem and increasing the manufacturer's profits.

The authors find that at equilibrium the manufacturer will price in the direct channel so as to drive all customers to the retail channel. In this case, the direct channel exists not to sell product, but as a mechanism for controlling the independent retailer's price. At the same time, there always exist circumstances under which the retailer and the manufacturer are both made better off as a result. This is possible because a reduction in the wholesale price accompanies the use of a direct channel. Hence, multi-channel approaches need not suffer from channel conflict.

Kumar and Ruan (2002)

In Kumar and Ruan (2002) the manufacturer contemplates augmenting the retail channel with a direct online channel.

The manufacturer moves first in choosing the unit wholesale price w to charge the retailer, and the direct sales price p_m if relevant. The retailer carries the manufacturer's product, as well as a substitute product that earns an exogenously specified profit margin k. In addition to retail price p_r for the manufacturer's brand, the retailer decides on the level of service (merchandising support) to provide the two products. The retailer incurs no direct costs for this service, but the total amount that can be provided is constrained (if one product enjoys "high" service, the other is left with "low").

Consumers are one of two types: retailer store loyal (segment size α_u) or manufacturer brand loyal (segment size α_s). Store loyal consumers buy only through the retail channel, with the choice of product determined by the service provided by the retailer: for a service level of s, $\alpha_u s$ store loyal consumers purchase the manufacturer's brand regardless of p_r and the remaining $\alpha_u(1-s)$ purchase the substitute product. Brand loyal consumers will purchase only the manufacturer's product, provided that its price is less than their reservation price r. If the product is available in both channels, a fraction $F=0.5+\beta(p_r - p_m)$ of brand loyal consumers will buy from the direct channel, where β measures the price sensitivity of brand loyal customers. The assumed form of F implies that some of these customers value the direct shopping experience enough to pay a premium over the prevailing retail store price.

As the retailer favors the product earning the higher margin, a manufacturer can buy a high level of retailer service for its product with a sufficiently low w. The manufacturer's total cost of achieving this goal can be reduced by adding a direct channel, which diverts some brand loyal customers away from the retailer (hence reducing the number of units impacted by w). Even with this reduction in sales the retailer may still be better off with the reduced w. The direct channel also affords the manufacturer a way to price discriminate between store loyal customers (who never buy direct at any price) and brand loyal customers (who never buy the substitute product), charging the latter a higher price in some cases. The authors provide scenarios in which the manufacturer benefits from opening a direct channel alongside its retail channel, and scenarios in which the direct channel benefits the retailer. However, the question of whether both outcomes can occur simultaneously is not addressed.

Rhee and Park (2000)

Rhee and Park (2000) examine the distribution strategy question for a scenario similar to that in Chiang *et al.* (2003) and Kumar and Ruan (2002).

The manufacturer moves first in choosing the unit wholesale price w to charge the retailer, and the direct sales price p_m if relevant. (The manufacturer is constrained by a no-arbitrage condition that forbids setting w higher than p_m, since otherwise the retailer would obtain its product from the direct channel. The analysis ultimately concludes that the manufacturer should set $p_m = w$ when using the hybrid system, meaning that customers buying direct can do so at exactly the wholesale price.) In addition to choosing its retail price p_r, the retailer provides S dollars worth of service with each unit, where the service is of a form that cannot be delivered to customers through an online interaction. Examples include personalized shopping assistance and convenient refunds/replacements. The direct channel is viewed as providing no service. The same cost is incurred per unit delivered in either channel, and delivery costs are ignored.

Individual end customers, who are each free to purchase up to one unit from their preferred channel, have reservation prices that are uniform on $[0,V]$, where V represents the market potential. A proportion m of the market is "service-sensitive," meaning that these consumers value the retailer service at $t\sqrt{S}$, while the other $(1 - - m)$ derives no value from service; t characterizes the degree of customer heterogeneity. From these assumptions, the induced demand for each channel can be derived as a function of the decisions p_m, p_r, and S.

The main findings are conditions on the parameters (m, t, and V) under which the manufacturer will favor each channel structure. For instance, a hybrid system is preferable when customers are similar across segments in their valuation of retail services, i.e., when m is small or when t^2/V is not too large. (This is at odds with the traditional wisdom that the main objective of a hybrid system is to increase coverage of a heterogeneous market.) On its own the retailer will price too high and serve too few customers; the manufacturer uses the competitive influence of a direct channel to lower retail prices, but still needs the retail channel to reach service-sensitive customers. Hence, this paper argues that a hybrid approach offers a manufacturer an implicit mechanism for exerting control over the retailer's price. This does not necessarily hurt the retailer's profits.

Rhee (2001)

Rhee (2001) generalizes Rhee and Park (2000) by explicitly modeling a competitive impetus for a manufacturer to add a direct channel alongside its established retail channel: to respond to a direct-sales-only challenger (usually Internet-based). A motivating example is Compaq's 1998 introduction of the Compaq.com online store, largely viewed as a response to the likes of Dell Computer.

Manufacturer M sells either exclusively via a retailer (charging the retailer a wholesale price of w per unit) or through a hybrid network (requiring the setting of a direct price p_m). Manufacturer D is an online-only direct marketer, selling its product for p_d. M and D are given identical cost structures, and the model excludes any inventory costs, scale economies in shipping or handling, or distribution costs. As in Rhee and Park (2000), the retailer sets a retail price of p_r and provides S dollars worth of service per unit, while no service is provided in any direct channel. M and D move simultaneously (setting w and p_m, and p_d, respectively), while M's retailer is a follower (in setting p_r and S).

All channels compete for the same set of customers, who are distributed along an interval $[0,l]$. M and D are located at opposite ends of the interval, and customers incur linear travel/search costs to obtain the product. A fraction m of the population values retail service at $t_1\sqrt{S}$ while the remaining $(1--m)$ assigns a valuation of $t_2\sqrt{S}$. An assumption $t_1 > t_2$ renders the first segment more service-sensitive, and the development immediately sets $t_2 = 0$. t_1 is then relabeled as t, which has the same meaning as in Rhee and Park (2000). These assumptions allow computation of the segment of each customer type that goes to each seller under each channel arrangement.

Conditions under which M will favor each arrangement are obtained as a function of t, m, and l. One key finding is that M can increase market coverage (sales) by supplementing its retail distribution with an online store. However, this intensifies the price competition with D, which in turn lowers the wholesale price necessary to appease the retailer, ultimately decreasing M's total profits. The authors offer this as an explanation of the financial ineffectiveness of the Compaq online store initiative, and a rationalization of the decision by manufacturer Levi Strauss to terminate its direct online sales activities.

Hendershott and Zhang (2001)

Hendershott and Zhang (2001) take a distinctive approach to modeling the hybrid channel arrangement in which a manufacturer can sell directly or through intermediaries. In contrast to the other models discussed in this section, the number of firms in their intermediary echelon is not discrete.

The monopolist manufacturer, which has infinite capacity and no operating costs, sets the wholesale price in the intermediated channel. Intermediaries are uniformly distributed on a unit interval, differentiated by their per-item transaction cost k_I. Each intermediary determines first whether to enter. On entering, an intermediary chooses a price p_I to charge and buys an amount equal to the expected demand in each search period. N^I is the equilibrium "number" of intermediaries (actually an interval size) who will be in the market.

Consumers have heterogeneous valuations for the good, represented by willingness to pay v. Each consumer purchases at most one item, and the population has a uniform distribution of v on a unit interval. Consumers have an expectation of the equilibrium distribution of prices $F(p_I)$, but individual intermediaries' prices are revealed only through search. This discovery takes time, delaying the consumer's benefit from purchase. This is modeled by applying a discount factor $\beta \in (0,1)$ to the value of the good. Based on a recursive value function, a consumer can calculate a reservation price r (as a function of v and $F(p_I)$) that characterizes the search policy: conduct search until finding a price lower than r.

Discounting creates a correlation between consumers' values and their search costs, i.e., consumers with higher valuations of the good will have higher reservation prices and hence engage in less search. The manufacturer may also sell in the direct channel at a price of p_M, incurring a transaction cost of k_M (either positive or negative) to provide the same services as the intermediaries. Direct purchases are discounted according to a factor $\beta_M > \beta$. Although search and discounting of value combine to inject time-sensitivity into the consumer's preferences, the sellers of product compete exclusively by setting prices.

Distribution strategy is evaluated not just in terms of the manufacturer's priorities, but also with respect to consumer surplus and social welfare. The analysis provides conditions under which either single-channel option will be used, and when the manufacturer will pursue the hybrid approach. The disintermediated market structure is used only when direct selling is more efficient than any of the intermediaries (i.e., $k_M > 0$). The manufacturer will use the intermediated channel for two reasons. First, the correlation between consumers' values and search costs due to discounting allows price discrimination through a higher direct price. Second, the intermediaries may provide a channel with an advantage in transactions costs. However, the drawbacks are that intermediaries mark up the price, and intermediated sales involve additional search, which delays those sales. Adding a direct channel alongside the intermediated channel increases the manufacturer's demand and profit by attracting the highest-value consumers from the intermediaries. Consumers are also made better off. However, this makes the intermediaries strictly worse off, as fewer of them will exist, markups will be reduced, and demand and profit will both drop. This argues that channel conflict is inevitable in dual-channel systems, although the analysis suggests that enough benefit can be generated to fund side payments that would appease the intermediaries.

A generalization of the basic model considers the case in which n symmetric manufacturers engage in Cournot competition in selling to the intermediaries and directly to consumers. The main conclusion is that the impact of adding direct sales is not sensitive to the assumption of a solitary upstream firm.

Tsay and Agrawal (2001)

Tsay and Agrawal (2001) consider a single manufacturer whose end customer market is sensitive to both price and sales effort. The manufacturer must decide whether to sell through an independent reseller, a direct channel[10], or a hybrid of the two, given each channel's relative supply chain and marketing cost structures, and the tendency of some customers to research the product in one channel but make the purchase in the other. The manufacturer also considers ways to improve the performance of the distribution system, such as revisiting the division of labor in the system.

The manufacturer sets the sales effort and selling price for the direct channel, while the reseller controls these decisions in its channel. Regardless of the choice of channel strategy, the demand in each channel is an increasing function of the sales effort within that channel. When both channels are open, each one's demand is increasing in the effort of the other as well, with magnitudes captured by parameters b_r and b_d. Hence, sales effort exhibits a positive externality across channels.

The reseller's cost of providing sales effort level s is represented as $\eta_r s^2/2$, whereas the manufacturer would have to spend $\eta_d s^2/2$ to achieve the same effect in the direct channel. Two supply chain unit cost parameters, c_d and c_r, are used to distinguish between the production/logistics methods needed to deliver direct to end customers vs. to an intermediary.

In this model, system inefficiency results from two factors. One is double marginalization within the reseller channel. Another is the failure of each channel to fully perceive the positive externality that its sales efforts can have on the other channel.

Contrary to expectation, the addition of a direct channel alongside a reseller channel is not necessarily detrimental to the reseller. In fact, to retain some of the reseller's selling effort the manufacturer will reduce the wholesale price, and in some cases this can make both parties better off. There can be a net system-wide efficiency gain to share because the wholesale price reduction can counteract double marginalization. In general, the desirability to each party of the distribution options depends on how the channels compare in terms of both supply chain efficiency and marketing capability, and none of the distribution strategies examined is universally preferred by either party. In fact, there are circumstances in which the conventional expectation is reversed: the manufacturer favors using only the reseller, but the reseller prefers that the manufacturer open a direct channel in parallel.

The analysis suggests that revisiting the wholesale pricing terms can improve the overall efficiency of a dual-channel system. However, the greatest improvements are realized when the pricing is premised on the reseller's sales effort, which may be difficult or impossible to monitor in practice. Fortunately, certain schemes observed in industry do not have this requirement. These include paying the reseller a commission for diverting all customers toward the direct channel, or conceding the demand fulfillment function entirely to the reseller. Such schemes could in fact be more profitable for both parties in that they achieve a division of labor according to each channel's competitive advantage: customers are *obtained* using the most cost-effective combination of channel efforts (as determined by η_r, η_d, b_r, and b_d) and *served* using the most cost-effective method (which is determined by the relative magnitudes of c_r and c_d).

Peleg and Lee (2002)

Peleg and Lee (2002) is distinctive in that the activities of the different sales channels are staggered in time. There is a primary market attended to by a traditional retail channel. The manufacturer may later emerge as a competitor to the retailers in a secondary market.

The manufacturer produces at unit cost c, and wholesales at price p_1, which is exogenous. Each of n identical retailers faces a stochastic primary demand in an isolated market, and purchases quantity Q from the manufacturer to sell at fixed price r_1. Value can only be recovered from overstocked product by opening a secondary market, which provides access to a different, more price-sensitive market segment. The manufacturer may view this market as an additional revenue source and then intervene by offering additional units for direct sales (of the amount by which initial production exceeds nQ). Transportation and handling costs required to move inventory from primary stores to secondary customers are ignored, so all units in the secondary market are treated equally.

Total demand in the secondary market is $D_2 = n[a-(a/b)p_2]$, where all units will be sold at price p_2. The equilibrium p_2 is the one at which supply and demand are equal. The retailers' concern is the downward price pressure resulting from the manufacturer's injection of additional supply.

Each retailer's overstock is dependent on its random primary demand, so p_2 will also be stochastic. However, for very large n (an assumed property of Internet-based markets), the Law of Large Numbers affords a limiting approximation of total retail overstock, and by extension a tractable form for the limiting value of the equilibrium p_2.

Opening the secondary market will always improve the profitability of the retailers despite the prospect of manufacturer intervention, but might hurt the manufacturer's profitability as well as supply chain performance in the primary market. In addition, it will not always be in the manufacturer's best interest to intervene in the secondary market. This is because the retailers may in turn reduce their initial orders in anticipation of the low secondary market price, and p_2 might end up below c. Compared to the scenario in which the secondary market is theirs exclusively, the retailers are always made worse off by manufacturer intervention,. However, their expected profits will still be higher than in the absence of the secondary market. Therefore there may be cases in which the retailers and manufacturer all benefit from the secondary market.

3.2 Manufacturer-owned channel contains physical stores

Ahn et al. (2002)

Ahn *et al.* (2002) consider competition between independent retailers and manufacturer-owned stores, where the parties compete in selling price and the manufacturer stores are in remote locations (which is common for discount factory outlets). The design of this distribution approach reflects intent to mitigate retailer concerns about channel conflict. (Retail stores tend to be in larger metropolitan locations while discount outlets are typically placed 60-80 miles away, where the real estate may also be cheaper.)

In this model, the manufacturer's product is sold in two spatially separated markets: market 1 contains an independent retailer and is K times the size of market 2, which contains the discount outlet. The manufacturer sets the wholesale price (p_0), the retailer then sets retail price (p_1), and finally the manufacturer sets the outlet price (p_2). Production cost is normalized to zero, and storefront i incurs a marginal cost of c_i.

Each potential consumer in each market has a reservation price that is uniform in [0,1], and incurs a fixed transportation cost T for buying from the other market. A distinctive feature is that price competition is only unidirectional: travel costs and the assumption that retail prices must be higher than outlet prices mean that some customers in the retail market might be willing to travel to get the outlet deal, but not vice versa. Hence the retailer's demand, $D_1(p_1) = K(1--p_1)$, is declared to be independent of p_2. The outlet demand is $D_2(p_1, p_2) = 1 -- p_2 + 1_{\{p_2+T \leq p_1\}} \cdot [K(p_1 - p_2 - T)]$. The analysis focuses on the manufacturer's perspective, in particular the usage of the channel options.

The manufacturer is found to have three equilibrium strategies: (1) "elimination" (set the wholesale price so high that the retailer decides not to carry the good), (2) "monopoly" (keep both channels, but price in the outlet so that no consumers cross regions, preserving a local monopoly for each store), and (3) "competitive" (price so that both channels carry the product, and some retail consumers cross regions to buy the from factory store). As for when each will occur, the authors note that as the transportation cost T is lowered, the monopoly strategy collapses. If the retailer's marginal cost c_1 is also sufficiently high, the manufacturer will choose to eliminate the retailer. Hence, as the locational advantage of the retailer diminishes, a retailer with higher marginal operating costs becomes more vulnerable. Conversely, as T or c_2 becomes larger, diminishing the manufacturer's cost advantage for selling through its outlet, a monopoly strategy becomes more likely. As long as c_1 is sufficiently low, the retailer will be used as a means of price discrimination: region 1 customers who are willing to pay a higher price will in fact be charged a higher price. The authors also consider a more general case that acknowledges the retailer's marketing contribution: when the retailer is removed, market 2 demand K is replaced by $K\alpha$ for some $\alpha < 1$ when the retailer is removed. α is an exogenous parameter, in contrast to the "effort" or "service" retailer decision variables seen in some papers. The analysis draws no conclusions about how the retailer behaves differently (i.e., pricing) when the manufacturer opens the outlet channel, hence is silent on the issue of channel conflict.

Bell et al. (2002)

Bell *et al.* (2002) propose the term "partial forward integration" to describe a manufacturer's use of a company-owned channel (involving physical storefronts) alongside a retail channel. In contrast to Ahn *et al.* (2002), they focus on the specific case where the company store and the retailers are colocated, such as in the same mall. (They also apply their findings to the case in which all stores are on the Internet, hence again are in the same "location.") For example, Polo Ralph Lauren, DKNY, Liz Claiborne, and Armani are among a number of apparel manufacturers who operate company stores in malls where independent retailers such as Macy's and Nordstrom carry these brands. A key research objective is to determine why a manufacturer would forward integrate when adding a company store would not extend total market coverage geographically. The authors note that this might be a way for the manufacturer to reach a different customer segment (where segmentation is along a dimension other than location, such as "brand loyalty"), to benchmark the performance of individual retailers, or to invest in marketing effort to provide price support for other independent retailers. The latter motive is the focus of the investigation.

The manufacturer sets a wholesale price w, and has a marginal cost of production c (normalized to zero). Each of n stores (indexed by s) affects demand via price p_s and "marketing effort" e_s (which captures retail value-add through activities such as salesperson support, retailer advertising, and in-store displays). The cost of providing effort is quadratic: $C(e) = e^2$. r_s is the unit retailing cost for independent retailer s, while the manufacturer incurs a unit retailing cost of r_m in its own store.

The demand faced by each of n stores is a linear function of all prices and effort levels, with the form

$q_s = 1 - p_s + [\theta/(n-1)] \Sigma_{s' \neq s} (p_{s'} - p_s) + e_s + [\beta/(n-1)] \Sigma_{s' \neq s} e_s$

θ is own-price responsiveness, and β measures the spillover effect. The latter acknowledges that a customer can benefit from a retailer's effort without necessarily purchasing from that retailer. This establishes effort as a "public good," with the accompanying potential for free-riding.

The base case has the manufacturer selling through three independent retailers, where the manufacturer first sets w and then the retailers maximize their individual profits by choosing their p_s and e_s. In the partially integrated case, the manufacturer also sets price and effort in the company store before the two independent retailers make their decisions simultaneously. The authors normalize r_s to zero for the independent retailers, while setting r_m to whatever value will lead to equal market share for all three stores in equilibrium. The rationale for this is to rule out the market share motive for opening a company store, and the bulk of the results are premised on this condition. The positive value of r_m also gives the independent retailers an efficiency advantage that justifies their existence.

Under the equal-market-share condition, the company store charges the highest price, followed by an independent retailer competing against the company store, and then the independent retailers in the base case with no company stores. The same ordering applies to marketing effort. So, left on their own the retailers will under-invest in marketing due to horizontal free riding. The marketing effort of the company store reduces the emphasis on price-based competition. Hence, the partial integration allows the manufacturer a form of retail price maintenance, except without relying on explicit arrangements that might violate antitrust laws. This finding may contradict a retailer's concerns about being undercut in price, because the channel that ostensibly enjoys an advantage in wholesale cost also contributes effort. The positive externality that drives customers to the retail stores is what creates the possibility that retailers can benefit from the competition presented by a direct channel.

The basic results are shown to extend to other cases, such as when the company store is a new addition rather than a replacement for an existing independent retailer, when the total demand does not increase on adding new stores, and when the company store has a base demand that differs from those of the retailers.

Boyaci (2001)

In Boyaci (2001), a manufacturer sells through a self-owned distribution channel as well as a competing retail channel[11]. This model is distinctive in considering stochastic demand, in which inventory is the basis of interaction between channels.

The manufacturer produces at unit cost c, and the retailer purchases at wholesale price $w \geq c$. All sales occur at the same exogenous price r. Random variables D_m and D_r describe "first-choice demand" in the direct and retailer channels, respectively. A fraction α of customers who encounter a stockout in their preferred channel will subsequently search the other channel before walking away. This dynamic is what induces interdependence between the two channels. The resulting objective functions have newsvendor form (with product substitution) and the two firms simultaneously choose order-up-to levels. The analysis proceeds to identify the equilibrium stocking levels and profits of both parties for any given w. w is treated not as a decision variable, but as a parameter for sensitivity analysis.

A combination of analytical and numerical investigations indicates that both channels will tend to overstock due to the channel interaction, and also examines the impact of changes in w and α. However, the parties' preferences for channel structure are not studied. Channel coordination is considered, with attention paid to the coordinating properties of various contracts: price-only, buyback, vendor-managed inventory (in which the manufacturer controls the retailer stock level and the retailer chooses the wholesale price), penalty (where the retailer pays a unit penalty for missed sales), and target rebate (where the manufacturer offers the retailer a rebate for every unit by which sales exceed a target level). Price-only, buyback, and vendor-managed inventory contracts are determined to be incapable of coordinating the system, as they are too simple to overcome the simultaneous influence of horizontal and vertical distortion of incentives. When designed appropriately, the penalty and target rebate forms have coordinating properties, although they face some implementation barriers.

3.3 Discussion

Traditional reasons for a manufacturer to install a captive channel alongside an independent retail channel include reaching a different market segment and achieving price discrimination. These play key roles in several of the papers described in this section, including Ahn *et al.* (2001), Rhee (2001), Hendershott and Zhang (2001), Kumar and Ruan (2002), and Peleg and Lee (2002).

A number of the models discover that the manufacturer-owner channel can offer an indirect means to influence the retailer's behavior (such as price) where more explicit control might be outlawed. However, there is some disagreement about the direction of pressure on the retail price. Bell *et al.* (2002) find "partial forward integration" to provide price support, while Rhee and Park (2000) and Chiang *et al.* (2003) see the direct channel as a way to keep prices low by combating double marginalization. The root explanation for this divergence is not obvious due to substantial differences in the assumptions of the three models. However, it is worth noting that the formulation of Bell *et al.* (2002) includes marketing effort that has positive spillover effects on all channels, while the service described in Rhee and Park (2000) has no externality across channels and Chiang *et al.* (2003) consider no non-price control variables at all.

Another recurring conclusion is that all firms involved can conceivably prefer a hybrid system to any single-channel alternative. A rationale offered by Tsay and Agrawal (2001) is that sales effort in a direct channel can be used to drive traffic to whichever channel is most efficient at fulfillment, possibly creating enough gains to share. This would suggest that retailer concerns about channel conflict might be unfounded. Unfortunately, this finding is not robust across the different model formulations, as summarized in Table 13.1.

A potentially significant caveat arises from the progression from Rhee and Park (2000) to Rhee (2001). Rhee and Park (2000) report that adding a direct channel alongside a retail channel can increase the manufacturer's profit. However, Rhee (2001) rules this out when the manufacturer faces competition from another direct seller, as the challenger weakens the manufacturer's pricing power. This raises the possibility that the research findings may depend on the manufacturer being a monopolist. Such an assumption is made in all these papers with the exception of Hendershott and Zhang (2001), whose qualitative conclusions are equally true of a symmetric set of manufacturers engaged in Cournot competition. This is a substantive source of controversy that invites further research.

4. Research Opportunities

A number of factors, including recent developments in Internet-based commerce, information and material handling technologies, and the transformation of shipping economics driven by the growth of the third-party logistics industry, have led many manufacturers to establish their own channels for reaching end customers. This may put such companies in direct competition with their existing reseller partners. The potential channel conflict has momentous implications for distribution strategy. In this chapter, we have reviewed the model-based literature on this topic.

Our review has led us to conclude that there are significant opportunities to enhance the extant literature. Although some may be particularly challenging, we state a number of these here with hopes of inspiring the community of researchers.

We provide recommendations along two broad categories. Section 4.1 discusses some limitations in existing model representations of the various channel types, especially in capturing the differences between online and bricks-and-mortar channels. Some perspectives on the analytical appraisal of alternative channel strategies are offered in Section 4.2. Section 4.3 presents our closing remarks.

4.1 Representing channel characteristics

To adequately represent disparate channel types within a single model requires capturing the ways in which they differ. These include differences in the "bundle of attributes" the customers actually get when they buy, the cost impact to the channel of satisfying demand, and the methods used to structure the terms of each sale. Also, the customers' preferences regarding any channel differences must be appropriately comprehended. The existing literature has approached these in a fairly parsimonious way (as mentioned in Section 3), yet the analysis has already been found to be challenging.

Much has been written about the diversity in possible channel types (cf. Stern *et al.* 1996), with the distinctive attributes of online channels receiving special attention in the last few years (e.g., Alba *et al.* 1997, Peterson *et al.* 1997, Kenney 1999). These bodies of work provide much more detail than we can or should replicate here. Instead, we will briefly mention a few of the more salient issues, and comment on the implications of incorporating these into analytical models of multi-channel distribution systems.

4.1.1   Pricing.   Existing models tend to assume that all channels take the same, fairly traditional approach to the pricing dimension of the transaction: (1) each seller posts a price; (2) potential customers decide whether to buy the product at all, and if so, from which seller. For tractability, most models consider just a single time period, so that every seller controls only a single price (if price is a control variable at all) and all units sold by each seller are at that same price. However, this paradigm might not adequately represent how new channel types approach pricing. For example:

- When comparing direct and bricks-and-mortar channels, differences in pricing strategies may be revealed by modeling the individual components that form the buyer's total out-of-pocket expense. One component of this is shipping and handling (S&H). Some sellers believe that consumers might not evaluate S&H fees as critically (or even at all) when comparing prices. Also, many comparison-shopping services on the Internet report selling price only, and do not handle S&H costs well since these vary by customer location. As a result, some direct sellers may price aggressively to attract customer attention while relying on S&H as a profit center. Some also use the S&H schedule to implement a volume discount, a simple example of which is offering free shipping with a minimum purchase. In comparison, bricks-and-mortar sellers lack this additional instrument for collecting money from customers and influencing buyer behavior. (Many of the same effects can certainly be achieved through other types of price promotions or policies, some of which may not enjoy the same cultural acceptance as S&H fees.)

- Dynamic pricing becomes much more feasible when selling online, as there are no physical price tags or display labels to manually update. Dell Computer is known to alter its prices on a frequent basis to balance demand and supply (McWilliams 2001).

- An online channel's ability to identify individual customers raises the prospect of personalized pricing, although public acceptance of the practice remains an obstacle[12].

- Auctions have always provided sellers a way to pursue the highest possible price for an item, and perhaps to offer multiples of the same product at different prices (if auctioned separately). The growing reach and market acceptance of online auction-enablers such as eBay have led an increasing number of firms to consider adding an auction-based channel for regular sales (as opposed to the occasional liquidation of distressed inventory[13]) and for reaching individuals (as opposed to other firms) (Brown 2002).

The Internet has enabled sellers to make innovative use of traditional instruments such as volume pricing. Mercata.com and others enabled unaffiliated individuals to spontaneously pool their buying power to become eligible for the bulk discounts[14]. Anand and Aron (2002) evaluate this selling approach in a context without intermediaries or multiple channels.

Modeling these types of pricing strategies would most likely require some combination of more time periods, more price variables and all the ancillary demand elasticities and cross-elasticities, and more detailed models of consumer behavior and preferences.

4.1.2 Non-financial aspects of the purchase.

A purchase is more than an exchange of dollars for product, and different channels can provide distinctive experiences for the buyer even if selling the same physical product at the same price. Many facets of this issue have been studied extensively in the marketing and consumer behavior literatures, but not in the competitive context of a multi-channel model. Two commonly mentioned examples are:

- A number of experiential distinctions are especially pronounced when comparing direct and bricks-and-mortar channels. These include visual and tactile interaction with the product[15], the method of obtaining product information (e.g., face to face, over a phone line, or through a computer screen), the delay between purchase and receipt, and even the crowds that often define bricks-and-mortar shopping (which can either be a positive or a negative depending on individual preferences). Naturally, these types of qualitative factors have been difficult to model in any detail.

- Ease of returning product is a major consideration when customers decide how to buy. Bricks-and-mortar channels enjoy an advantage in this category, as most virtual channels do not reimburse return freight charges except for defective products (and even then, the customer must still endure the hassle of shipping and then ensuring the proper crediting of a refund). Existing models offer guidance on how a seller might design a customer return policy to accommodate legitimate customer concerns while guarding against opportunistic behavior (e.g., Davis et al. 1995, Hess et al. 1996, Chu et al. 1998, Davis et al. 1998). However, they do not consider the return policy in a competitive context.

Modeling the competitive consequences of non-financial dimensions can be challenging. Unlike a seller's price, which ultimately is only relevant to either the buyer or seller when a purchase is made from that seller, some of the non-price factors can be decoupled from the actual purchase of the item. That is, there are cases in which one channel delivers value to consumers but is uncompensated for this because the consumers ultimately spend their money in a different channel. An obvious example is when a consumer researches a product using the resources of a retail store, but makes the actual purchase online. On the other hand, a direct seller may invest heavily in generating demand for the product only to have customers make the purchase at a local store (usually for immediate gratification and convenient return privileges)[16]. Either way, a firm is choosing investment levels in actions that may influence collective demand but which might not provide an appropriate individual payback. This is a major contributor to concerns about channel conflict. It is not clear when the net externality of such activities will be positive (i.e., some new customers also spill over to the other channel) and hence subject to free-riding, or negative (i.e., cannibalizing) in which case the non-price factor is more a competitive weapon like price. Hence, both possibilities need to be modeled for adequate realism[17]. At the very least this suggests the limitations of considering just one type of "service" or "sales effort" as many models do. Existing research has not done an adequate job of handling both in the same model, likely because of the additional dimensionality this would entail.

4.1.3 Non-price product attributes.
The prevailing modeling framework has competing channels selling the same product, or at least closely substitutable variants. In reality, key attributes of the product may depend significantly on the channel type.

A direct channel may be selling the same good as a bricks-and-mortar channel, but in a different form. This is especially likely for information goods such as software or music, which are purchased on tangible storage media at retail stores but may be downloaded in purely digital form online[18]. Or, a direct channel and a bricks-and-mortar channel may be selling fundamentally different products. Customization is often a key dimension of the value proposition. By self-selection, customers of direct channels are willing to tolerate some delay in obtaining the product. This creates the prospect of shifting from make-to-stock mode (which dominates traditional retail) to make-to-order mode. This option has always existed for all forms of mail-order, but the Internet has provided an automated and efficient mechanism for extracting and even influencing customer preferences, and communicating these from the customer to the factory. Furthermore, recent manufacturing and supply chain process innovations have facilitated time compression and efficiency in production and logistics. Together these have given direct sellers a viable way to meet individual customer preferences better than traditional retailers can, and at reasonable prices. Dell Computer is the most prominent practitioner of this business model.

When selling the same product in different forms or fundamentally different products, the competing channels may exhibit sharply divergent economic properties, in areas such as inventory management. Customer preferences are also likely to vary substantially across the channels.

4.1.4 Product assortment.

Assortment planning is a fundamental core competence and competitive differentiator in traditional retail. The decision process attempts to take into account simultaneously a multitude of factors, some more quantifiable than others (cf. Smith and Agrawal 2000, Mahajan and van Ryzin 2001, Rajaram and Tang 2001). As a result, it may be reasonable for a given retailer's stores in two neighboring shopping centers to carry strikingly different product offerings. (Whether the stores actually do so is a management decision.) The topic is far too complex to address here, so we will simply highlight some factors that might lead different channels to carry different assortments.

- Manufacturers, selling either direct or through company stores, are generally not likely to offer as broad an assortment as retailers (although a manufacturer might offer a much greater selection within its own brands, and might offer new products earlier). This is mostly attributable to the differing business objectives of manufacturers and retailers.

- Products whose attributes are difficult to convey virtually[19] (hence have more of an experiential requirement, as described earlier) are more likely to be included in bricks-and-mortar assortments.

- Shipping costs and constraints are another influence. Home Depot's website states, "Due to UPS shipping restrictions, not all products that are available in our stores can be delivered. Large items and heavy items are the most notable exceptions. For our full selection of products, please visit your local store."

- Product sales volume can shape assortment strategy, although occasionally in contrasting ways. Perhaps seeking the advantages of focus, Macy's only offers its store bestsellers through Macys.com. Alternatively, a direct channel is unconstrained by the need to carry inventory for presentation purposes and also can pool demand for each SKU across a larger set of customers, so can more feasibly offer less-popular items such as extreme sizes of clothing. This is true of the size selection offered online by The Gap Inc.

- Just as with pricing, selling online allows dynamic updating of the assortment to achieve a better match between supply and demand. Again, Dell is known to alter its website offerings frequently. A related strategy is to customize the assortment presented to each individual based on past purchase behavior or stated preferences.

Product assortment is a means by which channel members can signal their concerns about channel conflict. A retailer may express dissatisfaction with a manufacturer's direct sales efforts by adding competing products, dropping that manufacturer's product altogether, or at least positioning the product in a less favorable context (e.g., by reducing the number of sizes/colors/styles carried, or discontinuing other items that are natural complements). To avoid such situations, some manufacturers will create entirely different product lines (or at least project that illusion through product names and model numbers) for the direct channel to obfuscate price comparisons or to target different segments, or may release products to different channels in phases.

Studying assortment decisions fundamentally requires a multi-product perspective that includes substitution effects. This has posed difficulties even in a traditional retail context without incorporating competition. As such, product line strategy for multi-channel settings presents a very open research opportunity.

4.1.5 Operational costs.
Alternative channels often represent fundamentally distinct operating models. The standard approach of using a single, exogenous unit-cost parameter per channel may obscure salient economic factors such as:

- Different channels may diverge in the fixed costs required for creation and ongoing operation.

- A number of costs are likely to be nonlinear. These include the costs of providing customer service or sales effort, and a seller's S&H expenditures.

- The costs experienced in different channels are not independent. For instance, in reality there are often scale economies in shipping goods to stores. In such a case, any reduction in volume (such as due to the diversion of some demand to a direct channel) will increase the average cost per unit shipped to the stores.

These issues can likely be represented in a mathematically straightforward way, but as always the challenge is tractability of analysis. We note that existing multi-channel models have inherited many of the cost-related simplifications that are common in models of a single channel or even a single firm.

4.1.6 Demand uncertainty. One motive for using a direct channel is to get better visibility into end-customer demand levels and/or product preferences. In a multi-channel strategy this could provide a vehicle for evaluating the quality of judgment and level of honesty exhibited by a retailer's forecasting and ordering behavior. However, to model this would require that a retailer have more information than a manufacturer who sells only through that retailer.

Most of the works that model competitive interactions between channel members assume entirely deterministic environments, so that improved information has no value whatsoever. Among models that include uncertainty, the vast majority assumes common knowledge, in which case the channel choice does not affect the manufacturer's state of information. Experience thus far indicates that asymmetric information is difficult to model even in dyadic models with an extremely stylized representation of uncertainty.

4.1.7 Alternative types of competition between channels.
Existing research has focused on conflict that arises when different channels compete for demand, primarily via price. However, retailers are increasingly able to detect when a manufacturer undercuts them, especially online. There is enough awareness of this irritant that many manufacturers deliberately price above the prevailing retail "street price" (cf. Machlis 1998(a,b), Bannon 2000). Yet channel conflict can still arise from a perception that a manufacturer might be favoring its own channels when allocating scarce products. Independent retailers carrying Apple Computer's hot-selling new iMac computer made this accusation in early 2002. Even though prices were the same across channels, the product seemed to be more readily available through Apple's own online and bricks-and-mortar stores (Wilcox 2002). Modeling this phenomenon may require explicit consideration of inventory, as well as allocation practices (which are often based on factors that can be difficult to quantify, such as gaining goodwill with strategically important partners).

4.1.8 Alternative types of multi-channel strategies. The basic distinctions among the channel types covered are (1) intermediated vs. manufacturer-owned; and (2) online vs. bricks-and-mortar. Multi-channel models typically represent some combination of the following possibilities:

a direct sales channel (usually Internet-based), a company store channel, or a channel containing a single intermediary level (usually bricks-and-mortar). However, current environments are characterized by a proliferation of distribution network types that are more complex, even ignoring the multitude of nuances mentioned previously.

One development is that many bricks-and-mortar retailers are adding their own online channels (Tessler 2000). So a manufacturer that chooses to distribute only through, say, Walmart may yet obtain an Internet sales presence (albeit still an intermediated one) through Walmart.com. Here any channel conflict would be between Walmart's stores and online divisions[20]. More generally, a modern manufacturer may simultaneously manage a variety of its own channels (e.g., online and company stores), sell through independent bricks-and-mortar retailers that also have online channels (e.g., Walmart, Macy's, Costco, etc.), and also sell through independent online retailers (e.g., Amazon.com).

Revisiting the division of labor of the channel functions can form new channel types. One approach is to "unbundle" a product offering, and then assign different channels exclusive rights to the various components. For example, the PC industry has migrated towards a model of integrated distribution, where manufacturers and their channel partners (such as value-added resellers, distributors, etc.) combine their strengths to provide a total solution composed of products and services. In Compaq's recent PartnerDirect program, computing products were ordered directly from Compaq but integration and complex configuration support were left to third-parties such as Tech Data Corp. and Ingram Micro Inc. (Zarley and Darrow 1999). Dong and Lee (2002) model a variety of intermediated channel structures observed in the PC industry.

Practices such as drop-shipping (Netessine and Rudi 2001(a,b)) separate the selling task from the physical or financial ownership of material, so that the seller becomes essentially an order-taker. For example, a book ordered from Amazon.com might travel directly from the book distributor to the end customer, with activity initiated only after Amazon has assured payment from the customer's credit card company. Monitors in Dell's orders are shipped directly from Sony's warehouses to the end customer without ever passing through Dell-owned facilities. A related development is exemplified by the option for Amazon.com's customers to pick up certain purchases (and make returns later if necessary) at a local Circuit City store, which was introduced in late 2001.

The dissolution of traditional divisions of labor raises new questions for managers and researchers. Which party should perform each channel function? How should the individual parties be compensated for their roles? (cf. Frazier 1999) A poorly designed strategy could suffer from even more channel conflict than its predecessor.

4.2 Evaluating distribution strategies

Section 4.1 outlined some channel characteristics and activities that might merit additional attention. Independently of whether these are modeled, there may be some need to revisit the approach to evaluating and comparing channel strategies. These reflect the assumptions about how the firms make decisions and interact.

4.2.1 Company objectives.

The notion of an optimal strategy is obviously sensitive to the choice of objective function. Like most economic models, the multi-channel literature has tended to assume that all firms seek to maximize some form of (expected) gross profit. While this may be the ultimate aim for most companies, certain legitimate channel strategies are not tied directly to this, at least not in the short run.

As is true of other areas of management, some channel decisions appear to trade immediate profit for some less tangible goal. A manufacturer may sell direct out of a desire to maintain "ownership" of the relationship with the customer, so that the customer is loyal to the product brand rather than to the retailer. Similarly, manufacturers may add direct channels primarily for advertising or to educate the market, even at an operating loss. In response to the initial hype about eCommerce, some firms quickly created Internet stores as a way to learn about online sales and to avoid being left behind. Some manufacturers simply distrust intermediaries or any other outside parties, perhaps due to difficulties in monitoring and controlling their behavior.

There may also be a time dimension to a company's goals, and hence the appropriate distribution strategy. An evolution may occur over a product's life cycle[21]. Early on, the prime concern might be market share rather than profit. Major retailers could be used at this stage for their ability to provide product exposure. As the product becomes established, the manufacturer might grow less interested in sharing the profit margin with an intermediary, so might sell exclusively direct or in a multi-channel arrangement (channel conflict notwithstanding). Toward the product's end of life, retailers might no longer be interested in carrying the product, leaving an exclusive direct channel as an option. Certain types of products, such as replacement parts, might be sold direct for years at a loss as a manufacturer's form of customer service. A similar evolution might occur over the life cycle of the manufacturer rather than the product. In either case, a major modeling challenge is the inherently multi-period nature of the setting.

Especially in turbulent environments, a multi-channel strategy can be a form of diversification by which a manufacturer protects itself against the failure of any individual channel (Anderson *et al.* 1997, Balasubramanian and Peterson 2000). This argues for an objective function that incorporates a tradeoff between profit and risk.

4.2.2 Channel power structure. To fall within the scope of this review, a model must represent the perspectives of multiple independent decision makers. The selection of the distribution strategy obviously depends on which party is making the decision, which is a statement of relative power. Indeed, this will determine how channel conflict will be ultimately resolved. Assumptions about the relative power of the parties are also embedded in the decision structure of the model.

The majority of the literature puts the manufacturer in a position of dominance, proposing the channel structure and taking Stackelberg leadership for various price and non-price decisions. Although the retailer might prefer some structures to others, it is almost always compelled to accept any arrangement under which a non-negative (expected) profit is attainable. In some cases this assumed allocation of power might be purely an artifact of the manufacturer being the common link between the two channels used, and is challenged by some empirical evidence (cf. Anderson *et al.* 1997). Shaffer (2001) argues that retailers are gaining power, and that bargaining models are necessary to see what channel contracts will result and how the gains will be split.

4.3 Concluding remarks

Existing literature on modeling coordination and conflict in multi-channel distribution systems has just begun to address some key issues, but opportunities for further research abound. Especially timely questions surround the use of nontraditional ways of selling and creative divisions of labor that have been enabled by information technologies. We hope that our review will provide a starting point for researchers who wish to contribute to this body of literature.

Acknowledgments

We would like to thank the following for feedback on versions of this manuscript: Krishnan Anand, David Bell, Kevin Chiang, Chan Choi, Vinayak Deshpande, Hau Lee, Serguei Netessine, Barchi Peleg, Greg Shaffer.

Notes

1. In general, this decision entails a determination of the number of levels in the distribution network, the number of outlets within each level, and other variables such as pricing, inventory levels, service levels, etc. The traditional marketing literature refers to these as *distribution strategy, distribution intensity,* and *distribution management*, respectively (Corstjens and Doyle 1979).

2. In this chapter, intermediaries should be understood to be largely independent of the manufacturers, as issues of conflict are only relevant when intermediaries are motivated by their own objectives. Intermediaries take various names depending on the context, including reseller, wholesaler, distributor, retailer etc. Our focus will be restricted to intermediaries that play a sales role, as opposed to, say, serving purely as a communication channel.

3. The use of print catalog or Internet sales does not necessarily constitute a direct channel, as in many cases this still relies on reseller intermediation. An obvious example is any manufacturer whose product is carried by Amazon.com.

4. In this chapter, the discussion surrounding this term applies to any "virtual" mode of distribution in which the seller provides no means for face-to-face contact with customers. In this context, print catalog and Internet sales are merely variants that differ in the medium through which customers obtain product information and communicate their desires. Indeed, increasingly rare is the virtual seller that does not accommodate nearly every available communication option. However, in describing the literature we will use the specific terminology used by the researchers, who in most cases draw motivation from Internet-based scenarios.

5. In marketing and other related disciplines, the term "channel conflict" has long been used to describe any tensions within or across channels (cf. Stern *et al.* 1996). Our usage is more specific than this.

6. There are two classes of Robinson-Patman violations (cf. American Bar Association Antitrust Section (1992)). One is termed "primary line price discrimination." An example is when manufacturer 1 claims that manufacturer 2 offered discriminatory prices to retailers. For this suit to succeed, manufacturer 1 must prove that it has been harmed and that competition has been injured. This generally entails proving that manufacturer 2's prices are predatory and below-cost (most courts apply a marginal-cost test). This form of Robinson-Patman violation is not relevant with a single manufacturer. The second class is "secondary line price discrimination." An example of this would be retailer 1 suing a manufacturer for giving a lower price to retailer 2. For this suit to succeed, retailer 1 must prove it has been harmed and competition has been injured. The latter can only occur when the retailers compete in the same market.

7. We note the possibility that the interaction across the independently managed channels has no basis for conflict. For instance, Seifert and Thonemann (2001) and Seifert *et al.* (2002) propose transshipping overstocked retail inventory to cover stockouts in the manufacturer's direct channel. But because their direct channel is assumed to serve a market entirely unavailable to the independent retailers, the retailers have no reason not to cooperate with this proposal as long as the shipping and handling costs are reimbursed.

8. From this point forward we will follow the general convention in the analytical literature of referring to a resale intermediary as "retailer" even if certain individual papers might not use this language. This is purely for clarity. Using the more general terms of "intermediary" or "reseller" is problematic when the need arises to describe multiple ones. For instance, a manufacturer that uses "multiple intermediaries" could be reselling through several in parallel within a single echelon, a series of echelons, or a combination. The term "retailer" avoids this ambiguity because of its terminal nature. Also, in many model formulations the business issues used to characterize the intermediary are evocative of traditional retail.

This restriction to two levels does rule out some plausible multi-channel scenarios. For example, the manufacturer may start selling through wholesalers who in turn sell to existing retailers, thereby creating a new channel with three echelons (manufacturer-wholesaler-retailer) instead of only two (manufacturer-retailer). In the analytical literature we have reviewed, this option has been considered only in cases in which the entire network was under the full control of the manufacturer, thereby preventing any channel conflict (cf. Cohen *et al.* 1995).

Note that attention is focused squarely on a retailer's function as a resale intermediary, whereas some of the horizontal competition frameworks treat retailers as the primary source of product (by excluding the retailers' procurement activities from the model scope). Such models blur the distinction between the terms "manufacturer" and "retailer."

9. A frequent complaint from bricks-and-mortar sellers is that even with parity in selling prices, their competitors selling from out-of-state are perceived to be cheaper due to the current practices around the collection of sales tax.

10. This model does not restrict the manufacturer-owned channel to be either direct sales or a manufacturer outlet. We have included this paper in the direct sales section because the formulation allows that channel to have an operating cost structure that differs substantively from that of the reseller channel.

11. Although the type of manufacturer channel is not explicitly stated, the similarity in the structural treatment of the two channels (identical selling price and same customer behavior) is more suggestive of a manufacturer outlet rather than direct sales.

12. Amazon.com terminated its Fall 2000 experiment with this strategy due to customer complaints after the practice was publicized (Rosencrance 2000).

13. The secondary market of Peleg and Lee (2002) is described as an auction, but is used only for the disposal of surplus. The modeled mechanism results in all units being sold at the same price to all interested buyers.

14. Mercata.com invited any online visitor to join a buying group for a particular product, with the price of the product dropping with each new member. The final price would be known only at the end of the joining period, but that price would be available to all group members regardless of when they joined. This was premised to accelerate demand creation since potential buyers would join even if the current price exceeded their maximum willingness to pay, as long as they could expect additional price declines. Such collective behavior would in turn make the desired price a self-fulfilling prophecy. Moreover, this method gives group members incentive to recruit additional buyers from their personal networks, effectively turning customers into a volunteer sales force. (The notion of customers providing sales effort, which is an example of what some term "viral marketing," is certainly outside the scope of any model of which we are aware.) Mercata.com ceased operations in January 2001, in part because manufacturers were reluctant to sell this way for fear of angering their traditional retail partners (Fowler 2002). This does not necessarily invalidate the sales model, but certainly underscores our thesis about the importance of managing channel conflict.

15. Goods are sometimes categorized into "search" and "experience" goods. Features of a search good can be evaluated from externally provided information, whereas experience goods need to be personally inspected (cf. Peterson et al. 1997).

16. This is an issue when the retailer is independent. However, sellers that control all their channels can exploit this concern by allowing direct purchases to be returned at the physical stores. The Gap Inc. is one firm using this approach.

17. A first step might be to categorize service or effort activities into "transactional" and "informational" types. Transactional service describes those aspects that are not meaningful to a customer who does not complete the purchase transaction in that channel. Examples include a fast checkout process or a lenient return policy. Informational service, such as activities that educate customers about the brand or product, can be "consumed" separately from the product. Informational service is obviously the type that is susceptible to free riding.

18. Reinhardt and Levesque (2001) discuss a product's level of "intangibility," which measures the feasibility of not only selling, but also distributing the product online.

19. For instance, a piece of clothing might be purchased based on the fit, the feel of the fabric, and the look of the color in natural light.

20. Being part of the same firm does not preclude conflict, as the channels may be controlled by divisions with individual incentives. Exacerbating this possibility, recently some online divisions have even been spun off into independent corporate entities to create a different culture and perhaps an opportunity to go to the capital markets for funding (e.g., Walmart.com, BarnesandNoble.com, Bluelight.com). However, many of these are in the process of being reabsorbed into the parent company.

21. As noted in Section 2.1, Lafontaine and Kaufmann (1994) examine the evolution of franchising strategy over a firm's life cycle.

References

Ahn, H., I. Duenyas, and R.Q. Zhang, "Price Competition Between Independent Retailers And Manufacturer-Owned Stores," Working Paper, University of California at Berkeley, 2002.

Alba, J., J. Lynch, B. Weitz, C. Janiszewski, R. Lutz, A. Sawyer, and S. Wood, "Interactive Home Shopping: Consumer, Retailer, And Manufacturing Incentives To Participate In Electronic Marketplaces," *Journal of Marketing*, **61**, July (1997), 38-53.

American Bar Association Antitrust Section, *Antitrust Law Developments, 3rd Edition* (1992), 401-450.

Anand, K.S. and R. Aron, "Group-Buying On The Web: A Comparison of Price Discovery Mechanisms," Working Paper, OPIM Department, The Wharton School, University of Pennsylvania, 2002.

Anderson, E.A., G.S. Day, and V.K. Rangan, "Strategic Channel Design," *Sloan Management Review*, **38**, Summer (1997), 59-69.

Artle, R. and S. Berglund, "A Note On Manufacturers' Choice Of Distribution Channels," *Management Science*, **5**, 2 (1959), 460-471.

Axsater, S., "Continuous Review Policies For Multi-Level Inventory Systems With Stochastic Demand," in S.C. Graves, A.H.G. Rinnooy Kan, and P.H. Zipkin (Eds.), *Handbooks in Operations Research and Management Science*, Volume 4 (*Logistics of Production and Inventory*), Elsevier Science Publishing Company B.V., Amsterdam, The Netherlands 1993, 175-197.

Balasubramanian, S., "Mail Versus Mall: A Strategic Analysis Of Competition Between Direct Marketers And Conventional Retailers," *Marketing Science*, **17**, 3 (1998), 181-195.

Balasubramanian, S. and R.A. Peterson, "Channel Portfolio Management: Rationale, Implications And Implementation," Working Paper, Department of Marketing, University of Texas at Austin, 2000.

Balasubramanian, S., R.A. Peterson, and S. Jarvenpaa, "Exploring The Implications Of M-Commerce For Markets And Marketing," *Journal of the Academy of Marketing Science*, forthcoming, 2002.

Balderston, F.E., "Communication Networks In Intermediate Markets," *Management Science*, **4**, 2 (1958), 154-171.

Bali, V., S. Callander, K. Chen, and J. Ledyard, "Contracting Between A Retailer And A Supplier," Working Paper, Michigan State University, 2001.

Baligh, H.H. and L.E. Richartz, "An Analysis Of Vertical Market Structures," *Management Science*, **10**, 4 (1964), 667-689.

Bannon, L., "Selling Barbie Online May Pit Mattel Vs. Stores," *Wall Street Journal*, November 17 (2000), B1.

Bell, D.R., Y. Wang, and V. Padmanabhan, "An Explanation For Partial Forward Integration: Why Manufacturers Become Marketers," Working Paper, The Wharton School, University of Pennsylvania, 2002.

Bernstein, F. and A. Federgruen, "A General Equilibrium Model For Decentralized Supply Chains With Price And Service Competition," Working Paper, Fuqua School, Duke University, 2001.

Blumenfield, D.E, L.D. Burns, J.D. Diltz, and C.F. Deganzo, "Analyzing Tradeoffs Between Transportation, Production And Inventory Costs On Freight Networks," *Transportation Research*, **19b** (1985), 361-380.

Boyaci, T., "Manufacturer-Retailer Competition And Coordination In A Dual Distribution System," Working Paper, McGill University, 2001.

Brickley, J.A. and F.H. Dark, "The Choice Of Organizational Form: The Case Of Franchising," *Journal of Financial Economics*, **18** (1987), 401-420.

Brickley, J.A., F.H. Dark, and M.S. Weibach, "An Agency Perspective On Franchising," *Financial Management*, **20** (1991(a)), 27-35.

Brickley, J.A., F.H. Dark, and M.S. Weibach, "The Economic Effects Of Franchise Termination Laws," *Journal of Law and Economics*, **34**, 4 (1991(b)), 101-132.

Brooker, K., "E-Rivals Seem To Have Home Depot Awfully Nervous," *Fortune*, **140**, August 16 (1999), 28-29.

Brown, E., "How Can A Dot-Com Be This Hot?," *Fortune*, January 21 (2002), 78-84.

Bucklin, C.B., P.A. Thomas-Graham, and E.A. Webster, "Channel Conflict: When Is It Dangerous?," *The McKinsey Quarterly*, June 22 (1997), 36.

Cachon, G.P., "Competitive Supply Chain Inventory Management," in S. Tayur, R. Ganeshan, and M. Magazine (Eds.), *Quantitative Models for Supply Chain Management*, Kluwer Academic Publishers, Norwell, MA, 1999, 111-146.

Cachon, G.P., "Supply Chain Coordination With Contracts," to appear in S. Graves and T. de Kok (Eds.), *Handbooks in OR and MS: Supply Chain Management,* North-Holland, 2002.

Cachon, G.P. and M.A. Lariviere, "Capacity Allocation Using Past Sales: When To Turn-and-Earn," *Management Science*, **45**, 5 (1999(a)), 685-703.

Cachon, G.P. and M.A. Lariviere, "Capacity Choice And Allocation: Strategic Behavior And Supply Chain Performance," *Management Science*, **45**, 8 (1999(b)), 1091-1108.

Cattani, K., W. Gilland, and J.M. Swaminathan, "Coordinating Internet And Traditional Channels," Working Paper OTIM-2002-04, Kenan-Flagler Business School, UNC Chapel Hill, 2002.

Caves, R.E. and W.F. Murphy, "Franchising: Firms, Markets And Intangible Assets," *Southern Economic Journal*, **42**, 4 (1976), 572-586.

Chen, F., A. Federgruen, and Y. Zheng, "Coordination Mechanisms For A Distribution System With One Supplier And Multiple Retailers," *Management Science*, **47**, 5 (2001), 693-708.

Chen, Y., G. Iyer, and V. Padmanabhan, "Referral Infomediaries," *Marketing Science*, **21**, 4 (2002), 412-434.

Chiang, W.K., D. Chhajed, and J.D. Hess, "Direct Marketing, Indirect Profits: A Strategic Analysis Of Dual-Channel Supply-Chain Design," *Management Science*, **49**, 1 (2003), 1-20.

Chiang, W.K. and G.E. Monahan, "The Impact Of The Web-Based Direct Channel On Supply Chain Flexibility In A Two-Echelon Inventory System," Working Paper, University of Maryland, Baltimore County, 2002.

Choi, S.C., "Price Competition In A Channel Structure With Common Retailer," *Marketing Science*, **10**, 4 (1991), 271-296.

Choi, S.C., "Price Competition In A Duopoly Common Retailer Channel," *Journal of Retailing*, **72**, 2 (1996), 117-134.

Choi, S.C., "Expanding To Direct Channel: Market Coverage As Entry Barrier," Working Paper, Rutgers University, 2002. Forthcoming, *Journal of Interactive Marketing*.

Chu, W., E. Gerstner, and J.D. Hess, "Managing Dissatisfaction: How To Decrease Customer Opportunism By Partial Refunds," *Journal of Service Research*, 1, 2 (1998), 140-155.

Clark, A.J. and H. Scarf, "Optimal Policies For A Multiechelon Inventory Problem," *Management Science*, **6** (1960), 475-490.

Cohen, M.A., N. Agrawal, V. Agrawal, and A. Raman, "Analysis Of Distribution Strategies In The Industrial Paper And Plastics Industry," *Operations Research*, **43**,1 (1995), 6-18.

Cohen, M.A., P. Kleindorfer, and H.L. Lee, "OPTIMIZER: A Multi-Echelon Inventory System For Service Logistics Management," *Interfaces*, **20,** 1 (1990), 65-82.

Cohen, M.A., Y. Zheng, and Y. Wang, "Identifying Opportunities For Improving Teradyne's Service-Parts Logistics System," *Interfaces*, **29**, 4 (1999), 1-18.

Collinger, T., "Lines Separating Sales Channels Blur: Manufacturers, Direct Sellers, Retailers Invade Each Others' Turf," *Advertising Age*, March 30 (1998), 34.

Corbett, C.J. and X. de Groote, "A Supplier's Optimal Quantity Discount Policy Under Asymmetric Information," *Management Science,* **46**, 3 (2000), 444-450.

Corbett, C.J., and U.S. Karmarkar, "Competition And Structure In Serial Supply Chains With Deterministic Demand, " *Management Science*, **47**, 7 (2001), 966-978.

Corstjens, M. and P. Doyle, "Channel Optimization In Complex Marketing Systems," *Management Science*, **25**, 10 (1979), 1014-1025.

Coughlan, A.T., "Competition And Cooperation In Marketing Channel Choice: Theory And Application," *Marketing Science*, **4**, 2 (1985), 110-129.

Coughlan, A.T. and B. Wernerfelt, "On Credible Delegation By Oligopolists: A Discussion Of Distribution Channel Management," *Management Science*, **35**, 2 (1989), 226-239.

Davis, S, E. Gerstner, and M. Hagerty, "Money Back Guarantees In Retailing: Matching Products To Consumer Tastes," *Journal of Retailing*, **71**, 1 (1995), 7-22.

Davis, S, M. Hagerty, and E. Gerstner, "Return Policies And Optimal Level Of 'Hassle,'" *Journal of Economics and Business*, **50**, 5 (1998), 445-460.

Deshpande, V. and L.B. Schwarz, "Optimal Capacity Allocation In Decentralized Supply Chains," Working Paper, Krannert School, Purdue University, 2002.

Dnes, A.W., "The Economic Analysis Of Franchise Contracts," *The Journal of Institutional and Theoretical Economics*, **152** (1996), 297-324.

Dong, L. and H.L. Lee, "Efficient Supply Chain Structures For Personal Computers," in J.S. Song and D.D. Yao (Eds.). *Supply Chain Structures: Coordination, Information, and Optimization (Volume 42 of International Series in Operations Research & Management Science)*, Kluwer Academic Publishers, Norwell, MA, 2001, 9-46.

Druehl, C. and E. Porteus, "Price Competition Between An Internet Firm And A Bricks And Mortar Firm," Working Paper, Graduate School of Business, Stanford University, 2001.

Federgruen, A., "Centralized Planning Models For Multi-Echelon Inventory Systems Under Uncertainty" in S.C. Graves, A.H.G. Rinnooy Kan, and P.H. Zipkin, (Eds.). *Handbooks in Operations Research and Management Science*, Volume 4, (Logistics of Production and Inventory), Elsevier Science Publishing Company B.V., Amsterdam, The Netherlands 1993, 133-173

Fowler, G.A., "Where Are They Now?," *Wall Street Journal*, June 10 (2002), R13.

Fransoo, J.C., M.J.F. Wouters, and T.G. de Kok, "Multi-Echelon Multi-Company Inventory Planning With Limited Information Exchange," *Journal of the Operational Research Society*, **52** (2001), 830-838.

Frazier, G.L., "Organizing And Managing Channels Of Distribution," *Journal of the Academy of Marketing Science*, **27**, 2 (1999), 226-241.

Gallini, N. T. and N.A. Lutz, "Dual Distribution And Royalty Fees In Franchising," *Journal of Law, Economics and Organization*, **8** (1992), 471-501.

Ghosh, S., "Making Business Sense Of The Internet," *Harvard Business Review*, **76**, 2 (1998), 126-135.

Gupta, S., "Coordination Incentives In Competing Supply Chains With Knowledge Spillovers," Working Paper, University of Michigan, 2001.

Gupta, S. and R. Loulou, "Process Innovation, Product Differentiation, And Channel Structure: Strategic Incentives In A Duopoly," *Marketing Science*, **17**, 4 (1998), 301–316.

Ha, A.Y., "Supplier-Buyer Contracting: Asymmetric Cost Information And The Cut-Off Level Policy For Buyer Participation," *Naval Research Logistics*, **48**, 1 (2001), 41-64.

Hendershott, T. and J. Zhang, "A Model Of Direct And Intermediated Sales," Working Paper, University of California at Berkeley and University of Rochester, 2001.

Hess, J.D., W. Chu, and E. Gerstner, "Controlling Product Returns In Direct Marketing," *Marketing Letters*, **7**, 4 (1996), 307-317.

Huang, W. and J.M. Swaminathan, "Pricing On Traditional And Internet Channels Under Monopoly And Duopoly: Analysis And Bounds," Working Paper, Kenan-Flagler Business School, UNC Chapel Hill, 2003.

Ingene, C.A. and M.E. Parry, "Coordination And Manufacturer Profit Maximization: The Multiple Retailer Channel," *Journal of Retailing*, **71**, 2 (1995(a)), 129-151.

Ingene, C.A. and M.E. Parry, "Channel Coordination When Retailers Compete," *Marketing Science*, **14**, 4 (1995(b)), 360-377.

Ingene, C.A. and M.E. Parry, "Manufacturer-Optimal Wholesale Pricing When Retailers Compete," *Marketing Letters*, **9**, 1 (1998), 65-77.

Ingene, C.A. and M.E. Parry, "Is Channel Coordination All It Is Cracked Up To Be?," *Journal of Retailing*, **76**, 4 (2000), 511-547.

Iyer, G., "Coordinating Channels Under Price And Non-price Competition," *Marketing Science*, **17**, 4 (1998), 338-355.

Jaikumar, R. and V.K. Rangan, "Price Discounting In Multi-Echelon Distribution Systems," *Engineering Costs and Production Economics*, **19**, 1-3 (1990), 341-349.

Katz, M.L., "Vertical Contractual Relations," in R. Schmalensee and R.D. Willig (Eds.). *Handbook of Industrial Organization*, Elsevier Science Publishers, Amsterdam, The Netherlands, 1989, Vol. 1, 655-721.

Keeney, R.L., "The Value Of Internet Commerce To The Customer," *Management Science*, **45**, 4 (1999), 533-542.

Klein, B., "Transaction Cost Determinants Of 'Unfair' Contractual Arrangement," *American Economic Review*, **70**, 5 (1980), 356-362.

Klein, B. and L.F. Saft, "The Law And Economics Of Franchise Tying Contracts," *Journal of Law and Economics*, **28**, 5 (1985), 345-361.

Kolay, S., G. Shaffer, and J.A. Ordover, "All-Units Discounts In Retail Contracts," Working Paper, University of Rochester, 2002.

Krishnan, H., R. Kapuscinski, and D.A. Butz, "Coordinating Contracts For Decentralized Channels With Retailer Promotional Effort," Working Paper, University of Michigan, 2001.

Kumar, N. and R. Ruan, "On Strategic Pricing And Complementing The Retail Channel With A Direct Internet Channel," Working Paper, University of Texas at Dallas, 2002.

Lafontaine, F., "Agency Theory And Franchising: Some Empirical Results," *The RAND Journal of Economics*, **23**, 2 (1992), 263-283.

Lafontaine, F. and P.J. Kaufmann, "The Evolution Of Ownership Patterns In Franchise Systems," *Journal of Retailing*, **70**, 2 (1994), 97-113.

Lal, R., "Improving Channel Coordination Through Franchising," *Marketing Science*, **9**, 4 (1990), 299-318.

Lal, R. and M. Sarvary, "When And How Is The Internet Likely To Decrease Price Competition," *Marketing Science*, **18**, 4 (1999), 485-503.

Lariviere, M.A., "Supply Chain Contracting And Coordination With Stochastic Demand," in S. Tayur, R. Ganeshan and M. Magazine (Eds.), *Quantitative Models for Supply Chain Management*, Kluwer Academic Publishers, Norwell, MA, 1999, 233-268.

Lee, H.L., V. Padmanabhan, and S. Whang, "Information Distortion In A Supply Chain: The Bullwhip Effect," *Management Science*, **43**, 4 (1997), 546-558.

Lippman, S.A. and K.F. McCardle, "The Competitive Newsboy," *Operations Research*, **45**, 1 (1997), 54-65.

Machlis, S., "Channel Conflicts Stall Web Sales," *Computerworld*, February 16 (1998(a)), 2.

Machlis, S., "Going Online, Lauder Remembers Retailers," *Computerworld*, July 6 (1998(b)), 79.

Machlis, S., "Beauty Product Sites Facing Channel Clash," *Computerworld*, November 9 (1998(c)), 24.

Mahajan, S. and G. van Ryzin, "Stocking Retail Assortments Under Dynamic Consumer Substitution," *Operations Research*, **49**, 3 (2001), 334-351.

Martin, R.E., "Franchising And Risk Management," *American Economic Review*, **78**, 12 (1998), 954-968.

Marx, L.M. and G. Shaffer, "Predatory Accommodation: Below-Cost Pricing Without Exclusion In Intermediate Goods Markets," *Rand Journal of Economics*, **30**, 1 (1999), 22-43.

Marx, L.M. and G. Shaffer, "Bargaining Power In Sequential Contracting," Working Paper, University of Rochester, 2001(a).

Marx, L.M. and G. Shaffer, "Opportunism And Nondiscrimination Clauses," Working Paper, University of Rochester, 2001(b).

Marx, L.M. and G. Shaffer, "Rent-Shifting And Efficiency In Sequential Contracting," Working Paper, University of Rochester, 2001(c).

Mathewson, G.F. and R.A. Winter, "An Economic Theory Of Vertical Restraints," *Rand Journal of Economics*, **15**, 1 (1984), 27-38.

McGuire, T.W. and R. Staelin, "An Industry Equilibrium Analysis Of Downstream Vertical Integration," *Marketing Science*, **2**, 2 (1983), 161-191.

McGuire, T.W. and R. Staelin, "Channel Efficiency, Incentive Compatibility, Transfer Pricing, And Market Structure: An Equilibrium Analysis Of Channel Relationships," *Research in Marketing*, **8** (1986), 181-223.

McIntyre, S.J., "How To Reap Profits And Avoid Pitfalls When A Catalog Is Only Part Of Your Business, Pt. 2," *Direct Marketing*, June 1 (1997), 32.

McWilliams, G., "Mimicking Dell, Compaq To Sell Its PCs Directly," *The Wall Street Journal*, November 11 (1998), B1.

McWilliams, G., "Lean Machine: How Dell Fine-Tunes Its PC Pricing To Gain Edge In A Slow Market," *The Wall Street Journal*, June 8 (2001), A1.

Minkler, A.P., "An Empirical Analysis Of A Firm's Decision To Franchise," *Economic Letters*, **34** (1990), 77-82.

Minkler, A.P., "Why Firms Franchise: A Search Cost Theory," *Journal of Institutional and Theoretical Economics*, **148** (1992), 240-259.

Moorthy, K.S., "Strategic Decentralization In Channels," *Marketing Science*, **7**, 4 (1988), 335-355.

Muckstadt, J.A. and R.O. Roundy, "Analysis Of Multistage Production Systems," in S.C. Graves, A.H.G. Rinnooy Kan, and P.H. Zipkin (Eds.). *Handbooks in Operations Research and Management Science*, Vol. 4 (Logistics of Production and Inventory), Elsevier Science Publishing Company B.V., Amsterdam, The Netherlands, 1993, 59-131.

Nasireti, R., "IBM Plans To Sell Some Gear Directly To Fight Its Rivals," *The Wall Street Journal*, June 5 (1998).

Netessine, S. and N. Rudi, "Supply Chain Structures On The Internet: Marketing-Operations Coordination Under Drop-Shipping," Working Paper, The Wharton School, University of Pennsylvania, 2001(a).

Netessine, S. and N. Rudi, "Supply Chain Choice On The Internet," Working Paper, The Wharton School, University of Pennsylvania, 2001(b).

Norton, S.W., "An Empirical Look At Franchising As An Organizational Form," *Journal of Business*, **61**, 4 (1988), 197-218.

O'Brien, D.P. and G. Shaffer, "On The Dampening-Of-Competition Effect Of Exclusive Dealing," *Journal of Industrial Economics*, **41**, 2 (1993), 215-221.

O'Brien, D.P. and G. Shaffer, "The Welfare Effects Of Forbidding Discriminatory Discounts: A Secondary-Line Analysis Of The Robinson-Patman Act," *Journal of Law, Economics, and Organization*, **10**, 2 (1994), 296-318.

O'Brien, D.P. and G. Shaffer, "Nonlinear Supply Contracts, Exclusive Dealing, And Equilibrium Market Foreclosure," *Journal of Economics & Management Strategy*, **6**, 4 (1997), 755-785.

Oxenfeldt, A.R. and A.O. Kelly, "Will Franchise Systems Ultimately Become Wholly-Owned Chains?," *Journal of Retailing*, **44** (1969), 69-87.

Padmanabhan, V. and I.P.L. Png, "Manufacturer's Returns Policies And Retail Competition," *Marketing Science*, **16**, 1 (1997), 81-94.

Parlar, M., "Game Theoretic Analysis Of The Substitutable Product Inventory Problem With Random Demands," *Naval Research Logistics*, **35** (1988), 397-409.

Pasternack, B.A., "Optimal Pricing And Returns Policies For Perishable Commodities," *Marketing Science*, **4**, 2 (1985), 166-176.

Peleg, B. and H.L. Lee, "Secondary Markets For Product Diversion With Potential Manufacturer's Intervention," Working Paper, Department of Management Science and Engineering, Stanford University, 2002.

Perry, M.K. and R.H. Porter, "Can Resale Price Maintenance And Franchise Fees Correct Sub-Optimal Levels Of Retail Service," *International Journal of Industrial Organization*, **8**, 1 (1990), 115-141.

Peterson, R., S. Balasubramanian, and B.J. Bronnenberg, "Exploring The Implications Of The Internet For Consumer Marketing," *Journal of the Academy of Marketing Science*, 25, 4 (1997), 329-346.

Purohit, D., "Dual Distribution Channels: The Competition Between Rental Agencies And Dealers," *Marketing Science*, **16**, 3 (1997), 228-245.

Rajaram, K., and C.S. Tang, "The Impact Of Product Substitution On Retail Merchandising," *European Journal of Operations Research*, **135**, 3 (2001), 582-601.

Raju, J., R. Sethuraman, and S. Dhar, "The Introduction And Performance Of Store Brands," *Management Science*, **41** (1995), 957-978.

Rangan, V.K. and R. Jaikumar, "Integrating Distribution Strategy And Tactics: A Model And An Application," *Management Science*, **37**, 11 (November 1991), 1377-1389.

Reinhardt, G. and M. Levesque, "Virtual Versus Bricks-and-Mortar Retailing," Working Paper, Department of Management, DePaul University, 2001.

Rhee, B., "A Hybrid Channel System In Competition With Net-Only Direct Marketers," Working Paper, HKUST, 2001.

Rhee, B. and S.Y. Park, "Online Stores As A New Direct Channel And Emerging Hybrid Channel System," Working Paper, HKUST, 2000.

Rosencrance, L., "Amazon Charging Different Prices On Some DVDs," *Computerworld*, September 05 (2000).

Schmid, J., "Reaching Into Retail: Can The Catalog And Retail Marketing Channels Coexist?," *Catalog Age*, January (1999), 59-62.

Scott, F.A., "Franchising Vs. Company Ownership As A Decision Variable Of The Firm," *Review of Industrial Organization,* **10** (1995), 69-81.

Seifert, R. W. and U.W. Thonemann, "Relaxing Channel Separation - Integrating A Virtual Store Into The Supply Chain," Working Paper, IMD – International Institute for Management Development, Lausanne, Switzerland, 2001.

Seifert, R.W., U.W. Thonemann, and S.A. Rockhold, "Integrating Direct And Indirect Sales Channels Under Decentralized Decision Making," Working Paper, IMD – International Institute for Management Development, Lausanne, Switzerland, 2002.

Shaffer, G., "Slotting Allowances And Resale Price Maintenance: A Comparison Of Facilitating Practices," *Rand Journal of Economics*, 22, 1 (1991), 120-135.

Shaffer, G., "Bargaining In Distribution Channels With Multiproduct Retailers," Working Paper, University of Rochester, 2001.

Shaffer, G. and F. Zettelmeyer. "When Good News About Your Rival Is Good For You: The Effect Of Third-Party Information On The Division Of Profit In A Multi-Product Distribution Channel," *Marketing Science*, 21, 3 (2002), 272-293.

Shapiro, C., "Theories Of Oligopoly Behavior," in R. Schmalensee and R.D. Willig (Eds.), *Handbook of Industrial Organization*, Volume 1, Elsevier Science Publishers B.V., New York, NY, 1989, 329-414.

Smith, S.A. and N. Agrawal, "Management Of Multi-Item Retail Inventory Systems With Demand Substitution," *Operations Research*, 48, 1 (2000), 50-64.

Stern, L.W., A.I. El-Ansary, and A.T. Coughlan, *Marketing Channels*, 5th edition, Prentice Hall, Upper Saddle River, New Jersey, 1996.

Tessler, J., "Hybrid's Here: Retailers Finding A Mix Of Stores, Ads, And Net Best Way To Reach Customers," *San Jose Mercury News*, December 24 (2000), G1-2.

Tsay, A.A. and N. Agrawal, "Channel Dynamics Under Price And Service Competition," *Manufacturing & Service Operations Management*, 2, 4 (2000), 372-391.

Tsay, A.A. and N. Agrawal, "Manufacturer And Reseller Perspectives On Channel Conflict And Coordination In Multiple-Channel Distribution," Working Paper, Santa Clara University, 2001.

Tsay, A.A., S. Nahmias. and N. Agrawal, "Modeling Supply Chain Contracts: A Review." in S. Tayur, and R. Ganeshan, and M. Magazine (Eds.), *Quantitative Models for Supply Chain Management*, Kluwer Academic Publishers, Norwell, MA, 1999, 299-336.

Wilcox, J., "Is Apple Stocking Its Own Shelves First?," *CNETnews.com*, March 13 (2002).

Winter, R., "Vertical Control And Price Versus Nonprice Competition," *Quarterly Journal of Economics*, 108, 1 (1993), 61-76.

Zarley, C. and B. Darrow, "Industry Moves To Integrated Distribution," *Computer Reseller News*, April 5 (1999), 1.

Zettelmeyer, F., "Expanding To The Internet: Pricing And Communications Strategies When Firms Compete On Multiple Channels," *Journal of Marketing Research*, 37, 3 (2000), 292-308.

David Simchi-Levi, S. David Wu, and Z. Max Shen (Eds.)
Handbook of Quantitative Supply Chain Analysis:
Modeling in the E-Business Era
©2004 Kluwer Academic Publishers

Chapter 14

SUPPLY CHAIN STRUCTURES ON THE INTERNET

and the role of marketing-operations interaction [1]

Serguei Netessine
University of Pennsylvania
Philadelphia, PA 19104
netessine@wharton.upenn.edu

Nils Rudi
University of Rochester
Rochester, NY 14627
rudi@simon.rochester.edu

Keywords: Internet, supply chain, competition, Nash, Stackelberg, marketing, advertising

1. Introduction

Spun.com, a small CD/DVD Internet retailer, has about 200,000 CD titles listed on its web site. Surprisingly, the company does not hold/own any inventory of CDs. Instead, the company partnered with the wholesaler Alliance Entertainment Corp. (AEC), which stocks CDs and ships them directly to Spun.com's customers with Spun.com labels on the packages. In this way, the retailer avoided an estimated inventory investment of $8M Forbes (2000), since it only paid the distributor for sold products. AEC calls this distribution system "Consumer Direct Fulfillment." According to the company's web site, "... using AEC as a fulfillment partner gives you more time and resources to focus on attracting more consumers to your store ...". The list of retailers practicing such forms of Internet business includes Zappos.com (Inbound Logistics

2000), Cyberian Outpost (Computer Reseller News 1999) and many others.

Drop-shipping is defined in marketing literature as "a marketing function where physical possession of goods sold bypasses a middleman, while title flows through all those concerned. The function of drop-shipping involves both the middleman who initiates the drop ship order and the stocking entity that provides drop-shipping services by filling the order for the middleman" (Scheel 1990). Clearly, the above example of Spun.com fits this description. Drop-shipping is different from many of the supply chain structures previously described in the literature in which the wholesaler is involved in the retailer's inventory management. It differs from the traditional consignment agreements in which the retailer holds (but does not own) inventory and decides what the stocking policy should be – under drop-shipping the stocking policy is entirely controlled by the wholesaler. Drop-shipping is close to but different from Vendor Managed Inventory (VMI), since the retailer does not deal with inventories and hence does not incur any inventory-related costs. At the same time, the wholesaler does not have direct access to the retailer's store where she could "rent" space and organize it in a way that influences demand according to the wholesaler's preferences (as is often the case under VMI). Drop-shipping also differs from outsourcing of inventory management, since under outsourcing the retailer usually still influences stocking quantities for each product.

Prior to the Internet, the practice of drop-shipping was mainly restricted to two settings. For large transactions of industrial goods, the wholesaler might have the manufacturer make the shipment directly to the retailer (and in some cases directly to the end customer). This is typically beneficial for shipments that in themselves achieve sufficient economies of scale, making the wholesaler act primarily as a market-maker. The second use of drop-shipping, which is more relevant to our setting, occurs when a catalog company has the wholesaler drop-ship the product directly to the end customer. This practice, however, has had very limited success, mainly due to problems in the integration and timeliness of information between the business partners, as well as high transaction costs. As a result, even the catalog companies using drop-shipping only use it for bulky and high cost items (see Catalog Age 1997). Hence, many marketing books see drop-shipping as having limited potential (see literature review in Scheel 1990 page 7). With the Internet, however, real time data-integration is readily available at a low cost. The combination of the physical concept of drop-shipping with the information integration made possible by the Internet resolves the problems that previously limited the adoption of drop-shipping. A survey of Internet retailers (Eretailing World 2000) indicates that 30.6% of Internet-only retailers use drop-shipping as the primary way to fulfill orders, while only 5.1% of multi-channel retailers primarily rely on drop-shipping. In a recent survey of Supply Chain Management in e-commerce applications Johnson and Whang (2002) cite several cases written about companies utilizing drop-shipping (or virtual supply chains).

One of the major differences between selling goods on the Internet and through the conventional brick-and-mortar retailer is the separation of physical goods flow and information flow. In a physical store, a customer selects a product and pays for it at the same time and place that she physically receives the product. On the Internet this is not the case. A customer on the Internet cannot observe from where the product is dispatched. Further, Internet customers (similar to mail-order catalog customers) do not expect an immediate delivery of the product. Together, this allows the retailer and the wholesaler to adopt the drop-shipping agreement efficiently at a low cost. Agreements of this type benefit the retailer by eliminating inventory holding costs and overall up-front capital required to start the company. The wholesaler increases her involvement in the supply chain and hence can potentially demand a higher wholesale price, thus capturing more profits (as evidenced by empirical data in Scheel 1990 page 42). Further, supply chain benefits occur due to risk pooling if the wholesaler performs drop-shipping for multiple retailers. Finally, each party can concentrate its resources: the retailer on customer acquisition and the wholesaler on fulfillment.

Despite several clearly attractive features, drop-shipping introduces new inefficiencies into the supply chain. Under drop-shipping, the wholesaler keeps the decision rights related to stocking policies, while the retailer's main task in the supply chain is customer acquisition. This separation of marketing and operations functions results in inefficiencies, some of which have been the subject of discussion in the literature on marketing-operations coordination. Many questions arise in such a situation: will the supply chain performance under the drop-shipping structure be better than under a traditional structure in which the retailer holds inventory? Further, is it preferable for both the retailer and the wholesaler to engage in this sort of agreement? Can drop-shipping agreements lead to system-optimal performance, and if not, what form of contract can coordinate the supply chain?

To the best of our knowledge, this chapter represents the first formal model of a drop-shipping supply chain, the practice where the retailer acquires customers while the wholesaler is responsible for fulfillment and takes inventory risk. We compare traditional supply chain structures in which the retailer both takes inventory risk and acquires customers with supply chain structures employing drop-shipping agreements. Our focus is on the supply chains in which both the retailer and the wholesaler are present. We concentrate on two cost aspects of the distribution channel: marketing (customer acquisition) and operational (inventory). In the case of Internet retailers, there exists vast evidence that these two cost components constitute a dominant portion of the company's budget. Since e-tailers do not have a physical presence that would attract customers by physical location or strong brand name, customer acquisition becomes a major issue. Online-only retailers spend about twice as much of their budget on customer acquisition as do multi-channel retailers Computer World (2000). At the early stages of Internet development, the typical marketing budget of an Internet-only retailer was 40.5% of sales, while the marketing budget for a multi-channel retailer was 21.4% Eretailing World (2000). A survey of online retailers conducted in 2000 shows that the catalog-based companies spend on average $11 to acquire a customer, compared to the $32 spent by a physical store and $82 by an e-tailer Computer World (2000). Although after the Internet bubble advertising spendings have gone down, Internet retailers still spend much more on advertising than traditional retailers (see Latcovich and Smith (2001). Furthermore, empirical studies have shown that advertising is used by the Internet retailers to signal quality and since Internet shoppers respond much more to advertising than to low prices (Latcovich and Smith 2001). These findings support our use of advertising as a key aspect of the Internet retailing. On the other hand, to be able to compete with traditional retailers, Internet companies typically offer extensive product variety and a high service level that in turn requires a large inventory investment. In addition, at the present time Internet retailing is in its early stages of development, and the demand for products is highly uncertain; hence, large inventories must be carried to maintain high service levels.

To better understand the wholesaler-retailer interaction and inventory risk allocation in drop-shipping supply chains, we focus on a simple model with a wholesaler and a single retailer. Note that for cases where the retailer buys directly from the manufacturer the wholesaler in our model is simply replaced by the manufacturer without any effect on other results. Three main models are analyzed and compared in this chapter: a traditional vertically integrated channel (Model I for "Integrated"), a traditional vertically disintegrated channel (Model T for "Traditional") and a drop-shipping channel (Model D for "Drop-shipping"). The solution for the drop-shipping channel further depends on the channel power (where power is modeled as being the first mover in a Stackelberg game). Hence, within the drop-shipping model we further consider three sub-models, in which either the wholesaler or the retailer is a Stackelberg leader or where each party has an equal decision power (i.e. moves simultaneously) and the solution is a Nash equilibrium. First, we demonstrate that a unique competitive equilibrium exists in each model, and we find optimal inventory and customer acquisition spending for each channel in analytical form.

We find that, for the drop-shipping channel, marketing-operations misalignment results from the fact that these functions are managed by two different firms. In addition, double marginalization is present in both the traditional and drop-shipping structures. Under identical problem parameters, we analytically compare the models in terms of decision variables and profits and show that the drop-shipping models, as well as the traditional model, always lead to underspending on customer acquisition in addition to understocking. We characterize the situations where drop-shipping supply chains outperform traditional supply chains and our analysis indicates the importance of the effects of channel power on supply chain performance. We then show that a price-only contract, a revenue sharing contract, or a returns/purchase commitment contract cannot coordinate the supply chain when customer acquisition costs are present. For the traditional and drop-shipping supply chains we propose a contract that combines a compensation plan that is linear in carried over inventory with subsidized advertising, and show that this contract induces coordination.

Our main interest in this chapter is to get a better understanding of inventory risk allocation issues in drop-shipping supply chains as well as the impact of power distribution between the channel entities. To keep focus and to not diffuse economic insights, we do not explicitly reflect other issues encountered in e-commerce fulfillment, since many are hard to include in a formal model. Among these issues are: possible differences in transportation costs and responsiveness, coordination issues arising when multiple wholesalers are needed to fulfill a single order, and the rationing of inventory when a wholesaler serves multiple retailers. These and other issues are treated qualitatively in Randall et al. (2002).

The rest of the chapter is organized as follows. In the next section, we provide a survey of the relevant literature. Section 3 outlines the notation and modeling assumptions. In Section 4, all three models are presented and in Section 5, we show that none of the channel coordination mechanisms described in the literature can achieve a first-best solution. Optimal coordinating contracts are also proposed in this section. Section 6 contains numerical experiments, and in Section 7 we wrapup the chapter with a discussion of managerial insights and conclusions.

2. Literature survey

The practice of drop-shipping has been described qualitatively in the marketing literature (see Scheel 1990 and references therein, page 7), but, to the best of our knowledge, its distinct features, i.e., the wholesaler taking inventory risk while the retailer, a separate firm, acquiring customers, have never been formally modeled and analyzed previously [2]. In a majority of marketing textbooks, the qualitative analysis of drop-shipping is limited to one paragraph. Further, the literature on drop-shipping does not raise the issue of marketing-operations misalignment and power in the supply chain. With the exception of Scheel (1990), all the sources we cite either ignore inventory or assume that inventory is held by the retailer.

Operations management has a wealth of literature that deals with the inventory aspects of the supply chain, but ignores marketing expenses like customer acquisition costs (see, for example, Tayur et al. 1998). At the same time, the marketing literature tends to deal with customer acquisition costs in the form of advertising and sales support (see, for example, Lilien et al. 1992), but ignores operational issues. For example, Chu and Desai (1995) model a supply chain with a retailer and a wholesaler, both investing into improvement of customer satisfaction which stimulates demand. However, they do not explicitly consider inventory risk.

This work belongs to the recent stream of research dealing with the alignment of marketing and operations incentives (see Shapiro 1977 and Montgomery and Hausman 1986 for discussions of some of the problems that arise from marketing-operations misalignment). Our model differs from the previous literature in several ways. First, in our drop-shipping model operations and marketing functions are performed by two independently owned and operated companies that make their decisions competitively. Since marketing and operational functions are performed by separate companies and we do not consider information asymmetry, the problem we consider is different from sales agent compensation. Second, inventory risk in the drop-shipping model is with the wholesaler, resulting in channel dynamics that differ from the existing literature. Finally, the marketing function is customer acquisition that affects demand for the product, while the sales price is exogenous. The majority of papers in this area assume that the marketing function is to set the sales price or promotion level and manufacturing makes production decisions within one company (see, for example, Sagomonian and Tang 1993 and references therein). De Groote (1989) considers product line choice as a marketing function. Eliashberg and Steinberg (1987) give an excellent summary of a number of papers modeling operations/marketing interfaces, but none of the papers they cite model uncertain demand.

Porteus and Wang (1991) model a company where the principal specifies compensation plans for one manufacturing and multiple marketing managers within the same company. Each product is exclusively assigned to a marketing manager, who stimulates the product's demand by effort. The manufacturing manager can affect capacity available to produce all these products. Although this paper is somewhat similar to our chapter in terms of structure of the problem and the issues considered, the principal-agent framework used by Porteus and Wang makes differences in motivation, structure and modeling. Porteus and Wang (1991) also cite a number of relevant papers, but none of these papers model customer acquisition spending or advertising as a marketing decision variable. Balcer (1980) models coordination of advertising and inventory decisions within one company and focuses on a dynamic nonstationary model in which demand is influenced by the level of goodwill. Balcer (1983) further extends this model by assuming that the advertising effect lasts for more than one period. Gerchak and Parlar (1987) look at a single-period model in which demand is a specific function of the marketing effort. None of the latter three papers consider supply chain issues, i.e., the interaction between several firms.

For traditional supply chains only, some papers model situations that in certain aspects are related to ours. Cachon and Lariviere (2000), among others, model a situation where the retailer both takes inventory risk and influences demand by exerting effort. They assume that the effort cannot be contracted upon, and hence they do not find a contract that coordinates the supply chain. While this assumption is reasonable in the problem setting they use, on the Internet it is possible to contract upon at least some forms of customer acquisition spending. For example, if the retailer pays for advertising based on the volume of click-through and purchase by the customers, then this type of customer acquisition can easily be independently verified and contracted upon. In the last section, we will comment more specifically on the viability of this type of contract and on practical ways to implement such contracts. Narayanan and Raman (1998) look at a problem that in some ways is the opposite of drop-shipping, i.e., where the retailer cannot affect demand distribution but the manufacturer can. In their paper, the manufacturer's effort is incorporated analogously to the way we use customer acquisition spending by the retailer. Most of their analysis, however, is done with some quite restrictive assumptions about the functional forms of the problem parameters and decision power in the channel. They demonstrate that vertical disintegration of the channel leads to suboptimal performance but do not derive an optimal coordination contract that would mitigate this problem without side-payments. Cachon (2003) studies a model that is closely related to ours: either the retailer, the wholesaler, or both may own inventory. He focuses on Pareto-improving wholesale price contracts while we take wholesale prices as exogenous. For the majority of the analysis, advertising effort is not considered and channel power is not an issue in his work.

Relevant work on supply chain coordination in traditional supply chains includes penalty contracts (Lariviere 1999) and returns contracts (Pasternak 1985 and Kandel 1996). See also Cachon (2002) for a survey of contracting literature in Supply Chain Management. Jeuland and Shugan (1983) model marketing effort for the problem of distribution channel coordination under deterministic demand, but ignore inventory issues. The marketing literature has widely addressed a phenomena of subsidized advertising in the context of franchising agreements (see Michael 1999 and references therein). The only work we are aware of that addresses subsidized customer acquisition or advertising between two firms that are not bound by a franchising agreement is Berger (1983), Berger (1972) and Berger and Magliozzi (1992), with inventory issues ignored. Corbett and DeCroix (2000) consider a shared savings contract, a problem that is different from ours but is similar in mathematical structure: in our problem, customer acquisition expenses affect the demand; in their problem, use-reduction effort affects the consumption of indirect materials. In their problem, however, there is no inventory involved, and both the upstream supplier and the downstream buyer exert effort of the same nature. Hence, issues related to marketing-operations interaction do not arise.

There is a wealth of marketing and economics literature addressing Internet-related issues, though, to the best of our knowledge, the practice of drop-shipping has never been mentioned. The most relevant work is Hoffman and Novak (2000), who describe customer acquisition models on the Internet. Finally, the only relevant operations management paper addressing supply chain structures on the Internet of which we are aware is Van Mieghem and Chopra (2000), in which the authors qualitatively address the choice of e-business for a given supply chain (without considering drop-shipping).

3. Notation and modeling assumptions

We model a supply chain with two echelons: wholesaler and retailer. Only a single product and a single firm in each echelon is considered. Demand for the product in each time period is uncertain and mean demand depends on the amount of customer acquisition spending by the retailer in this period (by customer acquisition we imply the total cost a company spends on advertising and marketing promotions). This is an extension of the standard marketing models that usually assume deterministic demand. Since our focus is the wholesaler-retailer interaction, we assume that the unit price is exogenously given, which is a departure from the traditional literature on the marketing-operations interface in which the price is a primary marketing decision variable. This assumption might be reasonable for Internet retail since advertising seem to influence Internet shoppers more than low prices (see Latcovich and Smith 2001 for discussion and many examples). Further, all the problem parameters are identical for the retailer and the wholesaler and known to everyone (there is no information asymmetry). We use a multi-period framework with lost sale representing a penalty for understocking through lost margin. Left-over inventory is carried over to the next period incurring holding cost. The following notation is used throughout the chapter (superscript t indicates relevant time period, vectors are underlined):

x^t beginning inventory before placing the order in period t,

Q^t order-up-to quantity in period t, $Q^t \geq x^t$,

r unit revenue,

w unit wholesale price,

c unit cost,

β discounting factor, $\beta \in [0, 1)$,

h unit holding cost per period,

A^t customer acquisition spending (includes all types of related marketing activities),

$D^t(A^t)$ demand (random variable), parameterized by the customer acquisition spending,

$f_D(\cdot)$ probability density function of the demand,

Π_r, Π_w, Π discounted infinite-horizon expected profit function of retailer, wholesaler, and total supply chain, correspondingly.

Superscripts I, T and D will denote vertically integrated, vertically disintegrated (traditional), and drop-shipping supply chains, correspondingly. Further, we will consider three drop-shipping models. In Model DW the wholesaler has channel power, in Model DR the retailer has channel power, and in Model DN the players have equal power. We also assume the following quite general form of the demand distribution:

Assumption 1. Demand distribution has the following form: $D^t(A^t) = \theta(A^t) + \varepsilon^t$, where $\theta(A^t)$ is a real-valued function and ε^t are i.i.d. random variables such that $D^t(0) \geq 0$.

Assumption 2. Expected demand is increasing in customer acquisition spending and it is always profitable to spend a non-zero amount on customer acquisition:

$$\frac{dED^t(A^t)}{dA^t} \geq 0, \quad \lim_{A^t \to 0} \frac{dED^t(A^t)}{dA^t} = \infty.$$

Assumption 3. Expected demand is diminishingly concave (i.e., it is "flattening out" as A approaches infinity) in customer acquisition spending:

$$\frac{d^2 ED^t(A^t)}{d(A^t)^2} \leq 0, \frac{d^3 ED^t(A^t)}{d(A^t)^3} \leq 0.$$

The form of the demand function specified in Assumption 1 is used extensively in the operations and economics literature (see Petruzzi and Dada 1999 and references therein). Under this assumption, only the mean demand depends on A^t, and the uncertainty is captured by an error term ε^t. The first part of Assumption 2 is standard in marketing models (see page 265 in Lilien et al. 1992). We add the conditions that guarantee the existence of the non-degenerate (interior) solution, which is needed for the comparisons of the models. Assumption 3 is supported by empirical evidence from the marketing literature (see, for example, Simon and Arndt 1980 and Aaker and Carman 1982).

4. Supply chain models without coordination

In this section we will assume that the retailer and the wholesaler are employing a contractual agreement with a fixed transfer price. This might be a result of outside competition or other arrangements existing in the industry. The objective is to maximize infinite-horizon discounted expected profit. If a product is out of stock, the sale is lost. Unsold inventory is carried over to the next period incurring holding cost. The lost sale assumption leads to the following inventory balance equation in all models:

$$x^{t+1} = \left(Q^t - D^t\left(A^t\right)\right)^+.$$

4.1 Model I - vertically integrated supply chain

The wholesaler and the retailer are vertically integrated. At the beginning of each period, the product is purchased at a fixed unit cost c and sold to the customers at a fixed unit price r. The integrated retailer-wholesaler is the sole decision maker who chooses both the stocking quantity and the customer acquisition spending. Using a standard setup (see Heyman and Sobel 1984), the integrated firm will seek to maximize the following objective function:

$$\Pi^I(\underline{A},\underline{Q},x^1) = \sum_{t=1}^{\infty} \beta^{t-1} E[r \min(D^t(A^t), Q^t) - h(Q^t - D^t(A^t))^+ - c(Q^t - x^t) - A^t].$$

The profit contribution in each period consists of the following four terms: revenue generated by the units sold (limited by demand and supply), holding cost for the units carried over (positive difference of supply and demand), purchasing cost of inventory (difference between order-up-to level and starting inventory), and customer acquisition spending. By standard manipulation, we can re-write the optimization problem as follows:

$$\max_{Q^t \geq x^t, A^t \,\forall t} \Pi^I\left(\underline{A}, \underline{Q}, x^1\right) = \max_{Q^t \geq x^t, A^t \,\forall t} \left[cx^1 + \sum_{t=1}^{\infty} \beta^{t-1} \pi^{I,t}\left(Q^t, A^t, x^t\right)\right],$$

where $\pi^{I,t}$ is the expectation of the single-period objective function defined as follows:

$$\pi^{I,t}\left(Q^t, A^t, x^t\right) = E\left[(r + h - c\beta)\min\left(D^t(A^t), Q^t\right) - (h + c(1-\beta))Q^t - A^t\right].$$

PROPOSITION 1. *There exists a unique pair* (Q^I, A^I) *optimizing the single-period objective function without starting inventory (i.e., unconstrained)* $\pi^I = \pi^{I,t}(Q^t, A^t, 0)$, *characterized by the following system of equations:*

$$\Pr\left(D\left(A^I\right) < Q^I\right) = \frac{r-c}{r+h-\beta c}, \quad (14.1)$$

$$\left.\frac{dED(A)}{dA}\right|_{A^I} = \frac{1}{r-c}. \quad (14.2)$$

Then, for any $x^1 \leq Q^I$, *the solution* (Q^I, A^I) *is a unique stationary solution to the infinite-horizon problem.*

PROOF: Consider the single-period unconstrained objective function π^I. To demonstrate uniqueness of the solution we will show concavity of the objective function, or equivalently we will verify that the diagonal elements of the Hessian matrix of the objective function are negative and that the determinant of the Hessian is positive. The first derivatives are

$$\frac{\partial \pi^I}{\partial Q} = (r + h - \beta c)\Pr(D(A) > Q) - (h + c(1-\beta)), \quad (14.3)$$

$$\frac{\partial \pi^I}{\partial A} = (r + h - \beta c)\Pr(D(A) < Q)\frac{dED(A)}{dA} - 1. \quad (14.4)$$

The second derivatives are

$$\frac{\partial^2 \pi^I}{\partial Q^2} = -(r + h - \beta c)f_{D(A)}(Q) < 0,$$

$$\frac{\partial^2 \pi^I}{\partial A^2} = -(r + h - \beta c)f_{D(A)}(Q)\left(\frac{dED(A)}{dA}\right)^2$$
$$+ (r + h - \beta c)\Pr(D(A) < Q)\frac{d^2 ED(A)}{dA^2} < 0,$$

$$\frac{\partial^2 \pi^I}{\partial Q \partial A} = (r + h - \beta c)f_{D(A)}(Q)\frac{dED(A)}{dA} > 0.$$

The diagonal elements are clearly negative by Assumptions 2 and 3. Positivity of the determinant is equivalent to the following condition:

$$\frac{\partial^2 \pi^I}{\partial Q^2} \frac{\partial^2 \pi^I}{\partial A^2} > \frac{\partial^2 \pi^I}{\partial Q \partial A} \frac{\partial^2 \pi^I}{\partial A \partial Q}.$$

This can be expanded as follows:

$$- f_{D(A)}(Q) \times [-f_{D(A)}(Q)(\frac{dED(A)}{dA})^2 + \Pr(D(A) < Q)\frac{d^2ED(A)}{dA^2}]$$
$$> [f_{D(A)}(Q)\frac{dED(A)}{dA}]^2.$$

After collecting similar terms we obtain

$$-f_{D(A)}(Q)\Pr(D(A) < Q)\frac{d^2ED(A)}{dA^2} > 0.$$

This is true by Assumption 3, which completes the proof of uniqueness. To obtain the system of optimality conditions, it is convenient to substitute the first optimality condition into the second. Finally, following Section 3.1 in Heyman and Sobel (1982), the solution to the problem is myopic in nature. Furthermore, if initial inventory is sufficiently low for our problem, i.e. $x^1 \leq Q^I$, the solution is stationary (i.e. the unconstrained solution is feasible). This completes the proof.

4.2 Model T - traditional supply chain

The wholesaler buys the product at a fixed unit cost c and sells it to the retailer at a fixed unit wholesale price w. The retailer holds inventory and sells it to the customers at a fixed unit price r. The retailer here is the sole decision maker who decides on both the stocking quantity and the customer acquisition spending. Similar to model I,

$$\max_{Q^t \geq x^t, A^t \, \forall t} \Pi_r^T \left(\underline{A}, \underline{Q}, x^1\right) = \max_{Q^t \geq x^t, A^t \, \forall t} \left[wx^1 + \sum_{t=1}^{\infty} \beta^{t-1} \pi_r^{T,t}\left(Q^t, A^t, x^t\right) \right],$$

where $\pi_r^{T,t}$ is the expectation of the retailer's single-period objective function defined as follows:

$$\pi_r^{T,t}\left(Q^t, A^t, x^t\right) \qquad (14.5)$$
$$= E\left[(r + h - w\beta)\min\left(D^t(A^t), Q^t\right) - (h + w(1 - \beta))Q^t - A^t\right].$$

The wholesaler's expected profit is

$$\Pi_w^T\left(\underline{A}, \underline{Q}, x^1\right) = -(w - c)x^1 + \sum_{t=1}^{\infty} \beta^{t-1} \pi_w^T\left(Q^t, A^t, x^t\right),$$

where $\pi_w^{T,t}$ is the expectation of the wholesaler's single-period expected profit:

$$\pi_w^{T,t}\left(Q^t, A^t, x^t\right) = (w-c)\, E\left[(1-\beta)Q^t + \beta \min\left(D^t\left(A^t\right), Q^t\right)\right]. \tag{14.6}$$

PROPOSITION 2. *There exists a unique pair* $\left(Q^T, A^T\right)$ *optimizing the retailer's single-period objective function without starting inventory (i.e., unconstrained)* $\pi_r^T = \pi_r^{T,t}\left(Q^t, A^t, 0\right)$, *characterized by the following system of equations:*

$$\Pr\left(D\left(A^T\right) < Q^T\right) = \frac{r-w}{r+h-w\beta}, \tag{14.7}$$

$$\left.\frac{dED(A)}{dA}\right|_{A^T} = \frac{1}{r-w}. \tag{14.8}$$

Then, for any $x^1 \leq Q^T$, *the solution* $\left(Q^T, A^T\right)$ *is a unique stationary solution to the infinite-horizon problem (i.e. the unconstrained solution is feasible).*

PROOF: Similar to Proposition 1.

Note that although in this model the retailer is the sole decision maker and therefore seems to possess some power in the chain, he also bears all the inventory-related risk. In addition, customer acquisition spending incurred by the retailer not only benefits him but also benefits the wholesaler. As we will demonstrate later, this leads to the suboptimal performance of the channel.

4.3 Model D – drop-shipping

The wholesaler buys the product at unit cost c, holds inventory, and ships the product directly to the customer upon the retailer's request. The retailer acquires customers and makes sales. He pays unit wholesale price w per closed sale to the wholesaler, receives fixed unit revenue r from the customer, and does not hold inventory. Note at this point that under drop-shipping, contracts between the wholesaler and the retailer are not limited to the transfer pricing agreements. Since the product is never physically transferred to the retailer, it is often natural to consider a contract where the revenue is split between the retailer and the wholesaler in proportions λ and $1 - \lambda$, and there is no need to establish a wholesale price. For example, Scheel (1990), page 17, indicates that in the drop-shipping business the dominant practice is for the wholesaler to give the retailer a discount over the suggested retail price. We do not consider such agreements since the analysis is identical to the transfer price contracts with $w = (1-\lambda)r$.

Under a drop-shipping arrangement, the channel members will make their decisions strategically and hence a game-theoretic situation arises. Similar to

Cachon and Zipkin (1990), we consider games where each player chooses a stationary policy: the retailer's strategy is A and the wholesaler's strategy is order-up-to quantity Q. Three situations arise: either the wholesaler or the retailer might have negotiation power in the supply chain and act as Stackelberg leaders, or it is possible that the players have equal power and therefore the solution is in the form of a Nash equilibrium. We will consider all three situations.

Denote the best response function of the retailer by $R_r(Q)$ and the best response function of the wholesaler by $R_w(A)$, both defined for zero initial inventory. At this point we have not demonstrated uniqueness or even existence of the equilibrium. It helps, however, to visualize the problem first. We begin by presenting the game graphically (see Figure 14.1, parameters are taken from the example in Section 6 with $r = 8$). The point (A^{DN}, Q^{DN}) is a Nash

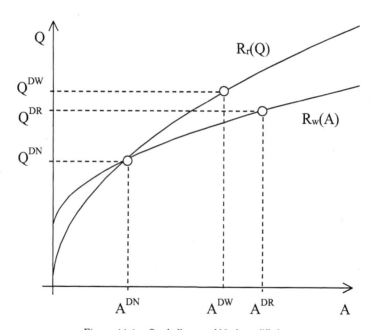

Figure 14.1. Stackelberg and Nash equilibria

equilibrium that is located on the intersection of the best-response curves. A Stackelberg equilibrium with the wholesaler as a leader (A^{DW}, Q^{DW}) is located on the retailer's best-response curve, and a Stackelberg equilibrium with the retailer as a leader (A^{DR}, Q^{DR}) is located on the wholesaler's response curve. We begin by characterizing the best response curves analytically. For a given Q, the retailer's optimization problem is

$$\max_{A} \Pi_r^D (A, Q) = \max_{A} \left[\sum_{t=1}^{\infty} \beta^{t-1} \pi_r^D (Q, A) \right],$$

where π_r^D is the expectation of the retailer's single-period objective function defined as follows:

$$\pi_r^D (A, Q) = E\left[(r - w) \min(D(A), Q) - A\right]. \quad (14.9)$$

Note that the objective function is concave, and hence the retailer's first-order condition characterizes the unique best response. For stationary policies, it is sufficient to consider the best response functions in the single-period game. The slope of the retailer's best-response function is found by implicit differentiation as follows:

$$\frac{dR_r(Q)}{dQ} = -\frac{\frac{\partial^2 \pi_r^D}{\partial A \partial Q}}{\frac{\partial^2 \pi_r^D}{\partial A^2}} = \frac{f_D(Q) \frac{dED(A)}{dA}}{f_D(Q) \left(\frac{dED(A)}{dA}\right)^2 - \Pr(D(A) < Q) \frac{d^2 ED(A)}{dA^2}} > 0, \quad (14.10)$$

where the second term of the denominator is positive by Assumption 3. Positivity of the slope means that the retailer's customer acquisition spending is increasing in the quantity that the wholesaler stocks. However, due to the complexity of the expression for the slope, we are unable to verify either concavity or convexity of the best response function. Fortunately, this is not essential for any of the later results. Note that Figure 14.1 shows the retailer's best-response as a convex function, which is the case for the specific problem parameters used. We will now characterize the wholesaler's best response. For a given A, the wholesaler's optimization problem is,

$$\max_{Q \geq x^1} \Pi_w^D (A, Q, x^1) = \max_{Q \geq x^1} \left[cx^1 + \sum_{t=1}^{\infty} \beta^{t-1} \pi_w^D (Q, A) \right],$$

where π_w^D is the expectation of the retailer's single-period objective function defined as follows:

$$\pi_w^D (A, Q) = E\left[(w + h - c\beta) \min(D(A), Q) - (h + c(1 - \beta)) Q\right]. \quad (14.11)$$

Again the objective function is concave, and the wholesaler's first-order condition characterizes the unique best response. The slope of the wholesaler's best response function is found by implicit differentiation as follows:

$$\frac{dR_w(A)}{dA} = -\frac{\frac{\partial^2 \pi_w^D}{\partial Q \partial A}}{\frac{\partial^2 \pi_w^D}{\partial Q^2}} = \frac{dE(D)}{dA} > 0. \quad (14.12)$$

We see that the stocking quantity of the wholesaler is increasing in the customer acquisition spending by the retailer. This time we can also establish the sign of the second derivative

$$\frac{d^2 R_w(A)}{dA^2} = \frac{d^2 E(D)}{dA^2} < 0,$$

and it follows that the wholesaler's best response function is concave. We are then ready to demonstrate that, when the retailer and the wholesaler have equal power in the channel, there will be a unique Nash equilibrium.

PROPOSITION 3. *There exists a unique Nash equilibrium pair (Q^{DN}, A^{DN}) in the single-period drop-shipping game without starting inventory that is characterized by the following system of equations:*

$$\Pr\left(D\left(A^{DN}\right) < Q^{DN}\right) = \frac{w - c}{w + h - \beta c}, \quad (14.13)$$

$$\left.\frac{dED(A)}{dA}\right|_{A^{DN}} = \frac{1}{r - w} \frac{w + h - \beta c}{w - c}. \quad (14.14)$$

Then, for any $x^1 \leq Q^{DN}$, the solution (Q^{DN}, A^{DN}) is a unique Nash equilibrium of the multi-period game.

PROOF: Existence of the equilibrium follows from the concavity of the retailer's and the wholesaler's objective functions (see Moulin 1986 page 114). To show uniqueness it is sufficient to notice that the optimality conditions can be solved sequentially, i.e., condition (14.14) can be solved uniquely for A^{DN} and then condition (14.13) can be solved uniquely for Q^{DN}.

We now consider the problem in which the wholesaler acts as a Stackelberg leader and offers the retailer a "take-it-or-leave-it" contract that specifies a quantity of merchandise that the wholesaler is willing to stock.

PROPOSITION 4. *Define:*

$$\xi(A^{DW}, Q^{DW}) = \min\left[\frac{w + h - \beta c}{w - c}, 1 - \frac{dED(A)}{dA} \times \left.\frac{\partial R_r(Q)}{\partial Q}\right|_{A^{DW}, Q^{DW}}\right].$$

For ε having an Increasing Failure Rate (IFR) distribution, there exists a unique Stackelberg equilibrium solution pair (Q^{DW}, A^{DW}) in the single-period game with a powerful wholesaler and without starting inventory that is characterized by the following system of equations:

$$\Pr(D(A^{DW}) < Q^{DW}) = \frac{w - c}{\xi(A^{DW}, Q^{DW})(w + h - \beta c)}, \quad (14.15)$$

$$\left.\frac{dED(A)}{dA}\right|_{A^{DW}} = \frac{1}{r - w} \frac{\xi(A^{DW}, Q^{DW})(w + h - \beta c)}{w - c}. \quad (14.16)$$

Then, for any $x^1 \leq Q^{DW}$, the solution (Q^{DW}, A^{DW}) is the unique Stackelberg equilibrium of the multi-period game.

PROOF: The retailer acts second by solving (14.9). The wholesaler takes the retailer's best response function into account and solves the following problem:

$$\max_Q \pi_w^D(Q) = \max_Q E\left[(w+h-\beta c)\min(D(R_r(Q)), Q) - (h+c(1-\beta))Q\right].$$

Since the retailer's best response function is single-valued, the Stackelberg equilibrium exists. Further, we will show that the second derivative of the wholesaler's objective function is negative (i.e., the objective function is concave), and so the Stackelberg equilibrium is unique. The first derivative of the wholesaler's objective function is

$$\begin{aligned}
\frac{d\pi_w^D}{dQ} &= \frac{\partial \pi_w^D}{\partial Q} + \frac{\partial \pi_w^D}{\partial A}\frac{dR_r(Q)}{dQ} \\
&= (w+h-\beta c)\Pr(D(R_r(Q)) > Q) \\
&\quad + (w+h-\beta c)\Pr(D(R_r(Q)) < Q)\frac{dED(A)}{dA}\frac{dR_r(Q)}{dQ} \\
&\quad - (h+c(1-\beta)) \\
&= (w-c) - (w+h-\beta c)\Pr(D(R_r(Q)) < Q)\left(1 - \frac{dED(A)}{dA}\frac{dR_r(Q)}{dQ}\right).
\end{aligned}$$

The second derivative is:

$$\begin{aligned}
\frac{d^2\pi_w^D}{dQ^2} &= -(w+h-\beta c)\frac{d\Pr(D(R_r(Q)) < Q)}{dQ}\left(1 - \frac{dED(A)}{dA}\frac{dR_r(Q)}{dQ}\right) \\
&\quad + (w+h-\beta c)\Pr(D(R_r(Q)) < Q)\frac{d\left(\frac{dED(A)}{dA}\frac{dR_r(Q)}{dQ}\right)}{dQ}.
\end{aligned}$$

Note first that

$$1 - \frac{dED(A)}{dA}\frac{dR_r(Q)}{dQ} = 1 - \frac{f_D(Q)\left(\frac{dED(A)}{dA}\right)^2}{f_D(Q)\left(\frac{dED(A)}{dA}\right)^2 - \Pr(D(A) < Q)\frac{d^2ED(A)}{dA^2}} > 0.$$

From the second derivative, consider the term:

$$\frac{d\Pr(D(R_r(Q)) < Q)}{dQ} = f_D(Q)\left(1 - \frac{dED}{dA}\frac{dR_r(Q)}{dQ}\right) > 0.$$

Next, consider the term:

$$\frac{d\left(\frac{dED(A)}{dA}\frac{\partial R_r(Q)}{\partial Q}\right)}{dQ} = \frac{d\left(\frac{1}{1-\frac{\Pr(D(A)<Q)}{f_D(Q)}\times\frac{d^2ED(A)}{dA^2}/\left(\frac{dED(A)}{dA}\right)^2}\right)}{dQ}$$

$$= \frac{\frac{d}{dQ}\left(\frac{\Pr(D(A)<Q)}{f_D(Q)}\right)\frac{d^2ED(A)}{dA^2}/\left(\frac{dED(A)}{dA}\right)^2 + \frac{\Pr(D(A)<Q)}{f_D(Q)}\frac{d}{dQ}\left(\frac{d^2ED(A)}{dA^2}/\left(\frac{dED(A)}{dA}\right)^2\right)}{\left(1-\frac{\Pr(D(A)<Q)}{f_D(Q)}\times\frac{d^2ED(A)}{dA^2}/\left(\frac{dED(A)}{dA}\right)^2\right)^2},$$

where

$$\frac{d}{dQ}\frac{\Pr(D(A)<Q)}{f_D(Q)} = \frac{d}{dQ}\frac{\Pr(\varepsilon<Q-\theta(A))}{f_\varepsilon(Q-\theta(A))} > 0,$$

since ε has an IFR distribution, and finally

$$\frac{d}{dQ}\left(\frac{d^2ED(A)}{dA^2}/\left(\frac{dED(A)}{dA}\right)^2\right)$$

$$= \frac{\frac{d^3ED(A)}{dA^3}\left(\frac{dED(A)}{dA}\right)^2 - 2\left(\frac{d^2ED(A)}{dA^2}\right)^2\frac{dED(A)}{dA}}{\left(\frac{dED(A)}{dA}\right)^4}\frac{dR_r(Q)}{dQ} < 0,$$

by Assumptions 2 and 3 so that $d\left(\frac{dED(A)}{dA}\frac{\partial R_r(Q)}{\partial Q}\right)/dQ < 0$ and the result follows.

We would like to point out that the IFR assumption on ε is very non-restrictive since the IFR family includes just about any commonly used demand distribution.

Suppose now that the retailer acts as a Stackelberg leader and offers the wholesaler a "take-it-or-leave-it" contract that specifies an amount of money the retailer is willing to spend on customer acquisition.

PROPOSITION 5. *There is a unique Stackelberg equilibrium solution pair (Q^{DR}, A^{DR}) in the single-period game with a powerful retailer and without starting inventory that is characterized by the following system of equations:*

$$\Pr(D(A^{DR}) < Q^{DR}) = \frac{w-c}{w+h-\beta c}, \quad (14.17)$$

$$\left.\frac{dED(A)}{dA}\right|_{A^{DR}} = \frac{1}{r-w}. \quad (14.18)$$

Then, for any $x^1 \leq Q^{DR}$, the solution (Q^{DR}, A^{DR}) is a unique Stackelberg equilibrium of the multi-period game.

PROOF: The wholesaler acts second by solving (14.11). The retailer takes into account the wholesaler's best response function and solves the following problem:

$$\max_A \pi_r^D(A) = \max_A E\left[(r-w)\min(D(A), R_w(A)) - A\right].$$

Since the wholesaler's best response function is single-valued, the Stackelberg equilibrium exists. The first derivative is

$$\begin{aligned}\frac{d\pi_r^D}{dA} &= \frac{\partial \pi_r^D}{\partial A} + \frac{\partial \pi_r^D}{\partial Q}\frac{dR_r(A)}{dA} \\ &= (r-w)\Pr(D<Q)\frac{dE(D)}{dA} \\ &\quad + (r-w)\Pr(D>Q)\frac{dR_w(A)}{dA} - 1 = (r-w)\frac{dE(D)}{dA} - 1,\end{aligned}$$

and the second derivative is

$$\frac{\partial^2 \pi_r^D}{\partial A^2} = (r-w)\frac{d^2 E(D)}{dA^2} < 0.$$

The retailer's objective function is clearly concave, and the uniqueness of the Stackelberg equilibrium follows. Finally, the optimality conditions are found by equating the first derivatives to zero.

Note that in Model D, as opposed to Model T, the wholesaler bears all the inventory-related risk. The retailer still incurs all the customer acquisition costs that will benefit not only him, but also the wholesaler. Hence, none of the players has an incentive to behave system-optimally, as we will show later.

4.4 Comparative analysis of the stationary policies

The interpretation of the first optimality condition (for Q) in each model is a standard one for the newsvendor-type models: equating the marginal cost of stocking an extra unit of the product with the marginal benefit. The second optimality condition (for A) has a similar interpretation in the marketing literature.

OBSERVATION 1. *Denote by $\eta(A)$ the elasticity of expected demand w.r.t customer acquisition spending. Formally:*

$$\eta(A) = \frac{dED(A)}{dA} \bigg/ \frac{ED(A)}{A}.$$

Then the optimality condition that defines customer acquisition for all Models can be re-written as follows:

$$\eta(A^I)\frac{r-c}{r} = \frac{A^I}{rED(A^I)},$$

$$\eta(A^T)\frac{r-w}{r} = \frac{A^T}{rED(A^T)},$$

$$\eta(A^{DW})\frac{w-c}{\xi(w+h-\beta c)}\frac{r-w}{r} = \frac{A^{DW}}{rED(A^{DW})},$$

$$\eta(A^{DR})\frac{r-w}{r} = \frac{A^{DR}}{rED(A^{DR})},$$

$$\eta(A^{DN})\frac{w-c}{w+h-\beta c}\frac{r-w}{r} = \frac{A^{DN}}{rED(A^{DN})}.$$

Each optimality condition is interpreted as follows: the ratio of the total customer acquisition spending to the total expected revenue (right-hand side) is equal to the demand elasticity times the retailer's relative marginal profit (left-hand side which is revenue minus marginal cost divided by the revenue).

The result of Observation 1 parallels a result frequently encountered in the marketing literature (see, for example, page 571 in Lilien et al. 1992).

OBSERVATION 2. *In all models, customer acquisition spending and stocking quantity are strategic complements.*

PROOF: A sufficient condition for strategic complementarity is positivity of the cross-partial derivative (see Moorthy 1993) of the objective function, which can be easily verified.

For identical parameters and no initial inventory we can perform analytical comparisons of the models. Since all the solutions are stationary, it is sufficient to compare single-period profits. The following table summarizes the optimality conditions of the five models:

Model I			$\Pr\left(D\left(A^I\right) < Q^I\right) = \frac{r-c}{r+h-\beta c}$	$\frac{dED(A)}{dA} = \frac{1}{r-c}$
Model T			$\Pr\left(D\left(A^T\right) < Q^T\right) = \frac{r-w}{r+h-\beta w}$	$\frac{dED(A)}{dA} = \frac{1}{r-w}$
	Powerful wholesaler		$\Pr\left(D\left(A^{DW}\right) < Q^{DW}\right) = \frac{w-c}{\xi(w+h-\beta c)}$	$\frac{dED(A)}{dA} = \frac{1}{r-w}\frac{\xi(w+h-\beta c)}{w-c}$
Model D	Powerful retailer		$\Pr\left(D\left(A^{DR}\right) < Q^{DR}\right) = \frac{w-c}{w+h-\beta c}$	$\frac{dED(A)}{dA} = \frac{1}{r-w}$
	Nash equilibrium		$\Pr\left(D\left(A^{DN}\right) < Q^{DN}\right) = \frac{w-c}{w+h-\beta c}$	$\frac{dED(A)}{dA} = \frac{1}{r-w}\frac{w+h-\beta c}{w-c}$

The next Proposition summarizes the comparative behavior of the models under the same wholesale price.

PROPOSITION 6. *Suppose that in all five models the retail price, the wholesale price, the unit product cost, and the demand distribution are identical. Then the following characterizations hold:*

a) **Customer acquisition:** $A^I \geq A^T = A^{DR} \geq A^{DW} \geq A^{DN}$, *the customer acquisition spendings are highest in Model I and lowest in Model D.*

b) **Stocking quantities:** $Q^I \geq Q^T$ and $Q^I \geq Q^{DR} \geq Q^{DN}$. If $(w + h - \beta c) \leq \sqrt{(h + c(1 - \beta))(r + h - \beta c)}$ then $Q^T \geq Q^{DR} > Q^{DN}$, otherwise $Q^{DR} > Q^T$ and $Q^{DR} > Q^{DN}$. *Further, among the three drop-shipping models, Model DN always has the lowest stocking quantity,* $Q^{DW} \geq Q^{DN}$ and $Q^{DR} \geq Q^{DN}$.

c) **Retailer's profits:** $\pi_r^{DR} \geq \pi_r^{DN}, \pi_r^{DW} \geq \pi_r^{DN}$, *the retailer makes the lowest profits under the Nash equilibrium.*

d) **Wholesaler's profits:** $\pi_w^{DR} \geq \pi_w^{DW} \geq \pi_w^{DN}$.

e) **System profits:** *For* $(w + h - \beta c) \leq \sqrt{(h + c(1 - \beta))(r + h - \beta c)}$, $\pi^I \geq \pi^T \geq \pi^{DR}$, *otherwise* $\pi^I \geq \pi^{DR} \geq \pi^T$. *Further, it is always true that* $\pi^{DW} \geq \pi^{DN}$ *and* $\pi^{DR} \geq \pi^{DN}$.

PROOF: Results $a)$, $b)$, and $c)$ are obtained by a pair-wise comparison of the first-order conditions and employing the fact that $dED(A)/dA$ is decreasing in A. Results $\pi_r^{DR} \geq \pi_r^{DN}$ and $\pi_w^{DW} \geq \pi_w^{DN}$, i.e. the Stackelberg leader makes more profits than in a Nash equilibrium, are standard for Stackelberg games (see Simaan and Cruz 1976). The other results in $c)$ and $d)$ are obtained as follows:

$$\pi_r^{DN} = \pi_r^D(A^{DN}, Q^{DN}) \leq \pi_r^D(A^{DN}, Q^{DW}) \leq \pi_r^D(A^{DW}, Q^{DW}) = \pi_r^{DW},$$

where the first inequality follows from the observation that in Model D the retailer's profit is increasing in Q for a fixed A, and the second inequality holds since A^{DW} is an optimal response to Q^{DW}. Similarly

$$\pi_w^{DN} = \pi_w^D(A^{DN}, Q^{DN}) \leq \pi_w^D(A^{DW}, Q^{DN}) \leq \pi_w^D(A^{DW}, Q^{DW}) = \pi_w^{DW},$$

and also

$$\pi_w^{DW} = \pi_w^D(A^{DW}, Q^{DW}) \leq \pi_w^D(A^{DR}, Q^{DW}) \leq \pi_w^D(A^{DR}, Q^{DR}) = \pi_w^{DR}.$$

Finally, $e)$ follows from the fact that the system profit is jointly concave in A and Q, combined with the results in $c)$ and $d)$.

From part $a)$ of Proposition 6, we see that vertical disintegration leads to underspending on customer acquisition by the retailer, and in drop-shipping models the retailer underspends more than in the traditional model due to the misalignment of marketing and operations functions. It is interesting to note that the drop-shipping model performs the worst when players have equal power. The customer acquisition spending is closest to the system optimum when the retailer has channel power. The first part of this finding, $A^I \geq A^T$, that vertical disintegration leads to underspending on marketing effort, is similar to the result obtained by Jeuland and Shugan (1983). They, however, ignore inventory issues and consider deterministic price-dependent demand.

The intuition behind part $b)$ of the proposition is that not only does vertical disintegration lead to putting too little effort into customer acquisition in Models T and D, but also it leads to understocking. This is an effect caused by the double marginalization that was described in the economics, marketing, and operations literatures. We also see that, for moderate to high wholesale price, the drop-shipping model with a powerful retailer always leads to ordering more than in the traditional model. The model with powerful wholesaler is not particularly transparent to the analysis, due to the presence of ξ, which depends on problem parameters in a non-trivial way. Note, however, that by studying Figure 14.1 we can see that Q^{DW} can be above or below Q^{DR}, depending on the curvature of the best response functions. The Nash equilibrium again results in the lowest (i.e. the worst) stocking quantity.

Parts $c)$ and $d)$ state that in Model DN both players are worse off than in the other two drop-shipping models. The intuition behind this result is as follows: note that both best response functions are monotonically increasing, and that the integrated solution consequently has higher optimal decision values than in the Nash equilibrium. It is well known that a Stackelberg leader is better off than under the Nash equilibrium. Hence, the Stackelberg solution will be within the rectangular area formed by the Nash equilibrium and the integrated solution (see Figure 14.2). Finally, since the follower's profit is increasing in the leader's decision variable, the result follows. Also, the wholesaler prefers the model with a powerful retailer over the other two. This makes Model DR a potential candidate for the best of the three drop-shipping arrangements.

Finally, part $e)$ summarizes the most important comparative findings. We see that for a relatively small w there is no hope that either Model DR or DN will outperform the traditional model. However, for moderate to high wholesale prices, Model DR always outperforms the traditional model. Moreover, there is hope that even Model DN outperforms the traditional model for high wholesale prices. Note also that in the condition $(w + h - \beta c) > \sqrt{(h + c(1 - \beta))(r + h - \beta c)}$, the threshold value $\sqrt{(h + c(1 - \beta))(r + h - \beta c)}$ is closer to $h + c(1 - \beta)$ than to $r + h - \beta c$, and hence this condition accounts for more than 50% of the possible values of $w + h - \beta c$.

5. Supply chain coordination

In the previous section, we demonstrated that the solutions of Models T and D are generally different from the system-optimal solution. In addition to the double marginalization effect, drop-shipping is also plagued by marketing-operations misalignment. Can we come up with a mechanism that will induce coordination? As the next observation shows, a price-only contract is not sufficient.

OBSERVATION 3. *In Model T, the sole price-only contract that induces coordination has $w^T = c$ in which the retailer captures all profits, and in Model D there is no such contract.*

Not only are the price-only contracts inefficient, but also none of the other currently known contracts can coordinate the supply chain when customer acquisition expenses are considered.

OBSERVATION 4. *None of the following contracts – returns, quantity-flexibility, penalty (all as described by Lariviere 1999), revenue sharing (as described by Cachon and Lariviere 2000), or quantity discount (as described by Jeuland and Shugan 1983) – can coordinate the supply chains in Models T or D.*

To our knowledge, the only known contracts that work here are quantity forcing and franchising (Lariviere 1999). These contracts, however, might be difficult to implement in practice, as was noted in the literature. Clearly, in order to coordinate the supply chains considered here, we need a new form of contract. In Model T, we need a mechanism that will allocate a part of inventory risk to the wholesaler and also make the wholesaler bear some part of the marketing expenses. Pasternak (1985) and Kandel (1996) show that a returns contract in which the wholesaler offers partial credit for returned merchandise achieves supply chain coordination in the absence of marketing aspects. In our model this mechanism corresponds to the wholesaler offering the retailer a compensation proportional to the inventory carried over in each period (somewhat similar to holding cost subsidy in Cachon and Zipkin 1999). To achieve coordination, we add the notion of subsidized advertising similar to Chu and Desai (1995) who use the term "customer satisfaction assistance".

PROPOSITION 7. *The following contract achieves supply chain coordination in Model T: the wholesaler sponsors a portion of the retailer's customer acquisition expenses $a = (w - c)/(r - c)$, and at the same time offers the retailer a compensation $b = a(r(1 - \beta) + h)$ for each unit of inventory carried* over in each period. Under this contract, the retailer and the wholesaler split the total profit in proportions $1 - a$ and a, respectively.

PROOF: Under the coordinating contract, the wholesaler's objective function is

$$\pi_w^T = E\left[(1-\beta)(w-c)Q + \beta(w-c)\min(D,Q) - b(Q-D)^+ - aA\right].$$

By substituting the proposed coordinating parameters and using the fact that $(Q-D)^+ = Q - \min(Q,D)$, we get

$$\begin{aligned}\pi_w^T &= aE[(1-\beta)(r-c)Q + \beta(r-c)\min(D,Q)\\&\quad - (r(1-\beta)+h)(Q-\min(Q,D)) - A]\\&= aE\left[(r+h-\beta c)\min(D,Q) - (h+c(1-\beta))Q - A\right].\end{aligned}$$

Clearly, the objective function in the brackets is the integrated supply chain's profit. Similarly, the retailer's objective function is

$$\pi_r^T = E\left[(r+h-w\beta)\min(D(A),Q) - (h+w(1-\beta))Q - A\right].$$

Using the same technique we get

$$\pi_r^T = (1-a)E\left[(r+h-c\beta)\min(D(A),Q) - (h+c(1-\beta))Q - A\right].$$

Hence, the retailer will choose the system optimal decisions. This completes the proof.

Interestingly, the proportion of profit captured by the wholesaler is equal to the proportion of the customer acquisition costs she sponsors. If the wholesaler can influence the wholesale price w, she should choose it in combination with a according to the coordinating contract, so that a is as high as possible. This finding has an interesting implication for business practices on the Internet. Currently, many e-tailers are plagued by huge marketing expenditures, while our finding demonstrates that wholesalers should consider subsidizing a significant proportion of these expenditures to their own benefit. Of course, the higher the subsidized advertising, the higher the wholesale price. In practice this would mean that a retailer without sufficient funds for customer acquisition should seek a contract with a wholesaler who would be willing to subsidize a large portion of advertising in exchange for a higher wholesale price.

A different but somewhat similar contract works for Model D. We need a contract that would allocate some inventory-related risk to the retailer, while at the same time allocating some marketing expenses to the wholesaler. One such contract would be for the retailer to compensate the wholesaler for each unit of the inventory carried over while the wholesaler subsidizes a portion of customer acquisition expenses.

PROPOSITION 8. *The following contract achieves supply chain coordination in Model D: the wholesaler sponsors a proportion of the retailer's customer acquisition expenses* $a = (w - c)/(r - c)$, *and at the same time the retailer partially compensates the wholesaler for all unsold merchandise in the amount* $p = (h + c(1 - \beta))(1 - a)$ *per unit. Under this contract, the retailer and the wholesaler split total profit in proportions* $1 - a$ *and* a, *respectively.*

PROOF: Similar to Proposition 7.

The insights here are similar: the wholesaler gets a proportion of profits that is equal to the proportion of the customer acquisition expenses she sponsors. It follows that the wholesaler should strive to sponsor a relatively large portion of the retailer's customer acquisition expenses in combination with charging a higher wholesale price and receiving lower inventory compensation according to the coordinating contract. Note that both optimal contracts do not depend on the demand distribution, as was noted by Pasternak (1985) for pure returns contracts. This leads us to another observation:

OBSERVATION 5. *Propositions 7 and 8 hold for any demand distributions, including the ones that do not satisfy Assumptions 1-3.*

The last observation shows that the contracts we propose are robust, as has been repeatedly shown for returns contracts. We can do some comparison of the optimal contracts for Models T and D. First, note that in both cases the wholesaler sponsors the same proportion of the retailer's marketing costs (provided that the wholesale price is the same). This is a convenient property since, in this way, if the wholesaler works with both the traditional and the drop-shipping retailers, she does not need to discriminate among them. Discrimination in terms of subsidized customer acquisition costs might invoke some undesirable consequences due to legal limitations, since such discrimination might be considered as preferential treatment for some retailers. It is, however, possible that the wholesale prices in Models T and D will be different. Since the wholesaler in Model D has more involvement in channel functions and takes inventory risk, she is likely to demand a higher wholesale price. Scheel (1990), page 42, provides empirical data that indicates 10-20% higher wholesale prices for drop-shipping vs. conventional distribution established by the wholesalers to cover their extra costs.

OBSERVATION 6. *Suppose* $w^T < w^D$. *Then in coordinated supply chains,* $a^T < a^D$ *and* $\pi_w^T < \pi_w^D$, *the wholesaler sponsors more of the customer acqui-*

sition costs and has higher expected profit in the drop-shipping model.

Drop-shipping requires a certain investment from the wholesaler, since she should be able to handle small shipments directly to the customer. In addition, there are costs associated with taking inventory risk. This would lead the wholesaler to demand an increase in the wholesale price, resulting in, as the last observation shows, capturing more profits, which is beneficial in the long run.

6. Numerical experiments

To illustrate the models and gain additional insights into the differences among the alternative supply chain structures, we will now consider a specific form of the demand distribution and a specific form of its dependence on customer acquisition expenses[3]. Let the random term of the demand follow a uniform distribution, $U[0, \Delta]$. Further, let $\theta(A) = \sqrt{A}$. For numerical experiments, we will assume that $c = 5$, $h = 1$, $\beta = 0.95$, and $\Delta = 1$. The variable of interest is, of course, the wholesale price. This also allows insight into cases where the wholesale price of the drop-shipping supply chains is different from the wholesale price of the traditional supply chain. In what follows, we assume that the price-only contract is used. We analyze two scenarios: $r = 8$ and 12, to illustrate the situations with low and high margins. Note that the optimal profits and decision variables can be obtained and compared in closed-form. First, we look at the optimal customer acquisition expense (Figure 14.2) and stocking quantity (Figure 14.3).

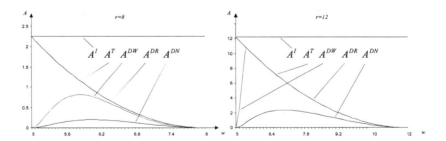

Figure 14.2. Optimal customer acquisition spending

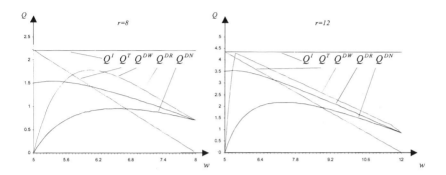

Figure 14.3. Optimal stocking quantity

Note the strange form of A^{DW} that is a result of highly non-linear term $\xi\left(A^{DW}, Q^{DW}\right)$ in the optimality condition. As we can see from both figures, in Model DN the retailer severely underspends on the customer acquisition and understocks. When margins are high, drop-shipping models with a Stackelberg leader closely resemble the traditional model. We will see this phenomenon again later. Finally, Model DW with the powerful wholesaler appears to perform somewhere in between the other two drop-shipping models, approaching Model DN for low margins and Model DR for high margins. We will now consider the retailer's profits (Figure 14.4).

In terms of the retailer's profit, the drop-shipping models outperform the traditional vertically disintegrated model for a wide range of parameters. The retailer (not surprisingly) is generally better off in Model DR where he has negotiation power. But even when the negotiation power is with the wholesaler (Model DW), the retailer's profit is higher than when the solution is a Nash equilibrium (Model DN), as we have demonstrated analytically. Any drop-shipping model dominates the traditional model for moderate to high wholesale prices. We consider the wholesaler's profit next (Figure 14.5).

Negotiation power does not do the wholesaler much good: she is generally better off when the retailer acts as a leader. The wholesaler's expected profit is virtually the same in the traditional channel structure T as in the drop-shipping

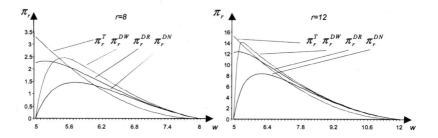

Figure 14.4. Retailer's profit

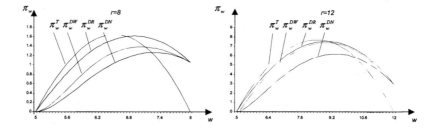

Figure 14.5. Wholesaler's profit

structure with powerful retailer DR. When the wholesale price is relatively high, then both players prefer Model DR over Model T – this gives indications for when (i.e. in terms of wholesale price) this model is preferable. The Nash equilibrium is again the least attractive to anyone. Model DW performs somewhere in between Models DR and DN. We then look at the system profits (Figure 14.6).

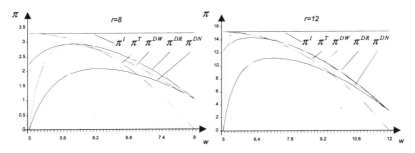

Figure 14.6. System profit

The drop-shipping model with the powerful retailer proves to be superior under moderate to high wholesale prices, as we demonstrated earlier. Under a very high wholesale price, any drop-shipping model outperforms the traditional model. Among the three drop-shipping models, Model DN is consistently the worst, and Model DR is consistently the best.

Finally, we will illustrate the coordinating contracts. Assume that $c = 5$, $r = 8$, and $\Delta = 1$. We denote profits for the retailer and the wholesaler resulting from the coordinating contract by π_r^C and π_w^C, correspondingly. As we noted before, a returns contract for the traditional supply chain and a penalty contract for the drop-shipping models splits profits in the same proportions. Figure 14.7 presents retailer's and wholesaler's profits with and without coordination for all models.

Observe that, in our example, under the coordinating contract, the retailer always makes more profit than without the coordinating contract. This is not true in general: under the low unit-revenue and high variability the retailer may be better off without the contract for high wholesale prices. The wholesaler, however, might be better off without coordination for low wholesale prices. Thus, Pareto optimality of the coordinating contract depends on the problem parameters.

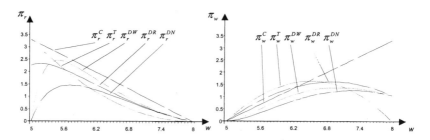

Figure 14.7. Retailer's and wholesaler's profits with and without coordination

7. Conclusions and discussion

Drop-shipping is a novel way of doing business on the Internet that has been inherited from catalog businesses and has already gained tremendous popularity. Since under drop-shipping the wholesaler takes inventory risk and performs fulfillment, the retailers can focus on customer acquisition. Currently, large retailers are able to capture a major part of the supply chain profit. Adopting drop-shipping will give wholesalers an opportunity to change this unbalance by increasing their involvement into the supply chain operations. Simultaneously, small retailers will benefit since barriers to entry in the form of large up-front inventory investment will diminish.

We find that drop-shipping introduces a conflict between marketing and operations functions that results in inefficiencies in the form of simultaneous understocking and spending too little on customer acquisition. As a result, both the retailer and the wholesaler should choose drop-shipping over the traditional contract only when the wholesale price is moderate to high. For example, if supply chain members decide to move from a traditional way of doing business to drop-shipping, one should consider a simultaneous wholesale price increase. This finding is consistent with the fact that under drop-shipping the wholesaler carries inventory-related risk and therefore should be able to increase the wholesale price; it is also consistent with the existing practice in catalog drop-shipping. Drop-shipping is more attractive when the retailer has the channel's power and can exercise it over the wholesaler. When channel power is equal, the retailer and the wholesaler arrive at a unique competitive equilibrium solution that significantly degrades system performance and usually does not benefit anyone.

We find simple contracts that achieve coordination in both the traditional supply chain and the supply chain with drop-shipping. According to these contracts, in the traditional channel the wholesaler subsidizes a portion of customer acquisition expenses as well as compensates the retailer for inventory carried over. In the case of drop-shipping inventory compensation goes from the retailer to the wholesaler. If the wholesaler can choose the wholesale price, the proportion of the customer acquisition expenses to subsidize, and the inventory compensation, an arbitrary split of profits can be achieved. In any case, the proportion of profits that the wholesaler captures coincides with the proportion of customer acquisition costs she subsidizes. Therefore, the higher the subsidy, the higher the wholesale price and the higher the wholesaler's profits.

One may wonder if a contract that specifies a proportion of subsidized customer acquisition expenses is enforceable. We have observed that some forms of subsidized marketing expenses are already in use by wholesalers. Alliance Entertainment Corp., a wholesaler that has implemented drop-shipping agreements, is one example. AEC publishes electronically an "All Media Guide" that is available to retailers working with AEC. According to AEC, the purpose of this database is to "... guide the consumer to make an intelligent purchasing decision and learn more about music and video." In order to publish this catalog, AEC employs about 600 professional and free-lance writers who create the content. Why would the wholesaler get involved into this completely different form of business? The "All Media Guide" is a form of subsidized customer acquisition expense. Spun.com, a retailer working with AEC according to the drop-shipping agreement, pays a basic weekly price of $1500 for access to the database, whereas it would cost Spun.com about $20M to create its own contents Forbes (2000).

Other methods of sponsoring customer acquisition costs are possible. Many companies now provide tools that register how many visitors saw the advertisement, how many interacted with it, and how many clicked through and made the purchase. Companies providing this kind of service include AdKnowledge, DoubleClick, MatchLogic, and others. By using these tools, both the retailer and the wholesaler can observe the impact of customer acquisition expenditures and contract upon it. As Scheel (1990), page 89, describes, in the practice of catalog drop-shipping it is conventional for the wholesaler to provide "... free photos, graphics, catalog sheets, color separations or other advertising aids or allowances." This practice seem to indicate that some sort of subsidized customer acquisition exists in drop-shipping.

Our model is an effort to introduce and understand the supply chain issues that arise under a drop-shipping supply chain structure with emphasis on the inventory risk allocation, supply chain interaction, and issues of channel power. Many extensions to our model are possible. The risk pooling effect when one wholesaler supplies several retailers will make drop-shipping even more appealing than with a single retailer as described in this chapter (see Netessine and Rudi 2002 for the analysis of this issue). It is, however, very encouraging to see that even without the risk pooling effect, drop-shipping in many cases outperforms the traditional supply chain structure. With multiple retailers, it is even possible that the system profit under drop-shipping can exceed the profit of a vertically integrated supply chain since the integrated channel does not enjoy the benefits of pooling. Finally, as Scheel (1990) suggests, retailers can carry the most popular products in inventory and drop-ship the rest directly from the wholesaler. This dual-sourcing problem raises many interesting questions for further research: which products should be stocked vs drop-shipped etc.

Notes

1. This is an invited chapter for the book "Supply Chain Analysis in the eBusiness Era" edited by David Simchi-Levi, S. David Wu and Zuo-Jun (Max) Shen, to be published by Kluwer. http://www.ise.ufl.edu/shen/handbook/. The authors are grateful to Gerard Cachon, Preyas Desai, Paul Kleindorfer and Martin Lariviere for helpful suggestions that significantly improved the chapter. The chapter also benefited from seminar discussions at the following universities: University of Rochester, Dartmouth College, Emory University, INSEAD, Georgia Institute of Technology, Cornell University, Northwestern University, University of Chicago, New York University, Columbia University, Carnegie Mellon University, Washington University at St. Louis, Duke University, University of Pennsylvania and University of Utah.

2. The exceptions are follow-up papers by Netessine and Rudi 2002 where multiple retailers are modeled and by Randall et al. 2003 that focuses on the empirical analysis of Internet companies. These papers build upon the analysis in this chapter and do not analyze marketing aspects

3. A link to an interactive web site for numerical experiments is provided at www.nilsrudi.com.

References

The ABCs of drop-shipping. *Catalog Age*, October 1997, 115.

Cheap tricks. *Forbes*, February 21, 2000, 116.

Online retailers less merry in '00. *Computer World*, August 14, 2000, 1.

Customer acquisition costs. *Computer World*, August 21, 2000, 48.

Cyberian Outpost taps Tech Data for online store. *Computer Reseller News*, October 11, 1999, 67.

Returns happen: reverse logistics online. *Inbound Logistics*, February 2000, 22-29.

The state of eretailing 2000. The supplement to *Eretailing World*, March 2000.

Aaker S. A. and J. M. Carman. 1982. Are you overadvertising? *Journal of Advertising Research*, Vol.22, no.4, 57-70.

Balcer, Y. 1980. Partially controlled demand and inventory control: an additive model. *Naval Research Logistics Quarterly*, Vol.27, 273-288.

Balcer, Y. 1983. Optimal advertising and inventory control of perishable goods. *Naval Research Logistics Quarterly*, Vol.30, 609-625.

Berger, P. D. 1972. Vertical cooperative advertising ventures. *Journal of Marketing Research*, Vol.XI, 309-312.

Berger, P. D. 1973. Statistical analysis of cooperative advertising models. *Operational Research Quarterly*, Vol.24, 207-216.

Berger, P. D. and T. Magliozzi. 1992. Optimal co-operative advertising decisions in direct-mail operations. *Journal of the Operational Research Society*, Vol.43, No.11, 1079-1086.

Cachon, G. P. 2002. Supply Chain coordination with contracts. Working Paper, University of Pennsylvania.

Cachon, G. P. 2003. The allocation of inventory risk in a supply chain: push, pull and advanced purchase discount contracts. Working Paper, University of Pennsylvania.

Cachon, G. P. and M. A. Lariviere. 2000. Supply chain coordination with revenue-sharing contracts: strengths and limitations. Forthcoming, *Management Science*.

Cahon, G. and P. Zipkin. 1999. Competitive and cooperative inventory policies in a two-stage supply chain. *Management Science*, Vol.45, 936-953.

Chu, W. and P. S. Desai. 1995. Channel coordination mechanisms for customer satisfaction. *Marketing Science*, Vol.14, 343-359.

Corbett, C. and G. A. DeCroix. 2001. Shared-Savings Contracts for Indirect Materials in Supply Chains: Channel Profits and Environmental Impacts. *Management Science*, Vol.47, No.7, 881-893.

De Groote, X. 1989. Flexibility and marketing/manufacturing coordination. *International Journal of Production Economics*, Vol.36, No.2, 153-169.

Eliashberg, J. and R. Steinberg. 1987. Marketing-production decisions in an industrial channel of distribution. *Management Science*, Vol.33, No.8, 981-1000.

Gerchak, Y. and M. Parlar. 1987. A single period inventory problem with partially controllable demand. *Computers and Operations Research*, Vol.14, No.1, 1-9.

Heyman, D. P. and M. J. Sobel. 1984, Stochastic models in operations research, Volume II: Stochastic Optimization. McGraw-Hill.

Hoffman, D. L. and T. P. Novak. 2000. How to acquire customers on the web. *Harvard Business Review*, Vol.78, No.3, 179-183.

Jeuland, A. P. and S. M. Shugan. 1983. Managing channel profits. *Marketing Science*, Vol.2, No.3, 239-272.

Johnson, E. M. and S. Whang. 2002. E-business and Supply Chain Management: an overview and framework. *Production and Operations Management*, Vo.11, No.4, 413-423.

Kandel, E. 1996. The right to return. *Journal of Law and Economics*, Vol.XXXIX, 326-356.

Lariviere, M. 1999. Supply chain contracting and coordination with stochastic demand. Chapter 8 in Quantitative Models for Supply Chain Management. S. Tayur, R. Ganeshan and M. Magazine, eds. Kluwer Academic Publishers.

Latcovich, S. and H. Smith. 2001. Pricing, sunk costs, and market structure online: evidence from book retailing. *Oxford Review of Economic Policy*, Vol.17, No.2, 217-234.

Lilien G. L., P. Kotler and K. S. Moorthy. 1992. Marketing Models. Prentice Hall.

Michael, S. 1999. Do franchised chains advertise enough? *Journal of Retailing*, Vol.75, 461-478.

Montgomery, D. and W. Hausman. 1986. Managing the marketing manufacturing interface. *Gestion 2000: Management and Perspective*, Vol.5, 69-85.

Moorthy, K. S. 1993. Competitive marketing strategies: game-theoretic models. In "Handbooks in operations research and management science: Marketing", J. Eliashberg and G. L. Lilien, eds.

Moulin, H. 1986. Game theory for the social sciences. New York University Press.

Narayanan, V. G. and A. Raman. 1998. Assignment of stocking decision rights under incomplete contracting. Harvard University working paper.

Netessine, S. and N. Rudi. 2002. Supply Chain choice on the Internet. Working Paper, University of Pennsylvania.

Pasternak, B. A. 1985. Optimal pricing and return policies for perishable commodities. *Marketing Science*, Vol.4, No.2, 166-176.

Petruzzi N. C. and M. Dada. 1999. Pricing and the newsvendor problem: a review with extensions. *Operations Research*, Vol.47, No.2, 183-194.

Porteus, E. L. and S. Wang. 1991. On manufacturing/marketing incentives. *Management Science*, Vol.37, No.9, 1166-1181.

Randall, T., S. Netessine and N. Rudi. 2002. Should you take the virtual fulfillment path?. *Supply Chain Management Review*, November-December, 54-58.

Randall, T., S. Netessine and N. Rudi. 2002. Empirical examination of the role of inventory ownership in the Internet retailing. Working Paper, University of Utah.

Sagomonian A. G., C. S. Tang. 1993. A modeling framework for coordinating promotion and production decision within a firm. *Management Science*, Vol.39, No.2, 191-203.

Scheel, N. T. 1990. Drop Shipping as a Marketing Function. Greenwood Publishing Group.

Simaan, M. and J. B. Cruz, Jr. 1976. On the Stackelberg strategy in nonzero-sum games. In "Multicriteria decision making and differential games", G. Leitmann, editor. Plenum Press, NY.

Simon, J. and J. Arndt. 1980. The shape of advertising function. *Journal of Advertising Research*, Vol.20, 11-28.

Shapiro, B. 1977. Can marketing and manufacturing coexist? *Harvard Business Review*, Vol.55, 104-114.

Tayur, S, R. Ganeshan and M. Magazine. 1998. Quantitative Models for Supply Chain Management. Kluwer Academic Publishers.

Van Mieghem, J. and S. Chopra. 2000. What e-business is right for your supply chain? *Supply Chain Management Review*, Vol.4, No.3, 32-40.

David Simchi-Levi, S. David Wu, and Z. Max Shen (Eds.)
Handbook of Quantitative Supply Chain Analysis:
Modeling in the E-Business Era
©2004 Kluwer Academic Publishers

Chapter 15

COORDINATING TRADITIONAL AND INTERNET SUPPLY CHAINS

Kyle D. Cattani
The University of North Carolina at Chapel Hill

Wendell G. Gilland
The University of North Carolina at Chapel Hill

Jayashankar M. Swaminathan
The University of North Carolina at Chapel Hill

Keywords: Channel coordination; pricing; retail channel; Internet channel; supply chain management; electronic commerce; mathematical modeling; game theory

1. Introduction

The Internet has provided traditional manufacturers and retailers a new avenue to conduct their business. On one hand, utilizing the Internet channel potentially could increase the market for the firm and, due to synergies involved, reduce the costs of operations. On the other hand, a new channel threatens existing channel relationships through possible cannibalization. This chapter explores recent research on coordination opportunities that arise for firms that participate in both traditional channels as well as internet channels. Three areas of coordination are discussed: procurement, pricing, and the backend operations of distribution and fulfillment.

1.1 Overview of Research

In its most parsimonious form, a supply chain consists of suppliers, manufacturers, retailers and customers who manage the bi-directional flows of goods, information and money. The opportunity for coordination arises at each interface. As indicated in Figure 15.1, we have labeled these three coordination opportunities procurement, distribution, and pricing. In this chapter, we identify and discuss research that has been conducted in each area. For a detailed overview of research on electronic supply chains see Swaminathan and Tayur (2003). Although the total research is these areas is quite voluminous, we focus on work that explicitly considers the role of the internet in conjunction with traditional supply chain relationships. We highlight opportunities for improved performance that arise from incorporating the internet into an existing supply chain. Figure 15.1 indicates the research papers that meet our criteria and are discussed in detail in this chapter. For ease of exposition, we cover these areas roughly chronologically; first describing the creation of a procurement policy, then discussing coordinated pricing practices, and finally addressing the opportunities for managing distribution after the product has been purchased.

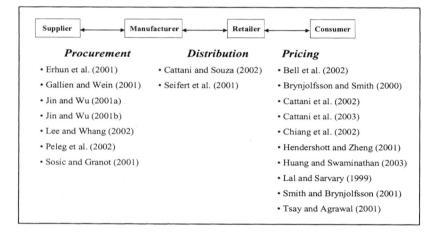

Figure 15.1. Map of Research

1.2 Procurement

An important change that has occurred due to e-business is in the area of supplier relationships and procurement in firms. The electronic business paradigm has created immense opportunity for firms to consolidate their buying processes. Many firms have benefited by adopting such *e-procurement* systems for procuring parts in the bill of materials as well as indirect goods and services. In fact, the first success stories in the area of e-procurement came from firms such as Ariba (www.ariba.com) and Commerce One (www.commerceone.com) that primarily dealt with indirect goods and services. The initial idea in these systems was that when procurement is consolidated under a single roof, the firm benefits from price reductions (a direct result of quantity discount). E-procurement systems provided an ability to bring all the qualified suppliers to the desktop of individual employees while at the same time making it easier and faster to complete the procurement process. Subsequently, more advanced models for procurement have evolved that utilize auctions to determine which firm wins a procurement contract. Each contract is open to bids in an electronic auction house and qualified suppliers are allowed to participate. Although auctions have been used for industrial contracts before, they were rather cumbersome to execute and thus were utilized only for very large and important contracts. E-business firms such as FreeMarkets (www.freemarkets.com, a business to business third party auctions firm) have made it much easier to conduct these auctions and as a result it is becoming more common for firms to use these auctions for procurement. This poses new research issues such as: should a firm use such auctions for all its components/suppliers; what is the long-term effect of such auctions on supplier relationships; how can firms use both traditional and auction mechanisms to hedge against risks in a more efficient manner; and how can a third party such as FreeMarkets make sure that capacities of suppliers are taken into account in these auctions so that the contract is executable.

A related phenomenon in the supplier management area is the formation of industry wide supply chain consortia such as E2Open (high tech) and Covisint (automobile). The motivation behind the formation of these consortia (also sometimes called marketplaces) is that they provide liquidity to inventory and capacity present in the extended supply chain as well as make it possible for the supply chain partners to get involved in long term collaborative planning and design efforts. For example, a manufacturer with excess inventory could salvage it in the marketplace. Similarly, buyers have an ability to conduct auctions for industrial parts as well as procure capacity options from the supply base. There are a number of new research issues that have evolved as a result of the above changes. For example, how could one quantify the benefits of joining such a consortia for a firm?; how many firms of the same type are likely to stay in a consortia?; how could the different firms use the liquidity and options in the marketplace to improve their operational performance? In section 2, we will discuss research models used to analyze the issues of procurement and supplier management.

1.3 Pricing

Using the Internet to sell products directly to customers provides several opportunities to affect the interactions and performance of an industry's supply chain. Among the choices that are most noticeable to the customer, as well as important for all participants in the supply chain, is the pricing decision that must be made at each node in the supply chain. We find that interesting and significant contributions can be gleaned from both prescriptive and descriptive research. Pricing in a supply chain that involves the Internet is affected by several important issues. The internet channel typically provides an alternative to an existing channel, so pricing decisions within the internet channel often must be made in the context of a multi-channel scenario. Customers will typically view the internet channel as an imperfect substitute for a traditional channel, meaning that differences in customer expectations or demographics must factor into any pricing decisions. From the supply side, the costs associated with an internet channel are likely to vary from those of the existing channel(s), further complicating the analysis required to determine prices in the new channel. When compared to traditional channels, knowledge of consumer behavior on the Internet is relatively scant; and due to the evolving nature of the Internet, one would expect consumer behavior to change over time. All these factors complicate the pricing decision, which can be one of the most important determinants in the ultimate success of companies as they manage the inevitable inclusion of the Internet as an integral part of their industry's supply chain.

Because this chapter focuses on coordination opportunities that arise between traditional and internet supply chains, we assume that products are being offered to customers through both traditional and internet channels. The degree to which a firm decides to vertically or horizontally integrate will determine the range of opportunities they have to coordinate different facets of the supply chain. Figure 15.2 displays four configurations that we will use to organize our discussion of research that discusses pricing and distribution issues for traditional and internet channels.

Imagine that we start from a standard abbreviated supply chain structure with a manufacturer selling products to a retailer, who in turns sells products to a customer through a traditional channel, such as a retail store or an industrial products distributorship. In model 1 of Figure 15.2, we consider the scenario when a company that is independent of both the manufacturer and the traditional retailer launches an internet-based alternative to the traditional channel (e.g., Amazon.com in books and each of the subsequent product categories in which they have chosen to compete). The second model we discuss, bricks and clicks, involves the traditional retailer opening an internet channel to provide customers multiple options for purchasing their products (e.g., The Gap extending their brand on-line, with a third-party manufacturer selling products

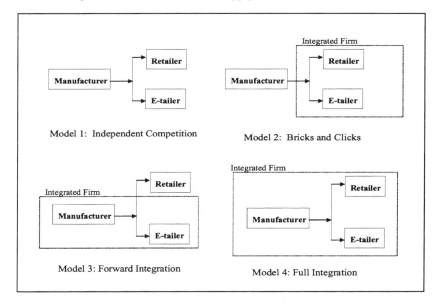

Figure 15.2. Internet Supply Chain Models

to The Gap to fill demand in both channels). In model 3, we depict a manufacturer who has elected to use the capabilities of the Internet to directly reach the customers of his product, forward integrating to provide an alternative channel that is under his direct control (e.g., Nike selling shoes and apparel on their web-site, in competition with traditional sporting goods retailers). Finally, we consider a fully integrated firm that controls the manufacturing function as well as end-customer relationships through both traditional retail channels as well as an internet channel.

1.4 Distribution / Fulfillment

For a manufacturer that has forward integrated to provide direct product sales (models 3 and 4 of Figure 15.1), the opportunity exists to coordinate distribution and fulfillment. Research taking a traditional approach to channel coordination (which seeks to eliminate double marginalization) is thoroughly reviewed in Lariviere (1999) and Tsay, et al. (1999). In this chapter, we focus attention on research that assumes the existence of both a traditional channel and an internet channel, and exploits the differences between the two channels to manage the supply chain more effectively.

A key feature of non-digital products purchased on-line is that the consumer has indicated a willingness to wait before actually receiving the product he/she has bought. In section 4, we summarize a couple of papers that have identified ways in which a forward-integrated or fully-integrated supply chain can use this information to improve performance.

2. Procurement

One of the major effects of the Internet on supply chain practices is in the area of procurement. Firms now utilize the Internet not only to diversify the supply base and hedge the risk but also to obtain lower costs through auctions. Several new firms such as FreeMarkets have built their business model around business-to-business supply auctions. Several other firms like Ariba and CommerceOne have added auction capabilities to their product suite. In this section we present several papers that address important issues related to supply chain management

At the strategic level, firms need to decide whether they should have long-term contracts with a few fixed suppliers or utilize auctions and a dynamic supplier base to reduce their costs. In particular, firms need to understand under what circumstances is it beneficial to have (1) long term relationships; (2) auction based short term relationships; or (3) a combination of (1) and (2). Another important decision is whether a firm should have one supplier or multiple suppliers and how that choice may depend on repetition in purchase. In the following sections, we discuss papers that study – (1) coordination of traditional and internet procurement; (2) formation of electronic consortia and their impact on procurement strategies.

2.1 Coordinating Traditional and Internet Procurement

Peleg, Lee, and Hausman (2002)

Peleg, et al. (2002) develop a two-period model with uncertain demand which considers the potential discounts that an exclusive long-term (strategic) supplier may be able to offer due to learning and compare that to the savings that might be generated through an auction. The strategic partner guarantees a price of p per unit in the first period and $p(1 - \Delta)$ in the second period where $0 < \Delta < 1$. They attribute this price decrease to the long term learning effects of having an exclusive supplier. In an auction based setting, the firm still pays a per unit price p in the first period but conducts an auction in the second period. The firm only knows the distribution of the minimum price (which is assumed to be independent of the quantity bought) that could be obtained as a result of the auction. In the combined strategy, the firm decides to use the exclusive supplier for the most part (by having a contract to order a minimum amount in the second period) and orders any remaining amount from the auction.

Let $H^i(x)$ denote the total expected costs for periods 1 and 2 under strategy i, given that the inventory level at the beginning of period 1 is brought up to x. Let Q_j denote the order quantity at the beginning of period j, D_j the demand in period j, and $\Phi_j(\bullet)$ and $\phi_j(\bullet)$ the cdf and pdf of the demand in period j, respectively. p is used to denote the unit price in period 1, Δ the percentage discount offered for period 2, and p_2 (with expected value $\bar{p}_2$) the price in period 2 (with pdf $f(\bullet)$ and cdf $F(\bullet)$), if an auction is conducted. Denote the unit per period shortage cost with π, the unit per period holding cost in period j with h_j, and the expected holding and shortage cost with $L_j(x)$.

Under the assumption of a strategic partnership with guaranteed prices p and $p(1-\Delta)$ in periods 1 and 2, respectively, the total expected cost for the two periods given Q_1 is:
$H^1(Q_1) = pQ_1 + L_1(Q_1) + E_{D_1}[G^1(Q_1 - D_1)]$, where
$G^1(Q_1 - D_1) = \min_{Q_2}\{p(1-\Delta)Q_2 + L_2(Q_1 + Q_2 - D_1)\}$.
The optimal order quantities Q_j^* are shown to satisfy:
$Q_2^* = [K_1 - (Q_1^* - D_1)]^+$ and
$\Phi_1(Q_1^*) = \frac{\pi - p\Delta}{\pi + h_1} - \frac{\pi + h_2}{\pi + h_1} \int_{D_1=0}^{Q_1^* - K_1} \phi_2(Q_1^* - D_1)\Phi_1(D_1)dD_1$, where
$K_1 = \Phi_2^{-1}\left(\frac{\pi - p(1-\Delta)}{\pi + h_2}\right)$.

Under the assumptions of a pure auction strategy, the buyer conducts an auction at the beginning of the second period and places an order with the lowest bidder at price p_2. Given Q_1, then the total expected cost for the two-period problem is

$$H^2(Q_1) = pQ_1 + L_1(Q_1) + E_{D_1}[G^2(Q_1 - D_1)],$$

where $G^2(Q_1 - D_1) = \min_{Q_2}\{E_{p_2}[p_2Q_2 + L_2(Q_1 + Q_2 - D_1)]\}$.
For this case, the optimal order quantities Q_j^* are shown to satisfy:
$Q_2^* = [K_2 - (Q_1^* - D_1)]^+$ and
$\Phi_1(Q_1^*) = \frac{\pi - (p - \bar{p}_2)}{\pi + h_1} - \frac{\pi + h_2}{\pi + h_1} \int_{D_1=0}^{Q_1^* - K_2} \phi_2(Q_1^* - D_1)\Phi_1(D_1)dD_1$, where
$K_2 = \Phi_2^{-1}\left(\frac{\pi - \bar{p}_2}{\pi + h_2}\right)$.

In the combined strategy the buyer forms an alliance with the first-period supplier such that the unit price in the second period is guaranteed to be $p(1-\Delta)$. To get this price guarantee, the buyer commits to purchasing a minimum quantity $\bar{Q}_2$. If additional units are required, the buyer can freely decide whether to buy from this supplier or to get the needed units in an auction at price p_2. Let n denote the number of units purchased in the second period at price $\min\{p_2, p(1-\Delta)\}$. Then the total expected costs for the two-period problem are

$H^3(Q_1) = pQ_1 + L_1(Q_1) + p(1-\Delta)\overline{Q_2} + E_{D_1}[G^3(Q_1 - D_1)]$, where
$G^3(Q_1 - D_1) = \min_{Q_2}\{E_{p_2}[\min\{p_2, p(1-\Delta)\}Q_2 + L_2(Q_1 + \bar{Q}_2 + Q_2 - D_1)]\}$.

For this case, the optimal order quantities Q_j^* satisfy:
$Q_2^* = [K_3 - (Q_1^* - D_1) - \bar{Q}_2]^+$ and

$$\Phi_1(Q_1^*) = \frac{\pi - p\Delta - \Gamma_{p_2}(p(1-\Delta))}{\pi + h_1}$$
$$- \frac{\pi + h_2}{\pi + h_1} \int_{D_1=0}^{Q_1^* + \bar{Q}_2 - K_3} \phi_2(Q_1^* + \bar{Q}_2 - D_1)\Phi_1(D_1)dD_1$$

where $K_3 = \Phi_2^{-1}\left(\frac{\pi - p(1-\Delta) + \Gamma_{p_2}(p(1-\Delta))}{\pi + h_2}\right)$ and

$$\Gamma_{p_2}(p(1-\Delta)) = \int_{p_2=0}^{p(1-\Delta)} [p(1-\Delta) - p_2]f(p_2)dp_2 = \int_{p_2=0}^{p(1-\Delta)} F(p_2)dp_2.$$

They show that there exists a Δ beyond which it is optimal for the buyer to prefer a strategic long-term partnership as opposed to utilizing an auction in the second period. Further, they show that depending on the distribution of price obtained from the auction, the combined strategy may be superior or inferior to a pure auction strategy. Finally, they explore the effect of increasing the supplier base (in the auction) by modeling the benefits obtained (due to potential lower prices) in the distribution of price obtained from the auction and conclude that the optimal number of suppliers is independent of the first period mean demand and could decrease or increase with the variance of demand in the first period. They also show that the optimal number of suppliers increase with the second period mean demand.

Erhun, Keskinocak, and Tayur (2001)

Erhun, et al. (2001) study a version of the combined strategy of Peleg, et al. (2002) in a more detailed setting. Their study is motivated by the automotive supply chain where a tier-1 manufacturer such as Ingersol-Rand may have to reserve capacity from their supplier worldwidepm.com which is a coalition of small powder metal technology manufacturers. Since the coalition is a powerful entity, the manufacturer may need to develop contracting mechanisms such as capacity reservation (which is similar in spirit to the minimum order assumptions of) Peleg, et al. (2002). Erhun, et al. (2001) study the strategic interaction and questions of participation (from the suppliers), profit sharing and impact on end customers. They analyze a two period model where the downward sloping demand could be in one of two states. This uncertainty in the first period is resolved in the second period (where demand becomes deterministic). As opposed to standard two period models, they assume that demand only occurs in the second period and that the first period is only for capacity reservation and demand resolution.

Let β_1 and β_2^i denote the capacity prices charged by the supplier (decision variables) in periods 1 and 2, when demand state is i ($i = 1$: high; $i = 2$: low). Assume a linear inverse demand function $p(Z^i) = a_i - b_i Z^i$, where $p(Z^i)$ is the product price when Z^i is the quantity produced by the manufacturer in demand state i. The manufacturer decides upon his production quantity Z^i, and upon the amounts of capacity to procure in each period (Q_1, Q_2^i) when demand is in state i. Π^S and Π^M denote the supplier and manufacturer profits, respectively. The solution to the two-period sequential pricing model is obtained by backward induction as follows. In the second period, after the supplier announces the capacity price β_2^i, the manufacturer sets Z^i and p_2 in order to maximize profits given demand state i and the first period's capacity procurement Q_1, i.e. $\Pi^M(\beta_2^i, Q_1) = \max Z^i(a_i - b_i Z^i) - \beta_2^i Q_2^i$ s.t. $0 \leq Z^i \leq Q_1 + Q_2^i$. The solution to this maximization problem determines the manufacturers reaction function as $Q_2^i = ((a_i - \beta_2^i)/(2b_i) - Q_1)^+$. Anticipating this response, the supplier determines the capacity price β_2^i to maximize his profit given demand state i and the manufacturer's first period capacity procurement Q_1, i.e. $\Pi^S(Q_1) = \max \beta_2^i Q_2^i = \max \beta_2^i \left((a_i - \beta_2^i)/(2b_i) - Q_1\right)^+$. The solution to this maximization problem determines the supplier's reaction function to Q_1 as $\beta_2^i = \frac{a_i}{2} - b_i Q_1$ if $Q_1 \leq \frac{a_i}{2b_i}$ and $\beta_2^i = 0$ otherwise. Anticipating the response functions of supplier and manufacturer, the manufacturer chooses the optimal procurement quantity Q_1 in the first period. Knowing this optimal procurement, the supplier sets β_1 to maximize profits. Under the above setting they demonstrate the existence of a unique subgame perfect Nash Equilibrium under various models. They show that having this additional negotiation period where demand is actually not realized, but demand resolution occurs and capacity is reserved, helps in improving the profits of the supplier, the manufacturer and the supply chain as a whole. Under uncertain demand and limited capacity at the supplier, they show that the additional period is beneficial but the impact on profits is complicated and depends on the level of capacity available.

Gallien and Wein (2001)

Motivated by FreeMarkets, Gallien and Wein (2001) study the design of smart markets for industrial procurement. They study a multi-item procurement auction mechanism for supply environments with capacity constraints. This enables them to model rational behavior of suppliers in terms of their responses when they have limited capacity. In particular, they consider a manufacturer interested in purchasing m different components (indexed by j) from a set of n possible suppliers (indexed by i). The desired quantity of each component j is fixed and denoted by Q_j. To this end, the manufacturer designs an auction to minimize the total cost of procurement across all the components taking into account the capacities of the various suppliers. The suppliers' capacity limitations are described by a linear model assuming a total amount K_i of production resource and amounts a_{ij} of this resource needed to produce one unit of each component type j. In each round, each supplier i gives a price bid b_{ij} for each component j. The assumption here is that the price bids decrease as the rounds progress. Further, at the end of each round, the auctioneer provides each of the suppliers information about allocations x_{ij} to them as well as the best bids for the next round that would maximize the profits of the supplier given all other suppliers hold on to their old bid. Allocations in round t are determined by means of the following LP (the "allocation engine"):

$$\underset{x_{ij}}{Min} \sum_{i=1}^{n} \sum_{j=1}^{m} b_{ij}(t) x_{ij}$$

$$s.t. \sum_{j=1}^{m} a_{ij} x_{ij} \leq K_i, \forall i$$

$$\sum_{i=1}^{n} x_{ij} = Q_j, \forall j$$

$$x_{ij} \geq 0, \forall (i,j)$$

The auction stops when the new set of bids from all the suppliers is identical to the previous round. In the terminal bidding round T, the bids $\mathbf{b}(T) = \{b_{ij}(T)\}$ are related to the allocations $\{x_{ij}(T)\}$ through the above LP. Having a fixed unit production cost C_{ij}, bidder i's payoff function is described by $\Pi^i(\mathbf{b}(T)) = \sum_{j=1}^{m} (b_{ij}(T) - C_{ij}) x_{ij}(T)$. Under the above conditions, the authors show that such an auction does converge and provides a bound for the profits of the manufacturer. Further, for special cases with two suppliers they provide insights on the impact of initial bids.

Jin and Wu (2001a)

While an auction typically serves as a price-determination mechanism, Jin and Wu (2001a) show that it also could serve as a coordination mechanism for the supply chain. Analyzing different auction schemes under both complete and asymmetric information assumptions in a two-supplier one-buyer framework, they demonstrate that different forms of auction and market mechanisms change the nature of supplier competition, thus the buyer-supplier interaction. In a wholesale price auction, the buyer announces the quantity Q and the two suppliers bid wholesale prices to maximize expected profits. In a catalog auction, the buyer determines the order quantity based on the given suppliers' price catalogs. In a two-part contract auction, additional side payments are introduced for channel coordination. Jin and Wu (2001a) assume an inverse linear demand function $p(Q) = a - bQ$, where $p(Q)$ is the product price when the quantity Q is sold. In the scenarios involving information asymmetry, the suppliers' marginal production costs as well as the intercept of the demand function parameter a are private knowledge to each supplier and the buyer, respectively, and common credible prior beliefs with respect to marginal costs are assumed (in terms of probability density functions). They show that channel coordination can be achieved in the two-part contract auction where the buyer announces a price-sensitive order function and the suppliers compete in an ascending bid side payment auction, if the market intermediary exerts effort to reduce the extent of information asymmetry, while restricting the buyer's profit on the side payments. Using the insights from the two-supplier one-buyer analysis they rank market schemes by their impact on expected channel efficiency, expected profitability for the buyer, expected profitability for the winning supplier, and expected commission revenue for the market maker.

Lee and Whang (2002)

Lee and Whang (2002) investigate the impact of a secondary market where resellers can buy and sell excess inventories. Motivated by HP's TradingHubs.com, they develop a two-period model with a single manufacturer and many resellers. The resellers order products from the manufacturer in the first period and are allowed to trade inventories among themselves in the second period. Consider n resellers that buy a product from the manufacturer at price p_1 (in period 1) and sell it at an exogenously given price π to end consumers. The salvage value for leftover inventory is assumed to be zero and the manufacturer's unit production cost is assumed to be constant and is normalized to zero. At the beginning of the first season, reseller i purchases a quantity Q_i and experiences sales D_i^0, which are independently and identically distributed with cdf $\Phi_0(\bullet)$. Without a secondary market, this is the standard newsvendor problem and the optimal order quantity for reseller i is given by $Q_i^0 = \Phi_0^{-1}\left((\pi - p_1)/(\pi)\right)$. To introduce the notion of a secondary market, Lee and Whang (2002) separate the sales season into two periods 1 and 2. As before, reseller i purchases a quantity Q_i from the manufacturer in the first period. First period sales are D_i^1 (i.i.d. with cdf $\Phi_1(\bullet)$). At the end of the first period, the second market is opened and resellers trade units at an equilibrium price p_2, which is determined to clear markets. Transaction costs are assumed to zero. Given p_2, every reseller chooses a stock level q_j to face the second period sales D_i^2 (i.i.d. with cdf $\Phi_2(\bullet)$). First and second period demand distributions are assumed to be independent and known to all parties. To guarantee comparability to the scenario without secondary market, it is assumed that $\Phi_0(\bullet)$ is the convolution of

$\Phi_1(\bullet)$ and $\Phi_2(\bullet)$. They solve for symmetric (subgame-perfect) equilibria by means of backward induction. In the second period the each reseller faces a newsvendor problem and $q_i^* = q^* = \Phi_2^{-1}((\pi - p_2)/(\pi))$ (owing to symmetry of the resellers). Given positive first period order quantities $\mathbf{Q} = \{Q_i\}$ (where $Q_i = Q, \forall i$, owing to symmetry) and sales $\mathbf{D}^1 = \{D_i^1\}$, they determine the equilibrium price p_2 and show that $\hat{p}_2 = \lim_{n \to \infty} p_2 = \pi[1 - \Phi_2(\int_0^Q \Phi_1(x)dx)]$. They argue that the number of resellers n in an internet-based market is sufficiently large to justify the analysis of limits. Then they show that $\hat{p}_2$ is decreasing in Q, while q^* is increasing in Q; as each reseller increases the initial purchase amount from the manufacturer, the secondary market price will go down, and this will encourage each reseller to carry more stock for the second period. Let $p_2^*(\mathbf{Q})$ and $q^*(\mathbf{Q})$, respectively, denote the equilibrium market price and the equilibrium stock level for the second period, when resellers make the first period ordering decisions $\mathbf{Q}$. With these responses functions and the symmetry assumption, they show that (for a sufficiently large number of resellers n), the first period order quantity Q^*, the second-period equilibrium price p_2^*, and the second-period stock level q^* satisfy the following simultaneous equations: $\Phi_1(Q^*) = (\pi - p_1)/(\pi - p_2^*)$, $p_2^* = \pi[1 - \Phi_2(\int_0^{Q^*} \Phi_1(x)dx)]$, and $q^* = \Phi_2^{-1}((\pi - p_2^*)/(\pi))$. They show that this second-period equilibrium price p_2^* is always smaller than the original (first period) purchase price. This is because a unit available in the second period has already lost the chance of being sold in the second period, such that it should have a smaller expected revenue. They show that there are two types of effects created due to the secondary market. First, there is a quantity effect (sales by the manufacturer) and then there is an allocation effect (supply chain performance). They show that the quantity effect is indeterminate; i.e., the total sales volume for the manufacturer may increase or decrease, depending on the critical fractile. In the traditional newsvendor problem, the optimal order quantity increases with the critical fractile. With a secondary market, however, this logic is diluted. They show that the reseller orders more than the newsvendor solution if the critical fractile is small, but less if the critical fractile is large. Consequently, a secondary market serves both as a hedge against excess inventory for a small critical fractile and a hedge against stockouts for a large critical fractile case. Controlling for the quantity effect, the existence of a secondary market always improves allocation efficiency. However, the sum of these two effects is unclear in that the welfare of the supply chain may or may not increase as a result of the secondary market. Finally, they present potential strategies for the manufacturer to increase sales in the presence of the secondary market.

2.2 Formation of Consortia

A related phenomenon in the area of procurement and supplier relationships has been industry wide consortia where multiple buyers and suppliers within an industry join and conduct business. Better transactional efficiency has been highlighted as the key benefit of these consortia such as Covisint (automotive) and Converge (hi-tech). The dynamics of these entities are not very well understood and pose several important questions. For example, one could expect that having multiple suppliers on the same platform is likely to reduce prices for the buyer, but it is likely to benefit other buyers in the consortium as well. Thus, it is not clear if one should join the consortium in the first place. In fact, one of the reasons for Dell deciding not to join either of the high tech consortia, Converge or e2open, could be that they do not want to open up their supply chain processes to competitors.

Sosic and Granot (2001)

Sosic and Granot (2001) study the benefits of consortia using a stylized model with three firms whose products have a certain degree of substitutability and explore conditions under which the formation of three-member and two-member alliances is optimal. In particular, they consider a deterministic linear model for demand that depends on the prices charged by the firm and its competitors as well as the degree of substitutability across the products. They assume that all retailers have full information and that when alliances are formed, the procurement cost (or the price paid by the firm) strictly decreases and that the sum of the reductions from a firm joining a two-firm coalition with each of the other firms is greater than the reduction obtained when all three firms are in the alliance. With these decreases in costs, they derive conditions under which joining a three- or two-firm alliance is beneficial for a firm. They identify conditions under which a firm may prefer to be in a two-firm alliance, despite the fact it might lead to lower profits, to prevent the other firms from forming an independent two-firm alliance. They also provide several other insights on the stability of these alliances.

Jin and Wu (2001b)

Jin and Wu (2001b) study the formation of supplier coalitions in the context of a buyer-centric procurement market. The buyer completely specifies an order (through quantity Q and reservation price $R(Q)$) before the auction starts. Let $C_i(Q)$ denote the cost function of supplier $i \in S$, where S is the set of all suppliers in the market. Each supplier computes pricing based on his own cost structure relative to the announced specifications and in a second-price seal-bid simultaneous descending auction, the buyer can reject any offer exceeding the reservation price. This auction scheme is incentive compatible and bidding the true cost is known to be a weakly dominant strategy. Without joining any coalition, the expected profit of supplier i is

$$E[\Pi^i(Q)] = \int_0^\infty [\min\{R(Q), C_j(Q) | j \in S/\{i\}\} - C_i(Q)]^+ f(Q) dQ.$$

Suppliers may form alliances under coalition mechanisms specified by the auctioneer. A coalition mechanism Ω specifies memberships in coalitions (coalition generation) and how profits are distributed among members. Once the profit distribution scheme is properly defined, the coalition generation can be derived from each player's incentives, which can be determined by comparing expected profits across possible coalitions. Building on the foundations of core games and bidding-rings, they explore the idea of "managed collusion" which provides a means to enhancing bidder profitability. They identify basic requirements for a valid coalition mechanism, including characteristics such as individual rationality, welfare compatibility, maintaining competition and financial balance. They show that a coalition mechanism could be constructed such that a stable coalition structure could be formed while the buyer does not lose the advantage from supplier competition. They propose a profit distribution scheme among members in the supplier coalition and show that the proposed scheme provides proper incentives such that (1) the best strategy for a coalition member is to comply with the coalition agreement, and (2) bidding the true cost is the best strategy so long as the bids are uniformly distributed and the bidder's cost is above a certain threshold. They also investigate the stable coalition structure under the proposed mechanism and show that under symmetric information there exists one unique strongly stable coalition structure.

3. Pricing

In this section we highlight research that addresses pricing issues under each of the four possible supply chain structures introduced in Figure 15.2.

3.1 Independent Competition

The Internet has provided customers (both industrial and consumer) with a new way to interact with their supply chain. Electronic commerce has been widely hailed as a revolution that will permanently transform the landscape of customer/supplier relationships. Although the realities of the dot-com bust have tempered the rhetoric used to describe the changes brought forth by electronic commerce, the Internet will continue to serve an important role in business during the 21st century. During the nascent period of electronic commerce, independent e-tailers, funded by a heavy flow of money from venture capitalists, played a dominant role in the development of internet-based sales to end consumers. For many product categories, independent e-tailers continue to play a vital role in their industry's supply chains. In this section, we focus our attention on the research that investigates pricing issues that arise when, in the most simplified form, a manufacturer, traditional retailer and internet-based retailer each exist as independent entities.

One of the most successful and well-known internet retailers is Amazon.com. In less than 10 years of existence, they have become a retailing giant with 2002 revenue of nearly $4 billion, coming entirely from online sales. They neither manufacture the products they sell, nor maintain a traditional retail channel. They participate in a supply chain that includes product manufacturers (Random House, HarperCollins, Sony, etc.) that are separate entities from the traditional retailers (Barnes & Noble, Borders, Waldenbooks, etc.) against which Amazon competes. The pricing decisions made by new, online retailers will have an important impact on the performance of all the members of the supply chain.

The pricing strategy employed by many independent online retailers is captured by Baker, et al. (2001). "Two very different approaches to pricing - neither optimal - have dominated the sale of goods and services through the Internet. Many start-ups have offered untenably low prices in the rush to capture first-mover advantages. Many incumbents, by contrast, have simply transferred their off-line prices onto the Internet."[1] Although the e-tailing shakeout of the past couple years has proved the use of the word "untenably" to be correct, there exist many factors that make it reasonable to imagine a profitable, independent e-tailer offering prices that are lower than his competitors in traditional channels. Traditional retailers make a significant investment in bricks and mortar storefronts, an expense that online retailers can largely avoid. Inventory management also provides an opportunity for an e-tailer to reduce expenses since they can pool inventory in a centralized location, rather than maintain separate inventory stocks at each retail store. In addition, internet customers provide a large degree of self-service by browsing the product offerings and placing an order without the assistance of any employees paid by the retailer. From a competitive perspective, the fact that consumers could theoretically switch from one e-tailer to another at the click of a mouse should also lead to lower prices because the relatively low switching costs would fuel greater price competition. One e-tailer, Buy.com, has even announced a strategy of pricing products at wholesale cost, thereby guaranteeing the company zero profits on each transaction, expecting to make their profit on advertising (Sgoutas (2000)).

One important advantage that the Internet provides in the pricing arena is the ability to rapidly and cost-effectively change prices. By simply manipulating a field in a database, internet prices can be immediately and automatically updated, without having to re-label any products or instruct any sales associates. Such flexibility can provide two important benefits. Prices as a whole can be adjusted, even by a very small amount, to reflect changing demand conditions, which might be triggered by macroeconomic factors or a move by a competitor. A second, more interesting opportunity, lies in the potential ability of internet retailers to price discriminate (i.e., charge different prices to different customers based on their willingness to pay). Taking advantage of the ability to "know" the customer before quoting a price, however, has proved a thorny issue for e-tailers. Partially due to the fact information can be communicated easily, customers are often aware of the prices paid for a product by other customers, and have strong feelings regarding the equity of price discrimination schemes. In a well-publicized incident, Amazon.com was testing price sensitivity by altering the price charged for a DVD to different customers. When some details of the test were pieced together by Amazon.com customers on an internet chat board, they created a backlash that resulted in severe negative publicity for the company (Hamilton (2001)).

Brynjolfsson and Smith (2000)

Although much of the discussion about the impact of internet channels on pricing has been anecdotal in nature, Brynjolfsson and Smith (2000) use a dataset of over 8,500 price observations collected over a period of 15 months to rigorously compare prices across internet and conventional retailers. They selected two product categories, books and CD's, with homogeneous products, and tracked the prices of 20 titles in each category at 12 different retailers (four pure e-tailers, four conventional-only retailers, and four retailers that sold products through both traditional channels and the web). Of the 20 titles in each category, half were selected from the best seller lists and the other half were randomly selected from a set of titles generally available at conventional retailers.

For both product categories, Brynjolfsson & Smith found that average posted internet prices are significantly lower (0.001 level of statistical significance) than conventional prices. For books, the Internet offered a price discount of 15.5% over the traditional channel, and for CD's the discount was 16.1%. The researchers also analyze the full price (including sales tax, shipping and handling, and transportation charges) and find again that the internet price is significantly lower than the price in conventional channels (by 9% and 13% for books and CDs, respectively).

Another aspect of Brynjolfsson and Smith's study examines the size of price changes in internet and conventional channels. They hypothesize that the lower cost of changing prices on the web will lead to less sizeable price changes by internet retailers. The average price change in conventional channels was $2.37 and $2.98 (for books and CDs respectively) as compared to $1.84 and $1.47 on the Internet, a difference that is significant at the .001 level for both product categories. Histograms of the price changes show that for both books and CDs, price changes smaller than $1.00 were much more frequent on the Internet than in conventional channels.

Finally, Brynjolfsson and Smith analyze price dispersion in both the conventional and internet channels. Classical economic theory suggests that price dispersion will arise when products are heterogeneous, customers face search costs, or consumers lack full information. The products in this study were selected to virtually eliminate product heterogeneity. Since the web is assumed to reduce search costs and make pricing information much more transparent, Brynjolfsson and Smith hypothesized that price dispersion would be much less on the Internet as compared to traditional channels. Price dispersion was measured in several different ways, but under each method the prices on the web were surprisingly diverse. For books, price dispersion was actually greater on the web in 8 of 12 months ($p < 0.01$), whereas price dispersion for CDs was somewhat greater in the conventional channel (statistically significant in only 5 of 12 months). They also find that the internet retailers with the lowest prices do not have the largest sales. They conclude that the surprisingly large dispersion of prices on the Internet is due to one of two sources of retailer heterogeneity: customer awareness or branding and trust.

Smith and Brynjolfsson (2001)

To further study consumer behavior when faced with a variety of prices on the internet channel, the authors conducted a second study analyzing choices made by customers using internet shopbots (Smith and Brynjolfsson (2001)). A shopbot is an internet tool that automatically compares prices across several internet retailers, thereby reducing search costs. They analyze almost 40,000 searches for a book title, monitoring the array of prices (both item price and total cost), and the internet retailer that the customer eventually visited. Smith and Brynjolfsson find that a large degree of price dispersion exists (lowest cost item is, on average, 28% below the mean price) and that a majority of customers do not choose the lowest priced item, despite the fact that the customers using an internet shopbot are likely to be more price sensitive than the general population. Among customers that do not choose the lowest priced item, the average price of the selected offer is 20.4% higher than the lowest priced selection. They use a multinomial logit model to assess the importance of various factors in the consumer decision making process, and find that consumers are willing to pay $1.72 more for an item from the "big three" internet booksellers (Amazon, Barnes & Noble, and Borders), with Amazon having a $2.49 price advantage over generic retailers and $1.30 advantage over their two biggest competitors. Smith and Brynjolfsson suggest that this result may be caused by consumers using brand name to signal reliability in service quality and other non-contractible characteristics.

3.2 Bricks and Clicks

As the dot-com shakeout continues, many of the pure e-tailing companies have disappeared. In most product categories, a bricks-and-clicks structure, where one company has both an internet and a traditional presence, is becoming the dominant form of internet participation. Large, traditional retailers such as Wal-Mart, Staples and Best Buy have websites that are selling their products in significant volume. These companies now have two distinct channels with which to reach their customers, and face significant challenges in effectively coordinating this two-pronged offering.

For many companies, the natural inclination is to price products on the web at the same price as their traditional retailers (Baker, et al. (2001), McIntyre and Perlman (2000)). An obvious benefit of maintaining consistent prices is that it facilitates centralized product decision making, as well as eliminates consumer feelings of confusion and inequity. Pursuing such a strategy, however, overlooks the opportunity to profit from differences that exist in the Internet and traditional marketplaces. Evidence exists of companies charging both higher and lower prices on their internet channel. Competition amongst on-line drugstores has led to prices at the internet sites of large drugstore chains, such as CVS, Walgreen and Eckard, to be 10% to 30% below the prices in their own traditional stores (Johannes (2000)). Marriott hotels, on the other hand, says the average price of rooms booked online is higher than those booked offline (Schlesinger (1999)).

Academic research is beginning to address the issue of how a bricks-and-clicks retailer should optimally set prices in each of their two channels. A conceptual model put forward by Balasubramanian and Peterson (2002) argues for an approach called "Channel Portfolio Management" that seeks to manage all of a firm's assets that interface with the marketplace in a coordinated fashion. They identify pricing as an important component of the marketing mix decisions that should be determined in a coordinated manner to optimally manage an entity that participates in multiple channels. A few authors have focused specifically on setting prices when competing in both an internet and traditional channel.

Lal and Sarvary (1999)

To address the issue of when and how the Internet is likely to decrease price competition, Lal and Sarvary (1999) utilize an analytic model that assumes demand comes from consumers who gather information on two types of product attributes: digital (that can be easily transmitted via the web) and nondigital (where physical inspection of the product is required). Consumers choose between two brands, but are only familiar with the nondigital attributes of the product they have most recently purchased. Consumers wishing to gather product information face search costs to uncover more information about either product, with the assumption being that digital information must be gathered at the time of each purchase.

Customers are assumed to be equally satisfied by the digital attributes of each product, but are in two distinct segments, of equal size, with respect to the nondigital attributes (such as the fit of a pair of jeans). Current customers of firm 1 are familiar with that firm's nondigital attributes, and have a reservation price R for buying their product. Firm 2's product will give them more utility $(R+f)$ with probability q and less utility $(R-f)$ with probability $1-q$, with q assumed to be less than $1/2$. To determine whether the competing product offers more utility, the customer must visit the firm's store, incurring a search cost. When a product is purchased over the Internet, it is not possible to determine the nondigital attributes of the product before making the purchase decision.

Two different scenarios are modeled: i) distribution through stores only, and ii) distribution through both stores and the Internet. The cost of selling a product through a store and the Internet is represented by c_T and c_I, respectively. A search cost of k_1 is incurred to visit the first store, and an additional cost of k_2 is incurred to visit the second store, with $k_2 < k_1$ to represent the fact that once a shopping trip is initiated, the cost of visiting a second store is less than the cost of beginning the trip and visiting the first store. No a priori assumption is made with respect to the relative magnitudes of c_T and c_I.

In the first scenario, the authors analyze the set of decisions facing the consumer and determine that there are three possible equilibria: i) consumers buy the familiar brand without searching, ii) consumers search to evaluate the unfamiliar brand and buy the familiar brand if there is a bad search outcome (i.e., the unfamiliar brand provides less utility than the familiar brand), and iii) consumers search, but do not buy anything if there is a bad search outcome. The third equilibrium, however, is proved to never exist. The prices charged are $R - k_1$ in the first equilibrium and $R - k_2$ in the second equilibrium.

When analyzing the second scenario, the authors assume that a company charges the same price in the internet and traditional channel. The fact that consumers can purchase products from the Internet allows consumers with a bad search outcome to go home and order their familiar product over the Internet. This results in two possible equilibria for this scenario: i) consumers do not search and buy the familiar product over the Internet, and ii) consumers search and buy the familiar product over the Internet in the case of a bad search outcome. The price in each equilibrium is R.

Under the set of assumptions used in this model, the introduction of an internet channel leads to a higher price than was charged with only the traditional channel. For some parameter values, the internet channel causes customers to abandon searching in favor of ordering the familiar product over the web, thus the Internet can lead to higher prices and a reduction in search activity. They also show that under certain conditions it is profitable for a company to introduce an internet channel even when the costs of that channel are higher than the traditional channel ($c_I > c_T$). The authors relax the assumption of equal prices in each channel to some extent by analyzing the case when a delivery charge, δ, is added to the price charged to internet customers, and show that their results hold under this assumption.

Cattani, Gilland, and Swaminathan (2003)

Cattani, et al. (2003) use a consumer choice model to determine the optimal prices that a firm should charge when they compete in both a traditional channel and an internet channel. They allow the firm to set different prices in each channel and analyze the effect of different factors on the relative prices in the two channels. Their results provide some guidance to firms attempting to establish an effective management structure for multi-channel competition.

Individual customers are modeled using a utility function for the product sold by the firm. A customer's utility for the product is decreasing in the price of the product and the effort to purchase the product. They assume that the scaled effort required to purchase a product is distributed uniformly between 0 and α_T for the traditional channel (representing factors such as distance from the store), and between 0 and α_I for the internet channel (representing factors such as internet connection speed and comfort with making purchases on the Internet). The traditional and internet efforts of a customer are assumed to be independent. A linear customer utility function common in the marketing literature is used: the utility a customer j receives from a product in channel x is

$$U_x^j = R - P_x - E_x^j$$

where $x \in \{T, I\}$ for traditional and internet channels, respectively, R = the customer's reservation price, P_x = the price in channel x, and E_x^j = the scaled effort customer j must expend to purchase the product from channel x. The relative values of α_T and α_I determine which channel is better suited to the customer population overall, but there will always be specific customers with a strong preference for each channel.

A customer will choose to purchase through the traditional channel or internet channel (or not purchase at all) depending on the prices in the two channels and the scaled effort they must exert to purchase from each channel. 15.3 depicts the market share that is garnered by each channel depending on the prices that are charged, where $I_1 + I_2$ represents the market share for the internet channel and $T_1 + T_2$ represents the market share for the traditional channel. The resulting profitability for a pair of prices can be easily calculated given the cost of selling products in each channel, c_I and c_T.

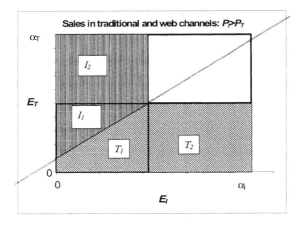

Figure 15.3. Customer Channel Preference Based on Distances ($P_I > P_T$)

Starting from a base case, where profits are maximized when the firm sells products solely through a traditional channel, the authors analyze several potential strategies for adding an internet channel. They first assume that the price in the traditional channel is left unchanged. The internet price can be set to maximize profits on the internet channel (such as would be likely if the internet channel and traditional channel were operated as two independent divisions), or to maximize total profits across both channels (as would result if both channels were centrally managed). In both scenarios, the internet price is higher than the traditional price when the cost of selling on the internet channel exceeds a certain threshold. The threshold is higher when the internet price is set to maximize total profits. When the cost of operating in each channel is identical ($c_I = c_T$), the internet price is higher than the traditional price when the internet price is set to maximize total profits, and is lower than the traditional price when the objective is to maximize just the internet profits.

A second strategy would be to allow prices in both the traditional channel and the internet channel to differ from the original price. When prices in both channels are matched (but not necessarily the same as the original price in the traditional channel), as many firms do in practice, the resulting price will be higher than the original price when the two channels' costs are the same. When different prices are allowed in each channel, the relationship between the prices in each channel is dependent upon the relative consumer acceptance of the Internet (characterized by the relationship between α_T and α_I in the model). Under the assumption that costs are the same in each channel, the internet price is higher than the price in the traditional channel when $\alpha_I > \alpha_T$ (i.e., the Internet is considered less convenient than the traditional channel), and the internet price is lower than the traditional price when $\alpha_I < \alpha_T$. This result suggests that as consumers become more comfortable with internet shopping, the optimal price charged by bricks-and clicks retailers will decrease relative to the price charged in the traditional channel.

As retailers consider adding an internet channel, the researchers find that it is important not to ignore profits from the existing channel in the determination of the web-channel price. In particular, the strategy that encourages the web channel to act entrepreneurially and maximize its own profits, does so at the expense of the existing channel and often leaves total profits (across both channels) far from optimal. In general, the strategy of holding traditional prices steady, but choosing web prices to maximize profits over both channels, seems to perform reasonably well based on computational experiments. This might be comforting to managers who wish to experiment with a web channel while not altering the existing channel. The strategy of pricing both channels equivalently performs reasonably well, especially when the costs in each channel are nearly identical. In practice, many companies have opted to match prices in the internet channel to the traditional channel prices, but when costs differ in the two channels, the overall profits from such a strategy can be significantly lower than the optimal profits that would arise from choosing the optimal prices for each channel.

Huang and Swaminathan (2003)

Motivated by the results in Cattani, et al. (2003), Huang and Swaminathan (2003) consider a linear downward sloping demand curve for the internet and traditional markets where the channels have a degree of substitution. In their model, demand (d_T and d_I) is given as follows:

$$d_T = uS \left(1 - \frac{\beta}{1-\theta}p_T + \frac{\beta\theta}{1-\theta}p_I\right)$$

$$d_I = (1-u)S \left(1 + \frac{\beta\theta}{1-\theta}p_T - \frac{\beta}{1-\theta}p_I\right)$$

where $0 \leq u \leq 1$, $0 \leq \theta < 1$, and β and S are positive.

In the above equations S represents the total market potential and β is the price elasticity factor. The parameters u and θ capture two different aspects of differentiation: the absolute difference in demand and the substitutability of the end products as reflected by the cross elasticity. Changes in u alter the relative product preferences in a way that preserve own- and cross-price elasticity, although the rates of change of quantities with respect to price are affected. The parameter θ, in contrast, affects the substitutability of the two products in terms of changes in prices. More specifically, θ is the ratio of the rate of change of quantity with respect to the price in the other channel to the rate of change of quantity with respect to price in the own channel. Using the above demand functions, they analytically prove that profits under the identical pricing strategy, as well as the strategy of holding traditional prices steady but choosing web prices to maximize profits over both channels, are no more than 4% away from the optimal level of profits under mild conditions. Further, they show that when there is competition between two players in the market, a retailer that uses both the internet and traditional channels will always price products higher than a pure internet retailer.

3.3 Forward Integration

For many companies that have traditionally sold their products through conventional retailers, the Internet provides an opportunity for them to directly reach their end customers. Many aspects of a direct channel are very appealing to the manufacturer: price control over their products, reduced competition from alternative vendors, stronger customer relationship, and more accurate indication of customer preferences. The Internet provides the manufacturer with the option of creating a direct channel at a fraction of the cost of establishing a bricks-and-mortar retail presence. Nike, the athletic footwear and apparel manufacturer, felt "the real opportunity for nike.com lay in defining a new, more profitable way of selling products to its loyal consumers" (McIntyre and Perlman (2000)). Creating an internet channel, however, often places the manufacturer in direct competition with their downstream supply chain partners. One important indication of the level of competitive threat posed by forward integration of manufacturers is the price they decide to charge for products purchased over the Internet. Thus pricing plays a role in determining the manufacturer's near-term profitability, as well as the degree of channel conflict they create. Not surprisingly, manufacturers that create an internet channel view their pricing choices as an important strategic decision. Nike, for example, hoped that by using full retail pricing on their website, they would reduce their traditional retailers concerns over unfair competition.

In another chapter of this book, Tsay and Agrawal discuss in detail much of the research that addresses channel conflict in multi-channel distribution systems. The pricing conclusions of this stream of research will be summarized here; the reader is encouraged to consult chapter 13 for a more detailed coverage of this subject.

Chiang, Chhajed and Hess (2002)

Chiang, et al. (2002) use a parameter, θ, with $0 < \theta < 1$, to measure customers' acceptance of products sold over the Internet. A consumer who places value v on the product when purchased through the traditional channel would value the product θv if it were purchased through the internet channel.

Each consumer's surplus equals the value they place on the product minus the price they pay, and a consumer uses the channel that gives them the highest value (provided it is non-negative). Prior to the introduction of the internet channel, the independent manufacturer and retailer face the classic problem of double marginalization, which causes supply chain profits to fall below their optimal level. Chiang et. al. find that a threshold value of θ exists, which separates the impact of the manufacturer's introduction of an internet channel. When $\theta \leq \hat{\theta}$, the internet channel is sufficiently unappealing to the customers that it has no effect on the marketplace. For $\theta > \hat{\theta}$, the introduction of the direct channel impacts the behavior of the traditional retailer, resulting in a new equilibrium even though no products are actually sold through the direct channel. In this new equilibrium, both the wholesale price and the retail price are lower than they were in the original case. For $\hat{\theta} < \theta < \bar{\theta}$, the profits of the retailer will increase, whereas for $\theta > \bar{\theta}$ the retailer's profits decline. The profits of the manufacturer increase for all cases where the equilibrium changes, as the lower price charged by the retailer eliminates some of the effects of double marginalization.

Hendershott and Zhang (2001)

Hendershott and Zhang (2001) model consumers with different product valuations who can purchase a product from one of many traditional retailers, or directly from the manufacturer. The retailers have different cost functions, and thus charge different prices. The consumers can only determine prices through a time-consuming search, which reduces the utility they experience when they eventually purchase the product. They find that the equilibrium market structure depends on the manufacturer's transaction cost for selling over the Internet, c_I. When c_I is sufficiently low, the internet channel will dominate the traditional channel and serve all customers. When c_I is sufficiently large, the manufacturer will choose not to utilize the internet channel. For intermediate values of c_I, a hybrid equilibrium will result, with both internet and traditional channels proving viable. In the hybrid equilibrium, the price charged by each traditional retailer is lower than the price charged by the same retailer before the introduction of the internet channel. The level of the internet price relative to the price of the traditional retailers depends on the costs of each retailer and the cost of selling over the Internet. The wholesale price charged by the manufacturer to the traditional retailers is the same before and after the introduction of the internet channel.

Tsay and Agrawal (2001)

Tsay and Agrawal (2001) present a model where a traditional and internet retailer compete against each other for customers based on both the price they charge and effort they put forth. A price increase by one competitor drives some customers out of the market and drives some customers to the other retailer; an increase in effort by one competitor causes demand for both retailers to increase. Supply chain costs of c_T and c_I are incurred by the traditional and internet retailer, respectively. Tsay and Agrawal find that the wholesale price charged by the manufacturer to the traditional retailer always decreases with the addition of an internet channel, but that the effect on the price charged by the retailer is difficult to characterize.

Bell, Wang, and Padmanabhan (2002)

Bell, et al. (2002) also analyze a model where each retailer's demand is negatively affected as their own price increases, and positively affected by increases in the price of other retailers, their own marketing effort, and the effort of other retailers. They consider competition between three retailers: either all traditional retailers, or two traditional retailers and one retailer that results from the forward integration of the manufacturer. Contrary to the findings in most other studies, Bell, et al. find that the prices charged in both the internet channel and the traditional channel are unequivocally higher than the price charged in the traditional channel prior to the introduction of the internet channel. They also find that the price charged by the manufacturer in the internet channel is higher than the price charged in the traditional channel (after the introduction of the internet channel). This result suggests that manufacturers who introduce a direct internet channel may reduce price competition, rather than increase it, even when they are acting optimally.

Cattani, Gilland, Heese, and Swaminathan (2002)

Cattani, et al. (2002) use the same model of consumer demand as in Cattani, et al. (2003) (see section 3.2) to study the effect of forward integration by the manufacturer on the equilibrium prices and profits. They assume that the manufacturer attempts to minimize channel conflict by setting their price to the customer equal to the price being charged by the retailer, who maximizes profits in the context of the competitive scenario. They consider three alternative strategies: i) maintain the wholesale price at the level prior to introduction of the internet channel, ii) set the wholesale price such that the price charged to the consumer will not change after the introduction of the internet channel, and iii) set the wholesale price to maximize manufacturer profits, given that the manufacturer will charge a price in the internet channel that matches the traditional channel price. Under the first strategy, the retail price will be higher after the introduction of the internet channel (with wholesale prices remaining constant by assumption), and the retailer's profits decline. Under the second strategy, the wholesale price is reduced sufficiently that the retailer's profits increase despite the reduction in market share. When the manufacturer maximizes profits, the resulting retail price is lower than the original price, provided that the cost of selling through the internet channel is not prohibitively large. When costs in the two channels are comparable, both the retailer and the manufacturer prefer (i.e., have higher profits) the third strategy.

3.4 Full Integration

Taking the notion of partial integration one step further, it would be possible to consider a manufacturer with a captive traditional channel that is considering the addition of an internet channel. This fully integrated organizational structure has received little attention in the literature, perhaps because it is relatively rare in business today. Certain branded fashion retailers (e.g., The Limited, Abercrombie & Fitch) or paint companies (e.g., Sherwin Williams) may provide the best examples of such a structure. Chiang, et al. (2002) focus some attention on the fully integrated model. They conclude that the retail price in the fully integrated model would remain the same (as compared to the retail price when there is no internet channel), and the internet price would be lower than the price in the traditional channel (they assume that the selling cost in the internet channel is less than in the traditional channel).

A fully integrated structure, with a manufacturer controlling both a traditional and internet channel, provides opportunity for further research in this area.

4. Distribution / Fulfilment

In addition to pricing and sourcing decisions, the introduction of a direct channel to complement an existing traditional channel can create opportunities to coordinate distribution and fulfillment through the channels to improve profits, especially for the more integrated firm structures such as models 3 and 4 of Figure 15.2. In particular, the willingness of customers to wait for delivery in the direct channel raises the possibility of using the direct channel as an outlet for leftover inventory in the traditional channel, or as a lower priority buffer that allows better service to the traditional channel.

In this section we provide an in-depth review of papers that cover these two possibilities, Seifert, et al. (2001) and Cattani and Souza (2002). Seifert, et al. (2001) considers the former: they explore the possibility of extending the cooperation between the manufacturer and the retailers to include the direct channel in existing distribution channels. This arises under model 3 of Figure 15.2. Cattani and Souza (2002) consider the latter possibility: they investigate inventory–rationing policies of interest to firms operating in a direct market channel, modeling a single product with two demand classes, where one class requests a lower order fulfillment time but pays a higher price. This arises under model 4 of Figure 15.2.

The related distribution and fulfillment literature includes traditional stochastic inventory comprehensively reviewed by Heyman and Sobel (1990).

4.1 Direct channel as outlet

Seifert, Thonemann, and Rockhold (2001)

Seifert, et al. (2001) assume that the manufacturer has a direct (internet) market that serves a different customer segment than the traditional retail channel (not owned by the manufacturer) and that demands in the two channels are thus independent. They model the inventory decisions for a supply chain comprised of a manufacturer and N retailers, as shown in Figure 15.4. They compare a dedicated supply chain to a cooperative supply chain. In a dedicated supply chain the inventory sent to each retailer is used exclusively to meet demand at that retailer while inventory at the internet store is used exclusively to meet demand at that store. In a cooperative supply chain, leftover inventory at the retailers is available to fill demand at the internet store.

The rationale for the model is that retail customers demand instantaneous availability and the retailer will experience lost sales if the product is not in stock. In contrast, internet customers are willing to wait. In particular, the short delay incurred by transshipping leftover product from a retailer to the customer does not result in lost sales in the internet channel, although the longer delay required to source the product from the manufacturer does result in lost sales in the internet channel (as it does in the traditional channel).

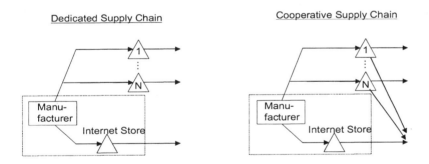

Figure 15.4. Dedicated and Cooperative Supply Chains

The problem is modeled as a stationary multi-period newsvendor problem with lost sales. Assuming a coordinated policy (where double marginalization has been eliminated through centralized decision making).[2] The decision to be optimized is the base stock level at the traditional retailers, S_{Ti}, with $i \in \{1,\ldots,N\}$ and at the internet store, S_I. Since each traditional retailer is identical, the subscript i can be dropped. Demand in each period has distribution $F_T(\cdot)$ at each traditional retailer and $F_I(\cdot)$ in the internet channel. The retail price p is the same in both channels but the unit production costs c_T and c_I, differ according to channel. There is a one-period discount factor α, holding costs by channel h_T and h_I, lost sales cost π (independent of channel), and a transshipment cost u, which is constrained to the range $(h_T - h_I) - \alpha(c_T - c_I) \leq u \leq \pi + h_T - \alpha c_T$, corresponding to the more interesting case where independent supply chains are attractive in spite of higher unit costs associated with demand met through transshipments.

Optimal inventory base stock levels in the traditional and internet channels for a dedicated supply chain are $S_T^* = F_T^{-1}((p + \pi - c_T)/(p + \pi + h_T - \alpha c_T))$ for each traditional retailer and $S_I^* = F_I^{-1}((p + \pi - c_I)/(p + \pi + h_I - \alpha c_I))$ for the internet channel. These levels are readily interpreted using the standard newsvendor analysis of underage and overage costs. The underage costs consist of lost margin plus lost sales cost while the overage costs are comprised of holding costs plus the financial cost of holding a unit in inventory for a period $(1-\alpha)c_T$ or $(1-\alpha)c_I$ for the two channels, respectively.

Optimal inventory base stock levels for a cooperative supply chain are
$$S_T^* = F_T^{-1}\left(\frac{p+\pi-c_T}{p+\pi+h_T-\alpha c_T} + \frac{p+\pi+h_T-(\alpha c_T+u)}{N(p+\pi+h_T-\alpha c_T)} q_T'(N, S_T^*, S_I^*)\right)$$
and $S_I^* = F_I^{-1}\left(\frac{p+\pi-c_I}{p+\pi+h_I-\alpha c_I} + \frac{p+\pi+h_T-(\alpha c_T+u)}{N(p+\pi+h_I-\alpha c_I)} q_I'(N, S_T^*, S_I^*)\right)$
where $q_T'(N, S_T^*, S_I^*) \geq 0$ and $q_I'(N, S_T^*, S_I^*) \leq 0$ are the partial derivatives of the expected number of units cross sold with respect to the base stock levels of the corresponding channel. In the cooperative supply chain, compared to the dedicated supply chain, the critical fractiles are adjusted to reflect the benefit of cross shipments which lower the cost of overage for the traditional retailers and decrease the cost of underage for the internet store.

Seifert, et al. (2001) also show that decentralized decision making in the cooperative supply chain can be coordinated through use of an ending-inventory subsidy of b and a transfer payment t for each unit that the retailer ships (at cost) to the virtual store in response to actual demand, both paid by the manufacturer to the retailer.

The following properties hold for coordinated supply chains:

1 The optimal base-stock level at a traditional retail store in the cooperative supply chain is higher than in the independent supply chain.

2 The optimal base-stock level at a traditional retail store in the cooperative supply chain decreases in N and approaches the optimal base-stock level in the independent supply chain as $N \to \infty$.

3 The optimal base-stock level at the internet store in the cooperative supply chain is lower than in the independent supply chain.

4. The optimal base-stock level at the internet store in the cooperative supply chain decreases in N.

5. The lower bound for the base stock level of the internet store as $N \to \infty$ is a newsvendor-type solution and if
$(u + \alpha c_T - c_I - h_T)/(u + \alpha(c_T - c_I) - (h_T - h_I)) \leq 0$, then the optimal base-stock level at the internet store will be zero for sufficiently large N. When this occurs, the benefit to the traditional retailer of having an outlet for excess inventory more than offsets the increased cost to the internet store of sourcing product indirectly through the traditional retailers.

While Seifert, et al. (2001) show the benefits of supply chain cooperation, they note that in the examples they have explored, the benefits of coordinating a supply chain are greater than the benefits of cooperating, indicating that supply-chain coordination (to eliminate double marginalization) probably should precede supply-chain cooperation. In any case, supply-chain coordination is a compelling motivation for supply chain integration.

In addition to the eased ability to coordinate channels in an integrated supply chain, the next section considers an example of other benefits possible if the supply chain is fully integrated.

4.2 Direct channel as Service buffer

Cattani and Souza (2002)

Cattani and Souza (2002) study rationing policies in which the firm either blocks or backlogs orders for lower priority customers when inventory drops below a certain level. Their model can be applied directly to this chapter's theme by considering two classes of customers to be alternate channels of the same firm, with the higher priority customers coming from the traditional retail channel and the lower priority customers from the internet channel. Given that the internet customers are willing to wait, rationing policies can be used to improve the overall performance of the firm compared to the alternative allocation of products to customer orders on a first-come, first-served (FCFS) basis.

The research is founded on the use of queuing models in an inventory system using methodologies similar to Cachon (1999) and Ha (1997). The model considers continuous-review inventory management for a single product type with two classes of customers, traditional (denoted with a T) and internet (I).

Demand for each class follows a Poisson process with rate λ_k, where k ∈ {T, I}; total demand has rate $\lambda = \lambda_T + \lambda_I$. Inventory is issued from a common stockpile of finished goods inventory (FGI); this stockpile is fed by a production system, or by a co–located supplier, which has exponentially distributed build times (for tractability) with rate μ. Inventory is held only in FGI and incurs a fixed holding cost rate h per unit inventory. Production stops after inventory reaches the level S — a base-stock policy is optimal for this supply chain in that it maximizes supply chain profit per unit time (see Beckmann (1961); Hadley and Whitin (1963)).

Customers from the traditional channel will purchase only if inventory is available. Lost-sale costs (goodwill) of π per unit are incurred. Internet customers will backlog at no cost as long as the wait is not excessive. This effect is approximated by limiting the backlog to K. The revenue, including shipping, charged to a customer of type k $\in$ {T, I} is r_k. Production cost per unit is c and there is an incremental channel cost of c_k.

The resulting margin due to a unit sale to a customer in channel k is $m_k = r_k - c - c_k$. Let $\rho_k = \lambda_k/\mu$ be the system's utilization regarding only type k customers; $\rho = \lambda/\mu$ the system's utilization; I be average inventory; D_k be expected sales rate for customers of type k; $L_k^= \lambda_k - D_k$ be the expected lost sales rate for customer type k, and B be the expected backorder rate for internet customers.

The profit rate for Policy j is a function of prices, costs and demand rates; it is determined by modeling inventory as a continuous time Markov chain, similarly to Cachon (1999) and is computed as $\prod = \sum_{k \in \{T,I\}} \{m_k D_k - \pi_k L_k\} - hI$.

The FCFS policy is analyzed as follows:

Policy 1: Ship orders as they arrive (FCFS), whether they are traditional retail orders or internet orders. When inventory drops to zero, backlog internet orders and ship them as inventory arrives.

For a given S, Policy 1 is a birth and death process, with states $\{-K, ..., -1, 0, 1, ..., S\}$. The death process rate is λ for states 1 through S, and λ_I for non-positive states. For a given S, the stationary probabilities are determined, and from them D_K, B, I, and Π.

If lost-sale costs from traditional retail orders are high, and if internet orders can be backlogged, then a FCFS policy such as Policy 1 may have lower profits compared to a policy that reserves some units for the retail customers by backlogging lower-margin orders when inventory levels are low.

Policy 2: If inventory drops to S_a or lower, ship only traditional retail orders; backlog internet orders and ship them after inventory returns to S_a.

The Markov chain associated with this policy is shown in Figure 15.5.

A state is defined as (i, u), where i is the inventory level and u is the backlogging level. The solution for the stationary probability distribution of the Markov chain in Figure 15.5 is calculated numerically by solving a system of equations as follows. Let Q denote the chain's rate matrix, p the matrix of $p_{i,u}$, and 0 the matrix of zeros. Then, $p_{i,u}$ is the solution to the following system of equations:

$$pQ = 0, \text{ and } \sum_{i=0}^{S} \sum_{u=0}^{K} p_{i,u}^{(2)} = 1.$$

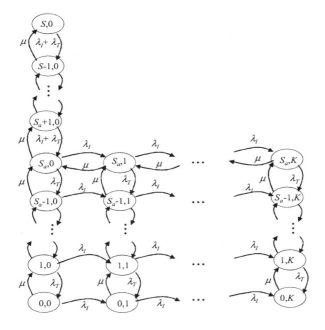

Figure 15.5. Markov Chain Associated with Policy 2

Then, $D_I^{(2)} = \lambda_I \left[\sum_{i=S_a+1}^{S} p_{i,0}^{(2)} + \sum_{i=0}^{S_a} \sum_{u=1}^{K-1} p_{i,u}^{(2)} \right]$, $D_T^{(2)} = \lambda_T \left(1 - \sum_{u=0}^{K} p_{0,u}^{(2)} \right)$, $I^{(2)} = \sum_{i=0}^{S} \sum_{u=0}^{K} i p_{i,u}^{(2)}$, and $B^{(2)} = \sum_{i=0}^{S_a} \sum_{u=1}^{K} u p_{i,u}^{(2)}$.

As an example, Figure 15.6 shows Policy 2 profit rate as a percentage of Policy 1 profit rate as retail orders vary as percent of total demand. The largest improvement in this example arises when there is substantial demand from both channels. If there are only a few retail customers, then only these few retail customers are able to capitalize on the presence of lower priority internet customers. On the other hand, if nearly all customers are retail, then there are few low-priority internet customers to act as a buffer for the retail customers.

Supply chains that are tightly integrated have rationing and shipment flexibility options that can capitalize on the customer's willingness to wait for delivery. As e-commerce matures and concerns over profitability continue to in-

Figure 15.6. Policy 2 Profit Rate versus Policy 1

crease, these options may become ever more vital in attaining and/or sustaining a competitive cost advantage.

5. Conclusion

In this chapter, we have discussed the opportunities for companies to harness the potential of the Internet by coordinating web-based activities with their traditional operations. We have focused on three critical supply chain elements: procurement, pricing and distribution/fulfillment. Most of the academic research in this area is quite recent, and hence much of the research we discuss is currently contained in working papers. We have selected to discuss several important papers that have come to our attention, but certainly can't claim to have produced an exhaustive survey of research in this nascent field. Like our colleagues in industry, academic researchers are still in the early stages of understanding the full impact of the Internet. We believe that issues of coordination between traditional and internet supply chains will continue to provide a fertile area for academic research.

Notes

1. The ramifications of the second strategy mentioned - companies setting identical prices in both online and offline channels - will be discussed later, when we focus on Bricks and Clicks competitors.
2. This section uses a more narrow definition of *channel coordination* than the rest of the chapter. Here channel coordination refers to efforts to eliminate double marginalization.

References

Baker, W., M. Marn and C. Zawada. 2001. Price smarter on the net. *Harvard Business Review.* **79** 122-127.

Balasubramanian, S. and R. A. Peterson. 2002. Channel portfolio management: Rationale, implications and implementation. *Working Paper.* The University of North Carolina, Chapel Hill, NC.

Beckmann, M. 1961. An inventory model for arbitrary interval and quantity distributions of demand. *Management Science.* **8** 35-57.

Bell, D. R., Y. Wang and V. Padmanabhan. 2002. An explanation for partial forward integration: Why manufacturers become marketers. *Working Paper.* Wharton School, University of Pennsylvania, Philadelphia, PA.

Brynjolfsson, E. and M. D. Smith. 2000. Frictionless commerce? A comparison of internet and conventional retailers. *Management Science.* **46** (4) 563-585.

Cachon, G. 1999. Competitive and cooperative inventory management in a two-echelon supply chain with lost sales. *Working Paper.* The Fuqua School of Business, Duke University, Durham, NC.

Cattani, K. D., W. Gilland, H. Heese and J. Swaminathan. 2002. Coordinating internet and traditional channels: The manufacturer's perspective. *Working Paper.* The Kenan-Flagler Business School, The University of North Carolina, Chapel Hill, NC.

Cattani, K. D. and G. C. Souza. 2002. Inventory rationing and shipment flexibility alternatives for direct market firms. *Production and Operations Management.* **11** (4) 441-457.

Cattani, K. D., W. G. Gilland and J. M. Swaminathan. 2003. Adding a direct channel? How autonomy of the direct channel affects prices and profits. *Working Paper.* The Kenan-Flagler Business School, University of North Carolina, Chapel Hill, NC.

Chiang, W. K., D. Chhajed and J. D. Hess. 2002. Direct marketing, indirect profits: A strategic analysis of dual-channel supply chain design. *Working Paper.* University of Illinois at Urbana-Champaign, Champaign, IL.

Erhun, F., P. Keskinokak and S. Tayur. 2001. Sequential procurement in a capacitated supply chain facing uncertain demand. *Working Paper.* GSIA, Carnegie Mellon University, Pittsburgh.

Gallien, J. and L. W. Wein. 2001. A smart market for industrial procurement with capacity constraints. *Working Paper.* MIT, Cambridge, MA.

Ha, A. Y. 1997. Inventory rationing in a make-to-stock production system with several demand classes and lost sales. *Management Science*. **43** (8) 1093-1103.

Hadley, G. and T. M. Whitin. 1963. *Analysis of inventory systems*. Prentice-Hall. Englewood Cliffs, N.J.

Hamilton, D. P. 2001. E-commerce (a special report): Overview — the price isn't right: Internet pricing has turned out to be a lot trickier than retailers expected. *Wall Street Journal*. (Feb 12, 2001) R.8.

Hedershott, T. and J. Zhang. 2001. A model of direct and intermediated sales. *Working Paper*. University of California, Berkeley, CA.

Heyman, D. P. and M. J. Sobel. 1990. *Stochastic models*. North-Holland. New York, NY

Huang, W. and J. M. Swaminathan. 2003. Pricing on traditional and internet channels under monopoly and duopoly: Analysis and bounds. *Working Paper*. The Kenan-Flagler Business School, University of North Carolina, Chapel Hill, NC.

Jin, M. and S. D. Wu. 2001a. Supply chain coordination in electronic markets: Auction and contracting mechanisms. *Working Paper*. Lehigh University, Bethlehem, PA.

Jin, M. and S. D. Wu. 2001b. Procurement auction with supplier coalitions: Validity requirements and mechanism design. *Working Paper*. Lehigh University, Bethlehem, PA.

Johannes, L. 2000. Competing online, drugstore chains virtually undersell themselves — their web sites charge less than their own stores, with some strings attached. *Wall Street Journal*. (Jan 10, 2000) B1.

Lal, R. and M. Sarvary. 1999. When and how is the internet likely to decrease price competition. *Marketing Science*. **18** (4) 485-503.

Lariviere, M. A. 1999. Supply chain contracting and coordination with stochastic demand. S. R. Tayur, R. Ganeshan and M. J. Magazine, eds. *Quantitative models for supply chain management*. Kluwer Academic Publishers, Norwell, MA, Chapter 8.

Lee, H. and S. Whang. 2002. The impact of the secondary market on the supply chain. *Management Science*. **48** (6) 719-731.

McIntyre, K. and E. Perlman. 2000. Nike - channel conflict. *Stanford GSB Case*. EC-9B.

Peleg, B., H. L. Lee and W. H. Hausman. 2002. Short-term e-procurement strategies versus long-term contracts. *Production and Operations Management*. **11** (4) 458-479.

Schlesinger, J. M. 1999. New e-conomy: If e-commerce helps kill inflation, why did prices just spike? *Wall Street Journal*. (October 18) A1.

Seifert, R. W., U. W. Thonemann and S. A. Rockhold. 2001. Integrating direct and indirect sales channels under decentralized decision-making. *Working Paper*. IMD, Lausanne, Switzerland.

Sgoutas, K. 2000. Pricing and branding on the internet. *Stanford GSB Case*. EC-8.

Smith, M. D. and E. Brynjolfsson. 2001. Consumer decision-making at an internet shopbot: Brand still matters. *The Journal of Industrial Economics*. **49** (4) 541-558.

Sosic, G. and D. Granot. 2001. Formation of alliances in internet-based supply exchanges. *Working Paper*. The University of British Columbia, Vancouver, Canada.

Swaminathan, J. M. and S. R. Tayur. 2003. Models for supply chains in e-business. *Forthcoming in Management Science*.

Tsay, A., S. Nahmias and N. Agrawal. 1999. Modeling supply chain contracts: A review. S. Tayur, R. Ganeshan and M. Magazine, eds. *Quantitative models for supply chain management*. Kluwer Academic Publishers, Norwell, MA, Chapter 10.

Tsay, A. A. and N. Agrawal. 2001. Manufacturer and reseller perspectives on channel conflict and coordination in multiple-channel distribution. *Working Paper*. Santa Clara University, Santa Clara, CA.

Part V

NETWORK DESIGN, IT, AND FINANCIAL SERVICES

David Simchi-Levi, S. David Wu, and Z. Max Shen (Eds.)
Handbook of Quantitative Supply Chain Analysis:
Modeling in the E-Business Era
©2004 Kluwer Academic Publishers

Chapter 16

USING A STRUCTURAL EQUATIONS MODELING APPROACH TO DESIGN AND MONITOR STRATEGIC INTERNATIONAL FACILITY NETWORKS

Panos Kouvelis
John M. Olin School of Business,
Washington University, St. Louis, MO 63130

Charles L. Munson
College of Business and Economics,
Washington State University, Pullman, WA 99164-4736

1. Introduction

1.1 A conceptual framework to Classify Global Network Structures

Supply chain management is no longer a local issue. As economies develop, foreign firms from more and more countries become competitive suppliers, and more and more countries become competitive locations for manufacturing. Transportation efficiencies have eased supply chain globalization. Furthermore, the internet has enhanced communication and coordination, simplifying the globalization of companies and enabling supply chain management across borders. Global companies, then, must develop strategic international facility network structures to handle the complex production and distribution of goods throughout the supply chain.

Structuring global manufacturing and distribution networks is a complicated decision making process. The typical inputs to such a process are composed of macro issues like which markets to serve and information about future macroeconomic conditions, as well as firm-specific issues like product mix decisions based on demand projections for different markets, transportation costs, and production costs. Given the above information, companies have to decide, among other things, where to locate factories, how to allocate production activities to the various facilities of the network, and how to manage the distribution of products (e.g., where to locate distribution facilities).

We propose a conceptual framework that classifies global network structures along three dimensions (Figure 16.1). 1. *Market focus* refers to the degree to which each marketing region produces subassemblies needed for that region. A *global* strategy generally has plants producing large fractions of worldwide demand, while a *regional* strategy generally segments production by region. 2. *Plant focus* refers to the degree to which different subassemblies are produced at the same facility. A *dedicated* strategy generally separates subassemblies into different plants, while a *flexible* strategy combines subassembly production at the same facilities. 3. *Network dispersion* refers to the number of different plants producing the same subassembly per marketing region that has any production at all. A *concentrated* strategy has a single plant for each subassembly type per region of production, while a *scattered* strategy has multiple.

Theoretically, a firm's network structure could fall into any one of the eight (implied) cells of Figure 16.1. For example, a *global-flexible-concentrated* strategy would have all subassemblies for the entire world's demand produced at one facility, while a *regional-dedicated-scattered* strategy would have each marketing region producing its own demand at multiple subassembly-specific facilities. We expect economies of scale, complexity costs of producing different subassemblies at the same plant, transportation costs, and tariffs to be the primary levers in determining a company's network structure. These hypotheses are intuitive and often implied in the operations management and international business literature. Our research enhances the level of understanding by first developing effective aggregate ways to measure "economies of scale," "complexity costs," and relevant "transportation" and "tariff" costs, and then explicitly describing via empirically derived equations the causal relationship of these measures with appropriately defined measures of "market focus," "plant focus," and "network dispersion." This allows us to validate the hypotheses and operationalize them into a set of tools, a set of structural equations, that can be used by the global manager to construct and monitor the firm's global facility network.

1.2 Literature Review

One of the early papers in this area, which presents a single-period model for simultaneously determining international plant location and financing decisions under uncertainty, is by Hodder and Dincer (1986). The model uses a

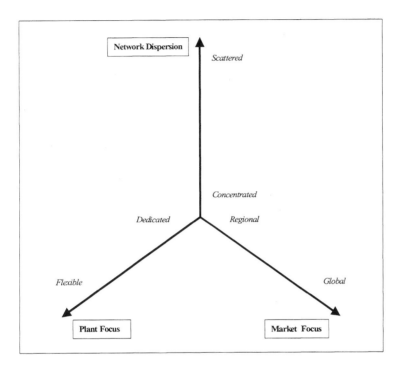

Figure 16.1. Global network structure classification along three dimensions

mean-variance approach in the objective function to incorporate price/exchange rate uncertainty and risk aversion in the location problem—an approach that builds upon earlier work of Jucker and Carlson (1976) and Hodder and Jucker (1985a, b). Aspects of subsidized government financing, tariffs, and taxation are not modeled.

Moving to more recent work that is closely related to ours (an exhaustive list of global location and sourcing papers can be found in Kouvelis (1998) and Cohen and Huchzermeier (1998)), Cohen and Lee (1989) develop an integer programming model for resource deployment decisions in a global manufacturing and distribution network for a U.S.-based manufacturer of personal computers. Huchzermeier and Cohen (1996) present a modeling framework that integrates the network flow and option valuation approaches to global supply chain modeling. The model maximizes discounted, expected, global after-tax profit for a multinational firm. The results in that paper provide a methodology for quantifying the risks and returns of flexible global manufacturing strategies. Their work clearly demonstrates how flexible facility networks with excess capacity can provide real options to hedge exchange rate fluctuations in the long term. Other related references to this type of work are Huchzermeier (1991, 1994).

Revelle and Laporte (1996) suggest that, even though plant location and configuration of production/distribution networks have been studied for many years, a number of important real-world issues have not received adequate attention. Even though there is ample evidence on the strong effects of financing and taxation factors on the structure of global manufacturing and distribution networks (see Bartmess and Cerny (1993) and MacCormack *et al.* (1994)), there is a lack of reasonably simple, yet comprehensive, models illustrating the effects of such factors on global location/network configuration decisions. Kouvelis *et al.* (2003) present one such attempt, and the mixed integer programming (MIP) model presented in the Appendix of this paper represents a related but different one.

Finally, Brush *et al.* (1999) report survey results analyzing actual international plant location decisions among large multinational firms, while various company-specific facility network design issues and decision support systems have been discussed in the applications-oriented literature—see the Breitman and Lucas (1987) work on General Motors, Davis (1993) work on Hewlett-Packard, Arntzen et al. (1995) work on Digital Equipment Corporation, and Cowie (1999) work on RCA.

Many of the models cited above provide more *tactical* solutions in the sense of describing specific production allocations for every possible facility in each country. Such models often either require a significant amount of detail, implying large models with long solutions times, or they must ignore important solution alternatives or aspects of the problem. Our structural equations modeling approach presented in Section 2 represents more of a *strategic* solution along the lines of Figure 16.1. A given network strategy would be expected to remain relatively stable over time, and managers would then desire simple monitoring tools to ensure that the current strategy remains reasonable. Detailed tactical models could be used to make more local decisions as local conditions change within a given country or region. For example, the closing of a single plant will not, in general, impact the overall strategic network structure. In addition, while we would expect production cost differences among countries to represent an important lever for tactical location decision within a network structure, we would not expect changes in those production cost differences to have a large impact on the nature of the network structure itself (see Section 3). Our approach is akin to the hierarchical production planning technique championed by Hax and Meal (1975), where long-term (strategic) decisions are made prior to and separately from short-term (tactical) ones.

1.3 Research Contributions

The main contributions of this paper are:
a) a conceptual framework classifying strategic global facility network structures along three dimensions that can be operationalized as a managerial tool to develop and monitor global networks;
b) a proposed approach for the development of structural equations that allows managers to classify their network structures via relatively simple calculations of primary levers, without the need to explicitly formulate and solve an MIP; and
c) an illustration of the effects of subsidized financing and taxation on the strategic network structure as well as the tactical allocation of production among countries.

Contributions (a) and (b) of our paper address the needs of global managers to have broad conceptual tools for developing effective rationalization strategies of their global facility networks in an ever-changing global environment. We abstract from detail to determine the fundamental variables that are relative to network design. Our numerical tests suggest that the formulas that we develop to represent economies of scale, complexity costs, transportation costs, and tariffs may be sufficient for managers to design and monitor their strategic network structures.

The rest of the chapter is organized as follows. In Section 2 we develop a structural equations model based on a relatively comprehensive MIP (Appendix). Normalized dependent variables are specified that act as proxies for a company's placement into one of the eight implied cells of Figure 16.1 via the calculation of just a few key independent variables. In Section 3 we perform sensitivity analysis on both tax rates and loan interest rates, illustrating their effects on the location decisions within the network, their effects on the network itself, and the penalty for not incorporating these important issues into a plant location model. Section 4 concludes the paper with a summary of our main results and some ideas for further research. In the Appendix we present an MIP formulation of the global facility network design problem that incorporates such issues as government subsidies in facility financing, trade tariffs and taxation into a multi-period model. We wish to emphasize that the structural equations modeling approach presented in the next section is not dependent upon the specific MIP presented in the Appendix. The approach should also work with other models in the literature, as long as they include the terms used in the independent and dependent variables of Exhibits 16.1 and 16.2.

2. Structural Equations Model

2.1 Problem Statement

A firm considers the issue of designing its global network of plants and distribution centers. The firm plans to produce a new product with sales expectations in many countries. The product is composed of many different subassemblies, but requires only one unit per subassembly (i.e., a two level product tree, with level 0 the final product and level 1 all its subassemblies). Subassemblies can be produced in some combination of facilities within the firm's market areas. In order to assemble the final product, distribution centers (DCs) can be located in various countries. To influence the firm's location decision, various governments are willing to grant to it tax incentives and loans with subsidized interest rates that reduce the facility costs. The firm wants to develop a network of plants and DCs that maximizes its discounted after-tax profit over the planning horizon for this product.

2.2 Motivation for the Structural Equations Modeling Approach

An MIP model such as that presented in the Appendix can be used to solve the global plant location problem just described. However, we now shift our emphasis towards identifying a few key measures that summarize the relevant cost input information from the MIP model. We illustrate how such aggregate measures can be used to effectively predict the nature of the optimal solution to the model (i.e., correctly place the network structure into one of the eight implied cells of the proposed conceptual framework) without having to solve it. In accomplishing this task we follow the global sensitivity analysis approach suggested by Wagner (1995). Using extensive experimental data generated from our MIP model, and through rigorous application of econometric techniques, we identified key summary measures (independent variables) that explain the observed key summary output variables of the model (dependent variables).

To use global sensitivity analysis (Wagner 1995), one first starts with a deterministic model for the problem at hand. The model is run multiple times for different values of the input variables. The model outputs are then regressed against the inputs to develop structural equations that can be used to predict outputs for any levels of the inputs, without re-running the original model. Global sensitivity analysis can be used to identify the most important inputs, to explain current solutions, and to predict solution change as key inputs change. In our paper, key inputs are incorporated into just four primary levers. The model outputs as incorporated into proxy measures of the structural classifications of Figure 16.1 are then regressed against the primary input levers. The regression results create our structural equations.

We see the value of the structural equations approach to a complex decision-making setting such as facility network design as twofold. At a primary level, the global sensitivity analysis of the MIP model we performed, and report in this paper, suggests four reasonably simple and explanatorily powerful cost measures on which managers can concentrate in order to understand the appropriate facility network structure for a specific environment. When those measures change, as a result of changes in product structure or macroeconomic or market demand data, managers can gauge the strategic impact without reformulating and rerunning the potentially very large MIP model. At a secondary level, if managers create structural equations for their products, then on an ongoing basis, as costs change the equations can be used to quickly determine if the firm's strategic network structure is sound or needs revisiting. If the structure should be shifted, the manager can proceed to make necessary adjustments in a two-phase process. At a first phase, use of the structural equations allows for determination of the desirable structure. Then at a second phase a detailed plant location model (similar in structure to our MIP model or even a more detailed model) can be used at that point because the network structure will be predetermined. This approach allows for decoupling of the strategic network decision and the actual plant location decision. The predetermination of the network structure at this phase simplifies the complexity of the MIP model (as many of the integer variables can be preset according to the imposed network structure) and thus allows the opportunity for more detailed operational issues to be entered in the remaining plant location/production allocation decision model. Further evidence is provided in Section 3 to validate the effectiveness of such an approach.

We should caution up front that while the *general* nature of the global sensitivity results described below for each measure on each of the dimensions will likely remain valid for a number of different products and industries, the *actual* structural equations that we derived will likely be product- and company-specific (as illustrated by the very different sets of equations derived for the three product types analyzed). Therefore, managers will need to develop their own structural equations by creating sample data runs from the MIP model that are more representative of their respective industrial setting.

2.3 Framework Measures (Independent Variables)

Based on intuition and implied in the operations management and international business literature, we propose four "framework measures" that we believe will have the largest impact on the type of facility network appropriate for a company: *economies of scale, complexity costs, transportation costs with tariffs,* and *transportation costs without tariffs*. We recommend two transportation measures because tariffs have an obvious impact on market focus, but they have a confounding effect on network dispersion according to our definitions. The particular structural equations derived from our experimental data with the MIP model included the first two measures and one of the second two.

Economies of Scale

$$EOS = \frac{\left[TF + \sum_{i=1}^{I} E_i + A\right] - \left[P - \sum_{i=1}^{I} V_i - G\ (Tf + a) - \sum_{i=1}^{I} e_i\right]}{WORLD}$$

Complexity Costs

$$COM = \frac{T(F + B_2) + A + 2(WORLD/MARK)G(Tf + a)}{2 \times [TF + A + (WORLD/MARK)G(Tf + a)]}$$

Transportation Costs with Tariffs

$$TRANS1 = \frac{MILE \times 1000 + TARIFF}{P - \sum_{i=1}^{I} V_i - v}$$

Transportation Costs without Tariffs

$$TRANS2 = \frac{MILE \times 1000}{P - \sum_{i=1}^{I} V_i - v}$$

Notes: Notation from Section 2 is used here, where the country indices have been eliminated and all costs and prices are based on a system-wide average. G = average space requirement per subassembly unit. $WORLD$ = average annual worldwide demand. $MARK$ = number of marketing regions. $MILE$ = transportation cost per mile for the final product (fixed shipping costs are excluded). $TARIFF$ = average tariff for shipping the final product outside of the marketing region.

Exhibit 16.1. Independent variables used for the structural equations model.

The framework measure calculations are shown in Exhibit 16.1. All four are dimensionless in nature. The formulas strike a balance between "richness" and "ease of use" while attempting to capture the "essence" of the four frameworks. The **economies-of-scale measure**, EOS, calculates a break-even volume as a percentage of annual worldwide demand, assuming all subassemblies produced at a single plant while ignoring complexity costs. The **complexity measure**, COM, calculates a ratio of the costs of producing two subassemblies at one plant vs. two subassemblies at two different plants. The **transportation measure with tariffs**, $TRANS1$, calculates the distance-related transportation costs plus tariffs, expressed as a percentage of gross margin. The **transportation measure without tariffs**, $TRANS2$, is the same as $TRANS1$, excluding the tariffs.

2.4 Proxies for the Global Network Structure Dimensions (Dependent Variables)

Exhibit 16.2 displays the calculations for proxies of the three network structure dimensions (the dependent variables in the structural equations model). Each measure is normalized to a range between 0 and 1. For example, a score of 0 on the market focus measure, $MARKET$, represents a pure regional strategy, 1 represents a pure global strategy, and 0.5 represents a "half-regional, half-global" strategy. The measures are computed from the output of the MIP model.

The **market focus measure**, $MARKET$, calculates the percentage of subassemblies imported from other regions, averaged over each region. A high value implies more of a global strategy. The **plant focus measure**, $PLANT$, is based on the average number of subassemblies per plant. A high value implies more of a flexible strategy. The **network dispersion measure**, DIS, calculates the average number of plants per subassembly, as a percentage of the number of possible plant locations per region, averaged over each region that has any subassembly production at all. A high value implies more of a scattered strategy.

2.5 Model Development

We used sample products to build the structural equations model. Parameters for the numerical studies are shown in Table 16.1 and Exhibits 16.3 and 16.4. The costs and, in fact, the products themselves are meant to be roughly representative of a particular type of industry and not specifically representative of any actual company data.

Market Focus

$$MARKET = \frac{1}{MARK-1}\sum_{REG}(1/I)\sum_{i=1}^{I}\left(1-Min\left\{1,\sum_{m\in REG}\sum_{q=1}^{N}\sum_{k=1}^{I}\sum_{t=1}^{T}X_{imqkt}\bigg/\sum_{k=1}^{N}\sum_{j\in REG}\sum_{t=1}^{T}x_{kjt}\right\}\right)$$

Plant Focus

$$PLANT = \frac{\left(\sum_{z=1}^{I}\sum_{m=1}^{N}\sum_{q=1}^{I}Y_{zmq}\bigg/\sum_{m=1}^{N}\sum_{q=1}^{I}Y_{1mq}\right)-1}{I-1}$$

Network Dispersion

$$DIS = \sum_{REG}\frac{1}{M(REG)-1} Max\left(0,\left\{\sum_{i=1}^{I}\sum_{m\in REG}\sum_{q=1}^{I}Z_{imq}\bigg/Max\left[1,\sum_{i=1}^{I}Min\left(1,\sum_{m\in REG}\sum_{q=1}^{I}Z_{imq}\right)\right]\right\}-1\right)\bigg/\left(\sum_{REG}Min\left\{1,\sum_{i=1}^{I}\sum_{m\in REG}\sum_{q=1}^{I}Z_{imq}\right\}\right)$$

Notes: Notation from Section 2 is used here. REG = a marketing region. $MARK$ = number of marketing regions. $M(REG)$ = number of production countries in region REG.

Exhibit 16.2. Dependent variables used for the structural equations model.

Table 16.1. Products (Industries) Used in the computational studies

Product (Industry)	No. of Sub-Assemblies	Baseline Variable Subassembly Cost		Baseline Variable DC Cost	Baseline Annual U.S. Demand	Transportation Costs	
						Fixed	Per Mile
Ceiling Fans	2	subpack	$5	$5	50,000	$1.25	$0.0025
		motor	$15				
Motorcycles	3	engine	$3,000	$500	20,000	$51.50	$0.75
		tires & seat	$150				
		other parts	$1,500				
Computers	3	CPU etc.	$150	$50	100,000	$8.50	$0.05
		monitor	$65				
		other parts	$160				

Solution times for the MIP in the Appendix can vary widely depending on the actual values of the parameters themselves. For example, a change of cer-

Europe	Asia
Ireland*	Taiwan*
Romania*	India*
United Kingdom	China
Germany	Thailand
France	Singapore
Sweden	South Korea
Portugal	Japan
Spain	
Norway	North America
Finland	Mexico*
Russia	Western U.S.*
The Netherlands	Eastern U.S.
Belgium	Canada
Denmark	
Switzerland	
Austria	
Italy	
Turkey	
Poland	

* Potential country of production

Exhibit 16.3. Markets used in the computational studies.

tain costs can increase a particular configuration's solution time from a few seconds to a few hours on a standard PC. For large problems, solution times may not only take many hours, but the program will often crash before an optimal solution can be obtained. In particular, we ran into difficulty trying to solve problems larger than the combination of nine countries of operation, thirty markets, four subassemblies, and five time periods.

As suggested in Wagner (1995), we used random data to vary model parameters and create a data set of experimental results. For our large study, we chose a problem with two subassemblies (ceiling fans), three marketing regions, six potential countries of operation, thirty markets, and five time periods. Low, medium, and high levels were chosen for four parameter sets: (1)

fixed and investment (base) costs (F_m, E_{im}, and A_m); (2) amount of tariff; (3) transportation cost per mile; and (4) complexity costs (B_{zm}). A full-factorial experiment of these combinations amounts to $3^4 = 81$ trials (Exhibit 16.5). We repeated this for 10 different random number seeds for a total of 810 trials to build the equations. We also created an 11th data set of 81 trials for validation purposes, as well as an additional validation data set of 81 trials with the same product but nine potential countries of production instead of six.

The effect of individual independent variables on the dependent variables can be determined by using a "one-at-a-time" approach. Following Wagner (1995), we regressed each dependent variable separately against a polynomial fit that allows terms x^Δ, where $\Delta = \{-6, ..., 5, 6\}$. We performed stepwise regression using SPSS® for Windows^{TM}, keeping the default alpha values of .05 to enter and .10 to remove. Table 16.2 shows the results of the fit on each dependent variable.

From these results we can claim that:

(a) the **transportation measure with tariffs**, *TRANS1*, has the most **explanatory power** for the **market focus**, *MARKET*, measure;

(b) **complexity costs** and **economies of scale** (as measured by *COM* and *EOS*, respectively) have the most **explanatory power** for the **plant focus**, *PLANT*, measure; and

(c) the **transportation measure without tariffs**, *TRANS2*, has the most **explanatory power** for the **network dispersion**, *DIS*, measure.

Even though these results were derived from the particular data sets we used, we expect that the general results hold across most reasonable product/market data sets representing most industrial settings. Further validation of this is provided in Section 2.7 for two other products.

Interaction effects can be explored by combining the independent variables into the same regression. In particular, we regressed 76 combinations of three major independent variables. Along with the 36 combinations from the "one-at-a-time" approach, we included 24 terms of the form $x^{\Delta 1}y^{\Delta 2}$, where $\Delta 1 = \{-2, -1, 1, 2\}$ and $\Delta 2 = \pm \Delta 1$; and we included 16 terms of the form $x^{\Delta 1}y^{\Delta 2}z^{\Delta 3}$, where $\Delta 1 = \{-2, -1, 1, 2\}$, $\Delta 2 = \pm \Delta 1$, and $\Delta 3 = \pm \Delta 2$. The regression results constitute the "structural equations." For our study, the combination of *EOS*, *COM*, and *TRANS1* fit *MARKET* the best (adjusted R^2 of .820), while the combination of *EOS*, *COM*, and *TRANS2* provided a better fit for the other two dependent variables (adjusted R^2 of .862 for *PLANT* and .833 for *DIS*). The structural equations are listed below.

$MARKET = 1.284 - .02462 TRANS1^{-1} + .00004163 TRANS1^{-2} + 33.546 (EOS \cdot COM)^2 - 28.688(EOS \cdot COM \cdot TRANS1)$
$- .0003153(EOS \cdot COM / TRANS1)^2 + 635.169(EOS \cdot TRANS1)^2$
$- .0001725(TRANS1 / EOS) - 997.638 TRANS1^5 - 8.509 TRANS1$
$+ 486.698 TRANS1^4 + 5.409(EOS \cdot COM) - 16.876(EOS \cdot TRANS1)$

Notes: $VARSUB(i) \equiv$ average variable cost for component i,
$VARDC \equiv$ average variable final assembly cost,
and $DEM \equiv$ average U.S. demand

Base demand for country $j = DEM \times$(country j=s population /
U.S. population)$\times$(country j=s per capita income / U.S. per capita income)
$D_{jt} = $ UNIFORM(0.75,1.25)$\times$(base demand for country j)
$\beta_n = 20\%$ for all countries
$\tau_n = $ actual tax rate in each country
$r_n = 34\%$ in each country (to encourage no financing in the base scenarios)
Mileage = actual mileage between countries.
$V_{im} = $ Uniform(0.8,1.1)$\times VARSUB(i)$
$v_k = $ Uniform(0.8,1.1)$\times VARDC$
$p_{imk} = 2 \times 1.1 \times VARSUB(i)$
$P_j = 2 \times [3_i \ p_{imk} + 1.13 VARDC]$
Line and DC capacities = maximum possible worldwide demand
$G_{im} = 1$ square foot for each subassembly
Plant capacity = 3_i ($G_{im} \times$ maximum possible worldwide demand)
$F_m = $ Uniform(0.8, 1.2)$\times 0.01 \times DEM \times 3_i$ [p_{imk} ! $VARSUB(i)$]
$f_m = $ Uniform(0.8, 1.2)$\times 0.05 \times 3_i$ [p_{imk} ! $VARSUB(i)$]
$U_k = $ Uniform(0.8, 1.2)$\times 0.01 \times DEM \times [P_j$!($VARDC + 3_i \ p_{imk}$)]
$u_k = $ Uniform(0.8, 1.2)$\times 0.05 \times [P_j$!($VARDC + 3_i \ p_{imk}$)]
$E_{im} = $ Uniform(0.01, 3)$\times 0.01 \times DEM \times [p_{imk}$! $VARSUB(i)$]
$e_{im} = $ Uniform(0.01, 3)$\times 0.05 \times [p_{imk}$! $VARSUB(i)$]
$A_m = $ Uniform(0.01, 3)$\times 0.01 \times DEM \times 3_i$ [p_{imk} ! $VARSUB(i)$]
$a_m = $ Uniform(0.01, 3)$\times 0.05 \times 3_i$ [p_{imk} ! $VARSUB(i)$]
$H_k = $ Uniform(0.01, 3)$\times 0.01 \times DEM \times [P_j$!($VARDC + 3_i \ p_{imk}$)]
$h_k = $ Uniform(0.01, 3)$\times 0.05 \times [P_j$!($VARDC + 3_i \ p_{imk}$)]
$B_{zm} = (z$ / total # of subassemblies)$\times$Uniform(0.8, 1.2)$\times 0.02 \times DEM \times 3_i$ [p_{imk} !$VARSUB(i)$]
Tariff between countries from different regions = 0.05 H P_j
Transportation cost = fixed transportation cost + transportation cost per mile $\times$ number of miles + tariff
Subassembly transportation cost fraction $\lambda_i = VARSUB(i)$ / ($3_i \ VARSUB(i) + VARDC$)
Depreciation = straight-line
$l_n = $ no limit

Exhibit 16.4. General input parameters used in the computational studies.

$PLANT = 12.902 + 12.036 COM^{-2} - .0000005751(EOS \cdot COM)^{-2} - .254 COM^{-6} - 2.494 COM - .001147(COM \cdot TRANS2)^{-1} + 749.921(EOS \cdot COM)^2 - 56.479 EOS - 21.333 COM^{-1} + 8.348E-14(EOS)^{-4} + 1.835E-13(EOS \cdot COM \cdot TRANS2)^{-2} + .000177[EOS / (COM \cdot TRANS2)]^2$

Parameter Set 1–Base Fixed and Investment Costs at Plants

Parameters: F_m, E_{im}, and A_m

Levels
Low   – 10% of Medium
Medium – See Exhibit 4
High   – 1000% of Medium

Parameter Set 2–Tariffs

Low   – 0%
Medium – 5%
High   – 20%

Parameter Set 3–Transportation Costs Per Mile

Low   – 5% of Medium
Medium – See Table 1
High   – 400% of Medium

Parameter Set 4–Complexity Costs B_{zm}

Low   – 20% of Medium
Medium – See Exhibit 4
High   – 500% of Medium

Exhibit 16.5. experimental design used in the computational studies.

Table 16.2. Adjusted R^2 Values for "One-at-a-Time" Regressions

Independent	Ceiling Fans			Motorcycles			Computers		
Variable	MARKET	PLANT	DIS	MARKET	PLANT	DIS	MARKET	PLANT	DIS
EOS	.016	.610	.159	.035	.713	.248	.000	.732	.129
COM	.006	.748	.088	.008	.745	.151	.000	.846	.088
TRANS1	.789	.005	.337	.763	.000	.209	.668	.000	.136
TRANS2	.423	.007	.462	.554	.000	.343	.488	.000	.537

$+ .000001682 TRANS2^{-2} - .009337[EOS / (COM \cdot TRANS2)]$
$- .0000001504(COM / TRANS2)^2$

$DIS = -.00155 + .0005559[TRANS2 / (EOS \cdot COM)]$
$- 3259.633(EOS \cdot COM \cdot TRANS2)^2 + 12783.277 TRANS2^4$
$- 1408.273(EOS \cdot TRANS2 / COM)^2 - 644156.8 TRANS^6 + 1.836\text{E-}11 EOS^{-3}$
$+ .0003994(COM \cdot TRANS2 / EOS) + 4.446(TRANS2 / COM)^2$

2.6 Model Validation

We validated the model by applying the above equations to a holdout sample of 81 trials. We also applied the equations to the same product structure with nine potential countries of production instead of six (adding Germany, Eastern U.S., and China). In addition to the adjusted R^2 numbers, we calculated the number of trials that were "correctly" classified according to the following scale. A trial is correctly classified under these conditions: (1) if estimate $\leq$ 0.25, then actual must be $\leq$ 0.5; (2) if $0.25 <$ estimate ≤ 0.5, then actual must be ≤ 0.75; (3) if $0.5 <$ estimate < 0.75, then actual must be ≥ 0.25; and (4) if estimate ≥ 0.75, then actual must be ≥ 0.5. This scale is somewhat arbitrary, but its purpose is simply to see if the structural equations place the network structure in the correct general area along all three dimensions. Table 16.3 shows the validation results for both sets of validation trials, as well as the original data set of 810 trials.

Table 16.3. Validation Results for the Structural Equations Model

Dependent Variable	Average Value	Range	Adjusted R^2	Correctly Classified
Holdout Sample of 81 Trials				
MARKET	.282	.0007–1	.800	100.0%
PLANT	.474	0–1	.818	97.5%
DIS	.160	0–1	.853	100.0%
Sample of 81 Trials with 9 Potential Countries of Production Instead of 6				
MARKET	.301	.0007–1	.793	97.5%
PLANT	.425	0–1	.811	96.3%
DIS	.086	0–.5	.494	100.0%
Original Data Set of 810 Trials				
MARKET	.247	0–1	.820	98.3%
PLANT	.400	0–1	.862	97.7%
DIS	.150	0–1	.833	100.0%

2.7 Other Products

While the *general* insights from the global sensitivity analysis results in Table 16.2 can likely be generalized to other products, the *actual* structural equations shown in Section 2.5 are specific to the product in question. Therefore, managers should develop equations for their particular products based on the approach described here. To reiterate, a primary benefit of creating structural equations after developing the MIP is that managers can use them to monitor their global network on an ongoing basis by focusing on only a few key summary measures.

As further experimentation, we tried 81 trials for two other products (motorcycles and computers) with three subassemblies and quite different cost structures. In general, the equations from Section 2.5 provided a very poor fit for these products. However, new equations developed specifically for these products result in similar fits to those described above for the ceiling fans. Table 16.2 shows the adjusted R^2 values for the individual regressions for the motorcycles and computers, while Table 16.4 lists the coefficients for the structural equations (note the dissimilarities compared to the ceiling fan equations). Table 16.5 shows the validation results for the motorcycles and computers.

3. Government Incentives: Tax Holidays and Subsidized Loans

Among the most important factors distinguishing the organization of global production activities from strictly domestic ones are the influences of various governmental policies, including tariffs and local content rules (Munson and Rosenblatt 1997). Moreover, governments often use special financing and taxation incentives to stimulate production investments in their respective country. While the framework measures described in Section 2 appear to have high explanatory power on a firm's network structure, actual location decisions within that structure will primarily depend upon differences among countries. In this section we perform sensitivity analysis on tax rates and loan interest rates to observe the effects on the global facility network structure.

For both the tax and interest rate analyses, we used eight experiments from the ceiling fan example described in Section 2 (Tables 16.6-16.9). The experiments were chosen because they represent a wide range of the possible global network structure cells from Figure 16.1 (see the initial levels (35% Mexican Tax Rate) of MARKET, PLANT, and DIS in Table 16.6). To represent government incentives, we lowered the applicable tax or loan rates under two conditions: (1) one country only with low tax (or loan interest) rate, and (2) one country from each region with low rates (resulting in 32 total data samples). We tabulated (1) the dimension proxy measures (*MARKET, PLANT,* and *DIS*), (2) the subassemblies produced in the countries with incentives, and (3)

Table 16.4. Structural Equation Coefficients for the Motorcycles and Computers (81 Trials Each)

	Motorcycles			Computers		
Term	MARKET	PLANT	DIS	MARKET	PLANT	DIS
Intercept	.906	!.430	!.008327	1.191	!2.978	7.996E-03
$TRANS1$	!7.202			!7.851		
$TRANS1^2$	16.332			12.796		
$TRANS1^4$	!22.089					
$TRANS1^{!3}$				!.02269		
$TRANS1^{!6}$				4.782E-16		
COM					1.340	
$COM^{!2}$					2.173	
$COM^{!3}$		.818				
$COM^{!6}$					!.140	
$(EOSACOM)^2$	44.412	187.427				
$(EOSATRANS1/COM)^2$	!35.693					
$(EOSACOM)^{!2}$		!1.136E-06			!2.257E-07	
$(EOS/COM)^2$		!99.427				
$(COM/EOS)^2$		!4.141E-07				
$(EOSACOM/TRANS1)^2$		5.784E-04				
$(EOS/COMATRANS1)$					!5.029E-02	
$(EOS/COMATRANS1)^2$					6.621E-04	
$(COMATRANS1)^{!2}$		!8.032E-07				
COM/EOS		.001067				
$TRANS2$			.772			
$TRANS2^2$						53.034
$TRANS2/EOS$			.002166			
$(TRANS2/COM)^2$						!11.575
$(TRANS2/EOSACOM)^2$						5.087E-06
$(EOSATRANS2)^2$			!286.913			
$(COMATRANS2)$						!1.564
$(EOSACOMATRANS2)^2$						!5817.426
$(COMATRANS2/EOS)^2$						1.803E-06

Table 16.5. Validation Results for the Structural Equations Model for the Motorcycles and Computers (81 Trials Each)

Dependent Variable	Average Value	Range	Adjusted R^2	Correctly Classified
Motorcycles				
MARKET	.320	.0028-1	.830	98.8%
PLANT	.440	0-1	.958	100.0%
DIS	.123	0-.6667	.940	100.0%
Computers				
MARKET	.226	.0001-1	.668	93.8%
PLANT	.429	0-1	.900	100.0%
DIS	.173	0-1	.971	100.0%

the optimal net present value. In addition, to illustrate the importance of incorporating taxes, depreciation, and financing in a plant location model, we solved the MIP under the varying tax and interest rates by using a more "traditional" objective function that maximizes total gross profit over the time horizon minus investment cost. (As with many location models, this "traditional" objective function does not calculate after-tax cash flows as our original MIP (Appendix) does (see, e.g., Love, Morris, and Wesolowsky (1988) and Francis, McGinnis, and White (1992)).) The resulting configuration was then applied to the objective function from the optimal NPV (our original MIP) model to determine the *actual* net present value of that configuration, and the results are compared with our original model's results. (In calculating the NPV for the "traditional" configuration, it was assumed that subsidized loans would be accepted and lower tax rates would be used, where applicable.)

Table 16.6. Tax Rate Sensitivity on Mexico's Tax Rate

	Optimal Solution Maximizing NPV					Maximizing Total Gross Profit Minus Investment Cost		
	Subs made in Mexico	MARKET	PLANT	DIS	Net Present Value	Subs made in Mexico	Net Present Value	% Deviation from Optimal
EXPERIMENT 1								
35% Mexican Tax Rate	303,954	0.002	0.000	1.000	18,501,230	284,048	18,452,120	0.27%
15% Mexican Tax Rate	579,988	0.001	0.000	0.667	19,391,310	284,048	19,199,740	0.99%
0% Mexican Tax Rate	579,988	0.001	0.000	0.667	20,533,120	284,048	19,760,460	3.76%
EXPERIMENT 2								
35% Mexican Tax Rate	303,954	0.002	1.000	0.667	18,194,560	284,048	18,101,540	0.51%
15% Mexican Tax Rate	579,988	0.001	1.000	0.333	19,115,440	284,048	18,837,000	1.46%
0% Mexican Tax Rate	579,988	0.001	1.000	0.333	20,248,130	284,048	19,388,600	4.24%
EXPERIMENT 3								
35% Mexican Tax Rate	0	1.000	1.000	0.000	24,688,820	0	24,189,270	2.02%
15% Mexican Tax Rate	0	1.000	1.000	0.000	26,953,770	0	24,189,270	10.26%
0% Mexican Tax Rate	1,661,448	1.000	1.000	0.000	30,917,790	0	24,189,270	21.76%
EXPERIMENT 4								
35% Mexican Tax Rate	0	0.001	1.000	0.000	16,245,130	0	16,092,760	0.94%
15% Mexican Tax Rate	579,988	0.001	1.000	0.000	17,288,980	0	16,092,760	6.92%
0% Mexican Tax Rate	579,988	0.001	1.000	0.000	18,343,260	0	16,092,760	12.27%
EXPERIMENT 5								
35% Mexican Tax Rate	0	1.000	0.000	0.000	24,752,710	0	24,253,160	2.02%
15% Mexican Tax Rate	0	1.000	0.000	0.000	27,017,670	0	24,253,160	10.23%
0% Mexican Tax Rate	1,642,390	0.984	0.000	0.000	30,989,870	0	24,253,160	21.74%
EXPERIMENT 6								
35% Mexican Tax Rate	286,853	0.006	0.000	0.000	22,935,730	287,351	22,514,180	1.84%
15% Mexican Tax Rate	590,592	0.009	0.000	0.000	24,955,060	287,351	22,672,450	9.15%
0% Mexican Tax Rate	592,552	0.011	0.000	0.000	26,622,700	287,351	22,791,140	14.39%
EXPERIMENT 7								
35% Mexican Tax Rate	0	1.000	0.000	0.500	24,564,810	0	24,398,530	0.68%
15% Mexican Tax Rate	821,755	1.000	0.000	0.000	25,819,480	0	24,398,530	5.50%
0% Mexican Tax Rate	1,612,344	1.000	0.000	0.500	30,479,100	0	24,398,530	19.95%
EXPERIMENT 8								
35% Mexican Tax Rate	0	0.500	0.667	0.250	22,265,560	0	21,374,100	4.00%
15% Mexican Tax Rate	587,648	0.007	0.750	0.167	23,720,790	0	21,374,100	9.89%
0% Mexican Tax Rate	643,150	0.047	0.750	0.167	25,321,730	0	21,374,100	15.59%

Table 16.7. . Tax Rate Sensitivity on the Tax Rates for Mexico, Romania, and India

	Optimal Solution Maximizing NPV						Maximizing Total Gross Profit Minus Investment Cost				
	Subasbls made in Mexico	Subasbls made in Romania	Subasbls made in India	MARKET	PLANT	DIS	Net Present Value	Subasbls made in Mexico	Subasbls made in Romania	Subasbls made in India	Net Present Value	% Deviation from Optimal
EXPERIMENT 1												
35%/25%/35%*	303,954	339,132	24,224	0.002	0.000	1.000	18,501,230	284,048	274,933	26,336	18,452,120	0.27%
15% Tax Rate	579,988	592,066	40,792	0.001	0.000	0.333	20,425,510	284,048	274,933	26,336	19,768,250	3.22%
0% Tax Rate	579,988	592,066	89,216	0.001	0.000	0.333	23,202,250	284,048	274,933	26,336	21,119,020	8.98%
EXPERIMENT 2												
35%/25%/35%*	303,954	339,132	0	0.002	1.000	0.667	18,194,560	284,048	274,933	26,336	18,101,540	0.51%
15% Tax Rate	579,988	592,066	40,792	0.001	1.000	0.333	20,142,850	284,048	274,933	26,336	19,384,170	3.77%
0% Tax Rate	579,988	592,066	89,216	0.001	1.000	0.333	22,889,490	284,048	274,933	26,336	20,704,660	9.55%
EXPERIMENT 3												
35%/25%/35%*	0	1,661,448	0	1.000	1.000	0.000	24,688,820	0	1,661,448	0	24,189,270	2.02%
15% Tax Rate	0	1,661,448	0	1.000	1.000	0.000	28,061,110	0	1,661,448	0	25,324,660	9.75%
0% Tax Rate	0	1,661,448	0	1.000	1.000	0.000	33,056,600	0	1,661,448	0	27,027,740	18.24%
EXPERIMENT 4												
35%/25%/35%*	0	592,066	0	0.001	1.000	0.000	16,245,130	0	0	0	16,092,760	0.94%
15% Tax Rate	579,988	592,066	0	0.001	1.000	0.000	18,243,260	0	0	0	16,092,760	11.79%
0% Tax Rate	579,988	592,066	0	0.001	0.500	0.000	20,254,740	0	0	0	16,092,760	20.55%
EXPERIMENT 5												
35%/25%/35%*	0	1,661,448	0	1.000	0.000	0.000	24,752,710	0	1,661,448	0	24,253,160	2.02%
15% Tax Rate	0	1,661,448	0	1.000	0.000	0.000	28,131,760	0	1,661,448	0	25,395,310	9.73%
0% Tax Rate	0	1,661,448	0	1.000	0.000	0.000	33,137,380	0	1,661,448	0	27,108,520	18.19%
EXPERIMENT 6												
35%/25%/35%*	286,853	599,278	244,232	0.006	0.000	0.000	22,935,730	287,351	599,172	243,734	22,514,180	1.84%
15% Tax Rate	574,702	599,278	487,468	0.006	0.000	0.000	27,086,860	287,351	599,172	243,734	23,235,590	14.22%
0% Tax Rate	574,702	599,278	487,468	0.006	0.000	0.001	31,988,050	287,351	599,172	243,734	24,084,620	24.71%
EXPERIMENT 7												
35%/25%/35%*	0	754,254	0	0.500	0.000	0.500	24,564,810	0	793,777	808,405	24,398,530	0.68%
15% Tax Rate	35,622	799,483	808,405	0.959	0.000	0.000	27,983,650	0	793,777	808,405	27,978,550	0.02%
0% Tax Rate	18,688	803,067	821,755	0.984	0.000	0.000	33,036,320	0	793,777	808,405	33,001,940	0.10%
EXPERIMENT 8												
35%/25%/35%*	0	1,172,518	0	0.500	0.667	0.250	22,265,560	0	607,013	0	21,374,100	4.00%
15% Tax Rate	576,704	611,824	425,484	0.040	0.750	0.167	24,975,150	0	607,013	0	22,584,220	9.57%
0% Tax Rate	576,704	622,552	425,484	0.040	0.750	0.167	29,498,620	0	607,013	0	24,399,410	17.29%

* 35% tax rate in Mexico, 25% tax rate in Romania, and 35% tax rate in India

Importantly, the global network structures did not, in general, change very much. To support this claim, for each dimension proxy measure (*MARKET*, *PLANT*, and *DIS*) we computed two measures of aggregation: (1) the absolute value of the deviation between the measure value at the initial rate (tax or interest) and the lowest possible rate (0%), averaged over all 32 samples, and (2) the percentage of the 32 samples having no deviation between the measure value at the initial rate (tax or interest) and the lowest possible rate (0%). The average absolute deviation and percentage of no deviations were .022 and 40.63% for *MARKET*, .055 and 84.38% for *PLANT*, and .169 and 56.25% for *DIS*, respectively. The general stability of the global network structures gives credence to the structural equations approach from Section 2 as a way to monitor the overall strategic configuration of the company. Under those circumstances, the actual plant location decisions within a region, based on country differences, can be decoupled from the global network decision.

As expected, government subsidies tended to significantly increase the investment in those countries. Low tax rates had a greater effect than low loan rates in terms of both net present values and movement of subassemblies. Furthermore, our results suggested that models ignoring tax rates when significant tax differences exist may perform very poorly. Specifically, the average deviation from optimal at 15% tax rates was 7.28% with a maximum of 14.22%, while at the tax-exempt levels the average deviation from optimal was 14.46% with a maximum of 24.71%.

Table 16.8. Loan Rate Sensitivity on Interest Rates in Ireland

	Optimal Solution Maximizing NPV					Maximizing Total Gross Profit Minus Investment Cost		
	Subs made in Ireland	MARKET	PLANT	DIS	Net Present Value	Subs made in Ireland	Net Present Value	% Deviation from Optimal
EXPERIMENT 1								
34% Interest Rate	254,744	0.002	0.000	1.000	18,501,230	321,733	18,452,120	0.27%
5% Interest Rate	294,754	0.004	0.000	1.000	18,656,340	321,733	18,636,600	0.11%
0% Interest Rate	597,722	0.004	0.000	0.667	19,165,670	321,733	18,997,760	0.88%
EXPERIMENT 2								
34% Interest Rate	254,744	0.002	1.000	0.667	18,194,560	321,733	18,101,540	0.51%
5% Interest Rate	292,227	0.004	1.000	0.667	18,361,000	321,733	18,297,130	0.35%
0% Interest Rate	597,722	0.004	1.000	0.333	18,957,100	321,733	18,680,030	1.46%
EXPERIMENT 3								
34% Interest Rate	0	1.000	1.000	0.000	24,688,820	0	24,189,270	2.02%
5% Interest Rate	0	1.000	1.000	0.000	24,688,820	0	24,553,270	0.55%
0% Interest Rate	1,661,448	1.000	1.000	0.000	25,607,840	0	25,265,840	1.34%
EXPERIMENT 4								
34% Interest Rate	0	0.001	1.000	0.000	16,245,130	597,722	16,092,760	0.94%
5% Interest Rate	596,928	0.004	1.000	0.000	16,546,080	597,722	16,544,800	0.01%
0% Interest Rate	597,722	0.004	1.000	0.000	17,429,720	597,722	17,429,720	0.00%
EXPERIMENT 5								
34% Interest Rate	0	1.000	0.000	0.000	24,752,710	0	24,253,160	2.02%
5% Interest Rate	0	1.000	0.000	0.000	24,752,710	0	24,617,160	0.55%
0% Interest Rate	1,661,448	1.000	0.000	0.000	25,633,290	0	25,329,740	1.18%
EXPERIMENT 6								
34% Interest Rate	0	0.006	0.000	0.000	22,935,730	0	22,514,180	1.84%
5% Interest Rate	0	0.006	0.000	0.000	23,058,330	0	22,648,520	1.78%
0% Interest Rate	599,854	0.012	0.000	0.000	23,430,140	0	22,911,500	2.21%
EXPERIMENT 7								
34% Interest Rate	67,501	1.000	0.000	0.500	24,564,810	0	24,398,530	0.68%
5% Interest Rate	821,755	1.000	0.000	0.000	24,684,110	0	24,398,530	1.16%
0% Interest Rate	849,733	0.977	0.000	0.000	24,938,460	0	24,398,530	2.17%
EXPERIMENT 8								
34% Interest Rate	16,972	0.500	0.667	0.250	22,265,560	4,465	21,374,100	4.00%
5% Interest Rate	38,761	0.500	1.000	0.500	22,271,440	4,465	21,375,880	4.02%
0% Interest Rate	869,885	0.500	0.667	0.250	22,481,570	4,465	21,379,380	4.90%

Table 16.9. Loan Rate Sensitivity on the Interest Rates for the U.S., Ireland, and Taiwan

	Optimal Solution Maximizing NPV						Maximizing Total Gross Profit Minus Investment Cost				% Deviation from Optimal
	Subasbls made in the U.S.	Subasbls made in Ireland	Subasbls made in Taiwan	MARKET	PLANT	DIS	Net Present Value	Subasbls made in the U.S.	Subasbls made in Ireland	Subasbls made in Taiwan	Net Present Value	
EXPERIMENT 1												
34% Interest Rate	274,224	254,744	465,170	0.002	0.000	1.000	18,501,230	292,134	321,733	462,264	18,452,120	0.27%
5% Interest Rate	274,224	291,702	465,098	0.002	0.000	1.000	18,921,270	292,134	321,733	462,264	18,902,790	0.10%
0% Interest Rate	579,988	592,860	465,630	0.000	0.000	0.333	19,978,820	292,134	321,733	462,264	19,773,940	1.03%
EXPERIMENT 2												
34% Interest Rate	274,224	254,744	489,394	0.002	1.000	0.667	18,194,560	292,134	321,733	462,264	18,101,540	0.51%
5% Interest Rate	274,224	289,175	488,600	0.002	1.000	0.667	18,659,550	292,134	321,733	462,264	18,590,760	0.37%
0% Interest Rate	579,988	592,860	488,600	0.000	1.000	0.000	19,917,270	292,134	321,733	462,264	19,536,860	1.91%
EXPERIMENT 3												
34% Interest Rate	0	0	0	1.000	1.000	0.000	24,688,820	0	0	0	24,189,270	2.02%
5% Interest Rate	0	0	830,724	1.000	0.000	0.000	24,991,980	0	0	0	24,553,270	1.76%
0% Interest Rate	0	830,724	830,724	1.000	0.000	0.000	25,835,710	0	0	0	25,265,840	2.21%
EXPERIMENT 4												
34% Interest Rate	579,988	0	489,394	0.001	1.000	0.000	16,245,130	575,126	597,722	488,600	16,092,760	0.94%
5% Interest Rate	575,126	596,928	489,394	0.004	1.000	0.000	17,255,060	575,126	597,722	488,600	17,253,780	0.01%
0% Interest Rate	579,988	592,860	488,600	0.000	1.000	0.000	19,510,290	575,126	597,722	488,600	19,497,030	0.07%
EXPERIMENT 5												
34% Interest Rate	0	0	0	1.000	0.000	0.000	24,752,710	0	0	0	24,253,160	2.02%
5% Interest Rate	0	0	830,724	1.000	0.000	0.000	25,088,900	0	0	0	24,617,160	1.88%
0% Interest Rate	0	830,724	830,724	1.000	0.000	0.000	25,871,720	0	0	0	25,329,740	2.09%
EXPERIMENT 6												
34% Interest Rate	286,853	0	244,232	0.006	0.000	0.000	22,935,730	287,563	0	243,628	22,514,180	1.84%
5% Interest Rate	284,144	0	495,688	0.010	0.000	0.000	23,239,840	287,563	0	243,628	22,932,020	1.32%
0% Interest Rate	574,702	597,140	589,606	0.005	0.000	0.000	24,110,480	287,563	0	243,628	23,735,070	1.56%
EXPERIMENT 7												
34% Interest Rate	0	67,501	821,755	1.000	0.000	0.500	24,564,810	13,350	0	27,978	24,398,530	0.68%
5% Interest Rate	0	821,755	821,755	1.000	0.000	0.000	24,860,830	13,350	0	27,978	24,423,500	1.76%
0% Interest Rate	0	821,755	821,755	1.000	0.000	0.000	25,445,280	13,350	0	27,978	24,471,980	3.83%
EXPERIMENT 8												
34% Interest Rate	0	16,972	454,020	0.500	0.667	0.250	22,265,560	583,454	4,465	448,578	21,374,100	4.00%
5% Interest Rate	0	38,761	459,623	0.500	1.000	0.500	22,361,810	583,454	4,465	448,578	21,655,300	3.16%
0% Interest Rate	189,780	652,755	501,300	0.339	0.750	0.167	22,769,890	583,454	4,465	448,578	22,188,530	2.55%

4. Conclusion

We have presented a modeling framework that provides decision support for global facility network decisions. Among other things, our detailed MIP model in the Appendix explicitly considers (1) economies of scale, (2) complexity costs, (3) transportation costs, (4) tariffs, (5) subsidized financing, and (6) taxation. Items (1) through (4) appear to have the largest impact on the *strategic* classification of the facility network according to our proposed conceptual framework. Furthermore, we demonstrate that once a network is classified, items (5) and (6) can have a significant impact on location decisions within the network. We also provide evidence suggesting that location models that include taxation and finance issues may provide significant savings compared to those that do not.

In Section 2 we presented a structural equations approach using global sensitivity analysis. Managers can use the model as a tool to help them make strategic network decisions by focusing on four reasonably simple, yet explanatorily powerful, cost measures. Managers can capture the essence of the strategic aggregated decision by using less detail than a traditional complete plant location model would necessitate. On an ongoing basis, as costs change the equations can be used as a tracking device to quickly verify the appropriateness of the current network structure. This can allow for a decoupling of the strategic network decision and the actual plant location decision for a given network structure.

We hope that the results of this work will stimulate further research into the observed location practice of firms in response to the many complicating factors of the global market environment. For example, the use of operational hedging to account for uncertainty could be incorporated into our models, perhaps by using a scenario approach with discrete probability distributions. Hopefully, future field research will use our facility network classification framework to understand network structure and location practices of firms in various industries. Research explaining differences of facility network structures within an industry will be useful in pointing out other primary levers of such decisions that are potentially not in our current models. Finally, we see an opportunity for consultants to develop industry-specific structural equations based on our approach.

Acknowledgments

We wish to thank Jianli Hu for her contributions to this work.

Appendix: MIP Model for Facility Location

Model Assumptions

- There are two facility levels: plants (producing subassemblies) and DCs (final assembly).

- The plants and DCs remain open throughout the finite planning horizon.

- The firm is a price taker in each market. All prices are quoted in local currency and then translated into common currency (say $) by using the real exchange rate.

- The market demands, selling prices, and transfer prices are independent of the structure of the facility network.

- At most one Distribution Center (DC) (assumed to have adequate capacity to cover any one country's demand) is allowed in each country. The

APPENDIX

rationale behind this assumption is that as shipment cost of a subassembly or final product within a country is assumed constant in our model, the opening of a second DC will result in additional fixed costs with no transportation cost savings. (Geographical large countries can easily be split into subcountries, as needed, to coincide with this assumption.)

- Any type of subassembly can be produced in at most one plant in any given country.

- Costs, prices, tax rates, interest rates, plant capacities, and discount rates remain constant over time. Demand and depreciation rates, however, may vary over time.

- Annual fixed costs and up-front investment costs are an increasing function of capacity and are modeled as a base cost plus a linear function of capacity.

- Capacities and costs are modeled for both plants and production lines within the plants. Production line variable investment and operating costs are linear functions of the number of subassemblies produced on those lines. Plant variable investment and operating costs are linear functions of the size of production lines (expressed as space requirements in square feet) installed in the plants.

Notation

Indices

i: subassembly index, $i = 1,..., I$;
n: country of operation index, $n = 1,..., N$;
k: country of locating a distribution center (DC) index, $k = 1,..., N$;
m: country of locating a plant index, $m = 1,..., N$;
q: subassembly plant index within a given country, $q = 1,..., I$;
j: country (market) index, $j = 1,..., J$;
t: time period, $t = 1,..., T$;
z: number of subassemblies produced at the same plant index, $z = 1,..., I$;

Parameters

V_{im}: unit variable production cost of subassembly i produced in country m;
v_k: unit variable assembly cost at a DC in country k;
F_m: annual fixed cost of operating a subassembly plant in country m (base cost);
f_m: annual fixed cost per square foot of operating a plant in country m (linear component);
B_{zm}: annual fixed cost of producing the zth subassembly in the same plant in country m, $z = 2,..., I$ (B_{zm} is increasing in z);
U_k: annual fixed cost of operating a DC in country k (base cost);

u_k: annual fixed cost of operating a DC in country k (linear component);

K_{im}: maximum capacity (in units) of a production line for subassembly i in country m;

C_m: maximum capacity (in sq. feet) of a subassembly plant built in country m;

c_k: maximum capacity (in units) of a DC built in country k;

G_{im}: space requirement (in square feet) per unit of a production line for subassembly i in m;

E_{im}: investment cost for equipment to build a production line for subassembly i in m (base investment);

e_{im}: investment cost per unit for equipment to build a production line for subassembly i in m (linear component);

A_m: investment cost for land and building to build a subassembly plant in country m (base investment);

a_m: investment cost per square foot for land and building to build a plant in country m (linear component);

H_k: investment cost to build a DC in country k (base investment);

h_k: investment cost per unit to build a DC in country k (linear component);

S_{kj}: cost to ship one unit of the final product from a DC in country k to market j (shipment costs include any assessed trade tariffs and other duties);

$\lambda_i S_{mk}$: cost to ship one unit of subassembly i from country m to country k (i.e., subassembly transportation cost is expressed as a fraction of the transportation cost of the final product, with λ_i the fraction for subassembly i);

D_{jt}: demand of the final product in country j in period t;

r_n: per-period interest rate on the loan in country n;

$R(r_n, T)$: loan payment factor given interest rate r and planning horizon T;

$\rho_t(r_n, T)$: interest calculation factor for period t given interest rate r and planning horizon T;

τ_n: marginal corporate income tax rate in country n;

P_j: sales price of the final product (translated into $ with real exchange rate) in country j;

p_{imk}: transfer price of subassembly i made in country m shipped to DC in country k;

d_{nt}: applicable depreciation rate in country n in period t (% per period);

β_n: discount rate of after tax cash flows in country n;

l_n: maximum loan that country n can give to the firm;

Decision Variables

X_{imqkt}: units of subassembly i made in country m at plant q and assembled in country k during t;

x_{kjt}: units assembled at a DC in country k and sold in market j during t;

L_n: the loan that the company will take from the country n government for investment in n;

APPENDIX

Z_{imq}: = 1 if a production line for subassembly i is built in country m at plant q, and 0 otherwise;

Y_{zmq}: = 1 if a zth subassembly is produced in country m at plant q, and 0 otherwise;

y_k: = 1 if a DC is operated in country k, and 0 otherwise;

W_{imq}: size (in units) of production line i built in country m at plant q;

Q_{mq}: size (in square feet) of plant q in country m;

w_k: size (in units) of a DC operated in country k.

The Objective Function

Revenue from units assembled and subassemblies produced in country n in period t:

$$\alpha_{nt} = \sum_{j=1}^{J} P_j\, x_{njt} + \sum_{i=1}^{I} \sum_{q=1}^{I} \sum_{k=1}^{N} p_{ink}\, X_{inqkt}.$$

Variable production costs and cost of goods sold in country n in period t:

$$\delta_{nt} = \sum_{j=1}^{J} v_n\, x_{njt} + \sum_{i=1}^{I} \sum_{m=1}^{N} \sum_{q=1}^{I} p_{imn}\, X_{imqnt} + \sum_{i=1}^{I} \sum_{q=1}^{I} \sum_{k=1}^{K} V_{in}\, X_{inqkt}.$$

Annual fixed costs of a DC and subassembly plants in country n:

$$\gamma_n = U_n\, y_n + u_n\, w_n + \sum_{q=1}^{I} \left(F_n\, Y_{1nq} + f_n\, Q_{nq} + \sum_{z=2}^{I} B_{zn}\, Y_{znq} \right).$$

Transportation costs from facilities (plants and DCs) located in country n in period t:

$$\eta_{nt} = \sum_{j=1}^{J} S_{nj}\, x_{njt} + \sum_{i=1}^{I} \sum_{q=1}^{I} \sum_{k=1}^{N} \lambda_i\, S_{nk}\, X_{inqkt}.$$

Loan payment in country n in period t: $\varphi_{nt} = L_n R(r_n, T)$

Loan interest payment in country n in period t: $\varphi'_{nt} = L_n \rho_t(r_n, T)$

Investment cost in country n:

$$\xi_n = H_n\, y_n + h_n\, w_n + \sum_{q=1}^{I} \left(A_n\, Y_{1nq} + a_n\, Q_{nq} \right) + \sum_{i=1}^{I} \sum_{q=1}^{I} \left(E_{in}\, Z_{inq} + e_{in}\, W_{inq} \right).$$

Depreciation expense in country n in period t: $\psi_{nt} = \xi_n d_{nt}$

Before-tax income in country n in period t:

$$\pi_{nt} = \alpha_{nt} - (\delta_{nt} + \gamma_n + \eta_{nt} + \varphi'_{nt} + \psi_{nt})$$

Corporate income tax paid in country n in period t: $\omega_{nt} = \pi_{nt}\tau_n$

Cash expenditures in fixed assets in year 0 that are not financed by external sources:

$$\mu = \sum_{n=1}^{N} (\xi_n - L_n).$$

So the objective function which maximizes the net present value is:

$$OBF = \max \left[\sum_{n=1}^{N} \sum_{t=1}^{T} (\pi_{nt} + \varphi'_{nt} + \psi_{nt} - \varphi_{nt} - \omega_{nt})/(1+\beta_n)^t \right] - \mu$$

The Set of Constraints

The Constraints are:

1) Demand Constraints: $\sum_{k=1}^{N} x_{kjt} \leq D_{jt} \quad \forall j, t$

2) Subassembly Line Capacity:

$$\sum_{k=1}^{N} X_{imqkt} \leq W_{imq} \quad \forall i, m, q, t$$

$$W_{imq} \leq K_{im} Z_{imq} \quad \forall i, m, q$$

$$\sum_{q=1}^{I} Z_{imq} \leq 1 \quad \forall i, m$$

$$\sum_{i=1}^{I} Z_{im(q+1)} \leq \sum_{i=1}^{I} Z_{imq} \quad \forall m, q < I(\text{for solution efficiency})$$

$$Z_{imq} = 0 \quad \forall i, m, q > i(\text{for solution efficiency})$$

3) Plant Capacity Constraints:

$$\sum_{i=1}^{I} G_{im} W_{imq} \leq Q_{mq} \quad \forall m, q$$

$$Q_{mq} \leq C_m Y_{1mq} \quad \forall m, q$$

APPENDIX

$$Y_{1m(q+1)} \leq Y_{1mq} \quad \forall m, q < I \text{(for solution efficiency)}$$

4) DC Capacity Constraints:

$$\sum_{j=1}^{J} x_{kjt} \leq w_k \quad \forall k, t$$

$$w_k \leq c_k y_k \quad \forall k$$

5) Conservation of Subassembly Flows $\sum_{m=1}^{N} \sum_{q=1}^{I} X_{imqkt} = \sum_{j=1}^{J} x_{kjt} \quad \forall i, k, t$

6) Loan Ceilings: $L_n \leq l_n \ \forall n$ and $L_n \leq \xi_n \ \forall n \ w_k \leq c_k y_k \quad \forall n$

7) Nonnegative Profit in Each Country in Each Period (assumption of convenience to avoid unnecessarily complicated tax calculations):

$$\pi_{nt} \geq 0 \quad \forall n, t$$

8) Count the Number of Different Subassemblies Produced at Each Plant:

$$\sum_{i=1}^{I} Z_{imq} = \sum_{z=1}^{I} Y_{zmq} \quad \forall m, q$$

$$Y_{(z+1)mq} \leq Y_{zmq} \quad \forall z \neq I, m, q$$

9) Specification of Decision Variables:
$X_{imqkt}, x_{kjt}, L_n, W_{imq}, Q_{mq}, w_k \geq 0$ for all indices
Z_{imq}, Y_{zmq}, y_k = binary for all indices

References

Arnzten, B.C., Brown, G.G., Harrison, T.P. and Trafton, L.L. (1995) Global supply chain management at Digital Equipment Corporation. *Interfaces*, **25** (1), 69-93.

Bartmess, A. and Cerny, K. (1993) Building competitive advantage through a global network of capabilities. *California Management Review*, 78-103.

Breitman, R.L. and Lucas, J.M. (1987) PLANETS: a modeling system for business planning. *Interfaces*, **17**, 94-106.

Brush, T.H., Maritan, C.A., and Karnani, A. (1999) The plant location decision in multinational manufacturing firms: an empirical analysis of international business and manufacturing strategy perspectives. *Production and Operations Management*, **8** (2), 109-132.

Cohen, M.A. and Huchzermeier, A. (1998) Global supply chain management: a survey of research and applications. In: *Quantitative Models for Supply Chain Management*, S. Tayur, R. Ganeshan, and M. Magazine (Eds.), Kluwer, Boston, 669-702.

Cohen, M.A. and Lee, H.L. (1989) Resource deployment analysis of global manufacturing and distribution networks. *Journal of Manufacturing and Operations Management*, 81-104.

Cowie, J. (1999) *Capital moves: RCA's seventy-year quest for cheap labor*, Cornell University Press, Ithaca and London.

Davis, T. (1993) Effective supply chain management. *Sloan Management Review*, 35-45.

Francis, R.L., McGinnis, Jr., L.F., and White, J.A. (1992) *Facility Layout and Location: An Analytical Approach, 2nd Ed.*, Prentice Hall, Englewood Cliffs, NJ.

Hax, A.C., and Meal, H.C. (1975) Hierarchical integration of production planning and scheduling, in *TIMS Studies in Management Science*, Vol. 1: *Logistics*, M. Geisler, Ed., New York: Elsevier.

Hodder, J.E. and Dincer, M.C. (1986) A multifactor model for international plant location and financing under uncertainty. *Computers and Operations Research*, **13** (5), 601-609.

Hodder, J.E. and Jucker, J.V. (1985a) A simple plant-location model for quantity setting firms subject to price uncertainty. *European Journal of Operational Research*, **21**, 39-46.

Hodder, J.E. and Jucker, J.V. (1985b) International plant location under price and exchange rate uncertainty. *Engineering and Production Economics*, **9**, 225-229.

Huchzermeier, A. (1991) Global manufacturing strategy planning under exchange rate uncertainty. Unpublished Ph.D. thesis, University of Pennsylvania, Philadelphia, PA.

Huchzermeier, A. (1994) Global supply chain network management under risk. Working paper, University of Chicago, Graduate School of Business, Chicago, Illinois.

Huchzermeier, A. and Cohen, M.A. (1996) Valuing operational flexibility under exchange rate risk. *Operations Research*, **44** (1), 100-113.

Jucker, J.V. and Carlson, R.C. (1976) The simple plant location problems under uncertainty. *Operations Research*, **24** (6), 1045-1055.

Kouvelis, P. (1998) Global sourcing strategies under exchange rate uncertainty. In: *Quantitative Models for Supply Chain Management*, S. Tayur, R. Ganeshan, and M. Magazine (Eds.), Kluwer, Boston, 669-702.

Kouvelis, P., Rosenblatt, M.J., and Munson, C.L. (2003) A mathematical programming model to global plant location problems: analysis and insights. *IIE Transactions*, forthcoming.

Love, R.F., Morris, J.G., and Wesolowsky, G.O. (1988) *Facilities Location: Models and Methods*, North-Holland Publishing Co., New York.

MacCormack, A.D., Newman, L.J. and Rosenfield, D. (1994) The new dynamics of global manufacturing site location. *Sloan Management Review*, Summer, 69-80.

Munson, C.L. and Rosenblatt, M.J. (1997) The impact of local content rules on global sourcing decisions. *Production and Operations Management*, **6** (3), 277-290.

Revelle, C.S. and Laporte, G. (1996) The plant location problem: new models and research projects. *Operations Research*, **44** (6), 864-874.

Wagner, H.M. (1995) Global sensitivity analysis. *Operations Research*, **43** (6), 948-969.

David Simchi-Levi, S. David Wu, and Z. Max Shen (Eds.)
Handbook of Quantitative Supply Chain Analysis:
Modeling in the E-Business Era
©2004 Kluwer Academic Publishers

Chapter 17

INTEGRATED PRODUCTION AND DISTRIBUTION OPERATIONS:

Taxonomy, Models, and Review

Zhi-Long Chen
Robert H. Smith School of Business,
University of Maryland, College Park, MD 20742

Keywords: Production distribution integration, literature review

1. Introduction

The supply chain of a typical product starts with material input, followed by production, and finally distribution of the end product to customers. The cost of a product includes not only the cost of factory resources to convert materials to a finished item but also the cost of resources to make the sale, deliver the product to customers, and service the customers. Consequently, in order to reduce cost, firms have to plan all the activities in the supply chain in a coordinated manner. It is well recognized that there is a greater opportunity for cost saving in managing supply chain coordination than in improving individual function areas.

Various types of coordination in a supply chain have been studied in the literature. We discuss the coordination of production and distribution in this chapter. Production and distribution operations can be decoupled if there is a sufficient amount of inventory between them. Many companies manage these two functions independently with little or no coordination. However, this leads to increased holding costs and longer lead times of products through the supply chain. Fierce competition in today's global market and heightened expectations of customers have forced companies to invest aggressively to reduce inventory levels across the supply chain on one hand and be more responsive to customers on the other. Reduced inventory results in closer linkages between production and distribution functions. Consequently, companies can realize cost savings and improve customer service by optimizing production and distribution operations in an integrated manner.

Many companies have adopted direct-sell e-business models as their way of doing business. In a direct-sell model, products are custom-made and delivered to customers within a very short lead time directly from the factory without the intermediate step of finished product inventory. Consequently, the production in the factory and the distribution from the factory to customers are closely linked. For example, in a typical computer direct-order system, there are hundreds of configurations available for a customer to choose from when ordering a computer. It is thus impractical to keep inventory of assembled and packaged computers with a particular configuration before the sale. As a result, assembly and packaging operations can only take place after customer orders come in. On the other hand, due to fierce competition in the computer industry, completed orders (i.e. assembled and packaged computers) are normally delivered to customers within a short lead time (e.g. two to three business days). Hence, production operations (i.e. assembly and packaging) and distribution operations (i.e. delivery of completed orders to customers) are linked together directly without any intermediate step. Thus, coordinated planning of these operations becomes critical in order to maintain a high level of customer service.

The interdependency between production and distribution operations, and the tradeoff between the costs associated with them can be illustrated intuitively by the following simple example. Consider a company producing multiple products for multiple customers. To save distribution cost, orders of closely located customers may have to be produced at similar times so that they can be consolidated for delivery right after they are produced. However, orders of closely located customers may require very different production setups, and producing them at similar times may incur a large production cost. Of course, in many cases, in addition to production and distribution, there are other factors such as inventory and capacity that also play important roles.

There are many different models in the literature involving joint considerations of production and distribution (possibly together with other functions such as inventory). The focus of this review is on tactical and operational level integration of production and distribution decisions. As such, we will not review strategic models that integrate design decisions in the supply chain such as location, plant capacity, and transportation channels. See Vidal and Goetschalckx (1997) and Owen and Daskin (1998) for reviews, and Jayaraman and Pirkul (2001), Dasci and Verter (2001), and Shen et al. (2003) for recent results in this area. We will only cover models that *explicitly* involve both production and distribution operations. There is a huge body of literature on the so-called *production-distribution* problems, models, networks, or systems. However, a major portion of this broad literature is on the following two classes of problems:

1. Problems that integrate inventory replenishment decisions across multiple stages of the supply chain. See, among others, Williams (1983), Muckstadt and Roundy (1993), Pyke and Cohen (1994), Bramel et al. (2000), and Boyaci and Gallego (2001).

2. Problems that integrate inventory and distribution decisions. See, among others, Burns et al. (1985), Speranza and Ukovich (1994), Chan et al. (1997), Bertazzi and Speranza (1999).

These problems either ignore or oversimplify production operations (e.g. assuming instantaneous production without production time or capacity consideration). We will not review them although many of the papers in this area have a title containing the term "production-distribution".

To distinguish the models we review here that involve explicitly both production and distribution operations from the models that do not, we call our models *explicit production-distribution* (EPD) models.

To our knowledge, no one has given a comprehensive review on EPD problems. Several existing surveys in the literature (Bhatnagar and Chandra 1993, Thomas and Griffin 1996, Sarmiento and Nagi 1999) have covered several papers on EPD problems. However, none of these surveys is dedicated to EPD models, and the EPD models covered in these reviews are only a subset of the models we will cover in this review. Our goal is to bring together all major EPD models, provide a comprehensive review on each of these models, and identify areas that require more research.

This chapter is organized as follows. In Section 2, we classify explicit production-distribution models into five different problem classes based on model assumptions and the nature of production and distribution decisions. Existing results on each of these problem classes are reviewed in Sections 3 through 7, and summarized in Tables 17.1 through 17.5. Finally, in Section 8, we conclude this chapter by providing some potential topics for future research.

2. Model Classification

We classify various EPD problems in the literature based on three different dimensions: (A) decision level, (B) integration structure, and (C) problem parameters, which are described in detail in the following.

(A) Decision Level

Since we are not going to cover strategic models, there are two types of EPD models we will discuss with respect to level of decisions:

(A1) Tactical EPD models – which mainly involve decisions such as: how much to produce and how much to ship in a time period, how long the production cycle/distribution cycle should be, how much inventory to keep, etc. We distinguish these decisions from higher-level ones such as facility location, capacity, and network structure. Also, they are different from the more detailed operational level decisions described below.

(A2) Operational EPD models – which mainly involve detailed scheduling level decisions such as: when and on which machine to process a job, when and by which vehicle to deliver a job, which route to take for a vehicle, etc. These are the most detailed decisions and at the level of individual jobs, individual processing machines, and individual delivery vehicles.

(B) Integration Structure

We classify the integration between production and distribution operations into the following three types of structure:

(B1) Integration of production and outbound transportation. This typically takes place in a two-stage supply chain involving manufacturers and customers. Products are delivered from manufacturers to customers after they are produced by manufacturers.

(B2) Integration of inbound transportation and production. This typically takes place in a two-stage supply chain involving suppliers and manufacturers. Suppliers supply raw materials or semi-finished products to manufacturers where final products are produced.

(B3) Integration of inbound transportation, production, and outbound transportation. This typically takes place in a supply chain with three or more stages involving suppliers, manufacturers, and customers. Raw materials are delivered from suppliers to manufacturers where final products are produced and delivered to customers.

(C) Problem Parameters

The difficulty of a problem is often determined by several key parameters. Everything else being equal, the difficulty of an EPD problem is mainly determined by two key parameters: length of planning horizon, and nature of demand. There are three variations on these parameters considered in the EPD literature:

(C1) One time period.

(C2) Infinite horizon with constant demand rate.

(C3) Finite horizon but with multiple time periods and dynamic demand.

Based on the three dimensions A, B, C of model characteristics described above, we classify EPD problems that have appeared in the literature into five problem classes as follows.

Class 1: Production-transportation Problems (A1, B1, C1)

Class 2: Joint Lot Sizing and Finished Product Delivery Problems (A1, B1, C2)

Class 3: Joint Raw Material Delivery and Lot Sizing Problems (A1, B2, C2)

Class 4: General Tactical Production-Distribution Problems (A1, B1 or B3, C1 or C3)

Class 5: Joint Job Processing and Finished Job delivery Problems (A2, B1, C3)

Note that the name "Production-Transportation" for Class 1 problems comes from the literature where these problems are commonly referred by this name. We name the other classes of problems to reflect their key characteristics.

3.   Production - Transportation Problems

A typical production-transportation problem can be described as follows. There are m production plants and n customers. A single product is produced at the plants and shipped from the plants to the customers. Each plant i has a capacity limit s_i, and the production cost is a concave function $f(x_1, x_2, ..., x_m)$ of the amounts produced at these plants $x_1, x_2, ..., x_m$. The transportation costs from the plants to the customers are linear; the unit transportation cost from plant i to customer j is c_{ij}. The problem involves a single time period only, and the demand of each customer j is known as d_j which must be satisfied. The problem seeks a production and distribution plan that minimizes total production and transportation cost. The problem can be formulated as the following mathematical program:

$$[PT] \min f(x_1, x_2, ..., x_m) + \sum_{i=1}^{m} \sum_{j=1}^{n} c_{ij} y_{ij} \quad (17.1)$$

$$\text{s.t.} \sum_{j=1}^{n} y_{ij} = x_i, \text{ for } i = 1, 2, ..., m \quad (17.2)$$

$$\sum_{i=1}^{m} y_{ij} = d_j, \text{ for } j = 1, 2, ..., n \quad (17.3)$$

$$x_i \leq s_i, \text{ for } i = 1, 2, ..., m \quad (17.4)$$

$$x_i, y_{ij} \geq 0, \text{ for } i = 1, 2, ..., m; j = 1, 2, ..., n \quad (17.5)$$

where x_i and y_{ij} are decision variables representing respectively the production amount at plant i and amount shipped from plant i to customer j. The objective function (17.1) representing the total cost including production and transportation is concave. Constraints (17.2), (17.3) and (17.4) are linear supply and demand constraints similar to those in the classical transportation problem. This is a concave minimization problem and hence not a convex program.

Most existing results for this problem class study either exactly the problem [PT] formulated above or its variations which differ mainly in number of plants (two plants, a fixed number of plants, or a general number of plants), and production cost function (general non-separable concave function, or separable concave function). The majority of the results are concerned with problems with a separable concave production cost function. Except one paper, all other papers assume that the demands of customers are deterministic and have to be satisfied.

We first review several results on computational complexity. Hochbaum and Hong (1996) show that the problem [PT] is NP-hard even if (i) the concave production cost function f is separable and symmetric, i.e. $f(x_1, x_2, ..., x_m) = g(x_1)+g(x_2)+ ... + g(x_m)$ for some concave function g; (ii) the transportation cost matrix has all elements equal to 0 or a constant. They also propose a polynomial time algorithm for the case when the number of plants m is fixed, the production cost f is non-separable and concave, and the transportation matrix has the Monge property. Tuy et al. (1993) give a strongly polynomial time algorithm for [PT] with two production plants (i.e. $m = 2$) and a general non-separable concave production cost function f. Tuy et al. (1996) extend this result by giving a strongly polynomial time algorithm for the same problem but with any fixed number of production plants.

Most other existing results are focused on solution algorithms. Due to the fact that the problem [PT] and its variations are non-convex, global optimization methodologies are required to find optimal solutions for this class of problems.

Youssef (1996) considers the problem [PT] with a special separable concave production cost function $f(x_1, x_2, ..., x_m) = \sum_{i=1}^{m} a_i x_i^{b_i}$ where $b_i \in (0, 1)$. The paper provides a numerical example using the tangent line approximation of Khumawala and Kelly (1974) to justify their claim that when there is a production economy of scale, there is a tendency toward centralized decision making.

Kuno and Utsunomiya (2000) consider the problem [PT] with a separable concave production cost function $f(x_1, x_2, ..., x_m) = g_1(x_1) + g_2(x_2) + ... + g_m(x_m)$ for concave functions $g_1, g_2, ..., g_m$. A Lagrangian based branch and bound algorithm is proposed and computational results are reported based on problems with up to 30 plants and 100 customers.

Kuno and Utsunomiya (1996) consider a model with a different structure on production plants than [PT]. There are a head production plant which can ship to all the customers, and a set of branch production plants, each of which can only supply a disjoint subset of customers. Two problems with this model are studied: (i) Production cost of the head production plant is not considered a part of the objective of the problem, but there is a capacity limit on that plant; (ii) Production cost of the head production plant is considered a part of the objective of the problem, but there is no capacity limit on that plant. In both problems, concave production cost function is separable. These problems are solved by dynamic programming which runs in time polynomial in the number of plants and customers and the total demand (i.e., it is in fact pseudo-polynomial).

Kuno and Utsunomiya (1997) study a problem with a more general structure than [PT] in which there are m-2 warehouses in addition to 2 production plants that can provide supply to the customers. There is no production and no cost involved in the warehouses. The production cost is non-separable and concave. The paper gives a pseudo-polynomial time primal-dual algorithm.

All of the papers reviewed above assume that the customer demand is deterministic and must be satisfied. Holmberg and Tuy (1999) consider a problem where the demand of each customer is stochastic and there is a convex penalty for unmet demand. The objective function includes a separable concave production cost, linear transportation cost, and the convex demand shortage penalty. The paper gives a branch and bound algorithm which can solve problems with 100 plants and 500 customers.

4. Joint Lot Sizing and Finished Product Delivery Problems

We first describe a *basic model* which will be used in this section as a reference point when we review results in this problem class. Most problems studied in this area have a similar structure and share many assumptions made in this basic model. A manufacturer produces one product on a single production line at a constant rate for a customer with a constant demand rate for the product which must be satisfied without backlog. At the manufacturer's end, there is a fixed setup cost for each production run, and a linear inventory holding cost. At the customer's end, there is a linear inventory holding cost. Between the manufacturer and the customer, there is a fixed delivery cost per order delivered from the manufacturer to the customer, regardless of the order size. The problem is to find a joint cyclic production and delivery schedule such that the total cost per unit time, including production setup costs, inventory costs at both the manufacturer and the customer, and transportation costs, is minimized over an infinite planning horizon.

As in this basic model, all the papers in this area assume that both production and demand rates are constant and the planning horizon is infinite, and consider all the costs in the system including production and inventory at the manufacturer, transportation from the manufacturer to the customer, and inventory at the customer. Most papers consider only one product, one production line, and one customer, and are concerned with finding an optimal solution from a given class of policies. There are three major classes of policies considered in the literature:

1 Production cycle length and delivery cycle length are identical, i.e. one lot of production is for one lot of delivery (this policy is called *lot-for-lot* in some papers).

2 There are an integer number of delivery cycles within each production cycle.

3 There are an integer number of production cycles within each delivery cycle.

For policy classes (a) and (c), it is commonly assumed that production quantity in each production cycle is identical and delivery quantity in each delivery cycle is also identical. As a result, production and delivery cycles are evenly spaced in time respectively. For policy (b), it is commonly assumed that production quantity is identical in each cycle; however there are two cases of delivery quantities considered in the literature: identical quantity or different quantities in different delivery cycles.

We review existing results separately mainly based on (i) closeness to the basic model described here and (ii) the policy used. The basic model with varying policies is studied respectively in the following sets of publications: (1) Banerjee (1986) under policy (a); (2) Goyal (1988) and Aderohunmu et al. (1995) under policy (b) with identical delivery quantities; and (3) Goyal (1995, 2000), Hill (1997, 1999), Goyal and Nebebe (2000) under policy (b) with non-identical delivery quantities. Except Goyal (1988) that requires that delivery can start only after the production batch is completed, all other results allow delivery to start before a production batch is completed. Among all the papers that use policy (b) with non-identical delivery quantities, Hill (1999) is the only paper that does not require delivery quantities to follow a pre-specified formula. Most of these publications use the terms "vendor" and "buyer" (instead of "manufacturer" and "customer" we use in the basic model) and address vendor-buyer coordination in their problem context. A review in the area of vendor-buyer coordination prior to 1989 can be found in Goyal and Gupta (1989).

Some variations of the basic model are also studied in the literature. Unlike the basic model and most models considered in this area where the objective is to minimize a total system-wide cost, Lu (1995) considers a model with the objective of minimizing the vendor's total cost subject to the maximum cost the buyer may be prepared to incur. Both the case with a single vendor and a single buyer and the case with a single vendor but multiple buyers under both policies (b) and (c) with identical delivery quantities are considered. The single-vendor-single-buyer problem is solved optimally by closed-form formulas, while a heuristic is proposed for the single-vendor-multiple-buyer problem.

Hahm and Yano (1992) consider an extension of the basic model in which there is a setup time in addition to a setup cost in each production run. They show that in an optimal solution, the production cycle length is an integer multiple of the delivery length. They formulate the problem as a nonlinear mixed integer program which is solved by a heuristic approach. They also consider the problem where the fixed delivery cost is associated with each shipment which has a limit vehicle capacity and hence the delivery cost per order is dependent on the order size. Benjamin (1989) studies the same problem except that the inventory cost at the manufacturer is calculated differently than Hahm and Yano (1992).

The model studied by Hahm and Yano (1992) is extended by Hahm and Yano (1995a, 1995b, 1995c) to include multiple products, each with a constant production rate and a constant demand rate. With multiple products to be produced on a single production line, the processing sequence of products in each production cycle becomes important. The first paper considers policy (a), and the second and third ones consider policy (b). The first two papers assume that within each production cycle, each product is produced exactly once. In the third paper, however, this assumption is relaxed, and it is assumed that within each production cycle, each product can be produced 2^u times for some integer u which is in general product dependent. The corresponding problems are formulated as nonlinear integer programs and solved by heuristics.

Models with multiple manufacturers or/and multiple customers are studied by Benjamin (1989), Blumenfeld et al (1985, 1991), and Hall (1996). Benjamin (1989) considers a model with multiple manufacturers and multiple customers where only a single product is involved. The problem is to determine production cycle at each manufacturer, and a delivery cycle for each transportation link between each manufacturer and each customer. It is formulated as a nonlinear program for which a heuristic solution procedure is designed. Blumenfeld et al (1985) consider various structures of delivery networks from manufacturers to customers including direct shipping, shipping via a consolidation terminal, and a combination of terminal and direct shipping. Problems with one or multiple manufacturers and one or multiple customers are considered.

Blumenfeld et al (1991) study a model with one manufacturer and multiple customers where the manufacturer produces multiple products, one for each customer. Each product is allowed to be produced multiple times within a production cycle. In the case when all the products are homogeneous (i.e. have identical parameters), the product cycle is identical for all the products, and identical number of production runs for each product within a production cycle, the authors derive the optimal production cycle length and optimal delivery cycle length under policy (b). They compare the case with the production-distribution coordination and the case without the coordination and conclude that the cost savings from coordination are between 15 to 40% based on a range of problem parameters they tested.

Hall (1996) considers various scenarios: one or multiple manufacturers, one or more customers, one or more machines at each manufacturer, and one or more products that can be processed by each machine. He derives cost formulas for many scenarios under policy (a).

All the problems reviewed above assume that all products produced are of acceptable quality and the production is at a constant rate. Khouja (2000) considers the same problem studied by Hahm and Yano (1995a) with the added model assumption that product quality deteriorates with lot sizes and/or decreased unit production times. The algorithm proposed by Hahm and Yano is modified to solve this extended problem.

5. Joint Raw Material Delivery and Lot Sizing Problems

The structure of this class of problems is symmetric to that of the problems reviewed in Section 4 in the following sense. They share most problem characteristics (e.g. infinite horizon, constant production and demand rates, batch production). The key difference is in their structure: the problems of Section 4 integrate production with outbound transportation (delivery of finished products to customers), while the problems in this section integrate inbound transportation (delivery of raw material from suppliers) and production.

We first describe a basic model which will be used as a reference point when we review existing results in this area. A manufacturer purchases raw material from a supplier, and uses the raw material to produce a final product, and ships a fixed quantity of the product to a customer periodically to satisfy the customer's constant demand. The size and timing of the shipments are specified by the customer. The manufacturer has a single production line with a constant production rate. There are four cost components in the system: (1) distribution cost for delivering raw material from the supplier to the manufacturer, (2) inventory cost for holding raw material at the manufacturer; (3) production cost at the manufacturer (there is a fixed setup cost for each production run), and (4) inventory cost for holding finished product at the manufacturer. The problem is to find a joint cyclic raw material delivery and production schedule such that the total cost per unit time is minimum over an infinite planning horizon.

As in this basic model, most papers consider only one type of raw material needed for production, one product, one production line, and one customer, and assume that the manufacturer delivers a fixed amount of the product to the customer periodically and the quantity and timing of the deliveries are prespecified. All the papers are only concerned with finding an optimal solution from a given class of policies. Three major policies considered in the literature are as follows (which are similar to the policies described in Section 4).

1. Raw material delivery cycle time and production cycle time are equal, i.e. lot-for-lot.

2. There are an integer number of equal-sized raw material delivery cycles within each production cycle.

3. There are an integer number of equal-sized production cycles within each raw material delivery cycle.

In addition, most papers assume that the production cycle time is an integer multiple of the pre-specified cycle time of finished product delivery. In this case, the production cycle decision reduces to finding such an integer multiplier.

Golhar and Sarker (1992) and Jamal and Sarker (1993) consider the basic model under policy (a) with the assumption that the conversion rate from raw material to final product is one to one. Two cases are considered: (i) *Imperfect matching* – production uptime and cycle time are not exact integer multiples of finished product delivery cycle; (ii) *Perfect matching*: the above numbers are integers. An iterative heuristic is used to solve the problem. Sarker and Parija (1994) consider exactly the same problem except that the conversion rate from raw material to final product is not assumed to be one to one. An exact algorithm is proposed.

Hill (1996) considers the basic model with a general raw material conversion rate under policies (b) and (c). It is assumed that the production cycle time is an integer multiple of the pre-specified cycle time of finished product delivery.

Extensions of the basic model with multiple types of raw materials needed for producing a single product are studied by Sarker and Parija (1996) and Sarker and Khan (1999, 2001). Sarker and Parija give a closed-form solution for the problem under policy (b) and with the assumption that the production cycle time is an integer multiple of the pre-specified finished product delivery cycle. Sarker and Khan propose a heuristic for the problem under policies (a) and (c) and the assumption that delivery of final product to the customer is carried out in the end of the whole production lot.

Parija and Sarker (1999) study another extension of the basic model where there are multiple customers and the delivery of the final product to each customer follows a pre-specified periodic schedule (which is customer dependent) like in all other problems reviewed earlier. A closed-form solution is obtained under policy (b).

6. General Tactical
PRODUCTION-DISTRIBUTION PROBLEMS

This class of problems are more general in both structure and parameters than the ones reviewed in the previous sections. Compared to the problems reviewed in Section 3 which involve only a single product and a single time period, the problems to be reviewed in this section involve multiple products or/and multiple time periods. Compared to the problems reviewed in Sections 4 and 5 which assume infinite planning horizon with a constant demand and seek optimal solutions among a given set of policies, the problems to be reviewed in this section have a finite horizon (one or multiple time periods) and dynamic demand over time (if there are multiple time periods), and seek optimal solutions among all feasible solutions. Furthermore, there are only two stages (manufacturer and customers, or suppliers and manufacturer) in all the problems reviewed earlier, whereas many problems to be reviewed here involve three or more stages (e.g. suppliers, manufacturers, warehouses, and customers).

Before reviewing existing results, we first give a typical model of this problem class that involves two stages (one manufacturer and some customers), multiple products, and multiple time periods. Many problems to be reviewed are either special cases or extensions of this model. A manufacturer produces m products to satisfy n customers' demand over T time periods. Customer k's demand ($k = 1, ..., n$) for product j ($j = 1, ..., m$) in period t ($t = 1, ..., T$) is d_{jkt} which is known in advance and must be satisfied without backlog. The parameters associated with the production at the manufacturer, inventory at the manufacturer and the customers, and the transportation between the manufacturer and the customers are as follows:

1. Production: The processing time required for producing one unit of product j, for $j = 1, ..., m$, is p_j, and the manufacturer has a total of B units of production time available in each time period. There is a setup cost s_j if product j is produced in a period.

2. Inventory: Both the manufacturer and the customers can hold inventory. Unit inventory holding cost and the initial inventory of product j at location k are h_{jk} and I_{jk0} respectively, where location $k = 0$ represents the manufacturer, and location $k = 1, ..., n$ represents the n customers respectively.

3. Transportation: Cost of each shipment from the manufacturer to customer k is c_k, and the capacity of each shipment is C.

The problem is to determine in each time period how much to produce at the manufacturer, how much to keep in inventory at the manufacturer and at each customer, and how much to ship from the manufacturer to each customer, so that the total cost including production setup, inventory, and transportation, is minimized.

Define the following decision variables

x_{jt} = amount of product j produced in period t

y_{jt} = 1 if the production facility is set up for product j in period t (i.e. $x_{jt} \xi$ 0), and 0 otherwise
r_{kt} = number of shipments to customer k in period t
q_{jkt} = amount of product j delivered to customer k in period t
I_{jkt} = inventory of product j at location k in period t
The above-described model can be formulated as the following mixed integer program.

$$[\text{GTPD}] \min \sum_{t=1}^{T} \sum_{j=1}^{m} s_j y_{jt} + \sum_{t=1}^{T} \sum_{j=1}^{m} \sum_{k=0}^{n} h_{jk} I_{jkt} + \sum_{t=1}^{T} \sum_{k=1}^{n} c_k r_{kt} \quad (17.6)$$

s.t. $\sum_{j=1}^{m} p_j x_{jt} \leq B$, for $t = 1, ..., T$ \hfill (17.7)

$x_{jt} \leq M y_{jt}$, for $j = 1, ..., m; t = 1, ..., T$ \hfill (17.8)

$I_{j0t} = I_{j0t-1} + x_{jt} - \sum_{k=1}^{n} q_{jkt}$, for $j = 1, ..., m; t = 1, ..., T$ \hfill (17.9)

$I_{jkt} = I_{jkt-1} + q_{jkt} - d_{jkt}$, for $j = 1, ..., m; k = 1, ..., n; t = 1, ..., T$ \hfill (17.10)

$C r_{kt} \geq \sum_{j=1}^{m} q_{jkt}$, for $k = 1, ..., n; t = 1, ..., T$ \hfill (17.11)

$y_{jt} \in \{0, 1\}, x_{jt} \geq 0$, for $j = 1, ..., m; t = 1, ..., T$ \hfill (17.12)

$q_{jkt} \geq 0, I_{jkt} \geq 0$, for $j = 1, ..., m; k = 1, ..., n; t = 1, ..., T$ \hfill (17.13)

$r_{kt} \geq 0$ and integer, for $k = 1, ..., n; t = 1, ..., T$ \hfill (17.14)

In this MIP formulation, constraints (17.7)-(17.8) represent production (capacity and setup), (17.9) represents the relation of production, inventory, and shipment at the manufacturer, (17.10) represents the relation of shipment, inventory, and demand at each customer, and (17.11) represents the relation between the number of shipments and the amount shipped.

All but one paper reviewed here address problems with multiple products, and many study problems with three or four stages (three stages, e.g. manufacturers – warehouses – customers; four stages, e.g. suppliers - manufacturers – warehouses – customers). Note that the problems reviewed in Section 3 all involve one product, one time period, and two stages, and hence are special cases of the problems reviewed here in terms of their structure. However, in the problems reviewed in Section 3, the production cost is a concave function of the production amount, whereas in the problems reviewed here the production cost is in most cases a linear function of the production amount or number of setups.

We review results on problems with one time period separately from results on problems with multiple time periods. Cohen and Lee (1988, 1989), Zuo et al. (1991), Chen and Wang (1997), and Sabri and Beamon (2000) consider problems with a single time period but multiple products. Each of these papers is reviewed in the following paragraphs.

Cohen and Lee (1988) consider a four-stage model in which demands for products are stochastic and the four stages involved are: multiple vendors – supplying raw materials to plants; multiple production plants; multiple DCs; and multiple customer zones. The problem is to determine ordering policies (lot sizes, reorder points, etc.) such that total system-wide cost is minimum subject to a certain level of customer service level. The problem is formulated as a complex mathematical program which is decomposed into four sub-models: material, production, inventory, and distribution. Four sub-models are solved sequentially in which the output of one sub-model is given as the input of the succeeding sub-model. The following questions are addressed: (a) How can production and distribution control policies be coordinated to achieve synergies in performance? (b) How do service level requirements for material input, work-in-process and finished goods availability affect costs, lead-times and flexibility?

Cohen and Lee (1989) study a four-stage model where demands for products are deterministic and do not have to be satisfied (but with a penalty if not satisfied). Both strategic decisions (location and capacity decisions associated with plants, DCs) and tactical decisions (how much to send from a vendor to a plant, from a plant to a DC, from a DC to a market region) are considered. International issues such as taxes in different countries and offset requirements (specifying that in order to gain entry into the government market segment of that country, a minimum level of manufacturing originating from that particular country must be carried out) are also considered in the model. The model is formulated as a mixed integer nonlinear program, and applied to a real-life case, a PC manufacturer. A hierarchical heuristic approach for solving the formulation is proposed.

Zuo et al. (1991) consider a real-life problem of seed corn production and distribution involving two stages with multiple production facilities and multiple sales regions. The production part has a special constraint that requires that each facility either produces 0 or produces more than a given minimum production amount. Demands for products are deterministic and have to be satisfied. Transportation cost is assumed to be a linear function of the amount shipped. The problem is formulated as an LP with some either-or constraints (production amount is either 0 or at least some minimum amount) which can be converted into a MIP. A heuristic approach is developed which first ignores the either-or constraints – resulting in an LP – and then adds integer constraints to the problem to enforce violated either-or constraints. It is reported that the annual cost savings provided by this study are about $10 million. Before this study, the company's production and distribution departments operated independently.

Chen and Wang (1997) study a three-stage (multiple suppliers, multiple factories, and multiple customers) problem motivated by a real-life steel production and distribution problem. Demands for products are deterministic. Each unit of a product sold earns a given revenue. Both purchasing cost of raw materials and transportation cost are linear. However, production cost consists of two components: a fixed cost and a variable cost depending on the production amount. The problem is formulated as an LP and solved directly using an LP software package.

Sabri and Beamon (2000) consider a four-stage (suppliers, plants, DCs, and customer zones) problem with both strategic (plant and DC locations) and tactical decisions. Demands for products are deterministic and have to be satisfied. There are fixed costs associated with DCs and transportation links between DCs and customer zones. Production cost is assumed to be linear. Two objectives are considered: (a) Total cost, (b) Volume flexibility (difference between plant capacity and its utilization, and difference between DC capacity and its utilization). The strategic sub-model of the problem is formulated as a multi-objective MIP. Two operational sub-models (suppliers, production) are formulated and solved as a non-linear programming problem. An overall iterative procedure is proposed which combines the strategic sub-model with the operational sub-models.

Now we review results on problems with multiple time periods. Such problems are studied by Haq (1991), Chandra and Fisher (1994), Arntzen et al. (1995), Barbarosoglu and Ozgur (1999), Dogan and Goetschalckx (1999), Fumero and Vercellis (1999), Mohamed (1999), Ozdamar and Yazgac (1999), and Dhaenens-Flipo and Finke (2001). Except for the problem considered by Haq (1991) which involves a single product, all the problems studied in the other papers involve multiple products. All these papers formulate and solve problems as mixed integer programs. The problems studied in these papers differ mainly with respect to the following structural parameters: (1) number of stages (two, three, or four); (2) whether demand has to be satisfied or backlog is allowed; and (3) whether each stage is allowed to hold inventory. Solution methodologies used in these papers may also be different. Each of these papers is reviewed in the following paragraphs.

Haq (1991) considers a 3-stage (production, warehouses, and retailers), 1-product problem where the production stage involves multiple serial sub-stages, however no transportation is involved between neighboring stages of production. Demand for the product is deterministic in each period and backlog is allowed but with a linear penalty cost. All the stages can hold inventory. There is a fixed and variable cost involved in each production stage. The transportation cost is a linear function of the amount shipped along each link. The objective is to minimize total production, inventory, and transportation cost plus backlog penalty. The problem is formulated as a MIP and applied to a real-life problem (manufacturing of urea fertilizer in India – involving 6 production stages, 12 warehouses, 6 periods). However, the paper doesn't mention how this MIP formulation is solved.

Chandra and Fisher (1994) consider a 2-stage, multi-product problem with a single production facility and multiple customers. Demand for each product in each period is deterministic and has to be satisfied without backlog. There is a setup cost for producing a product in each period. Inventory is allowed at both the plant and the customers. Transportation cost consists of a fixed part and a variable part which is determined by the routes of vehicles, and hence vehicle routing is one of the decisions of the problem. The problem is formulated as a MIP which is similar to the formulation [GTPD] above with some additional constraints for the vehicle routing part. The authors compare sequential (first production, then transportation) and integrated approaches. They make the following observations based on computational tests on randomly generated data sets on various parameters: (i) Value of production-distribution coordination (measured as average cost reduction achieved by the coordination) increases with production capacity, vehicle capacity, number of customers, number of products, and number of time periods; (ii) Value of coordination increases with relatively high distribution costs (fixed and variable) compared to production cost; and (iii) Cost reduction varies from 3% to 20%. In some cases (e.g. when vehicle capacity is small), there is no value of coordination because in this case all deliveries will be made full truckloads and hence no consolidation is necessary.

A similar problem to the one considered by Chandra and Fisher (1994) is studied by Fumero and Vercellis (1999). They assume that there is a limited number of vehicles available for product delivery in each time period, whereas no such assumption is made by Chandra and Fisher. They give a different MIP formulation than the one by Chandra and Fisher, and solve it by Lagrangean relaxation. As in Chandra and Fisher (1994), the integrated approach is compared to a decoupled approach in which the production part of the problem is solved first and then the distribution part is solved based on the given solution of production decisions. Similar observations to those obtained by Chandra and Fisher are reported about the value of coordination.

Arntzen et al. (1995) study a real-life problem encountered at Digital Equipment Corporation. It is a 2-stage, multi-product problem involving multiple production facilities and multiple customers. The demand for each product in each period is deterministic and has to be satisfied without backlog. The production part involves transformation of materials through multiple stages, and the production cost has a fixed part and a variable part. Inventory is only allowed at production facilities, and no inventory can be kept at the customers (i.e. customer demands are satisfied exactly at each time period). There are multiple transportation modes available and the transportation cost consists of a fixed part and a variable part. However, no routing decisions are involved.

Issues related to international operations such as duty drawback and duty relief are also considered. The objective of the problem is to minimize weighted total costs and cumulative production and distribution times. The problem is formulated as a MIP and solved using techniques of elastic constraints (which may be violated at a given linear penalty), and row factorization. It is reported that the results of this study saved DEC over $100 million.

Barbarosoglu and Ozgur (1999) consider a 3-stage, multi-product problem involving one plant, multiple DCs, and multiple customers. The demand for each product in each period is deterministic and has to be satisfied in JIT fashion (i.e. no inventory is allowed at the customers). There is a fixed setup cost if a product is produced in a time period. The transportation cost has a fixed part (which occurs if a nonzero amount of product is shipped along a given transportation link at a time period – regardless of how much shipped) and a variable part. However, no routing decisions are involved. The problem is formulated as a MIP and solved by Lagrangean relaxation which dualizes some constraints and decomposes the problem into two subproblems: production and distribution.

Dogan and Goetschalckx (1999) consider a 4-stage, multi-product problem which integrates strategic decisions such as facility and machines with tactical decisions. There are fixed costs associated with each facility and each machine at each production facility. The demand for each product in each period is deterministic and has to be satisfied without backlog. There are no production setup costs in the product level (instead there are fixed costs in the facility and machine level). There are no routing decisions and no fixed transportation cost involved. Replenishment frequency is given for each transportation link. The problem is formulated as a MIP and solved using Benders decomposition. It was applied to a real-life problem in the packaging industry and achieved 2% of cost savings ($8.3 million).

Mohamed (1999) considers a 2-stage, multi-product problem involving multiple production facilities and multiple markets possibly located in different nations. The demand for each product in each time period is deterministic and must be satisfied exactly in a JIT fashion (i.e. the markets do not hold inventory). Production capacity is a part of the decision. Capacity can change from one time period to another, but there is a capacity retaining cost and a capacity changing cost in each period for each facility. Linear production and transportation costs are considered. No routing decisions are involved. Exchange rates of the host countries of the facilities in each time period are also considered. The problem is formulated as a MIP. An example with changing exchange rates is used to illustrate the usefulness of the model.

Ozdamar and Yazgac (1999) investigate a 2-stage (one production facility and multiple warehouses), multi-product problem in which the demand for each product in each time period is deterministic and backlog is allowed. There are production setup times. However, the production cost is linear. Inventory is allowed only at the warehouses, and no inventory is held at the plant, i.e. whatever is produced in a time period is shipped to the warehouses in this period. Transportation cost is linear, and no routing decisions are involved. A hierarchical planning approach is proposed for the problem. In the aggregated model, production setup times are ignored to avoid binary variables, and time horizon is aggregated into a smaller number of periods. This results in an MIP model. In the disaggregated model, production setup times and all time periods are considered, which also results in an MIP formulation. An iterative constraint relaxation scheme is used to solve the problem. The results are applied to a real-life problem (production and distribution of liquid and powder detergents in Turkey).

Dhaenens-Flipo and Finke (2001) study a 3-stage, multi-product problem involving multiple plants (each with multiple parallel production lines), multiple warehouses, and multiple customers. The demand for each product in each time period is deterministic and must be satisfied in a JIT fashion. There are a fixed setup cost, variable cost, and setup time for producing a product on each production line. Due to long changeover times, it is restricted that only a small subset of products are allowed to be produced on each production line. Inventory is only allowed at the warehouses. Transportation cost is linear. The model is formulated as a MIP, applied to a real-life problem (the manufacturing of metal items), and solved directly by CPLEX.

In all the above-reviewed papers in this section, underlying problems are explicitly formulated mathematically (LP, MIP, etc.). There are also case studies reported in the literature that address real-life integrated production and distribution but give no mathematical formulations. Glover et al. (1979) consider an integration problem of production, distribution and inventory of Agrico Chemical Company. The problem involves both long-term (location of DCs, transportation equipment, inventory investment, etc.) and short-term decisions (allocation decisions, where/how much/when products should be produced). The project saved $18 million during the first 3 years of implementation. King and Love (1980) study coordination of sales forecasting, inventory control, production planning, and distribution decisions for Kelley-Springfield Tire Company, and their results brought an annual savings of $500,000. Martin et al. (1993) consider integration of production and distribution (inter-plant, from plants to customers) for Libbey-Owens-Ford Glass Company. They mentioned that they developed a LP model, but gave no details of their formulation. Their results achieved an annual cost savings of $2 million.

7. Joint Job Processing and Finished Job Delivery Problems

This class of problems involves detailed scheduling of production and distribution at the individual job level. We first describe a general model of this type. All the problems to be reviewed here can be viewed as special cases of this model. A set of n jobs $N = \{1, 2, ..., n\}$ is to be processed at a manufacturing facility which may consist of a single machine, a set of parallel machines, or a series of flow shop machines. After processing, jobs must be delivered to the corresponding customers which may be located at different locations. There are non-negligible transportation time and cost needed to go from one location to another. A number of delivery vehicles are available to deliver completed jobs from the manufacturing facility to customers. Each vehicle has a capacity limit and is initially located at the manufacturing facility. A job cannot be picked up for delivery until it is processed in the manufacturing facility. The delivery time of a job is defined as the time when it is delivered to the customer. The objective of the problem is to minimize a weighted sum of a job performance measure represented by a non-decreasing function of job delivery times, and the total transportation cost.

Most existing results consider one of the following two special cases of this general model: (i) transportation costs are assumed to be 0, and hence the objective is optimize a job performance only; (ii) transportation times are assumed to be 0, i.e. job delivery can be done instantaneously. Most papers do not consider routing decisions for delivery vehicles.

We first review papers that do not consider transportation costs. Potts (1980) and Hall and Shmoys (1989) consider a problem where there is a single machine in the manufacturing facility. It is assumed that a sufficient number of deliver vehicles are available such that each job once completed will be picked up for delivery immediately and separately without consolidation with other jobs. As a result, each delivery trip carries one job only and is a direct shipment from the manufacturing facility to the job's destination. The objective is to minimize the time when the last job is delivered. Potts proposes a heuristic and analyzes its worst-case error bound. Hall and Shmoys give an approximation algorithm with an arbitrary good performance. Woeginger (1994) studies a more general version of the problem where there are a set of parallel machines in the manufacturing facility. He gives some heuristics with constant worst-case error bound guarantee.

Zdrzalka (1991, 1995) and Woeginger (1998) investigate a similar problem except that jobs belong to different families and a setup time is required if a job from one family is processed immediately after a job from another family. Zdrzalka (1991) proposes heuristics and analyzes their worst-case error bounds for the case with a single machine in the manufacturing facility and unit setup times. Zdrzalka (1995) considers the same problem as Zdrzalka (1991) except that the setup times are not assumed to be unit. Heuristics are proposed and their worst-case error bounds are analyzed. Computational results are also reported to show the average performance of these algorithms based on a set of randomly generated instances. Woeginger (1998) studies the same problem as Zdrzalka (1995) and gives a polynomial-time approximation scheme for the problem.

Lee and Chen (2001) investigate various problems with combinations of the following parameters: (i) the manufacturing facility has a single machine, a set of parallel machines, or a set of flow-shop machines; (ii) there is either one delivery vehicle, or multiple but a fixed number of delivery vehicles; (iii) each delivery vehicle can deliver one or multiple jobs; (iv) the objective is to minimize the time when all the jobs are delivered, or the total delivery time of jobs. It is assumed that a vehicle can be used repeatedly, and that all the jobs belong to a single customer and hence like in the papers reviewed in the previous paragraphs no vehicle routing decisions are involved. Many problems are shown to be NP-hard, while some others are shown to be solvable by polynomial or pseudo-polynomial time dynamic programming algorithms.

Next we review papers that assume zero delivery time but include transportation costs as part of the objective. All these papers assume that the transportation cost is a function of the number of deliveries, independent of how many and what jobs form each delivery batch. Since zero delivery time is assumed, there are no vehicle routing decisions involved.

Cheng and Kahlbacher (1993), Cheng and Gordon (1994), and Cheng et al. (1996) consider different cases of the following problem. There is a single machine in the manufacturing facility and the objective is to minimize the sum of total weighted earliness and total delivery cost. The earliness of a job is defined as the difference between the delivery time of this job and the completion time of this job on the machine. The delivery cost is a general function of the number of delivery batches used. Various cases with respect to job processing times are considered in these papers. They give an NP-hardness proof for one case of the problem, and dynamic programming algorithms for some other cases. Cheng et al. (1997) consider the same problem except that it is assumed that there is a constant setup time before each batch on the machine. NP-hardness proofs, dynamic programming algorithms and heuristics are given for different cases of the problem.

Herrmann and Lee (1993), Chen (1996), and Yuan (1996) study different cases of a problem with a single machine, a common due date shared by all the jobs, and the objective of minimization of the sum of total weighted earliness and tardiness of jobs and total delivery cost. Herrmann and Lee consider the case with the common due date given in advance and jobs have an identical earliness weight and an identical tardiness weight. They give a pseudo-polynomial dynamic programming algorithm and discuss some special cases. Yuan (1996) studies the complexity of a slightly more general case where the earliness and tardiness weights are job dependent. He shows that the problem is strongly NP-hard. Chen considers the same case of the problem as Herrmann and Lee except that the common due date is not given but to be determined. He gives a polynomial time dynamic programming algorithm.

Among all the papers that assume zero delivery time, Wang and Cheng (2000) is the only one that considers a parallel-machine configuration in the manufacturing facility. The objective is to minimize the sum of total delivery times of jobs and total delivery cost. The problem is shown to be NP-hard, and dynamic programming algorithms are given for various cases of the problem.

Hall and Potts (2000) study scheduling problems with batch delivery in an arborescent supply chain where a supplier makes deliveries to several manufacturers, who also make deliveries to customers. Various problems from a supplier point of view, from a manufacturer point of view, or from an entire system point of view are considered. Each problem is either proved to be NP-hard or given a polynomial dynamic programming algorithm. They also discuss the possible benefits from coordinated decision making, as well as some mechanisms for achieving it.

None of the papers we have reviewed so far in this section consider vehicle routing decisions. To the best of our knowledge, there are only two papers (Hurter and Van Buer 1996, Van Buer et al. 1999) that do address routing decisions. These papers consider both transportation times and transportation costs. They consider the following joint production and distribution scheduling problem that arises in the newspaper industry. Every night, a number of different newspapers are first printed in a printing facility, and then immediately delivered to drop off points where they will be picked up by home delivery carriers in early morning. It requires a setup time to print a different type of newspaper on the same machine after one type of newspaper is printed. The demand for different types of newspapers at each drop off point is known and has to be delivered before a given deadline (e.g. 6:00am the next morning). The problem is to find a production schedule in the printing facility, a delivery route for each delivery van involved, and the amount of each type of newspaper to be carried by each delivery van so that the number of vans used is minimized.

If the demand from each drop off point for each type of newspaper is required to be processed as a single job without interruption, the above-described problem is similar to the general mode described in the beginning of this section. Hurter and Van Buer (1996) propose a heuristic with the assumption that each delivery van can only be used once. The heuristic is an iterative approach. In each iteration, it first determines how many vans and what routes to use, and then based on the given routing decisions finds a production schedule. Van

Buer et al. (1999) allow repeated use of the delivery vans and give some local search based heuristics for the problem.

8. Directions for Future Research

We have reviewed existing results on five different classes of EPD problems. These classes of problems address various practical situations in the EPD area and the existing results provide useful problem-solving tools in this area. Although a large amount of research has been done, this is still a relatively new area, as it can be seen that most of the results reviewed here were published in the last decade. The following five topics deserve more research:

1. EPD models with stochastic demand. Most existing results consider deterministic models where demand for a product is known in advance and has to be satisfied without backlog. However, many products in practice have a large variety and a short life cycle, and the market demand for such products is difficult to predict and has a high degree of uncertainty.

2. EPD models at the detailed scheduling level. As we mentioned in Section 1, in a direct-sell e-business model, production and distribution operations are linked directly without any intermediate step of inventory typically seen in a traditional business model. Consequently, joint consideration of production and distribution decisions at the detailed scheduling level becomes critical in order to achieve short delivery lead time at minimum cost. Unfortunately, few models we have reviewed study joint production-distribution decisions at the job scheduling level. All the scheduling models reviewed in Section 7 make some special assumptions such as no transportation time, no transportation cost, an infinite number of delivery vehicles or one delivery vehicle, etc. Much research is needed in this area to investigate problems with these assumptions relaxed so that the models are more applicable to real-world situations.

3. Value of coordination. A handful of papers (e.g. Chandra and Fisher 1994, Fumero and Vercellis 1999) have studied the value of coordination between production and distribution based on computational experiments on random test problems. It would be very interesting if one could quantify the value of coordination *theoretically* (e.g. worst-case or average-case analysis) for certain models. This could then be used to decide whether it is worth the effort to consider production and distribution jointly.

4. Mechanisms for coordination. None of the results reviewed here has addressed the fundamental question whether it is possible at all to inte-

grate production and distribution operations. All the results have implicitly assumed that the decision making is centralized, i.e. there is only one decision maker involved in the system who seeks to optimize the system-wide performance. However, as a recent industrial survey (Langley et al. 2001) revealed, a majority of companies are using or planning to use third party logistics services. This means that often times production and distribution operations are executed by different companies which seek to optimize their own objectives. In this case, it is critical to design coordination mechanisms and incentives so that the companies involved are willing to cooperate and production and distribution operations can be executed in an integrated manner.

5 Fast and robust solution algorithms. Most problems reviewed here are NP-hard and difficult to solve. Most existing algorithms for the first four classes of problems are optimization based (e.g. global optimization, MIP). For large-scale problems, such approaches typically require a large amount of computational time. For decisions that have to be made in a short time frame, fast heuristics are perhaps more useful. To guarantee certain level of solution quality, one can combine heuristics and optimization based approaches. Most existing results for the last class of problems are dedicated to complexity analysis. There is clearly a need to develop solution algorithms (both heuristics and optimization based approaches such as branch and bound) for these problems.

There are also topics that require further research within each individual problem class. For production-transportation problems, all the results reviewed consider one product and linear transportation costs. A more general model should address multiple products (which compete for limited production capacity) and a nonlinear (e.g. fixed charge plus variable) transportation cost structure. Note that in practice transportation cost is normally nonlinear.

For joint lot sizing and finished product delivery problems, all the results assume a constant demand rate and a constant production rate, and almost all the results are concerned with finding an optimal solution from a given class of policies. Research is needed for studying more general problems with a dynamic demand rate and finding optimal solutions. Also, few existing models consider vehicle routing decisions. Many companies use JIT manufacturing practice and require frequent deliveries of small orders. If there are multiple customers with different locations, it may be necessary to consolidate orders of different customers in order to save transportation cost. Thus it will be worthwhile to consider vehicle routing decisions as well for problems that involve multiple customers.

Extending from joint raw material delivery and lot sizing problems, one can consider a more general model involving three stages: raw material delivery, lot sizing, and finished product delivery. The existing results in this problem class consider mainly the first two stages. Such a three-stage model will unify the models discussed in Sections 4 and 5.

For general tactical production-distribution problems, few papers consider vehicle routing decisions. However, as mentioned earlier, due to the JIT practice, it may be necessary to consolidate orders and thus routing decisions become a part of the problem. Certainly with routing decisions, problems become much more difficult since a typical vehicle routing problem alone (without the production part) is already known to be very challenging. To solve such problems, it is important to exploit problem structure. It may be necessary to use heuristics (which can be coupled with an optimization approach) that can generate good solutions quickly.

Acknowledgments

This research was supported in part by the National Science Foundation under grant DMI-0196536. We thank Prof. David Wu and an anonymous referee for their constructive comments on an earlier version of this chapter.

References

Aderohunmu, R., A. Mobolurin, and R. Bryson, "Joint Vendor-Buyer Policy in JIT Manufacturing", *Journal of the Operational Research Society*, 46 (1995), 375-385.

Arntzen, B.C., G.G. Brown, T.P. Harrison, and L.L. Trafton, "Global Supply Chain Management at Digital Equipment Corporation", *Interfaces*, 25 (1995), 69 – 93.

Banerjee, A., "A Joint Economic-Lot-Size Model for Purchase and Vendor", *Decision Sciences*, 17 (1986), 292-311.

Barbarosoglu, G. and D. Ozgur, "Hierarchical Design of an Integrated Production and 2-Echelon Distribution System", *European Journal of Operational Research*, 118 (1999), 464 – 484.

Benjamin, J., "An Analysis of Inventory and Transportation Costs in a Constrained Network", *Transportation Science*, 23 (1989), 177-183.

Bertazzi, L. and M.G. Speranza, "Minimizing Logistics Costs in Multistage Supply Chains", *Naval Research Logistics*, 46 (1999), 399-417.

Bhatnagar, R. and P. Chandra, "Models for Multi-Plant Coordination", *European Journal of Operational Research*, 67 (1993), 141-160.

Blumenfeld, D.E., L.D. Burns and C.F. Daganzo, "Synchronizing Production and Transportation Schedules", *Transportation Research*, 25B (1991) 23-37.

Blumenfeld, D.E., L.D. Burns, J.D. Diltz, and C.F. Daganzo, "Analyzing Tradeoffs between Transportation, Inventory and Production Costs on Freight Networks", *Transportation Research*, 19B (1985), 361-380.

Boyaci, T. and G. Gallego, "Serial Production/Distribution Systems under Service Constraints", *Manufacturing & Service Operations Management*, 3 (2001), 43-50.

Bramel, J., S. Goyal, and P. Zipkin, "Coordination of Production/Distribution Networks with Unbalanced Leadtimes", *Operations Research*, 48 (2000), 570-577.

Burns, L.D., R.W. Hall, D.E. Blumenfeld, and C.F. Daganzo, "Distribution Strategies that Minimize Transportation and Inventory Costs", *Operations Research*, 33 (1985), 469-490.

Chan, L.M.A., A. Muriel, and D. Simchi-Levi, "Supply-chain management: Integrating inventory and transportation", Working Paper, (1997), Department of Industrial Engineering and Management Sciences, Northwestern University.

Chandra, P. and M.L. Fisher, "Coordination of Production and Distribution Planning", *European Journal of Operational Research*, 72 (1994), 503 – 517.

Chen, M. and W. Wang, "A Linear Programming Model for Integrated Steel Production and Distribution Planning", *International Journal of Operations & Production Management*, 17 (1997), 592 – 610.

Chen, Z.-L., "Scheduling and Common Due Date Assignment with Earliness-Tardiness Penalties and Batch Delivery Costs", *European Journal of Operational Research*, 93 (1996), 49-60.

Cheng, T.C.E. and V.S. Gordon, "On Batch Delivery Scheduling on a Single Machine", *Journal of the Operational Research Society*, 45 (1994), 1211-1215.

Cheng, T.C.E., V.S. Gordon, & M.Y. Kovalyov, "Single Machine Scheduling with Batch Deliveries", *European Journal of Operational Research*, 94 (1996), 277-283.

Cheng, T.C.E. and Kahlbacher, H.G., "Scheduling with Delivery and Earliness Penalties", *Asia-Pacific Journal of Operational Research*, 10 (1993), 145-152.

Cheng, T.C.E., M.Y. Kovalyov, and B.M.-T. Lin, "Single Machine Scheduling to Minimize Batch Delivery and Job Earliness Penalties", *SIAM Journal on Optimization*, 7 (1997), 547-559.

Cohen, M.A. and H.L. Lee, "Strategic Analysis of Integrated Production-Distribution Systems: Models and Methods", *Operations Research*, 36 (1988), 216-228.

Cohen, M.A. and H.L. Lee, "Resource Deployment Analysis of Global Manufacturing and Distribution Networks", *Journal of Manufacturing and Operations Management*, 2 (1989), 81-104.

Dasci, A. and V. Verter, "A Continuous Model for Production-Distribution System Design", *European Journal of Operational Research*, 129 (2001), 287-298.

Dhaenens-Flipo, C. and G. Finke, "An Integrated Model for an Industrial Production-Distribution Problem", *IIE Transactions*, 33 (2001), 705 - 715.

Dogan, K. and M. Goetschalckx, "A Primal Decomposition Method for the Integrated Design of Multi-Period Production-Distribution Systems", *IIE Transactions*, 31 (1999), 1027 – 1036.

Fumero, F. and C. Vercellis, "Synchronized Development of Production, Inventory, and Distribution Schedules", *Transportation Science*, 33 (1999), 330 – 340.

Glover, F. G. Jones, D. Karney, D. Klingman, and J. Mote, "An Integrated Production, Distribution, and Inventory Planning System", *Interfaces*, 9 (1979), 21 – 35.

Golhar, D.Y. and B.R. Sarker, "Economic Manufacturing Quantity in a Just-In-Time Delivery System", *International Journal of Production Research*, 30 (1992), 961-972.

Goyal, S.K., "A Joint Economic-Lot-Size Model for Purchase and Vendor: A Comment", *Decision Sciences*, 19 (1988), 236-241.

Goyal, S.K., "A One-Vendor Multi-Buyer Integrated Inventory Model: A Comment", *European Journal of Operational Research*, 82 (1995), 209-210.

Goyal, S.K., "On Improving the Single-Vendor Single Buyer Integrated Production Inventory Model with a Generalized Policy", *European Journal of Operational Research*, 125 (2000), 429-430.

Goyal, S.K. and Gupta, Y.P., "Integrated Inventory Models: The Buyer-Vendor Coordination", *European Journal of Operational Research*, 41 (1989), 261-269.

Goyal, S.K. and F. Nebebe, "Determination of Economic Production-Shipment Policy for a Single-Vendor-Single-Buyer System", *European Journal of Operational Research*, 121 (2000), 175-178.

Hahm, J. and C.A. Yano, "The Economic Lot and Delivery Scheduling Problem: the Single Item Case", *International Journal of Production Economics*, 28 (1992), 235-252.

Hahm, J. and C.A. Yano, "Economic Lot and Delivery Scheduling Problem: the Common Cycle Case", *IIE Transactions*, 27 (1995a), 113-125.

Hahm, J. and C.A. Yano, "Economic Lot and Delivery Scheduling Problem: the Nested Schedule Case", *IIE Transactions*, 27 (1995b), 126-139.

Hahm, J. and C.A. Yano, "Economic Lot and Delivery Scheduling Problem: Powers of Two Policies", *Transportation Science*, 29 (1995c), 222-241.

Hall, L. and D. Shmoys, "Approximation Schemes for Constrained Scheduling Problems" In *Proceedings of the 30th Annual Symposium on Foundations of Computer Science*, 1989, 134-140.

Hall, N.G. and C.N. Potts, "Supply Chain Scheduling: Batching and Delivery", 2000, to appear in *Operations Research*.

Hall, R.W., "On the Integration of Production and Distribution: Economic Order and Production Quantity Implications", *Transportation Research*, 30B (1996), 387-403.

Haq, A.N., "An Integrated Production-Inventory-Distribution Model for Manufacturing of Urea: A Case", *International Journal of Production Economics*, 39 (1991), 39-49.

Herrmann, J.W. and C.-Y. Lee, "On Scheduling to Minimize Earliness-Tardiness and Batch Delivery Costs with a Common Due Date", *European Journal of Operational Research*, 70 (1993), 272-288.

Hill, R.M., "The Single-Vendor Single-Buyer Integrated Production-Inventory Model with a Generalised Policy", *European Journal of Operational Research*, 97 (1997), 493-499.

Hill, R.M., "The Optimal Production and Shipment Policy for the Single-Vendor Single-Buyer Integrated Production-Inventory Problem", *International Journal of Production Research*, 37 (1999), 2463-2475.

Hill, R.M., "Optimizing a Production System with a Fixed Delivery Schedule", *Journal of the Operational Research Society*, 47 (1996), 954-960.

Hochbaum, D.S. and S.-P. Hong, "On the Complexity of the Production-Transportation Problem", *SIAM Journal on Optimization*, 6 (1996), 250-264.

Holmberg, K. and H. Tuy, "A Production-Transportation Problem with Stochastic Demand and Concave Production Costs", *Mathematical Programming*, 85 (1999), 157-179.

Hurter, A.P. and M.G. Van Buer, "The Newspaper Production/Distribution Problem", *Journal of Business Logistics*, 17 (1996), 85 – 107.

Jamal, A.M.M. and B.R. Sarker, "An Optimal Batch Size for a Production System Operating under a Just-In-Time Delivery System", *International Journal of Production Economics*, 32 (1993), 255-260.

Jayaraman, V. and H. Pirkul, "Planning and Coordination of Production and Distribution Facilities for Multiple Commodities", *European Journal of Operational Research*, 133 (2001), 394-408.

Khouja, M., "The Economic Lot and Delivery Scheduling Problem: Common Cycle, Rework, and Variable Production Rate", *IIE Transactions*, 32 (2000), 715-725.

Khumawala, B.H. and D.L. Kelly, "Warehouse Location with Concave Costs", *INFOR*, 12 (1974), 55-65.

King, R.H. and R.R. Love, Jr., "Coordinating Decisions for Increased Profits", *Interfaces*, 10 (1980), 4 – 19.

Kuno, T. and T. Utsunomiya, "A Lagrangian Based Branch-and-Bound Algorithm for Production-Transportation Problems", *Journal of Global Optimization*, 18 (2000), 59-73.

Kuno, T. and T. Utsunomiya, "A Decomposition Algorithm for Solving Certain Classes of Production-Transportation Problems with Concave Production Cost", *Journal of Global Optimization*, 8 (1996), 67-80.

Kuno, T. and T. Utsunomiya, "A Pseudo-Polynomial Primal-Dual Algorithm for Globally Solving a Production-Transportation Problem", *Journal of Global Optimization*, 11 (1997), 163-180.

Langley Jr., C.J., G.R. Allen, and G.R. Tyndall, "*Third Party Logistics Study: Results and Findings of the Sixth Annual Study*", 2001.

Lee, C.-Y. and Z.-L. Chen, "Machine Scheduling with Transportation Considerations", *Journal of Scheduling*, 4 (2001), 3-24.

Lu, L., "A One-Vendor Multi-Buyer Integrated Inventory Model", *European Journal of Operational Research*, 81 (1995), 312-323.

Martin, C.H., D.C. Dent, and J.C. Eckhart, "Integrated Production, Distribution, and Inventory Planning at Libbey-Owens-Ford", *Interfaces*, 23 (1993), 68-78.

Mohamed, Z.M., "An Integrated Production-Distribution Model for a Multi-National Company Operating under varying exchange rates", *International Journal of Production Economics*, 58 (1999), 81 – 92.

Muckstadt, J. and R. Roundy, "Analysis of Multistage Production Systems", In *Logistics of Production and Inventory, Handbooks in OR/MS*, Volume 4, S. Graves, A. Rinnooy Kan, and P. Zipkin (eds.), Elsevier, North-Holland, Amsterdam. 1993.

Owen, S.H. and M.S. Daskin, "Strategic Facility Location: A Review", *European Journal of Operational Research*, 111 (1998), 423-447.

Ozdamar, L. and T. Yazgac, "A Hierarchical Planning Approach for a Production-Distribution System", *International Journal of Production Research*, 37 (1999), 3759 – 3772.

Parija, G.R. and B.R. Sarker, "Operations Planning in a Supply Chain System with Fixed-Interval Deliveries of Finished Goods to Multiple Customers", *IIE Transactions*, 31 (1999), 1075-1082.

Potts, C.N. "Analysis of a Heuristic for One Machine Sequencing with Release Dates and Delivery Times", *Operations Research*, 28 (1980) 1436-1441.

Pyke, D.F. and M.A. Cohen, "Multiproduct Integrated Production-Distribution Systems", *European Journal of Operational Research*, 74 (1994), 18-49.

Rebaine, D. and V.A. Strusevich, "Two-Machine Open Shop Scheduling with Special Transportation Times", *Journal of the Operational Research Society*, 50 (1999), 756-764.

Sabri, E. and B.M. Beamon, "A Multi-Objective Approach to Simultaneous Strategic and Operational Planning in Supply Chain Design", *Omega*, 28 (2000), 581 – 598.

Sarker, B.R. and G.R. Parija, "An Optimal Batch Size for a Production System Operating under a Fixed-Quantity, Periodic Delivery Policy", *Journal of the Operational Research Society*, 45 (1994), 891-900.

Sarker, B.R. and G.R. Parija, "Optimal Batch Size and Raw Material Ordering Policy for a Production System with a Fixed-Interval, Lumpy Demand Delivery System", *European Journal of Operational Research*, 89 (1996), 593-608.

Sarker, R.A. and L. Khan, "An Optimal Batch Size for a Production System Operating under Periodic Delivery Policy", *Computer & Industrial Engineering*, 37 (1999), 711-730.

Sarker, R.A. and L. Khan, "An Optimal Batch Size under a Periodic Delivery Policy", *International Journal of Systems Science*, 32 (2001), 1089-1099.

Sarmiento, A.M. and R. Nagi, "A Review of Integrated Analysis of Production-Distribution Systems", *IIE Transactions*, 31 (1999), 1061-1074.

Shen, Z.-J. M., C. Coullard, and M.S. Daskin, "A Joint Location-Inventory Model", *Transportation Science*, 37 (2003), 40 – 55.

Speranza, M.G. and W. Ukovich, "Minimizing Transportation and Inventory Costs for Several Products on a Single Link", *Operations Research*, 42 (1994), 879-894.

Thomas, D.J. and P.M. Griffin, "Coordinated Supply Chain Management", *European Journal of Operational Research*, 94 (1996), 1-15.

Tuy, H., N.D. Dan, and S, Ghannadan, "Strongly Polynomial Time Algorithms for Certain Concave Minimization Problems on Networks", *Operations Research Letters*, 14 (1993), 99-109.

Tuy, H., S. Ghannadan, A. Migdalas, and P. Varbrand, "A Strongly Polynomial Algorithm for a Concave Production-Transportation Problem with a Fixed Number of Nonlinear Variables", *Mathematical Programming*, 72 (1996), 229-258.

Van Buer, M.G., D.L. Woodruff, and R.T. Olson, "Solving the Medium Newspaper Production/Distribution Problem", *European Journal of Operations Research*, 115 (1999), 237-253.

Vidal, C.J. and M. Goetschalckx, "Strategic Production-Distribution Models: A Critical Review with Emphasis on Global Supply Chain Models", *European Journal of Operational Research*, 98 (1997), 1-18.

Wang, G. and T.C.E. Cheng, "Parallel Machine Scheduling with Batch Delivery Costs", *International Journal of Production Economics*, 68 (2000), 177-183.

Williams, J.F., "A Hybrid Algorithm for Simultaneous Scheduling of Production and Distribution in Multi-Echelon Structures", *Management Science*, 29 (1983), 77 – 92.

Woeginger, G.J. "Heuristics for Parallel Machine Scheduling with Delivery Times", *Acta Informatica*, 31 (1994) 503-512.

Woeginger, G.J. "A Polynomial-Time Approximation Scheme for Single-Machine Sequencing with Delivery Times and Sequence-Independent Batch Set-Up Times", *Journal of Scheduling*, 1 (1998), 79 – 87.

Youssef, M.A., "An Iterative Procedure for Solving the Uncapacitated Production-Distribution Problem under Concave Cost Function", *International Journal of Operations & Production Management*, 16 (1996), 18-27.

Yuan, J., "A Note on the Complexity of Single-Machine Scheduling with a Common Due Date, Earliness-Tardiness, and Batch Delivery Costs", *European Journal of Operational Research*, 94 (1996), 203-205.

Zdrzalka, S. "Approximation Algorithms for Single-Machine Sequencing with Delivery Times and Unit Batch Set-Up Times", *European Journal of Operational Research*, 51 (1991), 199 – 209.

Zdrzalka, S. "Analysis of Approximation Algorithms for Single-Machine Scheduling with Delivery Times and Sequence Independent Setup Times", *European Journal of Operational Research*, 80 (1995), 371 - 380.

Zuo, M., W. Kuo, and K.L. McRoberts, "Application of Mathematical Programming to a Large-Scale Agricultural Production and Distribution System", *Journal of Operational Research Society*, 42 (1991), 639-648.

Table 17.1. Production-Transportation Problems

	Integration Structure	Planning Horizon	Nature of Demand	Production Cost Function	Other Characteristics
Hochbaum and Hong (1996), Youssef (1996), Kuno and Utsunomiya (2000)	Production & outbound transportation	One time period	Deterministic, must be satisfied	Concave, separable	
Kuno and Utsunomiya (1996)					Each plant can only supply a subset of customers
Hochbaum and Hong (1996), Tuy et al. (1996)				Concave, non-separable	
Tuy et al. (1993)					There are only two plants
Kuno and Utsunomiya (1997)					In addition to two plants, there are two warehouses where no production is involved
Holmberg & Tuy (1999)			Stochastic. Convex penalty cost for unmet demand	Concave, separable	

Table 17.2. Joint Lot Sizing and Finished Product Delivery Problems

	Integration Structure	Planning Horizon	Nature of Demand	Policy on Production & Delivery Cycles	Other Characteristics
Banerjee (1986)	Production & outbound transportation	Infinite horizon	Constant demand rate	Policy (a): Production cycle length equal to delivery cycle length	
Hahm and Yano (1995a), Khouja (2000)					Multiple products
Hall (1996)					Multiple manufacturers, multiple customers
Goyal (1988), Aderohunmu et al. (1995)				Policy (b): Integer number of delivery cycles within each production cycle. Identical delivery quantity for different delivery cycles.	
Lu (1995)					Multiple customers
Benjamin (1989), Hahm and Yano (1992)					Setup time
Blumenfeld et al (1985, 1991)					Multiple customers, multiple products
Hahm and Yano (1995b, 1995c)					Multiple products
Goyal (1995, 2000), Hill (1997), Goyal and Nebebe (2000)				Policy (b). Non-identical delivery quantities for different delivery cycles.	Require delivery quantities to follow a given formula
Hill (1999)					Do not require delivery quantities to follow a given formula
Lu (1995)				Policy (c): Integer number of production cycles within each delivery cycle. Identical delivery quantity for different delivery cycles.	Multiple customers

Table 17.3. Joint Raw Material Delivery and Lot Sizing Problems

	Integration Structure	Planning Horizon	Nature of Demand	Policy on Production & Delivery Cycles	Other Characteristics
Golhar and Sarker (1992), Jamal and Sarker (1993)	Inbound transportation & production	Infinite horizon	Constant demand rate	Raw material delivery cycle length equal to production cycle length	
Sarker and Parija (1994)					General raw material conversion rate
Sarker and Khan (1999, 2001)					Multiple types of raw materials
Hill (1996)				Integer number of equal-sized raw material delivery cycles within each production cycle	General raw material conversion rate. Production cycle time equal to integer multiple of finished product delivery cycle time.
Sarker and Parija (1996)					Multiple types of raw materials. Production cycle time equal to integer multiple of finished product delivery cycle time.
Parija and Sarker (1999)					Multiple customers
Hill (1996)				Integer number of equal-sized production cycles within each raw material delivery cycle	General raw material conversion rate. Production cycle time equal to integer multiple of finished product delivery cycle time.
Sarker and Khan (1999, 2001)					Multiple types of raw materials

Table 17.4. General Tactical Production-Distribution Problems

	Integration Structure	Planning Horizon	Number of Products	Nature of Demand	Other Characteristics
Zuo et al (1991)	Two stages: Production and outbound transportation	One time period	Multiple time periods	Deterministic, must be satisfied without backlog	
Arntzen et al. (1995)					International issues also considered
Chandra and Fisher (1994), Fumero and Vercellis (1999)					Vehicle routing decisions also considered
Mohamed (1999)					Demand satisfied in JIT fashion (i.e. no inventory allowed at the customers)
Ozdamar and Yazgac (1999)				Deterministic, backlog not allowed in the aggregate model	A hierarchical planning approach proposed
Chen and Wang (1997)	Three or more stages: Inbound transportation, production, and outbound transportation	One time period	Multiple products	Deterministic, does not have to be satisfied but with a penalty for demand shortage	
Cohen and Lee (1989)					Some strategic decisions also considered
Sabri and Beamon (2000)				Deterministic, must be satisfied without backlog	
Cohen and Lee (1988)				Stochastic demand	
Barbarosoglu and Ozgur (1999), Dhaenens-Flipo and Finke (2001)		Multiple time periods		Deterministic, must be satisfied without backlog	Demand satisfied in JIT fashion
Dogan and Goetschalckx (1999)					Some strategic decisions also considered
Haq (1991)			One product	Deterministic, backlog allowed with a penalty	

Table 17.5. Joint Job Processing and Finished Job Delivery Problems

	Integration Structure	Planning Horizon	Transportation Times/Costs	Machine Configuration	Other Characteristics
Potts (1980) Hall and Shmoys (1989) Zdrzalka (1991, 1995), Woeginger (1998)	Production and outbound transportation	Finite planning horizon	Transportation times considered. Transportation costs not considered	Single machine	Multiple families of jobs. Setup times required between different families of jobs.
Woeginger (1994)				Parallel machines	
Lee and Chen (2001)				Single machine, Parallel machines, Flow shop	One or multiple delivery vehicles, each with a limited capacity
Cheng and Kahlbacher (1993), Herrmann and Lee (1993), Cheng and Gordon (1994), Chen (1996), Cheng et al. (1996), Yuan (1996)			Transportation times not considered. Transportation costs considered	Single machine	
Cheng et al. (1997)					Setup times required before each batch of jobs
Wang and Cheng (2000)				Parallel machines	
Hall and Potts (2000)				Arborescent supply chain (One supplier, multiple manufacturers, multiple customers)	Problems studied from a supplier point of view, from a manufacturer point of view, or from an entire system point of view
Hurter and Van Buer (1996), Van Buer et al. (1999)			Both transportation times and costs considered	Single machine	Vehicle routing decisions considered

David Simchi-Levi, S. David Wu, and Z. Max Shen (Eds.)
Handbook of Quantitative Supply Chain Analysis:
Modeling in the E-Business Era
©2004 Kluwer Academic Publishers

Chapter 18

NEXT GENERATION ERP SYSTEMS: SCALABLE AND DECENTRALIZED PARADIGMS

Paul M. Griffin
School of Industrial & Systems Engineering,
Georgia Institute of Technology, Atlanta, GA 30332-0205

Christina R. Scherrer
School of Industrial & Systems Engineering,
Georgia Institute of Technology, Atlanta, GA 30332-0205

Keywords: Enterprise planning, coordination and collaboration, scalable systems, decentralized systems

1. Introduction

The Internet has opened new venues for companies to create flexible systems (or "supply webs") by offering high-speed communication and tight connectivity. This presents a tremendous potential to change the way companies distribute goods, communicate with suppliers and customers, and collaborate both within their company and with their business partners. Hence, businesses increasingly look for ways to use these new tools to achieve "collaboration" and "coordination" between entities within their company, as well as between entities within their supply chain.

The development and implementation of Enterprise Resource Planning (ERP) systems was one of the first steps towards collaboration and sharing information between business units in a firm. However, traditional ERP systems are quite limited in providing the necessary tools for increased coordination and collaboration. While most of these systems are very good in storing, organizing and maintaining data, they usually do not provide decision support to convert the data into meaningful information. Companies such as SAP (www.sap.com) have recently attempted to close this gap by introducing decision support tools such as APO (Advanced Planning and Optimization). ERP systems also tend to be inflexible and expensive and have very structured representation requirements for data and processes. Another important drawback is that they do not meet the needs of today's extended enterprises or supply webs. They tend to focus on processes internal to a firm, such as finance, manufacturing and human resources, but cannot support the needs of a "virtually integrated" company. Traditional ERP systems are not Internet-enabled; they do not provide any functionality for a company to get "connected" with its suppliers and customers.

More recently, many supply chain management software companies such as i2 (www.i2.com) and Manhattan Associates (www.manh.com) have moved from the purely supply chain optimization solutions including demand fulfillment and transportation planning to offerings that provide many of the same capabilities as ERP vendors. Detailed discussion of supply chain operations and offerings may be found in chapter 17 and in Sections III and IV of this book.

In this chapter we will discuss the development of enterprise planning systems, and the inadequacies of these systems in an e-business environment. In particular we will discuss the issues of scalability and decentralization. The layout of the Chapter is as follows. In Section 2 we provide a very brief overview of the history of ERP systems along with some of the benefits and drawbacks of these systems. In Section 3 we discuss the current functionality found in most ERP implementations. A discussion of implementation issues such as cost and time are discussed in Section 4. The topic of scalable enterprise systems is covered in Section 5 and decentralized systems in Section 6. Current enterprise issues such as ECM and ERP II are presented in Section 7. Conclusions and issues for next generation enterprise systems are given in Section 8.

Before continuing, we should first say a few words about the use of the name "enterprise resource planning". For several reasons (which we will discuss later) this term has fallen out of favor. In fact, many have argued (and as we discuss later, we do not disagree) that ERP is dead. Major ERP vendors such as SAP and Oracle (www.oracle.com) no longer refer to themselves as such. If you visit their web pages it is difficult [64] to find the term ERP anywhere, but instead a focus on phrases such as customer relationship management (CRM). According to AMR Research [64], Oracle's CEO, Larry Ellison, has not mentioned the term ERP in over three years. Although here have been many significant changes in enterprise software such as advances in optimization-based planning and web-enablement, the product marketed by vendors is still fundamentally ERP. At the risk of sounding archaic, we will use this term throughout the chapter. We will, however, discuss some of the more marketable terms for ERP in Section 7.

2. A Brief History

From an engineering perspective, the roots of ERP are planted firmly in the idea of material requirements planning (MRP). MRP was developed in the 1960s and is a rather simple accounting scheme that determines when to release work (planned order releases) into the production facility based on forecasted demand. The MRP system first defines a set of gross requirements for an end item by aggregating the forecast into time buckets. On hand inventory is netted against the gross requirements to define net requirements. A lot sizing procedure is then used to determine the appropriate planned order receipts to meet the net requirements. Next, planned order releases are determined by offsetting the planned order receipts by the leadtime (which is assumed to be constant). The key insight for MRP was the recognition that production for components that make up the end item should be based on production for the end item. Therefore the final step of MRP is to explode the bill of materials (BOM) so that the planned order releases for the end item become the gross requirements for the next level in the BOM. This set of steps is repeated through the BOM [51].

MRP became extremely popular in the 1960s and 1970s. However, it became apparent that there were several problems with its use. First and foremost was the assumption that leadtime is constant regardless of the lot size. As a result there was no check if the capacity was high enough to feasibly meet demand. In practice this tended to result in large levels of inventory and long leadtimes to enable companies to meet capacity requirements. Another problem with MRP was that it caused *system nervousness* [66]. Nervousness occurs when small changes to the input (e.g. gross requirements) lead to large changes in the output (i.e. planned order releases). The end result was that developing stable production plans could be quite difficult.

In the 1970s and 1980s, there was an attempt to correct the difficulties associated with MRP through the development of what became known as manufacturing resources planning (MRPII). While there is no real agreement about what MRPII means (since there have been so many different types of MRPII implementations), in essence the issue of capacity was addressed through the use of rough cut capacity planning (RCCP) and capacity requirements planning (CRP). RCCP uses a bill of resources as a rough check against the aggregate requirements. CRP determines capacity load profiles of predicted job completion times and checks them against projected plant capacity. Fixed leadtimes (now for a process center rather than the whole facility) were still incorporated into the analysis. Many MRPII implementations also focused on more real-time outputs including job dispatching and shop floor control, and integrated more closely with forecasting and demand management systems. In addition, several MRP II implementations include feedback loops among the strategic, tactical and operational planning levels. This feedback is helpful in reducing the system nervousness previous discussed for traditional MRP systems. Still at the core of MRPII implementations, however, was MRP. As a result, although MRPII helped to address some of the shortcomings of MRP, the end result was similar when implemented (i.e. high inventories and long production lead times). Some other reasons for the lack of overwhelming success of MRPII implementations include: i) inaccurate data from inventory and forecasts, ii) complexity of the implemented system, and iii) the focus on production without integrating to other business functions.

Throughout the 1980s and 1990s, several other types of systems with various acronyms were developed, including MRPIII and BRP (business requirements planning). The real change, however, came when MRPII was integrated with other business functions including human resource management, finance and accounting. Since this approach crossed most functions of the businesses, it was given the name enterprise resource planning (ERP). ERP became extremely popular through the 1990's. As evidence of the popularity, ERP software sales in the US grew from under $1B in 1993 to over $8B by 1999. Toward the end of the 1990s, many firms such as AMR Research predicted growth rates of over 35% per year and worldwide ERP sales of $52B by the end of 2002. Those lofty projections have been scaled back more recently, however. ERP sales fell by over 35% in 2001 and the Gartner Group forecasts negative growth through at least 2004.

There were several reasons for the explosion of popularity of ERP in the 1990s. Hopp and Spearman [39] argue that there were three developments that occurred preceding its development. First was the increased interest in supply chain management that forced firms to look outside of their current operations. Second was the popularity of business process reengineering (BPR). This caused many organizations to dramatically change their management structure and helped to foster the notion that ERP implementations would facilitate this change. Finally, there was tremendous growth in distributed processing. There are other reasons that should be mentioned as well. First, when a new technology is developed many firms purchase the technology without completely understanding the ramifications for fear that if they don't, their competition will use it as a competitive advantage against them. Second, many enterprises set requirements for their suppliers in order to facilitate transactions. For example, large suppliers to the "Big 3" automobile manufacturers were forced to purchase "compatible" systems. Third, many firms felt that ERP implementation would be an effective mechanism to deal with the potential Y2K threat, and ERP sales showed a spike in the late nineties as a result. Finally, ERP was simply oversold by both vendors and consultants who collected huge fees for implementation. As we will see in the next section, many of the perceived benefits of ERP were never realized.

Toward the end of the 1990's, the benefits of ERP systems seemed much less promising. First, there were numerous examples of ERP failures. For example, in 1996 FoxMeyer Drug Company filed for bankruptcy protection on the basis that its new SAP ERP implementation could not handle its transaction volume. They filed suit for $500 million against both SAP and Anderson Consulting (the firm hired to install the system). In 1997, Dell Computer stopped its SAP implementation after investing $30 million since they realized that the total investment would be over $130M. In 1999 Hershey Foods argued in its annual report that it lost over $200M in sales due to problems in order taking from its SAP ERP system. Recently an executive at Intel informed us that they scrapped an ERP implementation after investing over $400M.

Another reason for the steep decline in ERP sales was the perceived low return on investment (which we discuss in Section 4). In addition, as Spearman [65] has argued, ERP performs basic logistical functions using exactly the same logic as MRP. Instead of re-engineering the entire paradigm, ERP vendors have provided multiple add-ons in the form Advanced Planning Systems. It is also the case that ERP systems have been very complex and inflexible, and so in sum, they simply have not worked well. Recent advances in enterprise systems offerings have tried hard to address many of these issues.

While we are not aware of any hard data to support this assertion, many within the operations research community have argued that one of the reasons for the success of decision support models in supply chain management is the fact that ERP systems have made available the (previously unavailable) data that these models require. This is certainly quite possible. In addition, many "stand-alone" decision support systems make use of data from ERP systems, but are not integrated with the ERP system and not viewed as ERP decision support systems. This would therefore clearly be an indirect ERP benefit.

3. Current ERP Functionality

As with MRP and MRPII, there is no universal set of ERP functions. However, the fundamental characteristics of current ERP systems are: i) it is packaged software which integrates the majority of a business' processes, ii) it uses a data warehouse that allows "real time" access, and iii) it processes many of the firm's transactions and integrates them with planning. A typical ERP system contains three layers: the *production layer* which consists mainly of manufacturing, the *back-office layer* which includes such core functions as human resource management and accounting, and the *front-office layer* which comprises the customer and inter-business functions such as sales and marketing [67]. A data warehouse is used at the core of the system in order to streamline the flow of information through modular applications to support the business activities. A shared database helps to eliminate the need for repetitive data input or reformatting, and contrasts with traditional information systems where the data are typically spread across numerous separate systems, each housed in separate functions, business units, regions, factories or offices [45].

One of the most common ERP implementations is SAP's R/3. The current list of mySAP functions (www.sap.com) includes supply chain management, customer relationship management, mobile business, human resources, enterprise portals, exchanges, financials, technology, product lifecycle management, supplier relationship management, and business intelligence; quite a broad list. The basic components of an ERP configuration are the use of models, artifacts and processes (MAPs). A *model* represents the "world" encompassed by the system. An example of such a world would be the organizational structure. The *artifact* is an interface between an "inner" environment and an "outer" environment such as an invoice document, vendor list, or product list. The *process* is the activity and information flows necessary to accomplish a particular task (or task set).

As an example, consider the SAP R/3 order management process in Figure 18.1 [36]. In this figure, the process maps into multiple integrated SAP modules. If this had been a function-based legacy system, these would not have been integrated and the information would have had to be exchanged manually. In order to effectively use this approach, however, an organization must restructure its business processes to fit into the structure provided by the ERP vendor. This is an extremely time consuming and costly process.

As mentioned previously, it is difficult to distinguish now between many of the offerings of ERP vendors with supply chain optimization vendors such as i2 and Manhattan Associates. For instance, Manhattan Associates focuses on moving goods across the supply chain, and has three main applications: i) a Trading Partner Management application suite which synchronizes business processes across an organization's trading partner network by extending execution capabilities to their suppliers, hubs, carriers and customers, ii) a Transportation Management System which facilitates transportation bidding processes, transportation planning and transportation loading and routing, and iii) a Warehouse Management Suite which helps with workload planning and slotting. i2 has many similar offerings and has recently moved into the area of revenue and profit optimization (since the acquisition of Talus).

Traditional ERP systems were client/server based and consisted of a database system, communication protocols and a user interface framework. With the advent of the Internet, vendors went to web-based interfaces. This made it possible to access the system though any web-enabled device (since no code rests

Next Generation ERP Systems: Scalable and Decentralized Paradigms   753

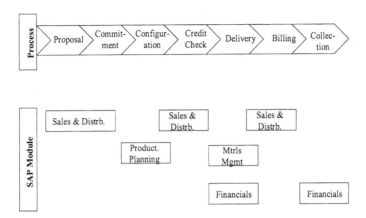

Figure 18.1. SAP R/3 Order Management Process

at the client) and reduced maintenance costs. More recently, personalized user interfaces (called *portals*) have been developed. Portals are web-based and organized by the roles of their users (such as quality manager or CIO) who tend to have very different informational and functional requirements. In addition, outward facing portals allow access to external partners (e.g. suppliers). Difficulties of implementing portals, however, include the lack of portal standards (discussed in Section 5) as well as the lack of administrative functions.

In order for ERP systems to be effective, they must be integrated with other components. There are three types of integration. *Horizontal integration* (in manufacturing applications) ties ERP to systems on the shop floor, typically through manufacturing execution systems (MES). *Functional integration* ties ERP with other functions such as engineering, marketing and human resources. Finally, *external integration* ties ERP with inter-organizational systems such as customers and suppliers. Enterprise application integration (EAI) packages (as opposed to internally developed middleware) are often used to provide these different levels of integration. EAI often makes use of extensible markup language (XML) data exchange and messaging and EAI products can be divided into three categories: i) *application integration* which is integration at the data level, ii) *business process integration* which supports business processes with application integration, and iii) *business community integration* which links the enterprise with suppliers and customers. This last form of integration has proven much more complicated than those within the firm, for reasons including security concerns, determining which internal systems need to be linked with which supplier/customer, standardization, etc.

For EAI systems to be effective, they must be able to fully understand the semantics from heterogeneous schemas. Currently, dealing with this issue is extremely time consuming and developers typically need to create transformation code manually (though textual languages and graphical transformation design tools help in this regard) [28]. Another challenge is that those schemas are constantly changing, sometimes rendering existing code unusable.

EAI is an important technology in terms of scalability. For instance, in Figure 18.2 [1], we see on the left what Gartner refers to as "application spaghetti" due to all of the point-to-point connections. As the number of applications increase, the number of these connections can grow exponentially. The EAI/hub (shown on the right of Figure 18.2) provides the infrastructure for more efficient connectivity. Scalability can also depend on the data. For example, Rensselaer's Enterprise Integration & Modeling Center has developed a "Metadatabase" for the purpose of addressing scalability and adaptability by focusing on the integration of LANs, WANs, and mainframes into a cohesive, enterprise-wide system. Their integration and modeling techniques are summarized in [40]. In addition, there is very active research on integration models and classifications based on different standards. A nice review of issues with respect to standards may be found in [42] and [43].

Of course EAI is far beyond the research stage as there are several major EAI vendors including Microsoft, SeeBeyond and TIBCO. In addition, this is a rapidly growing area and Gartner forecasts that approximately 30% for all future enterprise IT budgets will be used for EAI [1].

Some of the more important ERP functionalities recently have been in advanced planning and scheduling (APS) and customer relationship management (CRM) (and the related supplier relationship management (SRM)). APS is often built on the idea of finite capacity scheduling which differ from MRP methods that do not consider actual availability of production resources (e.g. it is assumed that material can be obtained in the specified lead-time). With finite capacity scheduling, many constraints from the resources, processes and materials can be included. There has been a broad range of approaches developed including rules-based logic, heuristics, mixed integer programming, neural networks and exponential smoothing. Although APS was primarily offered as a "bolt-on" to ERP systems from companies such as Red Pepper Software, there has been considerable consolidation in this area, and now most major ERP vendors offer it as an option.

The role of CRM is to manage the customer relationship from inquiry to sales and servicing and returns. CRM is typically divided into front-office applications (including sales force automation and customer profiling) and back-office applications (including call center management and service and repair).

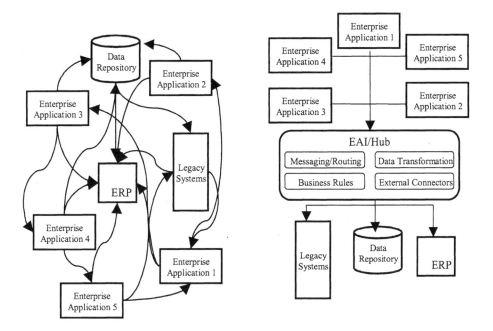

Figure 18.2. SAP R/3 Order Management Process

CRM connects the different systems for sales, marketing, customer service, and warranty departments and provides employees with a complete view of customer data. Integrating CRM with ERP allows additional benefits, such as providing real-time information on order status. An important function is the analytics on the customer information. CRM has been one of the more popular functionalities in recent years, and is offered by all of the major ERP vendors.

4. Implementation Issues

The motivation, cost, and effort of an ERP implementation can vary widely between firms. Mabert et. al [52] surveyed over 450 U.S. manufacturing firms and found that the most important company motivations for ERP implementation included[1]: replace legacy system (4.06), simplify and standardize systems (3.85), improve communication with suppliers and customers (3.55), gain strategic advantage (3.46), link to global activities (3.17) and pressure to keep up with competitors (2.99). The motivation, however, did vary by function. For instance, the most important motivations from the perspective of operational programs included: i) poor/ uncompetitive performance, ii) complex, ineffective business practices, iii) cost structures too high, iv) not responsive enough to customers, and v) business becoming global.

The cost of an ERP implementation can also vary widely and is a function of firm size. A conference board study of 186 SAP implementations in 2000 found:

- Companies with annual sales less than $1B paid between $5M and $20M for ERP implementation.

- Companies with annual sales between $1B and $10B paid between $6M and $150M for ERP implementation.

- Companies with annual sales between $10B and $25B paid between $30M and $600M for ERP implementation.

- Companies with annual sales greater than $25B paid between $150M and $800M for ERP implementation.

A Benchmarking Partners study of 500 executives from 300 firms (houns54.clearlake.ibm.com/solutions/erp/erppub.nsf/Files/BENCHIBM/$FILE/BENCHIBM.pdf) found that the cost breakdown for implementation was as follows: consulting services (38%), internal resources (21%), software (13%), hardware (11%), other (9%), and training (8%). AMR Research found similar figures of: reengineering (43%), data conversion (15%), training (15%), software (15%), and hardware (12%). Both surveys show that the ERP software is a fairly minor component of the total cost of implementation.

Implementation time varied quite a bit between firms. According to Mabert et. al [52], 34% of implementations took less than a year, 45% of implementations took 1 to 2 years, 12% of implementations took 2 to 3 years, and 9% of implementations took more than 3 years. In addition, it was estimated that the breakeven point occurred on average 40 months after implementation completion. However, a survey of executives by Boston Consulting Group (BCG) [15] showed that relatively few companies achieved significant value. In fact, they found:

- Only 33% of outcomes were viewed as positive

- Only 37% of companies could point to tangible financial impact

- Only 58% of companies finished on time and budget

- 20% of companies stated that they could have achieved the same value for less than half the cost.

BCG summarizes the reasons for less successful enterprise initiatives as:

- Companies responded to the hype of ERP before considering their strategies and business capabilities

- Companies were drawn to "big-bang" ERP projects as a means to reengineer or force change

- Companies allowed vendors and integrators to make decisions that served more to increase the vendors' and integrators' fees than to create true business capabilities and value for themselves.

AMR Research also reported a survey of supply chain management suites in August of 2001 [27]. The key findings were that basic supply chain planning (SCP) modules are the most widely implemented (i.e., demand planning 62%, supply planning 56%, and plant scheduling 31%). In addition, they stated that collaboration was oversold. Only 12% of customers had over 5% of their suppliers or customers feeding data into their systems, but on average, 78% of a company's products were being planned using the software (i.e., "pervasiveness within the four walls is impressive"). They also summarize that there was little demand for full SCP suites.

As mentioned earlier, CRM has been one of the key applications in recent years. In 2003, the Gartner Group surveyed 343 North American Firms [30]. They found the following:

- Organizations' continued tight control of expenses, which has cut (or frozen) spending on CRM initiatives, makes ROI an even more important factor in the purchasing process.

- Organizations do not deploy 41.9 % of the CRM licenses they buy.

- CRM sales deployments have begun to show positive with partner relationship management (PRM) implementations producing the highest ROI.

- 66% of respondents said they had received demonstrable ROI from PRM. The second highest rating was incentive compensation management systems, with 55% claiming verifiable ROI from the deployment of the application.

The Gartner group recommended that buying more than what you need in the short may seem like a good idea, but that it costs money in the long term. In addition, they argue that organizations should not confuse benefits with ROI.

In spite of this fairly negative picture, ERP did have some significant benefits. These are summarized in Figure 18.3 using data from AMR Research and the Gartner Group. Notice that the largest improvements came in the form of reducing financial transactions times. In addition, AMR Research projected in June of 2003 that after years of declining sales, the enterprise software market would rebound in 2003 [5].

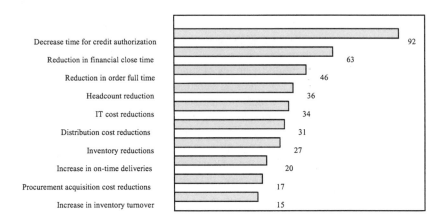

Figure 18.3. ERP Benefits

Banker et. al [10] performed an empirical analysis to determine what reasons motivate organizations to adopt ERP systems. They found that the organizations more likely to adopt these systems either view coordination as a critical function, use low-cost or low-price as their strategic advantage, are highly centralized, or have a high-up locus of control (i.e. follow a top-down focused financial strategy). In addition, they found that firms that use technology rather than marketing as a means of differentiating themselves from their competitors or face high uncertainty in output prices are also most likely to adopt ERP systems.

Much of the current focus for ERP implementations is collaboration. However, an AMR Research study showed that there is actually very little collaboration in current systems. The only significant collaboration (greater than 20% of firms) is in the area of demand forecasting. Even with demand forecasting, there is only significant collaboration internally and with suppliers. Demand forecasting collaboration with contract manufacturers, distributors, retailers, third party logistics, and customers was significantly less than 20%. Other areas of potential collaboration (including production planning, replenishment, logistics, product design and promotion) are all insignificant except among internal groups. There are several reasons why collaboration is not occurring, including security issues and lack of standardization. In addition, firms fear sharing information. One reason is that any internal problems (production, quality, etc.) become apparent to external partners. Additionally, firms are unwilling to share information that they feel gives them a competitive advantage, such as design data.

Key components for effective collaboration, and for effective enterprise systems in general, are scalability and decentralization. We discuss these issues in Section 5 and 6.

5. Scalability and ERP

Scalability can be defined as the ability of a system to consistently adapt to different situations, which could include different volumes of use, technology changes, different types of products or users, changing markets (especially to e-business), and different clients/business partners. The growth of e-business has put special emphasis on scalability as it refers to Internet (and intranet) related issues. In this environment, the users of an ERP system can vary significantly in size from just a few users to many thousands of users at one time. For effective scalability, we need to be able to increase the number of users indefinitely without needing a change in software or causing a decrease in the efficiency of the system.

Traditional ERP systems are not scalable. They require a high amount of domain-specific customization, and there is a high cost to implement the system, as well as to add new users. We will discuss these issues, and progress currently being made, in the next sections.

5.1 New Developments in Enterprise Scalability

In 1999, the National Science Foundation held a workshop to address the issue of scalability and enterprise systems (www.eng.nsf.gov/ses/Workshops/workshops.html). The workshop resulted in NSF sponsoring exploratory research in the area, with the objective to "foster the development of a science base for enterprise-wide business automation". As a result of the workshop, there have been several new research initiatives in this area (www.eng.usf.edu/nsf/conference/scalable/scalable_enterprises/ speakers_presentations.html). Summaries of several of the research efforts follow. An excellent summary of research in this area (in much more detail than the speaker presentations) is found in Prabhu, Kumar and Kamath [60]. In addition, a good reference for enterprise modeling techniques for scalability may be found in Kamath et al [47].

Kamath et al [45] developed a Distributed Integrated Modeling of Enterprises (DIME) framework for modeling of next generation enterprise systems. They created an XML-based markup language and combined it with both a set of Petri net-based representation constructs and a mapping scheme between the two. Adjusting the granularity of the processes and activities controls the size and complexity of the Petri net models. This scalable model uses elements of enterprise process modeling, distributed computing, modern accounting techniques, and engineering methods.

Kamath et al.'s [46] intent for the framework was a web-based environment in which enterprise users will define, modify, and analyze business processes using user-oriented modeling applications resident on Web clients. Petri net-based constructs will be used to translate business process models into formal models. This formal representation of the enterprise's business processes will be stored on enterprise servers. The Web-based clients will address the needs of diverse enterprise users, ranging from the managerial to the technical. Along with storing and manipulating the formal representations, server applications will also perform qualitative and quantitative analyses which can be used to support process improvement initiatives.

Their research focuses on scalability of the theoretical modeling formalism, the modeled enterprise system, and the Internet-based deployment environment, along with development of scalability metrics. The DIME framework is scalable in several ways. Users from different locations can work the model on independently, it isn't affected by the addition of a new sub-model or group of users, and new sub-models can easily be linked to the existing model. Additionally, the XML-based scalable representation layer allows members of the supply-chain to share and access the model as needed and the need to build new XML documents from scratch is reduced when the size and complexity of the enterprise system increase over time.

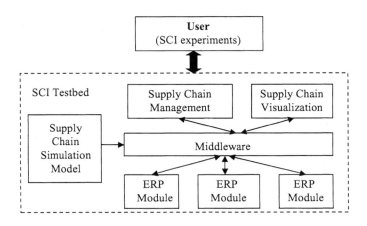

Figure 18.4. Scalable supply chain test bed [46]

Ball et al. [8] added to the research on scalability in ERP, specifically as ERP ties in to supply chain management (SCM) structures. They built a scalable supply chain test-bed that consists of the several distinct but integrated components: ERP, SCM, a simulation, middleware, and visualization (see Figure 18.4). In the model, each supplier and customer is assumed to have its own ERP system, but they are linked by common end products and use collaboration techniques in their supply chain. The SCM component integrates all of these ERP instantiations.

They include the user as an integral part of the supply chain control system in their research test-bed. To this end, the simulation component dynamically generates live data, such as customer orders, to the ERP and SCM components. Then, the user makes decisions about the ERP and SCM systems to manage the supply chain. The system periodically gives feedback to the user and the rest of the model. Ball et al. intend to investigate both the performance of supply chain software and human decision-making. For more information about the model and how it relates to available-to-promise (ATP) decision models, refer to chapter 11.

Their research objective is to use simulation results to design systems that have an overall system quality of service, individual client quality of service, and optimal performance with fair allocation of resources to clients (www.rhsmith.umd.edu:8001/ckim/nsf_sci.htm). Two specific components that Ball et. al investigate are aggregation of client requests and traffic management.

When there are a large number of client requests, a supply chain monitor can aggregate some requests to potentially improve overall system performance while increasing the latency of individual client requests. Further benefits can be obtained by processing these results as they arrive; since a server may provide replies to some clients earlier than if it delayed providing a response until after the entire answer was obtained.

When there are multiple replicated server sites, the supply chain monitor can use traffic management to direct requests to servers to balance workload at the server and to improve latency of the client requests. Similarly, when service is site specific, the monitor can use semantic knowledge about the sites, such as location or transportation costs, to direct requests to servers to balance the server workload while also meeting the specific semantic criteria that may be specified by clients.

Chung, Kwon and Pentland [22] developed an XML-based scalable enterprise systems that they call Manufacturing Integration and Design Automation System (MIDAS) which helps to support collaboration for a network of suppliers during prototype part design. The system that they developed has two parts: i) a formal specification of supported methodologies and tools (or syntactic layer), and ii) an environment that helps designers to construct and execute workflow (or execution layer that provides the detail on how each task should be executed based on the situational criteria). This layered approach is a key factor that helps the scalability of MIDAS. With regard to scalability, the authors point out that it is not only the quantity of parts that is important, but also the fact that the system must be able to adapt to a changing environment from such factors as new products, new processes, and changing market conditions. Although the authors have applied their system to prototyping services, they believe that the process is generalizable to a broader range of manufacturing industries.

Graves et al. [31] applied scalable technologies to the electronics industry by developing new methodologies for network-based design and agile manufacturing, an emerging environment among corporate enterprise systems. They developed the *Virtual Design Environment* (VDE), which is a distributed, heterogeneous information architecture with a Virtual Design Module (VDM) that supports design-manufacturing-supplier optimization through the use of intelligent agents and evolutionary computation. Distributed modules are also included for cost and cycle time that reside at the enterprise supplier sites and provide analysis support for the VDM optimization process, enhance scalability of the VDE system, and provide timely advice to designers.

They contributed to the theory of database management systems (DBMS) by using relations to provide a consistent modeling of structures within the database. Updates, insertions, and deletions are possible without the need for global restructuring. The distributed relational framework then forms the basis for developing an integrated approach to network-based decision systems that formally and efficiently link local relational databases into a global decision framework.

In this decision environment, network delays can present a major obstacle to efficient search. Their approach to reducing the network usage in a distributed computational environment belongs to a "class of distributed evolutionary computation that has been proposed in which the problem is solved by multiple cooperating algorithm components that concurrently explore subspaces of the feasible space, and achieve coevolutionary collaborative problem solving by exchanging partial results."

Prabhu and Masin (www.ie.psu.edu/cmei/ses_ws/abstract/prabhu.htm) have been working to build scalable distributed algorithms for a new class of adaptive manufacturing enterprises using distributed agent architecture over the Internet. These algorithms will be embedded in agents geographically distributed throughout the enterprise, with the goal of maintaining responsiveness and effectiveness to enable the next generation of scalable enterprises. The objective of these algorithms is to reconfigure work-in-process inventory levels and production schedules to adapt to changing market demands and supply-chain conditions.

Scalability will be achieved through distributed algorithms that are predictable and computationally efficient; distributed agent architectures that support growth in size and capability; and distributed clusters that provide rapid access to information using cost effective technologies. They intend to develop analytical models to predict the emergent behavior of such systems and stability and convergence properties of their computations. These models will also be used for determining computational complexity, communication requirements, and scalability of the algorithms. Distributed agent architecture will be developed in which clusters of similar agents will be identified based on task decomposition and task similarity.

We recognize that there are also other definitions of scalability, including feasibility regardless of the actual firm size. Right now ERP systems are only economically feasible for large firms. However, these large firms need to connect to the small firms that are part of their supply web if they are to achieve true collaboration and coordination. Section 6 will further discuss decentralization and outsourcing of enterprise systems. Additionally, for scalability, this communication must take place in a standardized manner, which we discuss below.

5.2 Standards in ERP

Originally, ERP systems were used only for intra-company processes. Despite the benefits offered by ERP systems, their potential was limited because they lacked the ability to communicate outside of the company—leading to a great deal of lost effort due to re-keying when sharing information with other companies. With the advent of EDI, ERP systems could communicate, albeit in a limited fashion, with business partners who were willing to make the investment to connect their systems. For the first time, business partners could exchange invoices, purchase orders, and other frequently used documents seamlessly without the need for manual data entry. However, with the growth of e-business, companies wanted to be able to communicate and conduct business in real time on their systems. It has become necessary for ERP data to be able to communicate better with other systems to allow business to occur at those rates.

Companies desiring to communicate with customers and suppliers electronically need to agree on a single format for that communication. To this end, the creation and adoption of standards has become an industry in itself.

XML (Extensible Markup Language) is an international standard for representing structured electronic data. The XML standard is a subset of SGML (Standard Generalized Markup Language) developed by the World Wide Web

Consortium (W3C; www.w3.org). In XML, Document Type Definitions, or DTDs, define a custom data structure and set of tags that describe the data contained in the document. XML provides a consistent way to represent the data in a structured, tagged format, and each defined document type provides a consistent way to describe a particular type of document, like an invoice. XML also provides a mechanism for validating data, even data in custom formats, enabling companies that exchange XML documents to efficiently check all incoming and outgoing data to ensure that it is well-formed. XML has emerged as a preferred method of communication between computing systems, especially over the World-Wide Web.

XML differs from EDI in one key respect, however: while EDI standards (primarily ANSI X12 in the United States and ISO's EDIFACT internationally) define the general structure of a document and further stipulate the fields that are used to create a standard set of business documents like invoices and purchase orders, XML provides only a way to structure the documents. The XML standard does not describe, for example, how to represent an invoice.

To provide standardization and a consistent means of communication, several collaborative groups in the computing industry have attempted to define the structure and data elements of common business documents. Two examples are xCBL (XML Common Business Library) and cXML (Commerce XML). More recently, standards organizations have also attempted to create a unified standard. In 2001, OASIS, the XML interoperability consortium, created UBL. In its initial form, UBL incorporated the documents in the xCBL standard.

UBL, xCBL, and cXML all amount to attempts to establish a common business language from industry to industry. By defining the data elements common to generic transaction types (e.g., invoice, purchase order, order response) and creating a set of open, extensible schemas that transcend industry barriers, the proponents of these systems hope to ultimately promote more widespread adoption of inter-industry standards, much as EDI did previously (see Figure 18.5).

In the current landscape, however, multiple standards are actually a hindrance to effective communication between companies. Multiple XML-based standards, two leading EDI standards, and multiple versions of each of those standards all create barriers for communication. To address the problem of multiple standards, several exchanges are offering hubs to manage data. They take whatever is communicated in, whether EDI, XML, or another message standard, and automatically translate it into the standard used by other trading partners. This enables communication or real time data interchange from your ERP system to their CRM, even if you are not using the same standard.

For example, Covisint (www.Covisint.com) was formed by the auto industry as a central hub where original equipment manufacturers (OEMs) and suppliers any size can communicate in a business environment using the same tools and

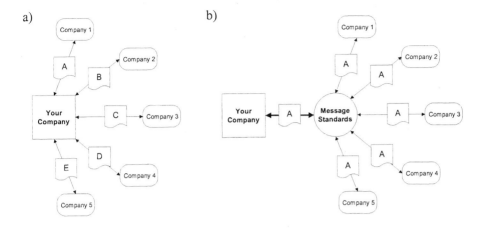

Figure 18.5. Without standards, each trading partner must agree on data and an exchange format for every transaction (left). With standards, trading partners only need to make these decisions for the first transaction. (right)

user interface. Their main standard is the ebXML message transport layer standard, and member companies are financially incented to conform to that standard, but Covisint continues to accept and transform messages with other messaging frameworks (www.covisint.com/about/ pressroom/pr/oagis_standards.shtml). Likewise, the forest products industry has Forest Express (www.ForestExpress.com), which exists as an independent business-to-business solutions provider for "enhancing existing trading relationships and promoting electronic commerce."

Traditional ERP providers are also getting involved in this market. For example, Baan (www.Baan.com) came up with iBaan Portal to provide end-users with access through standard internet browsers, to "all company information sources, whether it is stored in disparate data sources, behind different web addresses or within different applications, folders, documents, or messages." This allows companies to integrate third-party and legacy applications though XML, aiding in the progression to the B2B *e*-commerce environment [35].

While much communication is within an industry, many elements, such as shipping documents, are used to communicate between industries, and it is difficult, if not impossible, to do this electronically if the different industries are using different standards. According to one EDI provider, this has caused many companies (including Sears, Home Depot, Wal-Mart, Sam's Club, Bell Atlantic, Certified Grocers, BMG Music, Ford, and GM) to force business partners to be EDI capable (www.xebecdata.com/ forced.html). Standards generally aid in scalability, because they tend to allow interested users the opportunity to participate (with little cost) in *e*-business transactions, regardless of the company size or technical expertise. Since you can use this same standard with every client, it is very efficient. These systems are robust and can handle many users. They can be used to remove bottlenecks in the information infrastructure by providing a centralized, easily accessible, and flexible information model that can be instantly used by any internal or external user or application.

XML has a very low cost of entry, especially when compared to ERP systems. Products for working with XML are generally very inexpensive or free (www.cidx.org/cidx/faq/rationale.html). For this reason, it is easier for smaller-sized enterprises to connect this way.

Two important self-funded, non-profit organizations are RosettaNet (www.rosettanet.org) and CIDX (www.cidx.org). RosettaNet is "a consortium of major Information Technology, Electronic Components, Semiconductor Manufacturing and Telecommunications companies working to create and implement industry-wide, open e-business process standards. These standards form a common e-business language, aligning processes between supply chain partners on a global basis. RosettaNet is a subsidiary of the Uniform Code Council, Inc. (UCC)." Based on XML, RosettaNet e-business process standards help to automate tasks such as collaborative demand forecasting, order management, shipping and receiving, and invoicing and payments. The trade association CIDX (Chemical Industry Data Exchange) runs the Chem e-standards initiative that focuses on networking IT and e-business peers across industries. In June of 2003, SAP integrated RosettaNet and Chem e-standards into their product offering. An additional industry supported standard is ebXML (www.ebXML.org) which "is a modular suite of specifications that enables enterprises of any size and in any geographical location to conduct business over the Internet. Using ebXML, companies now have a standard method to exchange business messages, conduct trading relationships, communicate data in common terms and define and register business processes."

6. Decentralized ERP

As stated earlier, one of the criticisms of current ERP implementations is the fact that they are complex and inflexible. As a result, there has been interest in the development of decentralized strategies for enterprise systems. Decentralized approaches can occur at two levels. In the first approach, decentralized decisions are made within the enterprise by function (or responsibility center [51]). In order for this approach to be successful, the incentives of the decentralized functions must be aligned based on enterprise goals. This has been a very active area of research (examples include [18], [19], [37], [51]). A good survey can be found in [17], and it is discussed in detail in Chapters 2, 4, 9, 13,

15 and 16. From an ERP perspective (as mentioned in Section 5.2), efficient and effective standardization of information is crucial to coordination of decentralized functions. The second approach to decentralized ERP is to link functions beyond the traditional enterprise (i.e. extended supply chains) through formal mechanisms such as intermediaries or through information sharing. In the case of intermediaries, some decision support (such as matching buyers and sellers) is performed by the intermediary. This approach has started to gain some popularity in practice (though with very limited success) and is discussed in Section 6.1. An alternative approach, which has become quite popular, is outsourcing of ERP functions, and is discussed in Section 6.2.

6.1 *e*-Market Intermediaries

In recent years, we have seen a rapid growth of electronic markets (*e*-markets or exchanges) that bring buyers and suppliers together and enable many-to-many transactions. Depending on who owns and operates them, we can divide *e*-markets into two categories [49]: (1) Private *e*-markets that are owned by a company (commonly referred to as a private trading exchange (PTX)) or a group of companies (co-op or consortia exchanges (CTX)) who are either buyer(s) or seller(s) in the market, and (2) independent trading exchanges (ITX) that are owned and operated by an independent entity (i.e., not wholly/partially owned by a buyer or seller in the market). Examples of such *e*-markets are GE Plastics (www.geplastics.com), which is a PTX, Transplace (www.transplace.com), which is a consortium exchange formed by the merger of logistics business units of the six largest publicly-held truckload carriers, and Chemconnect (www.chemconnect.com), an ITX specializing in chemicals and plastics. The benefits of *e*-markets compared to traditional channels include reduced transaction costs and access to a larger base of potential buyers and suppliers.

Although *e*-markets began with great fanfare, they have not lived up, at least so far, to their advanced billing. This is especially true for ITXs. For example, at the start of 2001 there were over 160 independent transportation exchanges but by the end of that same year there were less than half of that number remaining. Two reasons why so many ITXs have failed are privacy concerns and, more importantly, the sellers' concerns about being compared solely on price. PTXs, on the other hand, have gained popularity as more companies strive to streamline their interactions with their supply chain partners. PTXs have the advantage of giving more control to the owner company and enabling information-sharing and other types of collaboration among participants. In order for ITXs and CTXs to be successful, they should offer more than decreased transaction costs or one-stop shopping convenience. All types of exchanges would benefit significantly from collaboration and decision-support tools for the trading process.

In many companies, purchasing is done by multiple functional divisions (or purchasing organizations within the company) that either act independently or have minimal interaction with each other. For example, until very recently purchasing was done locally by managers at each of the Dial Corp.'s sites. It was typical for a buyer at one facility to buy the same raw material as a buyer at another Dial plant from two different suppliers and at different prices. This approach was ineffective in taking advantage of Dial's volume and corporate-wide buying power [63]. Similarly, until 1997, purchasing at Siemens Medical Systems was done locally, where buyers at Siemens' ultrasound, electromedical, computer tomography, magnetic resonance imaging and angiography divisions independently bought the components and materials that their individual plants needed and rarely communicated with each other. There was no pooling of component demand for leveraging purposes [20]. Recently both companies have moved with great success towards centralized procurement, which allows collaboration among internal purchasing units.

Most research on interactions among participants of an enterprise system has been in the context of decentralized versus shared information. When the information is decentralized, studies are primarily on constructing different mechanisms to enable coordination in a two-stage setting and to eliminate inefficiencies stemming from factors such as double marginalization (e.g. Cachon and Zipkin [19] and Jin and Wu [41]). Detailed discussion of these topics may be found in Chapters 2, 4 and 17.

Since new technology has made the sharing of information possible at every stage, collaboration has arisen as a means for extracting hidden benefits of enterprise systems and supply chains. To achieve vertical collaboration relationships, the collaborative planning, forecasting and replenishment (CPFR) concept has been introduced by the Voluntary Inter-industry Commerce Standards (VICS) Association. The first pilot effort was undertaken by Wal-Mart and Warner-Lambert to explore the potential for retailers and manufacturers to collaborate on forecasts (www.cpfr.org). At this point, is being implemented by several major suppliers in network connections with retail partners [53]. In addition, there are many software packages that implement CPFR including packages from i2 Technologies, IPNet Solutions, and Manugistics. Very little research, however, has been done on collaboration under shared information. One example is Aviv [7] who studied the benefits of collaborative forecasting with respect to local forecasting. A detailed discussion of this topic is found in Chapter 10.

Some researchers have studied the problem of matching multiple buyers with multiple suppliers from a resource allocation perspective. Ledyard et. al. [50] test allocation mechanisms in decentralized markets with uncertain resources and indivisible demand. The results indicate that high efficiency could be obtained if coordination is enabled among buyers. A similar problem where multiple buyers and suppliers contract is studied by Kalagnanam and Davenport [44]. Their motivation is electronic markets in the paper and steel industries. In their study, they design the contracting process as a generalized assignment problem. The aim is to maximize the net surplus given the prices that sides are willing to pay. Detailed discussion may be found in Chapter 5.

A key function of e-market intermediaries is to match buyers and suppliers. Buyers often consist of functional divisions responsible for purchasing. One important question is when collaboration through intermediaries is beneficial. In practice, there are several different types of collaboration. For example, one model is when buyer divisions and suppliers trade through traditional sales channels, via one-to-one transactions. In this case, there does not exist any information flow or collaboration among the functional units of a buyer or among multiple buyers. However, advances in information technology and enterprise systems have increased the availability of real-time data. This in turn has led to increased levels of information sharing and collaboration between the divisions (or business units) of a company (e.g. through EDI). Before 1997, each division in Siemens Medical Systems had its own supplier, which significantly deteriorated buying power. Centralization of purchasing has saved 25% on material costs [20]. Chevron Corp. has begun purchasing greater volume of materials from fewer suppliers. It is aiming to cut 5% to 15% from annual expenditures by centralizing the procurement system and thus leveraging the volume buys [62]. In these examples, functional units of a buyer collaborate internally, allowing buyers to achieve economies of scope across multiple items due to consolidation in transportation.

A third case occurs when a third party intermediary model allows the participants to achieve benefits from both economies of scale and scope due to reduced fixed production and transportation costs. The total set of trades among the buyers and suppliers is a *matching*. Note that the connectivity requirements (and hence, related transaction or search costs) between buyers and suppliers decrease as the collaboration level increases.

Griffin et. al [32] discuss the benefits of collaboration along with the market conditions under which these benefits would be realized. They found that when there is tight capacity in the market and when potential benefits from economies of scope are high (i.e. when the fixed cost of transportation is high), the "full collaboration" model performs significantly better than other strategies in terms of total surplus obtained. These benefits are much more significant when benefits from economies of scale are also high (i.e. when both fixed manufacturing and transportation costs in high). Conversely, the extra benefits obtained by full collaboration are relatively low when the capacity is high and the fixed cost factors are low.

They also observed that internal collaboration performs very well provided that potential benefits from economies of scope are high. On the other hand, when potential benefits from economies of scale are high, buyer strategies with a "look-ahead" perform well. These are the strategies that consider potential future trades in the market by other buyers while contracting with a supplier.

It is clear from the analysis that intermediaries will be most beneficial in markets with a high fixed production cost and/or high fixed transportation cost. Process industries such as rubber, plastic, steel and paper are typically characterized by high fixed production costs. High fixed transportation costs can arise in industries that have to either manage their own distribution fleet or establish contracts with carriers that guarantee a minimum capacity in order to ensure adequate service levels (e.g. Ford Motor Company guarantees deliveries to their dealers every three days and hence has a high fixed transportation cost).

We should mention that auctions are another popular approach for matching buyers and sellers. This is discussed in detail in Chapters 5 and 7.

6.2 Outsourced ERP Systems

There has been a tremendous increase in outsourcing of ERP functions such as CRM and e-procurement in recent years. As an example, a recent AMR Research report stated that IT outsourcing made up approximately 15% of Accenture's portfolio in 2001, but is estimated to make up 45% of their portfolio in 2002. Similarly CSC's IT outsourcing accounts for 55% of revenue in 2002, up from 23% in 2001. It has also become the largest component of IBM Global Service's portfolio. One of the reasons for this increase is the maturity of the ERP market. Enterprise systems are installed in most large organizations and hence no longer provide a means to achieve a strategic advantage.

One common way to outsource ERP is through application service providers (ASPs). Typically the applications are run on the server of the provider and accessed through a web browser. The primary advantages of outsourcing ERP through ASPs are the lower cost for application and maintenance compared to in-house development, and the removal of much of the database and management tasks. A survey in *Infoworld* [4] found that early adapters saved 25 to 45 percent on data management alone. There are some significant risks, however. For example, an ASP typically offers a one-size-fits-all approach, which might not always fit the business scenario. In addition, the business is relying on the fact that the ASP will both stay in business and maintain the product features that the business has purchased. Finally, an ASP might (inadvertently or otherwise) disclose sensitive business information of one's company.

Application service providers are also offering applications that can be integrated with an existing ERP system. This is a partially outsourced model and reduces the business's dependency on the ASP. It can also provide some additional flexibility since it allows the firm to choose a mix of applications that fit their business. Finally, this partial outsourcing allows firms to keep

Next Generation ERP Systems: Scalable and Decentralized Paradigms 771

the control over their sensitive data. The key to the success of such a system is the integration of the ASP with the in-house ERP. There has been some development of software that analyzes the system (e.g. Rational Rose www.rational.com/products/rose/index.jsp), maps the applications and discovers missing data links. Unfortunately, these applications require a large amount of in-house expertise [1].

An example of a partially outsourced ERP is shown in Figure 18.6. In this case, existing enterprise systems at the corporate and regional headquarters are integrated with applications provided by ASPs. Coorporate data tends to be stored at multiple locations to reduce the risk of data failure. The custom and outsourced software is integrated over the internet by middleware. Any outsourced application can be accessed (as long as integration issues are resolved).

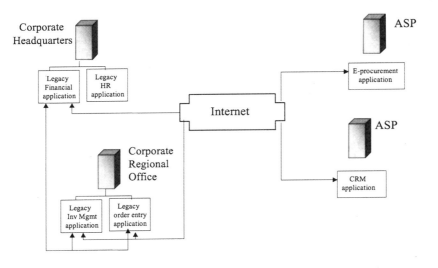

Figure 18.6. Partially outsourcing ERP through ASPs [1]

Another reason for the increase in outsourced ERP is saturation of the market. Most large firms already have some form of enterprise system, so the next target is small to medium sized enterprises (SMEs). This has already occurred to a large extent in Europe where there was over $4B in sales of enterprise systems to medium sized enterprises in 2000. In the US, over 96% of firms have less than 250 employees, and so there is a huge potential for sales.

In the past, SMEs could not consider purchasing ERP software due to the huge cost. Some vendors offer limited implementation methods that have a lower cost but only limited functionality (e.g. SAP Business One, designed for companies of less than 250 employees). Using ASPs, can be a lower priced alternative to achieve the functionality that the SME needs. Other reasons for SMEs outsourcing ERP functions through ASPs is pressure from large suppliers or customers in order to tie into their systems.

7.   Current Enterprise Issues: ERPII and ECM

As we mentioned in section 1, very few traditional ERP vendors actually call themselves ERP vendors anymore. Two of the more currently marketable terms are Enterprise Resource Management (ERM) and ERP II. The key influence for both ideas is the reality that the Internet is moving ERP software beyond just the enterprise. We highlight the similarities and differences between these two 'next-generation' enterprise systems below.

Gartner Group (www.gartner.com) coined the phrase ERP II in 2000, and defines it as a both business strategy and a set of collaborative operational and financial processes existing for use internally and beyond the enterprise [16],[34]. ERP II is accomplished by traditional ERP vendors integrating their legacy ERP systems with new "outward-facing" elements, such as B2B applications. ERP II is based on the idea of collaborative-commerce - recognizing that, as companies begin collaborating, their core ERP systems have to be redefined and extended to embrace the internet, new virtual supply chain models, customer relationship management (CRM) systems and the new business-to-business (B2B) and business-to-consumer (B2C) ecommerce models.

The goal of ERP II is to be able to have zero latency - give customers what they want, when they want it. In Gartner's view, systems with multiple vendor products patched together will develop bottlenecks and break points that will create delays in the fulfillment process, and to be successful will require deep levels of integration across a trading community. They believe users will still want the broad, cross-sector functionality of ERP, but they are also going to want deeply integrated industry-specific functionality.

Similarly, AMR Research (www.amrresearch.com) developed the concept of Enterprise Commerce Management (ECM) to attempt to unify the spectrum of enterprise applications and provide technical standards for software vendors to meet for ECM compliance [6], [57]. In this sense, it is not an application users have to buy or a software market in and of itself, but instead provides a blueprint for companies to maximize the benefit they receive from their enterprise providers. ECM is made up of 5 levels: information services, integration services, interaction services, exchange services, and collaboration services (see Figure 18.7, from their web site).

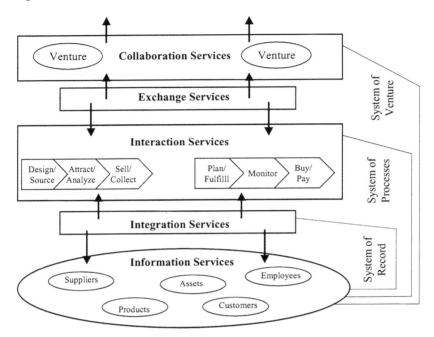

Figure 18.7. Enterprise Commerce Management

ECM attempts to help the many companies that are now functioning with ERP systems across multiple business units, disconnected e-commerce, supply chain, and customer applications, and a range of legacy systems still delivering functionality. The ECM blueprint provides a logical unified view of information, business rules, and presentation through the use of integration. This requires an investment in technology in seven key areas as defined by AMR to ensure business agility, cost-effectiveness (interoperability and the reuse of applications and technology) and competitive advantage (efficiencies with trading partners).

A significant portion of the change from ERP to ERP II or ECM involves adding supply chain management (SCM), supplier relationship management (SRM) and customer relationship management (CRM) practices to the core ERP model. In the more customer-centric environment of late, many companies have turned their focus to the efficiencies of managing their supply chains and the effectiveness of their customer-facing business processes. Companies need to take advantage of the option of connected supply chains to move com-

Pillar Name	Purpose
Private Trading Exchange (PTX)	Focal point for enterprise commerce with trading partners and customers
Data Analytic Model	Effectively measure business ventures
Industry-Standard Application Server	Simplifies inter-application collaboration
Independent Application Integration	Must be an intrinsic element within enterprise software frameworks
Business Process Management	For dynamically managing business ventures
Single Portal Framework	Required for consistency of content and community definition
Integrated Systems Management	Provides end-to-end service management of business processes and ventures

merce into the customer's context, to form ventures on an ad hoc basis, and deliver measurable results. This new way of doing business will require an enterprise to actively manage commerce. Many of the traditional ERP vendors have adopted fast-track programs to add this functionality, while SCM, SRM, and CRM vendors are using their current advantage from already possessing the functionality to sell their services as add-ons to customers' current ERP models.

It is clear that collaboration plays a major role in both systems. While a benefit of traditional ERP is to provide manufacturers a way to consolidate business applications within a common framework, today's economy requires adding functionality spanning their vertical supply chain to maintain competitiveness. Gartner intends to foster this evolution to a vertical industry focus through ERP II, while AMR Research uses the Private Trading Exchange (PTX).

While ECM and ERP II are similar in many ways, there are at least two key differences. First of all, Gartner doesn't identify a company as an ERP II vendor unless it provides traditional ERP capabilities, which has come under significant criticism [55] since these very ERP vendors have all but abandoned the ERP label as they recast themselves in the role of *e*-business providers. That leaves out vendors such as Siebel Systems, Ariba, and i2 who pioneered the B2B applications that spurred on the development of collaborative commerce. While Gartner recognizes that these vendors' systems could potentially be used as part of an ERP II solution, they are not characterized as ERP II vendors. Meanwhile, ECM recognizes that multi-vendor ECM systems are likely and recognizes the B2B software vendors as part of ECM. In their evaluation system of ECM vendors, vendors are classified as providers of ERP, Global

Process Management (GPM), *e*-business, or Platforms and are then evaluated in whatever categories they belong to. For example, PeopleSoft falls under ERP while Siebel is evaluated under GPM.

A second difference is that while ERP II expects companies to thoroughly integrate outward-facing applications (i.e. CRM, SRM, SCM) into back-office ERP systems, ECM supports an integration architecture that provides a layer of abstraction between individual ERP modules and B2B applications (Figure 18.6 from above). ECM promotes private trading exchanges as a way for companies to integrate with customers and suppliers without the costs and complexity of complete integration.

8. Conclusions

Enterprise systems have been effective in helping to integrate internal operations (though historically at a very high cost). The current focus on ERPII and ECM is also helping to address the tremendous inflexibility of traditional ERP implementations. This focus has also helped to take IT from behind the firewall of the corporation. That said, there are still many issues that need to be addressed for next generation enterprise systems to realize their potential.

One of the key internal issues of enterprise systems is the development of ERP techniques that allow the efficient collection and presentation of information that is specific to the user. These "data-mining" or "business intelligence" techniques have been actively pursued, and several products are available on the marker such as Microsoft's Data Analysis and Cognos' Finance 5. These tools are still rather primitive and this should continue to be one of the most important research areas for next-generation ERP systems.

Another key issue for enterprise systems is to help companies move from an internal focus to an external one. Of foremost importance is the development of technology and standards to integrate companies that are employing different information technology strategies (i.e. inter-enterprise integration). Currently, developing these systems is time consuming and costly, and the result is fairly inflexible. Continued PTX development can help in this regard. However, PTXs will only work if there is a core set of standards/applications. AMR Research has argued that there will ultimately be a few leading software provides that will form the core PTX platform. This will in turn lead to a network effect around this core (i.e. as more people use this core, the more value it has for users). We are quite skeptical that this will occur in the short term. However, even if this does occur, there is a more fundamental issue. If we look at what is currently being traded over PTXs, it tends to be simple commodities such as plastics and chemicals. The big advantage has been for achieving economies of scale for these goods.

Firms, however, seem unwilling to share strategic information that they perceive gives them a competitive advantage. It is easy to argue that firms need to open up and share their critical information in order to realize the gains for collaboration. However, there are strong business reasons for keeping much of the information internal. One problem is the potential for exploitable quasirents from partnerships (e.g. if a large investment were required by a firm in a partnership). In addition, cases where there are information asymmetries between parties can lead to moral hazard and therefore makes contractual monitoring difficult at best. Finally, sharing of strategic information can provide a temporary advantage at the expense of strategic control (see Fine [26] for a detailed discussion). Further work must be done to determine what information can be shared that provides a long-term benefit while allowing a firm to maintain its competitive advantages.

In addition, collaboration through information sharing by itself does not necessarily lead to improvement. For instance, one means of collaborative forecasting is through information sharing. If one party believes that there will be a significant change (e.g. demand spike), this can be shared with all other interested parties so that they can appropriately respond. However, if the forecast is wrong, then instead of one party overreacting, all parties overreact. It is quite possible, therefore, that certain forms of information sharing introduce variability into the system rather than reduce it. There are many research opportunities for developing robust techniques for collaborative information sharing.

Another important issue in the development of enterprise systems is defining the proper boundaries of the system. Many argue that the boundaries should increase without end because of the potential to reduce local optimization of subsystems. However, complexity increases as the boundary widens. At some point the marginal benefit is less than the marginal cost from complexity. Advances in scalability and reductions in the cost of transactions helps in this regard, but there are still boundaries. More importantly, businesses are not simple economic systems of transactions, but are driven by strategies whose end is to maximize shareholder wealth over the long term. As such, there are issues of control by the organizing firm. There are also antitrust issues from the Justice Department that must be considered. While these issues are well studied from the industry perspective in the field of Industrial Organization, there are many opportunities for building decision-based models to aid enterprise systems in boundary definition from engineering and business perspectives.

In order for multiple enterprises to collaborate around an idea (e.g. the notion of a virtual enterprise), decision tools must be designed that help identify and exploit competitive and comparative advantages of partnerships. Parker [57] gives an example of a Father's Day gift developed from the formation of a coalition of a TurtleWax product with a Hallmark card packaged by an outsourced firm and sold through an AutoZone store. While this is certainly a potential partnership, there are many issues that would need to be addressed. How is profit shared between the parties? Who organizes the coordination of products (or is this contracted out)? How is effective advertising performed? Answering these and other related research questions would make inter-enterprise collaboration a more feasible option.

Although the focus of this chapter has been on scalability and decentralization, we believe that many of the ERP functions should remain centralized; particularly internal ones. There will always be the tradeoff between the potential flexibility and reduced transactions costs of decentralization and the potential benefits of system-wide optimization from centralized ones. The tradeoffs will vary by type of business, environmental factors (such as regulations) and customer characteristics. For this reason we believe that there will continue to be a wide variety of ERP offerings. The deciding factor, as always, is how the software aligns with business strategy.

Notes

1. The scores given in parenthesis quantify the importance of each factor. The scale given is 1 for "not important" and 5 for "very important".

References

[1] Acharya, R., "EAI: A Business Perspective", *eAI Journal*, 37-44, April, 2003.

[2] AMR Research,

[3] Apicella, M. "Alternatives to the Traditional ASP Model", *Infoworld*, June 23, 2001.

[4] Apicella, M., "Playing Data Host", *Infoworld*, April 26, 2002.

[5] Associated Press, "AMR Research sees upturn in enterprise software market", June 2, 2003.

[6] Austvold, E. "The Seven Technology Pillars of Enterprise Commerce Management" *AMR Research Note*; www.amrresearch.com/free/0107etsstory1.asp, 2002.

[7] Aviv, Y., "The Effect of Collaborative Forecasting on Supply Chain Performance", *Management Science*, 47, 1326-1343, 2001.

[8] Ball, M.O., Boyson, S., Raschid, L. and V. Sambamurthy, "Scalable Supply Chain Infrastructures: Models and Analysis", in *Proceedings of the 2000 NSF Design & Manufacturing Research Conference*, Vancouver, 2000

[9] Ball, M.O., Boyson, S., Chen, C.-Y., Rabinovich, E., Raschid, L., Sambamurthy, V., Zadorozhny, V., and Z.Y. Zhao, "Scalable Supply Chain Infrastructures and Their Impact on Organizations", Proceedings of *2001 NSF Design, Service and Manufacturing Grantees and Research Conference,* Tampa, FL, 2001.

[10] Ball, M. O., Chen, C.-Y., Chen, M., Raschid, L., and Z.Y. Zhao, "Scalable Supply Chain Infrastructure: System Integration and Embedded Decision Models", Proceedings of *2002 NSF Design, Service and Manufacturing Grantees and Research Conference,* Puerto Rico, 2002.

[11] Banker, R.D., Janakiraman, S.N., Konstans, C. and S.A. Slaughter, "Determinants of ERP Adoption: An Empirical Analysis", woking paper, University of Texas at Dallas, www.utdallas.edu/dept/aim/ Working%20Papers/ 19_Final_ERP.pdf, 2002.

[12] Bingi, P., Sharma, M.K., and J.K. Godia, "Critical Issues Affecting ERP Implementation", *Information Systems Management*, 16, 7-14, 1999.

[13] Blum, A., Sandholm, T. and M. Zinkevich, "Online algorithms for clearing exchanges," working paper, Carnegie Melon, 2002.

[14] Bond, B., Genovese, Y., Miklovic, D., Rayner, N., Wood, N. and B. Zrimsek, "ERP is Dead – Long Live ERP II," Research Note, Gartner Group, October 4, 2000.

[15] Boston Consulting Group, "Getting Value from Enterprise Initiatives: A Survey of Executives", BCG Report, www.gcg.com, March 2000.

[16] Brant, K "Discrete Manufacturing: The Promise of the ERP II Model", 11 February 2002, *Gartner Research Note* #AV-15-4168.

[17] Cachon, G.P., "Supply Chain Coordination with Contracts", *Handbooks in Operations Research and Management Science: Supply Chain Management*, North-Holland, to appear.

[18] Cachon, G.P. and M. Fisher, "Supply Chain Inventory Management and the Value of Shared Information", *Management Science*, **46**, 1032-1048, 2000.

[19] Cachon, G.P. and P.H. Zipkin, "Competitive and Cooperative Inventory Policies in a Two-Stage Supply Chain", *Management Science*, **45**, 936-953, 1998.

[20] Carbone, J., "Siemens Medical Systems Rx: Supplier Leverage and Buyer/Supplier Involvement in Design", *Purchasing.com*, http://www.manufacturing.net/pur/index.asp?layout=articleWebzine &articleID=CA159215, 2001.

[21] Chandler, A.D., Scale and Scope: *The Dynamics of Industrial Organizations*, Harvard University Press, Cambridge, MA, 1990.

[22] Chung, M.J., Kwon, P. and B. Pentland, "Design and manufacturing process management in a netword of distributed suppliers", in *Scalable Enterprise Systems - An Introduction to Recent Advances*, Kluwer Academic Press, in press, 2003.

[23] Davenport, T.A., "Putting the Enterprise into Enterprise Systems", *Harvard Business Review*, **76**, 121-131, 1998.

[24] Eisenberg, R., "Business Process Management: The Next Generation of Software", *eAI Journal*, 28-25, June, 2003.

[25] Fan, M., Stallaert, J. and A.B. Whinston, "The Adoption and Design Methodologies of Component-Based Enterprise Systems", *European Journal of Information Systems*, **9**, 25-35, 2000.

[26] Fine, C., *Clockspeed: Winning Industry Control in the Age of Temporary Advantage*, Perseus Publishing, Boston, 1999.

[27] Fontanella, J., "AMR Research: The overselling of supply chain suites", October 9, http://www.plantservices.com/Web_First/ps.nsf/ ArticleID/DLOZ-53CSH4/, 2001.

[28] Fox, J., "Active Information Models for Data Transformation", e*AI Journal*, 26-30, May 2003.

[29] Gallien, J. and L. Wein, "Design and Analysis of a Smart Market for Industrial Procurement," working paper, MIT, 2002.

[30] Gartner Group, "Gartner CRM Survey of Sales Organizations Shows Partner Relationship Management as Having the Highest Return", April 29, http://www4.gartner.com/5_about/press_releases/ pr29apr2003c.jsp, 2003.

[31] Graves, R.J., Sanderson, A.C., and R. Subbu, "A Scalable e-Engineering Technology for Network-based Concurrent Engineering in Electronics" *ISPE/IEE/IFAC International Conference on CAD/CAM, Robotics & Factories of the Future*, Proceedings, 2002.

[32] Griffin, P., Keskinocak, P. and S. Savaaneril, "The Role of Market Intermediaries for Buyer Collaboration in Supply Chains", working paper, Georgia Institute of Technology, 2002.

[33] Gruser, J.-R., Raschid, L., Zadorozhny, V. and T. Zhan, "Learning Response Time for WebSources using Query Feedback and Application in Query Optimization,." *VLDB Journal, Special Issue on Databases and the Web*, Mendelzon, A. and Atzeni, P., editors, **9**, 18-37, 2000.

[34] Harrington, A.; "Gartner touts the ERP II vision", *VNUNet*, April, http://www.vnunet.com/Analysis/1115981, 2001.

[35] Harwood, S; "Baan makes a comeback: Launches an Array of Enterprise e-Tools", http://www.baan.com/home/bulletins/41840? version=1, January 2001.

[36] Hiquet, B.D. and A.F. Kelly, SAP R/3 Implementation Guide, Macmillan Technical Publishing, New York, 1999.

[37] Hirshleifer, J., "Internal Pricing and Decentralized Decisions", in *Management Controls: New Directions in Basic Research*, McGraw-Hill, New York, 1964.

[38] Holland, C.P. and B. Light, "A Critical Success Factors Model for ERP Implementation", *IEEE Software*, **16**, 30-36, 1999.

[39] Hopp, W.J. and M.L. Spearman, *Factory Physics*, 2nd Ed., Irwin McGraw-Hill, Boston, 2001.

[40] Hsu, C., *Enterprise Integration and Modeling: The Metadatabase Approach*, Kluwer Academic Publishers, Amsterdam, Holland and Boston, Mass, 1996.

[41] Jin, M. and D. Wu, "Supply Chain Contracting in Electronic Markets: Incentives and Coordination Mechanisms", working paper, Lehigh University, 2002.

[42] Juric, M.B., "J2EE as the platform for EAI: supporting open-standard technologies", *JAVA Development Journal*, 7(3), http://www.sys-con.com/java/article.cfm?id=1346, 2002.

[43] Juric, M.B. and I. Rozman, "J2EE for EAI: transactions, security, naming, performance, and Web services", *JAVA Development Journal*, 7(4), http://www.sys-con. com/java/article.cfm?id=1403, 2002.

[44] Kalagnanam, J., Devenport, A.J. and H.S. Lee, "Computational Aspects of Clearing Continuous Call Double Auctions with Assignment Constraints and Indivisible Demand", *Electronic Commerce Journal*, 1, 3-15, 2001.

[45] Kamath, M., Dalal, N.P., Kolarik, W.J., Lau, A.H., Sivaraman, E., Chaugule, A., Choudhury, S.R., Gupta, A. and R. Channahalli, "An Integrated Framework for Process Performance Modeling of Next-Generation Enterprise Systems: Design and Development Issues", in *Proceedings of the 2001 University Synergy Program International Conference*, 2001.

[46] Kamath, M., Dalal, N.P. and R. Chinnanchetty, 2002, "The application of XML-based markup languages in enterprise process modeling," *Industrial Engineering Research Conference*, Miami, CD-ROM Proceedings, 2002.

[47] Kamath, M., Dalal, N.P., Chaugule, A., Sivaraman, E. and W.J. Kolarik, "A review of enterprise modeling techniques," in *Scalable Enterprise Systems - An Introduction to Recent Advances*, Kluwer Academic Press, in press, 2003.

[48] Kekre, S., Mukhopadhyay, T and K. Srinivasa, "Modeling Impacts of Electronic Data Interchange Technology", in *Quantitative Models for Supply Chain Management*, Tayur et al., editors, Kluwer Academic Publishers, Boston, 1999.

[49] Keskinocak, P. and W. Elmaghraby, "Ownership in Digital Marketplaces", *European American Business Journal*, Winter, 71-74, 2000.

[50] Ledyard, J.O., Banks, J.S., and D.P. Porter, "Allocation of Unresponsive and Uncertain Resources", *RAND Journal of Economics*, **20**, 1-25, 1989.

[51] Lee, H.L. and S. Whang, "Decentralized Multi-Echelon Supply Chains: Incentives and Information", *Management Science*, **45**, 633-640, 1999.

[52] Mabert, V., Soni, A., and M.A. Venkataramanan, "Enterprise Resource Planning Survey of US Manufacturing Firms", *Production and Inventory Management Journal*, Spring, 52 – 58, 2000.

[53] McCarthy, J., "Starting a Supply Chain Revolution", InfoWorld, November 11, http://www.infoworld.com/article/02/ 11/01/ 021104 ctcpg_1.html? Template=/storypages/printfriendly.html, 2002

[54] McHugh, J., "Binge and Purge", Business 2.0, www.business2.com/articles/mag/print/0,1643,6580,FF.html, June 2000.

[55] Mello, A. "Battle of the labels: ERP II vs. ECM", *ZDNet*, Oct. 25, http://techupdate.zdnet.com/techupdate/stories/main/0,14179,2812226,00.html, 2001.

[56] O'Toole, A., "A Tectonic Shift for Integration?", *Business Integration Journal*, 17-20, July 2003.

[57] Parker, B., "Enterprise Commerce Management: The Blueprint for the Next Generation of Enterprise Systems" *AMR Research Note*, http://www.amrresearch.com/free/0106emsstory1.asp, 2002.

[58] Porteus, E. and S. Wang, "On Manufacturing/Marketing Incentives", *Management Science*, **37**, 1166-1181, 1991.

[59] Prabhu, V and M. Masin, http://www.ie.psu.edu/cmei/ses_ws/_abstract/prabhu.htm, presented at: *Scalable Enterprise Systems Workshop*, Institute of Industrial Engineers, Manufacturing Division, Dallas, 2001.

[60] Prabhu, V.V., Kumara, S.R.T. and M. Kamath (Editors), *Scalable Enterprise Systems – An Introduction to Recent Advances*, Kluwer Academic Press, in press, 2003.

[61] Presley, A.R., Sarkis, J. and D.H. Liles, "A Soft Systems Methodology Approach for Product and Process Innovation", *IEEE Transactions on Engineering Management*, **47**, 379-392, 2000.

[62] Reilly, C., "Chevron restructures to leverage its buying volumes", *Purchasing.com*, http://www.manufacturing.net/index.asp?layout=articleWebzine&articleid=CA149560, 2001.

[63] Reilly, C., "Specialists Leverage Surfactants, Plastics, Indirect and other Products", *Purchasing.com*, http://www.manufacturing.net /index.asp?layout=articleWebzine&articleid=CA191383, 2002.

[64] Richardson, B., "ERP: Dead or Alive?", *AMR Research Alert*, July 19, 2001.

[65] Spearman, M.L., "The Best of Times, The Worst of Times," International Consortium for Innovative Manufacturing, University of Wisconsin, Milwaukee, 2001.

[66] Vollman, T.E., Berry, W.L., and D.C. Whybark, *Manufacturing Planning and Control Systems*, 3rd Ed., Irwin Press, Burr Ridge, 1992.

[67] Wong, E. and L. Martin-Vega, "Scalable Enterprise Systems," Memorandum to NSF Design and Manufacturing Conference Attendees, February 5, 1999.

David Simchi-Levi, S. David Wu, and Z. Max Shen (Eds.)
Handbook of Quantitative Supply Chain Analysis:
Modeling in the E-Business Era
©2004 Kluwer Academic Publishers

Chapter 19

DELIVERING E-BANKING SERVICES: AN EMERGING INTERNET BUSINESS MODEL AND A CASE STUDY

Andreas C. Soteriou
University of Cyprus
Nicosia, Cyprus
basotir@ucy.ac.cy

Stavros A. Zenios
University of Cyprus
Nicosia, Cyprus
and The Financial Institutions Center, The Wharton School
Philadelphia, USA
zenioss@ucy.ac.cy

1. Introduction

The increasing presence of the Internet in everyone's life is changing the way business is done. The financial services industry is no exception. As of March 2000, 27.5 million individuals were cyberbanking—up from nine million a year before. The euphoria witnessed towards the end of last decade surrounding the use of the internet in service provision, was based primarily on the notion of "infinite scalability," that is the ability to serve increasing numbers of customers at low incremental costs. This notion justified high valuations of internet firms from venture capitalists. *e*-banking[1] within the information-based environment of financial services made infinite scalability appear even more promising compared to other types of *e*-commerce.

The often unrealistically optimistic projections regarding internet use, however, led to the dot.com shakeout that came with the dawn of the new millennium. In the US, at year end 2000 only 19% of the US commercial banks and savings institutions offered *e*-banking services, a number that shrank even further in 2001. According to a recent projection by the office of the Comptroller of the Currency, it is likely that almost half of all financial institutions have no plans to offer *e*-banking, ever.

As the dust settles from the dot.com crush, however, a number of success "survivor" stories make it clear that *e*-banking is here to stay. What is needed are new business models that suit the new business environment. The potential is there. The internet has become a part of everyday life with tens of millions online every day engaging in various internet activities, 50% of which include *e*-commerce. (See a special issue of American Scientist, 2001, and of Communications of the ACM, 2001, for trends and statistics.) According to a recent study by TowerGroup, there are about 16 million *e*-banking accounts in the US, 35% of which belong to eight banks, the top three being Bank of America, Wells Fargo and Wachovia (after acquisition by First Union). E*Trade Bank —perhaps the most successful and profitable *e*-bank today— has close to half a million online accounts, $7.7 billion in deposits, and $13 billion in assets. One of its primary assets, unique for an *e*-bank, includes a large ATM network of 10,000 stations, acquired through the acquisition of Card Capture Services. Only a few weeks after the terrorist attacks of September 11, 2001, Bank of America (BofA), proceeded aggressively with major *e*-banking programs. It launched an account aggregation program, where a user can, with a single sign-on, have no-fee access to all his or her BofA accounts as well as those at other institutions. The service, called My Bank of America, was rolled out nationwide in 2002[2]. On-line banking is rapidly adopted in Europe (with more than 20 million internet users logged onto finance and personal banking sites in May 2001 alone) with the trailblazer being Scandinavia where almost 60% of the online population in Sweden were visiting finance sites in May 2001 (Carter, 2001).

Over the last few years a number of internet business models have appeared in the literature, addressing various aspects of *e*-commerce. With very few exceptions, these do not target the idiosyncrasies of *e*-banking. This chapter builds upon the previous literature and presents an internet business model that focuses on *e*-banking and can provide the foundation for successful *e*-banking strategies. More specifically, we first review some of the changes of the world of financial services in the internet era that are relevant for the proliferation of *e*-banking services. We examine both supply and demand forces that shape the drive for internet-based financial services, paying particular attention to Italy—the country on which we later on base our case study. Based on the above, we next present a general business model specific to *e*-banking. Finally, we present as a case study a system for personal financial planning which has been deployed by some Italian banks and analyze some of its components from the point of view of the general business model.

Table 19.1. Traded financial assets by Italian households during 1997–2002 in billions of ITL. (Source: ISVAP, the board of regulators for Italian insurers.)

	1997	1998	1999	2000	2001	2002
Household total	944.853	1427.999	1781.996	2124.102	2488.154	2877.773
% of household's assets	23.6	31.4	34.6	38.3	41.9	44.8
Mutual funds	368.432	720.823	920.304	1077.360	1237.964	1386.519
Asset Management	375.465	542.205	673.500	781.300	880.450	956.970
Life and general insurance	165.000	202.300	257.400	329.600	433.400	574.000

2. The changing landscape of demand for financial services

The decline of the welfare state has created an increasing level of awareness by individuals that the well-being—current and future—of themselves and their families is more in their own hands and less in the hands of the state. This development has created a breed of consumers that demand anytime-anywhere delivery of quality financial services. At the same time we have witnessed an increased sophistication of the consumers in terms of the financial products they buy and the channels of service delivery they use. These trends are universal among developed economies, from the advanced and traditionally liberal economies of North America, to the increasingly deregulated economies of the European Union and pre-accession States, and the post-Communist countries.

The numbers are telling: In the 1980's almost 40% of the U.S. consumer financial assets were in Bank deposits. By 1996 bank deposits accounted for less than 20% of consumers' financial assets with mutual funds and insurance/pension funds absorbing the difference (Harker and Zenios, 2000, ch. 1). Similar trends are observed in Italy. The traded financial assets of Italian households more than doubled in the 5-year period from 1997, and the bulk of the increase was absorbed by mutual funds and asset management; see Table 19.1.

The increase in traded financial assets comes with increased diversification of the Italian household portfolio, similar to the one witnessed in the U.S. a decade earlier. Figure 19.1 shows a strong growth of mutual funds and equity shares at the expense of liquid assets and bonds. Today one third of the total revenues of the Italian banking industry is originated by asset management services.

These statistics reveal the *outcome* of a changing behavior on the part of consumers. What are the changing characteristics of the consumers, however, that bring about this new pattern of investment? The annual *Household Savings Outlook* (2001) provides important insights. First, the traditional distinction between *delegation* of asset management to a pension fund Board or an insurance firm directors by the majority of consumers, and *autonomy* in the

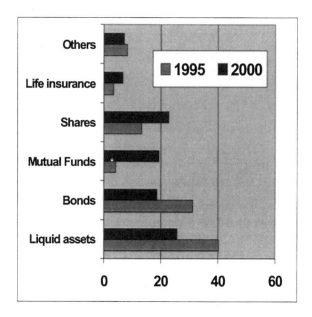

Figure 19.1. The evolution of Italian household portfolios

management of assets by wealthy investors, no longer appears to be valid. Both attitudes are present in the behaviorial patterns of private savers.

Second, the trend in behaviorial profiles is towards higher levels of autonomy, and there is an increased propensity towards innovative instruments as manifested in the data of Figure 19.1. The group of Italian households classified as "innovators" grew steadily from 6.7% in 1991 to 22.6% by 2001. Each percentage point increase added a further 200,000 households to this category. Today this segment numbers 4.3 million Italian households. Households in this category adopt a very professional approach to questions of finance. They are able—or at least they feel so—to manage their financial affairs, and they rely on integrated channels for doing so, using on-line information and conducting business by phone.

Third, an analysis of the influence of quantitative variables on the savings habits of households shows that awareness of financial indicators, and in particular the performance of managed asset returns, is influencing household behavior. Investors in older age groups are more aware of such indicators than the younger generations. The survey also reveals that the trend towards increased diversification of assets under management will continue unabated during the next three years. The investors' favorites are insurance and portfolio management. However, the survey was conducted just prior to the stalling of the world-wide bull markets so the projection of a continued favor towards portfolio management can be questioned.

The changes in attitudes towards innovative products came with a change in attitudes towards delivery channels. Data from a survey of households in the U.S. (Kennickel and Kwast, 1997) show that modern consumers also demand access to more than one delivery channel. While a personal visit to a bank branch remains the predominant way of doing business, a significant percentage of U.S. households use alternative channels such as phone, electronic transfer, ATM, PC-banking etc; see Figure 19.2. Italian households follow this trend, although with a delay of some years. In 2000 only 16% of households could recognize on-line brands. Today this number has grown to 56%. Brand recognition has been followed by use of the new channels as illustrated in Figure 19.3.

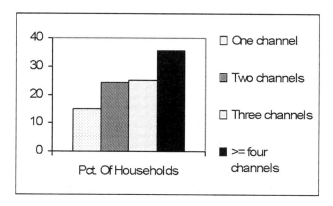

Figure 19.2. Percentage of U.S. households using alternative delivery channels. (Data from Kennickell and Kwast, 1997.)

3. The changing landscape of supply of financial services

Painting the changing landscape of the supply of financial services is a daunting task, better left to collections of works by scholars and practition-

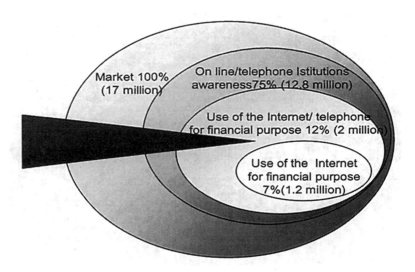

Figure 19.3. The use of new channels—on line and telephone—by Italian households. (Data from *Household Savings Outlook, March 2001*, a joint publication of Prometeia Calcolo and Eurisko.)

ers such as those found in Harker and Zenios (2000). We focus here on key changes that are relevant to the provision of asset allocation support to individuals through the World Wide Web.

Technology and the internet are gaining a growing role in the world of finance and investing. It is impossible to enumerate all those companies operating on the internet offering research, advise, brokerage operations and other important financial data, just *one click away*. Browsing the web can lead anyone, anytime, to security prices, company and market news, retirement plan consultants. Web sites are designed in a way that even financial novices are able to make decisions on which mutual fund to purchase, whether it is convenient to surrender their life insurance, or place an order to sell or buy a given stock. The *web-investor* is afforded the autonomy of deciding what is important and what is not, much as the institutional investor has been doing for years. The value added by the internet consists of spreading financial information, besides offering the possibility to interact immediately on the basis of the news just downloaded.

The market for direct distribution of financial products through the web is, however, a niche market. It is relatively modest in terms of the actual shares traded through the web compared to traditional channels. For instance, it is

E-Banking Services

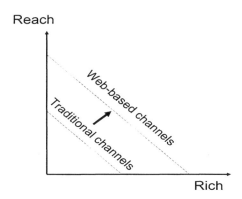

Figure 19.4. Rich communication channels and the reach of clients for a given fixed cost

estimated that in Italy only 500,000 investors out of 12 million potential users rely on the web for trading. This is consistent with another significant change on the supply side of financial services: the changeover from product sales, pure and simple, to the active management of customers' financial planning expectations and needs. In this respect the internet has considerable potential as a facilitator. It is one more channel to be used in the management of existing relationships established with customers through the traditional channels of banks, agencies and advisors. Financial institutions endeavor to provide multi-channel support, keeping up with the trends in Figure 19.2.

The shift towards multichannel distribution is due to both pull and push forces. The pull is coming from the changing demands of consumers and especially the younger generations. The push is coming from the suppliers of financial services who find in the web a rich medium that allows them to reach a wider client base. For a given fixed cost an inverse relation exists between the richness of a channel and its reach to clients. In providing financial services the web considerably reduces costs since no logistics are involved in the delivery of financial products. The frontier of *reach vs rich* is pushed out; see Figure 19.4.

The web is not used only to reach customers, it is also a valuable tool for the support of financial advisors. These are the internal clients of a financial service provider. Web-based services support the financial advisors so they can better serve their clients. This leads not only to customer loyalty towards the advisor, but also creates disincentives for advisors in switching firms (Roth and Jackson 1995). In addition, web-based services provide an alternative channel of communication between the firm and the customer, so that the customer-firm relationship is not under a monopoly control by the network of financial advisors.

The last points should not be underestimated. Indeed, they are key considerations in the adoption of any system. The advisors are one of the most valuable assets for a financial service provider. The firm needs to serve them well, but also to loosen their tight grip on the clients. Broker Stephen Sawtelle made front-page news in *The Wall Street Journal Europe* (August 29, 2001) when he left Wadell & Reed and a clash followed for control of his 2,800 clients. Mr Sawtelle was eventually allowed to keep 2,600 of his clients, while an arbitration panel ruled that the firm must pay $27.6 million in damages to its ex-broker. Mr Sawtelle's departure could not have be avoided even if Wadell & Reed had supported him by a web-based system—he was fired for "personality conflicts." However, the 2,800 clients would have been more autonomous in managing their assets directly, and the ensuing battle for their control would have been less disruptive.

4. An Emerging *e*-Banking Model

The basic elements of customer experience found in traditional *e*-commerce are illustrated in Figure 19.5. They include the following: i) *Navigation*, or the ability to access and move around the site, ii) *Information*, that is providing enough information (depth and breadth) to help customers make a purchase decision, iii) *Support*, i.e. providing customer support regarding various aspects of the product or service and be able to answer questions promptly, and iv) *Logistics*, that is, handling, packaging, and delivering the physical goods or service to the customer and arranging for payment.

An orchestrated execution of all four elements is expected to have a positive impact on customer loyalty and, in turn, on long term growth and profitability (Heskett et al., 1997). Each of the aforementioned elements can also affect each other. For example, if no adequate attention is paid on making the site easy to use (navigation), the result maybe an increase of the number of customers acquiring customer support. Information handling can greatly reduce both customer support and logistics costs. In turn, well executed logistics may further result in reduced customer support costs. Interestingly enough, even today most companies choose to build their competitive priorities around the first two elements: navigation and information. Unfortunately for them, these are also the easiest to imitate, thus eliminating the potential for building a lasting competitive advantage.

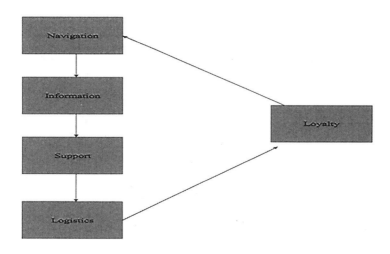

Figure 19.5. A traditional e-commerce framework

An *e*-banking model that builds on the traditional *e*-commerce dimensions is outlined in Figure 19.2. Navigation is still an important element in *e*-banking. An abundance of articles in the popular literature suggest elements to keep in mind when designing a web site (Wen et al., 2001). Customer oriented cyberscape design and various issues of *e*-quality pertaining to the tangible aspects of the site are the focus of many on-going research efforts, mostly from the marketing area (i.e. Parasuraman and Grewal, 2000).

Also important in *e*-banking is the interrelated dimension of information handling. Customizing, for example, the complex financial services requires continuously updated, relevant and complete information available to the right customers. The potential to truly create and manage a customized experience in the customer interface is further discussed by Wind (2001), who presents the concept of *customerization* in *e*-banking services. The concept of customerization moves beyond simply recognizing customers and setting up web pages with account information. Customerization creates a true customized experi-

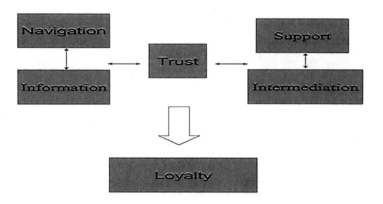

Figure 19.6. An e-banking framework

ence that integrates a number of elements including, but not limited to, personalized messages and custom-made or bundled products (Bakos and Brynjolfsson, 1999) with the customers' lives.

The wealth of information available through electronic transactions also creates a different type of strategic advantage. With *e*-transactions one can track down, among others, the data flow, amount, and source of transactions, which can in turn create a body of knowledge regarding customers and their needs. Most banks utilize such information today to further segment their customers, identify needs and build customer loyalty. Only last year, Fidelity Investments identified and directly contacted 250,000 "serial-callers" prompting them to use their web site and/or their automated system. Quite often such information is used to identify the revenue-generating customers. At First Bank in Baltimore, for example, moneymakers are given the option to click on a web icon to chat with a customer service representative. The rest of the customers do not even know about the existence of cush an option.

Although many believe that a successful site will result in minimal customer support requests, this has not been the experience of many successful internet firms (i.e. Everdream.com, Amazon.com, LandsEnd.com etc.), which have recognized the importance of customer support within *e*-service. As a result, such firms place increasing attention on their support function. According to a 2001 study by Gomez Inc., 84% of *e*-banking customers actively use and pursue customer support. Given the increasingly complex financial environment customer support—which in the case of financial services also includes financial advising—becomes crucial and can provide a major source of competitive advantage. The application we present in the following Section, is an internet based system adopted by italian banks, providing support for personal financial planning. This system also analyzes the risk of the portfolio in terms that are intuitive for a layperson and monitors its performance in achieving the target goals. As such, it provides additional support towards customer advising.

Given the primary absence of a tangible product in financial services, logistics is no longer a key element of the framework. It is well accepted that the dimension of logistics can be a major source of competitive advantage. Amazon, for example, the leading *e*-commerce in sales with nearly $3 billion of sales a year already, turns over its inventory—held in centralized warehouses—much faster than bricks-and-mortar retailers do. In *e*-banking, however, the dimension of logistics is replaced with financial intermediation and the management of risk (see also Zenios and Holmer, 1995). Logistics in financial services is not so much about the physical transfer of products, as is the transfer of risks. The integration with traditional and/or existing models for financial intermediation provides an additional challenge in *e*-banking, although little literature exists on the topic. Saatcioglou et al. (2001) present an interesting approach for developing proprietary indices—that focus on individual customer needs—and using the internet to provide customers with proprietary financial instruments on these instruments. These include a tool for selecting customized optimal portfolios and a bundle trading mechanism (see also Fan et al., 1998) to help establish and rebalance portfolios as needed.

The most important perhaps element in the customer experience in *e*-banking, is trust, which is gained by an orchestrated delivery of all the aforementioned elements of the framework. Surprisingly, despite all the advertised and promising features of *e*-banking being convenient 24 hours x 7 days a week, reliable and at least as secure with traditional banking, consumer demand is still not there. In a recent study on *e*-banking from Gomez Inc. perceived security was still among the primary reasons for "disusers." Electronic money transactions can be undermined by latent frauds and illegal access to private information. Issues of data confidentiality and privacy become crucial, especially in a risk-dominated cyberspace where transactions are conducted at a distance with limited contact with the service provider and uncertain knowledge of the outcome (Reicheld and Schefter, 2000).

Trust is, of course, a more general concept relating to the ability of the firm to deliver its promise (Berry, 1995). Thus, a well executed experience involving all the elements of navigation, information, support and financial advising and efficient financial intermediation, will result to higher levels of trust, as perceived by the customer.

The interrelationships among loyalty and trust have been well explored within the marketing and services management literatures (i.e. Berry, 1995; Berry and Parasuraman, 1991) and are now becoming the focus of numerous studies within the management information systems (i.e. Ba and Pavlou, 2002) and *e*-commerce literatures (Urban et al., 2000). Trust in turn leads to higher levels of customer loyalty. Trust is thought to be the cornerstone for successful and lasting relationships, as it determines the customers' future behavior and loyalty towards the business (Berry and Parasuraman, 1991).

Reicheld and Schefter (2000) also report that *e*-customers are less price sensitive as originally thought and more sensitive to trust. Consider for example, the Vanguard Group, the fastest growing mutual fund company over the past decade, with more than $ 500 billion in assets currently under management, having spent more than $ 100 million into the development of its website. It is not uncommon to see certain "high flier" funds flagged, bearing a word of caution from the CEO Jack Brennan, warning that recent returns may not be sustainable in the future. This is in contrast with most competitors who aggressively promote the returns of such "hot" funds. Despite the difficulty of accessing its site (need specialized sophisticated encryption technology software and a password that will only be mailed) Vanguard has gone overboard not to jeopardize the trust among the customer base on which they focus.

The primary antecedent of trust is customer loyalty, which has well been discussed in the literature as a driver of long term profitability and growth (Reicheld, 1996; Heskett et al., 1997). Today, the exploration of the construct in *e*-banking, including its drivers and antecedents, and its various sub-dimensions (i.e. "attitudinal" and "behavioral" loyalty) is at best in its infancy. Despite, however, the dearth of relevant literature specific to *e*-banking, *e*-loyalty is considered by both academics and practitioners as the most important driver of long term growth and profitability (Reichheld and Schefter, 2000).

5. The design of a web-based personal asset allocation system

In the Section we present a brief description of a web based personal financial system and discuss some of its elements as they relate to the *e*-banking model presented in the previous Section. The system was designed by Consiglio et al. (2002) and has been adopted by several Italian institutions. We also briefly discuss some key characteristics of a successful business plan for the implementation of the system.

Individuals have a variety of financial goals for which they must plan. A house, a car or other tangibles must be purchased, children education financed, retirement planned, and health care and other insurance covered. All of these requirements face the typical family, but they appear with varying intensity at different stages of one's life-cycle. The young parents are concerned mostly about children education, newlyweds for purchasing a home, and middle-age business executives for their retirement. Some personal asset allocation systems (e.g., the HOME Account Advisor^{TM}, see Maranas et al.,1997) advocate an integrative approach to financial planning taking into account all of the above targets. Others (e.g., FinancialEngines.com of Sharpe) focus on a single problem, namely that of retirement planning.

The system of Personal Financial Tools (PFT) we discuss provides support for each one of the goals facing a typical family by segmenting the family's planning problem into distinct sub-goals. The user specifies the financial planning problem by indicating the time horizon of the project, the target goal, and the current asset availability. This information is sufficient in calculating the target return that the individual expects. The system of PFT will then assist the user in structuring an asset allocation consistent with this target return and the appetite towards risk revealed by answering an on-line questionnaire.

For each user-specified goal PFT provides three interactive modules:

Personal asset allocation: This determines the *strategic* asset allocation decisions. Users are asked to specify their targets, their planning horizon, and the availability of funds. The users must also reveal their attitude towards risk. A scenario optimization model (see Consiglio et al., 2002; Consiglio et al., 2001) specifies an asset allocation plan that meets the target using the available endowment, and which is consistent with the user's risk profile.

Personal rating: This provides a data warehouse of financial indicators and ratings of mutual funds to assist users in *tactical* asset location. The Personal Rating tool provides a menu of mutual funds offered for sale by the institution, that are consistent with the broad asset categories in the client's strategic plan. The menu comes with ratings of the funds and other information relating to their past performance. The multitude of mutual funds available to investors makes the tool invaluable in assisting users. (For instance, the *Investment Funds* brochure of UBS offers more than 270 funds.)

Personal risk analyzer: This measures the portfolio risk and monitors the portfolio performance in achieving the target goals.

These three tools form part of an integrated interactive system that allows users to carry out game-of-life simulations, addressing both strategic and tactical

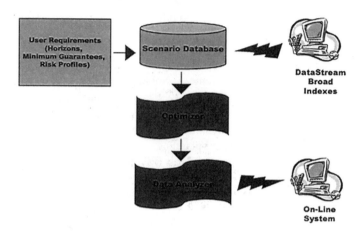

Figure 19.7. The off-line system: The optimization model is run every month for several combinations of risk profiles, horizons, and target portfolio growth rates. (Source: Consiglio et al., 2002.)

issues. The personal risk analyzer provides a control module to ensure that the developed strategy and its execution will meet the targets.

5.1 The integrated decision support system

The system is designed as a combination of an off-line module that runs the optimization, and an online module for product customization. The off-line module exploits the fact that large population segments are homogeneous, so that one could optimize for a range of planning horizons, financial targets, and risk preferences. Customization of a plan for an individual is then extrapolated from the pool of optimized plans.

The scenario optimization model is run off-line and the results are stored in a *solution database*, see Figure 19.7. The online system (Figure 19.8) interacts with the user and, for a given risk profile, horizon and final goal, interpolates the optimal portfolio from the available solutions in the database.

The online system is interfaced through a set of *web pages*. The user's inquiry is first analyzed by an *expert system* that maps the risk profile to the proper risk aversion parameter, and the minimum growth rate is calculated. These data are passed on to the *interpolation module* that consults the off-line

E-Banking Services

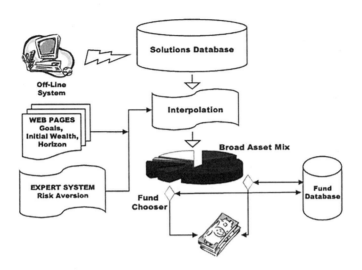

Figure 19.8. The online system: The user interacts through the web pages and the specific enquiry is mapped on to the solution database and matched with specific products from the fund database. (Source: Consiglio et al., 2002.)

system through the database of solutions, and determines the strategic asset allocation which is closer to the user requirement. The broad asset allocation is then mapped to a set of mutual funds which can be purchased by the investor. A *fund chooser* takes care of this task by showing to the user a set of mutual funds, offered for sale by the institution, from the broad asset classes selected by the optimizer. A database of available funds is maintained by each institution providing the online service, and it reflects the institution's product portfolio.

5.2 Business plan for the deployment of the system

The development of the appropriate engine and the conceptualization of the web-based service was a significant step in building the system. However, its eventual success was critically dependent on the business plan. We attribute the successful adoption to two key characteristics of the business plan:

1 Concentrate the development to an *Application Service Provider*. Prometeia Calcolo[3] stuff, working closely with client institutions and the team

of academic consultants, designed a turn-key system relying on the off-line optimization model, with custom-made on-line system to support the idiosyncratic needs of each institution.

The off-line system constitutes the generic engine box and it is identical for all applications. The input data and the user interface were customized. Input data concern primarily the types of products offered for sale, and this is already part of the business strategy of the institution. Similarly, some institutions may have views on market trends and may wish to convey those to their clients in the specification of scenarios. More often than not, however, they rely on market expectations from other sources. The information required on the part of the consumer is also custom-made for each application, driven by the market segment at which the business is aiming. Similarly the requirement for the user-interface hinge upon the core business of the sellers and is well-understood by their marketing departments.

No particular expertise was expected on the part of the client institutions either in financial engineering or in web-based services. However, the clients' performance specifications were adhered too. Much as a new car buyer can be satisfied with a particular automobile without knowledge of the complex electronic controls running the engine under the hood, so the clients of Prometeia were satisfied with the services provided by the web-based system and the Personal Financial Tools, without being aware of the advanced technology behind the user-interface.

2 Adopt a b2b4c view of the web-based system: *Business-to-Business-for-Consumers*. In the provision of financial services there are two types of business involved. The product originators, such as investment Banks, and the distributors, such as retail banks, financial advisors and brokers. The system developed enhanced the integrations of the process from product origination to distribution to the consumer. Distributors have direct access to several product originators if they so choose, and they can even use the system to originate their own products. A product originator can also reach financial advisors of multiple distributors.

Not only b2b4c integrates better the existing service chain—originator-distributor-consumer—but also creates alternative processes at no extra cost (e.g., originator-consumer, distributor-multiple originators). These alternative processes may cater to different market segments. It is still too early in the use of the system to determine whether if any of these new processes become dominant.

In our business plan we took the view that the developed system is one more channel of delivery of services to be added to existing channels. This is consistent with the behavior of consumers who rely on more than one delivery channels as highlighted in Figure 19.2. Both distributors and originators could achieve the same results with traditional delivery channels, but the rich web-based channel enhances their reach; recall Figure 19.4. Perhaps this enhancement is better understood in observing that product originators can now reach directly the consumer without the need for an elaborate network of sales intermediaries.

The system resulting from the combination of technology and business plans adds value at two levels. At a *basic level* it supports the provision of personal asset allocation advise to consumers. At an *advanced level* it supports the complex needs of financial advisors in serving their clients, designing customized portfolios, and dealing with the product originators. Depending on the business plan of each client institution the system would add value at either one, or at both levels.

Four banks are now using the system and have benefited from its adoption. The benefits vary, depending on each bank's niche market. For instance, www.Comdirect.com advertises itself as "Europe's leading online broker". It was created in 1995 as a direct banking subsidiary of Commerzbank AG to offer clients a complete range of direct brokerage services. Within five years in the business, ComDirect became one of Europe's leading online brokers with the most heavily frequented financial website and over 631,000 clients (over 595,000 of them direct brokerage clients; data as of June 30th, 2001). ComDirect takes pride in offering a whole range of professional information and analysis tools to help clients in their direct trading. Clients order via the Internet on the basis of solid information provided either by ComDirect or by other information providers. In the first trimester of 2000 more than 8 million orders were placed through 60 million visits to the company's site. ComDirect offers services trough .com subsidiaries operating in Germany, France, the UK, Austria and Italy each specializing to some extend to the conditions of the local market and complying with local regulations. The personal financial tools were first deployed by the Italian subsidiary of the company, and are supporting a client base of 10,000 during the first year if its operation.

A second user is a subsidiary of one of the oldest and largest banks in Italy going back to the Italian renaissance in the 1400's. The subsidiary has a network of 1,500 financial advisors that support tens of thousands of clients in planning their personal investments. This Bank uses the web-based system at the basic level, adding one more channel of delivery of services to their clients. The Bank also uses the system for advanced support to the financial advisors enhancing their access to product originators. The use of the system for both basic and advanced support poses a challenge in ensuring that consistency is preserved between the recommendations of the advisors and the recommendations of the system. In part this issue is resolved by careful design of the web-based system specifications, so that its recommendations are consistent with the general recommendations offered by the advisors. The Bank management also encourages the use of the system by the advisors in order to create awareness of new products or market outlooks, and to pinpoint any inconsistencies before they are noticed by the consumer and undermine its confidence in the advise he or she receives.

The third user is a new virtual Bank set up recently by one of the major Banks in Northern Italy with the widest country-wide reach. The web-based system has been adopted as part of their core-business and is an integral part of the services offered by this virtual bank. The Bank is completely independent from the Northern Italian bank and is, in a sense, competitor to the parent company. The view adopted by the new bank is that virtual banking should not focus only on the electronic provision of traditional banking services — deposits, credit cards, account payments etc.— but must offer new services in a seamless environment. The web-based personal asset allocation system provides a service not offered by most retail Banks which integrates naturally with the core business of electronic banking. The Bank uses the web-based system at the basic level offering one more service to its clients. In doing so it must be noted that it competes with the parent company.

The fourth user is a private Bank serving the needs of a small client base of high net-wealth individuals. The system is used for advanced support to financial advisors thus enhancing their access to product originators. For their sophisticated client base it is important that financial advisors have advanced support to design custom-made products as the client needs evolve and market conditions change. The system also enhances the quality of the working environment for the financial advisors and enhances the already excellent retention rates that this institution enjoys, and its clients expect.

5.3 Bringing it altogether: Lessons from the implementation of the system

The basic premise of the emerging *e*-banking model presented in Section 4, is that a successful implementation of any *e*-banking system can only be guaranteed by an orchestrated integration of all the dimensions of Figure 19.6. The way management handles information, navigation, support and financial intermediation affect perceptions of trust, which in turn is a primary driver for customer loyalty in *e*-banking services. For the case of the personal financial planning tool we presented, issues of navigation and information become crucial especially when considering the on-line module system, which parallels a traditional *e*-banking series of web pages. The system of personal financial tools developed does not impose the information requirements that an integrative approach to financial planning does, resulting in a system that is easier to navigate. High information requirements by the user would have a negative impact on the system, since that would require a certain level of sophistication from the user, as discussed above. Also crucial for its success is the availability of relevant, and complete information regarding the database of available funds and financial indicators maintained at each institution. Clearly, the way this information is handled can greatly impact perceptions of trust. This by itself is, however, not enough as the availability of *current* information must also be communicated to the user.

In fact, the first effort to deploy the personal financial planning tool on the web came through the internet service providers (ISPs). A number of ISPs had originally expressed a keen interest in the on-line system. However, the system would add value to the service provider only as one more web site advertised through the portals. The value to the general client base of the service providers was falling short of the full potential of the system, which is geared towards the innovator household. Indeed, working with internet service providers we could have provided only superior navigation and information capabilities, and those would be the easiest to imitate. This approach did not go far, as it simply overemphasizes the first two dimensions of the model, namely those of navigation and information. In addition, it is likely that ISPs are not very familiar with financial services, which would create a number of problems regarding the dimensions of support and intermediation, resulting in subsequent decrease of perceptions of trust.

Presentations to senior management of financial service providers were met with eventual acceptance. The development of operational systems was carried out over a period of around 6 months. A team of two from Prometeia staff designed jointly with the clients the navigation requirements. A team of six at Prometeia handled the back-office operations—data management, software development, on-line and off-line system customization and testing. Finally, at each customer's site a team of web-integrators would design and implement the web site and integrate it with the back office systems, in order to provide known products of the bank.

This approach ensured a balanced attention not just to issues of navigation and information alone, but also to those of support and intermediation. In our case, the off-line system responsible for the optimization and the generation of the solution database, as described in the previous Section, provides the risk management mechanism that the system depends upon. The b2b4c approach adopted further strengthens the support function, as existing service chains are now better integrated and better financial advising is available to the consumer. This integration of the four key dimensions of Figure 19.6 is crucial towards trust and loyalty building.

The acceptance of the system was achieved after several obstacles were first overcome. We mention them, in no particular order and without wrapping them with attempts of deeper understanding: The deployment of the system would require significant organizational resources; It would conflict with the role of the financial advisors which is to sell, but not to offer any advise; It would hamper the efficiency of financial advisors who would be spending much more time with each client custom-designing a portfolio, than selling from off-the-shelf products; it will undermine the firm's contacts with clients. These objections were eventually overcome, addressed or side-stepped either due to the features of the technology, or the commitments of the client's top management.

6. Concluding Remarks

Although some scepticism still exists regarding the future of e-banking, most recent studies reveal a growth of both internet usage and e-banking adoption. This growth, even though not as explosive as some might have expected at the turn of the milennium is there, and so is the potential of e-banking. It is now well accepted that e-banking can indeed lower operating costs and pass these savings onto consumers. A number of other advantages including faster service, mass customization through product bundling etc., have been reported in the literature.

In this chapter we present a conceptual model that can help managers improve their e-banking performance. According to a recent study by Gomez Inc., consumers are only moderately satisfied with e-banking. What we need are models that take into consideration the idiosyncrasies of e-banking and help improve customer loyalty. Although attention to each of the elements is important, our model suggests that it is the orchestrated delivery of the various elements of the framework that can indeed create competitive advantage.

Notes

1. The term "e-banking" for the purposes of this article encompasses all those banking services provided over the internet either by traditional brick-and-mortar banks or by e-banks (virtual or direct banks).

2. The My Bank of America service is available to customers whose primary accounts are located outside the states of Washington, Oregon, or Idaho.

3. Prometeia Calcolo S.r.l. is a limited liability company established in 1981 to carry out economic research and analysis, and provides consulting services to major industrial companies, insurance companies and banks, as well as government agencies in Italy.

References

American Scientist, Special Issue on Internet Usage in the New Millenium, 89(5).

Ba, S. and P. Pavlou. Evidence on the effect of trust building technology in electronic markets: Price premiums and buyer behavior. *MIS Quarterly*, 26(3).

Bakos, Y. and E. K. Brynjolfsson. Bundling information goods: Pricing, profits and efficiency. *Management Science*, 45(12):1613–1630, 1999.

Berry, L. L. *On Great Service*. The Free Press, New York, 1995.

Berry, L. L. and A. Parasuraman. *Marketing Services: Competing Through Quality*. The Free Press, New York, 1991.

Carter, B. Scandinavia leads in spread of e-banking Across Europe. *Jupiter MMXI, NEW MEDIA AGE*, 16, 2001.

Communications of the ACM, Special Issue on E-Banking, 44(6).

Consiglio, A., F. Cocco and S.A. Zenios, www.Personal_Asset_Allocation, *Interfaces*, (to appear).

Consiglio, A., F. Cocco, and S.A. Zenios, "The Value of Integrative Risk Management for Insurance Products with Guarantees", *Journal of Risk Finance*, 6–16, Spring 2001.

Fan, M., J. Stallaert, and A. B. Whinston. A web-based financial trading system. *IEEE Computer*, 64–70, 1998.

Harker, P.T. and S.A. Zenios, editors. *Performance of Financial Institutions: Efficiency, Innovation, Regulations*. Cambridge University Press, Cambridge, England, 2000.

Heskett, J.L., W. E. Sasser Jr. and L. A. Schlesinger. *The Service Profit Chain*. The Free Press, New York, New York, 1997.

Holmer, M. R. and S. A. Zenios. The productivity of financial intermediation and the technology of financial product management. *Operations Research*, 43(6):970–983, 1995.

Household Savings Outlook. Eurisko and Prometeia Calcolo Report, Italy, 2001.

Kennickell, A. B. and M.L. Kwast. Who uses electronic banking? Results from the 1995 survey of consumer finances. Working paper, Division of Research and Statistics, Board of the Governors of the Federal Reserve System, Washington, DC, 1997.

Maranas, C. D., I.P. Androulakis, C.A. Floudas, A.J. Berger, and J.M. Mulvey. Solving long-term financial planning problems via global optimization. *Journal of Economic Dynamics and Control*, 21(8-9):1405–1426, 1997.

Parasuraman, A. and D. Grewal. The impact of technology on the quality-value-loyalty chain: a research agenda. *Journal of the Academy of Marketing Science*, 28(1):168–175, 2000.

Reichheld, F. *The Loyalty Effect: The Hidden Force Behind Growth, Profits, and Lasting Value*. Harvard Business School Press, Boston, MA, 1996.

Reichheld, F. F. and P. Schefter. E-loyalty. *Harvard Business Review*, 78(4):105–112, 2000.

Roth, A. V. and W. E. Jackson III. Strategic determinants of service quality and performance: Evidence from the banking industry. *Management Science*, 41:1720–1733, 1995.

Saatcioglou, K. Stallaert, J. and Whinston, A. B. Design of a financial portal. *Communications of the ACM*, 44(6):33–38, 2001.

Urban, L., F. Sultan, and W. J. Quals. Placing trust at the center of your intrenet strategy. *Sloan Management Review*, 42(1):39–48, 2000.

Wen, J. H. G. H. Chen, G. H. Hwang. A dynamic model of customer loyalty to sustain competitive advantage on the web. *European Management Journal*, 20(3):299–309, 2001.

Wind, Y. The challenge of "customerization" in financial services. *Communications of the ACM*, 44(6):39–44, 2001.

Index

σ-value, 48
Account aggregation, 784
Account management, 316, 318
Adaptive conjoint analysis, 300
Adaptive inventory policies, 399
Advanced planning and scheduling, 754
Adverse selection, 69, 71–72, 110
After-sales service, 317
Agency theory, 4
Agents are substitutes condition, 263
Alliance, 607, 637, 649, 655
Allocation of production, 685
Allocative efficiency, 159, 200
Alternating-offer bargaining, 69, 80, 90
Amazon.com, 302, 312, 314, 336, 378, 593, 596–597, 646, 657–658, 792
American Marketing Association, 295
Analysis, 535
APO, 454–455, 748
Apple Computer, 592
Apple Remote Desktop, 318
Application service providers, 770
APS, 754
Arbitration, 70, 106
ARIMA, 409
ARMA, 417, 425
Armani, 583
Ascending auction, 160, 187, 191, 264–266, 272, 276–279, 281–282
Asked price, 73, 75–76, 78, 87, 91, 103–105, 108
ASPs, 770–772
Assembly, 41, 315, 451–454, 473, 478, 712
Asset allocation, 788, 794–795, 797, 799–800
Assortment planning, 590
Asymmetric, 14
Asymmetric information, 71, 90, 93, 109
Asymmetric Information, 592, 652
Asymptotic analysis, 535
Auction, 2, 4–5, 13–14, 58, 67, 143–148, 154–155, 157, 159–160, 162–164, 166, 168–169, 172–173, 175–176, 181–198, 200–201, 203–204, 214–224, 227–228, 230–232, 238, 240–242, 247–251, 254–255, 257–258, 262, 264–266, 276–279, 281, 284–285, 307, 588,
597, 645, 648–653, 655
  ascending-price, 145, 147, 159–160, 162, 181–183, 187, 191
  combinatorial auction, 160, 162, 165, 168–169, 175, 181, 183–189, 193–195, 201, 204
  combinatorial exchange, 175, 180, 201
  double auction, 172, 197–198, 204
  dutch, 143, 204
  english, 143, 160, 162–163, 166, 194, 204
  exchange, 144, 154, 173–175, 180, 198–199, 201, 447, 588, 630, 683–684, 727, 753, 763–767, 773–774
  first-price, 193–194, 204
  forward, 144, 169, 189, 194–195, 348, 381, 470, 583, 585, 596, 647–648, 660, 665, 667
  multi-unit, 145, 160, 166, 185, 189, 191, 194, 197
  multiattribute, 144, 155, 160, 164–165, 176, 178, 192–193, 197
  procurement, 143–144, 147, 169–170, 172, 182, 186, 192, 194–197, 375, 449, 453, 557, 596, 643–645, 648, 651–652, 655, 674, 768–770
  reverse, 144, 166, 169, 176, 194–196, 200, 357, 573
  second-price, 163, 193, 204, 655
Auto-regressive process, 400, 408, 414, 425
Available to promise (ATP), 449, 761
  Advanced ATP, 448, 454–455
  ATP quantity, 450, 455, 462, 470
  Conventional ATP, 449–451, 454, 461
Average earliness, 491
Average lateness, 491
Average tardiness, 491
B2B, 4–5, 213–217, 219, 231–233, 238–240, 453, 456–457, 559, 765, 772, 774–775
B2B e-marketplaces, 214
Backlog, 45, 236, 297, 308, 672, 717, 722, 725–727, 732
Balanced shop, 506
Bargaining power, 80–81, 88–90, 107–108
Bargaining Theory, 67
Bargaining with incomplete information, 70, 76, 101, 106
Barnes and Noble, 301, 319

Barriers to entry, 636
Base-stock policy, 411, 416, 418–419, 421, 430, 433, 439, 672
Bass Ale, 559
Batching, 456, 460, 474, 477–479
Bayesian framework, 412
Bayesian games, 52, 57–58
Belief, 53
Bertrand competition, 369
Best response functions, 17–18, 22, 29, 32, 35, 621
Best response mapping, 21, 26
Bid price, 73, 75–76, 78, 87, 89, 91, 96, 102–105
Bidding-ring, 656
Bidding decision support, 227
Bidding language, 146, 164–166, 168–169, 175, 179, 181, 184, 186, 199, 201
Bids, 145, 159–161, 163–168, 170–171, 173–176, 178, 182, 186–194, 196–199, 204, 645, 652, 656
  all-or-nothing, 145, 164, 166, 171, 194
  bundle, 145, 158, 165, 168, 171, 173–176, 185–189, 195–196, 204, 380, 587, 793
  divisible, 164, 166, 174, 178
Biform games, 48, 50–52
Bilateral bargaining, 69–70, 76–78, 80, 84, 87, 89, 93, 97, 99–101, 106–108, 110
Bilateral monopoly, 567, 569
Bill of materials, 645
Bill of Materials, 645
Bluelight.com, 597
Boeing, 304
Bottleneck shop, 506
Boundary equilibria, 34
BPR, 750
Branch and bound, 259, 716–717, 733
Branch and cut, 259
Break-even volume, 689
Breakdown probability, 79, 81, 86, 89
Bricks and clicks, 646, 660, 675
Bricks and mortar, 657
Brouwer, 21
Budget balanceness, 103
Bullwhip effect, 400
Bundles, 5, 222, 241, 247–248, 258–259, 264–265, 269, 274, 277, 285
Burning money, 55
Business process reengineering, 750
Business rules, 164, 169–170, 178, 194–195
ByAllAccounts, 319
Calculus of variations, 46
Call center, 310, 317, 455, 754
Cannibalization, 575, 643
Capacity, 6, 16, 38, 45, 48, 52–55, 57–58, 131, 168, 196–197, 213–214, 217, 221–222, 225, 228, 231–232, 234, 239, 241, 260, 285–286, 296, 308–309, 317–318, 339, 344, 346–348, 357–360, 362–363, 366–367, 370–372, 379–380, 394, 400–401, 424, 442, 449, 452, 454, 458, 461, 464–465, 467–468, 473–474, 478, 573, 578, 613, 645, 650–652, 684, 712–716, 719, 722–727, 729, 733, 749, 754, 769–770
Capacity investment, 337, 347, 366
Catalog companies, 608
Catalog sales, 558
Ceiling fans, 696
Central control, 567
Central server job shop, 506
Centralized decision making, 716
Channel conflict, 559–562, 575, 579, 582, 586, 589, 591–597, 665, 667
Channel coordination, 368, 568, 573, 585, 611, 614, 647, 653, 675
Channel power, 595, 610–611, 613, 615, 628, 637–638
Channels, 337–338, 340, 452, 457, 461, 463, 559–561, 565–566, 571–573, 575–576, 578, 580–582, 584–593, 595–597, 643, 646–647, 657–660, 662–666, 668–671, 673, 675, 712, 767, 769, 785–789, 798–799
Characteristic form, 49, 52
Characteristic function, 49–50
Choice-boards, 302–303, 306
Circuit City, 7, 594
Clearance, 350–352, 355–356, 362
Clinique, 559
Closed-loop, 46, 48
Coalition, 16, 49–50, 650, 655–656, 776
Coalition of players, 49
Cobb-Douglas function, 537
Collaboration, 6, 397, 402, 442, 747, 757–759, 761, 763, 767–770, 773–774, 776–777
Collaborative forecasting, 397–398, 401–402, 406–407, 413–414, 418, 429, 433, 441, 768, 776
Collusion, 222–224, 232, 241, 265, 276, 656
Column generation, 260
Combinations, 5, 165, 177, 247–248, 258, 287, 374, 408, 440, 458, 467, 692, 730, 796
Combinatorial allocation problem, 158, 185
Combinatorial auction problem, 185, 189
Combinatorial auctions, 4–5, 160, 165, 169, 181, 183, 186, 188–189, 193–194, 217, 241, 248, 257, 277, 284
Common due date, 730–731
Common due date problem, 500
Common due dates, 490
Common prior, 250–251, 255
Common prior assumption, 251
Compact strategy space, 22
Compaq, 49, 378, 559, 577–578, 593
Competition, 7, 17, 21, 29, 45, 48, 50, 59, 118, 149, 216, 224, 226–227, 237–241, 254, 264–265, 282, 284, 298, 302, 306–307, 309, 311, 319,

Index

337, 340, 367, 369–373, 377–380, 559–560, 565, 568–570, 572–575, 578–579, 582, 584, 586, 591–592, 596, 616, 647, 652, 656–657, 660, 662, 665, 667, 711–712, 750
Competitive analysis, 494
Competitive equilibrium, 146–147, 157–161, 187–188, 190–191, 194, 197, 371, 610
Competitive ratio, 494
Competitive scenario, 667
Complementarity, 24, 626
Complete information, 4, 16, 40, 69–70, 73–77, 85, 89–90, 102, 106, 108, 110, 148, 154, 157, 182, 241, 372, 374, 395, 791
Complexity, 145
Complexity costs, 682, 685, 687, 689, 692, 701
Complexity of communication, 278
Complexity
　communication, 149, 179–182, 185, 201, 571, 596, 681, 747, 752, 755, 763–765, 789–790
　implementation, 146, 153, 159–160, 164, 179–181, 183–184, 190, 201, 455–456, 458–459, 463, 477, 558, 585, 728, 747–748, 750–751, 755–756, 772, 794, 800
　strategic, 179–180, 185, 194, 201, 355, 372, 378–380, 456, 557, 574, 626, 648–650, 665, 681, 684–685, 687, 699, 701–702, 712–713, 724–725, 727, 750, 755, 758, 770, 776, 792, 795, 797
　valuation, 147–148, 150–151, 153–154, 156–157, 160, 169, 176, 179, 181–183, 186, 193–194, 201, 204, 577–578, 683
Computational complexity, 171, 174, 179, 716
Computational studies, 690–691, 693–694
Computers, 299, 310, 336, 350, 378, 683, 696–697, 712
Concavity, 21, 24, 39, 259, 344, 364, 617, 621–622
Conceptual framework, 8, 681–682, 685–686, 701
Conjoint analysis, 299–301, 304, 320
Consignment agreements, 608
Consortia, 645, 648, 655, 767
　Formation of, 645, 648, 655
Constant demand, 714, 717, 719–720, 722, 733
Constant lead time, 486
Consumer behavior, 588, 646, 659
Consumption-effort problem, 122–123
Contracts, 42, 54, 56–57, 118, 121, 137, 170, 214, 216–217, 233, 248, 285–286, 312–313, 316, 456, 569, 585, 595, 611, 613–614, 619, 629, 631, 635, 637, 645, 648, 770
Contradiction, 28–29, 341
Control chart, 526
Cooperative, 3, 13, 48–49, 58, 559, 669–671
Cooperative game, 14, 16, 48–51
Cooperative sub-game, 51–52
Coordinating Traditional and Internet Procurement, 648

Coordination, 2, 5–7, 52, 118, 303, 313–314, 319, 336–337, 340, 350, 364, 367–369, 373, 380, 393–394, 401, 459, 562, 565–571, 573–574, 585, 595, 609, 611, 613–614, 616, 629, 631, 635–637, 643–644, 646, 648, 652–653, 671, 674–675, 681, 711, 718–719, 726, 728, 732–733, 747, 758, 763, 767–768, 777
Core, 48–52, 59, 144, 213, 262–263, 307, 318, 397, 406, 590, 656, 750–752, 772–773, 775, 798, 800
Core of the game, 49
Correspondence, 18
Cost analysis, 435
Cost assessment, 398, 435, 440
Cost minimization, 489
Costco, 593
Costless signal, 55
Cournot competition, 45, 369, 579, 586
Critical fractile, 654, 670
Critical ratio, 497
CRM, 748, 754–755, 757, 764, 770, 772–775
Customer acquisition, 7, 314, 568, 609–610, 612–618, 622, 624–633, 636–637
Customer acquisition costs, 611–612, 625, 630, 632, 637
Customer classes, 545
Customer complaints, 319, 597
Customer feedback, 318
Customer relationship management, 752, 772–773
Customer relationships, 456
　Partnership-based, 453
　Transaction-based, 452
Customer support, 316, 558, 790, 792–793
Cutting planes, 259
Cycle time, 536
Decentralized marketing-operations framework, 536
Decision power, 7, 610, 613
Decision support, 227, 233, 394, 454–455, 684, 748, 751, 767, 796
Decreasing best responses, 26
Delayed Strategies
　Delayed pricing, 346
　Delayed production, 346, 359
Delegated control, 118, 133, 138
Delivery, 296
Dell, 49, 137, 302, 310, 315, 336, 377–378, 452–453, 577, 587, 590–591, 593, 655, 751
Dell Computer Corporation, 452
　Two-stage Order promising, 452
Demand, 5–6, 9, 17, 22, 41–45, 50, 52–58, 130–131, 145, 157–158, 161, 166, 170, 172, 174, 176, 178, 185, 187, 190–191, 195, 214, 216–218, 221–222, 225–227, 232–240, 242, 265, 272, 274, 285–286, 296–297, 301, 307–308, 310–311, 315, 336–360, 362–375, 377–380, 394–419, 421–425, 427–433, 436,

439–443, 448, 454, 459, 461–470, 558, 561, 564, 566–569, 573, 575, 577–584, 586–589, 591–592, 597, 608–610, 612–617, 624–628, 631–632, 647–651, 653, 655, 658, 660, 664–665, 667–673, 681–682, 687, 689, 714–715, 717, 719–720, 722–723, 725–728, 731–733, 748–750, 757, 768, 776, 784–785, 787, 793
Demand forecast, 449, 460, 471, 758, 766
Demand Functional Form, 338
   Additive, 153, 165, 171, 174, 176–177, 186, 193, 339, 341–342, 345, 349, 375
Demand learning, 340, 373–374, 380
Demand management, 441
Demand models, 6, 33, 399, 402, 404, 414
Demand
   deterministic, 567
   Stochastic, 338–341, 343, 345–348, 352, 357, 359–360, 368, 374, 379, 467–468, 567, 581, 584, 669, 717, 724, 732
Dependent variables, 685
Depreciation, 697
Deterioration, 350
Determinant, 34, 37, 617–618
Deterministic demand, 338–339, 341, 344, 347, 350–351, 357, 567, 575, 614
Deterministic model, 45, 686, 732
Diagonal dominance, 33
Differential games, 14, 45–46, 48
Differential topology, 34
Direct-sell, 712
Direct-to-Customer (DTC), 336
Direct channel, 561, 575–581, 584, 586, 589–592, 594, 596, 665–666, 668–669, 671
Direct exchange, 68, 73, 87
Direct mechanism, 150, 181–183, 201, 249–250, 264, 278
Direct revelation mechanism, 149, 249–251
Direct sales, 558–560, 571, 573–577, 579, 581, 593, 597
Discount rate, 132, 136, 224, 299
Discretionary sales, 346
Disintermediation, 58, 68, 558
Disruptive technology, 299
Distribution, 7–8, 17, 22, 41, 50, 53, 134, 148, 151, 154, 156, 250, 254, 278–279, 285, 295–296, 298–299, 301, 312, 315–316, 318, 320, 336, 340, 343–345, 349, 354–355, 362–364, 367–369, 372, 374–375, 394, 396, 400, 402, 404–407, 409, 412, 415–416, 424–425, 427–428, 433, 436–438, 448–449, 455–456, 458–461, 463, 467, 470–472, 557–559, 561–567, 571–572, 578, 580, 582, 584, 587, 593–595, 607, 610–611, 613–615, 622, 624, 627, 631–632, 643–644, 646–648, 650, 656, 661, 665, 668–670, 672, 674, 682–683, 711–715, 719–721, 724–729, 731–734, 770, 788–789, 798
Distribution centers (DCs), 686
Distribution intensity, 596
Distribution management, 458, 596
Distribution networks, 571, 681, 684
Distribution strategy, 571, 573, 576, 579, 586, 594–596
DKNY, 583
Documentation, 316
Double auction, 102–103, 105
Double marginalization, 567, 575, 580, 629, 647, 666, 669, 671, 675
Drop-shipping, 7, 72, 314, 320, 568, 593, 608–615, 619, 622, 627–629, 631–633, 635–638
Drop-shipping channel, 7, 610–611
Dual-channel, 579–580
Due date, 485
Due date management, 486
Due date tightness, 498
Dynamic, 14, 503
Dynamic demand, 8, 722, 733
Dynamic games, 21, 40, 45
Dynamic pricing, 307, 336–337, 346–347, 351–352, 369, 373–374, 376–378, 588
Dynamic program, 411
Dynamic programming, 124–125, 127, 129, 132–133, 204, 415, 730–731, 40, 46, 716
E-banking, 9, 791–792
E-business models, 712
E-commerce, 296, 298–299, 306, 308, 310–313, 315, 319–321, 358, 376–377, 379, 452–453, 457, 559, 609, 611, 673, 772, 791
E-tailer, 610, 630, 656–658
EAI, 753–754
Earliest due date, 497
Earliest finish time (EFT), 497
Earliest operation, 497
Earliness, 730–731
EBay, 143, 588
EBusiness, 558
Echelon-based inventory systems, 432
Echelon-based Inventory systems, 432
Echelon-based inventory systems, 437
Echelons, 232, 238, 432, 460, 567, 596, 614
ECM, 748, 772–775
Econometric techniques, 686
Economic efficiency, 250
Economic Order Quantity (EOQ), 338, 347, 350
Economies of scale, 222, 608, 682, 685, 687, 692, 701, 769–770, 776
Edsel, 301
Efficiency, 558
Efficient auctions, 254
Eigenvalue, 32
Elasticity, 214, 233–235
Electronic commerce, 393, 656, 765
Electronic markets, 146, 194, 201, 377, 767

Index

Empty core, 49, 52
English auction, 160, 162–163, 194, 216, 264–265
Enriched information structures, 425
Enterprise application integration, 753
Enterprise commerce management, 772–773
Enterprise Resource Management, 772
Equilibrium, 13, 16–17, 19–21, 25–26, 28–30, 34–40, 42–44, 47, 51, 54–55, 58–59, 128–129, 146–149, 151–154, 157–161, 163, 169, 179–183, 185–188, 190–191, 193–194, 197–198, 200, 202, 204, 241–242, 251, 265, 272, 276, 279, 343, 362, 368–373, 573, 575, 578, 581–582, 584, 610, 620, 622–628, 633, 635–636, 651, 653–654, 661, 666–667
Equilibrium market, 654
Equilibrium
    Bertrand, 369–370, 372–373, 565
    Cournot, 369, 373, 565
    dominant strategy, 147, 151–152, 180, 191, 198, 656
    Nash, 151, 153, 157, 161, 180, 187–188, 191, 193, 198, 200, 202, 370, 372, 620
    Quasi Bertrand, 373
    Quasi Cournot, 373
Equivalent, 254
ERM, 772
ERP II, 748, 772–775
Estee Lauder, 559
Event tree, 83
Evolutionary design, 306
Ex post efficiency, 76
Excess inventory, 366, 645, 654, 671
Exchanges, 137, 172–173, 175, 197, 199, 201, 295, 752, 764, 767
Existence, 20
Existence of equilibrium, 21, 26, 369
Existence of the equilibrium, 620, 622
Experience goods, 597
Experimental design, 694
Experimental mechanism design, 201–202
Explanatory power, 399, 692, 696
Exposure problem, 274–276, 282, 284
Extended formulation, 188, 194–196, 267–269
Extensible Markup Language, 753, 763
Extensive, 4, 6, 15–16, 20–21, 30, 39, 169, 287, 312, 317, 610, 686
Facility location, 714
Facility networks, 684
FAQ, 318, 766
Feasible, 15, 17, 29, 41–42, 121, 126, 132, 135, 148, 151, 155, 158–160, 162–163, 172, 179–180, 184–185, 189, 195, 250, 252, 254–255, 260, 262, 269, 271–274, 280, 297, 310, 357, 364, 369, 472, 475, 558, 587, 618–619, 722, 762–763, 777
Federal Express, 558

Feedback, 5, 46–48, 145, 182, 186–187, 189, 193–194, 196, 221–227, 241, 301, 316, 318–319, 321, 595, 750, 761
Financial advisors, 790, 798–800, 802
Financial intermediation, 793–794, 800
Financial service provider, 790, 801
Financial services, 2, 8, 783–785, 787, 789, 791, 793, 801
Finite horizon, 8, 348, 714, 720, 722
First-best, 52, 121, 136, 611
First-mover advantage, 85
First come first serve, 497
Fixed costs, 239, 591, 725, 727
Fixed point, 19, 24–26, 29–30, 47
Fixed point theorems, 21
Fixed pricing, 358–359
Flexible strategy, 689
Flow allowance, 500
Flow shop, 506
Flow time, 491
Folk theorem, 42, 59
Forced compliance, 55, 401, 567
Ford, 248, 301, 728, 765, 770
Forecast evolution, 402, 404–407, 410, 414, 419, 435, 442
Forecasting, 2, 6, 237, 296, 313, 374, 394–395, 397–404, 406–407, 413–414, 418, 421–422, 424–429, 431–433, 435, 438–439, 442, 468, 472, 592, 728, 750, 758, 766, 768, 776
Forecasting capabilities, 441
Forward Integration, 585, 665, 667
Framework measures, 687, 696
Franchising, 562–564, 597, 614, 629
Fulfillment, 7, 237, 314, 449, 451, 453–455, 457–458, 474, 560, 568, 580, 585, 607, 609, 611, 636, 643, 647, 668–669, 674, 748, 772
Full Integration, 668
Game theory, 1, 3, 13–14, 46, 48–49, 58
Game Theory, 67
Game theory, 308
Game without time dependence, 42
General tactical production-distribution problems, 734
Global facility networks, 685
Global strategy, 689
Government subsidies, 8, 685, 699
Gross-substitutes condition, 259
Hamiltonian, 47–48
Hessian, 33–34, 37, 39, 617
Heuristic policies, 416, 421
Heuristics, 260, 346, 353, 355, 719, 729–730, 732–734, 754
Hierarchical production planning, 684
History-dependent compensation, 121–122
Home Depot, 285, 559, 765
Horizontal competition, 380, 560, 565, 568, 572–573

Horizontal integration, 753
Hotelling, 573
I2
   Allocated ATP (AATP), 454
   Rapid Optimized Integration for R/3, 454
   Rhythm module, 454
IBM, 49, 318, 378, 400, 480, 559, 756, 770
IFT, 32, 37–39
Immediacy, 71
Implicit collusion, 42
Implicit Function Theorem, 28, 32
Impossibility theory, 91
Imputation, 262
Incentive alignment, 442
Incentive compatibility, 119, 124–125, 127–129, 144, 251, 255, 260, 280
Incentive compatible, 75, 91–93, 98, 110, 149, 655
Incomplete contract, 107
Incomplete information, 14
Increasing best response functions, 26
Increasing Failure Rate, 622
Independent competition, 656
Independent variables, 8, 685–688, 692
Index theory approach, 34
Indirect mechanism, 145, 147, 150, 160, 164, 181–183, 186, 202, 249–250, 264, 278
Individual rationality, 119, 287
Individually rational, 75, 92, 96, 110
Inefficiency, 21, 567, 580
Infinite horizon, 132, 338, 345, 714, 720
Infinitely-repeated games, 42
Information, 13
Information asymmetry, 70–71, 73, 89, 91, 93, 106, 109, 612, 615, 653
Information pump, 300
Information revelation, 57, 185
Information sharing, 6, 297, 395–396, 399, 427–429, 441–442, 767, 769, 776
Informational Intermediary, 71
Innovation, 2, 9, 296, 298, 300, 305–306
Installation-based Inventory systems, 430
Installation-based inventory systems, 430, 436
Installation, 316–317, 437, 443
Integer programming, 5, 183, 260, 266, 683–684
Integration, 7–8, 25, 156, 231, 296, 315, 319, 336, 454, 463, 564, 583–585, 593, 608, 665, 667–668, 671, 712–714, 728, 753–754, 761, 771–775, 793, 800–801
Integration of inbound transportation and production, 714
Integration of production and outbound transportation, 714
Interdependent values, 278–279, 282
Intermediaries, 2–3, 5, 7, 146, 186, 558–562, 565, 567, 569–571, 573, 578–579, 588, 594, 596, 767, 769–770, 799
Intermediary, 67, 70, 73, 87, 109

Internet, 2, 5, 7, 9, 58, 119, 137, 143, 213–214, 219, 238, 240, 296–308, 311–312, 314–315, 317–321, 336, 356, 363, 376–377, 381, 557, 559–560, 565, 575, 577, 581, 583, 586–588, 590, 593–594, 596, 607–610, 613–614, 630, 636, 639, 644, 646–648, 654, 656–658, 672–674, 681, 747–748, 752, 759–760, 762, 765–766, 771–772, 783–784, 788–789, 792–793, 799, 801–803
Internet channel, 643, 646–647, 659–670
Internet commerce, 377
Internet retailing, 610
Internet store, 594, 669–671
Inventory-related risk, 619, 625, 630, 636
Inventory, 4–7, 22, 35, 38, 41–45, 47, 50, 52, 57, 118–119, 127, 129–133, 214–216, 218, 220, 232–233, 236, 238–239, 297, 299, 302, 307–308, 310, 312–316, 320, 336–357, 359–360, 362–370, 373–376, 379, 393–395, 397–400, 402–404, 406, 410–417, 419–422, 427–430, 432–440, 442, 448, 450, 453, 456, 458, 462–466, 468–471, 474, 476, 478, 565, 567–568, 571, 574, 578, 581, 584–585, 588, 590–592, 596, 607–608, 614–618, 620, 622, 624, 626, 628, 637, 645, 649, 653–654, 657, 668–672, 711–714, 717, 719–720, 722–728, 732, 749–750, 762, 793
Inventory control, 358, 462, 480
Inventory control policies, 402, 410, 440
Inventory investment, 348, 396, 607, 610, 636, 728
Inventory management, 17, 320, 374, 394, 399, 404, 407, 410, 413, 416, 421, 428, 432, 442, 449, 657
Inventory position, 403, 412–413, 416, 421, 424, 433
Inventory risk, 609, 612–613, 629, 631–632, 636
Inventory risk allocation, 610–611, 638
Iterative play, 30
Job characteristics, 543
Job shop scheduling, 471
   Due date assignment, 471
Joint forecasting, 71
Joint Job Processing and Finished Job delivery, 715
Joint job processing and finished job delivery, 729
Just-in-time, 71, 396, 453
Kakutani, 21
Kalman filter, 407–410, 415–416, 419, 421–422, 424, 426, 432, 436
Kinexus, 319
Kozmo.com, 313–314
Kramer's rule, 37
Labor assignment, 492
Lagged forecast errors, 438
Lagrange multiplier, 38, 273–274
Lagrangean relaxation, 272, 274, 727
Lanes, 225–226, 248, 258, 260, 264, 284–287
Lattice, 26, 39

Index

Lead-time demand, 398, 416, 419, 424, 432, 435–436
Lead time flexibility, 542
Lead times, 485
Leader, 40–41, 227, 620, 628, 633
Leadtime, 337, 351, 364–365, 369, 380, 472, 749
Leadtime differentiation, 365
Leadtime quoting, 308–311
Leadtimes, 308–310, 313, 365, 471, 749
Least remaining operation, 497
Levers, 682, 685–686, 702
Linear compensation systems, 119
Linear model, 652, 655
Linear regression model, 429
Linear state-space framework, 6, 402, 427
Liz Claiborne, 583
Loan interest rates, 696
Location decisions, 696, 701, 728, 795
Logistics, 182, 186, 217, 241, 248, 260, 284–285, 317, 393–394, 447, 459, 571, 573, 575, 580, 586, 590, 607, 733, 758, 767, 789–790, 793
Logistics auctions, 264, 284–285
Longest processing time, 497
Lost sales, 345, 358, 669–670, 672
Lot sizing, 242, 368, 465, 714–715, 717, 720, 733–734, 749
Loyalty and trust, 794
Macy's, 583, 591, 593
Maintenance, 137, 194, 220, 316–318, 567, 570, 584, 753, 770
Make-to-order, 352, 367, 370–372, 448, 458, 472, 485, 590
Make-to-stock, 352, 448–449, 458, 463, 485, 590
Manufacturer-owned channel, 558–560, 573–575, 582, 597
Manufacturer, 357–358, 365, 368, 457, 470, 473, 475, 478, 560–563, 565–573, 575–586, 590–597, 608, 610, 613, 645–647, 650–654, 657, 665–670, 683, 717–720, 722–724, 731
Manufacturer returns, 567
Manufacturing, 8, 118, 236, 241, 309, 315–317, 335–337, 339, 351, 365, 377, 380, 393, 397, 448, 452, 456, 458, 461, 466, 469, 563, 575, 590, 612–613, 647, 681, 683–684, 724–725, 728–731, 733, 748, 751, 753, 755, 761–762, 766, 769
Manufacturing resources planning, 749
Manugistics, 455, 768
Mapping, 18–19, 22, 26, 29–30, 32–35, 47, 448, 759
Marginal product, 193, 256, 261–263, 270–272, 367
Marginal revenue, 100–101
Markdown pricing, 352, 356
Market clearing, 77, 102–103, 105
Market focus, 8, 682, 687, 689, 692
Market share, 170, 371, 378, 380, 558, 584, 594, 662, 667
Market signals, 427
Marketing-operations interface, 296–299, 305, 308–309, 311, 315, 319–320, 614
Marketing-operations misalignment, 612
Marketing, 5, 16, 45, 48, 219, 230, 233, 241, 295–296, 301–302, 304, 306–307, 311–312, 314, 316–320, 337, 348, 356–357, 359, 367–368, 370, 373–374, 376, 381, 442, 461, 467, 558, 560, 562, 565, 573, 579–580, 588, 596–597, 608–612, 614–616, 625–627, 629–631, 636–637, 639, 660, 662, 682, 752–753, 755, 758, 791, 794, 798
Marketing effort, 574, 583–585, 613–614, 628, 667
Marketing region, 682, 691
Marketplace, 573, 645, 660, 666
Markov-modulated demand, 406, 417
Markov Decision Process, 118, 122, 408
Markov games, 43
Markov Perfect equilibrium, 43
Mass customization, 302, 802
Matching markets, 73, 97
Material requirements planning, 749
Materials, 449, 452, 458, 460–461, 464, 466, 471, 473, 558, 560, 571, 614, 711, 714, 721, 724–726, 749, 754, 768–769
  Component instances, 473, 475
  Incompatibility, 454, 473, 475–476
  Substitution, 454
Mathematical programming, 8
Matrix norms, 32
Mattel, 559
Maxtor Corporation, 453
Mechanism, 143–150, 152–154, 156, 158, 160, 163–164, 179–186, 188, 192–193, 196–198, 200–202, 374, 451, 468, 562, 566, 574–575, 577, 590, 597, 629, 651–652, 656, 750, 764, 793, 801
Mechanism design, 56, 70, 74–76, 92, 98, 144, 146–155, 157, 159, 179–182, 185–186, 201–204, 249–250
Mechanism
  direct, 145–147, 149–150, 153, 160, 181–183, 190, 193–194, 201–202, 378, 452, 455, 460, 558–561, 565, 569–571, 573–581, 584–594, 596–597, 607–608, 645, 647, 665–669, 671, 712, 719, 729, 732, 761, 788, 798–799, 803
  efficient, 146, 148–155, 157, 159–163, 183–184, 186–189, 191–192, 197–198, 201, 204, 336, 344, 346, 354, 376, 380, 574, 579, 585, 590, 645, 754, 762–763, 766–767, 775, 794
  indirect, 145–147, 150, 159–160, 164, 181–183, 186, 201–202, 585, 614, 645, 751
Mechanisms for coordination, 2, 732
Menu of contracts, 56–57
Mercata, 588, 597

Microsoft, 303, 305, 754, 775
Minimize the average due date, 491
Minimizing tardiness, 491
Minimum commitment, 56
Mixed integer programming, 474, 754
Mixed strategy, 16
Model validation, 695
Modeling, 683
Modeling framework, 118, 312, 589, 701
Modified Due Date, 497
Modular design, 302–303
Monotone comparative statics, 38
Motorcycles, 696
MRP, 310, 449, 749, 751, 754
MRPII, 749–751
Multi-channel distribution, 561, 572, 587, 595, 665
Multi-channel retailers, 610
Multi-lateral CPFR agreement, 442
Multi-machine setting, 503
Multi-period game, 41, 44, 622
Multi-period games, 43
Multi-period model, 369, 685
Multi-unit combinatorial auction, 258, 287
Multilateral bargaining, 107
Multilateral trade, 76–77, 97, 100–101, 106
Multinomial logit, 659
Multiple-sourcing, 176
Multiple equilibria, 20, 24, 35–36, 38
Multiple products, 6, 218, 337, 340, 347, 352, 357, 359, 362, 380, 428, 458, 473, 712, 719, 722–723, 725
Multiple time periods, 362, 567, 714, 722–723, 725
Multiplicative, 339, 341–342, 345, 349, 355, 358, 360, 375
Multiplicity of equilibria, 20, 42
Myopic policies, 416–417
Nash Bargaining Solution, 69
Nash equilibrium, 18
Nash Equilibrium, 18
Nash equilibrium, 20, 43, 82, 109, 128–129, 180, 187–188, 191, 193, 198, 200, 202, 272, 370, 372, 610, 620, 626–628, 632, 635, 651
National Shoe Association, 559
NE, 18–22, 25–26, 29–30, 33–35, 40–43, 46–49, 54, 56
NetOp, 318
Netscape, 303, 305
Network design, 2, 8, 684–685, 687
Network dispersion, 8, 682, 688–689, 692
Network structure, 8, 402, 681–687, 689, 695–696, 699, 702
Newsboy problem, 469
Newsvendor formula, 411
Newsvendor problem, 17, 46–47, 340–342, 375, 653–654
Nike, 559, 647, 665
Niketown, 559

Non-cooperative, 18, 48, 370, 372
Non-cooperative game, 3, 25, 49–52, 180
Non-cooperative static games, 14, 16, 58
Non-credible threats, 20
Non-decreasing best responses, 26
Non-empty core, 50, 52
Non-equilibrium outcome, 36
Non-existence of NE, 22, 24
Non-monotone best responses, 26
Non-stationary equilibria, 43
Non-transferable utility, 48
Nordstrom, 583
Normal form, 15–16, 20
NP-hard problem, 473
Nucleous, 48
Numerical studies, 689
NYSE, 102
Offline, 487
Online, 487
Online sales, 578, 594, 657
Open-loop, 46–48
Open-loop strategy, 46
Operation, 488
Operation due dates, 498
Operations incentives, 612
Operations/marketing interfaces, 612
Opportunity cost, 71, 73, 76, 90–93, 95–96, 98–104
Optimal auction design, 98
Optimal control theory, 46
Optimal mechanism, 56–57, 146, 149, 154–156, 201
Oracle, 197, 454–455, 748
 APS modules, 454
 Global ATP Server, 455
Order acceptance, 487
Order promising, 449–450, 452–453, 455, 457, 459–460, 467, 470–471, 474
Orders, 345, 358, 363, 365, 369–370, 447–453, 455–460, 462–463, 467, 469–475, 477–478, 558, 581, 593, 608, 649, 654, 671–673, 712, 733–734, 761, 763–764, 799
 Customer order requests, 449
 Order commitments, 451–452
 Order promising and fulfillment, 451, 453, 455, 461, 474
Outside options, 71, 75, 79–81, 85–86, 88–91, 93, 96, 108–109
Outsourcing, 545
Outsourcing of Inventory Management, 608
Package bidding, 276
Parallel machines, 729–730
Pareto frontier, 21
Pareto inferior, 21
Pareto optimal, 21, 635
Pareto optimal equilibrium, 36
Participation constraint, 119–121, 124–125, 128

Payoffs, 15–17, 20–22, 24–26, 38, 42–43, 46, 49–51, 151, 156, 158, 163
Penalty, 17, 55, 131, 133, 241, 274, 286, 342–344, 353, 371, 403, 410, 416, 435, 439, 476, 585, 614–615, 629, 635, 685, 717, 724–725, 727
Pension funds, 785
Perfect information, 56, 368
Performance-based compensation, 119
Performance criteria, 478
   Customer response time, 457, 478
   Fill rate, 465, 565
   Overall profit, 370, 474, 561, 664
   Tangible profit, 478–479
Performance
   Back-end logistics efficiency, 447
   Efficiency, 148–150, 152–153, 184, 197, 200, 336, 376–377, 447–448, 567, 580, 584, 590, 653–654, 759, 769, 802
   Flexibility, 145, 356, 365, 376, 448, 454, 457, 460, 473, 478, 558, 657, 673, 724–725, 771, 777
   Front-end customer satisfaction, 447
   Reliability, 371, 457, 659
   Responsiveness, 448, 583, 611, 762
Periodic review, 354
Planning, 448–449, 454–455, 458, 461–462, 464, 466, 474, 590, 645, 684, 686, 712, 717, 728, 733, 747–752, 754, 757–758, 768, 784, 789, 793, 795, 799–801
Planning horizon, 133, 234, 236, 410–412, 415, 421, 462, 475–477, 714, 717, 720, 722, 795–796
Planning
   advanced planning and scheduling (APS), 454
   Enterprise resource planning (ERP), 454, 750
Plant focus, 8, 682, 689, 692
Plant location, 682, 684–687, 697, 699, 702
Plants, 682, 686, 689, 715–717, 724–725, 728, 768
Players, 14–18, 20–22, 24, 26, 29–30, 32–43, 45–54, 58–59, 278, 615, 620, 625, 628, 635, 665
Policy coordination, 437
Polynomial fit, 692
Pooling equilibrium, 54–55
Portals, 752–753, 801
Positive externality, 569, 574, 580, 584
Positive information rent, 57
Post-sales service, 298, 316–319
Powerful retailer, 7
Powerful wholesaler, 7, 626, 628, 632
Preemption, 501
Preferences, 144–147, 152–153, 158, 161, 164–165, 180–184, 187, 336, 454, 460, 575, 579, 585, 587–588, 590–592, 608, 664–665, 796
   marginal-decreasing, 168, 190–191
Price-only contract, 611, 629, 632

Price, 143–147, 157, 159–166, 168, 171, 174–177, 181–183, 187–198, 204, 336–356, 358–360, 362–377, 379–380, 457, 468–469, 473, 565, 567–585, 587–589, 591–592, 595, 597, 609, 612–616, 618–619, 626–628, 630–632, 635–637, 645, 648–655, 657–668, 670, 682, 758, 767, 794
Price discrimination, 214–215, 219, 235, 238–240, 242, 367, 377–378, 574, 579, 583, 585, 596
Price elasticity of demand, 537
Price markdown, 341
Price protection, 567
Price schedule, 164, 166–167
Price smoothing, 343–344
Price
   competition, 149, 337, 340, 367, 369–373, 377–380, 559, 565, 568–570, 573–575, 578, 582, 584, 586, 591–592, 596, 616, 647, 652, 656–657, 660, 662, 665, 667, 711–712, 750
   Dispersion of prices, 659
   elasticity, 351, 376, 625–626, 664
   Equilibrium market, 666
Pricing, 2, 4–7, 48, 157, 199, 213–219, 227, 230–233, 236–241, 295, 297, 300, 307–309, 320, 335–337, 340–341, 343–344, 346–348, 350–356, 358–359, 362–370, 372–382, 468, 487, 558, 565, 570–571, 580, 583, 586–588, 591, 596, 619, 643–644, 646, 651, 655–660, 664–665, 668, 674
Pricing decisions, 537
Primal-dual, 147, 160–163, 188, 190–191, 194–196, 201, 717
Principal-agent, 4, 14, 118–119, 121–122, 124, 127–128, 130, 132–133, 137–138, 613
Priority dispatch policy, 494
Private information, 53, 57, 144, 147–149, 156, 192, 197, 249, 278–279, 568
Private trading exchange, 767, 774–775
Procurement cost, 197, 655
Product attributes, 228, 558, 589, 660
Product configuration, 466, 475
Product development, 297
Product Development, 297
Product development, 298–307, 319–320
Product life cycles, 306
Product returns, 316
Product support, 318, 453
Product variety, 296
Production-transportation problems, 714, 733
Production, 5–6, 8, 45, 48, 119, 129–133, 137–138, 194, 196–197, 221–222, 224, 234–236, 239, 242, 295–296, 298–299, 301, 303, 308–310, 312–313, 315–316, 320, 336–340, 343–348, 350–351, 356–360, 362–368, 370–374, 376, 379–381, 394, 397, 400–401, 448–461, 463–467, 469–471, 473–475, 477–478, 562–563, 566, 573, 580–583, 590, 612,

651–653, 670–672, 681–682, 684–685, 687, 689, 692, 695–696, 711–729, 731–734, 749–751, 754, 758–759, 763, 769–770
Production costs, 197, 344, 347, 356, 653, 670, 682, 770
Production environments, 363, 448
  High mix low volume, 448
  Short-term operational environment, 449
Production line, 465, 467, 717, 720, 728
Production strategies
  Assemble-to-order (ATO), 448, 458
  Configure-to-order (CTO), 458
  Make-to-order (MTO), 352, 448, 458
  Make-to-stock (MTS), 352, 448, 458
Profit-to-go, 345
Profit maximization, 489
Profit seeking, 73
Promotion, 343, 351, 356–358, 379, 612, 758
Proper equilibria, 20
Prototyping, 304–306, 762
Proxies, 685, 689
PTX, 767, 774–775
Pull, 451, 459–460, 469–470, 473–474, 477, 480, 789
  First-come-first-served policy, 470
  Pull-based ATP model, 459–460, 469–470, 473–474
  Pull control, 448
  Reserve level, 476
  Shortfall penalty, 476
Pure strategy, 16
Push-Pull, 448–450, 455, 459–460
  Push-pull boundary, 448–450, 460
  Push-pull Framework, 448
Push, 459–464, 467–468, 480, 789
Push
  Demand classes, 359, 459, 461–465, 467–469, 668
  Optimization-based allocation, 463
  Push-based ATP model, 459–462, 464, 468
  Push control, 448
  Rationing policies, 463, 668, 671
QAD, 454–455
  Supply Chain Optimizer, 454
Quality management, 311
Quantity-Flexibility, 629
Quantity discount, 567, 629
Quantity Flexibility, 567
Quantity forcing, 629
Quantity setting, 265
Quasi-concavity, 24, 26, 34, 41
Quasi-definiteness, 34
Quasi-linear valuation, 98
Queue length, 491
Queueing systems, 310
Quick response, 296, 393
Quoting, 365, 471, 473–474, 476, 478, 658

Rapid delivery, 313–314
Rationing, 463, 467, 611, 668, 671, 673
Regional strategy, 689
Regression, 228–229, 374, 430, 564, 686, 692
Relative explanation power, 430
Repair, 194, 297, 316–318, 754
Repeated games, 40, 42–43
Reputational effects, 42
Resellers, 593, 653–654
Reservation Price, 157, 171, 354, 356, 360, 576–577, 579, 582, 655
Reservation utility, 124–125, 127–129
Reserve capacity, 535
Residual bargaining, 50
Resource, 143–145, 147, 157, 186, 359, 380, 449, 452–454, 458–464, 468–469, 471–472, 564, 652, 683, 747–748, 750–751, 768
  Distribution resource, 458
  Factory resource, 711
  Material resource, 469
  Reservation, 157, 171, 354–356, 360, 467–469, 474, 476, 576–577, 579, 582, 650–651, 655, 661–662
  Resource availability, 449, 453, 458, 464, 468, 472
  Resource commonality, 458
Response function, 620–623, 625
Retail channel, 381, 575–577, 581, 585–586, 647, 657, 669, 671
Retail price, 342, 356, 368, 567, 574–578, 582, 584–585, 619, 627, 666–668, 670
Retail sales, 575
Retailer, 344, 354, 362, 367–369, 372, 374, 470, 565, 567–570, 572–578, 581–586, 590–592, 594–597, 607–610, 612–616, 618–638, 646, 657, 659–660, 665–671
  Bricks-and-clicks, 660
  bricks-and-mortar, 558, 565, 586–590, 592–593, 597, 665, 793
  Conventional, 376–377, 449, 462, 472, 565, 573, 580, 609, 631, 637, 658–659, 665
Returns, 319, 380, 451, 567, 593, 611, 614, 629, 631, 635, 672, 684, 754, 786, 794
Revelation principle, 56, 91–92, 149–150, 153–154, 159, 249–251
Revenue management, 336, 351, 379, 468
Revenue maximization, 5, 257
Reverse logistics, 29, 43, 319
Risk-attitude, 119
Risk-averse, 121, 123
Risk-neutral, 123, 137, 242
Risk neutral, 73
Risk pooling, 50, 566, 609, 638
SAA, 274, 277, 282–284
Safety-stock, 419
Safety-stock, 419

Index

Safety-stock, 420–421, 424, 434, 436–437, 439–440
Sales, 7, 44, 214–216, 218–220, 230, 233–234, 236–237, 239, 296–297, 306–307, 311–313, 316–319, 339, 343–347, 351–354, 358–361, 366–369, 372, 374–375, 378, 381, 393, 395–396, 441, 451–452, 455–457, 461–462, 558–560, 565, 567–569, 571, 574–579, 581, 584–585, 588–590, 593–594, 596–597, 610, 612, 619, 647, 653–654, 656–659, 669–670, 672, 686, 724, 728, 750–752, 754–757, 769, 771–772, 789, 793, 799
Sales effort, 45, 573, 580, 586, 591–592, 597
Sales tax, 597, 658
Salvage value, 340, 342, 653
SAP, 454–455, 748, 751–753, 755–756, 766, 772
Scattered strategy, 689
Scheduling, 487
Screening, 52, 56–58
Searching and matching, 71, 87
Second-best, 121, 127, 129, 136
Secondary market, 581–582, 597, 653–654
Self-service support, 318
Sensitivity analysis, 8, 350, 585, 685–687, 696, 702
Separating equilibria, 55
Sequential equilibrium, 20, 70, 90
Sequential moves, 40–41
Service, 4, 6, 8, 22, 45, 119, 133–137, 215–216, 220–221, 225, 248, 281–282, 295, 297, 305–306, 309–313, 316–318, 320–321, 363–364, 370–372, 377, 396, 398, 452–453, 460, 465, 474, 478, 558, 562, 564–565, 569, 571, 573–574, 576–578, 583, 585, 589, 591, 594, 597, 637, 657, 659, 668, 711–712, 724, 754–755, 761, 770, 774, 783–785, 790, 792–793, 797–798, 800–803
Service buffer, 671
Service guarantees, 504
Service leasing, 319
Service Level, 341–342, 359, 363–365, 380, 461–462, 576, 596, 610, 724, 770
Service level constraints, 490
Set-packing problem, 258
Setting auction, 265
Shapley value, 48, 50–51
Shared information, 768
Shipping and handling, 587, 596, 658
Shop utilization, 498
Shortest processing time, 497
Side payment, 579, 653
Signaling, 52–53, 55–58, 564
Signals, 55, 145, 279, 424, 427, 429–430, 432–434, 441, 564
Simulation, 304, 437, 440, 477, 494, 761
Simultaneous ascending auction, 274, 282
Single-sourcing, 145, 176, 192
Single crossing, 35

Single machine, 503
Single product, 130, 340, 362, 461, 463, 572, 614, 668, 671, 715, 717, 719, 721–722, 725–726
Single time period, 587, 715, 722–723
Slack (SLK SLK_o), 497
Slopes of the best response, 33
Smith's rule, 498
Sojourn times, 553
Sony, 593, 657
SPE, 82–87
Spectral radius, 32
SRM, 754, 773–775
Stability, 30, 448, 655, 699, 763
Stable, 33, 362, 409, 418, 423–424, 449, 656, 684, 749
Stable equilibria, 30
Stackelberg competition, 565
Stackelberg equilibrium, 40–41, 46, 48, 623, 625
Stackelberg follower, 40
Stackelberg game, 40–41, 610, 627
Stackelberg leader, 573, 595, 610, 620, 622, 624, 627–628, 632
Static, 14
Static and dynamic flowtime estimates, 554
Stationary Inventory policy, 43
Stationary inventory policy, 43
Stationary Inventory policy, 44
Stationary policies, 43, 621, 625
Statistical process control, 527
Stochastic demand, 339, 341, 343, 345, 348, 352, 357, 359, 369, 379, 567, 584, 732
Stochastic differential games, 45
Stochastic games, 41, 43–45, 58
Strategic decisions, 724, 727
Strategic network structure, 684–685, 687
Strategies, 6–7, 9, 14–17, 20–22, 24, 26, 36, 38–39, 43, 46, 49, 51, 54, 58–59, 124, 126, 128, 146–149, 151–152, 160–161, 179–180, 183, 185–186, 188–189, 193, 200, 202, 213–214, 241, 298–299, 302, 305, 311, 314, 335–337, 343, 346, 366, 369–370, 372, 376–377, 379–381, 424, 448, 455, 458, 460, 466, 559, 561, 565, 568, 571, 580, 582, 586–588, 593–594, 648, 654, 663, 667, 684–685, 756, 766, 769–770, 775–776, 784
Strategy, 9, 15–20, 22, 29, 32, 36, 39–40, 42, 46–48, 51, 59, 122–127, 129, 134, 136, 146–147, 149, 151–152, 161, 163, 169, 180–182, 185, 190–191, 193–194, 196, 198, 219, 226, 239, 251, 255–256, 265, 298–300, 302, 305–307, 312, 316, 346, 348–349, 369–370, 372–373, 376–378, 380, 448–449, 453, 457–458, 559, 569, 571, 573, 575–576, 579–580, 582, 586, 590–597, 620, 648–650, 656–657, 660, 663–665, 667–668, 675, 682, 684, 689, 758, 772, 777, 796, 798

Strategy space, 15, 21–22, 26, 30, 34, 38, 59, 180
Strategyproof, 152, 157, 162, 180, 184, 186, 199
  feasible strategyproofness, 185
Structural equations, 8, 684–689, 695–696, 699, 702
Subassemblies, 682, 686, 689, 691, 696, 700
Subgame perfect equilibrium, 82–84
Subsidized advertising, 611, 614, 629–630
Subsidized customer acquisition, 637–638
Subsidized financing, 8, 701
Subsidy, 629, 637, 670
Substitutability, 655, 664
Substitutable products, 538
Supermodular games, 24, 29, 36, 38–39
Supermodularity-preserving transformations, 25
Supermodularity, 24–26, 38–39
Supply chain, 1, 3–4, 6–9, 21, 29, 43, 52, 54, 57,
  127, 145, 149, 213, 219, 232, 234, 237–239,
  303, 310, 312–313, 336–337, 365, 367–369,
  380, 393–396, 398–403, 406, 408, 410, 421,
  423, 427–430, 432–433, 435–443, 449,
  453–456, 458, 460, 470–471, 558–559,
  566–567, 573, 579–581, 590, 608–616, 618,
  620, 629–632, 635–638, 644–648, 650–652,
  654–657, 665–667, 669–674, 681, 683,
  711–714, 731, 747–748, 752, 760–761,
  766–768, 772–774
Supply chain coordination, 2, 5–6, 42, 67, 108, 340,
  367, 373, 459, 614, 629
Supply chain integrator, 72–73, 109
Supply Chain Intermediation, 67, 70, 73, 76–77
Supply chain management, 2–3, 14, 53, 58, 227,
  397, 427, 448, 454, 470, 609, 614, 648, 681,
  748, 750–751, 757, 761, 773
Supply chain performance, 399, 581, 609, 654
Supply chain
  cooperation, 668, 671
  coordination, 336–337, 340, 350, 364, 367–369,
    373, 380, 562, 565–571, 573–574, 585, 595,
    609, 611, 613–614, 616, 629, 631, 635–637,
    643–644, 646, 648, 652–653, 671, 674–675,
    681, 711, 719, 726, 728, 732–733, 747, 758,
    763, 767–769, 777
  Dedicated, 570, 669–670, 682, 713, 733
  performance, 184, 201–202, 365, 367, 369, 376,
    447, 462, 472, 478, 560–563, 566–567, 575,
    580–581, 583, 609, 611, 613, 619, 636,
    644–646, 648, 654, 657, 671, 724, 729–730,
    733, 755, 761, 786, 793, 795, 798, 802
  structure, 145–146, 155, 158, 164–165, 169–171,
    173–174, 177, 182–183, 196–197, 339, 342,
    344–345, 369, 373, 375, 459–460, 561–562,
    565, 569–571, 573, 575, 577, 579, 585, 587,
    595, 597, 609, 613–614, 635, 638, 646,
    655–656, 660, 662, 666, 668, 682, 684, 687,
    695–696, 713–714, 716–717, 720–721, 723,
    733–734, 750, 752, 764

Symmetric, 19, 22, 157, 204, 370, 372, 579, 586,
  654, 656, 716, 720
Symmetric equilibrium, 36
System-optimal performance, 21, 609
System surplus, 73–74, 79–81, 84, 86, 106, 108
Tactical decisions, 724–725, 727
Tardiness, 730–731
Target pricing, 336
Tariffs, 8, 682–683, 685, 687, 689, 692, 696, 701
Tarski, 21, 24–26
Tax incentives, 686
Taxes, 697, 724
Tech Data Corp., 593
Telephone support, 316–317
The Gap, 57, 366, 381, 597, 646
The Right Start, 319
Threshold problem, 276–277, 282
Throughput, 536
Timbuktu, 318
Time-dependent multi-period games, 43
Time-discounted gain, 69, 79
Time-series framework, 399
Toshiba Corporation, 451, 473
Total quality management (TQM), 311
Traditional and Internet channels, 565, 646
Traditional supply chain, 609, 611, 613, 618, 635, 637–638, 644
Transaction cost, 68–69, 71, 77–79, 87–89, 106
Transactional Efficiency, 655
Transfer payment, 670
Transferable utility cooperative games, 48
Transition probabilities, 43, 123
Transportation costs, 8, 681–682, 685, 715, 727,
  729–731, 733, 761, 769–770
Transportation costs without tariffs, 687
Trigger strategies, 42–43
Truncated LPT (LPTT), 497
Trust agent, 71
Trusted institution, 72
Truthful bidding, 182, 198, 255
Two-stage supply chain, 6, 399–400, 402, 427–428,
  435, 714
Type, 14, 41, 43, 53–58, 137, 144, 147–157, 166,
  171, 174, 179, 181–184, 196, 201, 204, 214,
  216, 221, 229, 233, 238–239, 242, 250–252,
  254–255, 258, 263, 265, 287, 307, 310, 344,
  350, 355–357, 360, 363, 371–372, 396–397,
  399, 403, 406, 408, 410, 412, 414–416,
  421–423, 425–426, 428, 433, 437, 440, 453,
  458, 460, 464, 466–468, 471, 473, 475,
  558–562, 565–567, 569–570, 578, 589, 597,
  609, 613, 625, 645, 652, 671–672, 682, 684,
  687, 689, 720, 729, 731, 764, 777, 792
UMTS-Auction, 281–282
Unbalanced shop, 506
Uncertain demand, 341, 372, 375, 612, 651
Underage and overage costs, 670

Index 817

Unidimensional strategies, 15, 36
Unique best response functions, 24
Unique fixed point, 29–30
Unique Nash equilibrium, 21, 371, 622
Unique NE, 24, 29–30, 34, 50
Unique Stackelberg equilibrium, 622–624
Uniqueness of equilibrium, 3, 29, 33
United Parcel Services, 558
Univalent mapping argument, 33–34
Upgrades, 314, 316, 318
UPS, 226–227, 315, 344, 590, 657
Urbanfetch, 314
User toolkits, 303
User training, 316
Utility vector, 49
Value of coordination, 726, 732
Variety, 3, 6, 25, 45, 200, 224, 296, 298, 302, 306–307, 312–315, 318, 338, 344, 351, 393, 397, 403, 410, 429, 448, 456, 458, 463, 466, 562, 566, 593, 610, 659, 732, 777, 795
VCG, 5, 152–154, 159–161, 163, 181–193, 198–199, 201, 204, 255–257, 260–264, 271
Vehicle routing, 264, 314, 320, 726, 730–731, 733–734
Vendor-buyer coordination, 718
Vendor Managed Inventory, 118
Vertical competition, 560
Vertically disintegrated channel, 610
Vertically integrated channel, 570, 610
Vertically integrated supply chain, 7, 616, 638
Vickrey-Clarke-Groves, 152

Vickrey-Clarke-Groves mechanism, 255
Virtual prototypes, 304–306
Virtual supply chains, 609
Virtual valuation, 96, 100
Virtual willingness to pay, 91, 96, 98, 100–101
Volume discount, 145, 164, 167, 171, 194, 587
Waiting time, 491
Wal-Mart, 248, 393, 395–396, 660, 765, 768
Warranties, 316
Web-based conjoint analysis, 300
Web-channel price, 664
Web-investor, 788
Webvan, 314
Wholesale price, 41–42, 242, 368, 568, 573, 575–578, 580, 582–585, 609, 613, 615, 618–619, 626–628, 630–633, 635–637, 652–653, 666–667
Wholesaler-Retailer interaction, 610, 614
Willingness to pay, 73, 76, 90, 93, 95–100, 103
Winner-determination, 166, 186, 257
Winner-determination problem, 164, 169, 174–175, 177, 183, 185, 187, 257
Winner determination, 258
Winner determination problem, 166, 169–170, 178
Winning supplier, 170–171, 195, 653
Workload-dependent due date assignment rules, 552
Worst case, 535
XML, 753, 759–761, 763–766
Yahoo!, 303, 305
Yodlee, 319
Zero latency, 772

Early Titles in the
INTERNATIONAL SERIES IN
OPERATIONS RESEARCH & MANAGEMENT SCIENCE
Frederick S. Hillier, Series Editor, *Stanford University*

Saigal/ *A MODERN APPROACH TO LINEAR PROGRAMMING*
Nagurney/ *PROJECTED DYNAMICAL SYSTEMS & VARIATIONAL INEQUALITIES WITH APPLICATIONS*
Padberg & Rijal/ *LOCATION, SCHEDULING, DESIGN AND INTEGER PROGRAMMING*
Vanderbei/ *LINEAR PROGRAMMING*
Jaiswal/ *MILITARY OPERATIONS RESEARCH*
Gal & Greenberg/ *ADVANCES IN SENSITIVITY ANALYSIS & PARAMETRIC PROGRAMMING*
Prabhu/ *FOUNDATIONS OF QUEUEING THEORY*
Fang, Rajasekera & Tsao/ *ENTROPY OPTIMIZATION & MATHEMATICAL PROGRAMMING*
Yu/ *OR IN THE AIRLINE INDUSTRY*
Ho & Tang/ *PRODUCT VARIETY MANAGEMENT*
El-Taha & Stidham/ *SAMPLE-PATH ANALYSIS OF QUEUEING SYSTEMS*
Miettinen/ *NONLINEAR MULTIOBJECTIVE OPTIMIZATION*
Chao & Huntington/ *DESIGNING COMPETITIVE ELECTRICITY MARKETS*
Weglarz/ *PROJECT SCHEDULING: RECENT TRENDS & RESULTS*
Sahin & Polatoglu/ *QUALITY, WARRANTY AND PREVENTIVE MAINTENANCE*
Tavares/ *ADVANCES MODELS FOR PROJECT MANAGEMENT*
Tayur, Ganeshan & Magazine/ *QUANTITATIVE MODELS FOR SUPPLY CHAIN MANAGEMENT*
Weyant, J./ *ENERGY AND ENVIRONMENTAL POLICY MODELING*
Shanthikumar, J.G. & Sumita, U./ *APPLIED PROBABILITY AND STOCHASTIC PROCESSES*
Liu, B. & Esogbue, A.O./ *DECISION CRITERIA AND OPTIMAL INVENTORY PROCESSES*
Gal, T., Stewart, T.J., Hanne, T. / *MULTICRITERIA DECISION MAKING: Advances in MCDM Models, Algorithms, Theory, and Applications*
Fox, B.L. / *STRATEGIES FOR QUASI-MONTE CARLO*
Hall, R.W. / *HANDBOOK OF TRANSPORTATION SCIENCE*
Grassman, W.K. / *COMPUTATIONAL PROBABILITY*
Pomerol, J-C. & Barba-Romero, S. / *MULTICRITERION DECISION IN MANAGEMENT*
Axsäter, S. / *INVENTORY CONTROL*
Wolkowicz, H., Saigal, R., & Vandenberghe, L. / *HANDBOOK OF SEMI-DEFINITE PROGRAMMING: Theory, Algorithms, and Applications*
Hobbs, B.F. & Meier, P. / *ENERGY DECISIONS AND THE ENVIRONMENT: A Guide to the Use of Multicriteria Methods*
Dar-El, E. / *HUMAN LEARNING: From Learning Curves to Learning Organizations*
Armstrong, J.S. / *PRINCIPLES OF FORECASTING: A Handbook for Researchers and Practitioners*
Balsamo, S., Personé, V., & Onvural, R./ *ANALYSIS OF QUEUEING NETWORKS WITH BLOCKING*
Bouyssou, D. et al. / *EVALUATION AND DECISION MODELS: A Critical Perspective*
Hanne, T. / *INTELLIGENT STRATEGIES FOR META MULTIPLE CRITERIA DECISION MAKING*
Saaty, T. & Vargas, L. / *MODELS, METHODS, CONCEPTS and APPLICATIONS OF THE ANALYTIC HIERARCHY PROCESS*
Chatterjee, K. & Samuelson, W. / *GAME THEORY AND BUSINESS APPLICATIONS*
Hobbs, B. et al. / *THE NEXT GENERATION OF ELECTRIC POWER UNIT COMMITMENT MODELS*
Vanderbei, R.J. / *LINEAR PROGRAMMING: Foundations and Extensions, 2nd Ed.*
Kimms, A. / *MATHEMATICAL PROGRAMMING AND FINANCIAL OBJECTIVES FOR SCHEDULING PROJECTS*
Baptiste, P., Le Pape, C. & Nuijten, W. / *CONSTRAINT-BASED SCHEDULING*
Feinberg, E. & Shwartz, A. / *HANDBOOK OF MARKOV DECISION PROCESSES: Methods and Applications*

* *A list of the more recent publications in the series is at the front of the book* *